D0811891

LOURDES COLLEGE LIBRARY
3 0379 1001 3974 2

JERRY S. FAUGHN — KARL F. KUHN

Department of Physics
Eastern Kentucky University
Richmond, Kentucky

1976

W. B. SAUNDERS COMPANY
Philadelphia • London • Toronto

SAUNDERS GOLDEN SUNBURST SERIES

Physics for people who think they don't like physics

W. B. Saunders Company: West Washington Square
Philadelphia, PA 19105

12 Dyott Street
London, WC1A 1DB

833 Oxford Street
Toronto, Ontario M8Z 5T9, Canada

Physics for People Who Think They Don't Like Physics ISBN 0-7216-3582-2

 Made in the United States of America. Press of W. B. Saunders Company. Library of Congress catalog card number 75-5047.

Last digit is the print number: 9 8 7 6 5 4 3 2 1

PREFACE

To the Teacher

Physics for People Who Think They Don't Like Physics is, as the name implies, intended for a course for nonscience majors; it is conceptual, nonmathematical, and sufficient for a two-semester course. Such a description fits a multitude of books now available. How, then, is this one different?

First, the authors have attempted to write in an informal, conversational style, avoiding pompous, stilted language. In support of this informality, a number of cartoon-style drawings appear in the margins, and bits of humor are scattered throughout the text material. At first glance these may appear to be simple, corny jokes, but a more careful examination will reveal that that is precisely what they are. We find that though science is serious business, few scientists go around frowning.

We have sought to minimize the use of the internal jargon of physics, realizing that few of the readers will be scientists and that in any case it is the concepts rather than the vocabulary of physics which are important.

The order of presentation of topics is basically traditional, beginning with mechanics. However, the chapters on "pure" mechanics have been kept as short as possible, and in an effort to increase student interest through this section, chapters on space travel and extraterrestrial life have been included. In addition, we realize that many instructors wish to cover the material in a nontraditional order or simply to omit certain topics. Thus, we have tried to make each of the six units as independent of the others as possible. This has resulted in some repetition, but we feel that review of important concepts never hurts.

Integrated into the text are a number of home experiments called "take-along-do-its" that require only readily available equipment. It is hoped that these exercises will help to clarify difficult ideas and that they will emphasize the reality of the situation under discussion.

The objectives listed at the end of each chapter are taken directly from the material of the chapter and can either be used by the student as a study guide or assigned by the instructor as questions. The questions themselves often go beyond the material of the chapter in order to encourage the student to apply the concepts to other situations, to relate ideas from various sections of physics, or simply to struggle with questions which have no simple answers.

To the Student

We advise all students to read the previous section, "To the Teacher." This section, however, has a curse on it such that any instructor reading it will grow hair on his/her eyeballs.

As that preceding section should indicate, capturing and maintaining your interest was a primary concern in writing this text. Physics is simply the study of nature, and physicists have fun in their work, so there is no reason why nonphysicists can't have fun studying physics.

Before starting your study, we must sound a warning: physics must be studied differently from the way some subjects are studied. Emphasis must be on *understanding* the concepts presented, rather than on memorizing facts. Thus, speed-reading is of little value. In fact, you should probably stop after reading each paragraph—or perhaps each sentence—and ask yourself if the ideas make sense and if they tie in with your experience and with what you have learned before.

In addition, understanding does not come quickly, so cramming for exams is the wrong way to study physics. Just as a football player who does all his practicing the night before a game will be practically useless to his team in the game, your cramming will leave you with a bunch of memorized terms but with no real understanding of the concepts involved.

Another analogy with sports comes to mind. We hope that physics will be fun for you, but the real fun will be of the type experienced by a ballplayer (or a pianist): this comes from satisfaction in being good at what you are trying to do—in this case, in understanding a part of nature. So have fun. We hope that by the time you are halfway through your course, the title of this book will no longer apply to you.

Acknowledgments

We are indebted to many people for their help in preparing this book. Deserving of special thanks for their reviews and hundreds of helpful suggestions are D. R. Bedding (University of Connecticut, Waterbury), R. M. Cotts (Cornell University), R. C. Davidson (University of Maryland), H. K. Schurmann (Temple University), and R. E. Simpson (University of New Hampshire). We are also grateful to Ann

Wesley and Lorainne Foley, who bore the major responsibility for typing and retyping the manuscript. Without the encouragement, expertise, and patience of our good friends at W. B. Saunders Company, this work would not have been possible.

Our heartfelt gratitude is extended to our families, Mary Ann, Laura, and David; and Sharon, Karyn, Kim, Karl, Keith, and Kevin, who "understood" and encouraged us throughout. Also, we must acknowledge our appreciation for the enthusiasm, encouragement, and helpful suggestions of the students who used the preliminary version of this text.

CONTENTS

SECTION V WAVE MOTION—EMPHASIS LIGHT

SECTION VI $E = MC^2$ AND ALL THAT

SECTION ONE

0 INTRODUCTION

Drawing by S. Harris, © 1973. The New Yorker Magazine, Inc.

Every respectable textbook in elementary physics begins its introduction with a definition of what physics is. The trouble is that the authors cannot state what physics is in a single chapter. Defining physics for you is really what this entire book is about. But surely there is a clear, concise definition of physics. Yes; in fact, there are several. Possibly the best known one is as follows: "Physics is what a physicist does." At first glance this seems trivial, but it really isn't. It indicates a truth about the field of physics: the range of subject matter is vast. Just about anything the physicist decides to apply his methods to can be called physics, and he has applied his methods to everything from particles which make up atoms to the galaxies which make up the universe. Perhaps, then, the *method* employed by the physicist is what determines what physics is and what it is not. Maybe. Partly.

The method used by the physicist is basically the method used by all of science. In fact, it is called the scientific method. As generally given, it consists of five steps:

(1) Recognize a problem.
(2) Guess ("hypothesize" is the more elegant word) an answer.
(3) Predict a result based upon the hypothesis.

(4) Devise and perform an experiment to check the prediction.
(5) Develop a theory which links the confirmed hypothesis to previously existing knowledge.

The scientific method, however, is not a magical prescription to solve a problem. You can't apply the scientific method by saying, "First let's do step one. . . . O.K., now it's time for step two. . . ." The steps given above are, however, recognizable in the development of scientific theory. In practice, they often overlap, and the basic ideas of hypothesizing and testing predictions are present in all true science.

Figure 0–1 illustrates the steps of the scientific method. They are shown as a circle because the method does not stop after a theory is developed. Instead, a theory often introduces an unforeseen problem, and the cycle starts again. In addition, there are often shortcuts (illustrated by dashed arrows) between different stages of the method.

GRASS GROWING

Since you are very likely to be a little more familiar with biology than with physics at this point, we will illustrate the steps of the scientific method with an example from that field. You will have ample opportunity in future chapters to see the method in operation in physics situations.

A classic experiment in elementary biological science is one which involves growing beans. But since we don't care for beans, we'll change it slightly and grow grass. The experimenter wishes to determine the factors that influence the growth of grass from seed (the problem). He decides that the amount of water in the soil is likely to be an important factor. (He is thereby *hypothesizing* that water influences

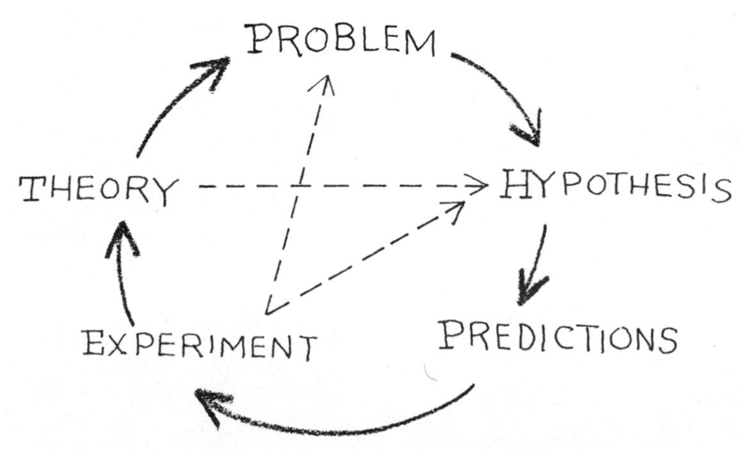

FIGURE 0–1. The steps of the scientific method.

growth.) Based on this very reasonable "guess," he predicts that if he puts a few grass seeds in each of a number of pots and adds different amounts of water to each, the growth will vary from pot to pot.

The decision as to how to set up and perform the experiment is critical if the experimenter is to give his hypothesis a fair test. Problems arise because there may well be other factors which also influence grass growth, and he must not ignore them. If he puts one pot on a windowsill, for example, and another in a closet, his results probably will not be reliable. He must be careful to *control* every variable which might possibly affect growth. Must all pots be the same size? The same color? The careful researcher tries to keep every variable the same except the one he is testing.[1] Although the control of variables cannot be listed as one of the steps in the scientific method, it is an important feature of all scientific experimentation.

So our curious scientist finds that there is a "best" amount of water for his grass. Too little or too much inhibits growth. He will then try to relate this discovery to what he already knows about grass growing and what causes seed to sprout, and he may develop a *theory* of the effect of water on grass growing. It is likely, though, that in doing the experiment, other questions will arise: will the amount of water which worked best in this case also work best under other circumstances—in the closet, for instance? Thus, in this cycle through the scientific method, the *theory* step might not occur, but instead the cycle would start again with a hypothesis.

Many of the activities of science—and particularly of physics—cannot so easily be classified into the five steps of the scientific method. This method, however, is common to all experimental sciences, including physics, chemistry, geology, biology, the social sciences, and all subdivisions and combinations thereof.

THE MANY SCIENCES

To identify what separates physics from the other sciences is not easy. It differs from the biological and social sciences in that it deals primarily with nonliving matter, but so do the other physical sciences, geology and chemistry. Within these sciences there are many overlapping areas, with

---

[1]In practice, he must use quite a number of seeds for each case, because if he uses only two seeds per pot, for example, the two seeds in a certain pot may both be "bad seeds." He uses many seeds in each pot and thus can assume that the number of bad ones in each is about the same.

the physicist having much to say about the geologist's rocks and the chemist's chemicals—and the geologist and chemist contributing to what the physicist calls his own.

To see what the physicist calls his own, one may look at the table of contents of a number of physics texts or at the physics section of a college catalog. Following is a partial list of topics whose names are somewhat self-explanatory:

Electricity	Sound
Magnetism	Heat
Atomic Physics	Light
Nuclear Physics	Astrophysics

The names of other areas, such as quantum mechanics, relativity, and solid state physics may mean little to anyone who has not studied these subjects.

The subject matter of physics is indeed hard to define. At the risk of offending nonphysicists, the authors are tempted to say that chemists study chemicals, geologists study rocks, psychologists study people, biologists study living things, and physicists study everything else.

THE UNITY OF NATURE

The difficulty in defining the subject matter of each science arises simply from the fact that these distinctions are man-made, while nature is united. One of the ideas we hope you will get from your study of physics is of the overall simplicity and unity—and beauty—found in all of nature. If such unity exists, it is no wonder that we have "hybrid" sciences such as biophysics, physical chemistry, and geophysics. Man has artificially divided a unified natural world.

But some division is necessary if we are to understand nature. We see the natural world as composed of a myriad of vastly different things. For example, there seems to be no connection between the study of how things fall to the earth when released and how the moon travels across the sky. So some people, in the past, concentrated their studies on falling rocks, while others studied heavenly bodies. Eventually it became apparent that falling rocks and heavenly bodies have much in common. But it was only by learning about each separately that the connection was finally seen.

Similarly, as we learn more about the human body, we find more and more areas where the biologist must consult chemistry and physics in order to better understand that very complicated creation.

Nature is a unit, and as we learn more about it, we see the unity more clearly all the time. It would seem, then, that there should be less need for the various scientific disciplines as the years go by. There is a complication, however: the more we learn, the more we realize how much we don't know. For every question we answer, we discover more questions. Example: the discovery that the atom is composed of parts answered some questions about the atom, but it revealed fantastic questions about the pieces that make up the atom. Such unanswered questions are what makes physics fun.

SCIENTISTS AND CHILDREN

Questions. Until you have spent some time with a four year old child, you have not experienced questions. Young children are naturally curious. They seek answers to their questions not only by asking adults but also by experimenting. They are scientists—with somewhat haphazard methods, perhaps, but scientists nonetheless. Somewhere in our educational process, however, too many people learn not to ask so many questions of the world. Their curiosity is drowned, or at least submerged, by the time they reach high school, in an ocean of memorized and regurgitated answers. If this curiosity about nature were absent in everyone, there would be no science. The scientist retains some of the child's ability to look at the world with questioning eyes. And he enjoys the search for answers just as the child does.

Perhaps if you approach your study of physics with the idea of getting the answers to some questions about nature, you'll experience some of this fun too.

OBJECTIVES

After your study of the introductory chapter, you should be able to:

1. Explain why there is a problem in clearly defining what physics is.
2. List the steps involved in the scientific method and use a diagram to illustrate its cyclic nature.
3. Perform an experiment and use it to demonstrate the steps of the scientific method.
4. Explain what is meant by "controlling the variables" and give an example.
5. List at least five of the subdivisions of the discipline of physics.
6. Explain why the dividing lines between the sciences are necessarily fuzzy.
7. Give an example of answering one question in physics only to find more questions.

1 KEEP MOVIN'

Man has always sought, as he looked at nature, to explain the many different things he sees and to find order in what he experiences. The manner used to seek this orderliness depends upon his perception of himself and of the world about him. In this chapter, we will look at some very basic concepts in nature: force and motion. Some of the ways we choose to look at these will be part of our everyday experience. Others will be alien.

SPEED

Of all the terms common to the world of physics, speed must rank as one of the oldest. Mankind has always been obsessed with the desire to get from hither to yon and to do so in the shortest possible time. Whether it's an ancient Babylonian wife anxiously watching her sundial in anticipation of her husband's arrival by camel, or a latter-day urban commuter, speed is of the essence.

We use the term "speed" daily as we glance at the speedometer on our car and note that we are going 50 miles per hour, for example. This 50 miles per hour which we read tells us that if we should travel for one hour at that speed, we would cover 50 miles. The expression, *miles per hour,* gives away the secret to the measurement of speed: we simply determine how far we go and divide by the time it takes to go that distance. There is nothing magical about miles per

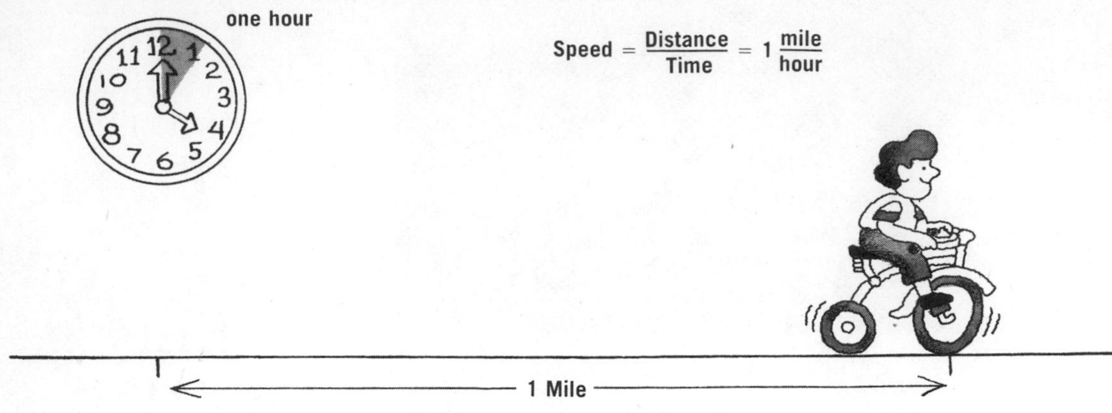

FIGURE 1–1. An example of speed.

hour. Speed can as well be expressed in feet per second, meters per second,[1] centimeters per second—even furlongs per fortnight—just as long as you express the measurement as a distance divided by the time used to cover that distance.

---

take-along-do-it[2]

Try estimating a few speeds. Roll a pencil across your desk. How many inches did it cover in one second? The answer gives the speed in inches per second. (Why is it more meaningful to use inches per second in this case than miles per hour?) Look out a window and observe a stroller walking by. Estimate his speed. (Is there a convenient distance and time scale to use here?) Repeat for a passing car.

ACCELERATION

As you travel down the highway in your nonpolluting Econo-Sport car, you normally do not travel long distances at a constant speed. Traffic lights, the need to adjust the radio, a nearby state police cruiser, and other factors require you to constantly alter your speed. A change in speed is referred to as an *acceleration.* But how is such a change measured? When you first enter your car, its speed is zero, and as you move from your parking place, its speed increases—the car accelerates. *Speed tells you how your distance changes with time; acceleration tells you how your speed changes with time.* As an example of the computation

[1]For an explanation of the metric system of measurement, refer to Appendix II.
[2]Take-home experiments (called take-along-do-its or TADI's) will be suggested throughout the book. They can be performed with a minimum of equipment, and they should be done. It is always better to see something than just to read about it. You will find that you will better understand physics if you spend a few minutes "doing" instead of reading.

FIGURE 1–2. An acceleration.

of acceleration, consider Figure 1–2. A car starts from rest and accelerates to a speed of 20 miles per hour in two seconds. The change in speed during the two seconds is 20 miles per hour. The acceleration, then, is 20 miles per hour per *two* seconds, which means that on the average the car changed speed by 10 miles per hour each second. The acceleration is best stated as 10 miles per hour per second. Further thought will reveal that if you had continued one more second at this acceleration, you would reach a speed of 30 miles per hour, and in another second, 40 miles per hour.

Try your hand at calculating the acceleration indicated in Figure 1–3. We suggest this because by actually calculating an acceleration or two you should get a better feel for what acceleration is.[3]

The distinction between acceleration and speed is an important one. When we say that a car has a great acceleration, it does not necessarily follow that it has a great speed. One car might go from rest to 20 miles per hour very quickly, as did the one in the first example. It would thus have a great acceleration, but a low speed. Another car might change from 60 to 70 miles per hour and take quite a while to do it—10 seconds, perhaps. It would have a small acceleration (one mile per hour per second) but a high speed.

[3]The turtle's acceleration = 2 inches per second per second.

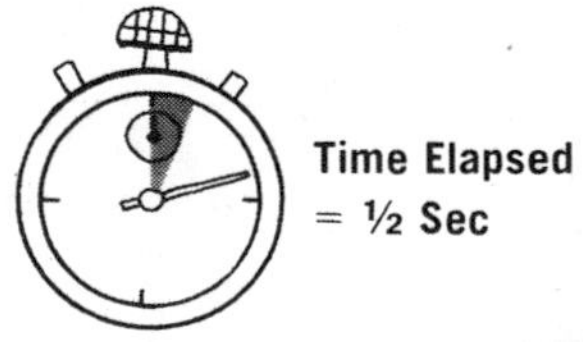

FIGURE 1–3. A do-it-yourself acceleration.

Acceleration Due to Gravity

We will see in Chapter 3 that all objects, heavy or light, fall in the same way. A brick and a tennis ball, when released, will hit the ground at the same time. It should be obvious that if they are dropped from rest and hit at the same time, their motions are alike in all ways as they fall. Because they are both accelerated, they must fall with the same acceleration. (We are ignoring the effect of air resistance, which would cause the fall times to be different, as in the case of a sheet of paper and a brick.)

It was Galileo Galilei who first gave a plausible argument for the theory that all objects fall with the same acceleration in a vacuum. His writings indicate that he did a number of experiments with accelerating objects and, in fact, first defined acceleration in the manner we did in the last section. His method was to roll balls down a slight incline and to measure the distance they moved in successive time intervals. The purpose of the incline was to sufficiently slow down their movement to enable him to take accurate measurements.[4] He then increased the incline to a steeper angle and repeated the experiment. Since a falling body can be considered to be the same as a ball rolling down a vertical incline, he was able to draw conclusions about freely falling bodies even though accurate measurement of such motion was impossible.

Galileo's experiments were important for two reasons: (1) they suggested that objects fall with the same acceleration (which remains constant as the objects fall), but more importantly, (2) they introduced the process of experimentation as a necessary part of scientific investigation. Prior to his time, investigation into natural phenomena had been considered primarily an exercise of the mind. Experimentation was not considered to be of major importance. Logical arguments were developed as to why and how something happens, and if no inconsistencies in the logic were apparent, the arguments were expounded as truths. If the credentials of the originators of an argument were good enough, it was "Damn the torpedoes"; they were quoted as authorities on the subject, and the authority behind a theory was perhaps the major determining factor in the acceptance or rejection of the theory. Today we like to think that it is experimental evidence which is the "proof of the pudding" in science, but we still retain some of the ancient reliance on authority. If an *expert* is quoted, we sometimes are unscientific in that we hesitate to question his hypothesis.

---

[4]In Galileo's time (1564–1642) the pendulum clock had not yet been invented. He used a water clock, which is an egg timer–like device by which time is measured by how much water has flowed through a small opening in the bottom of a bucket.

Galileo suggested that objects fall with the same acceleration everywhere, but actually the force of gravity is slightly different at different places on the earth, and thus the acceleration which a freely falling object experiences varies slightly at different locations. Measurements show, however, that acceleration is approximately 32 feet (or 9.8 meters) per second per second[5] everywhere on earth. In other words, any falling object will accelerate in such a way that its speed changes at a rate of 32 ft/sec every second. A ball dropped from rest off a high building (Fig. 1–4) will thus be traveling at a speed of 32 ft/sec after one second has passed. Let another second pass and its speed will increase to 64 ft/sec, then in yet another second to 96 ft/sec, and so forth.

If the ball is thrown upward (Fig. 1–5), the acceleration due to gravity remains the same, but its effect is now to *slow down* the object at a rate of 32 ft/sec every second. If the ball has a speed of 96 ft/sec at the instant it leaves the hand, one second later it will be traveling upward at only 64 ft/sec. In another second, its speed will have declined to 32 ft/sec, and finally, in another second, it will stop. It then comes back down as if it had been dropped from rest at the top.

In Figure 1–6 an object is thrown upward from the edge of a tall building. Verify that its speed will be as indicated at the times shown. Note also that at the top of its path, although the speed is zero for an instant, the acceleration is not zero. This may seem strange, but it must follow that if the object is at one time moving upward and at a later time moving downward, it must at some time in between have stopped its motion. This happens for an infinitesimally short time. At that time its speed must be *changing*, however, so its acceleration is not zero; it is in fact 32 feet/sec/sec just as during the rest of the object's flight.

Galileo Galilei (1564–1642)

Galileo stands out as perhaps the dominant figure in leading the world of physics into the modern era. He constructed the first workable telescope in 1609, with which he discovered mountains on the moon, the satellites of Jupiter, the rings of Saturn, and spots on the sun. His observations also convinced him of the correctness of the Copernican theory that the earth is not the center of the universe, but instead is a planet moving about the sun. He attempted to convince the Church of this idea but ran into troubles with religious leaders during the Inquisition. His work with motion is particularly well known, and because of his leadership, experimentation became an important part of man's search for knowledge.

NEWTON'S LAW OF INERTIA

Imagine the following experiment: take a shoe and slide it across a shag rug. What happens? It stops—and very quickly—after you release it. The ancient Greeks chose to concentrate on this aspect of matter. They said that the natural state of matter was for it to be stopped. When an object is set into motion (as was the shoe above) it naturally comes to a stop. But take this same shoe and slide it across a smooth, highly polished floor. What happens now? Again, it comes to a stop, but not nearly as soon.

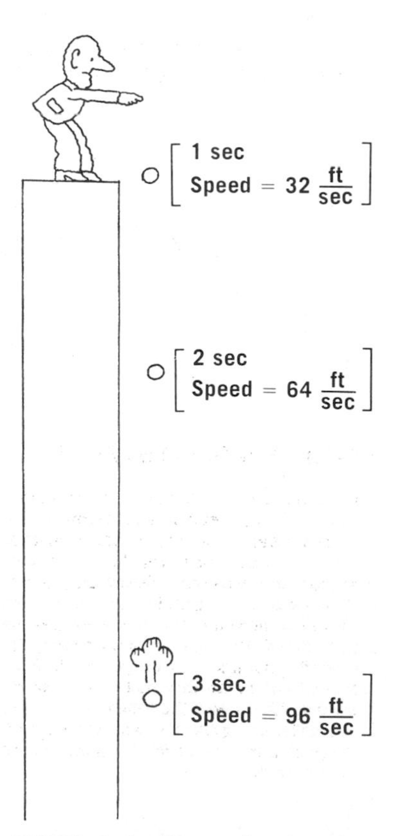

FIGURE 1–4. The acceleration of a falling ball. (If the fellow leans any farther, he will fall with the same acceleration.)

[5] This "per second per second" may sound odd. But we have simply expressed speed as feet per second rather than miles per hour, so change in speed divided by time becomes "feet per second per second."

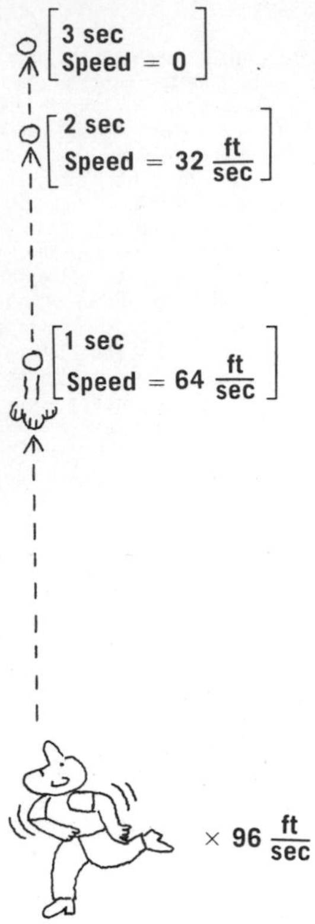

FIGURE 1–5. Acceleration (or deceleration) of an object thrown upward.

In the sixteenth and seventeenth centuries, Galileo looked at motion and concluded that man was considering it from an unproductive viewpoint. He said that the natural state of matter is to remain in motion once it is set in motion. Objects do not "naturally" stop; they "naturally" continue to move. Does this make sense? As we look around us, every object we see that moves eventually stops. But why does the shoe on a rug come to a stop more quickly than the shoe on the polished floor? Something slows down the shoe on the carpet; we call that something *friction.* Consider what would happen if we slide a shoe on an even smoother, slicker floor and imagine that we have a perfectly smooth shoe sole. Could it be that it would slide forever? If no friction were present, it would! "Ah," you say, "but that is an imaginary case, and it does not exist in actual practice on earth." This is true. But we will see that if we consider this imaginary case, we can learn a lot about motion and then later we can add friction—a complicating effect—and come to an understanding of what happens in a real-life situation.

This new idea about motion was formalized by Isaac Newton (1642–1727) and can be stated in part as follows: An object at rest tends to remain at rest and an object in motion tends to remain in motion at the same speed and in a straight line. The name we give this tendency of matter to remain in the state of motion in which it finds itself is *inertia.*

Assume that you are a passenger in a fast-moving automobile when the driver slams on the brakes (Fig. 1–7). You find yourself thrown forward because your body has inertia. Before the emergency, you were traveling at a certain speed, and your body desires to continue traveling at that speed. And that's exactly what it is going to do until something stops it. A force exerted on your body by a seat belt—or, if bad goes to worse, by the windshield—will do the job. This observation allows us to amend our original statement concerning the motion of objects as follows: *An object at rest will remain at rest and an object in motion will remain in motion at the same speed and in a straight line unless a force acts on it to change the motion.*

As another example of the law of inertia, consider two rocks made of the same material and alike in all respects except that one is the size of a bowling ball and the other the size of a ping-pong ball (Fig. 1–8). What would happen if you tried to drive them with a golf club? Both are at rest, and their inertia causes them to prefer to stay that way. When you hit them with the golf club, you are applying the force required to set them into motion, but the small rock may go a considerable distance while the larger one will barely move. The larger one has more inertia than the

Isaac Newton (1642–1727)

Born on Christmas day 3 days before Galileo died, Newton was perhaps the most brilliant man of science who ever lived. As a boy, however, he was a poor student and was more interested in mechanical devices than in study. During an 18-month period in his early twenties, he formulated the law of gravitation, invented calculus, and proposed theories of light and color. He was buried in Westminster Abbey with the following epitaph: "Mortals, congratulate yourselves that so great a man lived for the honor of the human race."

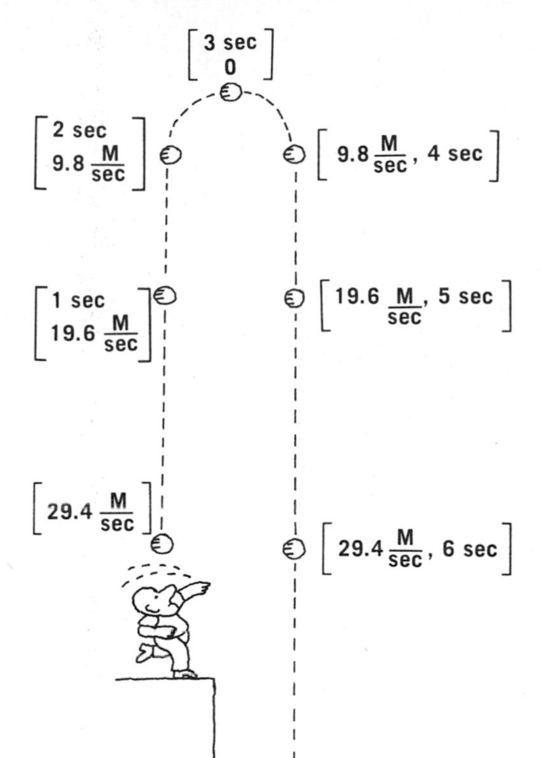

FIGURE 1-6. Check the figures for this accelerating object. (If you try this throw, you'll find that the ball does not curve at the top as shown. It will fall back on your head.)

FIGURE 1-7. This might happen if the windshield does not exert enough force to stop the unfortunate passenger.

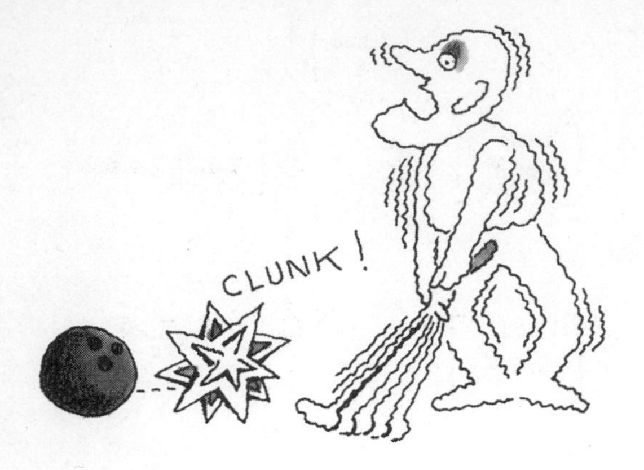

FIGURE 1–8. The result of application of a golf club to a bowling ball.

smaller one. Thus different objects may have different amounts of inertia.

WEIGHT AND MASS

Mass is a term used to describe how much inertia an object has. A very massive object will require a large force to start it in motion or to stop it once it is started. Civil War soldiers found this out with disastrous results. A soldier would see a cannonball rolling leisurely across a field and would attempt to stop it by placing his foot in its way. But because of the ball's deceptive speed and large mass it still required a large force to stop it quickly, and the result was often a broken leg.

Mass is a word that is often used synonymously with weight, but the two are very different! *The weight of an object is the force with which the earth pulls on it.* At the surface of the earth, a very massive object is also very heavy because the earth pulls on it very strongly. Imagine now that you take this same object hundreds of thousands of miles into outer space. At this location, the pull of the earth on the object has weakened greatly. We say that its weight is much less, but what about its mass? To determine whether or not its mass has decreased, imagine that you grab it and shake it. You would find that some effort would be required to set it in motion and then to stop it in each cycle of your shake. This indicates that it *still* has mass. If you are not yet convinced, imagine that a fellow space traveler throws a brick at you. If it has mass, you will have to exert a force on it to stop it. You *will* have to exert this force.

These hypothetical situations serve to point out that weight and mass are distinctly different quantities. If you travel an infinite distance away from any planet or star, you would find that the weight of an object becomes zero. Its mass, however, remains exactly the same as it did on earth.

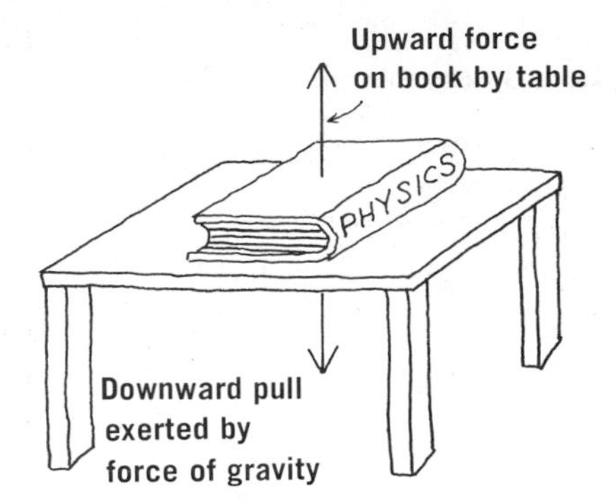

FIGURE 1–9. The forces on a book as it sits on a table.

The way in which mass is measured is discussed in Appendix II.

NET FORCE

Newton's law of inertia has been stated as follows: An object at rest will remain at rest and an object in motion will remain in motion at the same speed and in a straight line unless a force acts on it to change the motion. So stated, the law has sufficed for the examples given. There is one point left to clarify: What is meant by a force which "acts on it to change the motion"? We have so far considered that only one force was acting on the object. What happens if, instead, two or more forces act simultaneously?

Think about a book sitting on your desk. It is motionless, but there are, in fact, forces acting on it. One is the force of gravity—the book's weight—pulling it down. This force does not accelerate the book, however, because the table exerts an exactly equal force *up* on the book (Fig. 1–9). The two forces cancel, so there is no net force on your book.

Figure 1–10 shows two bank robbers trying to remove a safe by sliding it along the floor. Each exerts a force of 100 pounds on the safe, but, as common experience tells you,

FIGURE 1–10. Two not-so-bright crooks at work.

the safe will not move because again the net force is zero. Thus, simply exerting a force is not sufficient to start an object moving.

If one of the robbers reduces his push to 40 pounds (Fig. 1–11*A*), the safe will move to the left. But it will move in exactly the same way as if the crook on the left had not pushed at all and the one on the right had pushed with a force of only 60 pounds. In this case, and in all others, it is the net force that is important.

FRICTION

Starting and Sliding Friction

We can now move closer to a real-life situation by including friction. In essence, friction can be treated as just another force, but *frictional forces always act in a direction such that they oppose motion.* For example, if an object slides to the left as in Figure 1–12, the friction force is directed toward the right. Complications arise when one tries to predict the strength of the force of friction in a given circumstance. Factors such as the conditions of the surfaces involved and the speed of the objects must be taken into consideration.

In Figure 1–13*A*, a shoe rests on a rough surface. In order to slide the shoe, a force must be exerted. This force is needed not only because the shoe has inertia but also because the shoe, in effect, "sticks" to the floor. Figure 1–13*B* represents a greatly magnified surface of a shoe sole in contact with a floor. Note that little bumps on each surface have settled against one another, effectively holding the shoe in place. To slide the shoe, you must either raise it over the bumps or else break them off. Either possibility requires a force, which may be relatively large if the resting

FIGURE 1–11. Both crooks exert a force in *A*, but the same result can be achieved by one, as shown in *B*.

FIGURE 1-12. A frictional force stops the ballplayer.

object is heavy. Once an object is set into motion, the bumps of one surface do not have time to fully settle into the holes in the other, and the friction force is less than when resting. The irregularities of each surface must still glide across one another, however, and so friction is still present.

The model which we use to explain friction forces also explains why friction is reduced after rough objects have been rubbed back and forth against one another. This wears down the irregularities, resulting in smooth surfaces. Friction is also reduced by lubricating the surfaces with oil. The film of oil separates the surfaces so that there is no intimate contact between the irregularities of each.

Let's return to the two crooks pushing on the safe. When last seen, one was pushing with a force of 100 pounds toward the left and the other with 40 pounds toward the right. We concluded that the effect was the same as if only one man pushed—with 60 pounds of force. Now we'll include the effect of friction. Suppose the weight of the safe and the surfaces involved are such that there is a frictional force of 50 pounds. Figure 1-14 represents this force by an arrow near the sliding surfaces. You should be able to verify that the net force is now 10 pounds toward the left. Thus the safe moves as if a single force of 10 pounds were acting on it.

FIGURE 1-13. You recognize *A*. *B* represents a greatly magnified view of the sole resting on the floor.

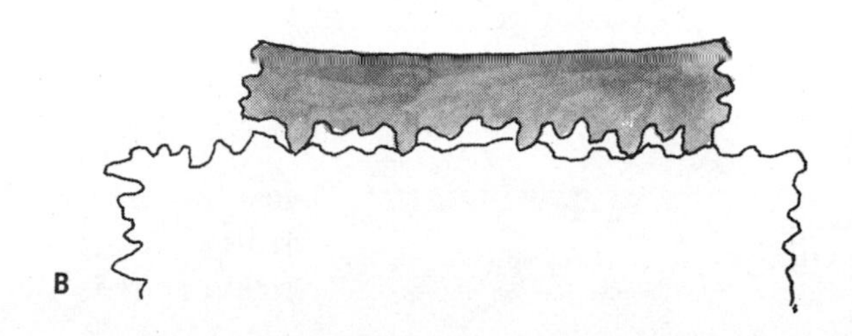

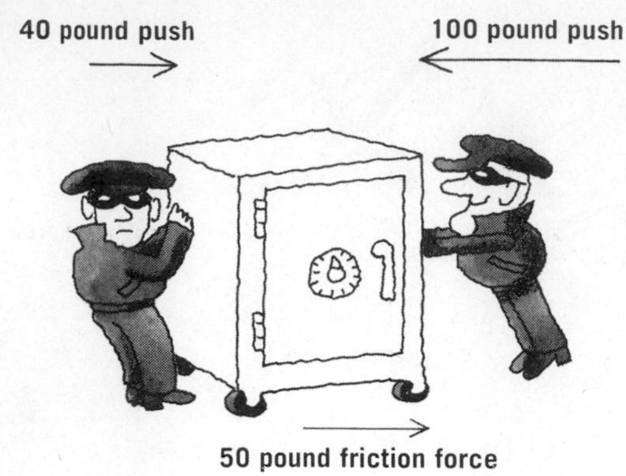

FIGURE 1–14. When the 50-pound frictional force is considered, we find that the net force is 10 pounds.

Rolling Friction

There is also friction when you roll a ball across the floor. Friction here is quite a bit less than when *sliding* one object across another. In the case of rolling, friction is caused by actual indentation of the rolling object and the flat surface (Fig. 1–15). Indentation of the flat surface forces the ball to continually roll *uphill*, while indentation of the ball means that you are continually trying to roll a "flat tire." In both cases the motion is impeded, and the impeding force is called the force of rolling friction.

Rolling friction can be reduced by using surfaces which are very hard and do not deform easily. Thus there is less rolling friction between a car's tires and the road if the tires are inflated to a high pressure. The steel wheels of a train roll on steel tracks, and because both surfaces are very hard, less friction is present than if a rubber-tired truck were to carry the load of a freight car.

When the wheel was first invented, it served as a method of greatly reducing friction, but as shown in Figure 1–16*A*, although wheels seem to roll around an axle, they are actually *sliding* around the axle. So, although the heavy wagon is no longer sliding on the ground, sliding friction is still present. Such friction can be reduced by greasing the axle, or by inserting roller bearings between the wheel and axle (Fig. 1–16*B*). The use of bearings further reduces frictional force, and such bearings are used in almost all machines. Figure 1–17 is a photo of bearings (and their holder) from a bicycle wheel.

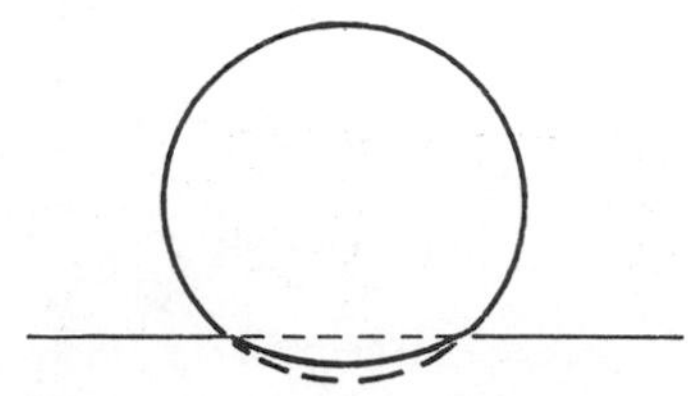

FIGURE 1–15. The mashing of the two surfaces involved as an object rolls.

Reducing Friction with Air

We slide a book across a table and it quickly stops. The law of inertia says that it would move forever in a straight line if no forces acted on it, but friction is always there to prevent this occurrence. An interesting device called an air

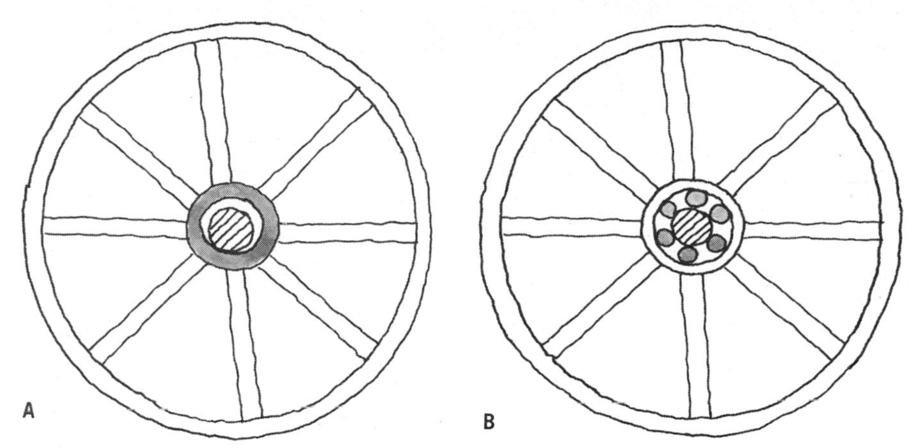

FIGURE 1–16. *A,* A wheel which must *slide* on the axle. *B,* The use of bearings to eliminate this sliding.

table was developed recently that helps to reduce this frictional resistance force.

The surface of an air table is perforated with hundreds of tiny holes spaced about an inch apart. A blower forces air through these holes, much like water being sprayed from a drinking fountain. A small, lightweight plastic puck placed on such a table never actually touches the surface. Instead, it is supported on a cushion of air (Fig. 1–18). As the puck slides across the table, the only frictional force involved is between it and the layer of air that supports it. This frictional force is considerably lower than that which would occur if the surfaces of the puck and table actually made contact, and as a result, the puck now slides for a considerably longer time than it normally would. It still does not go on forever, because friction is not reduced completely to zero. Figure 1–19 shows an air table used for recreational purposes. The players are actively engaged in a game of air hockey. Your recreation building may have one of these machines. Take a break from your studies, drop a quarter in the slot, feel the air flowing out of the holes, slide the puck, and marvel at the law of inertia. At least marvel until the puck goes into your goal and you lose the point (and your quarter).

Streamlining

The presence of friction is the only reason that an engine is needed to keep a car moving on a horizontal road. The law of inertia tells us that in the absence of a net force, a car will continue to move at the same speed once we get it moving. But because a frictional force does exist, we must supply a force to compensate for it. If the engine supplies a force exactly equal to the force of friction, a car will con-

FIGURE 1–17. Bearings from a bicycle wheel. (Courtesy of Jim Shepherd.)

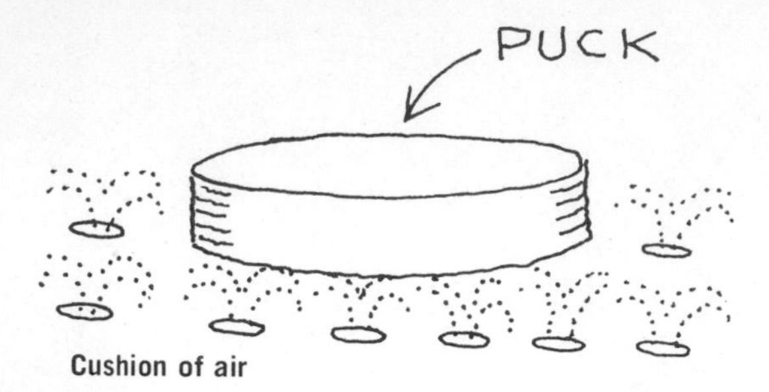

FIGURE 1–18. Air flows from holes in an air table and supports the puck above the table.

tinue at a constant speed (on a horizontal surface), because in this case the two forces cancel out one another, leaving no overall force on the car. Because of its inertia, then, the car continues to move.

At low speeds, the primary source of friction is the tires and the road. At higher speeds, another source of friction becomes important—air friction, or air resistance. As a car moves, it must push aside the air in front and allow the air to fill in the space behind it. As it moves faster, the force is greater, because the air must be moved more quickly.

Air friction can be reduced by changing the shape of the object moving through the air. Figure 1–20A shows air

FIGURE 1–19. The air-hockey puck is held up by air coming from holes in the table.

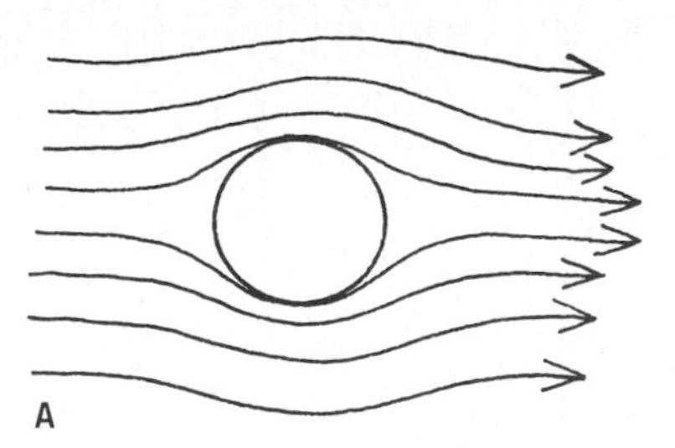

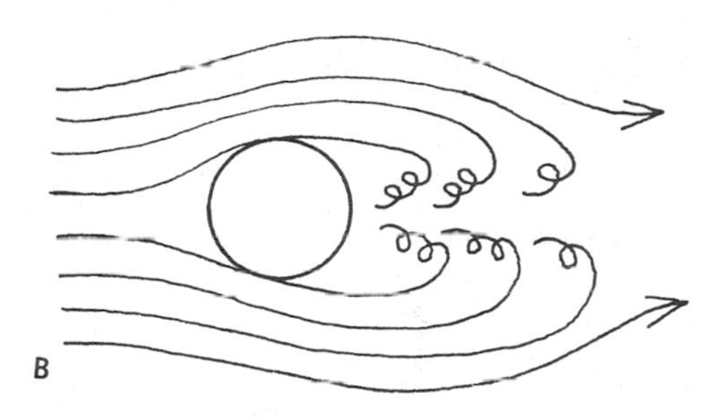

FIGURE 1–20. *A*, Air moves past the ball at a low speed. *B*, Speed is greater, and turbulence results.

flowing past a circular object moving at low speed; Figure 1–20*B* shows the air flow when the object moves at a high speed. In this case, the air does not move smoothly around the object; instead, there is a turbulence of the air behind the object. This turbulence is the major cause of air resistance at high speeds.

The lines drawn in Figure 1–20 represent the flow of the air stream, and when we change the shape of an object to reduce the turbulence, we say that we are streamlining the object. Figure 1–21 shows an object of about the same size as the one in Figure 1–20, but this object is shaped somewhat like a teardrop, and even at a speed which caused turbulent flow around the circular object, there is smooth airflow around this streamlined object.

take-along-do-it

When riding in a car, hold your hand out the window and feel the force of the air on your hand. Change the shape of your hand by making a fist, and note the difference in air resistance. Twist your hand so that air hits it at different angles and again note the different force exerted by the air.

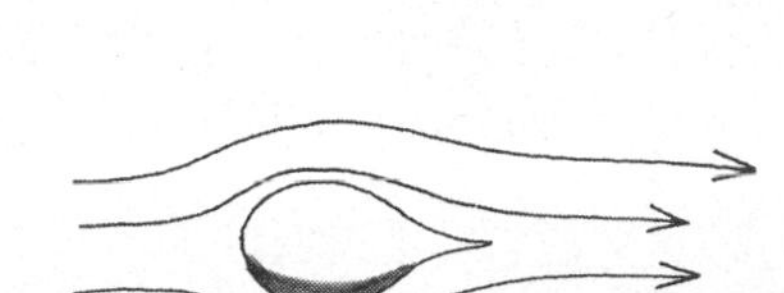

FIGURE 1–21. Air flowing at a fairly great speed around a streamlined object.

A study of how shape affects the amount of air resistance can be performed with a wind tunnel (Fig. 1–22). A fan pushes the air through the tunnel, and a support holding the object in the airflow is connected to a scale that meas-

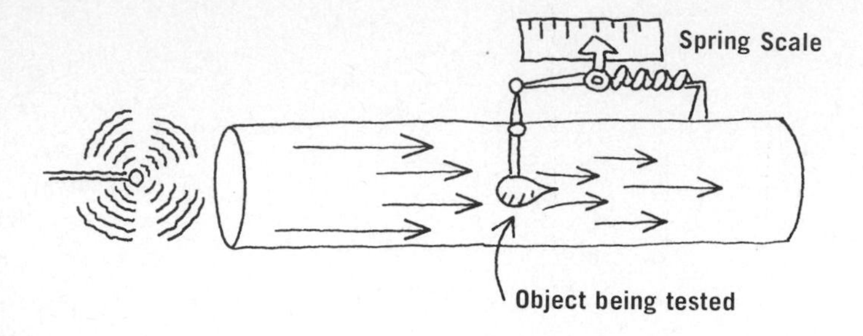

FIGURE 1–22. A basic wind tunnel.

ures the force exerted on the object by the air. A well-streamlined object will cause less force. It may be surprising to learn that it is the *rear* of an object which is most important in streamlining. Thus, the teardrop-shaped object shown in the wind tunnel has less resistance in the position shown than if it were turned around so that the pointed end faced into the wind.

TERMINAL SPEED

When an object falls from a great height, its weight acts simply as a force to accelerate the object. As the object falls, it continues to speed up because its weight is forever acting downward on it. But as it gains speed, air resistance enters the picture; the greater the speed, the greater the air resistance. Finally the object attains a speed at which the frictional force of air resistance equals the weight of the object. Remember that friction is always opposed to the direction of motion and is simply a force, just as weight is a force. Thus, when the force of air resistance is equal to the force of weight, the two forces cancel one another and the object behaves as if there were no forces on it—it continues at the same speed. (So says the law of inertia.) Therefore, in practice, an object falling from a great height reaches a terminal or maximum speed (Fig. 1–23).

The terminal speed of an object depends upon its weight and shape. A leaf falling from a tree has little weight and is not streamlined, so not much speed is needed for air resistance to equal weight. Its terminal speed is low, and it flutters down slowly. A person falling from an airplane, on the other hand, has quite a bit more weight and can change his streamlining by changing the position of his arms and legs as he falls. Thus, a person in a diving position as he falls from a plane will have a higher terminal speed than a person falling spread-eagle. We therefore advise our readers, when such a situation occurs, to spread their arms and legs wide. In fact, open up your coat and use it as an air foil to catch the wind. In this manner, your terminal speed will be

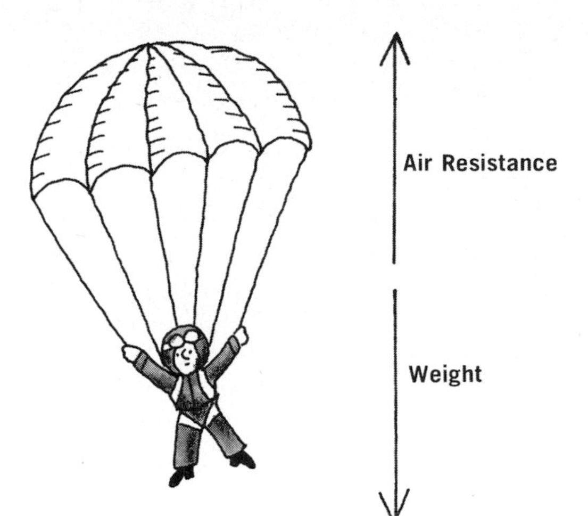

FIGURE 1–23. A parachutist reaches terminal speed when air resistance equals his (and his parachute's) weight.

less than if you take a prayerful position with your hands pressed together in front of your chest. Incidentally, the terminal speed of a person falling through the air can be varied from about 100 to 120 miles (160 to 190 kilometers) per hour—enough to cause trouble no matter what the position of the hands.

The fact that an object reaches a terminal speed when falling serves to lessen the destructive effects of rain and hail. Raindrops quickly reach a terminal speed of 7 to 15 miles per hour, depending upon their mass. If it were not for air resistance, raindrops would reach speeds great enough to produce disastrous results. Their effect would be like a rain of bullets directed toward the earth—yet another reason why it is nice to be surrounded by a blanket of air.

NEWTON'S SECOND LAW

The law of inertia is called Newton's First Law. It states that a net force is needed to alter the state of motion of an object. In this section, we investigate the factors which determine the exact effect of a given force. The results will be stated as Newton's Second Law.

Imagine a brick placed on a table. A force is exerted on it by pushing as shown in Figure 1–24*A*. The force causes the object to change from being at rest to being in motion, and hence, there is an acceleration of the brick. Assume that the force exerted and the resulting acceleration are measured. Now place a second identical brick on the first and push again. If you push with the same force this time, the acceleration will be half as great (Fig. 1–24*B*). Add a third brick, push again with the same force, and the acceleration

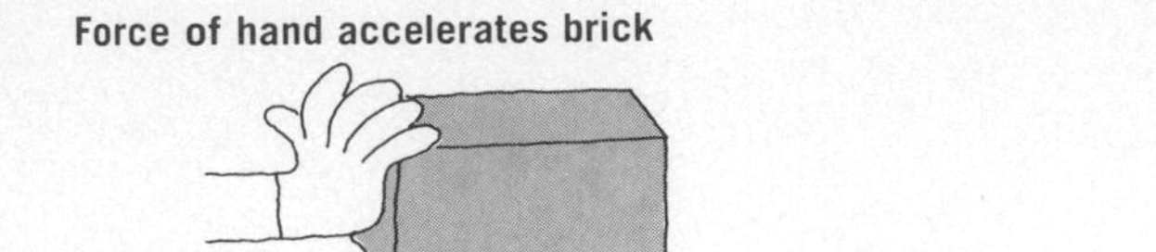

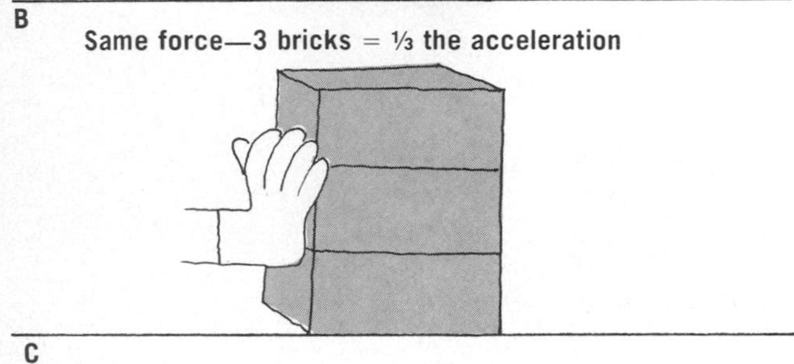

FIGURE 1–24. A "thought" experiment indicating the change in acceleration with mass when force remains the same.

decreases to one third of the value when one brick alone is involved.[6] The results are summarized by the relationship:

$$\text{Force} = \text{Mass} \times \text{Acceleration}$$

In this equation the force which appears is the net force acting on the object. The equation corresponds to the results of our hypothetical experiment with the bricks, for it indicates

[6]We are again ignoring friction for the sake of simplification. To include it we would need only to speak of net forces on the bricks. Or the bricks can be thought of as being on an air table.

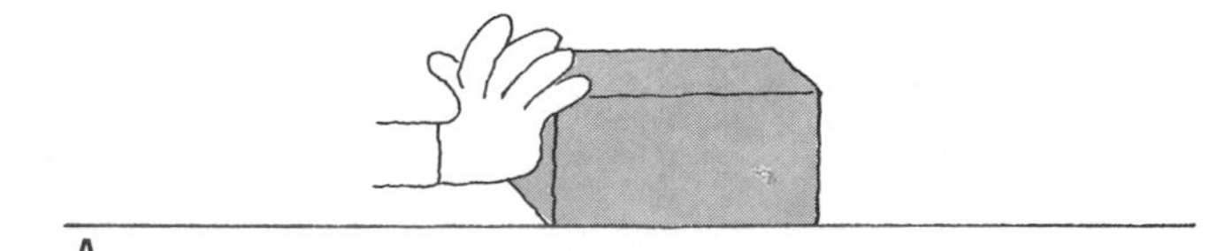

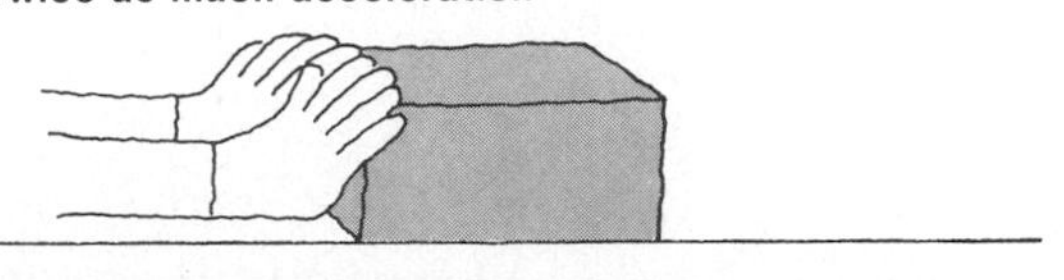

FIGURE 1–25. The effect of different forces applied to the same mass.

that if the same net force is applied to two different masses, the acceleration will depend only upon the masses of the objects. If one object is twice as massive as another, its acceleration will be one half as great; three times as massive, one third the acceleration.

Consider the experiment illustrated in Figure 1–25A. An object of a given mass is pushed by a certain force, and its acceleration is measured. Now repeat this experiment (Fig. 1–25*B*) with the same brick, but this time push twice as hard. Newton's Second Law, as expressed by the equation above, applies to this case also. In this instance, the mass is the same for both experiments, but the force is not. The equation tells us that if the force is doubled, the acceleration must be doubled. Figure 1–26 shows a similar set-up. Use Newton's law to verify that the correct answer is given.

VECTORS

There are a large number of physical quantities that require more than a simple statement of "how much" in order to tell you all that you need to know about them. To completely describe these quantities, called vector quantities, a statement of "what direction" is also required. For example, suppose someone tells you that he walked 30 yards, stopped, and then walked 20 more yards. How far from his starting point would he then be? If you answer, "Fifty yards," you are not necessarily right. You assumed that both walks were in the same direction, but that is an unwarranted assumption. The 20-yard walk may have been in the opposite direction from the 30-yard walk, in which case the distance-from-start is only 10 yards. Or suppose that the person is on a football field (Fig. 1–27). He walks 30 yards down the field and then 20 yards toward the sideline. His distance-

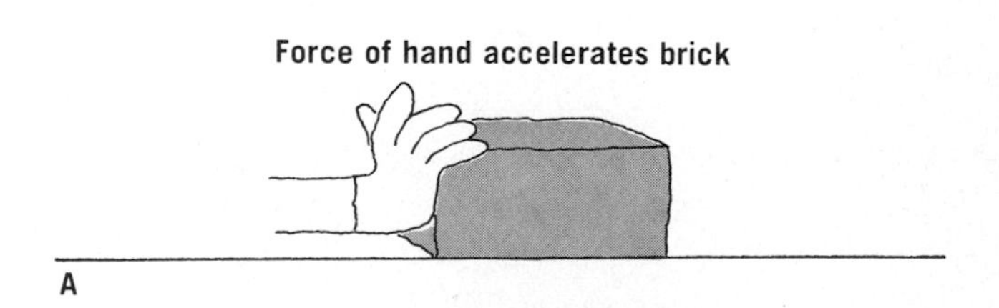

FIGURE 1–26. Convince yourself that this is reasonable.

from-start is about 36 yards. In the second part of the figure, his two walks are shown as two arrows and the final distance as a third (dashed) arrow. The length of this third arrow on the football field is about 36 yards.

What does this fellow who is reliving past football glories have to do with forces? Quite a bit, because, like the distance walked by the man, force is also a vector quantity, and it can be analyzed in the same way. To show this, let's return to our two robbers still trying to move the safe. Now one is pulling forward with a force of 30 pounds; the other is pulling sideways with a force of 20 pounds.

To analyze this situation and determine the net force on the safe, refer to Figure 1–28. In part *B*, arrows (vectors) have been drawn to represent the forces involved. To concentrate only on the vectors involved, the safe is represented by a dot. Note that the "30 pound arrow" is $1\frac{1}{2}$ times as long as the "20 pound arrow." In part *C*, the "20 pound arrow" has been moved to the head of the "30 pound arrow." (It was moved without changing its direction, though.) Notice that the arrows now appear just as did the arrows for the football field walker. And they tell the same story. In *D*, a dashed arrow is used to show the final net force which results from the original two. Perhaps not so surprisingly, this force is 36

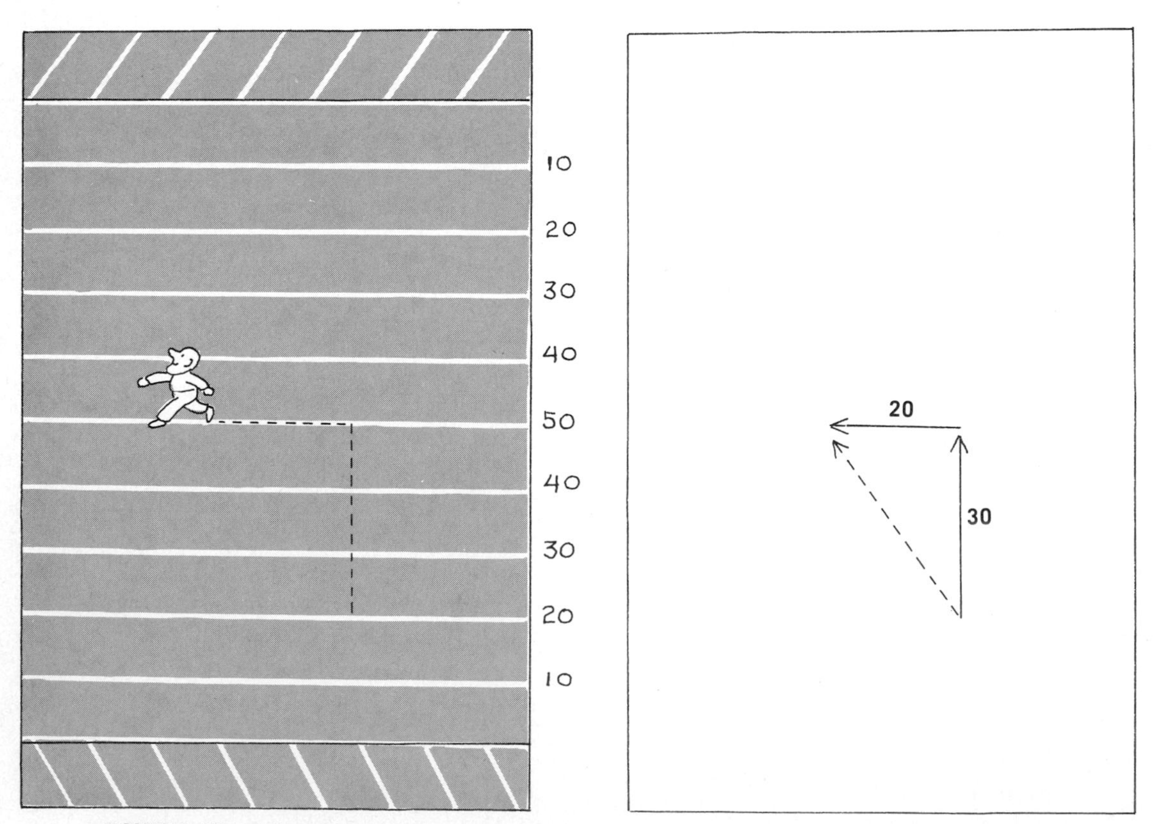

FIGURE 1–27. A football-field walker who walks 50 yards and winds up 36 yards from where he started.

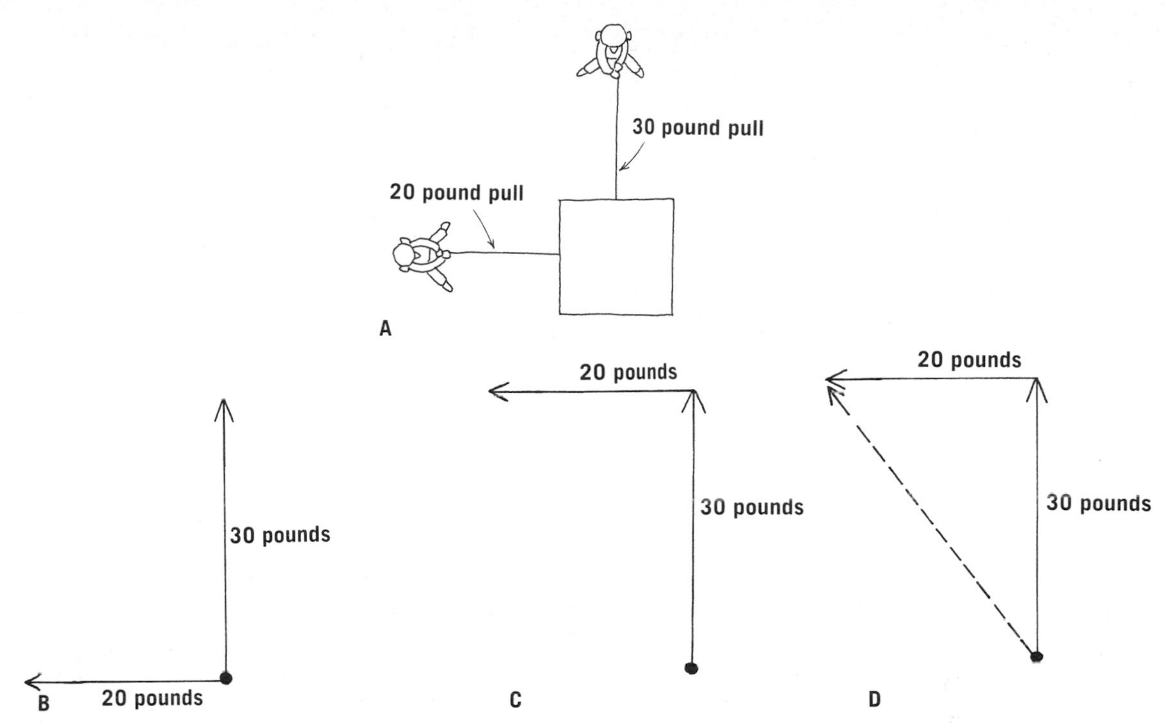

FIGURE 1–28. Vectors applied to the crooks-and-safe problem. The two forces in *A* have the same effect as the force represented by the dashed arrow in *D*.

pounds. Thus the overall effect of the pulling by the two robbers is exactly the same as if one were pulling with 36 pounds in the direction shown and the other doing nothing (Fig. 1–29).

Figure 1–30 shows another example of vector addition of forces. Examine it step by step to see that it follows the same procedure as the ones above.

The beauty of the idea of vector addition is that it can be used for many different quantities. We have considered only distance and force thus far. In both cases we work with the vectors only, without necessarily considering whether the problem involves distance or force. For example, when vectors of lengths 20 and 30 are perpendicular to one another, it is found that they total 36. If the vectors represent pounds, the result is also in pounds; if the first two are distance, the third is distance too.

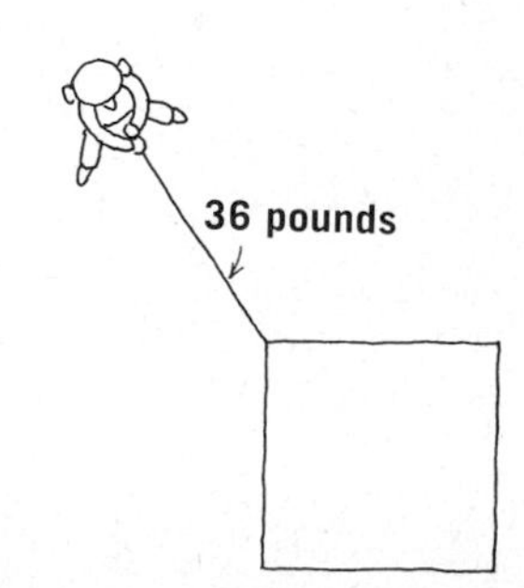

FIGURE 1–29. Here's that one fellow pulling with 36 pounds. He replaces both of the others and achieves the same result.

Vectors and Newton's Second Law

Because distance-from-start is a vector quantity, velocity and acceleration are likewise vectors. (They are derived by combining distance with time, which is not a vector because it cannot be assigned a direction in space.) Newton's Second Law relates force, mass, and acceleration. The first and last of these are vector quantities, and the law itself thereby has a vector nature. It's simple enough: the acceleration pro-

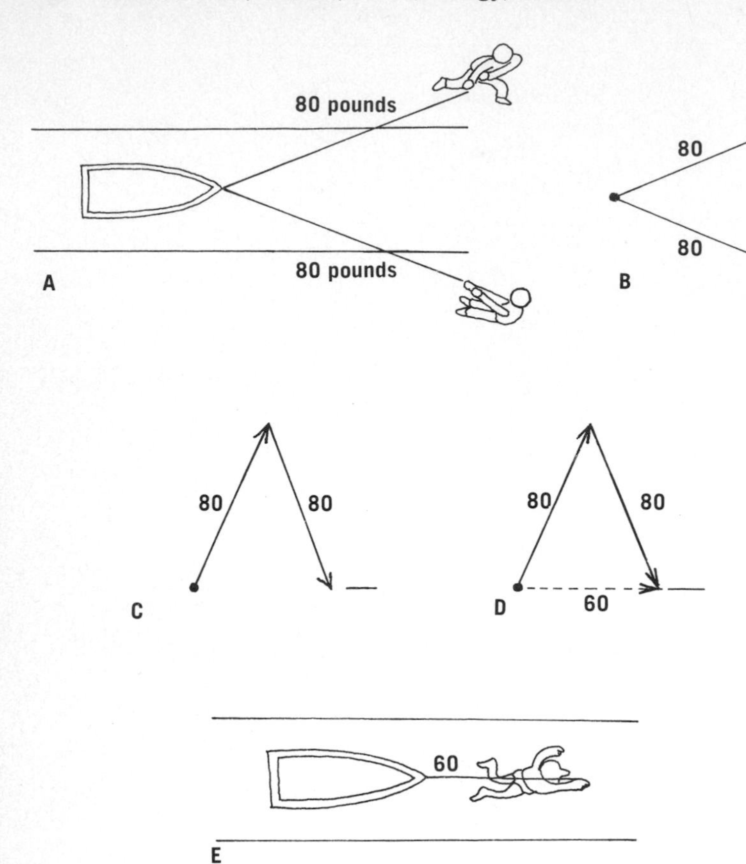

FIGURE 1–30. A check-it-yourself problem. *A*, Two men pulling on a boat, each pulling with a force of 80 pounds. *B*, The two forces represented by vectors. *C*, The lower vector is moved (while keeping the same direction). *D*, The net resultant force is drawn in. It is only 60 pounds because most of the men's force was used against one another. *E*, The effect on the boat is the same as if one man were swimming ahead, pulling with 60 pounds of force.

duced by the net force is always in the direction of that force. Note that it is the *acceleration* and not the *velocity* that is in the direction of the force.

Roll a marble or any round object along a flat table top, and then gently slap it toward the side with a stick, as shown in Figure 1–31. Note the direction taken by the marble. If you hit it perpendicular to its direction of travel, did it move along that perpendicular direction? Or did it move somewhat as in Figure 1–32*B*?

Although the angle at which the ball in the TADI rebounds from your slap will depend upon how fast it was moving beforehand and how hard you slap it, it could not have acted as in Figure 1–32A.[7] Thus the *velocity* after the force was applied is not necessarily in the direction of the force.

The ball, however, did *change* its velocity *toward* the direction of the slap. Its acceleration, therefore, was in the

[7]If it appeared to do so, you probably hit it from the front rather than the side, or else you hit it so hard that it *very nearly* took the path of part *A*.

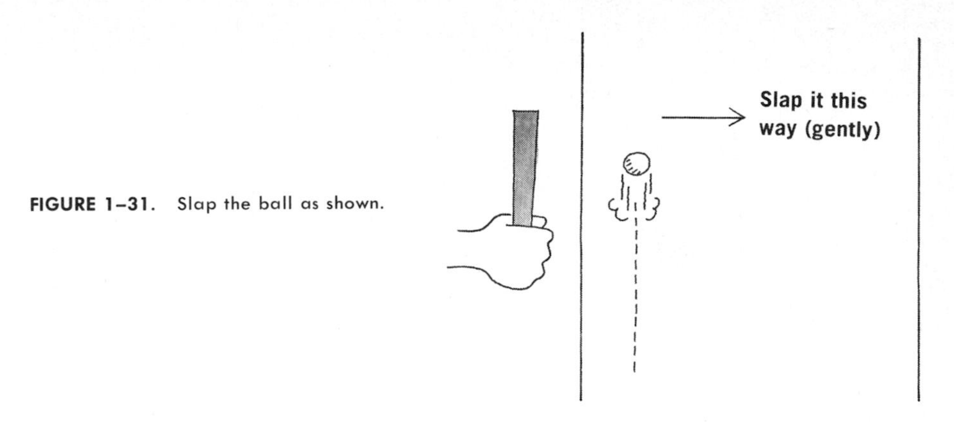

FIGURE 1–31. Slap the ball as shown.

direction of the slap. We will see another example of a difference between direction of the velocity vector and the acceleration vector in the next section.

CIRCULAR MOTION

The motion we have discussed so far has largely been motion involving straight lines. Except for the example in the last section, our objects have accelerated and moved in a single direction. We now examine what happens if, instead, they move around a circular path. As an indication of what to expect, consider a young coed driving down a straight street with her boyfriend in the middle of the seat, his head on her shoulder. A sharp left turn around a corner sends the boy toward the door (Fig. 1–33). Why? Could it be bad breath? No! It's all due to inertia. The boy was in motion in a straight line, and because of his inertia, he tends not to alter this condition—and he *will not* unless a force acts on him. The result is a slide across the seat until something exerts a force strong enough to cause him to turn with the car: this something is the door. If the turn had not been very sharp, friction between his pants and the seat would have

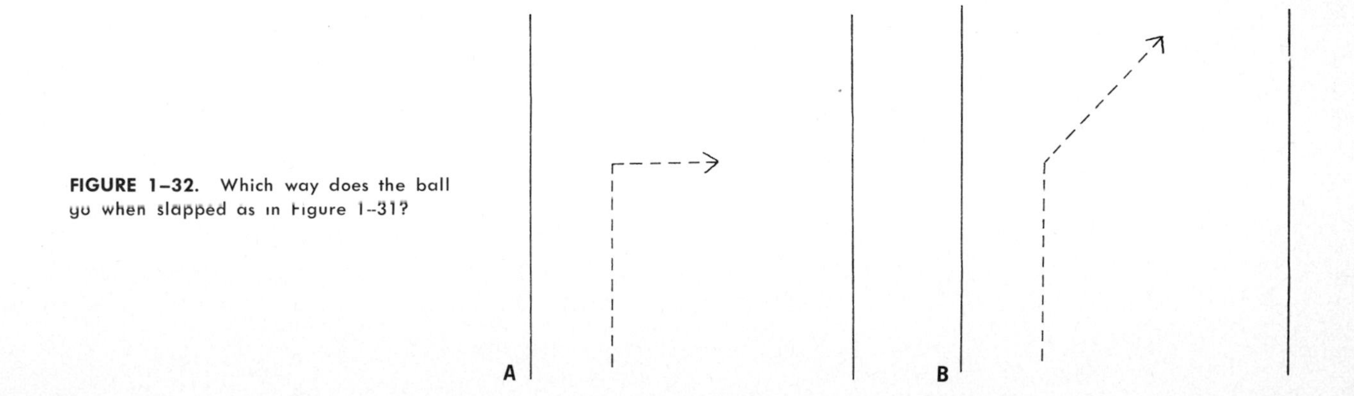

FIGURE 1–32. Which way does the ball go when slapped as in Figure 1–31?

FIGURE 1–33. A sharp turn results in a slide across the seat.

been enough to produce the force to turn the boy with the car.[8]

Circular motion is a prolonged case of the above. Instead of a single sharp turn, it consists of a continuous series of turns which cause an object to follow a circular path. A rock tied to a string and whirled in a circular path as in Figure 1–34 is an example. At every instant, the rock is trying to move in a straight line, but because of the force exerted on it by the string, its path is continually altered to move in a circle.

Several stages of the motion of the rock as seen from above are shown in Figure 1–35A, and the directions of the forces exerted on the rock by the string are indicated. It

[8]This example might also cause one to form a law stated as follows: Love is a sharp turn to the right. Can you use the law of inertia to explain this statement? (Which, incidentally, would not be true in the United Kingdom—or the Virgin Islands.)

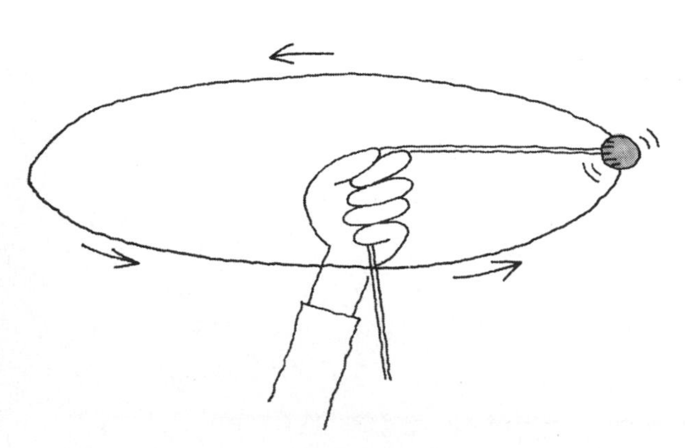

FIGURE 1–34. When you whirl a rock on a string, you can feel that a force is needed to keep it from going straight.

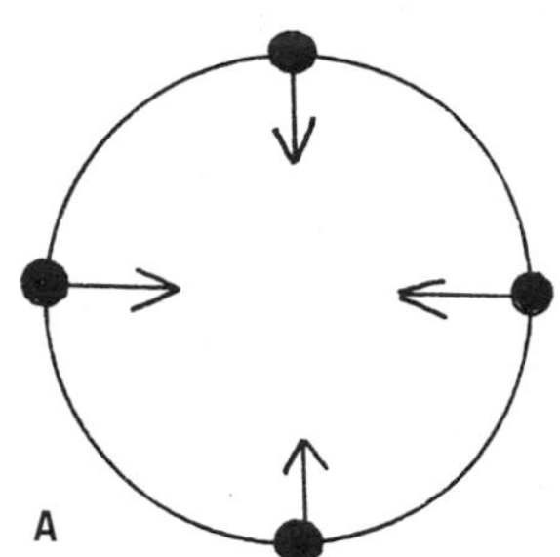
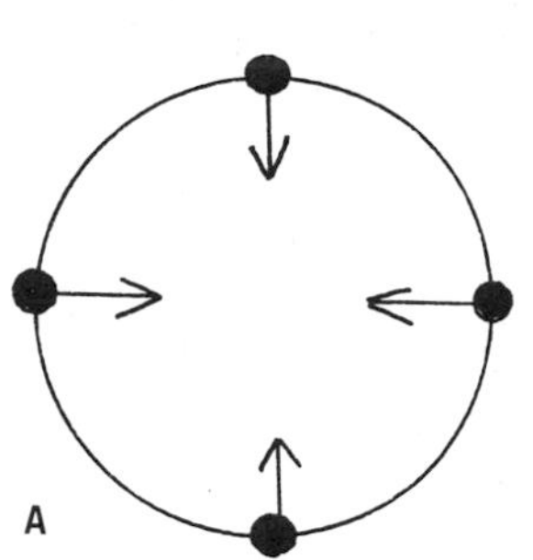

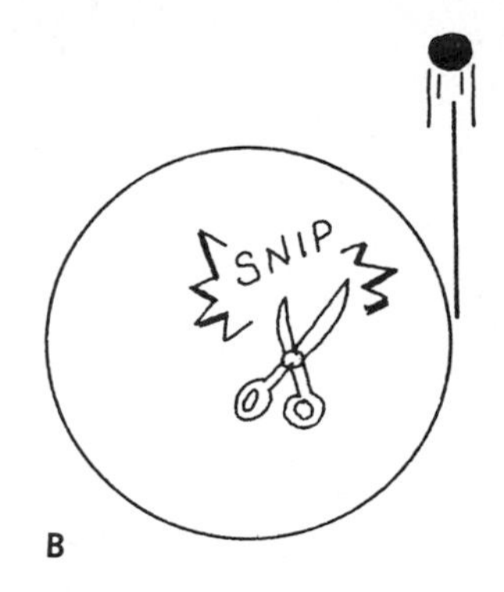

FIGURE 1–35. *A*, Direction of the centripetal force; *B* indicates what happens when the string no longer exerts this force.

should be noted that the force is always directed toward the center of the circular path. This force is called a *centripetal force.* If the string is cut, this centripetal force vanishes, and the rock no longer has the necessary force acting on it to cause circular motion. It then flies off in a straight line (Fig. 1–36).

All objects which move in a circular path have a centripetal force acting on them. The moon, as it circles the earth, is forced to follow its path because of the gravitational attraction of the earth for it; this attraction is the centripetal force holding the moon in orbit. The gravitational attraction of the sun for the planets likewise produces their orbital motion.

Does Newton's Second Law apply to an object moving in a circular path? The law states that if a net force acts on an object, then it has an acceleration in the direction of that

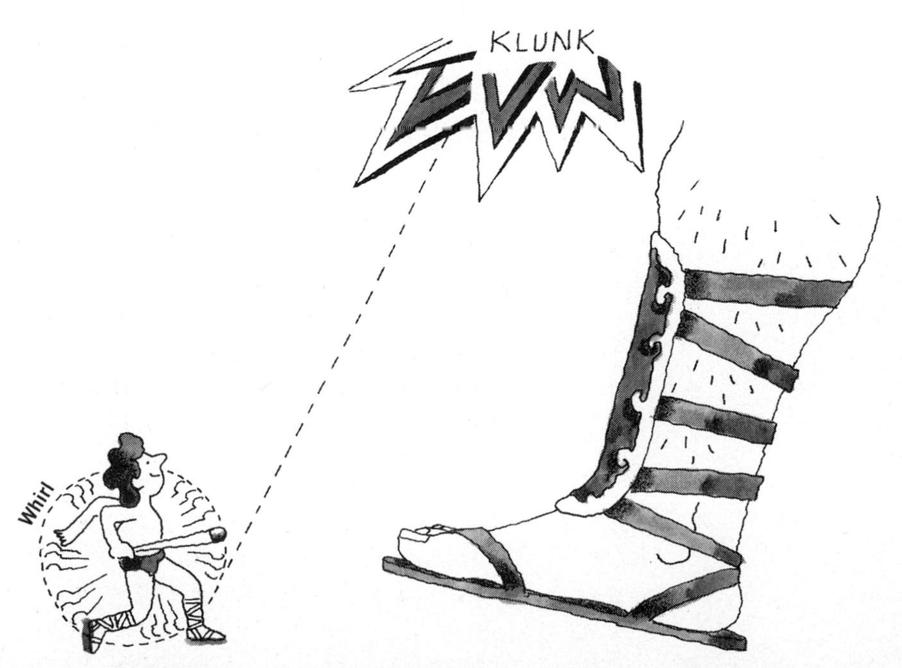

FIGURE 1–36. A practical use of the phenomenon depicted in Figure 1–35.

force. The centripetal force acts toward the center of the circular path, and if the Second Law is to be obeyed, there must be an acceleration also directed toward the center. We have defined acceleration as the amount that the speed changes divided by the time used to make the change. But what if our object moves in a circular path with *constant* speed? It seems that a paradox has arisen. But we can accommodate the case of circular motion by a slight modification in the definition of acceleration. The terms "speed" and "velocity" are often used interchangeably, but there is a distinction between the two in physics. Speed requires only one piece of information to define it completely, and that is "how much." To say that a car is going 10 miles per hour tells you the speed. Velocity requires not only "how much" but also "in what direction." In short, velocity is a vector. The velocity of a car is specified if you are told it is going 10 miles per hour *north.* As a result of this new distinction, we can reformulate our definition of acceleration to be the *change in velocity divided by the time used to make the change.* With this as our definition, we see that an acceleration can be created in two ways. If a car speeds up, it is accelerated because its "how much" is changing; if a car moves in a circular path, it is likewise accelerated because its "what direction" is continually changing. This observation applies not only to circular motion but to any motion in which the direction changes.

Newton's Second Law is obeyed, then. The centripetal force is toward the center of the circle, and the *change in velocity*—the acceleration—is also toward the center. It may seem odd, but it is true, that although the rotating object accelerates toward center, it never gets any closer to center as long as it remains in circular motion.

SCIENTIFIC DEFINITIONS

Perhaps this wheeling and dealing with definitions bothered you. (If not, maybe you should read it again.) At first glance Newton's Second Law does not seem to apply to circular motion when speed remains constant. So what do scientists do? They define velocity to include direction as well as speed, so that the law is obeyed. To see that this procedure is on the up-and-up, consider what scientists are trying to do when they state a law of nature. They are seeking to relate some of the myriad of events of the physical world to one another; they are seeking connections between these events; and they are seeking a few simple statements which apply to all of the events. In the case of his Second Law, Newton could have stated it as we did earlier in the chapter, so that it applied to straight-line motion only. A "second" Second Law would have been needed for curving motion, and perhaps different laws for different types of curves. But by simply defining a term—velocity—to include speed and direction, Newton could state a simple, concise law which applies to every object in the entire universe.

Such defining of terms to simplify language (and thought) is not foreign to our everyday experience. When a certain device became popular some 30 years ago it could well have been referred to as "radio with a moving picture which people stare at for hours at a time." But instead of all that, we just call it "television."

PROJECTILE MOTION

A baseball hit by a bat, a bomb dropped from a plane, and a bullet shot from a gun are all examples of another class of motion called projectile motion. The path of a batted baseball is, at first glance, complicated. But nature has made such motion basically simple; it is only a combination of two separate motions—one horizontal and one vertical.

Suppose you are in a low-flying plane above a baseball park. You see a ball hit by the batter toward left field. From the plane, however, you are unable to tell how high the ball is flying; you can see only how fast it moves across the ground. If you took measurements of this horizontal speed of the ball, you would find that it moves across the ground at a constant speed. For example, you might see the ball pass over the third baseman (90 feet from home plate) one second after it is hit. Then in another second, it would move exactly 90 feet further, and in another second, another 90 feet (Fig. 1–37). The horizontal motion is thus as simple as can be: constant speed. Consider, now, the view of the ball seen by someone sitting in a good seat where he can look straight over home plate toward the third baseman (Fig. 1–38). Since he has a hard time telling how far the ball has been hit toward the outfield, the motion he sees is primarily an up-and-down one. And this motion is simple, too: it is exactly the same motion as that of a ball thrown straight up. Note the similarity between the motion in this figure and that in Figure 1–6.

The reason for this independence of the two motions is that the only force acting on the ball in flight is its weight. This force is always downward, and as a result, no forces are present to change the ball's speed along the horizontal direction. In the vertical direction, the ball is constantly changing its speed at a rate of 32 feet per second per second. The resulting motion is indicated in Figure 1–39. The velocity along the horizontal plane is represented by an arrow that maintains a constant length to indicate that its speed does

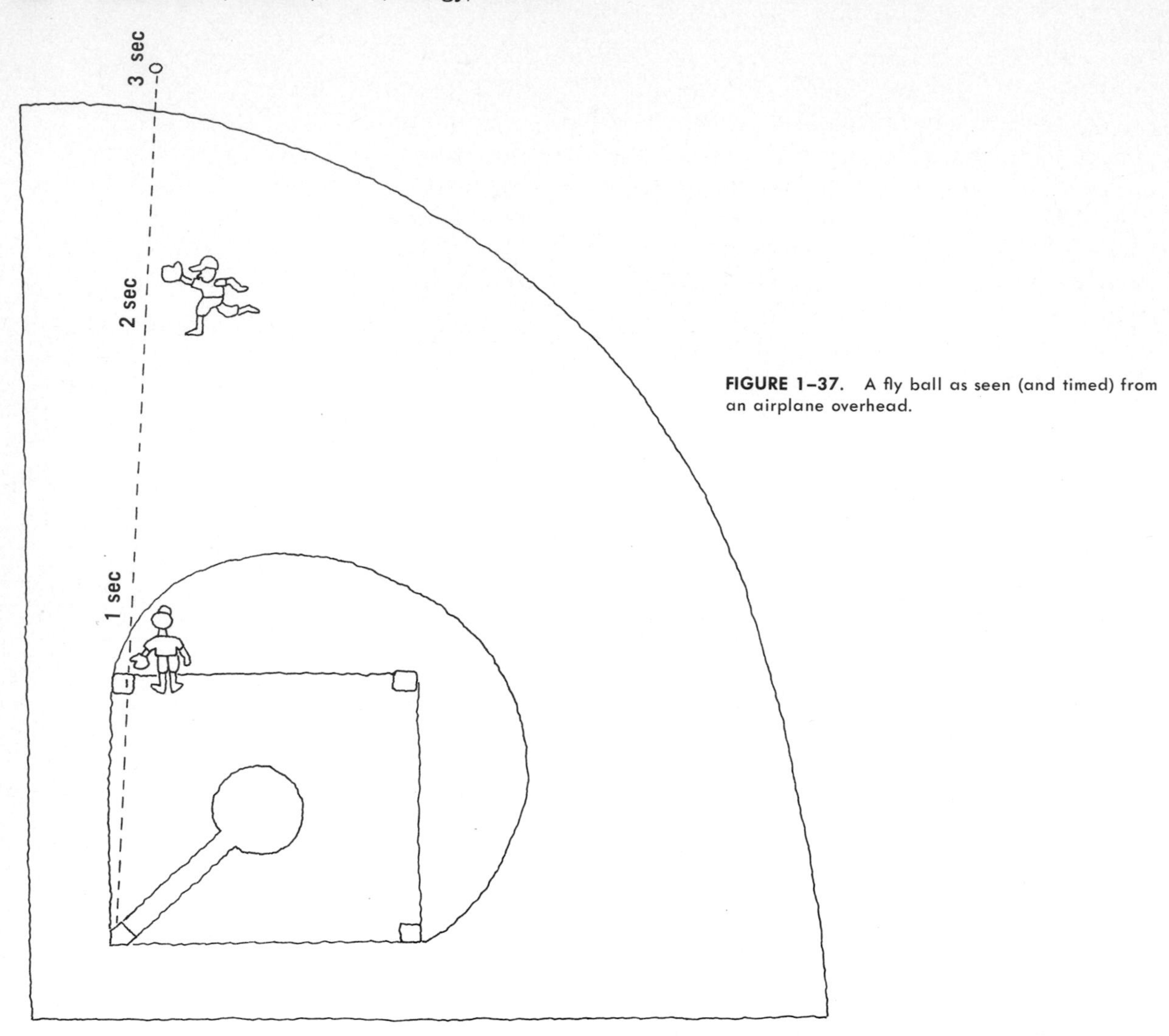

FIGURE 1–37. A fly ball as seen (and timed) from an airplane overhead.

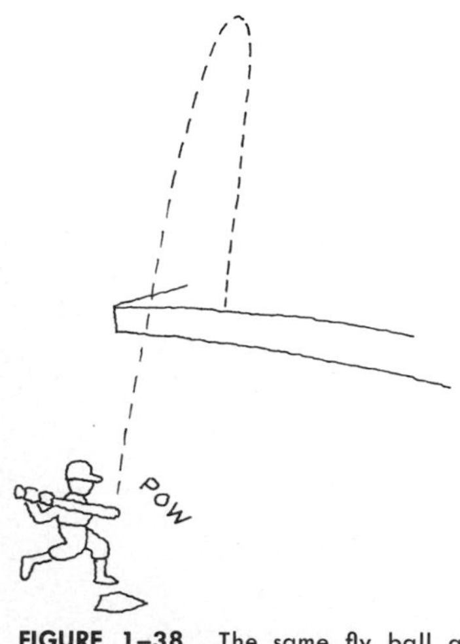

FIGURE 1–38. The same fly ball as seen from behind home plate.

not change. In the same way, the vertical arrows indicate that the ball starts with some great speed upward, which gradually decreases as the projectile rises, finally falling to zero at the highest point in its flight. It then accelerates downward, as a freely falling body, to finally strike the earth with the same speed it had when sent upward.

Of interest to gunners and punters is the maximum horizontal distance which can be obtained by a projectile. Assuming that every time a kicker boots a ball, it leaves his toe traveling at the same speed, he will find that the maximum distance is reached when the flight angle is 45 degrees (Fig. 1–40). Figure 1–40 also shows the paths which would result for angles both greater and less than 45 degrees.

A bomb released from an airplane is followed in its flight in Figure 1–41. When released, the bomb is traveling horizontally at the same speed as the plane. In the absence of air resistance, nothing alters this horizontal speed as it falls. Thus if the pilot does not alter his velocity, as by changing direction, the bomb will remain directly below him throughout its descent.

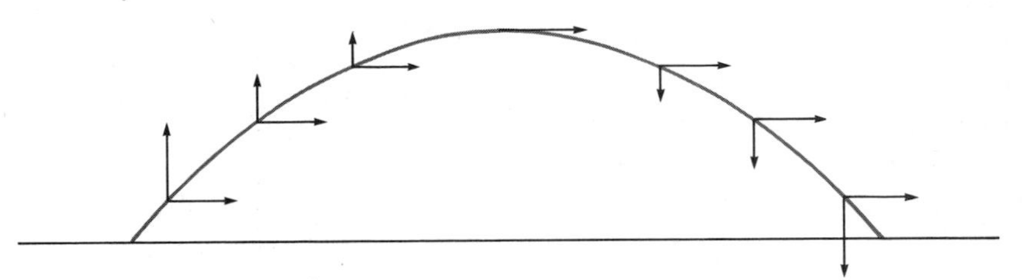

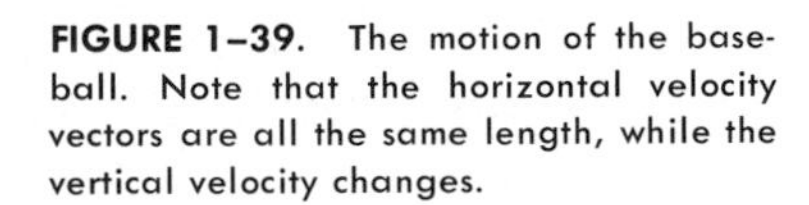

FIGURE 1–39. The motion of the baseball. Note that the horizontal velocity vectors are all the same length, while the vertical velocity changes.

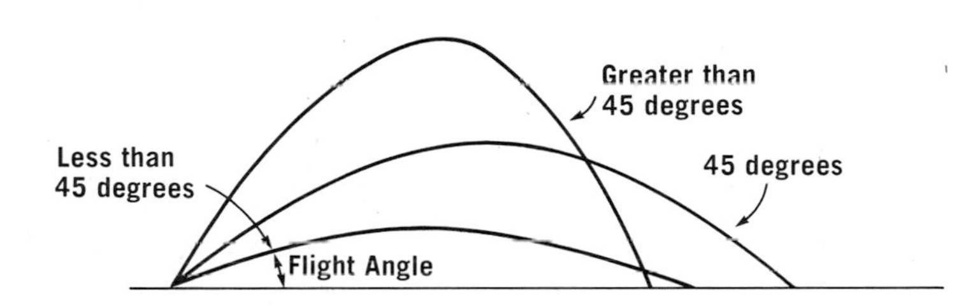

FIGURE 1–40. Maximum distance is obtained at a 45° projection angle (if friction is ignored).

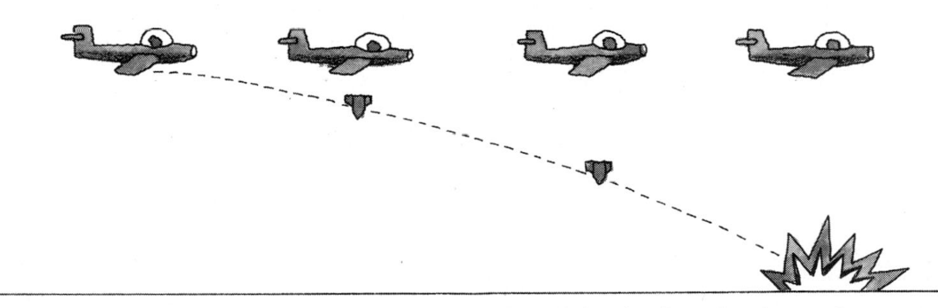

FIGURE 1–41. A bomb dropped from a plane would stay below the plane if it were not for air resistance.

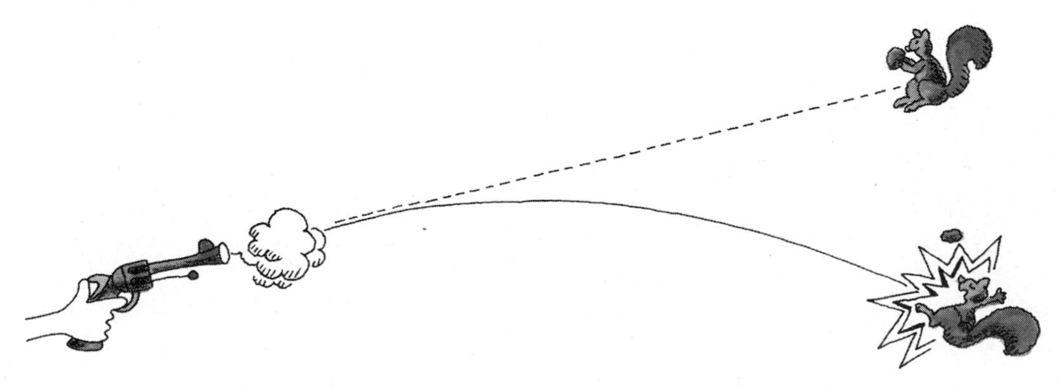

FIGURE 1–42. Luckily, the squirrel falls at exactly the same time the bullet leaves the gun.

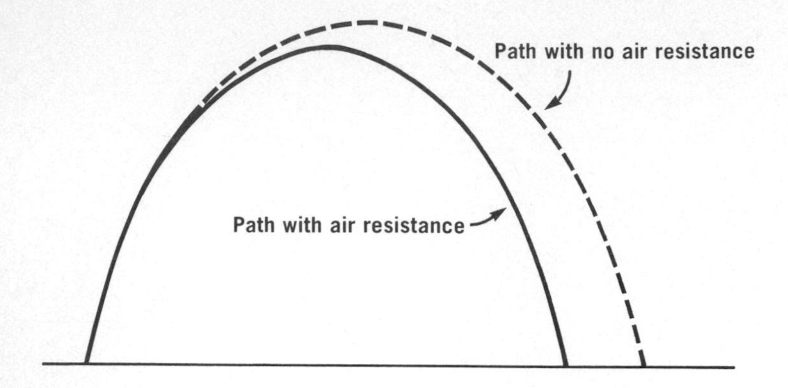

FIGURE 1–43. Air resistance results in less distance for a baseball.

In Figure 1–42 a squirrel in a tree notices a hunter taking aim. In order to avert impending disaster, he decides to escape by dropping straight down off the limb at the instant the gun is shot. Never having read a physics book, the squirrel is not aware that he and the bullet are pulled toward the earth with the same acceleration. The end result is summarized in the figure. The squirrel and bullet reach the same point in space at the same time; a very bad scene for the squirrel.

We have neglected the influence of air resistance throughout this discussion of projectile motion. The error introduced by this simplification is greater in some cases than in others. For instance, the bomb dropped from the plane is moving at a great speed and thus experiences a lot of air resistance. This results in a decrease of its forward speed so that it, in fact, lags behind the plane as it falls. A baseball's path is less altered by air resistance, but there is still an effect caused by the air. The solid line of Figure 1–43 more nearly shows the true path of a ball than does the dashed line, which is the path which would be taken by a ball if there were no air resistance.

OBJECTIVES

The authors hope that a study of Chapter 1 will enable you to:

1. Define speed and state four possible units in which it can be measured.
2. Define acceleration, state four possible units in which it can be measured, and give a numerical example of the meaning of the units.
3. Given the acceleration of gravity, calculate the speed of an object at any whole number of seconds after it is dropped from rest (ignoring friction).
4. Relate (and explain) an example in which speed is zero but acceleration is not zero.
5. State and explain the law of inertia, and discuss the

value of this principle as opposed to that of the Greek idea.

6. Distinguish clearly between the concepts of weight and mass, giving an example that illustrates the basic difference between them.
7. Explain the concept of *net* force, and how the force of friction enters into the determination of net force.
8. Sketch a microscopic view of two surfaces in contact and use this to explain both starting and sliding friction.
9. State the cause of rolling friction and use your explanation to show why rolling friction is greater in some cases than in others.
10. Explain from a microscopic viewpoint how bearings and lubricants reduce friction.
11. Describe the principle of the air table.
12. Discuss the streamlining of objects and its importance.
13. Explain the effect of air friction on a falling body and how it results in a terminal speed.
14. State Newton's Second Law and give illustrative examples involving: (a) a change of force, and (b) a change of mass.
15. Distinguish between vector and non-vector quantities, giving two examples of each.
16. Describe the vector nature of Newton's Second Law.
17. Explain circular motion by considering Newton's First and Second Laws.
18. Distinguish between speed and velocity, and discuss the reason for making such a distinction.
19. Explain projectile motion as a combination of two independent motions and give two examples.

QUESTIONS—CHAPTER 1

1. A rock dropped into a deep well falls for seven seconds before striking the water. How fast was it going when it hit?

2. If a rock is dropped from the top of a sailboat's mast, will it hit the deck at the same point whether or not the boat is in motion?

3. Why does mud fly off a rapidly turning wheel?

4. Racetracks are often banked at the turns. Why?

5. A pail of water can be whirled in a vertical path such that none is spilled. Why does the water stay in even when the pail is above your head?

6. In everyday language, speed and velocity mean the same thing. Of what value is the distinction which is made between them in physics?

7. Pigeons have wings much smaller than seagulls. Explain why pigeons must flap faster to stay aloft.

8. Why does the back window of a station wagon get dirty more easily than that of a sedan? Draw flow lines of the air past each at high speeds. Which normally has greater air resistance?

9. If gold were sold by *weight*, would you rather buy it in Denver or in New Orleans? If sold by mass, would it make any difference?

10. Suppose you see a film of a ball being thrown up and then caught. If air friction were not present, there would be absolutely no way to tell *from the path of the ball* whether the film is being run forward or backward. Since friction *is* always present, how could you detect in which direction the film is running?

11. Although the equal arm balance depends upon weight to operate, it measures mass. Explain. (See Appendix II for a discussion of the balance.)

12. When you round a curve in a car, you feel as if there is a force on you toward the outside of the turn. Some people have said that there is a force outward on you and that this force is equal to the centripetal force. Use the law of inertia to show that this cannot be the case.

13. Analyze the motion of a rock dropped in water in terms of its acceleration and speed as it falls.

14. Suppose a falling object has almost reached terminal speed. How do the forces of weight and air friction compare at that point?

15. Suppose you slap a rolling marble as in the TADI on page 22 and it has exactly the same speed before and after your slap. Is Newton's Second Law obeyed in this case? Explain.

16. Imagine that you attach a heavy object to one end of a spring and then whirl the spring and object in a horizontal circle (by holding the other end of the spring). Does the spring stretch? If so, why? Discuss in terms of centripetal force.

17. Can a force directed southward on an object ever cancel a force directed eastward?

18. At the end of its arc, the speed of a pendulum is zero. Is its acceleration also zero?

19. According to the *Guinness Book of World Records*,[9] the greatest altitude from which anyone has fallen without a parachute and survived is 21,980 feet. After falling this far a freely falling body would have attained a speed of about 800 miles per hour. Explain how Lt. I. M. Chisov of the USSR (the faller) could have survived. (He struck the edge of a snow-covered ravine, incidentally.)

20. It has been suggested that rotating cylinders about ten miles in length and five miles in diameter be placed in space for colonies. The purpose of the rotation is to simulate gravity for the inhabitants. Explain this concept.

---

[9]McWhirter, N., and McWhirter, R.: *Guinness Book of World Records.* New York, Bantam Books, 1974.

(Photo from Tennessee Valley Authority.)

2

A COUPLE OF CONSERVATION LAWS

In this chapter, we develop two ideas which are very important in physics: the conservation of energy, which is used throughout the book, as indeed, it is used throughout physics; and the conservation of momentum.

Energy in this chapter will be considered in its most basic sense rather than in the context of a world energy shortage, for example. The chapter is not impractical, however, because an understanding of the nature of energy will help you to understand problems associated with obtaining new sources of energy for use by man.

HOW DOES A PENDULUM KNOW?

Figure 2–1 is a multiple exposure photo[1] of a pendulum. It shows that the weight on the end reached the same height at both ends of its swing. The weight started from a certain height at the left and moved to the bottom of its arc. But

[1]Taken by shining a blinking strobe light on the pendulum as it swings.

FIGURE 2–1. A pendulum swinging through a long arc.

how did it "know" how high to rise at the other end? If we did this experiment a number of times, using pendulums of various lengths, we would find that they always return to the same height from which they left. (We neglect the effects of friction here.)

Figure 2–2 is a photo of the same pendulum shown in Figure 2–1, but it is now released from a lower point. As always, it goes to the same height from which it started. But notice how much closer together the images of the pendulum are at the bottom of the swing in the second photo than they are in the first, even though the rate at which the film was being exposed was the same in each of the two photos. We can conclude, then, that the pendulum was moving more slowly in the second case. Again, no surprise—a pendulum falling from a lower height naturally would not be expected to move as fast through its lowest point.

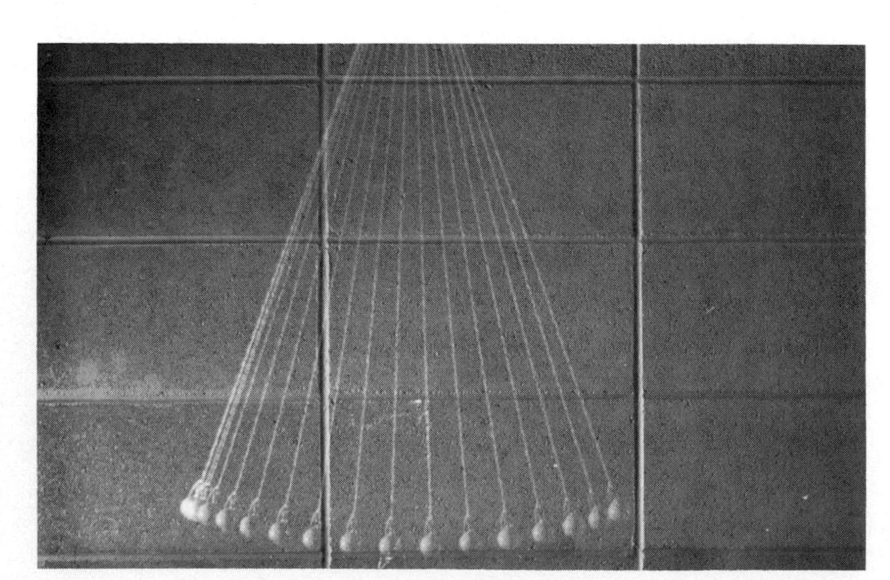

FIGURE 2–2. A pendulum swinging through a shorter arc. Time elapsed between exposures is the same as in Figure 2–1.

Therein lies the secret: the pendulum falling from the greater height goes faster at the bottom, and it then makes sense that it will go higher again before stopping. It knows what height it came from by how fast it is going at the bottom.

We don't really believe that pendulums "know"; we explain their actions by using the idea of energy. Before we can discuss energy, however, we must go to work.

WORK

You probably noticed in Chapter 1 that acceleration was given a somewhat more specific definition than is often the case in nonscientific usage. It was defined in such a way that it could be measured, because when we really want to learn more about something, we seek to measure it. Likewise with the concept of work. *Work*, in physics, *is defined as the product of force and the distance over which the force acts.* An example: you lift a five-pound book from the floor to a height of three feet. The force you exert is five pounds and the distance is three feet, so you have done 15 foot-pounds of work. The unit, foot-pounds, follows from the definition, which in equation form is

$$\text{Work} = \text{Force} \times \text{Distance}$$

Because we expressed force in pounds and distance in feet, work was expressed as the product of pounds and feet: pound-feet, or as is more commonly stated, foot-pounds.

Note that the definition is not completely at odds with the everyday definition of work. Most people would agree that lifting two books to a height of three feet requires twice as much work as lifting one book to that height, or that lifting a book twice as high takes twice as much work. The definition does, however, lead to some conclusions which seem odd. Consider holding a sofa at shoulder height for one hour. How much work is done? According to our definition, there is *no* work done *while it is being held there,* because the force is not being exerted over a distance—the sofa is not being lifted but is simply being held in place. Any force multiplied by zero distance is zero work.

ENERGY AND ITS CONSERVATION

The reason for the above digression from the swinging pendulum is that *energy is defined as the ability to do work.* The amount of energy an object has depends upon how much work it can do. We will find that energy can exist in several forms. Two of these, gravitational potential energy and kinetic energy, are important now; others will be added as we need them.

Gravitational potential energy is the energy possessed by an object because of its position. In Figure 2–3, a spurned lover holds a rock over the edge of a cliff. His intention is to do damage to the cad below who stole his love from him. In the position shown, we say that the rock has gravitational potential energy. To determine how much energy it has, consider the amount of work that was necessary to lift it to the top of the cliff. This is equal to the weight of the rock (the force needed to lift it) times the height of the cliff (the distance lifted). A heavier rock would have more gravitational potential energy on the cliff. Similarly, the present rock would have more gravitational potential energy on a higher cliff.

But energy was defined as the ability to do work, or to exert a force over a distance. And that is exactly what that rock will do when it hits the unsuspecting guy below—it will dent his head.

If no cliffs are handy, the spurned lover has another form of energy in his arsenal to aid him. In Figure 2–4, he is at the same height as the others, having just tossed a rock at them. In this case the energy of the rock is in the form of *kinetic energy:* the energy an object has because of its motion. (It obtained this energy because the thrower threw it by exerting a *force* on it as he swung his arm through a *distance.*) The amount of kinetic energy a moving object possesses depends upon two factors: the mass of the moving object and its speed. A heavy rock will produce a greater effect when tossed than a light one, and the faster a given rock is thrown, the greater is its effect. (In fact, the kinetic energy of

FIGURE 2–3. Both of the guys in the drawing have something on their minds. The rock has gravitational potential energy.

Figure 2–4. The rock now has kinetic energy.

an object depends on the square of the speed. This means that if a given rock is thrown and then another just like it is thrown twice as fast, the second has four times the kinetic energy. If tossed three times as fast, it has nine times the kinetic energy.)[2]

Energy can be transformed from one kind to another. A rock held high above the earth has potential energy,[3] but if the rock falls, this potential energy gradually decreases as its height above the ground becomes less. We find, however, that its speed, and hence its kinetic energy, is increasing. Finally, just before striking the surface of the earth, all of the rock's initial potential energy is completely converted to kinetic energy.

Figure 2–5 shows a circus spectacular wherein five identical divers follow one another off the high dive. Since all are identical, they have the same weight and thus the same potential energy when they are on the high platform. Suppose each weighs 200 pounds and the platform is 50 feet high. It therefore takes 10,000 foot-pounds of work to lift each to the top, so each has 10,000 foot-pounds of potential energy. The drawing shows how potential energy changes to kinetic energy as the divers near the water. Note that at every instance, the total of the kinetic energy and the potential energy remains the same. We say that *energy is conserved.* This means that *energy can be changed from one form to another, but it never disappears.*

This principle also applies to our discussion of the pendulum. At one of its end positions, it has potential energy, and if the pendulum is not pushed, it has no kinetic energy because its speed is zero. At its lowest position, some of this potential energy has been lost, but in its place we find an equivalent amount of kinetic energy. If this is true, the next time the pendulum comes to rest at the other

---

[2]For those of you inclined toward formulas, the formula for kinetic energy is

$$K.E. = \frac{1}{2} mv^2$$

where m is the mass and v is the speed.

[3]Often gravitational potential energy is referred to simply as potential energy. We will, however, encounter another type of potential energy later.

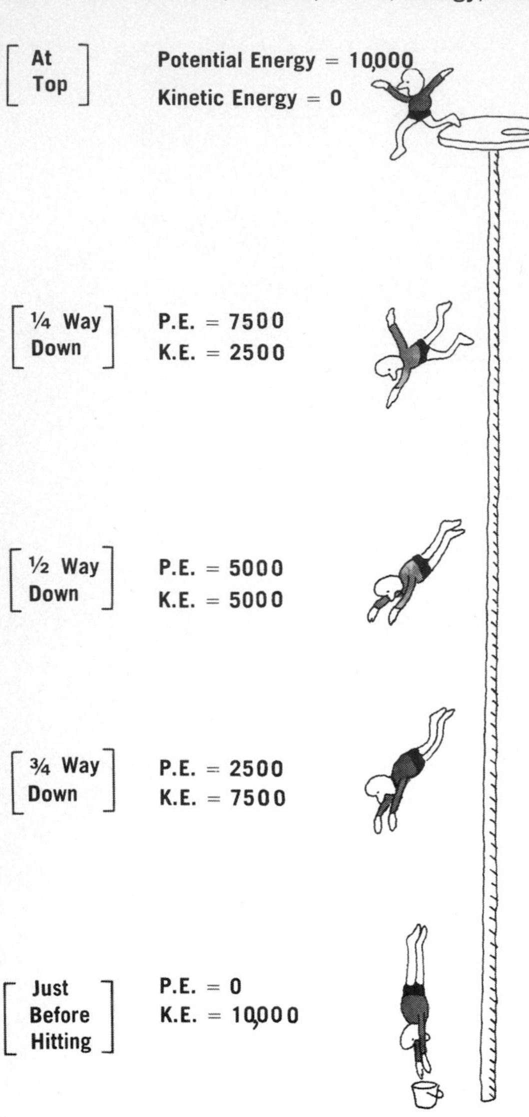

FIGURE 2–5. Calculate the total for each position of the diver.

end of its swing, it must have the same amount of potential energy as it had when it started its swing. (The kinetic energy is again zero, since the pendulum stops at the end.) In order to have the same potential energy, it must rise to the same height.

take-along-do-it

Construct a pendulum by hanging a bottle by a string from a door frame so that it can swing through the door. Swing the pendulum, and note that it swings to the same height on each end of its swing—no surprise there! While it is swinging, slap a broomstick or yardstick across the frame about halfway along the string's length (Fig. 2–6). Observe what happens as the string is caught by the stick. Notice carefully how high the pendulum rises on the other side.

Figure 2–7 is a multiple-exposure photo of a pendulum swinging past a stick as described in the preceding TADI.

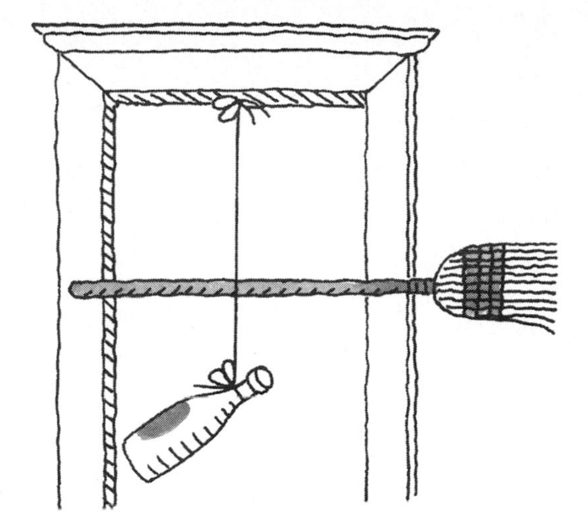

FIGURE 2–6. The broomstick alters the pendulum's swing.

The photo shows that the weight on the end reaches the same height at each end of the swing. The conservation of energy explains this case also. Suppose our pendulum had 25 units of potential energy at position A in Figure 2–8, and that when it reached position B it had only 10 units of potential energy. Thus according to the principle of conservation of energy, it must have 15 units of kinetic energy at B. As it swings past B, however, the string strikes the stick. It now changes its swing and rises to height C, where it stops. If energy is conserved, the 15 units of kinetic energy must change back to potential energy so that the pendulum again has 25 units of potential energy. It therefore must be at the same height at C as it was at A.

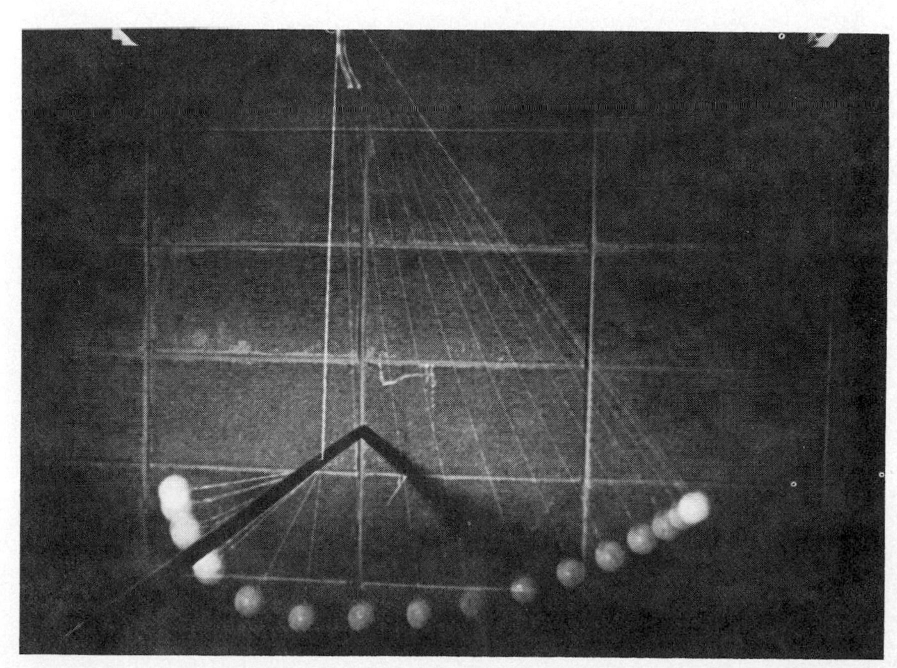

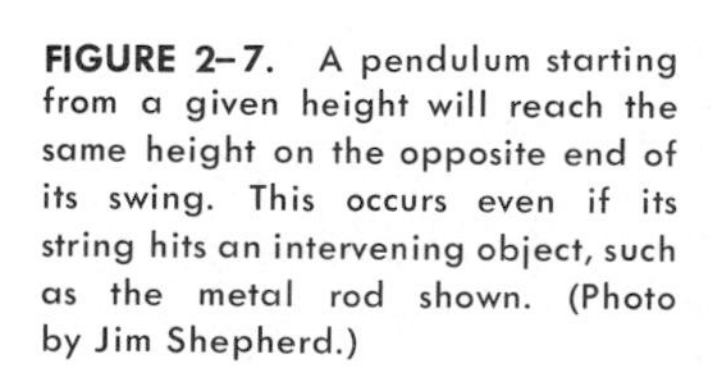

FIGURE 2–7. A pendulum starting from a given height will reach the same height on the opposite end of its swing. This occurs even if its string hits an intervening object, such as the metal rod shown. (Photo by Jim Shepherd.)

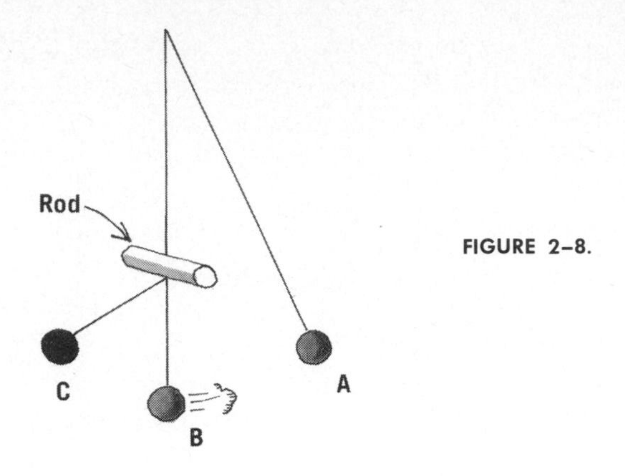

FIGURE 2–8.

The case of the pendulum is but one example of the application of the conservation of energy law. We stated previously that energy could be changed from one form to another but it never disappears (i.e., it is conserved). A more general—and useful—statement of the principle is as follows: *If no energy is transmitted into or out of a system, the total amount of energy within that system remains the same.* In the preceding example, the pendulum was the system. We could have added energy to the system by pushing hard on the pendulum, or we could have removed energy by slowing down the pendulum with our hands, but if no such energy is added to or taken from the pendulum, it will always keep the energy it had at the beginning.

Another Type of Energy

We hope that the reader has seen a difficulty with our discussion of energy. Consider what would happen if the pendulum continued to exchange kinetic and potential energy, and *never lost its energy.* It would continue forever! But pendulums do not do this. Either the principle of conservation of energy has failed in the very example with which we introduced it, or we have left out something. As you may suspect, the latter is the case.

In Chapter 1, we found it convenient to ignore, for a while, the effect of friction, and we have done so again. As a pendulum moves through the air, it experiences a frictional force. This is primarily the air friction against the swinging weight and string. Friction causes a decrease in the amount of energy, so that as a pendulum swings from an extreme position to the bottom position, the amount of kinetic energy gained is *not quite* equal to the amount of potential energy lost. A *little* more potential energy is lost than kinetic energy is gained. And as the pendulum moves up toward the other end of its swing, it does not gain back quite all of the potential energy it had initially. The energy, however, has not been *lost;* it has merely changed to a third form of energy (other than kinetic or potential).

In order to see how this third form of energy enters the picture, imagine the effect the object on the end of the string has on the molecules of the air (Fig. 2–9). A molecule just in front of the object is in a position similar to a golf ball as the golf club approaches. When the pendulum hits the molecule, it will knock it away just as a golf club knocks away the ball. Other molecules will be hit by this molecule as it flies away, and in general the billions of billions of molecules near the pendulum will gain speed from the stirring effect of the pendulum's motion. Different molecules are affected differently, but the important thing is that, on the average, the air molecules gain speed as a result of the swinging object.

Individual molecules have mass just as pendulums do, and thus a gain in speed by molecules means a gain in kinetic energy. This, then, is where the pendulum's energy goes: to the energy of the molecules. As the pendulum swings through the air, it stirs up air molecules and increases their kinetic energy. If one were able to measure the increase in energy of the molecules during one swing of the pendulum, one would find that it would be almost enough to account for the decrease in the pendulum's energy. Some of the pendulum's energy is also lost to friction at the point where the string is supported; this would account for the rest.

The law of conservation of energy has already been stated: if no energy is transmitted into or out of a system, the total amount of energy within that system remains the same. We saw that a pendulum's energy remains very nearly the same from one swing to the next, but not quite. The reason

FIGURE 2–9. The molecule is about to be hit by the pendulum.

is that the pendulum is not completely isolated from other things, and energy is transmitted from it to the air and to its support. So, rather than being an exception to the law of conservation of energy, the pendulum is actually a good example of it.

Bicycles. Anyone who has ridden a bicycle knows about the principle of conservation of energy, whether he calls it by that name or not. A bike rider who starts from the top of one hill, coasts through a valley, and on up the next hill, knows that he does not quite reach the height on the second hill that he started from on the first (Fig. 2–10). He is also aware that he coasts farther up the second hill if he leans down on his bike so that his body does not stir up as many molecules. Thus, just as occurred for the pendulum of the preceding section, it is air friction that is playing a major role in determining how far up the hill the bike will go. This streamlining effect is quite noticeable, and if you have never experienced it, we suggest that you get on a bike and start pedaling.

Not all of the energy lost by a cyclist is lost to air friction. Some is lost to friction of the wheels on the road. Such friction results in an energy increase for the molecules of the tires and the roadway, but, again, all the energy lost by the bike is accounted for in the energy gained by atoms and molecules. As we will see in a later chapter, when the atoms and molecules of an object gain kinetic energy, we perceive it as an increase in the temperature of the object. We call the energy transferred to the object "thermal energy." Thus, a bike tire becomes slightly warmer after high-speed travel because its molecules have increased in energy. (This effect is greater for car tires, of course, which can actually become hot enough to be uncomfortable to touch.)

The bicycle's energy losses due to friction are unwanted by the rider. When he applies his brakes, however, he is purposely transforming unwanted kinetic energy to another form (thermal energy of the brake system). Since he pedals hard to give that energy to his bike (and himself), the

FIGURE 2–10. Because of friction the bike rider does not reach the same height from which he started.

experienced rider hesitates to use his brakes. Their use represents energy lost—and, worst of all, it is energy he worked hard to transfer from his body to the bike. The thoughtful car driver, likewise, drives in such a way as to use his brakes as little as possible. The amount of gasoline saved can be considerable.

Energy Released to the Air. The fact that automobiles convert other types of energy to thermal energy of their tires is easy to verify: the tires heat up. But as the auto, the bicycle, and the pendulum move, they also increase the kinetic energy of the surrounding air. Therefore, the thermal energy of the air is increased, and its temperature is thus increased. Such an increase is almost impossible to detect, however. As a bike moves, molecules of air all along its path are affected. Thus, the energy is spread out over a wide area. In addition, that slightly warmer air mixes quickly with cooler air nearby, and detection of a temperature increase is unlikely. In like manner, a temperature increase of the roadway is practically undetectable because the energy is spread out so much. The tires, on the other hand, continue to receive energy as the bike rolls along, and after a while, the temperature increase becomes perceptible.

IS ENERGY CONSERVED, THEN? In the previous discussion we stated that energy is lost because of friction. It may appear that if it is *lost*, it is not *conserved*. It is true that the energy that is converted to thermal energy is of no more use to us. The bike rider cannot recover the energy he loses when he applies his brakes.[4] The energy still exists, however. If we were to measure the total amount of energy in its various forms, we would find that this total does not change: energy is conserved.

We have so far considered only three forms of energy: gravitational potential energy, kinetic energy, and thermal energy. There are many other kinds, among them chemical, electrical, light, and nuclear. When these kinds of energy are involved (such as chemical energy within the body changing to potential energy as one walks uphill), they must also be included in the calculation of total energy. But again, the rule holds: energy is conserved.

We began this chapter by speculating that a pendulum could *remember* the height from which it fell, but we moved on to give the scientist's outlook, which is based on the conservation of energy law. This may, at first, seem little improvement. We have simply defined three different types of energy and said that their sum was conserved in the pendulum system. If this conservation law were valid only for pendulums, it would indeed be hardly better than the

[4] If he gets his brakes hot enough, though, he might use them to warm his hands!

theory of remembering. But the value of the concept of conservation of energy is that it works in all known cases; it applies to birds, bananas, and planets.

This, then, is the value of the energy-view in explaining the pendulum: exactly the same ideas can be used to explain other mechanical devices, and, indeed, many other phenomena. In addition, since each type of energy is defined in a way such that it can be measured, one is able to measure energy at one instant and predict what it will be at another instant, even though it may change form. Knowing the height from which a pendulum or a bicycle starts, one can predict its kinetic energy—and therefore its speed—at some lower height. In this way, we have been able to make connections between such seemingly different events as the splitting of an atomic nucleus and the path of a comet through the heavens.

PERPETUAL MOTION

The phrase "perpetual motion" means different things to different people. To some it means simply forever-moving, but even that phrase is ambiguous—how long is forever? The earth will continue its motion around the sun for billions of years. Is that close enough to "forever"? Since a strict use of the word "forever" would put one on shaky ground in talking about any form of material object, one would probably agree that the earth's motion constitutes perpetual motion in this loose sense. And such perpetual motion is not restricted to planetary objects. We are now capable of constructing spinning devices so friction-free that they will continue in motion for thousands of years. (By the law of inertia, if friction were reduced completely to zero, such objects would continue to move "forever.")

This is not, however, the type of perpetual motion that people have spent countless years searching for. The reason is simple: such "perpetual motion" devices might make a good conversation piece but they are almost useless.[5] If one of these spinning objects were hooked up so that it provided energy for some worthwhile purpose, it would necessarily slow down in the very act of providing that energy. This happens because any force provided by the spinning object would act to slow down the object just as friction would.

The perpetual motion machine about which dreamers dream is one which would provide energy without slowing down and without needing an outside source of energy. Such a machine, if mounted in an automobile, would power

---

[5]They do find use in such instruments as gyroscopes, but this use is certainly not a significant breakthrough in the history of the human race.

the car without ever needing gas. As a matter of fact, small demonstration cars have been built which need no gas, oil, coal, or even an overnight electrical plug-in. They run on energy which comes from the sun and is converted to electrical energy by solar cells on top of the car. Although these cars may be of value, they are not examples of perpetual motion, because there is energy supplied to them—by the sun. And if you measure the solar energy absorbed, you would find it equal to the total of the various energies involved in running the car (potential, kinetic, electrical, and thermal).

No, this is not the perpetual motion that, if found, would solve a great number of the world's problems. We need a machine which would violate the law of conservation of energy! Let us investigate a couple of ingenious proposals and find out why a successful working model has never been built.[6]

Machine 1

The grand award for the most often proposed perpetual motion machine would probably go to one based upon the principle of the lever.

take-along-do-it

Place a couple of books at the edge of a table with a ruler or pencil pushed under them an inch or two and protruding over the table-edge. (See Fig. 2-11.) Push down on the ruler a couple of inches from the table and note how hard you need to push to lift the books. Then push down farther out toward the end of your lever. Quite a bit less force should be needed.

The lever principle explains why wrench handles must be long if one is to loosen very tight bolts; the farther from

[6]The US Patent Office has granted numerous patents for perpetual motion machines based upon applications with complete detailed drawings. Some years ago, though, the patent office began requiring working models of such a machine before a patent would be granted. Result: no patents granted for perpetual motion machines since that time.

FIGURE 2-11. Ruler used as lever.

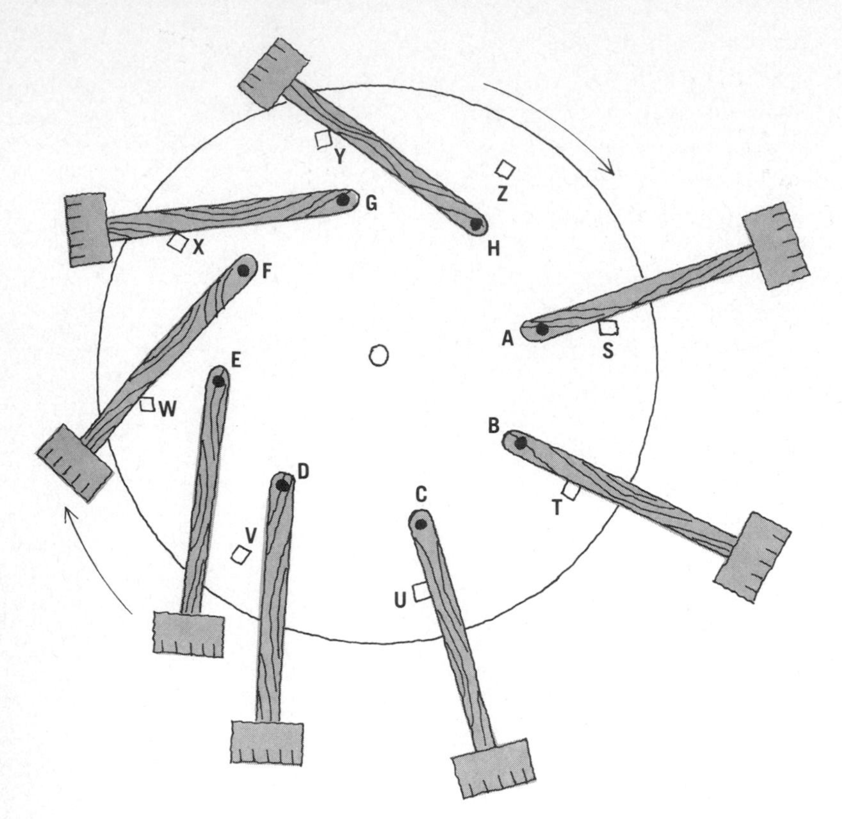

FIGURE 2–12. A perpetual-motion wheel. Note the greater distance from center of the hammer heads on the right. But also note the greater number of hammers on the left.

the point of rotation one pushes, the more turning-force (or *torque*) one achieves. An attempt to construct a perpetual motion machine which uses this principle is shown in Figure 2–12. A circular wheel has hammers hung from pegs located in a circle on the wheel (at *A*, *B*, *C*, *D* . . . *H*). The hammers can swing on these pegs but are stopped by other pegs (at *S*, *T*, *U* . . . *Z*). For example, if one begins to move the wheel clockwise as shown by the arrows, after a little turning the hammer swinging from *H* will swing across and be stopped by peg Z.

Enter the lever: notice that the heavy heads of the hammers on the right side of the wheel are farther from the wheel's center. A hammer on the right, therefore, exerts more turning-force on the wheel than a hammer on the left, whose head is nearer center. (The hammers on the left have a tendency to turn the wheel counterclockwise, while those on the right work to turn it clockwise.) The argument continues that these hammers on the right—since each produces more turning-force than its counterpart on the left—win the battle. The wheel takes off!

Before rushing off to build this contraption, let's consider it more carefully. One could argue that it can't work because it would violate the law of conservation of energy, but this misses the point. Those proposing the idea claim just that: they say it *does* violate the energy conservation principle. Others may argue that any slight gain in energy would be compensated for by a loss of energy via friction. This again misses the point. One could reduce friction to a very small value, and one could increase the energy-gain effect by using heavier hammers. No, the friction argument is not a substantial barrier in this age of technology.

The answer is really simple. Draw a line straight up and down through the center of the wheel. Count the number of hammers on each side of that line. More on the left, right? If we were to go to the trouble of doing some calculations, we could show that the extra turning-force produced by those "far-out" right-side hammers is exactly canceled by the oppositely-directed turning-force produced by the extra left-side hammers.[7] The wheel won't turn by itself!

Machine 2

The ingenuity of man never ceases, however, especially when success would mean a life of luxury. "Why not," the hopeful inventors ask, "use the idea that made machine 1 fail, to make another one work?" Figure 2–13 shows an end-

[7]It is interesting to note that some would-be inventors proposed such a wheel, predicting that it would go *counter*clockwise. They noticed the extra number of hammers on the left but failed to understand the principle of the lever at work on the right.

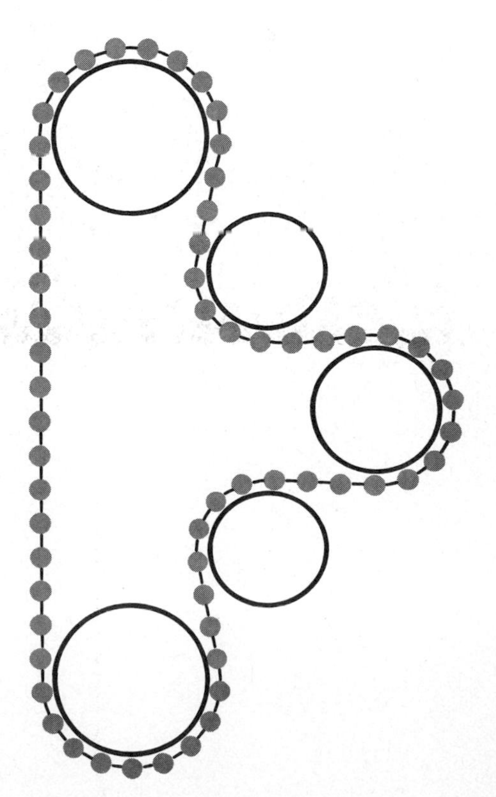

FIGURE 2–13. A second perpetual motion machine.

less chain (of the key-ring type) going around two big wheels and some small ones. Obviously, there are more links of the chain on the right side of the large wheels than on the left side. So she goes around of her own accord, right? Before building a model, one had better put a brake on it, right? Wrong and wrong again!

But where's the catch? Again, it is not friction, because we could sufficiently reduce friction to make it negligible. The catch is simply that those small wheels on the right are exerting force on the chain, too. There are more links on the right than on the left, and if we did a careful analysis we would find that the small wheels support exactly the weight of these links. Failure again—you can't beat the law of conservation of energy!

Admittedly, these examples of efforts to defeat the conservation of energy law are a little far-out and are insufficient to prove that there are no exceptions to the rule. But this law has been around for physicists to play with for a very long time, and in no case has it ever been violated. Investigations have ranged from the depths of space to the submicroscopic world, and it hasn't failed yet. You will find the law of conservation of energy to be a recurring theme throughout this book, and so you will have ample opportunity to see it in action. We picked the two preceding examples only to show some interesting directions in which people have turned in an attempt to defeat it. In case you feel you have a better plan—try it. We predict defeat.

CONSERVATION OF ENERGY ISN'T EVERYTHING

We'll go from perpetual motion to a game of billiards (or pool, depending upon where you play). If you've ever played this game, you know what to expect when a rolling ball hits a stationary one head on. The stationary ball takes off in the direction that the moving one was going, while the original ball stops. The struck ball also moves with very nearly the same speed as the original (Fig. 2–14).

The question is: why? Or we might ask: How did the second ball know how fast to go? Having studied the conservation of energy, one might draw upon knowledge of that law to explain the phenomenon. One would say that (1)

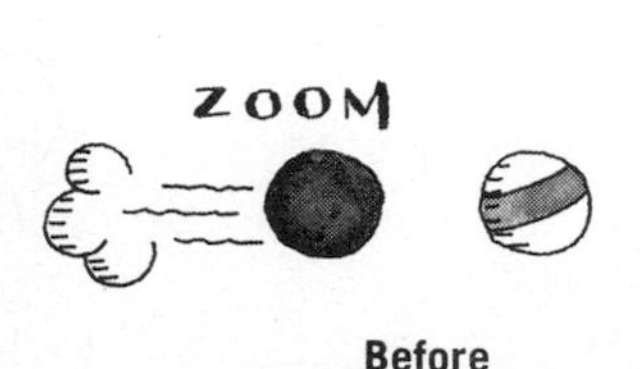

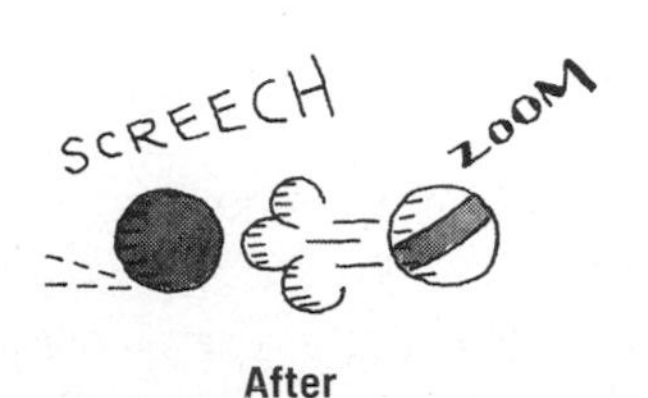

FIGURE 2–14. A head-on collision between pool balls.

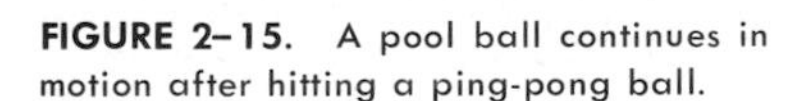

FIGURE 2–15. A pool ball continues in motion after hitting a ping-pong ball.

there was no change in the height above earth, so that no change in gravitational potential energy is involved; (2) not enough friction is involved in the collision to change kinetic energy to thermal energy; and (3) thus kinetic energy after collision is the same as before. Because the speed of the identical balls determines their kinetic energy, the speeds are therefore the same, one would say. And one would be right, as far as one went. But something else is involved.

To see this, consider the following. A ping-pong ball is inadvertently left on the pool table. Along comes a pool ball to collide head-on with the ping-pong ball. In this case, the ping-pong ball would again take off in the direction the pool ball was going, but *at a greater speed* (Fig. 2–15). In addition, the pool ball would continue on in that direction at a lesser speed. This should not be surprising, because we do not expect the little ping-pong ball to stop that massive pool ball.

Again, the law of conservation of energy can be applied, and again it would work: the total kinetic energies of the two balls after the collision would be equal to the initial kinetic energy of the pool ball (except for a slight loss caused by friction). But energy could have been conserved in any number of ways. Why didn't the pool ball stop completely and the ping-pong ball move off really fast? Such an event would conserve energy, because the law of conservation of energy requires only that their total energies be equal to the original energy; it does not specify how the energy is to be shared. If you do the experiment over and over, however, you will find that the energy is always shared in the same way—the pool ball never stops and gives all the energy to the ping-pong ball.

In order to understand the rules of this sharing of energy, we must look at the forces involved in the collision, and in fact, at the nature of forces in general.

Forces Come in Pairs

Forces never exist singly; they come in pairs. As you sit studying this book, you are exerting a force down on your chair. But the chair is also exerting a force back up on you. This simple force-pair will serve to point out the characteristics of all force-pairs: (1) the two forces are always equal, (2) they are in opposite directions, and (3) they are always exerted on different objects. These three rules are easy to apply to you and your chair, but let's consider a more complex event.

Suppose that you wish to express displeasure, in a subtle way, with an exam score. To do so, you punish your professor with a force of 10 pounds directed north on his south-facing nose. According to our law,[8] a second force arises. You exert a force of 10 pounds on the nose, but simultaneously, the nose exerts a 10-pound force on your fist (Fig. 2–16). The old adage that "it hurts me more than it does you," does not seem to apply, at least not to forces.

Does this example fit what we learned about forces in Chapter 1? The impact of the fist was great enough to alter the state of motion of the head, sending it into motion. The force of the nose on the fist also changed the speed of the fist, slowing it down. Thus, in each case, the force caused a change in speed (an acceleration) and this situation fits our two previous laws of force and motion.

The weight hanging from the end of a string in Figure 2–17*A* also illustrates the force-pair law. The book is prevented from falling because of the force which the string exerts on it. If the book weighs 10 pounds and is supported by the string, the string must exert a force upward of this amount (Fig. 2–17*B*). If the string exerts a force of 10 pounds on the book, the book must reciprocate. It exerts a force of 10 pounds on the string, and since this must be in the opposite direction, it is a downward force. Other pairs of forces exist. If the string is not moving, there cannot be a net force on it, and so something must exert a force to cancel this

---

[8]This force-pair law is called Newton's Third Law and is the last of Newton's triple-header.

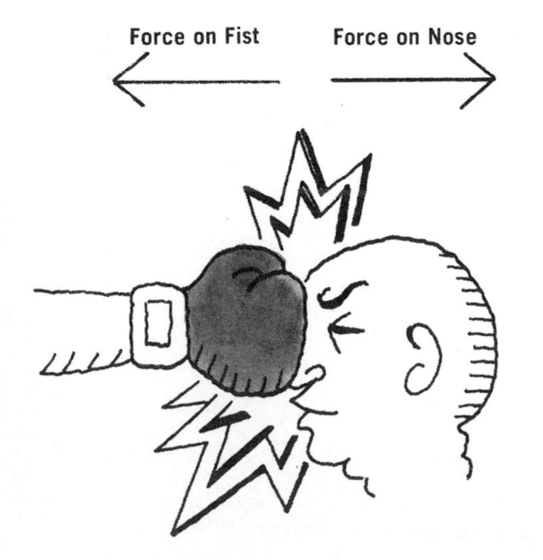

FIGURE 2–16. A force-pair. Note that the forces are equal and opposite.

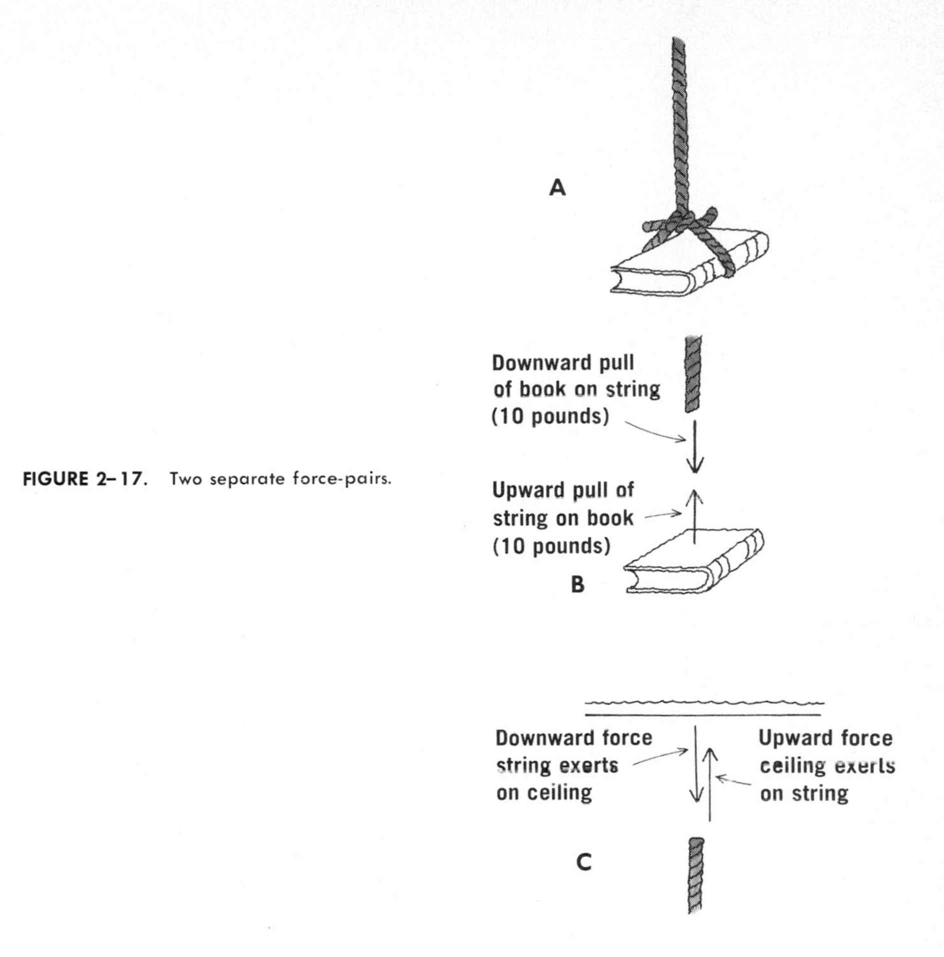

FIGURE 2–17. Two separate force-pairs.

downward 10-pound force. This something is the ceiling, which exerts an upward force of 10 pounds (Fig. 2–17*C*). The force opposite to this is the downward force which the string exerts on the ceiling.

All the forces seem to be taken care of quite well in this instance, except for one which you probably have already noticed. In case you didn't, let's emphasize its presence by cutting the string (Fig. 2–18). The book falls and accelerates, and so a net force must act on it. This force is its weight, but where is its oppositely directed counterpart? In order to answer this, you must recall that weight is defined as the force which the earth exerts on the book. The opposite force, then, is the force that the book exerts on the earth. This small force on the earth moves it in accordance with Newton's First and Second laws, but the earth's huge mass prevents a noticeable effect. We do not expect 10 pounds of net force on the earth to accelerate it much. And we're right, but it does accelerate the earth upward by a minute amount.

Let's look at another example. If you inflate a balloon and then release it, the stretched rubber of the balloon causes the air to be pushed out the opening. But when the balloon pushes on the air, the air pushes back on the balloon and causes it to take off (Fig. 2–19).

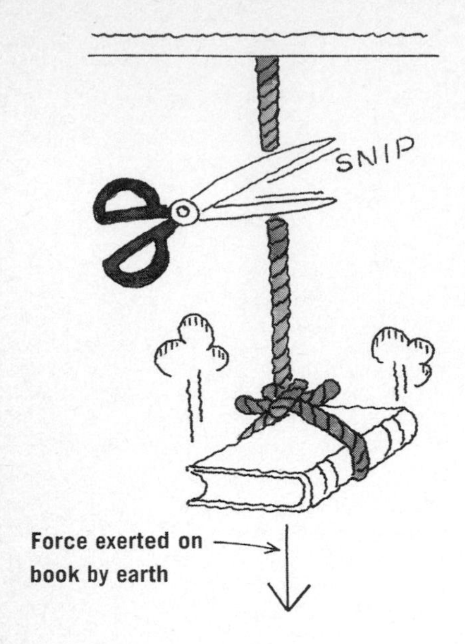

FIGURE 2–18. When the string is cut, the earth's force on the book is obvious. The other half of the pair is a force *upward* on the earth.

It was mentioned earlier that it is possible to reduce friction enough so that a spinning wheel could be made to spin almost indefinitely, but if you try to obtain energy from the spinning wheel, you slow it down. This is precisely because of the force-pair law. If you get the wheel to exert a force on something, that something necessarily exerts a force back on the wheel. And that force slows the wheel!

THE CONSERVATION OF MOMENTUM—APPLICATION ON THE BILLIARD TABLE

During the short time that a billiard ball is in contact with a struck ping-pong ball, they exert forces on each other. The force exerted on the pool ball is equal in strength and opposite in direction to the force that it exerts on the ping-pong ball—so says the force-pair law. Remembering how forces act on masses to cause acceleration, we see that because the forces are equal, there will be more acceleration of the less massive ball. The result is that the ping-pong

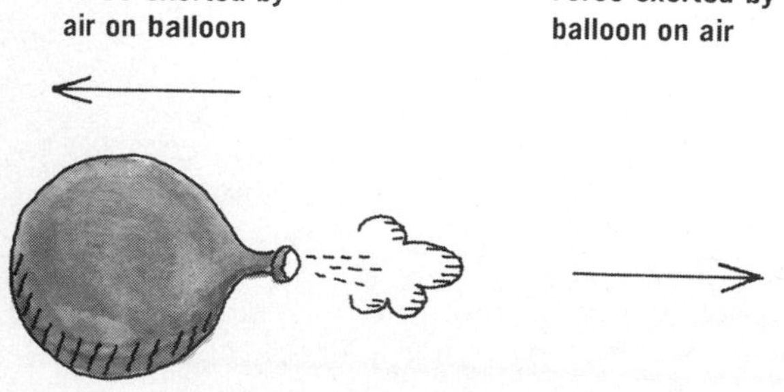

Force exerted by air on balloon

Force exerted by balloon on air

FIGURE 2–19. A force-pair is responsible for the balloon's zipping around.

ball speeds up more than the pool ball slows down. In fact, the change in speed depends upon the masses of the colliding objects. If a ping-pong ball has a mass $1/20$ of that of a pool ball, it changes its speed 20 times as much.

The results of the preceding experiment can be summarized in a formal statement of a conservation law—the conservation of momentum. Momentum is defined as the mass of an object times the velocity of that object. In equation form:

$$\text{Momentum} = \text{Mass} \times \text{Velocity}$$

From this we can see that a massive object moving slowly can have the same momentum as a less massive object moving rapidly. The conservation of momentum principle says that *the net momentum of a system before any event is equal to the net momentum of the system after the event.*

In the case of the pool table collision, the pool ball had a certain momentum that could be found by multiplying its mass times its velocity. The collision (or event) occurs such that the total of the momenta of the pool and ping-pong balls is the same as the original momentum of the pool ball (Fig. 2–20).

Thus we have two conservation laws: energy and momentum. Although the mathematical details are beyond the scope of this book, the application of both laws to any pool table collision (or any other kind of collision) determines exactly the speeds of the objects after the collision.

Examples of Momentum Conservation

Consider a man about to start a track meet by shooting a gun. As he prepares to shoot, the momentum of gun, bullet, and man is zero (because there is no motion). Bang! The bullet comes racing out of the barrel. It has gained momentum in a certain direction and this means that something else must have received momentum in the opposite direction if momentum is to be conserved (Fig. 2–21). It is the gun (and the shooter's arm) which gains this momentum

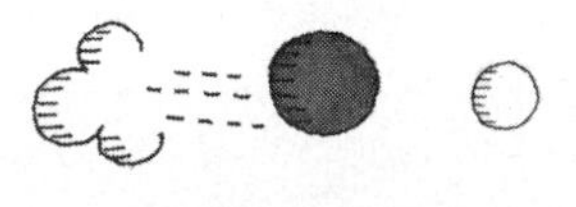

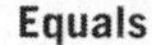

FIGURE 2–20. The pool ball vs. ping-pong ball event from a conservation of momentum viewpoint.

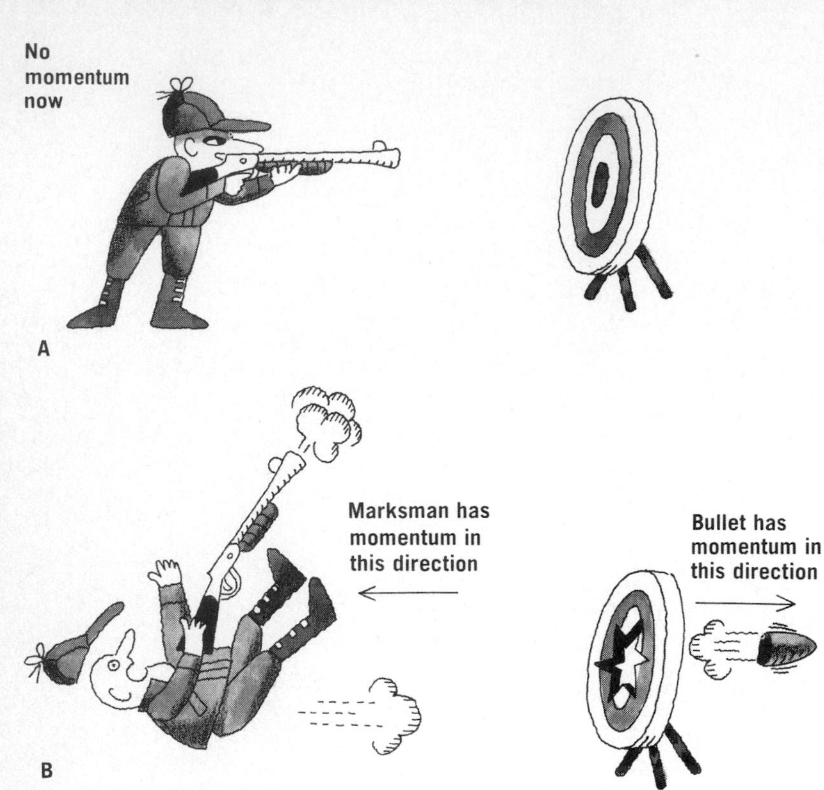

FIGURE 2–21. Rather than shooting at the track stars, the starter practices conservation of momentum by shooting at a target.

as the gun recoils.[9] A Volkswagen driver hears the shot and is so distracted that he fails to see another VW stopped at a traffic light (Fig. 2–22). Suppose that when the inevitable crash occurs, the two cars stick together. We now apply the law of conservation of momentum to predict the speed of the mass after collision. Momentum is mass times velocity, so the initial momentum is

$$1 \text{ VW Mass} \times 30 \text{ mph}$$

In part *B* after the collision, the mass is 2 VW masses. If the momentum is the same, then, the speed afterward must be 15 miles per hour because

$$2 \text{ VW mass} \times 15 \text{ mph} = 1 \text{ VW mass} \times 30 \text{ mph}$$

After the crash, there is twice as much mass moving but at half the speed.

While all this is happening, the track stars are taking off. When they do so, they gain momentum toward the finish line. What has momentum the other way? The whole earth does! This may be hard to believe, but in every measurable case, the law of conservation of momentum applies. Do we

[9]Note the pair of forces involved here. One force pushed the bullet. An equal force pushed back the gun.

expect it to suddenly fail simply because its effect cannot be measured by man? No. Because of the earth's large mass, the law predicts a very, very slow motion given to it—so slow that we could never expect to measure it. But we believe that nature is consistent, so that when a principle applies in every measurable case, it will apply in the immeasurable ones, too.

A Man With a Problem. Just how useful the law of conservation of momentum is will become apparent to you when you are in the position similar to that in which Mr. Abraham Maharba found himself on February 29, 1969. On that day, Mr. Maharba found it necessary to carry a large can of oil across a frozen lake in Saudi Arabia. The sun was shining and the ice had just begun to melt, leaving a very slippery glaze on top. As you might suspect, the poor man fell and spilled the entire five-gallon bucket of oil all around him. After checking his bruises, Mr. Maharba carefully got up and tried to walk off the lake, but the ice and oil had become so slippery that there was absolutely no friction between the ice and his shoes (Fig. 2–23).

Normally when we start walking, we move our legs so as to push backward on the ground, and because there is friction between our shoes and the ground, the ground exerts a force forward on our shoe (the other half of the force pair). But Mr. Maharba had no friction, so he was unable to push backward on the ice and allow it to push forward on him.

Not having studied physics (which naturally would have provided him with an answer), the luckless lake walker sat there pondering his dilemma. Soon, a dog on shore noticed

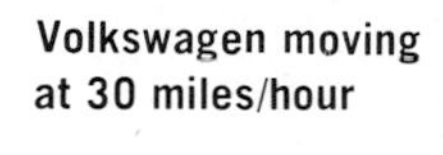

Volkswagen stopped
at traffic light
(no momentum)

A

Net Momentum = (1 VW mass) $\times$ $30\left(\frac{\text{miles}}{\text{hour}}\right)$

Two Volkswagens at 15 miles/hour

B

Net momentum after collision = (2 VW masses) $\times$ $15\left(\frac{\text{miles}}{\text{hour}}\right)$

FIGURE 2–22. A not-so-foreign example of momentum conservation.

FIGURE 2–23. How can the man get off the frictionless ice?

him and started to bark—laughing, no doubt. This was too much for Mr. Maharba's pride, and after cursing the dog a few times, he began to look for something to throw at the cur. He couldn't reach the oil can, so he took off a shoe and hurled it as hard as he could at the dog.

Although he missed the dog, his throwing act saved the day, for in throwing the shoe he exerted a force on it and gave it momentum toward the dog. Thus, the shoe exerted a force on him, giving him an equal amount of momentum in the opposite direction! The shoe's mass was small, but in his anger, Maharba gave it a great deal of speed. Thus, though his mass was large, he was given enough speed to get off the lake before he froze or starved. (Once he started moving, of course, he continued across the frictionless ice until he reached shore—across the lake from the dog.)

Glancing Collisions—Vectors Again

In the collision examples given previously, the collisions were always head-on. A pool ball arrived from a certain direction, and after colliding with another one, both balls went off in that same direction. However, anyone who has ever played pool knows that glancing collisions are much more common occurrences (Fig. 2–24). In all of our collisions, we have said that the conservation of momentum law is obeyed: the momentum before the collision is equal to the momentum after it. This law is still valid in glancing collisions, although perhaps not as obviously so.

We have seen that forces and velocities are vector quantities, and so is momentum. In order to find the net momen-

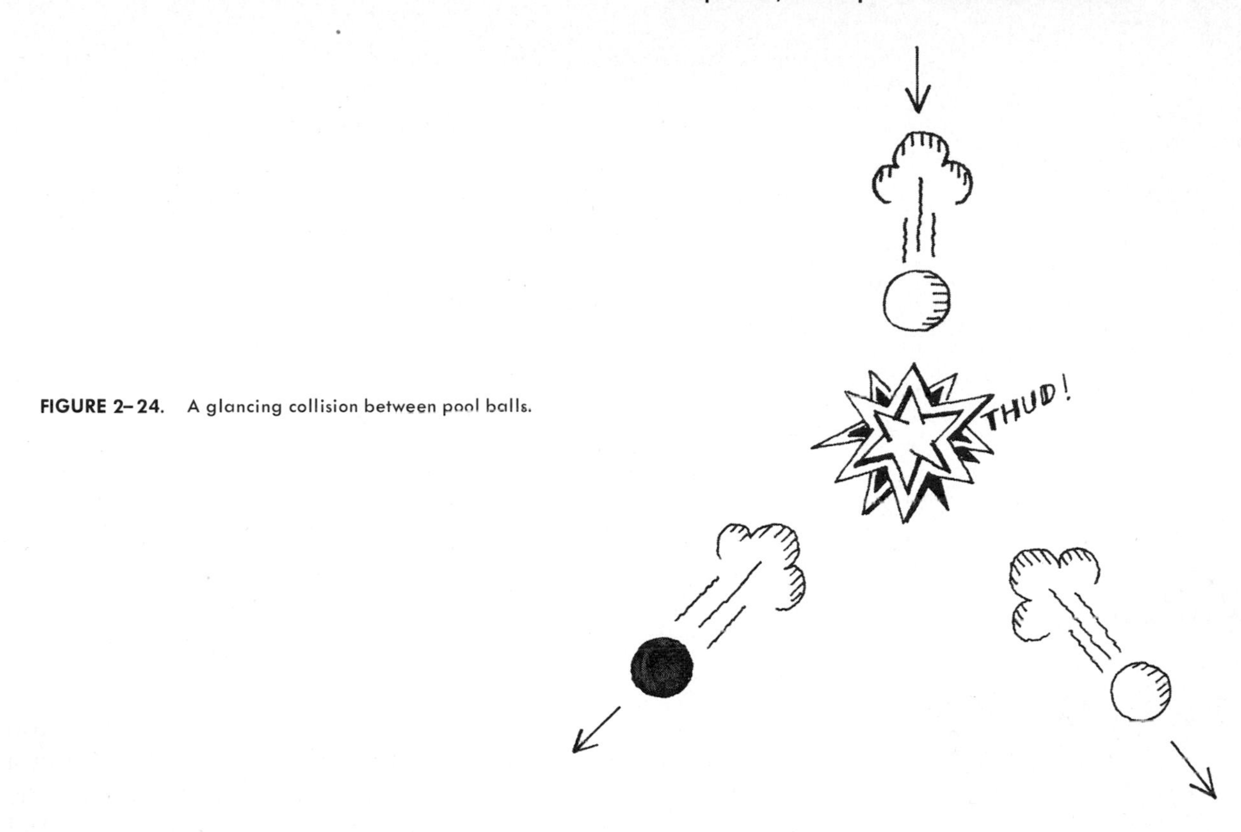

FIGURE 2–24. A glancing collision between pool balls.

tum of the balls after a collision, we must add the momenta of the balls in the manner that we always use to add vector quantities. In Figure 2–25A we see an oncoming ball whose momentum is indicated by a vector. After the collision, the two balls travel along the path shown in *B*. Also shown are the vectors which represent the momentum of each of the balls. We proceed now as we did when adding forces together (Fig. 2–25*C*). Again, the net momentum after the collision is equal to the net momentum before; the conservation of momentum law holds in *all* collisions.

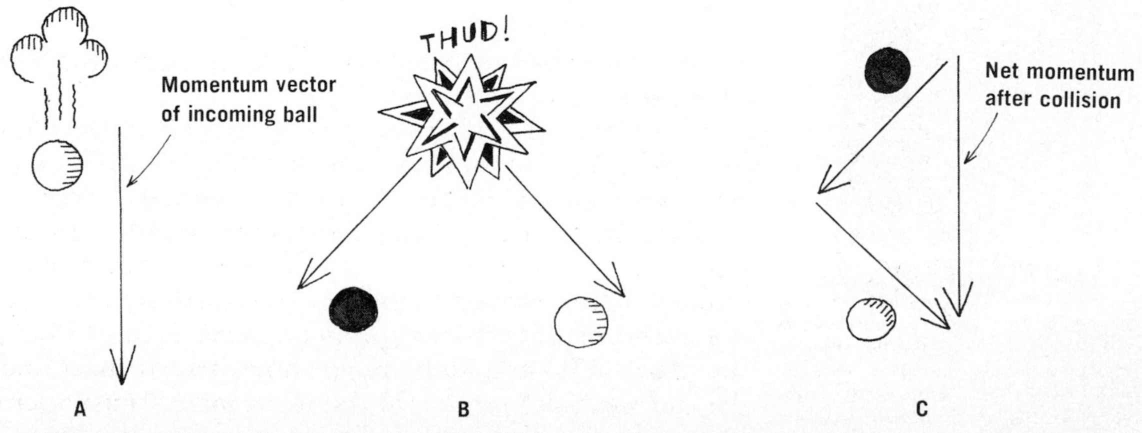

FIGURE 2–25. Note that the momentum after collision (C) is equal to the momentum of the white ball before collision (A).

ANGULAR MOMENTUM

Thus far we have applied the law of conservation of momentum only to objects moving in a straight line. We have not considered spinning objects, to which a principle similar to momentum conservation applies, namely, the conservation of angular momentum.

take-along-do-it

Tie a pencil to the end of a string and whirl it in a circle above your head. Stop moving your hand suddenly and note how the object gradually slows down until it finally hits you and stops. (This, of course, occurs because of friction and possible small motions of your hand which slow it down.)

Whirl the object again at a fairly great speed, but this time, when you stop whirling it, let the string wind up on your finger. (See Figure 2–27.) How does the rotational speed of the object change as the string winds up on your finger?

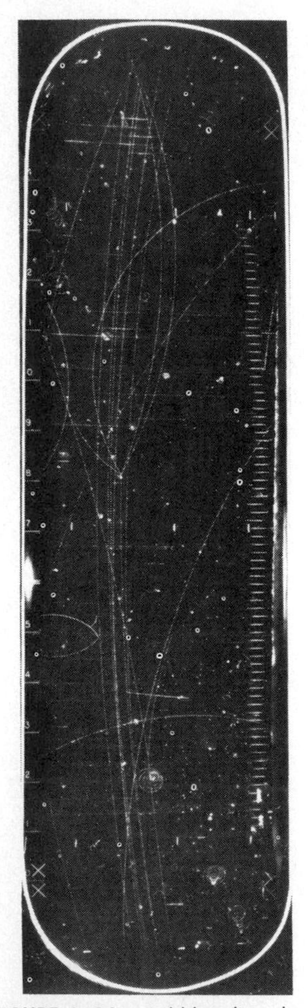

FIGURE 2–26. Bubbles show the path of submicroscopic particles passing through a bubble chamber. In each collision, momentum is conserved. (Courtesy of University of California Lawrence Berkeley Laboratory.)

We hope you noticed that the object circled your finger more quickly as the string shortened. For example, you might estimate that when the object whirled with the string at full length, it took 1/2 second to circle around, but then as the string wound around your finger, it took less and less time until it got around in 1/4 second.

This behavior of the pencil can be explained in terms of momentum conservation, but in this case, the momentum is *angular* momentum. The calculation of angular momentum can become quite a complex problem if the object that rotates has an unusual shape, but for the cases that we will consider, only a slight modification needs to be made to our expression for regular linear momentum considered previously. We must include a distance in our equation as:

$$\text{Angular momentum} = \text{Mass} \times \text{Radius} \times \text{Velocity}$$

These quantities—mass, radius, and velocity—are indicated in Figure 2–28.

Imagine yourself riding on the pencil in the preceding example. In addition, assume that the pencil is equipped with a speedometer (as are all good pencils). When swinging in a large circle, the speedometer would register a slow speed. As the string winds up on your finger, however, the radius of the circular path becomes smaller, and, therefore, because angular momentum is conserved, the velocity must increase. (The equation above shows that if mass and angular momentum remain the same, velocity must increase as the radius decreases.)

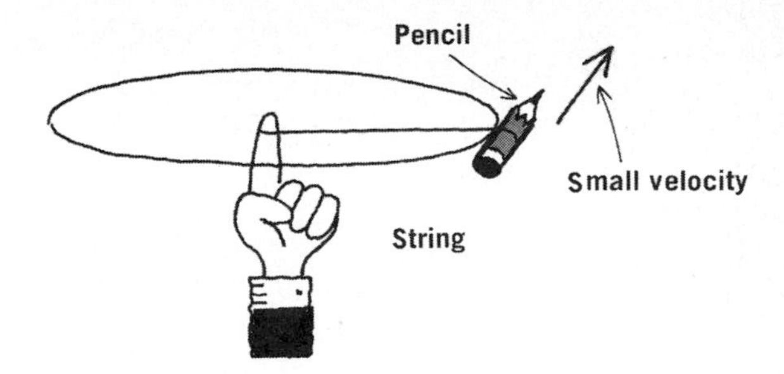

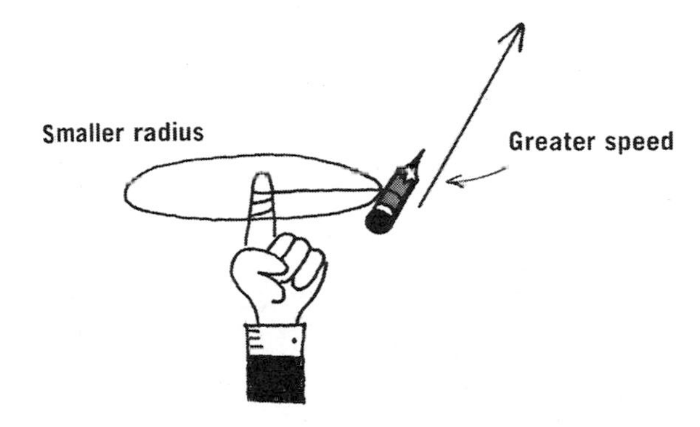

FIGURE 2–27. Note the change in circling time as the string winds around your finger.

Angular Momentum in Action

Suppose a skater starts whirling on the ice with both arms and perhaps one leg extended (Fig. 2–29*A*). Her limbs are moving in a circular path. She now pulls her limbs in toward her body, the center of the circle. This decrease in the radius of the circular path causes her arms and leg to go faster. As they go faster, the rest of her body must follow, and her spinning increases until finally the circumstances of Figure 2–29*C* prevail. If the skater wishes to increase the effect, she can hold weights in her hands. If she does, these whirling weights will help pull the rest of her body around as they increase their rotational speed. If you skate, try this.

If you can find a rotating chair (the type often used as an office desk chair) this TADI is well worth doing. Hold a couple of books in each hand. Sit on the chair, extend your arms, and let someone start spinning you slowly. Pull the books in toward your chest and prepare for a surprising sensation. If you are athletic enough to start with your legs extended, the effect is even greater.

CONCLUSION

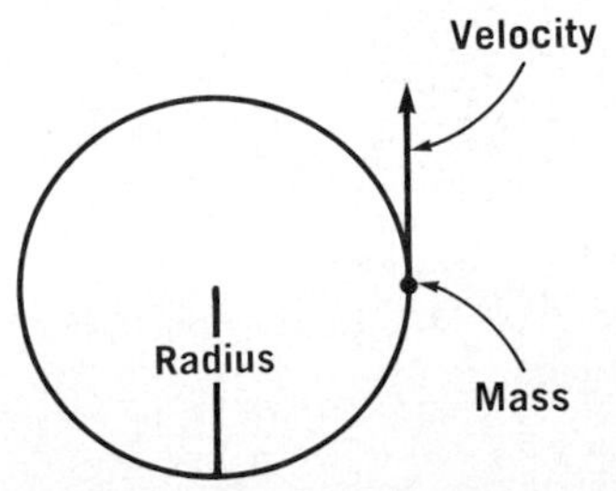

FIGURE 2–28. The three factors involved in angular momentum conservation.

We have covered, in the last two chapters, a small part of the subject which physicists call *mechanics*. Mechanics

FIGURE 2–29. A skater uses the law of conservation of momentum.

(along with astronomy) was the area of study where physics historically started, and many of the ideas in mechanics are useful throughout all of physical science. We will use them, particularly the conservation of energy, throughout this text.

Perhaps in this chapter some of the power of the two conservation laws was revealed: they are universally applicable. Nature obeys them in every instance.

OBJECTIVES

The student who understands the material of Chapter 2 will be able to:

1. Describe qualitatively the relationships between the heights of a pendulum at each end of its swing and its speed at the lowest point of its swing.
2. Relate the concept of conservation of energy to the motion of a pendulum, naming and explaining the two types of energy involved and describing qualitatively how the energy is distributed between the two at various points in the swing.
3. Define work and energy and describe their interdependence by the use of an example.
4. State and explain the law of conservation of energy.
5. Give three examples of the conservation of energy using the terms kinetic energy, potential energy, and thermal energy.
6. Describe thermal energy from a molecular viewpoint.
7. Use the conservation of energy law to explain the following:
   (a) Why a cyclist can coast farther by leaning forward on his bike.
   (b) Why the tires of a car heat up during travel.
   (c) Why a perpetual motion machine is impossible.
8. Explain the value of the scientist's energy-conservation approach in explaining nature.
9. Distinguish between an energy-producing perpetual motion machine and each of the following: a motor run by solar cells; electricity from hydroelectric plants; and a frictionless ever-spinning wheel.
10. Perform a demonstration to show the principle of the lever, describe (with a drawing) an attempted perpetual motion machine based upon this principle and then use the principle to explain why the machine will not work.
11. Explain, by the use of an example, why use of the law of conservation of energy is not sufficient to predict the outcome of a collision.
12. Describe and give at least five examples of the force-pair law (Newton's Third Law). In each example, indicate the object on which each force is exerted.

13. Define momentum and state the law of conservation of momentum, giving one example involving an explosion and one involving a collision.
14. Explain why the following two events are not violations of the conservation of momentum principle: a ball being released to fall freely, and a car accelerating from rest.
15. Define angular momentum, and perform a demonstration illustrating the law of conservation of angular momentum.
16. Use the concept of angular momentum to explain how a whirling skater can increase her speed of rotation.

QUESTIONS — CHAPTER 2

1. A snowball is thrown against a wall with a velocity great enough that the ball melts on impact. Discuss the energy changes which occur in this collision. Is the conservation of momentum principle violated?

2. If the forward momentum of a bullet is the same as the backward momentum of the gun, why isn't it as dangerous to be hit by the gun as by the bullet?

3. Why does a heavy gun kick less than a light one?

4. Discuss the energy changes that occur in a cycle in which water flows over a waterfall, evaporates, and returns to the head of the falls.

5. Use the conservation of angular momentum law to devise a hypothesis that explains why a cat always lands on its feet regardless of the position from which it is dropped.

6. The statement has often been made that rockets cannot work in outer space because there is no air present for their exhaust to push against. Why is this false?

7. As the moon circles the earth, it causes tides in the oceans through which the continents move as the earth turns. Thus there is friction slowing down the earth's rotation. Explain why this does not violate the law of conservation of energy.

8. Suppose you are standing on the sixth floor of a large hotel waiting for one of the ten elevators to take you *down* to street level. Every elevator, however, seems to be going *up*. State a conservation-type law for elevator-direction which allows you to retain hope of finding a "down" elevator.

9. Discuss the energy changes which occur as a ball is thrown straight up and then caught as it falls. Does the ball have the same speed when caught as it did immediately after being thrown? Consider the effect of thermal energy also.

10. Is the law of conservation of momentum violated during the action of Question 9? Defend your answer.

11. Discuss the use of battery-driven automobiles to conserve energy.

12. Rapid transit systems are being developed in which application of the brakes on the vehicle results in increasing the speed of flywheels mounted beneath the floor. These spinning flywheels are later used to accelerate the vehicle. Discuss this idea from an energy conservation point of view. Is an engine needed at all?

13. Identiy the force pairs in the following situations: a man takes a step; a snowball hits a girl in the back; a baseball player catches a batted ball; a gust of wind strikes a window.

14. Early in this century Dr. Robert Goddard proposed sending a rocket to the moon. Critics took the position that in a vacuum, such as exists between earth and moon, the gases emitted by the rocket would have nothing to push against to propel the rocket. According to *Scientific American* (Jan., 1975), Goddard placed a gun in a vacuum and fired a blank cartridge from it. (A blank cartridge fires only the hot gases of the burning gunpowder.) What happened when the gun was fired?

Reprinted with the permission of Sidney Harris.

3

GRAVITY AND WEIGHTLESSNESS —WHAT'S UP?

It is said that Isaac Newton was sitting under an apple tree and was inspired to formulate the law of gravity when an apple hit him on the head. Whether or not the idea hit him in just this way, it soon became obvious to him that not only did the apple fall to meet the earth but also that the earth, in its lumbering fashion, rose to meet the apple (Fig. 3–1). Indeed, all of the objects in the heavens respond to the principle of gravitation. The moon, the earth—every object in the universe—feels the gravitational tug of every other object. And, apparently, the great forces involved between the "everyday" heavenly bodies are not nearly the greatest gravitational forces in the universe. It is believed that there are objects in space that attract things so strongly that not even light can escape from them. (They're called black holes.)

On the other hand, we see astronauts tumbling around in space capsules, apparently without weight at all. In order to fully appreciate these phenomena, we will have to take a backwards look into history to see how our ideas concerning gravitation were born.

THE WEIGHT OF PEOPLE BEFORE 1600 AD

The story of the beginning of our modern view of weight and gravitation must start with a look at the medieval system of physical thought, which prevailed until about the

FIGURE 3–1. The earth rises to meet a falling apple—but not much.

seventeenth century. We learn of this system primarily from the works of the Greek philosopher Aristotle, who lived in the fourth century BC. Nearly 2000 years after his death, Aristotle's ideas of nature were incorporated into Christian theology, and they formed the basis for man's outlook on the physical universe until the seventeenth century.

Aristotle emphasized the concept of natural motion. He held that the fall of a heavy object toward the earth was its natural motion because "down" was the natural place for such objects. Since heavier objects sought "down" more than did lighter objects, it was believed that heavier objects would fall faster than lighter objects (Fig. 3–2). Let us quote Aristotle:

> A given weight moves a given distance in a given time; a weight which is heavier moves the same distance in less time. . . . For instance, if one weight is twice another it will take half as long over a given distance.

FIGURE 3–2. Aristotle held that a heavy object falls faster than a light one.

Drop a quarter and a piece of paper at the same time. The results will probably not contradict Aristotle to any great degree. Now try another experiment: Drop a quarter and a penny at the same time. Aristotle's theory probably did not fare so well in this experiment. A quarter weighs about twice as much as a penny, and Aristotle's theory predicts that it will reach the floor twice as fast. Do you distinguish any difference between their times of fall?

It is from conflicts such as the above that science advances. Clearly, experimental observation contradicts what is predicted from the theory. At times like this, one must ask whether the theory is being applied correctly, whether the measurements taken are accurate and are being interpreted correctly, and whether every experimental detail is being considered. If the conflict cannot be resolved by such questions, there is no alternative but to alter or discard the theory. A theory cannot be allowed to stand if it violates the result of valid observations.

So Aristotle was in trouble, and although he was not

Aristotle (384-322 BC)

Aristotle was the son of the king of Macedonia's personal physician. When he was about 18, he entered the Academy of Plato, where he studied and worked until Plato died, 20 years later. In about 342 BC, Aristotle became the tutor of young Alexander the Great. Around 334 BC, Aristotle founded his own school, but after Alexander died (323 BC) he was charged with "impiety" and was forced to flee for his life. Because of his writings in logic, ethics, politics, science, and metaphysics, Aristotle has been one of the most influential thinkers in western civilization.

FIGURE 3–3. Reprinted with the permission of Sidney Harris.

there to defend himself against the scientists of the sixteenth century who questioned his ideas, there were many influential people of the time who refused to accept the results of experiments. They simply could not allow their long-standing ideas about the physical universe to be challenged. That Aristotle's ideas were so bound up with the theology of the times further complicated the problem of changing people's views.

ISAAC NEWTON'S WEIGHT

Isaac Newton (1642–1727), in his book *Philosophiae Naturalis Principia Mathematica* (usually called simply *The Principia*), presented the theory of gravitation which is in general use today. This theory explained why falling objects, such as your quarter and penny, behave as they do, but it also explained phenomena far greater in scope. We will not try to describe Newton's reasoning in arriving at the theory, but will simply present it and look at what it means.

According to Newton's Law of Gravitation, *every object in the universe has an attraction for every other object in the universe with a force which we call gravitation.* One must ask if this statement fits one's experience. Hold a brick in each hand. If you can feel the bricks being attracted to one another, you have a rich imagination (Fig. 3–4). There is apparent conflict between theory and experience here, but this arises only because we have not yet stated how large a force we should expect. Theory states that the amount of force depends upon the masses of the two objects involved. In the case of two bricks about a yard apart, the force is predicted to be about 4×10^{-10} pounds—hardly measurable.

For an explanation of powers of ten notation, see Appendix I.

To obtain realistic forces, large masses are needed. The largest mass is that of the earth. It certainly produces a meas-

FIGURE 3–4. Can you feel the gravitational force between two bricks?

urable effect. Its large mass (about 6×10^{24} kilograms) attracts your body (of perhaps 75 kilograms) with a force of about 160 pounds. If the earth were the same size but had only half as much mass, you would weigh half as much (80 pounds). You'd not be skinnier; you would simply weigh less.

One other factor enters the theory: distance. As two masses move apart, the gravitational force between them becomes less, as you would probably expect. In fact, if the masses were twice as far apart, the force would be only 1/4 as great. Right now, you are about 4000 miles from the center of the earth. If you go out into space 4000 miles, you would be 8000 miles from center, and your attraction toward earth—your weight—would be 40 pounds instead of 160. Go out to 12,000 miles from earth's center (three times as far as now), and you will weigh 1/9 as much—about 18 pounds.[1] (See Figure 3–5.)

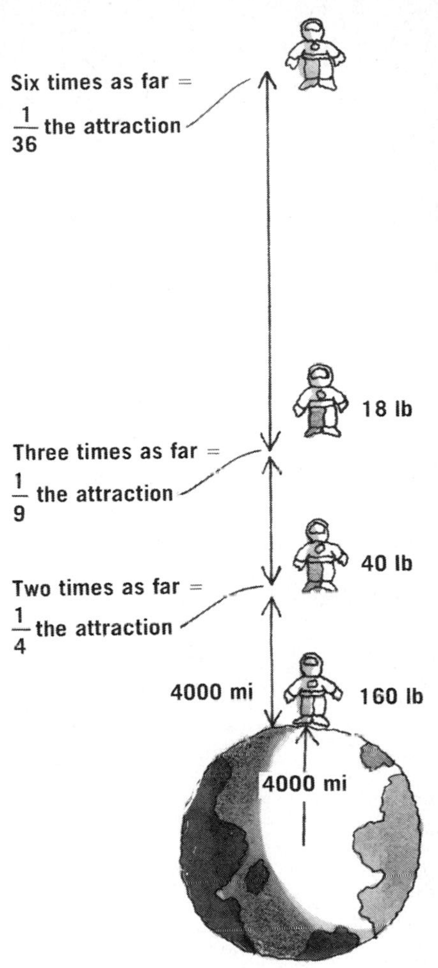

FIGURE 3–5. The weight of an object decreases as the square of the distance from the center of earth.

Astronauts in Orbit Are Not Weightless

Note the pattern in the preceding examples: two times as far—1/4 the force; three times as far—1/9 the force. In each case the force decreases as the square of the distance. This means that no matter how far you are from earth, a force is still there. At 24,000 miles (six times as far as now) you would weigh 1/36 as much—about four pounds. The further out you go, the smaller the gravitational force becomes, but a small remnant is always there. What about astronauts in orbit around the earth? Diagrams of these orbits are usually similar to Figure 3–6. The orbit appears to be far from the surface of the earth. But read the article about the orbiting spaceship, and you'll see that the object is about 100 to 200 miles above the surface, or 4100 to 4200 miles from center. This is hardly different from 4000 miles. If an astronaut who weighs 160 pounds on earth were in orbit 200 miles above us, the earth would still exert a gravitational force of about 145 pounds on him. This certainly is not weightlessness.

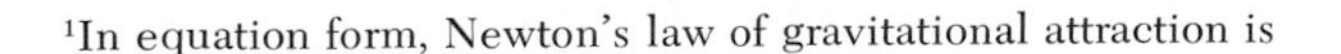

[1]In equation form, Newton's law of gravitational attraction is

$$F = G\frac{m_1 m_2}{d^2}$$

where G is a constant, m_1 and m_2 are the masses of the two attracting objects, and d is the distance between their centers. If the metric system of units is used, the constant G has the value

$$G = 6.67 \times 10^{-11} \frac{\text{newton} \cdot \text{m}^2}{\text{kilogram}^2}$$

In the metric system, forces are measured in newtons (1 newton = 0.225 pound). A stick of butter weighs approximately 1 newton.

FIGURE 3–6. The distance at which most spacecrafts orbit is usually greatly exaggerated in popular writing.

Television pictures from space vehicles, however, show astronauts turning flips in mid-air and generally acting weightless (Fig. 3–7). To see how this is possible, imagine yourself in an elevator 50 floors above the ground when the cable breaks. Imagine, also, that just as the cable breaks you drop two sacks of groceries. They will fall side by side, obeying the law of gravity. But remember, you are falling, too, and so is the elevator. (We must assume that air friction does not prevent the elevator from falling freely.) If quarters fall like pennies, as demonstrated in the last TADI, so should people and groceries and elevators. Thus, when you release the groceries, they will stay in the air in front of you; they will fall right along with you (Fig. 3–8). In addition, one other feature should be noted in your fall. This can best be explained with reference to a common experience. If you have ever descended in an elevator from the top of a tall structure (such as the Washington Monument), you probably have noticed that as the elevator begins its descent, the floor seems to drop out from under you. In other words, the elevator floor is not pressing upward against your feet very hard. The elevator floor would take the opposite point of view: it would say that you are not pressing down against it very hard. In the extreme case of an elevator falling completely freely, you would find the elevator floor not pressing against your feet at all. It would be falling at the same rate as you are.

Some people have the idea that a person would hit his head on the ceiling of the elevator when the cable breaks.

FIGURE 3–7. Edwin E. Aldrin, Apollo 11 lunar module pilot and the second man to walk on the moon, is shown getting in some weightlessness time on a KC-135 airplane. (Photo courtesy of NASA.)

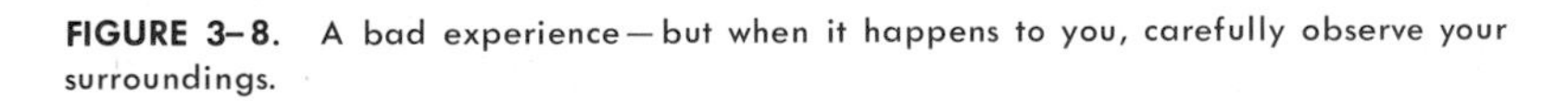

FIGURE 3–8. A bad experience—but when it happens to you, carefully observe your surroundings.

To illustrate that this is not true, put a coin in a bottle and drop the bottle. The coin stays on the bottom and falls with the bottle. This is what one would expect from our initial penny-quarter experiment. The coin and the bottle fall together, regardless of whether the coin is inside or outside the bottle.

You are now in an elevator, feeling no floor under you and with two bags of groceries "floating" in front of you.[2] If you were standing on a bathroom scale in the elevator, it would read zero (Fig. 3–9). You fall at the same rate that it does, so you do not push against it.

Are you then weightless? You certainly would feel as if you were, and the groceries would certainly appear to be weightless. But it is an *apparent* weightlessness. If weight is defined as the force of attraction that the earth exerts on your body, you certainly have weight, or you would not be falling. Astronauts are in the same situation as you in the elevator, as we see in the next section.

Newton Knew about Sputnik

When the USSR orbited the first artificial satellite in October, 1958, the triumph was primarily one of technology rather than of gravitational science. The basic gravitational theory of orbiting satellites was formulated by Isaac Newton 300 years earlier. In his writings he presented the following argument.

[2]They are not floating, of course. The word is used for lack of a more appropriate one. The sacks simply remain in front of you.

FIGURE 3–9. Why does the fellow "weigh" nothing?

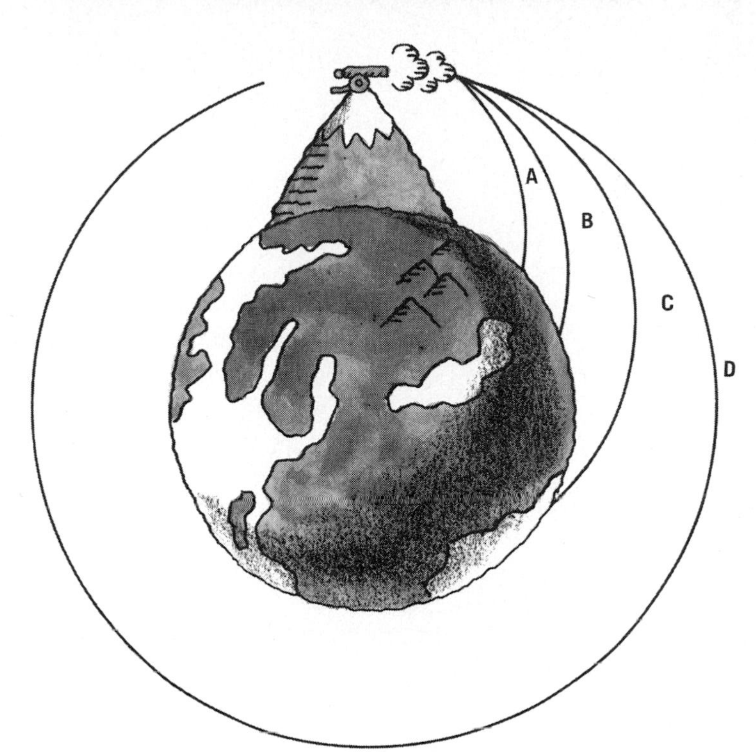

FIGURE 3–10. Newton drew a figure similar to this to illustrate how satellites orbit.

If a mountain could be found which extended above the earth's atmosphere, and a cannonball were fired from the summit, its path would be somewhat as shown in Figure 3–10, line A. Use a greater charge to shoot the cannonball more forcefully, and you get a path as in B; shoot harder yet and get C. With careful adjustment (and a powerful cannon) you should be able to shoot around the earth and hit yourself in the back of the head (line D). The cannonball will then go around again in the same path; it will be in orbit.

Even though there seems to be no obvious relationship, the situation described here is similar to that of the falling elevator. After the ball leaves the cannon, it falls toward the earth, and if you tried to ride the cannonball, you would be falling also, just as in the case of the elevator. Even in the extreme case of path D of Figure 3–10, you would be falling. The only reason you never hit the earth is because the curvature of the earth is such that it curves away from you by exactly the same amount that you are curving to meet it. And just as objects apparently floated in the falling elevator, they would do the same thing on the cannonball or in a space capsule. This effect is referred to as weightlessness, but it should actually be called "apparent weightlessness."

Since Newton's time we have not found that 100-mile high mountain; we had to devise large rockets to make Newton's idea a reality. And even though we mistakenly say that astronauts are "weightless," Newton would have understood.

BEYOND ORBIT

Thus far we have considered situations in which our satellite moved in perfectly circular orbits. In reality, orbits are elliptical.[3] This is the shape that the opening in a basketball hoop appears to have when viewed from the free throw line (Fig. 3–11). Thus the satellite comes closer to earth at one point in its orbit and then swings out away from earth again (Fig. 3–12). The explanation of the apparent weightlessness is exactly the same, however. The satellite is still always "falling." This is also the case on a trip to the moon.

take-along-do-it

Hold a penny and a quarter in one hand and throw them into the air. Throw carefully so that they leave your hand at the same time. Note whether or not they stay together all the way to the top of their flight and all the way down.

[3]In 1684 Edmund Halley (the astronomer for whom Halley's Comet is named) visited Newton and mentioned that a number of scientists were trying to calculate the path that an object would take due to a force (such as gravity) which decreases in proportion to the square of the distance. Newton immediately replied that he had calculated years earlier that it would be an elliptical orbit. He was, however, unable to find the paper on which he had derived this result. (He later reproduced the work, though.) It is curious that while the rest of the world was looking for the answer, the great Newton had already found it and lost it again.

FIGURE 3–11. The hoop appears to be an ellipse to the shooter.

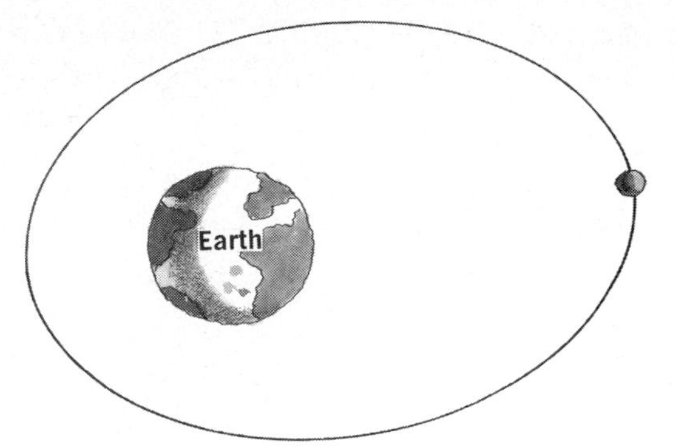

FIGURE 3–12. The elliptical orbit of a satellite.

An astronaut on the way to the moon is in somewhat the same situation as the coins above. After the rockets are deactivated, the space capsule and everything in it are coasting upward just as the penny and quarter. You presumed you were weightless in the previous example of the coasting elevator because you, the groceries, and the elevator all fell at exactly the same rate. The astronauts in the spaceship consider themselves weightless for the same reason. They are coasting right along with everything else in their spaceship. The only difference is that they move up instead of down as you did in the elevator. The astronauts are thus apparently weightless during that coasting period (Fig. 3–13). This has been demonstrated many times in television transmissions sent back showing objects such as flashlights floating by the lens of the camera.

Interesting Effects of Weightlessness

One of the early fears concerning man in space was that prolonged periods of weightlessness would result in adverse physiological and/or psychological effects. This has proven not to be a major problem on flights up to a few months in length. The astronaut finds, however, that many muscles used almost constantly on earth are used little in weightlessness. Thus, he must exercise to keep fit for return to earth. The exercises must, of course, be different from those we might practice on earth. Because he feels no force against the floor of the capsule, the astronaut must make use of springs and other mechanical devices to exercise. We do push-ups against gravity; he uses springs.

It is interesting to consider how some of our common games would be played in conditions of apparent weightlessness. Dice thrown to the floor would take one bounce and then continue in a straight line toward a wall, where another bounce could send them along another path (Fig. 3–14). Perhaps the players would simply read the dice each time they went by.

Flying paper airplanes proved to be fun on the first Skylab flight. To get a plane to glide was no problem; a

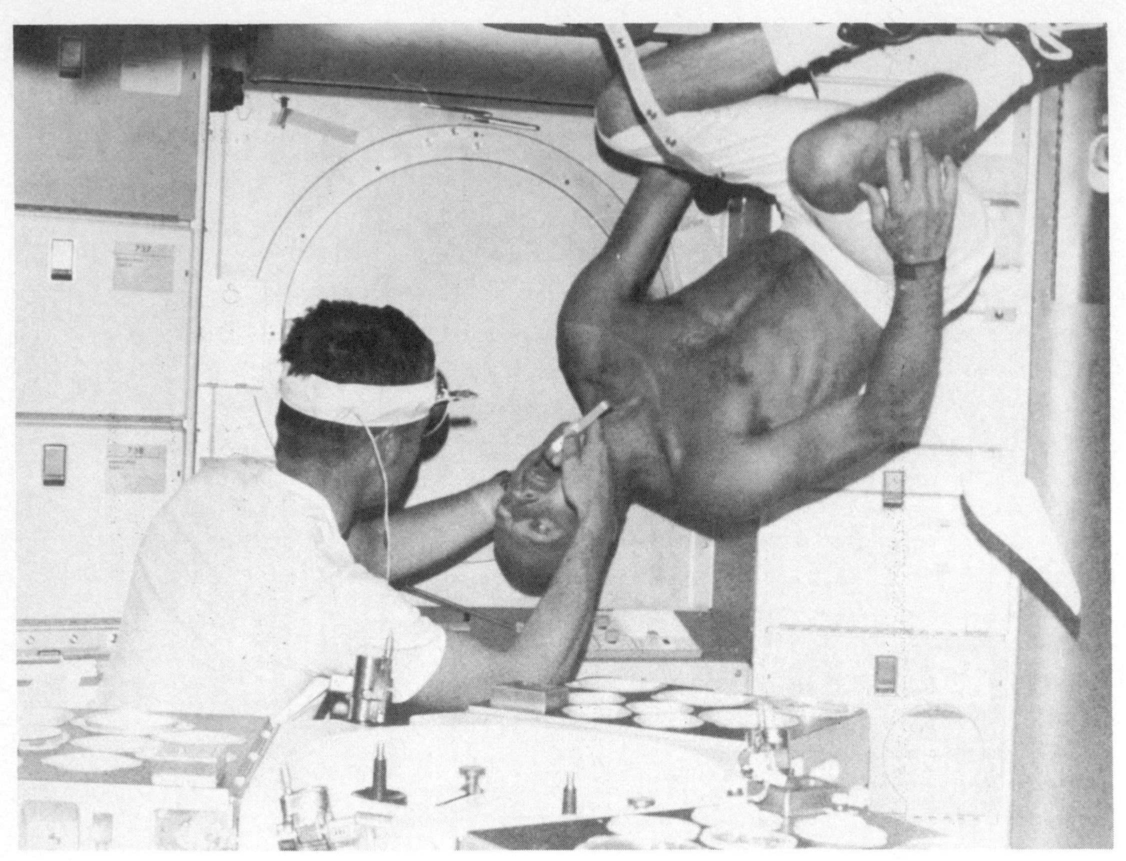

FIGURE 3–13. Scientist-astronaut Joseph P. Kerwin, Skylab 2 pilot, gives a physical examination to astronaut Charles Conrad, Jr. Under zero gravity conditions, Conrad is held in place by a restraint around his left leg. Note the floating pieces of paper at right. (Photo courtesy of NASA.)

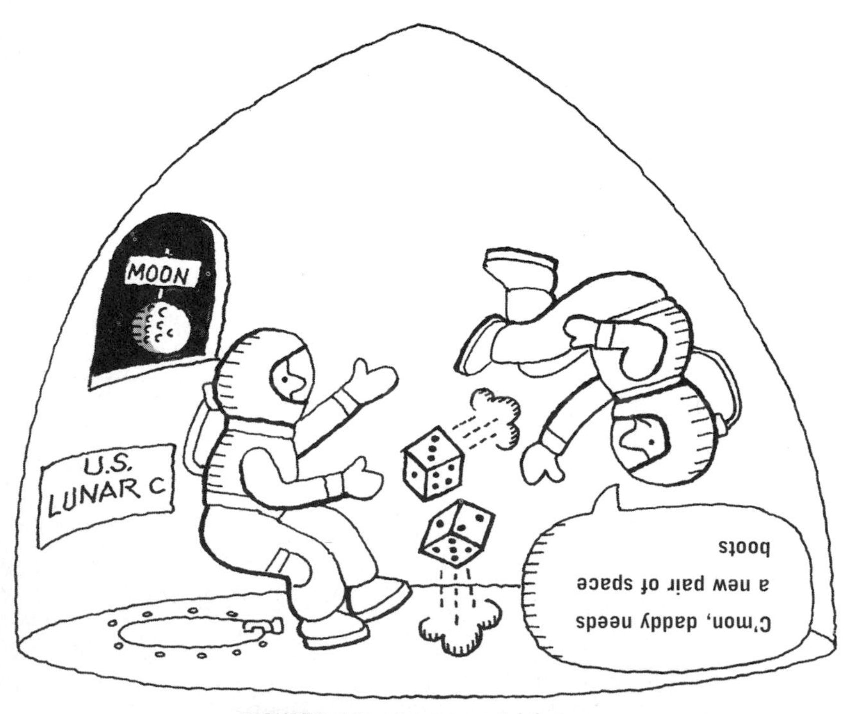

FIGURE 3–14. Apparent weightlessness.

paper wad or even a screwdriver glides straight across the spaceship cabin. A paper plane designed to do a loop-the-loop on earth would do a loop-the-loop-the-loop-the-loop-... in a spacecraft.

Gravity on a Larger Scale

The phenomenon of gravitation is far more important in the design of the universe than indicated by our discussion of orbiting satellites. Early in this chapter it was stated that according to the law of gravitation, every object in the universe has an attraction for every other object in the universe. "Every object" includes the moon as well as man-made satellites, and the moon is, in fact, in orbit around the earth just as is an orbiting satellite. It it weren't for the force of gravity, the moon would not stay near the earth; it would go off in a straight line toward who-knows-where.

As the moon orbits the earth, there is an attraction of the moon toward the earth and of the earth toward the moon—the attraction is a two-way affair. In fact, the moon most strongly attracts those parts of the earth which are closest to it, and least strongly attracts those parts farthest from it. This would be unimportant if the earth were truly a single, solid body. But the earth has large oceans, which can respond to the gravitational pull of the moon. The result is that the moon pulls harder on the water on the side of the earth nearest it than it does on the main body of the earth. The water thus bulges away from the earth as shown (in exaggerated form) in Figure 3–15. The main body of the earth is attracted more toward the moon than is the water on the op-

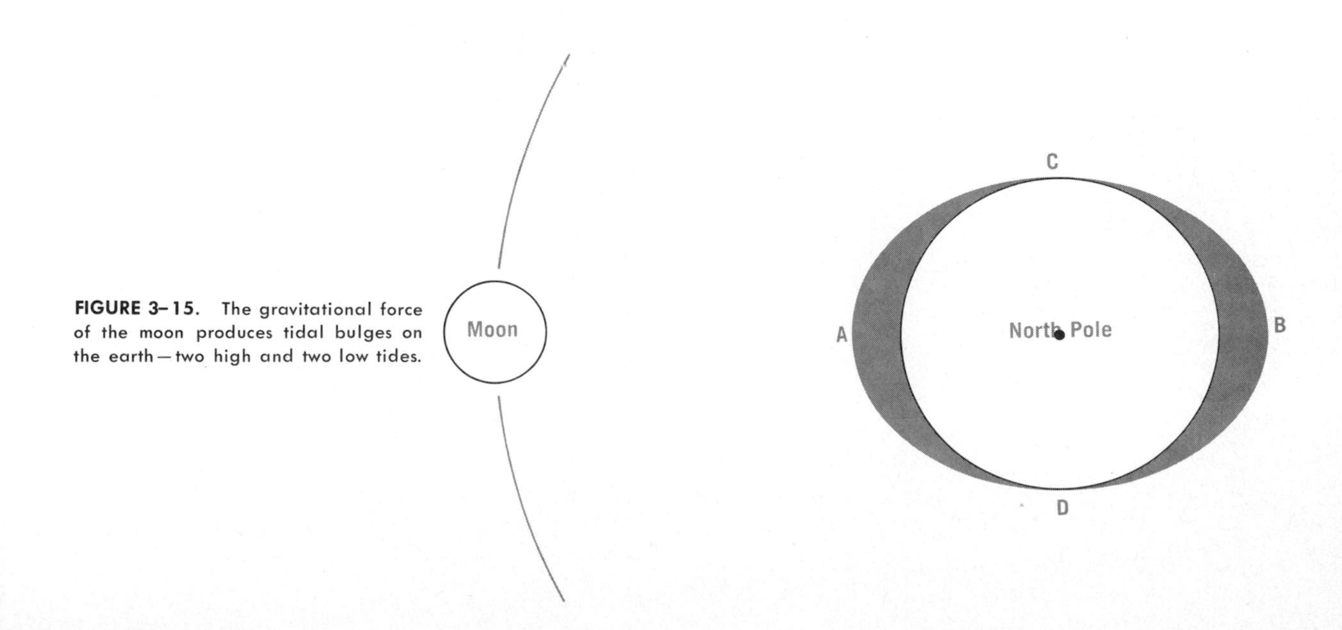

FIGURE 3–15. The gravitational force of the moon produces tidal bulges on the earth—two high and two low tides.

posite side from the moon, and thus, the water on the side opposite the moon bulges away from the earth. The earth is turning under these bulges of water. The effect seen by people on a seashore is that there is a high tide each time their seashore comes under a high-point of the water. Thus, the people at points A and B in Figure 3–15 are experiencing high tide while those at C and D experience low tide.

But the moon in orbit is still small stuff for the force of gravity. Along with the earth there are eight other planets orbiting the sun in elliptical orbits. In order, starting from the sun, these are Mercury, Venus, Earth, Mars, Jupiter, Saturn, Uranus, Neptune, and Pluto. Table 3–1 is a chart of the planets, showing their relative sizes, distances from the sun, and other data. Note the wide variation in mass of the planets, from 1/20 of the earth's mass to 318 times the earth's mass.

The earth is 93 million miles away from the sun. (This is an average distance; because of its elliptical orbit the distance varies somewhat.) This distance is such that it takes light, traveling at 186,000 miles per second, about eight minutes to get to earth from the sun. Yet the earth is fairly near the sun compared to Pluto—it takes light 5½ hours to make the trip from the sun to Pluto.

And what is the force that spans this great distance?—gravity, the same force that deigns to do such menial tasks here on earth as pulling pins to the floor.

BEYOND THE SOLAR SYSTEM

As astronomical distances go, the sun-to-Pluto distance is not much. Most of those tiny objects we see in the sky at night are other "suns" much like ours, and the closest one to us, Alpha Centauri, is 24 trillion miles away.[4] This distance is about average for distances between stars, and because it is so great, we use another unit of measurement to describe it. As we did above for earth and Pluto, we use the speed of light as a yardstick to speak of astronomical distances. Light takes 4.5 years to reach us from Alpha Centauri, so we say that Alpha Centauri is 4.5 light-years away. This new distance unit, then, is the distance light travels in one year (about 5.8 trillion miles).

Alpha Centauri is but one of about 100 billion stars that are clustered together into what we call the Milky Way galaxy. Figure 3–16 is an edge-on view of a galaxy similar to

[4]There may be closer stars, but if so they are not bright enough to be seen by us. There may also be nonluminous objects like our planets in the space around the solar system.

TABLE 3-1. PLANETARY DATA

	MERCURY	VENUS	EARTH	MARS	JUPITER	SATURN	URANUS	NEPTUNE	PLUTO
Average distance from sun (millions of miles)	36	67	93	142	483	887	1780	2790	3660
Average distance from sun relative to earth's distance	0.4	0.7	1.0	1.5	5.2	9.5	19.2	30.1	39.4
Revolution period (in earth years)	0.24	0.62	1.0	1.88	11.9	29.5	84.0	165	248
Mass relative to earth	0.05	0.82	1.0	0.11	318	95	14.5	17	?
Average diameter (in miles)	3000	7600	7930	4270	89,000	75,000	33,000	31,000	3000–4000
Number of known "moons"	0	0	1	2	13	10	5	2	0

ours and shows how the stars of the galaxy are located in a disk which bulges at the center.

If we were able to see our galaxy from above (or below, although such terminology is meaningless in this context because there is no *real* top or bottom) it would look like Figure 3–17. Notice the spiral arms in this photo. Each contains billions of stars. The arrows in Figures 3–16 and 3–17 represent the approximate position of our sun if these were photos of our own galaxy.

The entire Milky Way galaxy is spinning in space, and the force which holds the stars in this grouping is none other than gravity. Our sun is about 25,000 light-years from the center of the galaxy, and the outer edge is approximately 50,000 light-years from center. The force of gravity becomes weaker with increasing distance, but because the masses involved are so large, the force is still great enough at 50,000 light-years to hold stars in a whirling group.

Galactic distances, however, do not represent the limit of gravitational force. Our galaxy is one of about seventeen galaxies which are clustered together in this region of space. Distances between galaxies in this cluster of galaxies are of the order of a few million light-years. Although there are smaller clusters of galaxies, ours must be considered a small cluster, for the giant Hercules Cluster contains about 10,000

FIGURE 3–16. Although this is not the Milky Way Galaxy, our galaxy (when seen edge-on) looks much like this. The sun would be approximately at the position shown. (Hale Observatories.)

FIGURE 3–17. This is a photo of a spiral galaxy seen from above—or below (these terms are meaningless in this context). If it were the Milky Way, the sun would be at about the position indicated. (Hale Observatories.)

galaxies (Fig. 3–18). And as you might suspect, the force which holds together these tremendous aggregates of millions of billions of stars is gravity.

ESCAPE VELOCITY

When you watch a football game on a Saturday afternoon, you feel secure in your knowledge of what will happen when a punter boots the ball. It first goes up and then it comes down. That's how it is, was, and ever shall be—unless. . . . If the kicker someday should kick the ball such that at the instant it left his toe it was traveling upward at a rate of seven miles per second, we would find that we would not have to worry about a punt return, because the ball simply would not come down. This particular speed is called the escape velocity of an object from the earth. At this speed, any object, large or small, will escape from the earth to soar forever upward until captured by the gravitational attraction of some other planet or celestial body. At first thought it might seem that a heavy object might require a

FIGURE 3–18. A cluster of galaxies in Hercules. (Hale Observatories.)

greater initial speed to escape from the surface of the earth than a lighter one. But remember that the way objects fall (or move upward against gravity) is independent of the mass of the object. As a result, if a velocity of seven miles per second is enough to cause a tennis ball to escape from the earth, it is also enough to send a bowling ball (or a freight car) on its way.

The concept of escape velocity applies to objects as small as atoms and molecules. The earth's atmosphere is composed primarily of oxygen and nitrogen, and it contains practically no trace of hydrogen or helium, the lightest elements. All of the atoms of a gas travel in random motion at very high speeds, and the less massive atoms and molecules travel at the greatest speeds. In fact, the lighter gases go so fast that their speeds are greater than the critical speed of seven miles per second. Thus they have long ago left our comfortable home and are now wandering through space in search of a more hospitable gravitational field.

The moon has a weaker gravitational field than the earth. The escape velocity of an object on the moon is about 1.5 miles per second, and even the heavier gases have enough speed to escape the moon. This explains why we find a vacuum there; any gases which may be released near the moon quickly escape.

GRAVITY IN THE EXTREME—BLACK HOLES

The idea of black holes is one of the most fantastic concepts of recent times. To explain what they are—or what they might be—we must briefly describe the life of a star.

Stars are formed by gravitational forces acting to pull together gases and dust particles in space. Each molecule of gas and each dust particle in the universe exerts a gravitational pull on every other molecule and particle in the universe. In regions of space where a greater-than-average number of these particles exists, their mutual gravitational attraction results in their being pulled together. As they get closer together, the force becomes greater, and they fall together faster. Their speed finally becomes great enough that when collisions between particles occur, nuclear reactions result. It is these reactions that are the source of energy in the stars (as we will see in a later chapter), and the constant release of heat energy keeps the star from further gravitational collapse. In other words, there is a balance between gravity pulling in and force produced by heat energy pushing out. And so the star burns on.

Finally, at some point in its life, the fuel for the nuclear reaction runs out, and after the star has burned all of the fuel it is going to burn, the balance is destroyed. Gravitational force is still pulling in, but nothing is pushing out.

A balloon filled with air has a balance of sorts. The stretched rubber seeks to pull in and make the balloon smaller, while the compressed air inside seeks to push out (Fig. 3–19). But suppose the air disappears. Now only the stretched-rubber force is there. The balloon collapses inward. For the balloon, however, the collapse stops when the rubber compresses enough so that there is no more "stretch." But in the case of stars, as the parts come closer

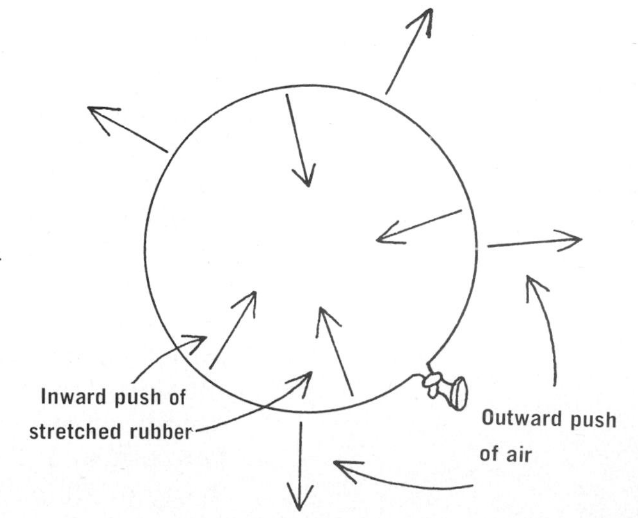

FIGURE 3–19. The equilibrium established in an inflated balloon.

together, the force responsible for the collapse (gravity) becomes *stronger,* not weaker.

Where does it all stop?—perhaps when the particles are in contact with one another. But the particles involved in the case of stars are the ultra-tiny nuclei of atoms. From what we know of these particles, it seems that they should be able to come together at least to the point where a star that was originally as big as our sun becomes a ball two miles in diameter. (Our sun now is about 80,000 miles across.) And at this point something really wild happens that gives this black hole its name.

Einstein, in his theory of relativity (Chap. 23), said that light has mass and is affected by the pull of gravity. This means that a flashlight beam shot across a room is pulled down by gravity just as is a bullet fired across the room. Impossible? Contrary to experience? Not necessarily. Remember that as you shoot a bullet across a room faster and faster, it drops less and less. Light travels at 186,000 miles per second—considerably faster than a speeding bullet. So it does not drop very much in going across the room, but it does drop. We will not go into the detail of the experiment, but this has been verified by measurements of light passing near the sun. There is good reason to believe that Einstein was right.

The gravitational force near a star compressed to a width of two miles would be tremendous. Thus the escape velocity on this star would likewise be tremendous. It would be greater than 186,000 miles per second, the speed of light. Any light that tried to escape would be pulled back in. Just as you cannot throw a ball away from the earth because you can't throw it at seven miles per second, the light emitted from the star could not get away, nor could any other kind of radiation.

Think of what such a star would look like. Think of nothing. No light, no radio waves—nothing comes from the star. In fact, any object or light passing near enough to the black hole would be sucked in.

We talk about such things, but do they exist? We don't know. As far as we know, there is no reason why they should not exist. And so astronomers look for them by looking for evidence of matter which is in the process of falling into them but is not so close that light cannot escape from it. At the time of publication, reports of observations were pouring in which seemed to confirm their existence, but there was not yet conclusive evidence.

Intuition tells us that black holes (if they exist) cannot go on collapsing forever. Yet we know of no force which would be able to stop their collapse . . . or maybe things *can* keep collapsing—perhaps that's how it all will end.

FIGURE 3–20. A black hole?

HOW SCIENTIFIC THEORIES AFFECT MANKIND'S OUTLOOK

Perhaps the most important aspect of Isaac Newton's work has been overlooked in this chapter. As was stated, the ancient philosophers had theories about the natural behavior of bodies on earth. They also had theories of the natural behavior of heavenly bodies. They observed the stars, moon, and sun apparently revolving around the earth and concluded that natural motion in the heavens was circular.[5] They also observed that nothing on earth has a natural circular motion. Thus they concluded that earth and heavens each had their own different laws, and the *uni*verse was not a *uni*ty at all. The entire concept of a heavens-and-earth universe was foreign to Aristotle's philosophy.

Newton, however, proposed that there is a gravitational force between any two masses in existence. The force is less at great distances, but if large masses are involved, the force is considerable. "Considerable" is not a strong enough word here, because it is the gravitational force acting between earth and moon that keeps the moon in its orbit. It is the gravitational force between earth and sun that keeps the earth in orbit (and in exactly the way that gravitational force keeps a satellite in orbit). As we saw in an earlier chapter, in the absence of forces, an object goes in a straight line without slowing down. Newton applied this concept to heavenly bodies as well as earthly ones. Thus without gravity, the moon would take off in a straight line and leave earth forever. Likewise, "Spaceship Earth" would leave its orbit around the sun and head off through space in a straight line.

Newton, then, showed that the heavens and the earth follow the same natural laws. In a sense, he invented the *uni*verse.

---

[5]This fits well into theological arguments. The heavens should be perfect, and what could be more perfect than the circle, which has no beginning and no end. The moon occupied a middle ground between earth and heaven, and, indeed, one could see imperfections on its surface.

OBJECTIVES

After completing this chapter you should be able to:

1. Explain Aristotle's concept of "natural" motion and how it differs from present notions.
2. Give an example of a case where theory and experiment conflict, and explain how such a discrepancy must finally be resolved.
3. Tell how the force of gravity changes with distance between objects.
4. Given a person's weight on the surface of the earth, calculate his weight at any multiple of 4000 miles above earth's surface.
5. Explain why gravitational forces between two handheld objects are not detectable.
6. Explain why astronauts are never truly weightless.
7. Explain what apparent weightlessness is and describe a situation in which such an effect may be felt on earth.
8. Describe and perform a simple demonstration which shows that people would not be pressed to the ceiling of a freely falling elevator.
9. Describe Newton's thought experiment which illustrated the possibility of orbiting a satellite.
10. Describe and perform an experiment which illustrates that even while coasting toward the moon, astronauts are in a state of apparent weightlessness.
11. Describe how prolonged apparent weightlessness affects man's muscles and how space-exercises must differ from earth-exercises.
12. Explain what controls the paths of the moon and the planets.
13. Explain why there are *two* tides each 24 hours.
14. List the planets in order from the sun, and state which is largest and which smallest.
15. Describe the appearance of the Milky Way galaxy and give the approximate location of the sun in the galaxy.
16. Describe the part played by the force of gravity in the structure of the galaxies.
17. Explain the special significance of the speed 7 miles per second, and what effect this has on the gases we find in our atmosphere.
18. Use the concept of escape velocity to explain why the moon has no atmosphere.
19. Explain why gravitational pull on light is not normally noticeable.
20. Describe what keeps a normal star, such as our sun, from collapsing under the force of gravity.
21. Explain why a star may finally begin to collapse.
22. Tell to what approximate size a star such as our sun must collapse in order to become a black hole.

23. Describe how Newton's law of gravity changed man's outlook on the world in which he lives.

QUESTIONS—CHAPTER 3

1. A quarter and a penny are allowed to fall to the floor, and the time of fall is measured for each. They are then glued together and dropped. Will the combination fall faster or slower than when separated? Argue your answer in two ways: first from the point of view of Aristotle, and second from that of Newton. Which one best fits the experiment?

2. If every body in the universe has a gravitational attraction for every other body, why do we not feel the pull of a ten-story office building on our car when we drive past? Can you develop a thought experiment that would allow this force to be measured?

3. When we run, we lean forward. From the standpoint of gravitational force, why is it better for a baseball player to slide head first rather than feet first?

4. When a rock is dropped, does the earth rise to meet it? Defend your answer.

5. Will a weight-conscious young physics student weigh more in Death Valley or in Denver?

6. What would happen to a rock if you dropped it into a hole drilled through the center of the earth?

7. A home builder drops two plumb lines in an attempt to get opposite sides of a building exactly parallel. Why will he not be able to do so? Is this important?

8. How far from the earth would you have to go in order to become truly weightless?

9. One might think that a raindrop falling from a cloud should keep gaining speed and achieve the destructive force of a bullet. What prevents a rainstorm from producing catastrophe?

10. A Cadillac and a Volkswagen both roll freely down a hill. Which one reaches the bottom first?

11. If the law of gravity should be repealed, what would happen to the moon?

12. Explain why there are *two* tidal bulges on the earth due to the moon.

13. Why does not the gravitational force between the stars of a galaxy pull the stars to the center and thus collapse the galaxy?

14. How would the escape velocity of objects on the surface of the planet Jupiter compare to their escape velocity on earth? Compare the escape velocity of objects on Mercury and on earth.

15. In what way can it be said that Newton "invented" the universe?

16. Describe a handball game in an environment of weightlessness.

17. The gravitational force exerted on the earth by the sun is much greater than that of the moon on the earth. Why is it, then, that the

moon plays a much more significant role in causing the tides than does the sun?

18. Why is it incorrect to speak of the *weight* of a planet?

19. Draw a diagram of the Milky Way, indicating the approximate position of the sun.

20. It is said that before Newton, man lived in a "duoverse." Explain.

Statue by Felix W. deWeldon honors Apollo astronauts on campus of Eastern Kentucky University.

4

THE HOW AND WHY OF SPACE TRAVEL

America's role in the race to space received its primary impetus from a pledge by President John F. Kennedy in 1961 to put a man on the moon by 1970. How we did it will be examined in this chapter. A more important question to many, however, is not how, but why. With many Americans besieged by hunger, poverty, and disease, were the moon landings worth the expense? Are continued efforts directed toward development of orbiting laboratories and planetary excursions worthy of additional expenditures?

The pros and cons of these issues will be considered here. It should be admitted at the beginning that no effort was made to be completely neutral, nor will both sides be given equal time. Most scientists feel that the space program has been beneficial and this biased viewpoint will be obvious.

THE 1866 TRIP TO THE MOON

Some have said that Jules Verne anticipated all the good science fiction plots and that the writers following him

Jules Verne (1828-1905)

Jules Verne, one of the first science fiction writers, started by writing plays and the lyrics for operas. He also wrote several historical novels, but he is best known for such works as A Journey to the Center of the Earth **(1864),** From the Earth to the Moon **(1865),** Around the Moon **(1870), and** Around the World in Eighty Days **(1873). Although his stories were popular when first published, he was far enough ahead of his time that some of them were made into movies during the 1950s, a half-century after his death. Not many future-based fantasies have survived their authors.**

have only added a few unique twists to his ideas. Our purpose here is not to debate this statement but instead to examine a portion of one of his famous works: *From the Earth to the Moon.* We ask you now to draw upon knowledge from the preceding chapters in order to pick out some of Verne's misconceptions about what would really happen on such a journey.

According to the story, a great hole was dug in the earth, and in it was placed a giant cannon. A bullet-shaped spaceship was fired from it, and after this initial thrust upward, the passengers (including some animals) coasted to the moon.

In the initial stage of the flight, the passengers felt themselves pressed to the floor on the earth side of their craft. As they sped upward, they found that this pressure decreased. (Recall how weight decreases as one goes further from the earth.) Nearing the moon, they reached a point where the gravitational attractions of the earth and moon became equal and opposite. At that point, they floated freely; they felt weightless. Closer still to the moon, they found its gravitational attraction becoming dominant, and they began to walk around on what had previously been the ceiling.

One fatality was registered during blast-off—a dog was killed. To dispose of him, they dumped him through a port in the side of the ship, placed there for just such emergencies. To their surprise, he remained their constant compan-

FIGURE 4–1. Photo of the earth from moon orbit, Apollo 16. (Photo courtesy of NASA.)

ion throughout the flight, floating alongside the spaceship—an unpleasant prospect indeed.

You should be able to point out a number of inconsistencies between Verne's story and the predictions based on the theory of gravitation, because as imaginative a writer as Jules Verne was, some of his physics concepts were inaccurate. We will point out a few of his errors in what follows. Why not attempt to find them before you read on?

THE WAY WE DO IT—BLAST-OFF

Fortunately, our astronauts are not "shot from guns." Instead, the principle of rocket action is used to propel them into space.

take-along-do-it

Blow up a balloon and release it. If it doesn't shoot around the room, you've got a strange balloon. To help understand why this happens,[1] there is a second part to this TADI. Pick up something heavy—a big book, or, better yet, a concrete block—and hold it at chest height. Now push it away from you—hard. What is the effect on you of pushing the object away? After doing this, you should understand why the man in Figure 4–2 had better be careful with that rock.

The explanation of this phenomenon lies in an often misunderstood principle: When you applied a force to push away the book, the book exerted an equal and opposite force

[1]Those who studied Chapter 2 already recognize this as an example of the force-pair law.

FIGURE 4–2. What happens when the man throws the rock?

on you, or, more specifically, on your hands. You can feel this force, and if you had not had experience with throwing things, you probably would have found that this force would have been enough to push you backward. As it was, however, you unconsciously anticipated the push backward and you probably leaned forward just slightly before you threw so that you would not lose your balance owing to the reaction force.

The balloon went forward for the same reason. The rubber of the balloon, being under tension, pushed the air out through the opening. But as it pushed on the air, the air pushed on it. So the balloon was forced forward.

A rocket works by exactly this same principle. The gases burning at the rear of the rocket expand and are forced out. When they are forced backward, the rocket is forced forward. A common misconception is that rockets depend for their operation upon the burning fuel having air outside the rocket against which to push. If this were true, rockets could not operate in a vacuum. However, because their operation depends only upon the fact that forces occur in pairs, rockets can operate in a vacuum. The rockets used by our astronauts today gain speed continuously over hundreds of miles. Even so, the astronauts feel pushed back into their couches with a fairly great force, just as you are pushed back in a car seat when you accelerate rapidly. If we attempted to shoot our astronauts out of a cannon and had to give them all that speed over a very short distance, we would have to scrape them out of their chairs afterward.

The damaging effects of such a blast-off can be understood by applying Newton's Second Law: force = mass × acceleration. An astronaut shot from a cannon would find his velocity changing quite rapidly, and thus the acceleration he experiences would be considerable. Examination of Newton's Second Law shows that a large force is necessary to produce such a large acceleration. The couch in which the astronaut rests exerts such a force.

As a slight digression, let's examine some other features of such a blast-off. Since the chair in which he rests is exerting a large force on the astronaut, he, in turn, is exerting a large force on it (the force-pair law in action). The result is that the astronaut is pushed back into the padding of the seat. Under normal circumstances, when we rest in a chair, our weight compresses the padding slightly. But in blast-off, the astronaut compresses the padding in the same way he would if his weight were, say, four times its normal value. A common expression has entered our language to describe the conditions encountered here. As we walk around on earth, we are attracted to the earth by a force equal to our weight; another way to express this is to say that we are experiencing a force of 1 g. In a rocket, astronauts experience g forces several times this value. A force of 2 g's means that he is pushing against the couch with a force equal to twice his weight. If a force of 5 g's is experienced for several seconds, the unfortunate victim usually loses consciousness. A force of 9 g's approaches the point at which the body is no longer able to withstand the push and the result is death by crushing.

It should be noted here that it is not necessary for astronauts to attain a speed of seven miles per second, the escape velocity from the earth. This number assumes that the spacecraft gets one push at ground level and that's all. Instead, our rockets receive a fairly constant thrust for several minutes to carry them into earth orbit. From there, another thrust of several minutes' duration carries them to a height such that the gravitational force has decreased to the point that escape is possible at a much lower speed.

THE WAY WE DO IT—TRAJECTORY

On all the missions to the moon, the spacecraft followed a flight path such as the one indicated by Figure 4–3. Initially, five huge engines thrust the rocket to a height of about 40 miles and to a speed of about 6000 miles per hour. At this point, a section (the first stage) of the rocket separates from the main body, and a second stage ignites. This stage carries the rocket to a height of about 120 miles and to a speed of about 15,500 miles per hour, at which time this section also drops away. A short burn by a third stage sends the vehicle into orbit. After sufficient time has elapsed to insure that all systems are working properly, the third stage is reignited (point A of Fig. 4–3). This thrust lasts for about five minutes and gives the rocket a speed of about 24,500 miles per hour; this speed must be attained in order to reach the moon. At the completion of this final burn, the third stage separates and the spacecraft is on its own.

Jules Verne's ship was given one big blast to send it off. It did not need to attain escape velocity, for the moon is not

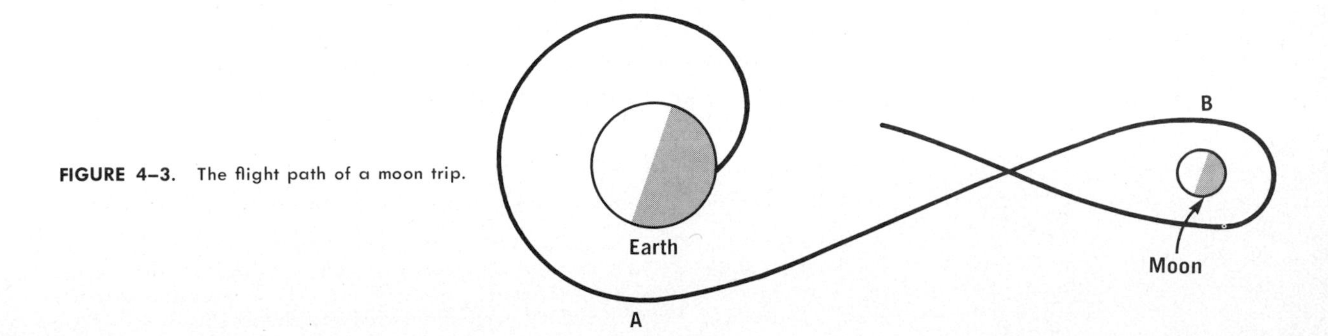

FIGURE 4–3. The flight path of a moon trip.

nearly beyond the pull of gravity of the earth. Nevertheless, quite a large speed is required to coast from earth to moon. Rockets impart this speed to our spaceships over hundreds of miles as the ships leave earth orbit for their trip to the moon. After this rocket burn, however, the flight consists of coasting.[2] Being retarded by the pull of earth's gravitation, the ship loses speed as it moves away from earth. Then, when it reaches the point in space where the moon's gravitational field is greater than earth's, it begins to gain speed, falling toward the moon. Contrary to Jules Verne's ideas, the astronauts experience no change as they pass this point. They feel weightless before reaching the point, at the point, and after leaving it.

At point *B* in Figure 4–3, braking rockets are fired to slow down the ship and put it into moon orbit. At a predetermined position, a shuttle craft carried along on the voyage is used to carry some intrepid souls to the moon's surface. Return to the earth is achieved by utilizing essentially the same steps in reverse, except the several rocket stages are not needed to escape the much weaker gravitational attraction of the moon.

MAN ON THE MOON

On July 20, 1969, Neil Armstrong became the first man to set foot on the moon. The kangaroo-hops of our bulky-suited astronauts on the moon became familiar sights on television screens in the late 1960s and early 1970s, and quite a bit of physics was illustrated in their motions.

The gravitational attraction of the moon at its surface is only about $1/6$ of the attraction on the earth's surface. Because the radius of the moon is smaller than that of the earth, it might seem that the gravitational attraction on the moon would be greater than on the earth (refer to the law of gravitational attraction). However, since the radius of the moon is smaller than that of the earth, its mass is also decreased, and it is this smaller mass which is the dominating factor. Because of this lesser attraction, when an astronaut jumps, it takes about $2\frac{1}{2}$ times as long for him to come back down as it would on earth. This accounts for the slow-motion appearance of moonwalks. Those bulky spacesuits and the life-support packs worn on the astronauts' backs would be very difficult to carry around on earth. They weight about 200 pounds on earth, but on the moon they weigh a mere 36 pounds. It is only the bulkiness of the spacesuits that prohibits the moonwalkers from making fan-

[2]On one of the Apollo flights, the astronauts were talking by radio to their families. One of their sons asked, "Who is driving up there?" The answer his dad gave was, "I guess Isaac Newton is doing most of the driving right now."

FIGURE 4–4. Astronaut Edgar D. Mitchell, Apollo 14 lunar module pilot, moves across the moon's surface. (Photo courtesy of NASA.)

tastic leaps. If they wore earth clothes on the moon, they could easily hop 8 to 10 feet off the ground. Stairs would hardly be necessary there.

In a moment of whimsy, Apollo 14 astronaut Alan Shepard pulled two golf balls from his pack and hit them with one of his tools. Not only is there less gravitational force to pull the ball back down to the surface, but in the vacuum of the moon, there is no air to slow the ball's progress. Likewise, there is no air to catch a spinning ball and cause it to curve away from where it was aimed. Golf pros who make their living correcting hooks and slices would be in need of a new occupation on the moon.

WHY GO INTO SPACE WHEN MAN NEEDS HELP ON EARTH?

The United States has spent quite a lot of money to explore the moon and "outer space" and continues to spend

FIGURE 4-5. Astronaut Harrison Schmitt explores the desolate lunarscape in the lunar "dune buggy." (Photo courtesy of NASA.)

large sums. Is it worth it? This is a big question and depends for its answer on our national priorities. To make such a judgment, it is necessary for us to look at some of the pros and cons of space travel.

First, it is expensive. The budget of the National Aeronautics and Space Administration (NASA) averaged nearly $4 billion per year during the late 1960s and early 1970s when man was exploring the moon. This is so much money that it is really beyond our comprehension. If you put down one dollar every second of every minute of every hour of every day of every year since the birth of Christ—about 1975 years—you would have a little over $60 billion. NASA spent about $28 billion on the Apollo program to get man to the moon and back. Many people question whether the money could not have been better spent. Perhaps. But let's look further into the United States budget. Table 4-1 shows the money allotted to various departments of our government in 1972. As the table indicates, our government spends a lot of money! So while NASA's budget represents a very great amount of money, it is only a drop in the bucket. (As another interesting insight into our priorities, during that same year, retail sales by American liquor stores totaled $9.2 billion.)

Still the question remains: Couldn't the $28 billion have been better spent, perhaps to solve our social problems? President Kennedy committed the nation to putting a man on the moon. Could he instead have chosen some social goal on which to spend the money? The answer is obviously "yes," but partisans of the space program argue that he could not have been confident of a positive result. They reason that the basic science needed for local space travel was known in 1960 and had been known since the latter part of the seventeenth century. (We pointed out in Chapter 3 that Isaac Newton formulated the basic gravitational theory of satellites.) The primary problem of putting a man on the moon was one of developing the necessary technology to accomplish the goal. In the case of our social problems, the situation is different. We simply do not know enough about people and how they act, either collectively or individually, to solve the problem of why we fight wars or how we can eliminate poverty. It is basic science which needs development here, and it is not a simple thing to hurry scientific development. Additional money will help, but no one can confidently say what will result from basic research and whether or not that research will lead to a solution to the problem at hand.

Opponents of the moon adventures can, of course, present persuasive counterarguments. If it truly is a more basic understanding of man and how he acts that is needed, then money should be spent to *try* to find some answer. This research may not lead to an immediate solution, but after all, Isaac Newton's basic research did not solve the space exploration problem for 300 years. Opponents argue that while no guarantees can be made of success, the effort is worthy of a try.

TABLE 4–1. 1972 BUDGET OF SELECTED GOVERNMENT AGENCIES*

AGENCY	BUDGET (BILLIONS OF DOLLARS)
Department of Defense	76.5
Pay to retired military personnel	3.9
Veterans' Administration	11.0
Health, Education and Welfare	71.7
Housing and Urban Development	3.4
President's Council on Youth	2.1
Department of Transportation	7.5
NASA	3.4

*Data from 1973 World Almanac. New York, Newspaper Enterprise Association, 1972.

Benefits from Previous Space Flights

The original motivation for our space program was primarily military. In October, 1958, the USSR launched the first Sputnik into orbit. The United States reaction was predictable, for the military implications of the conquest of space were obvious.[3] Such was the motivation behind our early space efforts, and, indeed, military reconnaissance satellites have been quite successful. By the late 1960s, however, the Apollo program had evolved from a military display into an effort at exploration of our environment. No military base was established on the moon.

Uncountable utilitarian applications of the technology from the Apollo program have become commonplace, including Teflon cookware, freeze-dried food, foam to refloat sunken ships, improved adhesives, subminiature semiconductors for electronics, improved plastic pipe materials, and waterproof thermal blankets. These, however, are almost a trivial return for $28 billion. But there are greater benefits: communication satellites for instant worldwide television transmission (costs are approximately $4000 per channel per year versus $25,000 for undersea cable); improved weather forecasting; improved navigation of ships and aircraft; better management of crop, wildlife, and timber resources through monitoring from space; detection of mineral deposits; forest-fire detection; and many advances in medicine (especially in the monitoring of body functions). It is difficult to put a price tag on such benefits, but it must be considerable. And there are benefits predicted but not yet fully realized: earthquake prediction and possible control, increased crop yields resulting from long-range weather forecasting, and new energy sources and distribution systems. The $28 billion suddenly seems like a little less money in the face of benefits such as these. And almost surely, benefits presently unforeseen will emerge as time goes by.[4]

Critics of the space program rightly point out that $28 billion would have gone far toward finding a cure for cancer. Even if the attempt had not been wholly successful, many unforeseen medical spin-offs could have been realized in this project also. Again, there is no black-and-white answer.

Perhaps, however, the greatest benefit from the American and Soviet space ventures does not lie in material benefits but in man's outlook upon himself and his planet.

---

[3]The idea of an orbiting satellite which has only to drop bombs over its targets is an oversimplification, however. From your study in the previous chapter, you know that a bomb released from a satellite in orbit would continue to orbit with the satellite. It would have to be taken out of orbit with retro-rockets just like any other satellite. Guidance would require the same sophisticated equipment as does the landing of a manned spacecraft.

[4]See Ordway, F. I., Adams, C. C., and Sharpe, M. R.: *Dividends From Space.* New York, Thomas Y. Crowell Co., 1971.

During the 1968 Christmas season, when the Apollo 8 astronauts were returning from orbiting the moon, they referred repeatedly to "Spaceship Earth." For centuries we have known that our planet is a small, almost insignificant spaceship in the vast universe, but we naturally tend to think of earth as a big, important object. The space ventures of the last decade have made us *realize* that the earth is indeed only a *spaceship.*

Since the dropping of the first atomic bomb in 1945, we have known that man has the power to destroy life on earth, but most people seemed to have faith that this could not happen, that somehow the earth and mankind were too important to be destroyed. After seeing pictures of earth from space and realizing that it is but a small object traveling through the near-emptiness of space, our outlook changed. We were forced to admit that our destruction would hardly matter in this immense universe—except to us.

The same three astronauts often used the expression "the good earth." Our planet seen from space is so much more beautiful than the desolation of Mars or the moon that the importance of preserving what we have here was made abundantly clear to us. It is no coincidence that the clamor for ecology programs began during the years of space exploration. The space program caused us to mature in our outlook toward our environment. It is ironic that the concern for preserving what we have has resulted in diminishing support for the space program, when it was the space program that spawned the ecology movement in the first place.

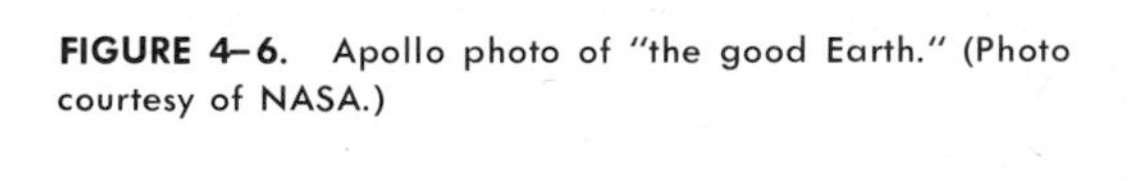

FIGURE 4–6. Apollo photo of "the good Earth." (Photo courtesy of NASA.)

The "Because It's There" Syndrome

A final reason for exploring space involves human nature itself: we are a curious and adventurous species. We climb mountains because they are there; we ask about the origin of mountains even though the only benefit of this knowledge is that it satisfies our curiosity. There are purely scientific reasons for going into space. We want to find out more about the earth, the moon, the solar system, and the universe; and space travel will help us toward this knowledge. For example, telescopes in space have the advantage of being above the earth's atmosphere, which does not allow some types of radiation to pass. These telescope observations from space will greatly advance the science of astronomy.

This quest for knowledge is often categorized as pure research. It has no immediate utilitarian benefits. The work of Isaac Newton would fit well into this category, and his work changed man's ideas about the universe. The "should we or shouldn't we" questions concerning space travel have largely subsided since the 1960s. There were returns from the money invested, and a large portion of the returns was connected with pure research, which shows no immediate useful benefits. Most of the readers of this text were never called upon to express their opinions about the advisability of this large investment. Many of you will say that it was wasteful and should not have been done. There will be similar issues to be decided in the future, and in the next chapter we will explore one possible question that might soon confront us.[5]

OBJECTIVES

A study of Chapter 4 should enable you to:

1. Describe Jules Verne's conception of a space flight from earth to moon, and point out errors in his ideas.
2. Explain the basic force law involved in rocket propulsion, and perform a simple demonstration of this effect.
3. Explain what is meant by g forces.
4. Explain why it is unnecessary for spaceships to achieve escape velocity from earth's surface in order to reach the moon.
5. Describe and explain the peculiar motions of astronauts walking on the moon.
6. Argue the pros and cons of the economic considerations of space flight.

---

[5]A second possibility (besides the material of the next chapter) is that of the colonization of space by man. In an article in *Physics Today,* (Sept., 1974), Gerald K. O'Neill presents a plan which would allow the human race to grow to 20,000 times its present number. And he suggests immediate implementation of the plan.

7. Discuss the difference between pure research and technological development as far as predictability of success, and show how this discussion entered the argument for or against the Apollo program.
8. List at least ten new products or technological advances resulting from the space program.
9. Discuss the effect that the space program has had on man's outlook on the universe.
10. Explain how the basically curious nature of man plays a part in questions of the value of research, and in particular space research.

QUESTIONS — CHAPTER 4

1. What part would air friction play in Jules Verne's method of space flight?

2. Will a rocket work in the air? in a vacuum? underwater?

3. How could a spaceship completely escape the gravitational pull of earth without ever attaining a speed of seven miles per second (the escape velocity from ground level)?

4. What difficulties due to different gravitational fields would an astronaut encounter as he tried to practice, here on earth, for moonwalks? Suggest possible (partial) solutions to these difficulties.

5. The population of the United States is somewhat over 200 million. On the average, how much did each person contribute to the total Apollo program?

6. Consult an encyclopedia to find additional practical benefits from the space program.

7. Do *you* believe the United States should have landed men on the moon? Defend your answer.

8. We pointed out some, but not all, of Jules Verne's misconceptions about space flight. Can you find others?

9. The reason that a balloon or rocket moves forward was explained using Newton's Third Law. An equivalent argument can be given which uses the principle of conservation of momentum. Try to apply this argument to explain the motion of a rocket.

10. Is there a contradiction in Jules Verne's story in that the dog floated alongside the ship but the people were pushed to the floor (and then the ceiling)? Explain. Why do you think Verne did this?

11. Suppose that on this planet the atmosphere were much thinner and the gravitational force much less. In what ways might humans have evolved differently to account for this different environment?

12. Answer Question 11 for the case of a heavier atmosphere and a stronger gravitational force.

13. Is attainment of escape velocity necessary in order for a rocket to reach the moon? Why or why not?

14. Suppose someday man builds airtight enclosures on the moon. How will houses within these enclosures differ from buildings on earth?

5

LIFE BEYOND THE SOLAR SYSTEM

The question of whether life—particularly intelligent life—exists anywhere but on earth has intrigued the human race for centuries. Many science fiction stories have been written about creatures from outer space. Most of these stories were bad science, and many of them were bad fiction, but the question is no less valid: are we alone? At some future date, perhaps our national priorities will lead us to attempt to answer this question. If contact with other civilizations is made, the advances in our technology and in pure science could be tremendous.

Let's investigate now some of the possibilities that may come before us at that future date. Let us also reformulate the question: Are we alone, and is it worth the money to find out?

ARE WE ALONE?

After much painstaking research, the authors have located a heretofore unknown scroll of heretofore unknown origin. We quote from it:

> The question of whether we are alone in the universe is an ancient one. Let us look at the facts. Life has evolved and is supported on our

planet by a great number of special circumstances. Temperatures here are perfect for life. Water and oxygen are plentiful. A delicately balanced amount of light and heat reaches us by day—neither too little nor too much. Our planet turns at just the correct speed to make day and night the right length. If the day were longer, the ground would heat too much before evening; if shorter, natural cycles would be upset.

Throughout our history all the great religions have ignored any natural life except our own. Who are we at this day and age to think that there is any chance that there is another planet revolving around some star out in space with all the right circumstances as exist on our own planet. And even if there were, there is no reason to suppose that the correct chemicals came together at the right time and under the right circumstances to form a beginning of life.

No, we must conclude that the chance of life existing elsewhere in the universe is so remote as to be ridiculous. The namuh race on the great planet Thrae of the system of star Nus is the only intelligent life in the universe. Certainly no life could exist on any planet that may possibly circle another star—such as star Sun, for example.

The above translation is admittedly from an obscure document, and some scholars have even had the audacity to claim that the text is a fake. The selection does, however, illustrate the fact that just as each person tends to consider himself special, each race probably considers itself unique. We used to think that the earth was the very center of the universe (and even that we lived at the center of the earth). We discovered that this was incorrect. We thought that the sun was special, only to learn that it is but an average star like billions of others. Can we realistically expect that our race is unique? As the unknown author of the above text wrote, let's look at the facts.

It is estimated that there are 2×10^{11} stars in the Milky Way galaxy, the sun being one of them. Roughly half of these stars are part of systems of two stars which revolve around one another, and these double star systems are not likely to have planets around them stable enough to support life. According to the present theory of the evolution of stars, however, most of the remaining single stars have planetary systems of some kind. So theory predicts that most stars have planets, but a healthy skepticism should lead you to ask whether there is evidence to support this hypothesis. Yes, there is. It must be admitted that no planet has ever been seen other than those circling the sun, and we know that none ever will be because of the limitations imposed by the great distances involved. The evidence for planets orbiting other stars depends upon a less direct observation and is based upon gravitational theory.

take-along-do-it

This could be a fun one. Two people are required, and if they are of opposite sex, there will be an advantage in that you will not get the two mixed up. The two people should face one another, hold hands, and whirl around. Note that if the two people are of about the same weight, they whirl around a point midway between them (Fig. 5–2). Each moves in a circle about that point. Now a small child is needed. Let the child take the

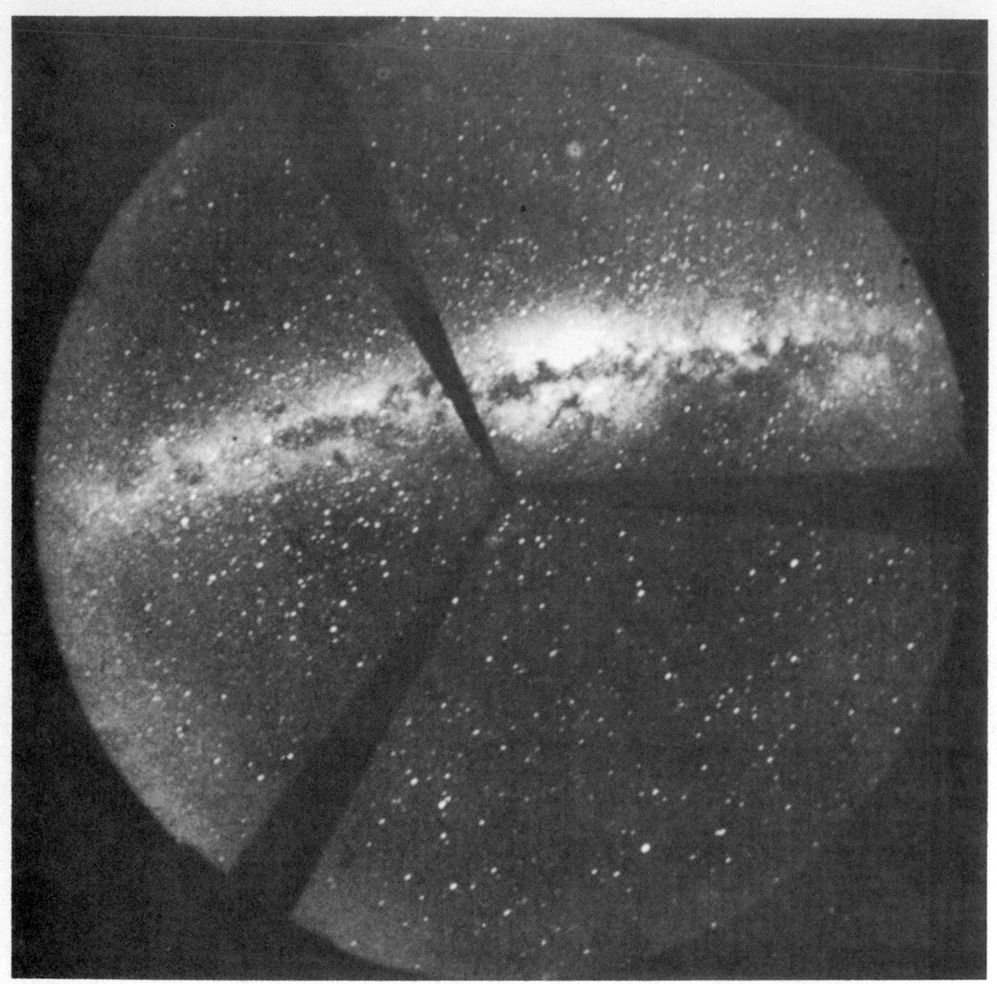

FIGURE 5–1. A wide angle view of the Milky Way. Supports for the camera obstruct part of the field of view. (Courtesy of A. D. Code and T. E. Houck.)

place of one of the whirlers. Note that the point about which the two turn is no longer midway between them but is much closer to the bigger person. This person moves in a very small circle while the child moves in a large one (Fig. 5–3).

FIGURE 5–2. Two people of the same mass rotate around a point midway between them.

If the preceding TADI is expanded to include a really big person and a really little one, the big person would act *almost* like the center of motion, with the little one covering a lot of distance to get around him (Fig. 5–4). Even in this extreme case, however, the big person would not be turning in the same way as he would if he were not whirling the small person. Planets and stars work in the same way (Fig. 5–5). As mentioned above, there are many systems of two stars which revolve around one another. If they are of the same mass, they both revolve around a center point. If one is larger, the two stars move around a point closer to the larger one.

Now suppose that by careful observation with a tele-

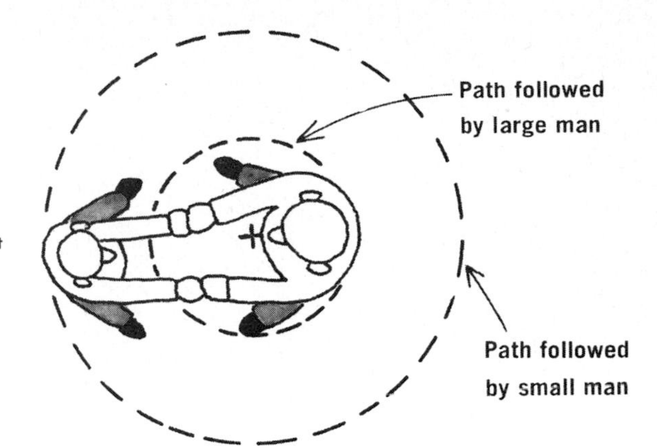

FIGURE 5–3. If one person is more massive than the other, the point around which the two revolve will be closer to the larger person.

scope we find a star that shows a slight wobble. We see it moving slightly to the left, then back to the right, and then back to the left again, each leg of its motion requiring the same amount of time. Although we cannot see another body rotating with the star, we can conclude that another body is there.

Such wobbling stars do exist. The first of these to be observed was Barnard's star. Its wobbling motion, observed over the last half-century, can be explained by hypothesizing that it has two massive planets revolving around it. Each of these has a mass far greater than the earth's mass, and one revolves around Barnard's star every 12 years, while the other completes a revolution every 26 years. A comparison can be made with our solar system. We have two large planets, Jupiter and Saturn, which have revolution periods of about 12 and 30 years respectively. In addition to these, of course, our system has a number of smaller planets. If we assume that our system was formed in a manner which is typical of the formation of planetary systems, we must conclude that Barnard's star is likely to have other, smaller planets circling it. Barnard's star was only the first of many to reveal its planetary system to us. Because the wobble of a star produced by a planet is so very small, we can expect to

Edward Emerson Barnard (1857–1923)

Edward Emerson Barnard was born of poverty-stricken parents in Nashville, Tennessee. As an amateur astronomer, he discovered a comet, and when he was 30 years old, he graduated from Vanderbilt University, where he had charge of the observatory. In 1892 he discovered a fifth satellite of Jupiter (Galileo had discovered the first four nearly 300 years earlier). It was Barnard who first detected the faster-than-usual motion of a particular star across the sky, and this star now bears his name: Barnard's star.

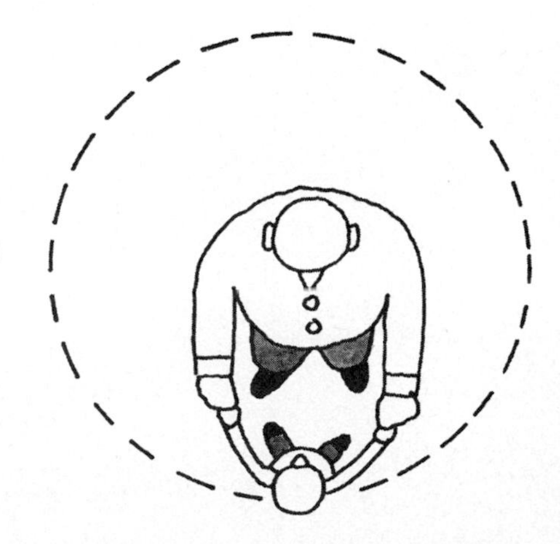

FIGURE 5–4. A giant and a child whirling. The giant turns almost as if the child were not there.

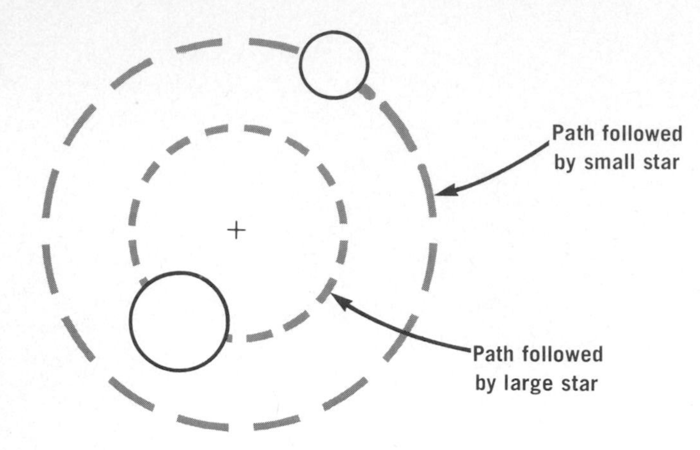

FIGURE 5–5. The two stars rotate around a point closer to the more massive one.

observe it for only the nearest stars. *And the wobble has been observed for almost every star which is close enough for such motion to be detectable.*

Thus the prediction resulting from the theory of stellar evolution is verified in this case. And we see that there is

"FROM ALL INDICATIONS, OURS IS ONE OF THE BIGGEST UNIVERSES THERE IS."

Reprinted with the permission of American Scientist/Sidney Harris.

FIGURE 5–6. A very massive planet will not support carbon-oxygen-water life.

good reason to believe that our solar system is not unique, but ordinary. However, the existence of a planetary system is not enough; not every planet will support life. Let us, for the moment, consider only forms of life which are based upon carbon, oxygen, and water, as is ours on earth. Other systems may conceivably be possible, but until more is known in this area, we must take the idea for what it is—an interesting conjecture.

Below are listed some of the requirements a planet must fulfill in order to support carbon-oxygen-water life:

(1) It must be massive enough to retain an atmosphere. Recall that the escape velocity for a planet depends upon the gravitational pull (and therefore the mass) of the planet. Earth's pull is not great enough to hold the lightest gases, hydrogen and helium, which move at speeds greater than the more massive gases, such as oxygen.

(2) It must have an average temperature which is not far below the freezing point of water (0° C, or 32° F) or too near the boiling point (100° C, or 212° F). If our earth were as close to the sun as is Venus, it would be too hot; if as far away as Mars, it would be too cold.

(3) It must contain organic molecules. Under the right circumstances, such molecules can be produced from a mixture of hydrogen, ammonia, methane, and water vapor by ultraviolet radiation and lightning.

FIGURE 5–7. A planet's temperature is an important consideration.

FIGURE 5–8. This planet is probably rotating too fast.

(4) It must have a rotational frequency such that its temperature extremes are not too great. A 100-hour day on earth would result in our all being cooked by day and then frozen by night.

(5) Its parent star must be of a temperature and mass such as to allow favorable life conditions. Too massive a star, for example, would cause tidal effects which would be catastrophic to life.

There are about 2×10^{11} stars in our galaxy, and approximately half of these have planetary systems. One fifth of the stars in the sky are of a suitable temperature and mass to support a life-planet, and about half of these should have a planet at a suitable temperature. Suppose that only one out of five of these is of suitable mass and rotational speed. We have thus far narrowed the number of possible life-stars in our galaxy to 2×10^{9}. What percentage of these have the correct chemical conditions to develop life? We know from the spectra of stars that the chemical composition of our sun is not unusual at all. Fifty per cent is a reasonable guess for planets with correct chemicals present.[1]

[1]Two meteorites (rocks hitting earth from outer space) have been found to contain amino acids, chemicals that played an important part in the evolution of life on earth.

FIGURE 5–9. A too-massive star causes disastrous tidal effects.

We are down to 10^9—one billion—possible life-sites, and now comes the most difficult question: What is the probability of life developing once these conditions are met? When conditions are different from those on earth, the probabilities may be vastly different. Life may be less probable under such conditions, but on the other hand, it may be *more* probable, and until we know more about organic evolution, the estimate of this probability will remain the weak link in our chain of calculations. If the probability is small, for example one in a billion, then we may be the only life around. Most "experts" who have considered the problem, however, believe that the probability is high, between one chance in ten and a sure thing. But to be on the safe side, let us say that only one in 100 planets with the right conditions actually develops life. We still have 10^7 planets with carbon-oxygen-water life on them.

Ten million planets with life on them! But note that we said *life*, not intelligent life. Life includes algae and moss. Algae and moss are interesting to a botanist, and they make wonderful pets, but they are hardly interesting dinner partners. Let us consider intelligent life, with whom communication may be possible.

INTELLIGENT LIFE

We are learning more all the time about the evolution of life on earth, from the earliest forms to man as he now exists. A number of pieces of the puzzle are still missing, however. In addition, we do not understand the process of evolution sufficiently to answer the question of how many alternate routes exist from basic life to intelligent life. It seems unlikely that man as we know him is the only possible intelligent life that could have developed.

The human race evolved as it did because of the particular set of circumstances that existed during the few billion years since the first life was formed on earth. If different circumstances had existed, we can suppose that intelligent life would have taken some form other than human. Consider the hypothetical being that may have formed instead of man on the earth. Such a being would necessarily have some characteristics in common with us, for the laws of nature would apply equally to it. Its size, for example, would have an upper limit, because a being of too great a bulk would not be able to feed itself adequately to provide energy for motion. It would have senses to perceive the external world, and although these might differ somewhat from ours, the range of wavelengths perceived by its "eyes" would probably correspond to those which pass freely through the atmosphere, that is, visible light and some infrared and ultravio-

"LIFE, YES- BUT AS FOR INTELLIGENT LIFE, I HAVE MY DOUBTS."

Reprinted with the permission of American Scientist/Sidney Harris.

FIGURE 5–10. A space being with "eyes" able to perceive radio waves.

let radiation.[2] Our atmosphere is transparent to radio waves, but in order to perceive them, a being's "eyes" would have to be the size of radio antennae, or several feet across (Fig. 5–10). So this is not likely.

Thus, it is unlikely that life on earth would have taken any radically different form than it has. But what about life on other planets? The gravitational force on another planet would be different from earth's, and creatures on a planet with a stronger gravitational pull would be smaller. On the other hand, creatures from a planet with a weak gravitational attraction could be larger. (Such creatures would have difficulty in coping with earth's gravity.) Life on other planets would differ from us in other ways also, depending upon the environment in which this life evolved.

---

[2]Infrared and ultraviolet are two among many types of waves referred to as electromagnetic waves. Other such waves include radio waves, visible light, x-rays, and gamma rays. We will explore these in greater detail in Chapter 16. For the moment, consider infrared rays in their more common context of "heat waves" and ultraviolet rays as the kind emitted by "black-lights."

What is the probability that intelligent beings exist on other planets somewhere in the universe? We are unable to say, but we concluded that there are probably 10 million planets in the Milky Way galaxy with some form of life. The reader is asked to choose some fraction of these on which life is likely to have evolved to produce intelligence. Surely not half of them. Is 1/10 too many? Who knows? Let's be conservative. Let's say that intelligent life has evolved on only one in 10,000. That still gives us 1000 planets in our galaxy with intelligent beings. Our galaxy is only one of billions of galaxies in the universe. Suppose each of them has 1000 homes for intelligence. That means trillions of planets with intelligent life!

Communications with "Them"

If "they" are out there, the obvious next step is to consider contacting them, or at least searching for signals from them. In 1960, a group of astronomers used the 85-foot radio telescope at Green Bank, West Virginia, for several months to listen for communications from beings in outer space. The telescope is normally used to receive radio waves emitted from stellar objects,[3] but during the search for life it was used to scan for signals which might vary in amplitude or frequency with some pattern, indicating the presence of a message. The search was unsuccessful, and the astronomers returned to their regular work.

In 1971, a group of scientists and engineers met for a summer to study methods of detecting extraterrestrial life.[4] The project was funded by the National Aeronautics and Space Administration (NASA) and was called Project Cyclops.[5] After reasoning somewhat as we did above to estimate the probability of the existence of intelligent extraterrestrial life, they went one step further in calculating probabilities. We decided that there are almost surely 1000 planets in our galaxy with intelligent life on them, but we did not consider the likelihood of such life being in the stage of its evolution when it would be sending out messages into space. Intelligent life—man—has existed on earth for 50,000 to 100,000 years. But radio communication has existed for less than 100 years. This is only 1/1000 of the time we have been here. The question of how long man will

[3]You may normally think of radio waves only in terms of communications, but actually, electromagnetic waves of radio wavelengths are randomly emitted from many starlike objects in space.

[4]Oliver, B. H., and Billingham, J.: *Project Cyclops: A Design Study of a System for Detecting Extraterrestrial Intelligent Life.* NASA CR 114445, 1971.

[5]Cyclops, in Greek mythology, was a member of a race of giants with only one eye, which was in the middle of the forehead.

remain in this communicative phase is a question we can only guess at. If our civilization should end in another 10 years and if it is typical of intelligent life, this would mean that we could expect only one out of 1000 of those intelligent-life planets to contain life in a communicative phase. Our 1000 possible planets would be reduced to one. And, if there is only one, or a few, planets out there from which we can expect communication, then the search for life in this galaxy is almost doomed before it begins.

But that is a very pessimistic view. Certainly our civilization will go on with at least today's technology for quite a while. The problem is that no one can guess the typical length of time a race spends in the communicative phase.

The Cyclops group took the optimistic approach. They proceeded on the assumption that a number of civilizations exist in this phase at the present time and concluded that many of them have probably contacted one another and are communicating at this very moment. Perhaps their signals pass right by the earth, and here we are, unaware of them. The Cyclops group suggested that we get busy and join the conversation. To do so, they made recommendations on a possible design for a system to detect any signals that may be reaching the earth.

A Listening System

These beings, if they are now communicating with one another or are trying to contact someone else in the universe, would be sending out electromagnetic waves. So it

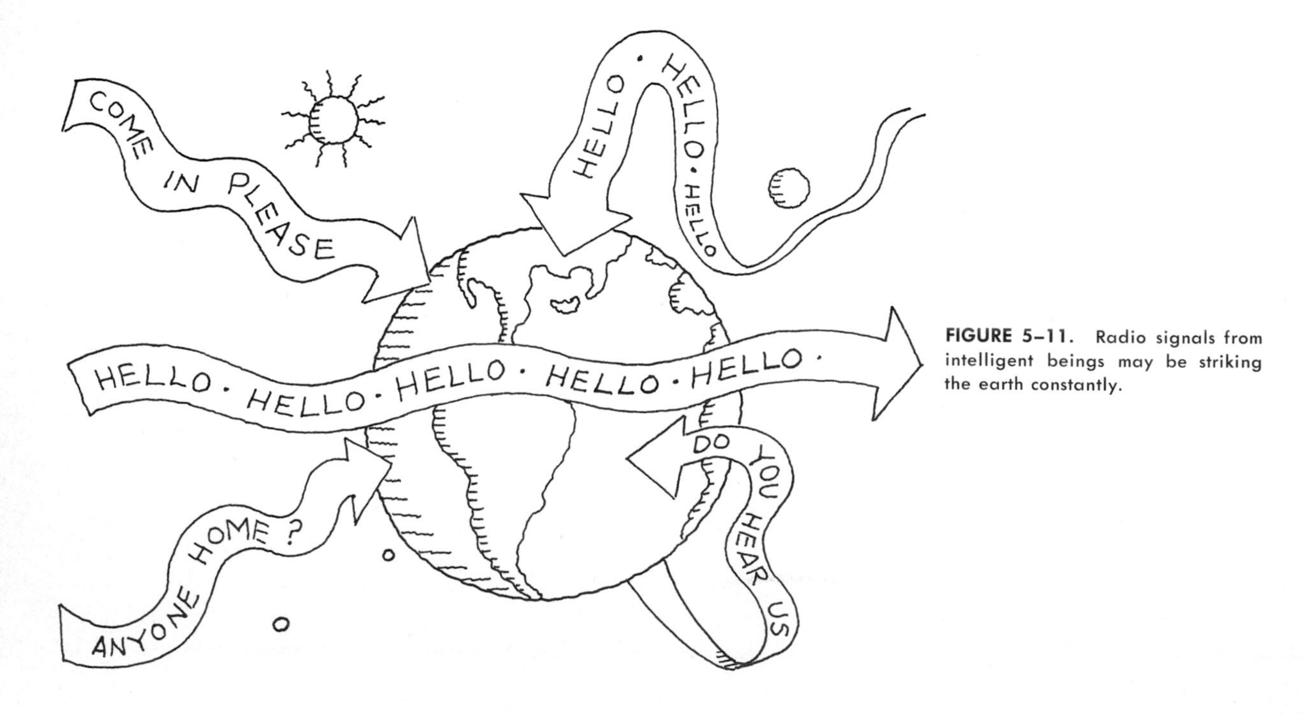

FIGURE 5–11. Radio signals from intelligent beings may be striking the earth constantly.

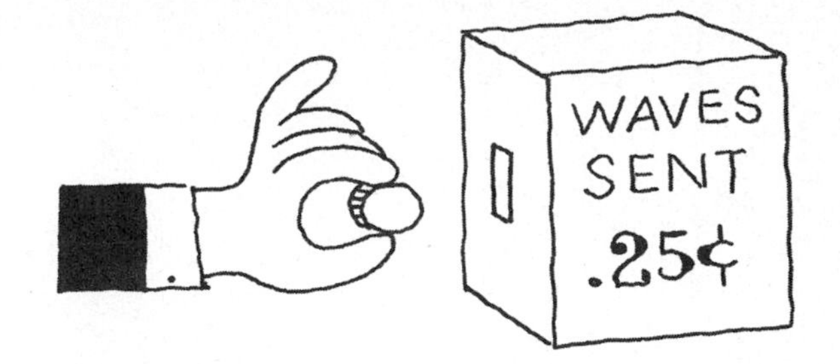

FIGURE 5–12. Economical?

may seem that all we need to do is to look for waves containing some type of message. But the electromagnetic spectrum ranges from the very short-wavelength gamma rays to very long radio waves. And different methods and equipment must be used to detect different parts of the spectrum. There are infrared detectors for infrared light and radio antennae for radio waves, while our eyes detect only visible light. So the Cyclops team first set out to determine where in the electromagnetic spectrum a search would have the best chance of success. A number of factors were considered in making this determination, including the following:

(1) The waves used must be economical to send.
(2) The waves must not be absorbed or deflected by material in space between the stars. (Outer space is not a perfect vacuum. It contains small amounts of gas, mainly hydrogen.)
(3) The waves must be easy to detect. This means that it must be easy to distinguish between them and waves coming from stars and other heavenly objects. Stars, for example, emit so much visible light that light signals would be drowned out.
(4) The waves must be able to be sent with a lot of power so that they have a very long range.

FIGURE 5–13. Easy to detect?

Considering these points, Cyclops suggested that there is a range of frequencies in the microwave region that is most likely to be chosen as a galaxy-wide communication frequency.[6] After calculating the size of the antenna which would be needed to detect a signal of reasonable power, they designed an antenna system. An artist's drawing of their proposed system is shown in Figure 5–14. Figure 5–15 shows a ground level view of this system of 1000 to 2000 individual bowl-shaped antennae covering an area about eight miles across. Such a system is estimated to cost in the neighborhood of $6 to 10 billion over a period of 10 to 15 years.

Sending Signals

The world's largest radio telescope is in a natural valley near Arecibo, Puerto Rico. The telescope's 1000-foot-diameter reflector is built right along the surface of the

[6]They suggested frequencies from 1.42×10^9 cycles/second to 1.66×10^9 cycles/second.

FIGURE 5–14. Artist's concept of high aerial view of the entire Cyclops system. Diameter of the antenna array is about 16 kilometers. (Courtesy of NASA.)

FIGURE 5–15. Artist's concept of low aerial view of a portion of the Cyclops system antenna array, showing the central control and processing building. (Courtesy of NASA.)

specially shaped valley. In 1974, the reflector was resurfaced, and at the dedication of the improved telescope on November 16 of that year, the most serious attempt to date was made to send signals to other beings in space. The signals were beamed toward a cluster of 300,000 stars (called Messier 13), which is some 24,000 light-years[7] away. If there are intelligent beings on some planet circling one of those stars, and if they are searching for signals, and if they answer our message, our great-×-1600-grandchildren may receive the reply.

The problem of deciding what message to send was not minor. Language differences here on earth probably do not compare to differences between earth languages and extraterrestrial languages. It was decided to send the message as only two symbols, similar to the dots and dashes of our Morse code, and to send a message containing 1679 symbols. This number is the product of 23 and 73 (and no other pair of numbers), and if the message is reassembled into a

[7]A light-year, the distance light travels in a year, is about 5.8×10^{12} miles (9.7×10^{12} km).

23 × 73 grid, the message of Figure 5–16 appears. We will not try to explain everything in the message, but the key lies in the numbers 1 through 10 written at the top as binary numbers. The message uses these to indicate the carbon-oxygen-water nature of life on earth, and the pattern of the DNA molecule is shown, as well as other information.

In sending the message, the 1679 symbols were repeated again and again, so that even if the alien civilization is unable to decipher the message, they will recognize it as being from intelligent life. And if they miss the first "Hello" from earth, the Arecibo telescope is fixed to automatically send the same message over and over whenever it is not being used in other astronomical work.

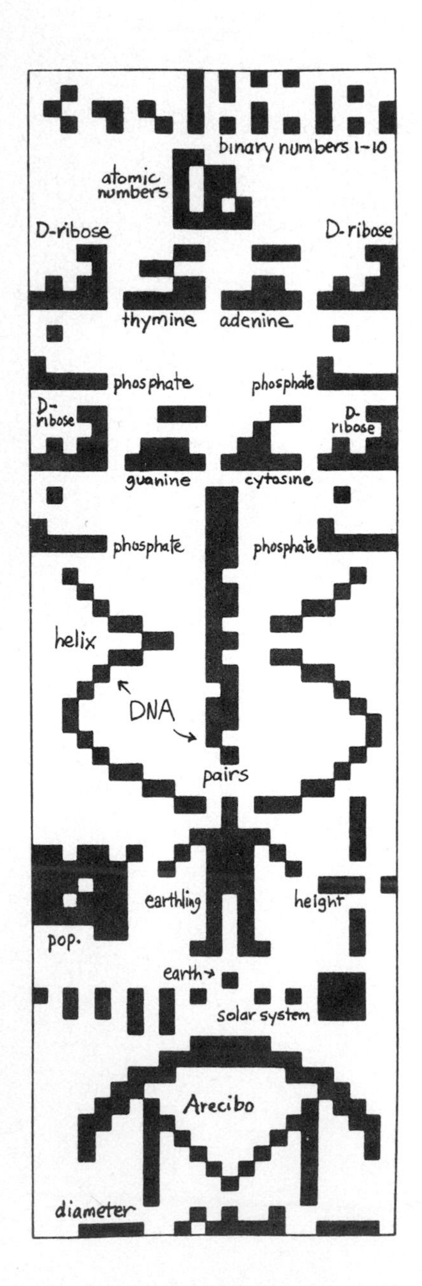

FIGURE 5–16. The message beamed into space by the Arecibo radio telescope.

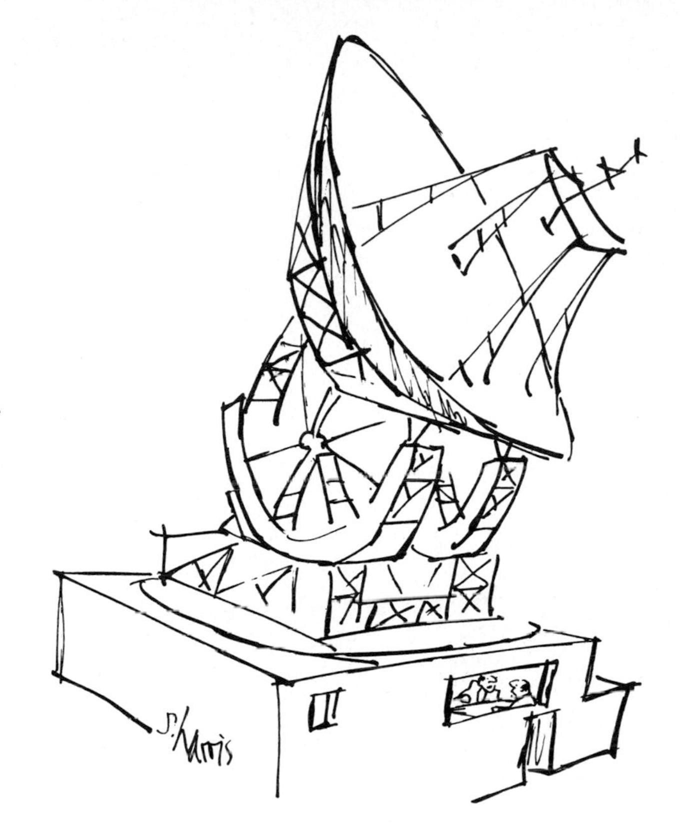

"As I understand it, they want an immediate answer. Only trouble is the message was sent out 3 million years ago."

Reprinted with the permission of American Scientist/Sidney Harris.

Should We Make Contact?

The question "Can we make contact?" is essentially a scientific and technical one. But difficult as it is to answer, the question "Should we?" is perhaps even more difficult. By doing so, we could be inviting destruction from outer space, a danger not involved in the attempt to detect waves sent by another civilization. Even if we choose only to listen, is mankind prepared for the expansion of outlook which would occur when we start getting messages from outer space? The prospect is both scary and exciting. Frank D. Drake, in introducing the report on Project Cyclops, writes:

> At this very minute, with almost absolute certainty, radio waves sent forth by other intelligent civilizations are falling on the earth. A telescope can be built that, pointed in the right place and tuned to the right frequency, could discover these waves. Someday, from somewhere out among the stars, will come the answers to many of the oldest, most important, and most exciting questions mankind has asked.[8]

Mankind has a history of not running from challenges. We have generally felt that the thrill of proceeding into the unknown outweighs the dangers involved. But are we ready? Are you?

---

[8]From Drake, F. D.: Intelligent Life in Space. New York, The Macmillan Co., 1967.

TRAVEL BEYOND THE SOLAR SYSTEM

We began this chapter by considering the probabilities of life beyond the earth and concluded that there is very probably intelligent life "out there." We considered communication, but what about travel? This question was saved for last because it is by far the most difficult. At present, the prospects are remote, "far out" perhaps.

The distance to the nearest star that is visible to the naked eye (Alpha Centauri) is 2.4×10^{13} miles, or 4.5 light-years, and this is a typical value for the distance between stars in our galaxy. So it takes $4\frac{1}{2}$ years for light to get here from the nearest star. Now, all electromagnetic waves travel at the speed of light, so consider how a conversation would go with someone near that star.[9] Suppose we start by asking, "How are you?" It would take nine years to get an answer!

Since it is unlikely that we would be lucky enough to find communicating life on a planet circling the very closest stars, we should figure that the nearest communicators may be 1000 light-years away. That's awfully far, so let's just imagine that Alpha Centauri has at least a hospitable planet on which we could land. Then let's consider traveling that far.

Suppose we try to use the method of chemical propulsion of rockets that is in use today, but imagine that the

[9]Actually, Alpha Centauri is one of a system of two or three stars orbiting one another. Any planet in the system would be too unstable to have supported the evolution of life.

FIGURE 5–17. Any intelligent life would be too distant for dialog.

rockets are made more powerful than at present. Still, the maximum conceivable speeds of such rockets would require more than 10,000 years to reach Alpha Centauri. So chemical propulsion is out.

Nuclear energy, though, provides far more energy than is possible by chemical means (as evidenced both by atomic bombs and by atomic power plants). At present, we do not have nuclear-powered rockets, but such rockets are within the realm of possibility with foreseeable advances in technology. Calculations of the maximum speeds obtainable with present knowledge of nuclear processes show that the fastest possible trip would still take 40 years—one way. Thus, it seems that technology of the foreseeable future does not make such a trip feasible. As a result, let's forget the near future and consider the maximum imaginable energy available in nuclear reactions.

As we will see in our study of the nucleus of the atom, nuclear energy is produced in a nuclear reaction by an accompanying loss of mass of the atom. Mass disappears and energy appears. The uranium fuel in an atomic bomb decreases in mass by about 0.01 per cent during the blast. This

FIGURE 5–18. The explosion of an A-bomb. (Courtesy of U.S. Air Force.)

means that if you start with exactly 100 pounds of fuel, after the reaction you have 99.99 pounds left. The 0.01 pound which disappeared was turned into heat, sound, light, and electromagnetic radiation. At present, we are unable to achieve a reaction in which more than about 0.01 per cent of the fuel is actually used. Clearly, if we could use even one per cent of the fuel, we could increase the energy produced to 100 times as much. But at present we are not even close to achieving such efficiency.

But we can still imagine a day when we will achieve total change of mass into energy, when we can cause *all* of the fuel to be used to boost our spacecraft toward Alpha Centauri. The Cyclops group calculated the fuel needs under such conditions to get to Alpha Centauri and back in 10 years. They based their calculations on a vehicle weighing 1000 tons, not counting the fuel. (One thousand tons is not as much as it may sound when you consider that it must support a crew for 10 years.) They concluded that the necessary fuel would weigh 33,000 tons and would cost at least $1 million billion. And to discover extraterrestrial life by such means might well require 1000 to 10,000 such trips. For economic reasons, then, electromagnetic waves seem to be the only hope of contact with "them."

UFO'S

Although speculation about life on other worlds has taken up much space in newspapers, magazines, radio, and television, perhaps nothing of this nature has caught the fancy of the American public quite so much as the description of sightings and alleged contacts with Unidentified Flying Objects, or "flying saucers." The name "flying saucer" is often applied because of their reported shape.

Sightings have been reported for an untold number of years. In 1947, the unanswered questions concerning the origin and aims of these phantom visitors caused the US Air Force to institute a formal study of them through a committee which, in 1951, became known as Project Bluebook. The efforts of this study were directed primarily toward investigating sightings and explaining the events, if possible, in terms of natural occurrences. For example, Jupiter, stars, airplanes, swamp gas, and weather balloons have found their place in explanations of UFO sightings. After a number of years of collecting data, the Air Force entered into agreement with the University of Colorado to formalize a report. Dr. Edward Condon, a respected physicist, was appointed director of this assignment, and the report, which was completed in 1968, bears his name.

The basic conclusion of the Condon Report was that

there is no substantial proof that flying saucers are attributable to anything other than natural phenomena. It suggests that there is very little scientific information of any value likely to be realized by additional study.

It is interesting to note that most of the reports have many features in common and that many of the reported sightings mention phenomena which are alien to the kind of physics we have discussed heretofore, or will discuss. For example, many observers report that the saucers can fly at tremendous speeds, but no sonic boom is ever reported.

The following is a description of an imaginary sighting which combines common features of reported sightings. Analyze it with respect to those observations that can be explained in terms of physics as we know it (or even by the laws of common sense) and those that go against the grain. For those outside the realm of scientific explanation, fantasize a *possible* explanation of how these creatures are circumventing our natural laws in order to achieve the observed results. If you are not sure about some, don't get upset. Either review what you have read previously, or wait until you have been through the remainder of the book to make your judgment.

(1) The UFO travels at very high speeds.
(2) It often makes right angle turns in flight.
(3) When it is in the vicinity of an automobile, the car's engine, radio, and headlights shut off. Diesel engines, however, continue to run. After it flies away, everything returns to normal.
(4) It leaves marks on the ground and scorches vegetation.
(5) An observer feels an electrical shock when the UFO is in close proximity.
(6) It can be sighted by radar.
(7) It sometimes rocks back and forth in flight.
(8) It can hover in midair.
(9) When relatively far away, it appears as a blue glow; at close range, it is red or orange.
(10) It is able to change its size and shape.
(11) It is usually sighted over power lines or regions in which there are deviations in the magnetic field of the earth.

Conclusion

The Condon Report concluded that no evidence was presented that UFO's are of extraterrestrial origin and that the Air Force is capable of investigating any further UFO sightings. There are many, though, who dispute these conclusions, among them reputable astronomers. These people claim that further investigation is warranted, and they say that our government should be funding such investigations. Even if UFO's are not of extraterrestrial origin, persuasive arguments suggest that there is life out there and that we should attempt to communicate with it. Such communication might be valuable to mankind, and it might not. Is there life out there, and should we spend the money to find out?

OBJECTIVES

After this chapter you should be able to:

1. Relate historical examples showing that man has a tendency to overestimate his uniqueness.
2. Describe double-star systems and demonstrate how the relative masses of the two stars affect their relative motions.
3. Explain how double-star systems are observed even when one star does not give off light.
4. Describe five requirements a planet would have to satisfy in order to support carbon-oxygen-water life.
5. Use the number of stars in our galaxy to estimate the probable number of life-sites.
6. Explain how our limited knowledge of evolution on earth affects our ability to determine the probability of extraterrestrial life.
7. Describe how gravitational forces and characteristics of the atmosphere affect the size and senses of beings on a planet.
8. Explain the effect of the length of a race's communicative phase on the probability of our receiving messages from extraterrestrial life.
9. Describe the Cyclops antenna system.
10. List and explain four factors that must enter into the decision of what wavelengths of radiation to use for interstellar communication.
11. Explain present and (probable) future limitations of interstellar travel, including considerations of nuclear energy.
12. List at least eight features that many UFO reports have in common.

QUESTIONS—CHAPTER 5

1. Write a short essay on the parallels between the way a child becomes less self-centered as he matures and the way mankind has come to realize that it is not the center of the universe.

2. The earth and moon form a rotating system of two astronomical objects. Does the earth remain stationary with respect to the moon's orbit? Refer to Chapter 3 to cite evidence for your answer.

3. How long do you think man's communicative phase will last?

4. Explain why there is a lower limit to the mass of a planet that permits it to support oxygen-based life.

5. The earth itself has regions with great temperature extremes. Yet life in some form has been found everywhere on earth. Discuss problems involved in trying to estimate temperature limits for carbon-oxygen-water life.

6. In what fraction of the cases of life existing somewhere do you predict evolution to intelligent life? Based on your prediction, how many sites in our galaxy should we expect to have intelligent life?

7. Describe the effect you would expect here on earth of receiving a message from a distant race.

8. Hypothesize an explanation for the common observation that UFOs cause regular car engines to stop but do not affect diesel engines.

9. Why would a planet revolving with two stars have less stable conditions than a planet with only one star?

10. The text stated that our great-×-1600-grandchildren might receive an answer from Messier 13. Why will it take so long?

11. Why did man's eyes evolve to perceive the particular set of wavelengths we see rather than other wavelengths?

12. It was stated that the Cyclops group assumed that communication is taking place now between other civilizations. Considering the distances involved, describe the nature of these "conversations."

13. In trying to deduce the likely number of planets supporting intelligent life, "pessimistic" estimates were made in this chapter, and the figure of 1000 intelligent races was reached. Repeat each step, but choose a more optimistic value in each case. How many life-sites for intelligence are estimated with these values?

14. Do you think that our government should support a search for extraterrestrial life? Write an essay in the form of a letter to your Congressman in which you tell him your views on the subject.

SECTION TWO

its FORMS and its PROPERTIES

6

IT'S ELEMENTARY

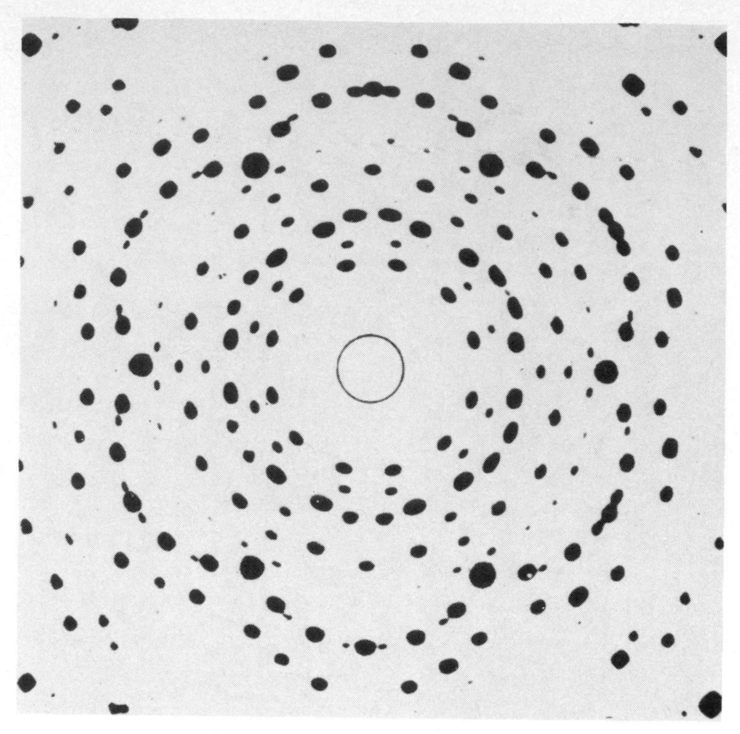

Courtesy of Eastman Kodak Company.

We have thus far studied about forces and motions and how these ideas apply to large-scale objects. Now we move to smaller things—smaller, in fact, than what is visible. We'll find that, with limitations, the laws which we have discussed apply on this scale also. These ideas and laws will be referred to in discussing such topics as the composition and behavior of solids, liquids, and gases. But first we must examine those elements which make up all of the material world. The elements, you know, are elementary.

THE ELEMENTS

Most of the substances we encounter in our everyday life can be decomposed chemically to other substances. Water, for example, can be broken up into two substances, hydrogen and oxygen. A small number of substances are incapable of being decomposed to more fundamental constituent parts. These are called *elements*. Hydrogen and oxygen are elements, as are carbon, iron, calcium, and gold. In all, there are 92 naturally occurring elements. Visits to the moon, telescopic observation of stellar objects, and other attempts to explore the vastness of space have not revealed the presence of any element other than the 92 found on earth. Besides these natural elements, however, there are several man-made elements, including such household names as americium, berkelium, californium, and fermium, which extend the list to 105.

Ninety-two seems to be a surprisingly small number of elements when one considers that all things in the universe, from shoestrings to brain cells, are composed of combinations of those few. For example, the human body is composed primarily of the elements carbon, hydrogen, oxygen, and nitrogen. Some ten years ago it was calculated that the net worth of the human body would be about $.97 if it were broken down into its elements and sold at the going rate. Since then, inflation has caused the price to soar to an estimated $3.26.

A Trip to Inner Space

In order to unlock the remarkable properties of these elements, let us use some magic from the classic *Alice in Wonderland.* Alice reportedly came upon a mysterious cake. When she took a bite, it caused her to shrink to a very small size. If we could possess one of these magic morsels and start to munch away, we would soon find ourselves shrinking into a microscopic domain of incredible wonder. An apparently smooth, well-polished floor would be revealed to have cracks which to our shrinking size would take on the vastness and grandeur of the Grand Canyon. Water, which normally seems so innocuous when we drink it, would be revealed to be teeming with marine life that would make prehistoric dinosaurs seem like kittens. The investigation of these creatures occupies the time of the biologist. Our present destination requires one more bite of the cake.

After that bite, we find ourselves in a situation in which the crack in the floor has given way to an apparent emptiness. No foothold remains, and we find ourselves falling, like Alice, into an apparently bottomless pit. Occasionally we see objects, somewhat like our solar system, floating around in an otherwise empty void. These are atoms, the ultimate building blocks which make up elements.

No self-respecting text would go beyond this point without attempting to give you some idea of the size of an atom. Let us, then, return to normal size and do so. Imagine that every atom in a thimbleful of water is suddenly changed to the size of one drop of lemonade. About 10^{23} atoms[1] are contained in a thimble of water, so we will have a lot of lemonade. In fact, we will have enough lemonade to cover the entire state of Texas to a depth of about 1000 miles—truly a Texas-size drink.

MOLECULES

If we had examined the structure of matter more closely as we shrank in our trip into fantasy, we would have found that often, two or more atoms are joined together.

The bond which holds atoms together is such that if the atoms get too far apart, they are pulled back together; and if they get too close, they are forced apart. The force acts much as if they were connected by a spring (Fig. 6–1). Combinations of two or more different atoms joined in this way are called *molecules*, and substances composed of molecules are called *compounds*. The most common molecule is that of water, which is made up of two hydrogen atoms linked to one oxygen atom. Another common (but dangerous) example is the poisonous gas carbon monoxide, composed of one carbon and one oxygen atom per molecule. In fact, most of the objects that you pick up each day have different kinds of atoms joined together into molecules which form the basic building block of the substance. In the case of elements, however, the basic block is the individual atom.

In discussing atoms, elements, and compounds it is convenient to adopt symbols to represent them. Hydrogen is symbolized by the letter H, helium by He, oxygen by O, and so forth. A compound such as water is written symbolically as H_2O. The "2" beside the symbol for hydrogen indicates that two hydrogen atoms are connected to the oxygen atom. Table 6–1 lists the elements along with the symbols used to identify them.

Vocabulary

It may be worthwhile to stop here and collect some of the terms used in discussing the elements. In science you'll often find everyday terms used with non-everyday meanings, so a review should help. One word of advice first: please don't consider this as a list of definitions to memorize. Try to understand the meanings of the definitions rather than to memorize the words.

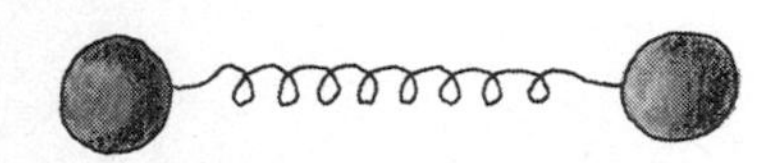

FIGURE 6–1. Atoms in solids act as if they were connected by springs.

[1]If somehow you skipped the preceding chapters and don't understand what this notation means, read Appendix I, which deals with scientific notation.

TABLE 6–1. THE ELEMENTS AND THEIR SYMBOLS

NUMBER	ELEMENT	SYMBOL	NUMBER	ELEMENT	SYMBOL
1.	hydrogen	H	54.	xenon	Xe
2.	helium	He	55.	cesium	Cs
3.	lithium	Li	56.	barium	Ba
4.	beryllium	Be	57.	lanthanum	La
5.	boron	B	58.	cerium	Ce
6.	carbon	C	59.	praseodymium	Pr
7.	nitrogen	N	60.	neodymium	Nd
8.	oxygen	O	61.	prometeum	Pm
9.	fluorine	F	62.	samarium	Sa
10.	neon	Ne	63.	europium	Eu
11.	sodium	Na	64.	gadolinium	Gd
12.	magnesium	Mg	65.	terbium	Tb
13.	aluminum	Al	66.	dysprosium	Dy
14.	silicon	Si	67.	holmium	Ho
15.	phosphorus	P	68.	erbium	Er
16.	sulfur	S	69.	thulium	Tm
17.	chlorine	Cl	70.	ytterbium	Yb
18.	argon	Ar	71.	lutecium	Lu
19.	potassium	K	72.	hafnium	Hf
20.	calcium	Ca	73.	tantalum	Ta
21.	scandium	Sc	74.	tungsten	W
22.	titanium	Ti	75.	rhenium	Re
23.	vanadium	V	76.	osmium	Os
24.	chromium	Cr	77.	iridium	Ir
25.	manganese	Mn	78.	platinum	Pt
26.	iron	Fe	79.	gold	Au
27.	cobalt	Co	80.	mercury	Hg
28.	nickel	Ni	81.	thallium	Tl
29.	copper	Cu	82.	lead	Pb
30.	zinc	Zn	83.	bismuth	Bi
31.	gallium	Ga	84.	polonium	Po
32.	germanium	Ge	85.	astatine	At
33.	arsenic	As	86.	radon	Rn
34.	selenium	Se	87.	francium	Fr
35.	bromine	Br	88.	radium	Ra
36.	krypton	Kr	89.	actinium	Ac
37.	rubidium	Rb	90.	thorium	Th
38.	strontium	Sr	91.	protractinium	Pa
39.	yttrium	Y	92.	uranium	U
40.	zirconium	Zr	93.	neptunium	Np
41.	niobium	Nb	94.	plutonium	Pu
42.	molybdenum	Mo	95.	americium	Am
43.	technetium	Tc	96.	curium	Cm
44.	ruthenium	Ru	97.	berkelium	Bk
45.	rhodium	Rh	98.	californium	Cf
46.	palladium	Pd	99.	einsteinium	Es
47.	silver	Ag	100.	fermium	Fm
48.	cadmium	Cd	101.	medelevium	Md
49.	indium	In	102.	nobelium	No
50.	tin	Sn	103.	lawrencium	Lw
51.	antimony	Sb	104.	rutherfordium	Rf
52.	tellurium	Te	105.	hahnium	Ha
53.	iodine	I			

Substance. A general term used here with no special meaning other than its everyday one: the matter of which a thing is composed.

Element. One of the 105 substances of which everything in the universe is composed.

Atom. The smallest possible chunk of an element which still is that element. Anything smaller would no longer have the properties of the original element.

Compound. A substance composed of more than one element bonded together as molecules.

Molecule. The smallest chunk of a compound which is still that compound. Molecules are composed of two or more atoms.

ATOMIC THEORY

A legitimate question to raise at this point is: How do we know that atoms exist? The idea that there is some final step in the division of matter, some ultimate building block, is not new. About 400 BC, Democritus, a Greek philosopher, proposed the idea. (He apparently had been persuaded in his belief by his teacher, Leucippus.) Plato and Aristotle later rejected the idea in favor of the belief that all matter is continuous and is composed of air, earth, fire, and water. In 1650, Gassendi, an Italian physicist, revived the atomic theory, and Isaac Newton became a convert to the idea.

John Dalton (1766–1844)

Dalton was born in Eaglesfield, England, the son of a weaver. He was educated in a Quaker school, and became a teacher there at the age of 12, upon the death of his teacher. At the age of 27, he taught at a college in Manchester, England. He is considered to be the father of present atomic theory, but his works included other fields of interest. He wrote a book about meteorology when he was 27 and also made the first study of color blindness.

The question of whether matter is continuous or discrete (individually distinct) was only a philosophical exercise prior to 1800. Then in 1806, an English schoolteacher, John Dalton, explained a well-known law of chemistry on the basis of the atomic theory. This was the first scientific evidence that atoms have a real existence. The law can be stated as follows: *In any pure compound the elements present are united in definite ratios by weight.* He used a compound called iron sulfide as his example. To form this compound, according to the above law, a specific weight of sulfur always combines with a specific weight of iron. Dalton argued that this occurs because one atom of sulfur (with its definite weight) always combines with one atom of iron with its definite weight. The resultant compound, made up of billions of these atoms, will always have a definite percentage, by weight, of these two atoms. If matter were continuous, Dalton reasoned, this definite ratio of weight would not always be found.

In the almost 200 years since Dalton's idea was presented, no contradictions have appeared. In fact, much additional information has lent support to it.

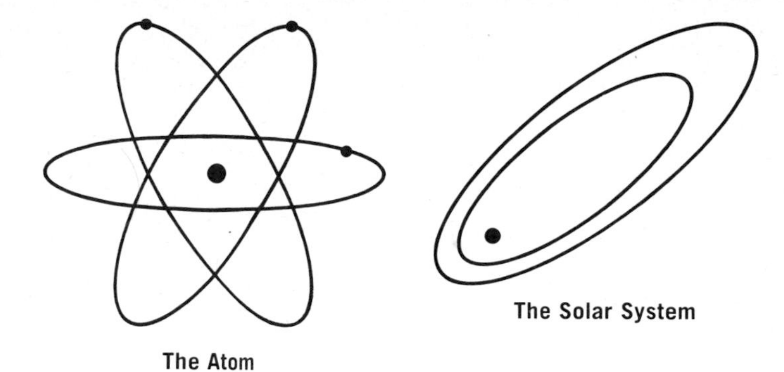

FIGURE 6–2. A comparison—the atom and the solar system.

WHAT DOES AN ATOM LOOK LIKE?

Back to Alice's cake. One more bite takes us spiraling down to the interior of the atom. We finally find ourselves at the center of the atom and, with a perfect three-point landing, we alight on the central core, the nucleus of the atom. Our journey started in our solar system and ends in a locale quite similar to it. The hub of our solar system is the sun, and orbiting around it are nine planets. The hub of our atom is the nucleus, and orbiting around it are electrons.

The solar system and the atom are compared in Figure 6–2. The atom differs from the solar system in that the solar system is almost entirely in one plane, while the atom has electrons traveling around it in many planes. These electrons are the carriers of negative electricity. Each one has a negative charge of an equal amount, and it is their motion through a wire which warms our homes, cooks our food, and lights our lights.

A closer examination of the nucleus on which we stand shows us that it is not a continuous surface. It, too, has building blocks, of which there are two primary kinds—protons and neutrons. The neutron is electrically neutral; the proton is nature's basic carrier of positive electricity. The proton and the neutron have almost the same mass; the neutron is slightly more massive. Both are great corpulent creatures compared to the electron, each being about 2000 times more massive than an electron.

You will see as you proceed through this text that this model of the atom did not materialize overnight. The model described above was proposed by Niels Bohr in 1913. For now, we will accept it and withhold until later our discussion of how and why he was led to this concept.

FIGURE 6–3. A representation of the helium nucleus with its two protons and two neutrons.

Distinction between Atoms

The primary feature that distinguishes one atom from another is the number of protons in the nucleus. This

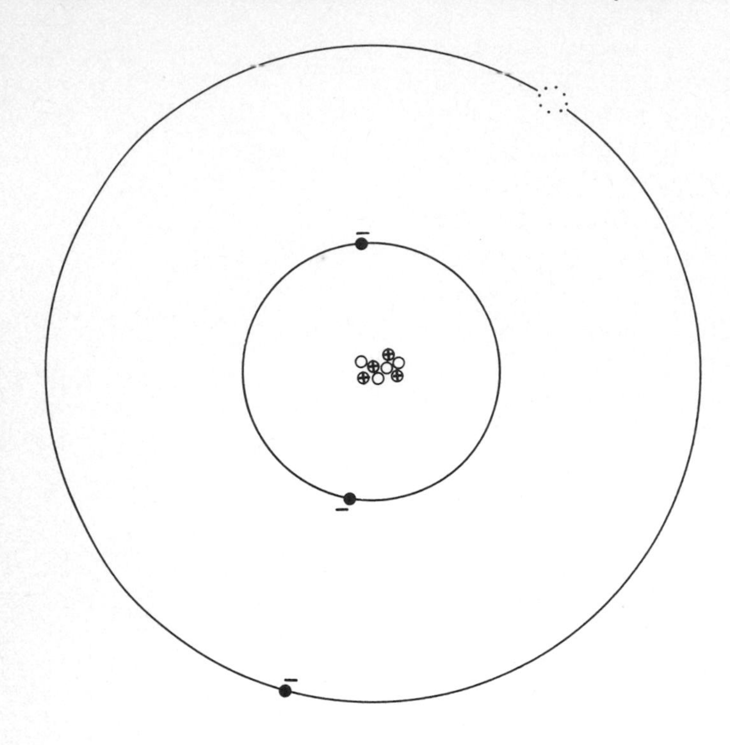

FIGURE 6–4. An electrically charged atom of beryllium. Note that it has four protons in its nucleus but that it has lost one electron.

number is called the *atomic number* and is listed in Table 6–1. Each hydrogen atom has one and only one proton in its nucleus, whereas each oxygen atom has eight, and only eight, protons in its center.

Looking out toward the electrons, we normally find as many of them as there are protons in the nucleus. This means that an atom normally does not have an overall electric charge, because for every positively charged proton there is a negatively charged electron which produces an effective cancellation. Occasionally, a collision or some other interaction occurs between an atom and another submicroscopic particle, producing an upheaval that can free one or more electrons from the atom. In this condition there are fewer electrons than protons, and the atom possesses a net positive charge. In this condition it is called an *ion* (Fig. 6–4). As long as it still has the same number of protons as before the interaction, it still is an atom of the same kind of element, however. Thus, though an oxygen atom may lose (temporarily) one or more of its eight electrons, it is still oxygen because it retains eight protons in the nucleus.

ELECTRON ORBITS

The simplest atom is hydrogen, with one proton in the nucleus and one electron circling around it. As the electron moves, sweeping out its orbit, it remains at a fixed distance from the nucleus (Fig. 6–5*A*). The next simplest atom is helium, with two protons and two electrons, both of which circle the nucleus at the same distance (Fig. 6–5*B*). The situation changes somewhat with the next element, lithium. In 1925, Wolfgang Pauli discovered a system according to which the electrons arrange themselves in orbits around the nucleus. It was found that this ordering system made the theory of atomic structure consistent with experimental observation.

Pauli's principle, sometimes called the *exclusion principle,* places a limit on the number of electrons which may orbit at any specific distance from the nucleus. The limit is two electrons in the innermost orbit, closest to the nucleus. Helium has two electrons orbiting the nucleus at this closest distance, and we say that this orbiting level is then filled. Lithium, the next element in the periodic table, has three electrons. One of these is excluded from orbiting with the other two and instead circles the nucleus in a higher orbit (Fig. 6–5*C*). The next element, beryllium, has two electrons at the first level and two at the second. The Pauli principle states that the second orbit would be filled when occupied by eight electrons. This occurs in the neon atom (Fig. 6–5*D*). As we go to elements with more electrons than neon, there is a need for a third orbit, and then a fourth, and so on.

Wolfgang Pauli (1900–1958)

Although he was born in Vienna, and died in Zurich, Pauli became an American citizen in 1946 and was a professor at the Institute for Advanced Study in Princeton, New Jersey, from 1935 to 1954. The exclusion principle was not his only contribution to science: among other things, he predicted (in 1931) the existence of the neutrino, a subatomic particle which was finally detected in 1956.

Once a new orbit is formed, the electrons in an inner orbit are largely excluded from participation in the normal day-to-day business of the atom. It is the electron or electrons in the outermost orbits which are responsible for chemical reactions, for the electrical properties of the atom, and for the production of light.

THE PERIODIC TABLE

The first efforts directed toward the determination of the actual weight of atoms relied on the law of definite proportion, which was used by Dalton in establishing his atomic theory. According to this method, the relative weights of the elements in a compound are assumed to be the relative weights of the atoms. By examining many compounds, it was possible to establish a list of the relative weight of several atoms, with hydrogen being the lightest.

In 1871, Dmitri Mendeleev, a Russian chemist, found some order among the 70 elements then known. He arranged the atoms in a table according to their atomic weights and their chemical similarities. Table 6–2 shows our present-day version of his work.

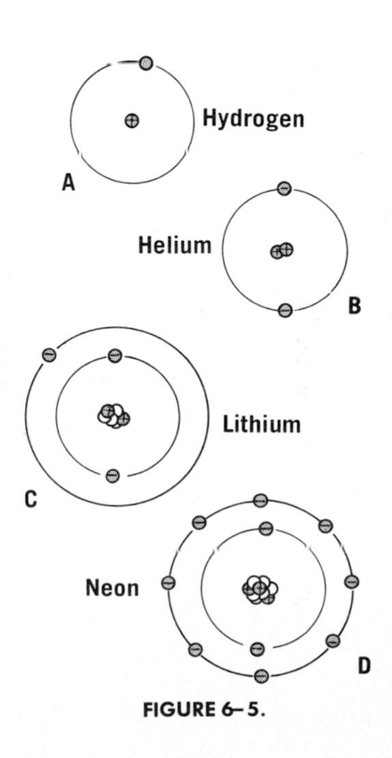

FIGURE 6–5.

TABLE 6-2. THE PERIODIC TABLE

1A	2A	3B	4B	5B	6B	7B	8B	8B	8B	1B	2B	3A	4A	5A	6A	7A	8A
																1 **H** 1.00797 ±0.00001	2 **He** 4.0026 ±0.00005
3 **Li** 6.939 ±0.0005	4 **Be** 9.0122 ±0.00005											5 **B** 10.811 ±0.003	6 **C** 12.01115 ±0.00005	7 **N** 14.0067 ±0.00005	8 **O** 15.9994 ±0.0001	9 **F** 18.9984 ±0.00005	10 **Ne** 20.183 ±0.0005
11 **Na** 22.9898 ±0.00005	12 **Mg** 24.312 ±0.0005											13 **Al** 26.9815 ±0.00005	14 **Si** 28.086 ±0.001	15 **P** 30.9738 ±0.00005	16 **S** 32.064 ±0.003	17 **Cl** 35.453 ±0.001	18 **Ar** 39.948 ±0.0005
19 **K** 39.102 ±0.0005	20 **Ca** 40.08 ±0.005	21 **Sc** 44.956 ±0.0005	22 **Ti** 47.90 ±0.005	23 **V** 50.942 ±0.0005	24 **Cr** 51.996 ±0.001	25 **Mn** 54.9380 ±0.00005	26 **Fe** 55.847 ±0.003	27 **Co** 58.9332 ±0.00005	28 **Ni** 58.71 ±0.005	29 **Cu** 63.54 ±0.005	30 **Zn** 65.37 ±0.005	31 **Ga** 69.72 ±0.005	32 **Ge** 72.59 ±0.005	33 **As** 74.9216 ±0.00005	34 **Se** 78.96 ±0.005	35 **Br** 79.909 ±0.002	36 **Kr** 83.80 ±0.005
37 **Rb** 85.47 ±0.005	38 **Sr** 87.62 ±0.005	39 **Y** 88.905 ±0.0005	40 **Zr** 91.22 ±0.005	41 **Nb** 92.906 ±0.0005	42 **Mo** 95.94 ±0.005	43 **Tc** (99)	44 **Ru** 101.07 ±0.005	45 **Rh** 102.905 ±0.0005	46 **Pd** 106.4 ±0.05	47 **Ag** 107.870 ±0.003	48 **Cd** 112.40 ±0.005	49 **In** 114.82 ±0.005	50 **Sn** 118.69 ±0.005	51 **Sb** 121.75 ±0.005	52 **Te** 127.60 ±0.005	53 **I** 126.9044 ±0.00005	54 **Xe** 131.30 ±0.005
55 **Cs** 132.905 ±0.0005	56 **Ba** 137.34 ±0.005	57 ***La** 138.91 ±0.005	72 **Hf** 178.49 ±0.005	73 **Ta** 180.948 ±0.0005	74 **W** 183.85 ±0.005	75 **Re** 186.2 ±0.05	76 **Os** 190.2 ±0.05	77 **Ir** 192.2 ±0.05	78 **Pt** 195.09 ±0.005	79 **Au** 196.967 ±0.0005	80 **Hg** 200.59 ±0.005	81 **Tl** 204.37 ±0.005	82 **Pb** 207.19 ±0.005	83 **Bi** 208.980 ±0.0005	84 **Po** (210)	85 **At** (210)	86 **Rn** (222)
87 **Fr** (223)	88 **Ra** 226.05	89 **†Ac** (227)	104 (257)	105													

*Lanthanum Series

58	59	60	61	62	63	64	65	66	67	68	69	70	71
Ce 140.12 ±0.005	**Pr** 140.907 ±0.0005	**Nd** 144.24 ±0.005	**Pm** (147)	**Sm** 150.35 ±0.005	**Eu** 151.96 ±0.005	**Gd** 157.25 ±0.005	**Tb** 158.924 ±0.0005	**Dy** 162.50 ±0.005	**Ho** 164.930 ±0.0005	**Er** 167.26 ±0.005	**Tm** 168.934 ±0.0005	**Yb** 173.04 ±0.005	**Lu** 174.97 ±0.005

†Actinium Series

90	91	92	93	94	95	96	97	98	99	100	101	102	103
Th 232.038 ±0.0005	**Pa** (231)	**U** 238.03 ±0.005	**Np** (237)	**Pu** (242)	**Am** (243)	**Cm** (247)	**Bk** (249)	**Cf** (249)	**Es** (254)	**Fm** (253)	**Md** (256)	**No** (253)	**Lr** (257)

Atomic Weights are based on C^{12}—12.0000 and Conform to the 1961 Values

Printed in U.S.A.

The number above the symbol for the element in this *periodic table* is the atomic number of that element. It corresponds to the number of protons in the nucleus of the atom. Note that there is an element for every atomic number from 1 to 105. If any new elements are discovered, they cannot have fewer than 105 protons in their nuclei; in other words, they must have an atomic number greater than 105.

It is interesting to note that when Mendeleev first proposed a table of this kind, many blanks were present. He boldly stated that the gaps were there only because these elements had not yet been discovered. By noting the column in which these missing elements should be located, he was able to predict roughly their chemical properties. Within 20 years these elements were, indeed, discovered.

Chemical Properties

The elements in the periodic table are arranged such that all those in a vertical column have similar chemical properties. For example, consider the elements in the last column—He (helium), Ne (neon), Ar (argon), Kr (krypton), Xe (xenon), and Rn (radon). The outstanding characteristic of these elements is that they do not normally become involved in chemical reactions; that is, they do not join with other atoms to form molecules.

It was noted earlier that the electrons in the outermost orbit of an atom determine whether or not an atom enters into chemical reactions. If we examine the elements above with regard to the Pauli principle, we see that each of them has a filled outer orbit. For example, helium's two electrons fill the first orbit; in neon, the second orbit is filled, and so forth. The conclusion that can be drawn from this situation is that when the outermost orbit is filled, the atom essentially ignores other atoms in the sense that it does not interact with them. Thus, Pauli's arrangement of electrons in their orbits ties in with the arrangement of the atoms in the periodic table.

Atomic Weight

The number below the symbol on the periodic table is the *atomic weight* of the element and is listed on a scale of relative weights, the basis of which will be discussed later. Note that on this weight scale, hydrogen has a weight very close to one, and in fact, most of the elements have weights very close to whole numbers. To see why some elements (such as chlorine, C1) do not follow this rule, and to learn the basis of the weight scale, you'll have to wait until Chapter 24, The Nucleus and Radioactivity. Right now you know everything you need to know in order to study solids, liquids, and gases.

OBJECTIVES

After a study of this short chapter, you should be able to:

1. Define the word "element" as it is used in science and distinguish elements from compounds.
2. Distinguish between a molecule and an atom.
3. Describe the planetary model of the atom, stating the names and locations of the three principal particles within it.
4. Explain, giving an example, how the exclusion principle accounts for the somewhat similar chemical behavior of different elements.
5. Define and give an example of atomic number and atomic weight.
6. Explain how elements are arranged in the periodic table.

QUESTIONS—CHAPTER 6

1. In what ways is the planetary model of the atom similar to the solar system? In what ways is it different?
2. What is "excluded" by the exclusion principle?
3. What is the name of the element having atomic number 73? What element has an atomic weight of 65.37?
4. Cobalt (element 27) and iridium (77) are both metals. What can be predicted concerning Rhodium (Rh)?
5. Why were the alchemists of old (who tried to change other elements into gold by chemical means) doomed to failure?
6. The chemical formula for common table salt is NaCl. Is this an element or a compound? Explain.
7. Imagine yourself to be an ace crime fighter disguised as a mild-mannered college student. Just before entering class you hear the news that the evil Dr. Atom has escaped from jail. Write a short essay describing the vile deed he tried to commit and how you caught him. Make use of as many actual properties of the atom as possible in your story.
8. If you could see atoms, how would you distinguish those of one element from those of another?
9. Draw a sketch of what the molecule Fe_2O_3 might look like. (Fe is the symbol for iron.)
10. Does an atom always have as many electrons as protons? Explain.
11. Sketch the electron arrangement in the shells of neon.
12. Microscopic particles of matter suspended in water can be observed to undergo a haphazard motion in which they dart first in one direction and then another. This type of motion, called Brownian motion, can be explained on the basis of the atomic theory of matter. What is your explanation?
13. Which of the following elements would you predict to have properties most like those of silver (Ag): copper (Cu), cadmium (Cd), or palladium (Pd)?

7 SOLID, LIQUID, AND GAS

Ordinary matter with which we come into contact exists in three states: solid, liquid, and gaseous, and all matter is made up of atoms. In most substances, each atom is joined with others to form a molecule. Water, for example, is made up of hydrogen atoms and oxygen atoms arranged in such a way that each molecule contains two hydrogen atoms and one oxygen atom. When you think of water, you probably think of it first in the liquid form because it is a liquid at room temperature. But when it becomes cold enough, water changes to a solid—ice. And when hot enough, it becomes water vapor, a gas. This freezing and vaporizing of water to change it to its various forms, however, does not change its chemical make-up. In each of the three forms, a water molecule contains two hydrogen atoms and one oxygen atom. This is true for all substances as they change from one form to another. Solid iron is melted in manufacturing processes at very high temperatures and poured into molds. Solid carbon dioxide (often called "dry ice" or "hot ice") changes directly from a solid to a gas when left in a warm room. But in every instance, the molecules of each substance are the same after the change as they were before. The change that takes place in a substance as it goes from solid to liquid occurs *between* rather than within individual molecules. We will consider each of the states of matter separately and see how simple changes in the relationship between molecules can so greatly influence the actions and appearances of a substance.

SOLIDS

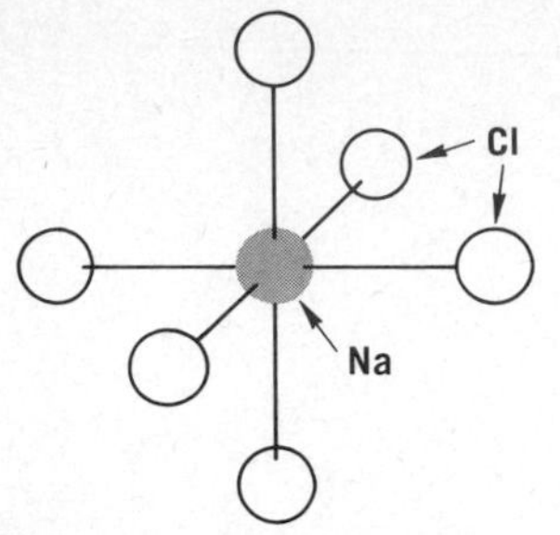

FIGURE 7–1. The sodium atom is surrounded by chlorine atoms and is thus bound in place. Each chlorine atom is likewise surrounded by sodium atoms.

In a solid, the atoms and molecules are held in fixed positions relative to one another. The force which holds them together is electrical and results from the fact that the atom is made up of electrically charged particles. Ordinary table salt is composed of sodium (Na) and chlorine (Cl), bound together as NaCl. In such a compound, the sodium in the salt has transferred one of its electrons to the chlorine. The sodium is left with a net positive charge and the chlorine with a net negative charge. In the solid state, the two types of atoms are positioned so that atoms of one kind are surrounded on all sides by atoms of the other kind. In Figure 7–1 a sodium atom (shown shaded) is completely surrounded by chlorine atoms. The sodium can go nowhere: it is bound in place by electrical forces between it and the chlorine. This, of course, is true for all of the atoms in the substance. The neat cubic pattern is shown in Figure 7–2. There are other more complex ways in which atoms can be arranged, and there are other types of bonds besides the one involving electron transfer, but explanations of these are also based on electrical properties of the atoms involved.

take-along-do-it

Sprinkle some table salt onto a dark surface so that you can examine it closely. A magnifying glass or a microscope will help greatly. Look for any regularity in the shape of the grains of salt. Sprinkle some sugar and compare the shape of its grains to that of the salt grains.

Although the salt grains had been tumbled over and over in handling before you examined them, perhaps their edges were not so rounded off that you were unable to see a pattern of 90-degree angles at the corners and rectangular (or square) sides. This tendency of salt grains to have 90-degree angles and rectangular faces results directly from the cubic arrangement of atoms. Imagine a chunk of salt made up of atoms arranged as in Figure 7–2. If this chunk is broken, it is natural for it to break along one of its layers, so that the remaining pieces have shapes similar to the regular molecular arrangement. Figure 7–3 is a photograph of a large salt crystal before and after breaking.

The molecular arrangement within a grain of sugar is not as simple as that of salt, and you were probably unable to see a repeating shape in the sugar grains. Water also forms a more complex molecular arrangement upon freezing. A pattern is still evident, however, in magnified pictures of snowflakes (Fig. 7–4). The symmetry of these flakes is testimony to an underlying pattern of molecular arrangement.

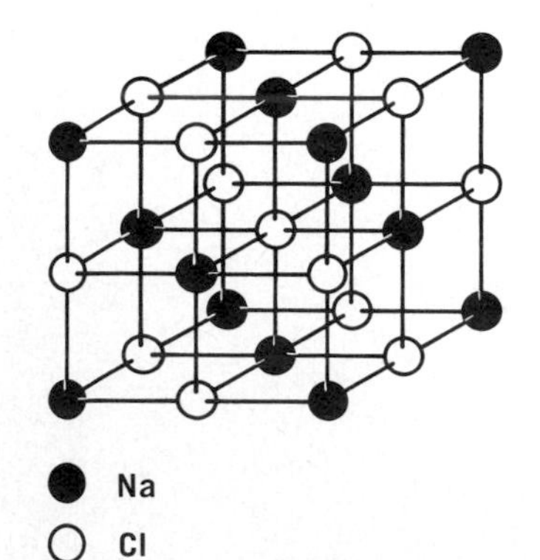

FIGURE 7–2. The cubic arrangement of sodium chloride.

FIGURE 7–3. A large salt crystal. *A*, Before cleaving; *B*, after cleaving. (Photo by Jim Shepherd.)

FIGURE 7–4. Snowflakes. (Photo courtesy of National Oceanic and Atmospheric Administration.)

Elasticity

The fact that solids have definite fixed shapes that resist being changed results directly from the forces between molecules within the solid. Some solids, such as rubber, distort easily but return fairly well to their original shapes. In these cases, the forces between atoms and molecules allow for stretching and twisting, but the atoms still are connected to one another and are pulled back to their original positions when the distorting force is released. This property is called *elasticity.*

---

take-along-do-it

Equipment needed: a couple of rubber bands, a number of coins, a spring, two paper cups, a ruler, and some string or wire. Hang your rubber band and spring side by side from some support—a clothes rod, for example. Punch holes under the rims of the cups and hang one from the spring and the other from the rubber band, as in Figure 7–5. Now add coins to the cups and note how much each coin stretches the rubber band and spring. Is there any regularity to the amount of stretch produced by each coin you add? For example, if the rubber band is stretched one centimeter by a single coin, will another coin stretch it another centimeter? Remove the weights one by one and note how well each cup returns to the position it occupied with a corresponding load when you were adding weights. Does the steel spring return more readily to its initial shape than the rubber band?

You probably found that not only was the steel spring more regular in its stretching (it probably stretched an equal amount with each weight) but it also returned to its original position more readily than the rubber band.

Some solids distort easily but have little tendency to return to their initial shapes. Examples are paper, putty, clay, and some types of wire. Such materials are *soft,* as is rubber, but are less *elastic* than rubber. All solids tend to lose their elasticity if left stretched for long periods of time. This is

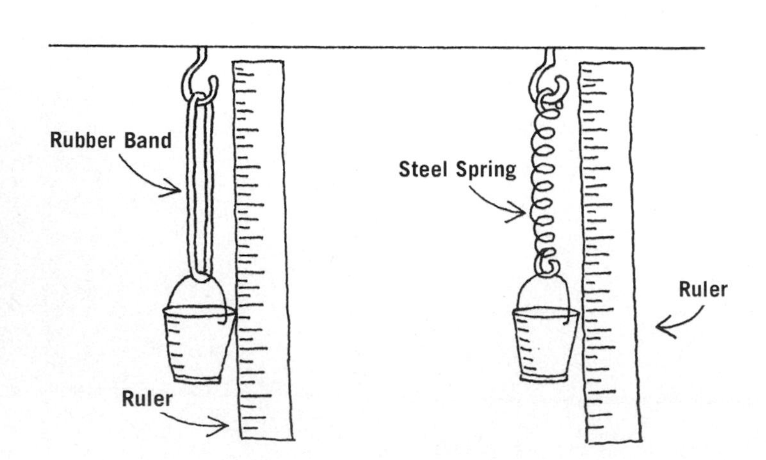

FIGURE 7–5. A set-up to investigate elasticity.

one reason why a violin or guitar gets out of tune when not used.

Other solids, such as the carbon in a pencil, are very brittle. They break rather than bend. All materials, however, distort at least a little when force is applied. The bricks and mortar supporting a large building bend enough so that a 100-story building sways at the top as much as a few feet in a strong wind. It is because of this property of elasticity that a force is applied on us by the floor upon which we stand. Consider a person standing on a stack of mattresses. What determines how far he sinks? He must sink until he compresses the mattresses enough to make them push back up with a force equal to his weight.[1] This is similar to the way the rubber band in the preceding TADI stretched until it exerted enough force to support the cup of weights.

If instead the person is standing on a piece of granite, we must not take it for granted that the stone is so hard that it does not distort. Figure 7–6 is a magnified (and simplified) drawing of molecules of a granite block compressed by the force of an ant's foot pushing down on them. They are compressed until the force pushing up on the ant equals his weight, assuming that all of his weight is on one foot. The drawing represents the forces as being caused by little springs. This turns out to be a rather good way to visualize forces between atoms in a solid. The springs at point A are completely relaxed, but when they are stretched such as at B, or compressed such as at C, they resist distortion and exert forces on the atoms to return them to their original positions. In actual practice, of course, electrical forces between atoms perform the functions of these imaginary springs.

Density

One of the obvious differences between different solids is what is often called weight but is better referred to as

[1]This is an example of Newton's Third Law, the force-pair law (Chap. 2). The man's force down on the mattresses must be equal to the mattresses' force up on the man.

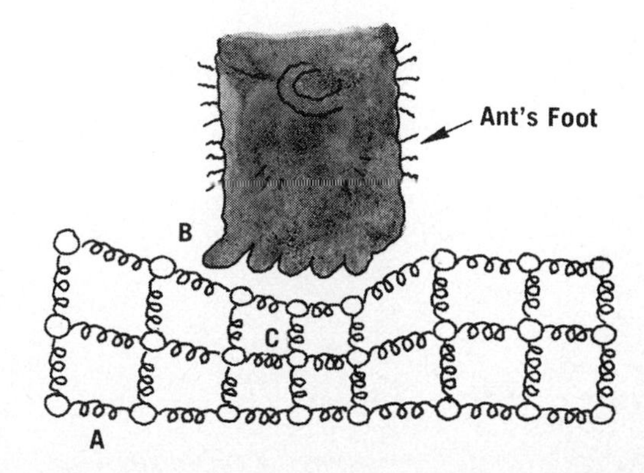

FIGURE 7–6. The ant's foot presses the atoms underneath it closer together.

density. To illustrate the difference between the two concepts we will ask the old trick question: Which weighs more, a pound of feathers or a pound of steel? The answer is, of course, that they both weigh the same—one pound. Although we sometimes might say that steel weighs more than feathers, what we really mean is that *considering equal volumes* of steel and feathers, the steel weighs more—it is therefore more dense.

Density is defined as the mass of a substance divided by its volume. Suppose one determines the mass of a block of pine wood to be 2000 grams and then measures the wood and computes its volume to be 4000 cubic centimeters. (Volume of a rectangular block is calculated by multiplying length by height by width—see Fig. 7–7. If each is measured in inches, this gives "inches × inches × inches" or cubic inches.) Computing the density:

$$\text{Density} = \frac{\text{Mass}}{\text{Volume}} = \frac{2000 \text{ grams}}{4000 \text{ cubic centimeters}}$$

$$= 1/2 \text{ gram/cubic centimeter}$$

This means that if we can assume that the pine wood is uniform throughout, we know that each cubic centimeter of the wood has a mass of 0.5 gram.

The usefulness of the concept of density is that for a material such as iron, one can state a density which applies to all iron. Iron has a density of 7.8 grams per cubic centimeter. Thus although one piece of iron—a small statue, for ex-example—may have a mass of 10 grams, and a large piece such as an iron anchor may have a mass of 500,000 grams, both have the same density. And although the statuette is not heavier than a pine tree, we can still compare densities and say that iron is much more dense than pine wood. We will find that the concept of density is even more relevant to the case of liquids and gases.

HEAT ENERGY AND TEMPERATURE

It was stated in Chapter 2 that energy can be added to or taken away from a body in the form of heat. This thermal energy, however, is not a new form of energy that is completely independent of those previously discussed.

Our model of a solid as presented in Figure 7–8A (with stationary atoms connected by springs) needs to be altered to include the idea of motion caused by the heat content of the object. In order to do this, we picture the atoms as being in a constant state of vibration. If we could see a single atom in a solid we would see it vibrating about a fixed location, as is the center atom of Figure 7–8*B*. *As a solid gains thermal*

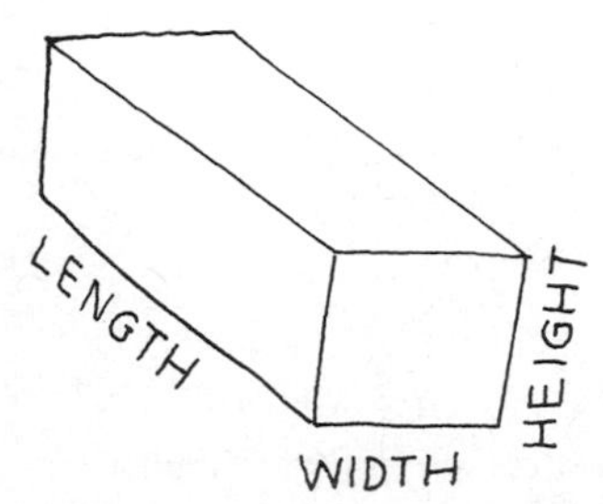

FIGURE 7–7. Multiply to get volume.

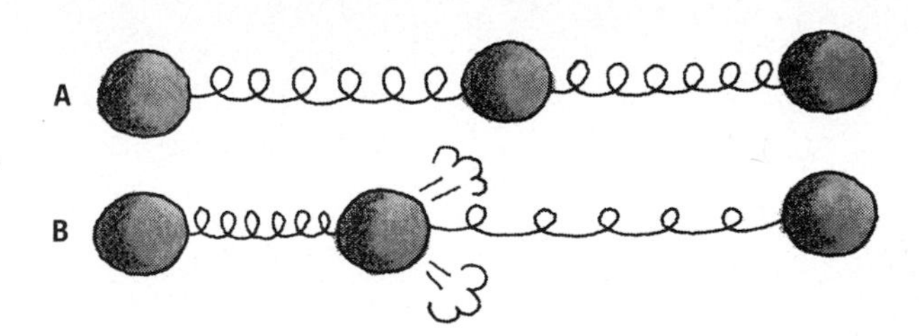

FIGURE 7–8. A, A solid; B, A more realistic representation because it shows an atom in motion.

energy, its particles vibrate with larger and larger amplitudes.

We perceive this increase in energy of the atoms of a substance as an increase in temperature. *If the average energy of the molecules is increased, the temperature is increased, and vice-versa.* Since it is the average energy of the molecules that is important, the same amount of heat energy added to different masses of the same material will not increase the temperature by the same amount. A measured amount of heat added to a 50-gallon drum of water may not result in a measurable temperature change. But add this same amount of heat to a small cup of water, and its temperature may rise by several degrees. Each of the many molecules in the large drum gains very little energy, but the average energy per molecule of the water in the cup increases quite a bit.

When two objects at different temperatures are placed in contact, thermal energy is always transferred from the hotter object to the colder one (Fig. 7–9). Figure 7–10 illustrates the mechanism by which this is accomplished. Part A shows two substances in contact, the molecules of the hot one (on the bottom) vibrating with large amplitudes. As the surface molecules of this hot substance collide[2] with the

[2]The molecules come in contact only in the sense that they get close enough so that they exert electrical forces on one another. Contact forces are basically electrical in nature.

FIGURE 7–9. Heat always flows from the hotter to the colder.

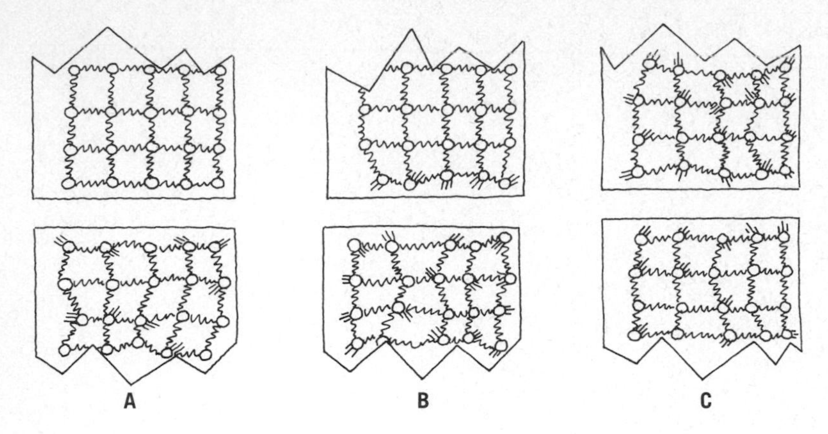

FIGURE 7–10. In *A*, the lower object is much hotter than the upper. *B*, Heat is transferred until both are at an intermediate temperature (*C*).

slower moving ones in the cold substance, they transfer some of their energy to them. Thus the molecules just above the surface begin to vibrate faster, but because they are connected by forces to others deeper within the substance, these molecules also gain kinetic energy. The heat is thus conducted from the hotter object to the cooler one and throughout the cooler one until both are at some intermediate temperature (Fig. 7–10*C*).

Gabriel Daniel Fahrenheit (1686–1736)

The inventor of the first accurate thermometer received his education in business and was by profession a manufacturer of meteorological instruments. Before his time, the gas thermometer of Galileo was in principal use, although alcohol thermometers were becoming more popular. It was Fahrenheit who first introduced the use of mercury in thermometers and who developed the scale which bears his name.

Measuring Temperature

Suppose that you ordered a thermometer from the Fly-by-Nite mail order house and found it to have no degree markings on it. When you exposed it to heat the mercury level rose as it should, so the thermometer seemed to work, but it still kept the temperature a secret. You would need to mark, or calibrate, the thermometer. Following are the directions for calibrating according to the Fahrenheit scale, which is most often used in this country.

(1) Place the thermometer in a mixture of ice and water. Mark the mercury height as 32°.
(2) Place the thermometer in boiling water[3] and mark the height as 212°.
(3) Make 180 divisions between these two marks, each representing one degree. Continue marking divisions of the same size below 32° and above 212° if you like.

Within the scientific community and throughout most of the world, a second scale is used: the Celsius scale. If you had desired to calibrate your thermometer according to this scale, you would follow the same procedure as for Fahren-

[3]To be exact, this should be pure water at standard atmospheric pressure. But an error of a degree or so won't matter here.

heit but would mark 0° as the freezing point, 100° as the boiling point, and 100 divisions between the two. Using this scale, you would find a comfortable room temperature to be about 22° C.

The coldest temperature attainable registers as about −273° C or −459° F. This temperature is referred to as absolute zero and is the lower limit on temperatures available to us. On earth a great deal of elaborate procedure must be followed in order to approach such a low temperature, but on the surface of the outermost planet of our solar system, Pluto, temperatures near this are a way of life, or rather, of nonlife. The sunlight reaching Pluto's surface is so scant that the planet's temperature is thought to remain constantly near −273° C.

In view of the fact that there exists an absolute minimum of temperature (absolute zero), another scale has been developed which uses absolute zero as its starting point. On the Kelvin scale, the freezing of water occurs at 273° K and the boiling of water at 373° K. The difference between the two temperatures is the same on this scale as on the Celsius scale. Thus the size of each degree is the same.

Melting

You may associate the word "melting" primarily with the process of ice changing into water, but the word applies equally to iron melting in the furnace of a steel mill. Although these events occur at different temperatures, they both can be understood by considering what happens as thermal energy is continually added to a substance, causing the molecules to vibrate faster and faster. At some point they will be vibrating so violently that the forces which had thus far restricted each of them to vibrate about a fixed position now break down. At this point, although an attractive force still holds the molecules near one another, they begin to move around each other, mixing freely.

Suppose people are dancing in a crowded ballroom of a transatlantic ship when the ship starts bobbing on the waves. To support themselves, everyone reaches out and grabs someone else. Then as the ship bobs and sways, each person can sway back and forth near the same position, but since his neighbors have a firm grip on him he cannot leave his area of the ballroom floor. The behavior of the people in this situation is analogous to the behavior of the molecules inside a solid.

But if the storm around the ship increases in violence, the people may sway so violently that they are unable to hold firmly to one another. Each person continues to grab at others to support himself, but the confusion is so great that he cannot hold on for long. The result is that the people begin to intermix on the floor. A person who started on the right side of the ship may find himself jostled about until he ends up on the left side. The motion of the people in the ballroom in this state resembles the motion of molecules within a liquid.

LIQUIDS

take-along-do-it

Put some cool water into one glass and about the same amount of hot water into another glass. (The use of styrofoam cups would be even better because styrofoam does not conduct heat well, and the water in each will not change temperature quickly.) Let each cup sit still for a minute or two until the water comes to rest. Then carefully put a single drop of ink into each cup. Do this in such a way as to disturb the water as little as possible. (A fountain pen or medicine dropper might help here.) Does the ink color stay where you put it, or does it spread throughout the water? In which glass of water does the color spread more rapidly? Try to devise a hypothesis to explain your observations.

There are probably a number of reasonable hypotheses which may explain the result of this TADI, but a good one is one which conforms to what we already know about liquids and temperature. The fact that molecules of a liquid move past one another to different parts of the liquid easily explains why the color spreads. It is as if all the dancers initially in one corner of the ship's ballroom had on green costumes. When the ship was rolling slowly enough that everyone held onto his neighbor, the green stayed in the corner. But in the more confusing situation where holds were torn loose, some green-costumed people would inadvertently get jostled around the room. After enough time passed, they would be spread all over the room. Such is the case with the ink molecules.

But you probably noticed that the ink spread faster in

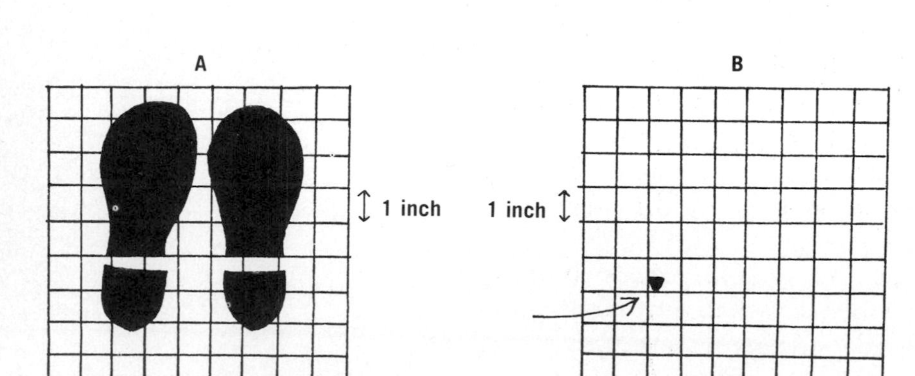

FIGURE 7–11. When standing in regular shoes, one's weight is distributed over a larger area than when standing on one spike heel.

the hot water. In hot water the molecules have more thermal energy; they are jostling around more, and so the ink molecules are spread around more rapidly.

FORCE AND PRESSURE

In Chapter 1 we examined the concept of force, and in this chapter we have spoken of forces between molecules and atoms. In everyday language, the terms "force" and "pressure" are often used interchangeably, but in physics we make an important distinction. Before defining pressure, let's give an example to show the need for a distinction between force and pressure.

In the 1960s, spike heels on women's shoes were fashionable. They caused some problems, however, on hardwood and tile floors because a woman could dent such floors by standing on one heel. The same woman standing in regular shoes caused no trouble, but somehow her 150 pounds did more damage when she stood on the spike heel. Although the force downward was the same in each case—150 pounds—the area on which she exerted the force was far different in the two cases. In regular shoes her weight was spread over about 25 square inches of floor (Fig. 7–11*A*). But standing on one heel caused the entire force to be exerted on only about 1/16 square inch (Fig. 7–11*B*).

To incorporate both ideas, force and area, into a single word we *define pressure to be equal to force divided by area.* Thus pressure is measured in pounds per square inch. The 150-pound woman exerted about six pounds per square inch standing on two regular shoes and more than 2000 pounds per square inch on one spike heel. (Use the definition to check our math.) Many floors will dent under such pressure.

Liquid Pressure

The concept of pressure is useful in working with liquids (and gases) because of a simple fact: the pressure exerted by a liquid at rest depends upon only two factors—the density of the liquid and the depth below its surface. We will look at the second factor first because it is easier to verify. If you dive to the bottom of a 10 or 12 foot swimming pool you will very likely feel an uncomfortable sensation of pressure in your ears. As you go deeper, the pressure is greater. To understand why, refer to Figure 7–12*A*. The drawing shows a square of tissue paper in a deep pool. Suppose the paper is exactly one square inch in area. If the water is still, it is easy to imagine that the paper will hover in the water wherever it is put. In the position shown, it is 10 inches below the surface. Above it there are 10 cubic inches

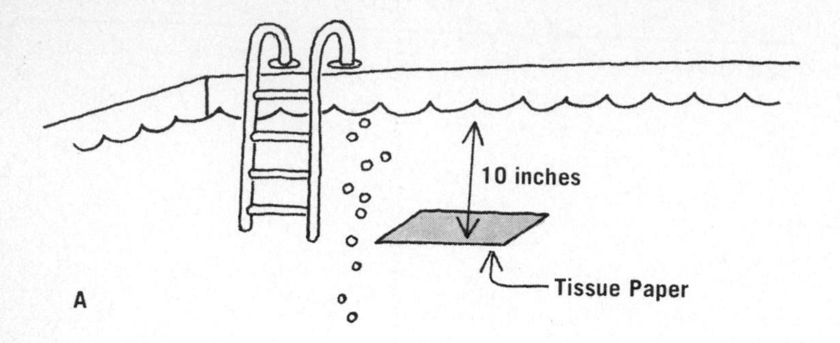

FIGURE 7–12. The piece of paper (1 sq in) has a force on its upper surface equal to the weight of water above it.

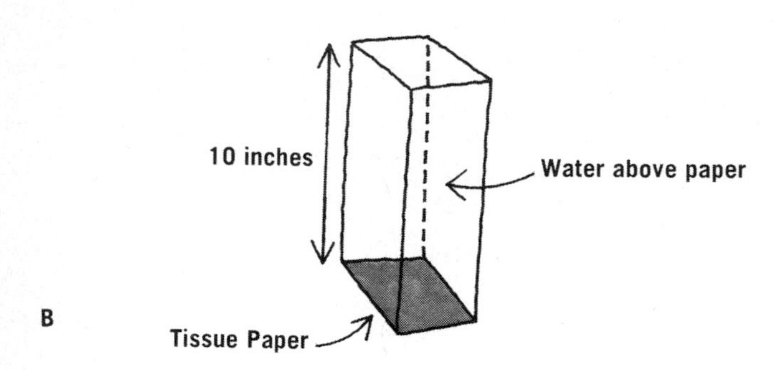

of water (Fig. 7–12*B*). Since the density of water is 0.036 pound per cubic inch[4] this means that a force of 0.36 pound is pushing down on the paper (0.036 pound/cubic inch × 10 cubic inches = 0.36 pound). If the tissue paper is stationary, however, the water below it must be exerting an equal force upward. The force on one square inch at a depth of 10 inches in water is thus 0.36 pound, or we can say that the pressure is 0.36 pound per square inch. If the tissue paper is moved to a depth of 20 inches, the same reasoning would show that the pressure is 0.72 pound per square inch.

Notice that our argument included the density of water in the calculation. If we had used a different substance, such as alcohol (which has a density of 0.029 pound/cubic inch), we would have found the pressure at 10 inches' depth to be 0.29 pound per square inch (check our math). In general, we can use the formula

$$\text{Pressure} = \text{Depth} \times \text{Density}$$

to calculate the pressure at any depth in a liquid.

---

[4]Notice that we are considering density to be *weight* divided by volume instead of *mass* divided by volume. Strictly speaking, we should call this "weight density."

The preceding equation explains many phenomena. For example, doesn't it seem fantastic that dikes can be built to hold back the entire ocean? When one considers how much water is in the oceans, it would seem that dikes would have to be unimaginably strong. The equation tells us, however, that the pressure on a dam depends only upon the depth of the dam below the surface of the water and not upon the length of the lake or ocean. Thus a 10-foot-high dam which will hold back the water in your local lake will *almost* do the job for a dike in Holland. Why "almost"? First, because salt water is slightly more dense than fresh water; one cubic inch of ocean water weighs 0.037 pound versus 0.036 pound for fresh water. Secondly, the waves and currents of the ocean exert additional forces. These two effects require a 10-foot-high ocean dike to be somewhat stronger than a 10-foot-high dam on a small lake, but the difference in size between the ocean and the lake has no direct effect on water pressure.

Submarines must be built to sustain pressures at great depths. At 100 feet, for example, the water pressure in salt water is about 44 pounds per square inch (or about 6300 pounds per square foot). Special deep-diving bathyscaphes have been built that are strong enough to withstand pressures of 16,000 pounds per square inch at 36,000 feet below the ocean surface.[5]

Direction of Pressure. The discussion of the forces acting on the square of tissue paper serves to indicate another property of pressure: *At a given depth the pressure exerted by a liquid is the same in all directions.* To see this, imagine the square of paper to be hovering at a certain depth below the surface. If it is hovering, Newton's Second Law states that no net forces are acting on it. Therefore, any upward push by the liquid must be canceled by a downward push, and any horizontal push to the left must be canceled by a horizontal push to the right.

Buoyancy and Floating

Imagine a tissue paper "box" with square sides one inch along each edge. The box encloses a volume of one cubic inch. Suppose further that the box (which is full of water) is below the level of water in a bowl. With what force is the water in the bowl pushing up to support the water in the box? Since the water in the box weighs 0.036 pound, this must be the force holding up the box of water (Fig. 7–13).

[5]An interesting book on this subject is: Shenton, Edward H.: *Diving for Science: The Story of Deep Submarines.* New York, W. W. Norton & Co., Inc., 1972.

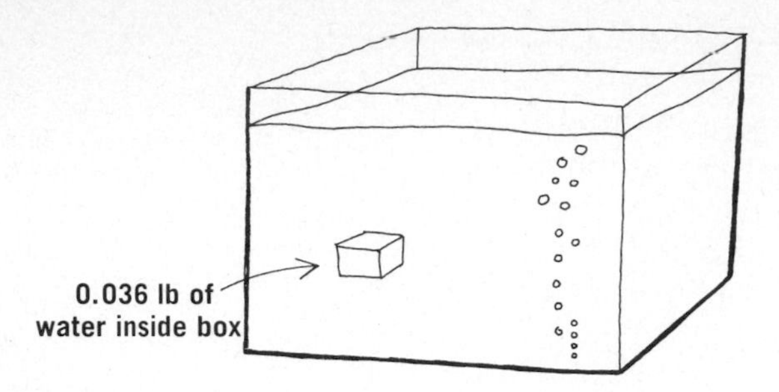

FIGURE 7–13. The cubic inch of water is held up by the water below it with a force equal to its weight.

Now suppose you replace the box with a piece of wood of exactly the same size. In this case, the water below the block will push up with exactly the same force as it did on the box. But one cubic inch of wood weighs about 0.020 pound, so we have its weight of 0.020 pound pulling down on the block and 0.036 pound pushing up. Result—the block rises. In fact, the block will continue to rise until part of it is above the surface. It will come to rest when it displaces (takes the place of) 0.020 pound of water (Fig. 7–14).

The preceding example can be generalized as follows: *The buoyant force on an object in a liquid is equal to the weight of the liquid displaced by the object.*[6] When the block was underwater, displacing 0.036 pound of water, 0.036 pound was the buoyant force on it. Figure 7–14*B* shows it floating at a depth at which it displaces 0.020 pound of water.

take-along-do-it

If you have access to a swimming pool, do this TADI in the pool; otherwise, a bathtub will work. Lift a brick or rock from the bottom of the pool and notice how heavy it feels. Then lift it out of the water and feel its weight. How do they compare?

[6]This law, which also applies to gases, is usually called Archimedes' principle. The story goes that Archimedes was sitting in his bathtub when he thought of the principle. He immediately jumped out and ran down the street naked yelling, "Eureka!" ("I have found it").

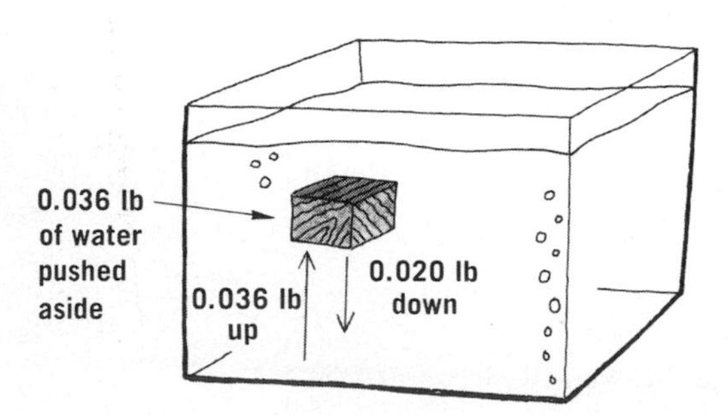

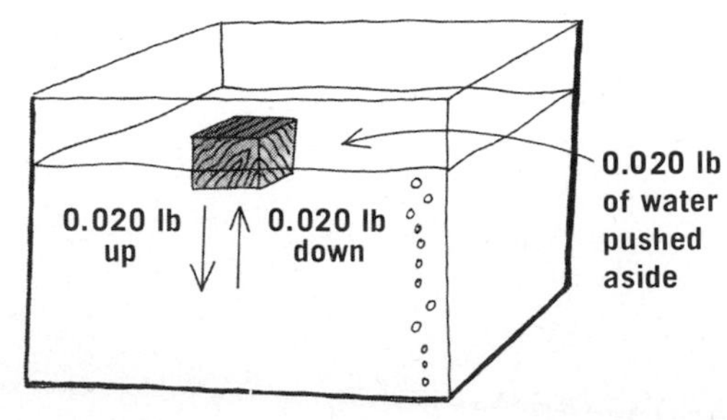

FIGURE 7–14. A 0.020 lb block of wood in water experiences a buoyant force equal to the weight of water displaced.

A brick has a volume of about 64 cubic inches (2 inches $\times$ 4 inches $\times$ 8 inches); when underwater, it displaces 2.3 pounds of water (0.036 lb/cubic inch $\times$ 64 cubic inches). The water buoys it up with a force of 2.3 pounds, which causes it to feel 2.3 pounds lighter when it is underwater.

Note that the brick is buoyed up with 2.3 pounds of force no matter at what depth it is located and therefore will apparently weigh the same at any depth in water. This is true because water is not compressible and so it has the same density at any depth;[7] thus the brick displaces 2.3 pounds of water at any depth.

Archimedes (about 287 BC–212 BC)

Born in Syracuse, Sicily, Archimedes (ar-ki-mee´-deez) was perhaps the greatest scientist of ancient times. The son of an astronomer, he was a relative and friend of Hiero II, king of Syracuse. Among his achievements are a number of mechanical inventions, a calculation of the value of "pi," and the buoyancy principle which today bears his name. Legend holds that Archimedes constructed lenses (or mirrors) to focus the sun's rays on enemy ships at sea, setting them afire, an example of the ancient use of science in warfare.

Comparing Densities

Application of the buoyancy principle makes it easy to determine whether something will float or sink in a liquid. If the object has a density greater than that of the liquid, it will sink; if its density is less than that of the liquid, it will float. The human body normally has a density just slightly less than water and will float with a very small part of it above water (Fig. 7–15). When a swimmer inhales, the volume of his body increases somewhat because his chest cavity expands, but since air has a very low density, the added air in his lungs adds very little to his weight. For this reason, many people float with their lungs full, but sink after exhaling. Bones have a greater density than fatty tissue, so a person with a large skeletal frame does not float as well as an overweight person with a smaller frame.

It was mentioned that salt water has a greater density than fresh water. Anyone who has played in the ocean knows how easy it is to float in salt water. The density of the Great Salt Lake in Utah is even greater than that of the ocean, and it is almost impossible for a person to completely sink in it.

take-along-do-it

Figure 7–16 shows a method which you can use to verify that liquids of high density provide a greater buoyant force than liquids of low density. Eggs, which have a slightly greater density than pure water, will sink when dropped into water. The density of an egg, however, is less than that of salt water.

[7] If water were compressible, the greater pressure below the surface would compress a given mass of water there into less volume, thus increasing its density.

FIGURE 7–15. Profile of floating person.

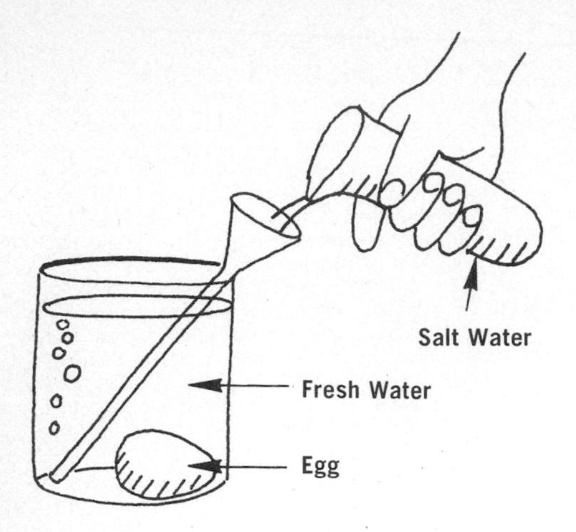

FIGURE 7–16. Carefully pour the salt solution to the *bottom* of the glass.

Use a long funnel to pour a salt solution to the bottom of the glass, beneath the pure water. The increasing density of the water will cause the egg to rise.

Mercury is a metal which is in liquid form at everyday temperatures. It is about 14 times denser than water, however. Figure 7–17 shows a steel ball floating on mercury.

Evaporation

Consider the molecules near the surface of a liquid. From time to time one of them may be hit hard enough by another molecule so that it is ejected from the liquid (Fig.

FIGURE 7–17. A steel ball floating in mercury. (Photo by Jim Shepherd.)

7–18). It flies out into the air and loses all contact with the liquid. This is the process which underlies the phenomenon of evaporation.

Evaporation also has an analogy in the ballroom of the rocking transatlantic ship. As the frightened travelers jostle about in the ballroom, occasionally one of them near a doorway happens to get shoved extra hard from behind. Through the door he goes, despite his frantic efforts to hold on to the people around him. He has become separated from his fellow dancers and corresponds to a freed molecule above a liquid. We call the dancer a goner, but we call the molecule a gas, or rather a gaseous molecule.

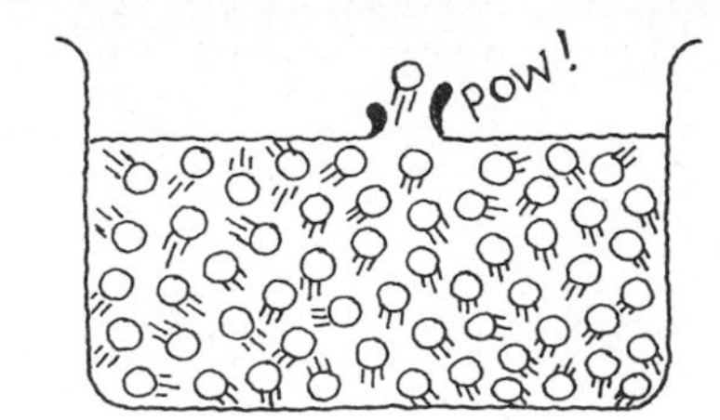

FIGURE 7–18. The process of evaporation.

The following occurrence could be presented as a TADI, but the phenomenon is so familiar there is no need for you to try it again. After swimming, you surely notice that your wet body feels cooler than it did when it was dry. The reason is not that the water in the pool is cold, for this cooling sensation is felt even after a hot bath. A deeper look at the process of evaporation provides an explanation.

Temperature is a measure of the *average* energy of the molecules of a substance. Note, however, that the molecules which escape by evaporation must have a greater-than-average energy, for it is only because a great amount of energy is given to a particular molecule that it is able to break free from its neighbors on the surface. But if the molecules leaving have a lot of energy, the ones left behind have, on the average, less energy than they had before; the remaining liquid is cooler. Thus evaporation cools the bather.

It is important to emphasize that it is not the fact that each evaporating molecule simply carries energy with it which results in the cooling process. The important point is that each leaving molecule carries a *greater-than-average* energy with it. The situation is analogous to a city which somehow is not attractive to people with high IQs. As children with a high IQ grow up, they leave town with greater frequency than people with lower IQ. The result is that the average intelligence (as measured by IQ) of the townspeople decreases as people leave.

FROM LIQUID TO GAS

We moved from solid to liquid by adding heat. As we continue to add heat to a liquid, the average energy of the molecules continues to increase. Finally, at some temperature they attain enough energy to break free of one another even below the surface of the liquid. When this happens, bubbles of vapor form in the liquid and rise to the surface. We call this phenomenon "boiling."

In a gas, the molecules have lost almost all contact with

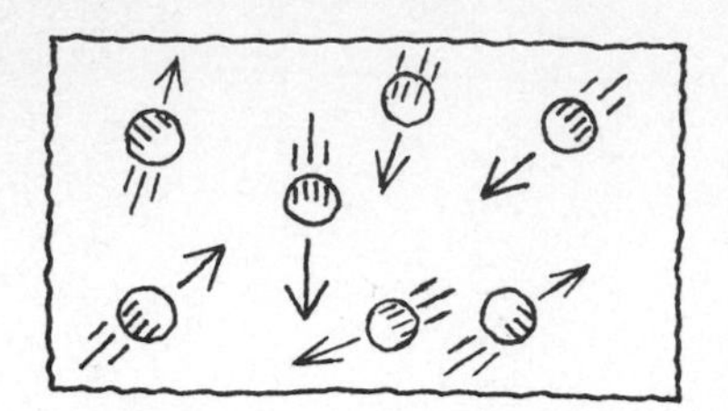

FIGURE 7–19. The molecules of a gas move freely in random directions.

one another. They fly around freely in space (it is not correct to say that they fly around in the air, because air itself is a gas composed of molecules which fly around in otherwise empty space). Figure 7–19 represents molecules of a gas moving in random directions. As they fly about, the molecules collide with one another, but other than that, they are almost independent of one another.

The most obvious property, besides its density, that differentiates a gas from a liquid is the fact that a gas expands to fill whatever container it is in, whereas a liquid redistributes itself to cover only the bottom of a container. In addition, the volume of a gas can be easily changed by compressing it. This can be demonstrated with a discarded hypodermic syringe. Hold your finger over the end so no gas can escape (remove the needle first), and push in the plunger to compress the gas inside. Try this with water in the syringe and you'll observe no compression; water resists compression.

Temperature also has a great effect on gas. For this reason it is dangerous to put a closed container of air in a fire. The heat will increase the pressure of the gas until the container explodes.

The pressure, volume, and temperature for a given amount of gas are related in a manner described by the equation.[8]

$$\frac{\text{Pressure} \times \text{Volume}}{\text{Temperature}} = \text{a constant}$$

In other words, regardless of the changes we make in the quantities pressure, volume and temperature, when they are multiplied and divided as above, we always get the same value – a constant. When we throw a closed bottle into a fire, the bottle does not change size, so the volume occupied by the gas does not change. We know that the temperature of the gas increases, however. Our equation states that if the temperature increases greatly, so must the pressure, until finally the bottle can no longer withstand the pressure: stand back!

If you slowly push the plunger into a hypodermic syringe filled with air, the temperature stays about the same, but the volume of the gas decreases. If the volume decreases by half, the equation says the pressure must double. That is why you have to push hard.

---

[8]The temperature used in this equation must be the absolute temperature, because only the absolute temperature scale starts at a real, meaningful zero point, and such a zero point is necessary if one is to talk of doubling and tripling the temperature. (If you measure your height from the top of a table next to which you are standing, it would be meaningless to say that someone else is 1.3 times as tall as you. You must measure from zero – the floor.)

Pressure of Gases

Fish living deep in the ocean exist in regions of great pressure. Yet they are unaware of it. But we land dwellers also live on the bottom of a great ocean—made of air. We, too, unknowingly exist under high pressures. Figure 7–20 shows that as we walk about, a force is exerted on us, caused by the weight of a column of air that extends to the "top" of the atmosphere. Measuring devices, which we will discuss in the next section, indicate that the pressure exerted on us is about 15 pounds per square inch.

A pressure of 15 pounds per square inch does not seem to be extremely large until we examine carefully what it means. Take a piece of chalk and draw a square anywhere on your body such that the square has sides one inch long (Fig. 7–21). The atmosphere pushes on that one square inch area with a force of 15 pounds. That still doesn't seem too bad until you consider how many such squares you could draw on your body. (Why don't you try it?) You will find that you can draw abut 2000 of them, and each one has a force pushing against it of 15 pounds. This means that over your entire body there is a total force of 30,000 pounds being exerted inward. Like Atlas of mythology fame, who supported the world on his shoulders, you support 30,000 pounds of air. The question is: how do you do it? The answer is, of course, that you were conceived, born, and grew in this environment. As a result, all of your body cavities and tissues are filled by air and liquids which push out with the same force with which the atmosphere pushes in. These two pushes cancel, and you are thus unaware of their existence unless. . . .

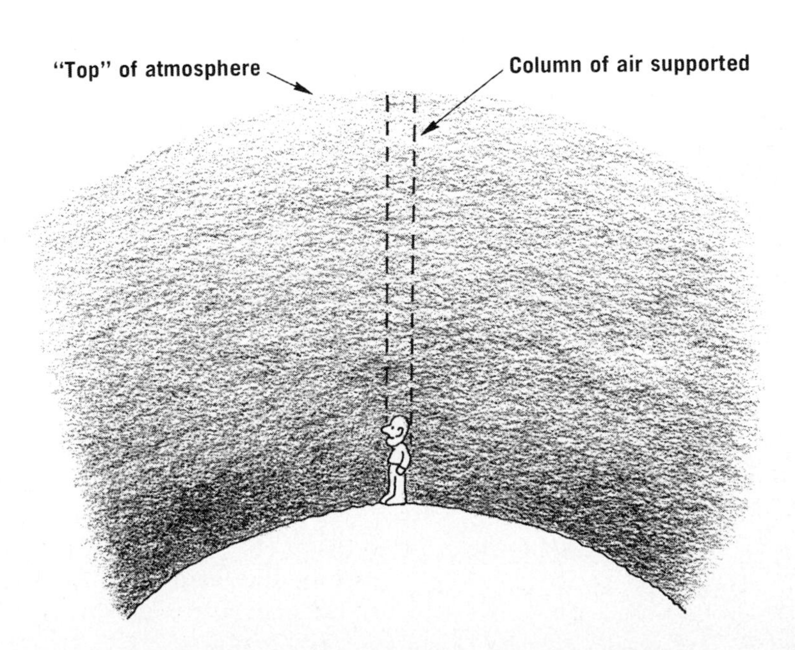

FIGURE 7–20. We must support the weight of the air above us.

FIGURE 7–21. Draw a square on your body.

As an indication of how you could be made aware of outside air pressure, consider the fate of a mouse in a vacuum chamber (a container from which the air can be removed rapidly). Without air, not only is the mouse unable to breathe, but if the air is removed quickly enough, the inward push of the air on the mouse's body drops to a very low value while the outward push of the air inside its body is not able to drop at the same rate. Result: boom—and Walt Disney stock drops 10 points.

We have examined some of the outward manifestations of pressure, but we have not considered its causes. In order to do so we must resort to molecular theory. Consider a gas confined in a container. The molecules move about freely and at high speeds. As a result, they make frequent collisions with the walls of the container (Fig. 7–22). Each collision is a tiny push, which, taken alone, would be insignificant, but billions of such pushes exert an outward push which seems constant and steady and which we call the pressure of the gas.

Barometers

Catch the next weather forecast on TV—the announcer turns to you and says: "The atmospheric pressure is at 29 inches and is falling." We just said that pressure is measured in pounds per square inch; he says it is measured in inches. Somebody is wrong, right? Not necessarily, if you understand how a barometer is used to measure pressure.

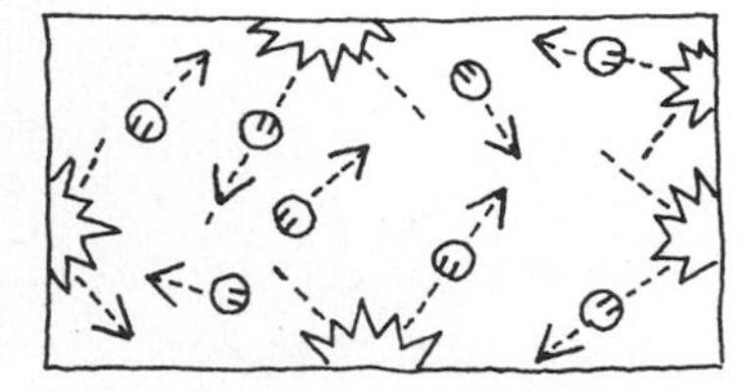

FIGURE 7–22. Gas pressure is the result of molecular collisions.

A barometer requires two pieces of equipment only: a glass tube, sealed at one end, and a container filled with mercury. All the air is removed from the tube, and the open end is inserted into the pool of mercury (Fig. 7–23A). The air of the atmosphere pushes down on the surface of the mercury, but inside the glass tube no air is present to

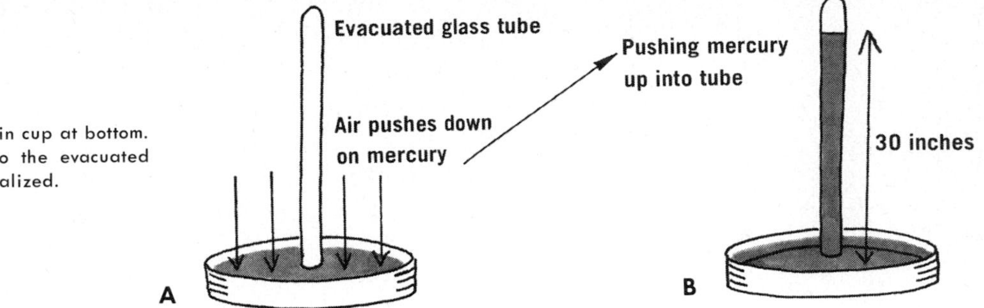

FIGURE 7–23. Mercury is in cup at bottom. Air pressure pushes up into the evacuated tube until pressures are equalized.

produce this downward push. The result is shown in Figure 7–23*B*. Mercury rises in the tube until the weight of the column of mercury pushes down with a pressure equal to the pressure with which the air pushes down on the rest of the mercury. Under normal atmospheric conditions the mercury rises to a height of about 30 inches (76 cm).[9] This provides a descriptive way for the weather forecaster to measure pressure.

Mercury is used in barometers because it has a low freezing point, but more importantly because of its great density. If water were used, the column in the glass tube would rise to about 34 feet before its weight equaled the push of the atmosphere.

As innocent as the barometer appears to be, it is interesting to note that one of its earliest users was almost tried for witchcraft. He used a water barometer constructed by running a long tube from a cistern in his cellar up through a hole in the roof of his house. Floating on the surface of the water, inside the tube, was a wooden figure of a man. Lower atmospheric pressure signaled by a falling barometer indicates bad weather, and the height of his barometer was such that on fair days the wooden figure would rise above roof level and survey the city, but on rainy days he would sink below roof level to hide in the attic. Such an enchanted figure and his inventor must surely be in league with the devil, and the barometer had to be hastily destroyed to prevent a lynching.

Buoyancy in Gases

Another property that gases have in common with liquids is that of buoyancy. The buoyancy principle applies to gas just as it does to liquid: *An object is buoyed up by a fluid (a gas or a liquid) with a force equal to the weight of the fluid displaced.*

[9] Mercury has a density of about 0.5 pound per cubic inch. Using the equation, pressure = depth × density: pressure = 30 inches × 0.5 lb/cubic inch = 15 lbs/cubic inch.

We walk around all day in a sea of air, but the human body is large enough to displace only a few ounces of air, so the buoyant force of air on the body is negligible. Some people have the misconception that a person is weightless in a vacuum. In fact, if the air were removed from around a person, a scale would indicate that he weighs *more,* because air would no longer be buoying him up with those few ounces.

Buoyant forces in air become important in the use of helium-filled balloons. At the same temperature and pressure, helium has a density less than air. If it is put into a light container, the helium-and-container combination still has a density less than that of air. Thus it rises in air because the buoyant force is greater than its weight. The air around the balloon becomes less dense as the balloon rises above the earth, however. Thus a helium balloon will rise only until its density is the same as that of the surrounding air.

BERNOULLI'S PRINCIPLE—AN ODD ONE

We discuss here a principle which explains many common phenomena seen when gases or liquids are in motion. We will start with a TADI which may surprise you.

FIGURE 7–24. The helium-filled Goodyear blimp is buoyed up by a force equal to the weight of air it displaces. (Photo by Goodyear.)

Cut a circle about 3 or 4 inches in diameter from a stiff piece of paper or cardboard, and push a thumbtack through the center of it. Rest the paper on a thread spool (as shown in Fig. 7–25A) so that the thumbtack sticks down into the hole in the spool and the paper cannot slide off. Now blow through the spool from the bottom to try to blow the paper off. If you refrain from blowing in sudden bursts, you will be unable to blow it off. In fact, if you turn the spool over and hold the paper until you start blowing, you won't even be able to blow the paper off downward.

take-along-do-it

It would seem that the air blowing through the hole would exert pressure on the paper to blow it away. Indeed, this effect does exist, but it is more than compensated for by another effect—one due to Bernoulli's principle. Bernoulli's principle can be stated as follows: *The pressure exerted by a fluid decreases when the gas or liquid is in motion.* This applies to the spool and paper as follows: the air you blow through the spool must move between the spool and paper in order to escape. The arrows in Figure 7–26 represent the airflow. According to Bernoulli's principle, there is less pressure on the cardboard due to the fast-moving air above the paper than due to the relatively stationary air below the paper. Thus, there is more air pressure upward on the paper than downward, and the paper is held up near the spool.

Bernoulli's principle, which was first stated in 1738 by the Swiss physicist Daniel Bernoulli, is not a self-standing basic principle of physics, but can be explained from more basic ideas. Consider the tube in Figure 7–27, which has a narrow section on its right and has water moving through it from left to right. It should be obvious that the water must move faster through the constricted part of the tube than through the larger part on the left. (This is similar to the way a river flows faster where it is narrow and slower where it is wide.) The three objects at A, B, and C are submarines carrying midget Martians. Submarine C is moving along with the water at a speed which is greater than the speed of Submarine A (because the water at C moves faster than the water at A). Submarine B is right at the point where it is moving into the constriction. It must, therefore, be speeding up. But if it is accelerating, an unbalanced force must be causing the ac-

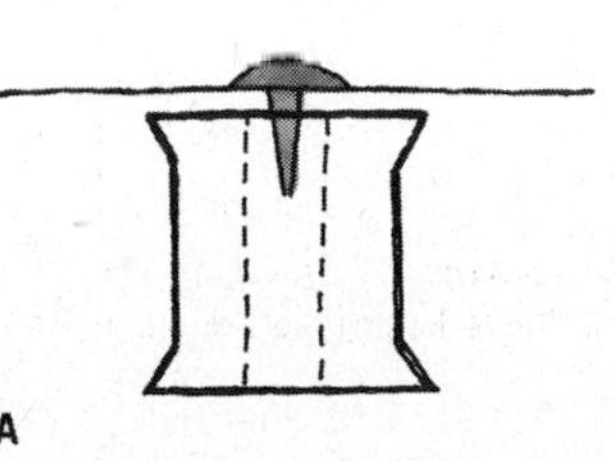

FIGURE 7–25. Try to blow the paper off the spool.

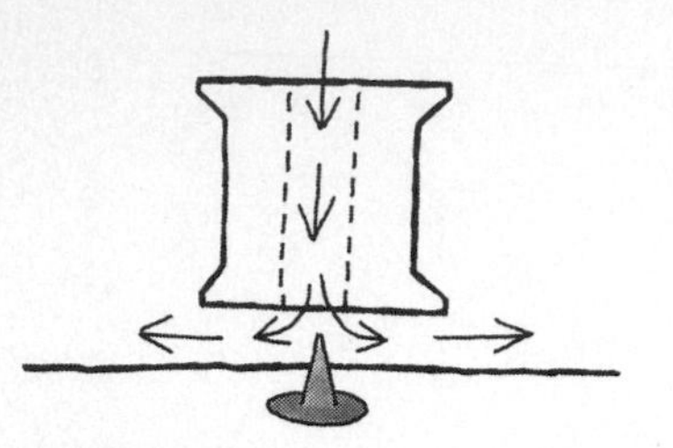

FIGURE 7–26. The moving air above the paper exerts less pressure than the still air below it.

celeration. (Remember Newton's Second Law.) This force *must* result from the fact that the water behind the submarine is exerting a greater pressure than the water in front of it. This same pressure difference causes the water to accelerate as it enters the constriction. It necessarily follows, therefore, that the pressure is greater where the water moves more slowly.

Applications

Figure 7–28 is a bird's-eye view of two ships moving toward the right. Notice the arrows representing water flowing by the ships. The water passing between the ships must be moving faster relative to the ships than the water on the other sides of the ships. This occurs because the water must move from the front of the ship to the back, and if some of it is forced through the constriction between the ships, it must move faster through this narrow space. The result is that there is less water pressure between the two ships than on the outside. This effect is actually experienced by ship captains; they must be careful not to steer too close to other ships lest old Bernoulli cause them to crash together.

take-along-do-it

(1) Hold a piece of paper along one edge and blow across the top of it. (See Fig. 7–29.) You should be able to predict the effect on the paper of the air moving across the top of the paper.

(2) Hold two pieces of paper together along one edge with the thumb and forefinger of each hand. Your hands should be several inches apart, as shown in Figure 7–30. Hold the paper between your slightly open lips and blow to produce a high-pitched squeak. By pulling the paper tauter, you can raise the pitch of the sound. Figure 7–31 explains this phenomenon. When the papers are together, your breath blows over and under them. Because the slight amount of air remaining between the sheets is at rest, it exerts more pressure than the air moving on the outer sides, so the sheets spread apart. Once apart, your breath blows between them, stopping the above effect. The tension from your hands pulls them together again. The vibration repeats hundreds of times per second, producing sound.

(3) Cut a soda straw in half. Lower one end into a glass of water, and use the other end to blow strongly across the top of the straw in the water. By blowing at the right position, you should be able to make the water rise in the straw until it reaches the top, whereupon the stream of air will spray it out in a blast. (It may help to pinch the far end of your blowing straw.) This principle is used in perfume atomizers and bug sprayers.

Blowing across the top of the vertical straw reduced the air pressure in the straw. The pressure of the atmosphere then pushed the liquid up that straw (Fig. 7–32).

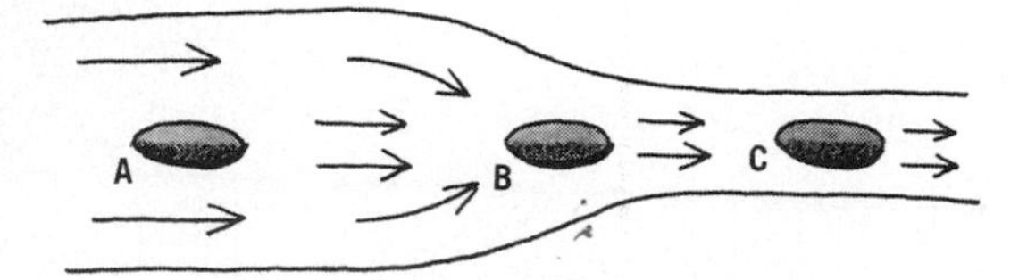

FIGURE 7–27. Submarine C moves faster than submarine A, so it must have been accelerated. An unbalanced force was needed to do this.

(4) Throw a very light ball through the air, flicking your wrist as you throw so that the ball spins from your hand. A styrofoam ball (such as those used in floral decorations) works best. A foam rubber Nerf ball also works well, or you can use a ping-pong ball. (It's best to roughen the surface of the ping-pong ball by rolling it first in glue and then in sand.) You should be able to throw a good curve with any of these balls.

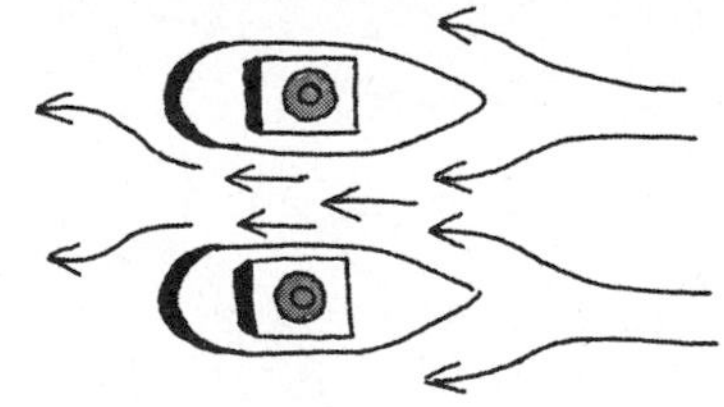

FIGURE 7–28. The water moves faster between the boats.

The curving ball is a less obvious application of Bernoulli's principle than the other examples. Figure 7–33 shows a spinning ball with air rushing by from left to right (which is what happens when a ball moves through the air from right to left). The spinning ball drags some air near its surface along with it, which is why we advised roughening the ping-pong ball. Note in the figure that this spinning air *aids* the air flowing over the top and retards the air across the bottom. Thus, the air across the top moves faster. *You* apply Bernoulli's principle.

FIGURE 7–29. Blow over the paper.

CONCLUSION

We have taken a brief look at some common properties of each of the three states of matter and have explained these properties in terms of the behavior of atoms and molecules. It is not immediately obvious, but given the right conditions, all substances can exist in any of these three states. For example, we normally think of iron as a solid, but at 1530° C, it melts and is a liquid. At 2450° C, it boils and becomes a gas. Helium, on the other hand, is a gas in our everyday experience, but under great pressure, at –268° C, it can be transformed into liquid, and at –271° C it can become solid. Table 7–1 shows the melting points and boiling points of several common substances.

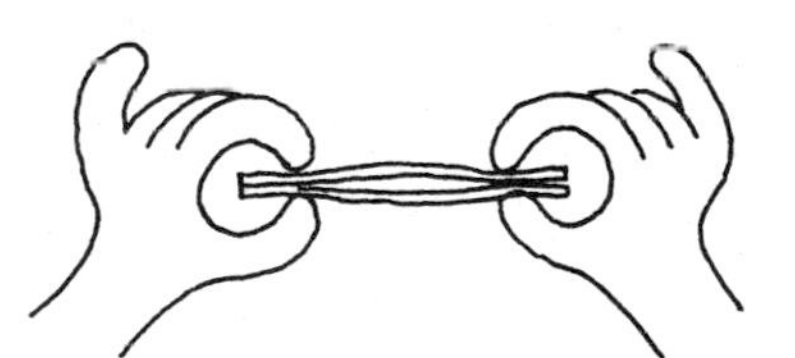

FIGURE 7–30. Hold two pieces of paper like this and blow.

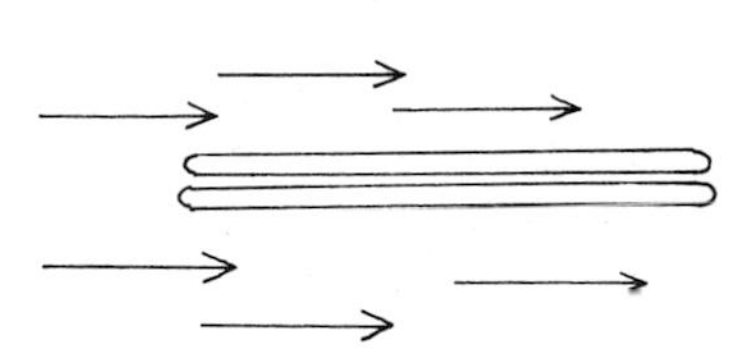

FIGURE 7–31. Air moving past the papers exerts less pressure than does the air between them.

TABLE 7–1. MELTING AND BOILING POINTS OF VARIOUS SUBSTANCES

	MELTING POINT (°C)	BOILING POINT (°C)
Aluminum	658	1800
Ethyl alcohol	–117	78
Gold	1063	2500
Hydrogen	–259	–252
Isopropyl alcohol	– 89	82
Lead	327	1525
Mercury	– 39	357
Oxygen	–218	–183
Water	0	100

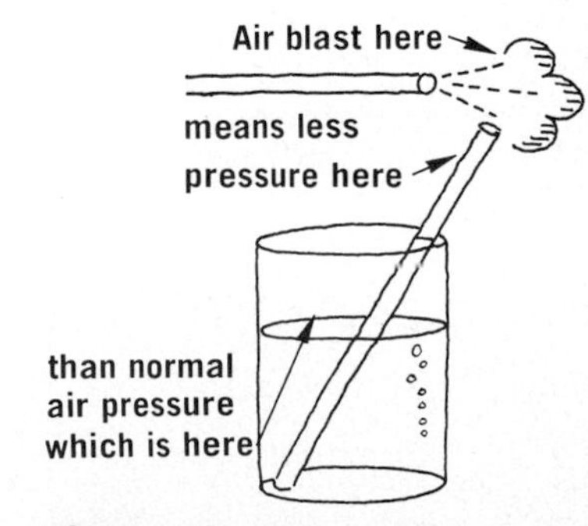

FIGURE 7–32. Blowing across the top of the vertical straw reduces the pressure in the straw.

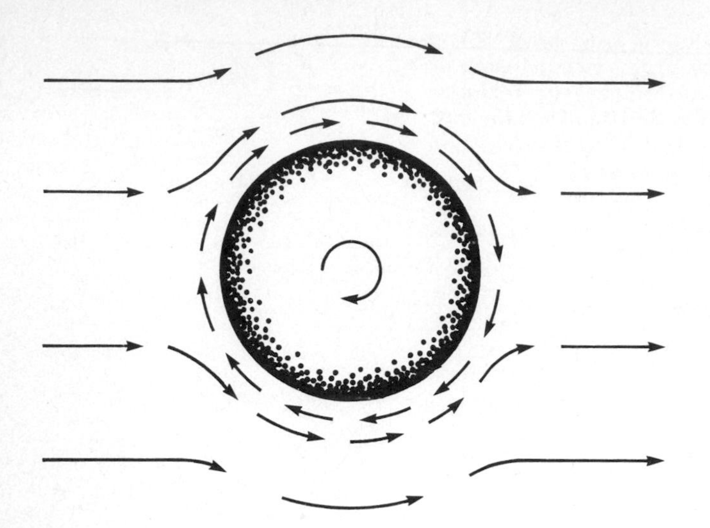

FIGURE 7–33. Air flowing over the top of the spinning ball moves faster, causing less pressure above than below it.

Notice that only a few of the substances listed—and in fact, only a relative few in existence—exist in all three states at temperatures one can ever expect to achieve in one's home. Water, of course, best fits this category; we freeze and boil it every day in our kitchens. Mercury, which is used in thermometers, freezes at –39° C. Since some chilly places on earth reach this temperature, mercury thermometers fail in these regions. Thermometers with alcohol (dyed red, normally) are often used because alcohol has a lower freezing point than mercury.

Table 7–1 serves to point out how limited is our view of the universe. We live on earth within a rather narrow range of temperature, and our tendency is to think that the entire universe is like the earth. In an earlier chapter, it was mentioned that we should avoid thinking that life elsewhere in the universe must necessarily be like life on earth. Data from space probes of Venus indicate that if you grew up there—which is highly unlikely—you would see lead melt at noon and hydrogen freeze at midnight. We earthlings, however, will go on thinking that lead is a solid and hydrogen a gas.

OBJECTIVES

After studying Chapter 7 you should be able to:

1. Explain in what way chemical reactions differ from changes between the states of solid, liquid, and gas.
2. Describe how atoms exist within a solid, and describe an observation which gives evidence of the orderly arrangement of atoms within a solid.
3. Define elasticity. Explain it from a molecular point of view, and explain how it is involved in contact forces between solids.

4. Define density; given the mass and volume of an object, calculate its density.
5. Describe how an increase in temperature affects the atoms within a solid.
6. Explain the process of melting from a molecular viewpoint.
7. Perform a demonstration which illustrates the difference between molecular motion in a cool liquid and a hot liquid.
8. Distinguish between force and pressure, and give an example illustrating the usefulness of the distinction.
9. State the factors that determine the pressure in a liquid, and given the density of a liquid, calculate the pressure at any depth below its surface.
10. Define buoyant force and list the factors that determine the buoyant forces of an object in a liquid.
11. Given the densities of various solids and liquids, predict what solids will float on what liquids.
12. Perform a demonstration in which the density of a liquid is increased so that a submerged body rises.
13. Describe evaporation from a molecular point of view and use the explanation to explain why evaporation results in cooling of a liquid.
14. Describe the behavior of molecules within a gas, describe how the pressure, volume, and temperature of a gas are related by an equation, and explain how molecular motion results in the phenomenon of gas pressure.
15. Explain how it is possible for animal life to live under great air and/or water pressure.
16. Describe the construction, use, and theory of a barometer.
17. Explain the phenomenon of buoyancy in a gas.
18. State Bernoulli's principle and describe and/or demonstrate at least five manifestations of the principle.
19. Show how Bernoulli's principle necessarily follows from Newton's laws of motion.
20. Discuss the chauvinism of earthlings which results from our life in a limited temperature environment.

QUESTIONS—CHAPTER 7

1. Why can you carry a person easily in a swimming pool when you could not carry her on land?

2. Nature has provided the body with an air-conditioning system that relies on perspiration and the process of evaporation. Explain how the act of perspiring results in cooling of your body.

3. In order to increase the rate of evaporation of a pan of water, one can warm it (even without boiling it). Explain, from a molecular point of view, why this works.

4. Steel is much more dense than water. How, then, do boats made of steel float?

FIGURE 7–34.

5. Figure 7–34 is a cross section of an airplane wing. Explain how Bernoulli's principle aids in giving "lift" to a plane. (Hint: note how the front of the wing divides the air so that the air going over the top has farther to go than the air going under.)

6. Consult an encyclopedia to explain how Bernoulli's principle applies to an automobile carburetor. (The principle is sometimes called the Venturi effect in this application.)

7. Use the distinction between force and pressure to explain why you can hold a stack of books on your head without pain, but if someone puts a small pebble between the bottom book and your head, the feeling is unpleasant.

8. Suppose a friend argues that when you swim in the ocean, there are miles and miles of water pushing in on your body and that your body should thus be squashed. What do you know about pressure in liquids which would remove his fears and get him in the water?

9. A helium-filled balloon will rise until its density becomes the same as that of the air. If a sealed submarine begins to sink, will it go all the way to the bottom of the ocean or will it stop when its density becomes the same as that of the surrounding water? (Hint: a liquid is practically incompressible. What does this say about its density?)

10. Does the volume of a gas increase or decrease if the pressure is raised without changing the temperature? Why? What happens to the volume if the temperature is raised without changing the pressure? Why?

11. What happens to the pressure of a gas if the temperature is raised without changing the volume? Why? What happens to the temperature if the volume is decreased without changing the pressure?

12. A fish rests on the bottom of a bucket of water while the bucket is being weighed. When he begins to swim around, does the weight change?

FIGURE 7–35. The ball is supported on a stream of air from the hose. Note that it is being held at an angle.

13. Will a ship ride higher in the water of an inland lake or in the ocean? Why?

14. How much force does the atmosphere exert on one square mile of land?

15. Use Bernoulli's principle to explain how you can support a ball on a stream of air as shown in Figure 7–35.

16. Bubbles of air rising in a liquid become larger as they approach the surface. Why?

17. If 125 tons were placed on the deck of the World War II battleship *North Carolina*, it would sink only one inch lower into the water. What is the cross-sectional area of the ship at water level?

18. You can freeze water by surrounding its container with ether (a substance which evaporates rapidly) and then blowing air over the ether with a fan. Explain why this occurs.

19. Why does a golf ball or tennis ball curve when hit so that it spins?

20. Explain why blowing over a piece of paper causes it to rise.

21. Note the size of the Goodyear blimp compared to the size of its passenger compartment (Fig. 7–24). Why must it be so large?

8

THE INTERACTION OF HEAT WITH MATTER

This photograph, taken from Skylab, shows a swirling solar eruption about 220 times the diameter of the earth, expanding outward into space at a speed of about one million miles per hour. (Photo courtesy of NASA.)

The preceding chapter, dealing with solids, liquids, and gases, indicated that the way in which one of these states of matter can be changed to another state is through the addition or subtraction of heat. Heat, however, is responsible for many changes in matter other than simple changes in state. In this chapter we will look at what heat is and what it does.

TEMPERATURE

One of the first sensations of childhood is that of the hotness or coldness of an object. This subjective "feel" for the temperature of a material is not sufficient, however, if we are to describe accurately and impartially just how cold or hot something is. In order to develop such a procedure we must use a property of a material that is affected by its temperature. Why don't you try one?

take-along-do-it

Blow up a rubber balloon and note its size. Then place it either in contact with a block of ice or in the freezer compartment of a refrigerator and note what happens to its size.

The relative size of the balloon before and after indicates a property that we can use in order to measure temperature. The balloon diminishes in size as the temperature

FIGURE 8–1. Cooling one down.

of the air inside decreases. In general, a change of size with a change of temperature is a property of all materials, regardless of whether they are solid, liquid, or gas. This property forms the basis for the most common of all temperature measuring devices: the mercury thermometer.

The Mercury Thermometer

A common mercury thermometer is illustrated in Figure 8–2. It consists of a glass tube (with a very thin inside diameter) from which all the air has been removed. This tube is connected to a small bulb filled with mercury When placed in contact with a hot object, both the mercury and the glass expand, but the increase in volume of the glass is not as great as the increase in volume of the mercury. The result is that the mercury gradually creeps up inside the stem. A scale engraved on the glass indicates the temperature according to how high the mercury rises inside the stem. (Recall the discussion in Chapter 7 of temperature scales.)

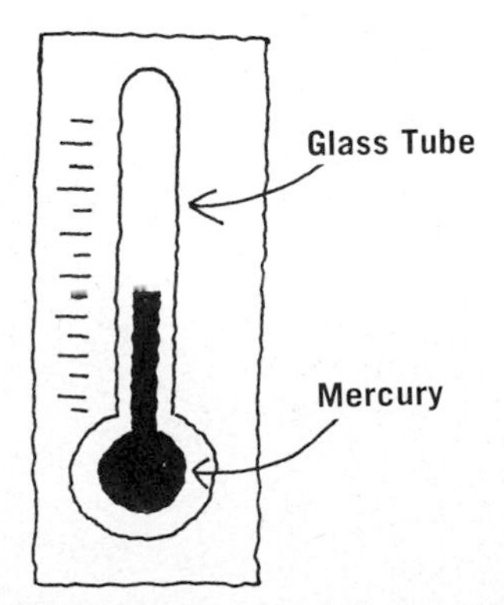

FIGURE 8–2. A mercury thermometer.

EXPANSION

With very few exceptions, the size of objects increases as their temperature increases. In the last chapter we saw that the addition of heat to an object causes the vibrational motion of the atoms to increase such that they swing back and forth with greater and greater amplitudes. Because of this increase in the amplitude of vibration, each atom effectively takes up more space than it did before heat was added, resulting in an expansion of the material. This expansion property of materials has been used to explain the rising mercury in thermometers. The thermometer also serves to point out another property of expansion: different materials expand differently (as evidenced by mercury expanding more than glass). In general, we find that most solids and liquids and all gases expand as heat is added, but the addition of the same amount of heat will cause a gas to expand most, a liquid next most, and a solid the least.

Rivets used in construction to hold together steel plates are cooled with dry ice so that their expansion upon warming firmly binds them in the plates. This is an example of the value of the expansion property, but the property has its disadvantages too. Imagine a long steel bridge built such that its ends are firmly anchored. It would do quite well as long as the temperature didn't change, but after a few degrees' rise, the bridge could take on a shape as shown in Figure 8–3. This might be a unique way to construct a drawbridge, but it is probably a bad prospect for an engineering design-of-the-year award. In order to avoid this difficulty, space is left at the ends of bridges to allow for expansion and contraction, and rocker supports like those in Figure 8–4 permit freedom of movement.

Poor quality dinner plates, especially thick ones, often crack when hot food is placed on them. This occurs because the top surface of the plate expands before heat has time to reach the lower surface, resulting in an uneven expansion. Some materials do not break upon uneven heating even though they may be brittle enough to break when dropped. One manufacturer of ceramic cookware once advertised its product with a photograph showing a pot that had one end in ice and the other over a flame (Fig. 8–5). Such materials do not break even under these extreme conditions for two reasons: (1) they expand very little upon heating, and

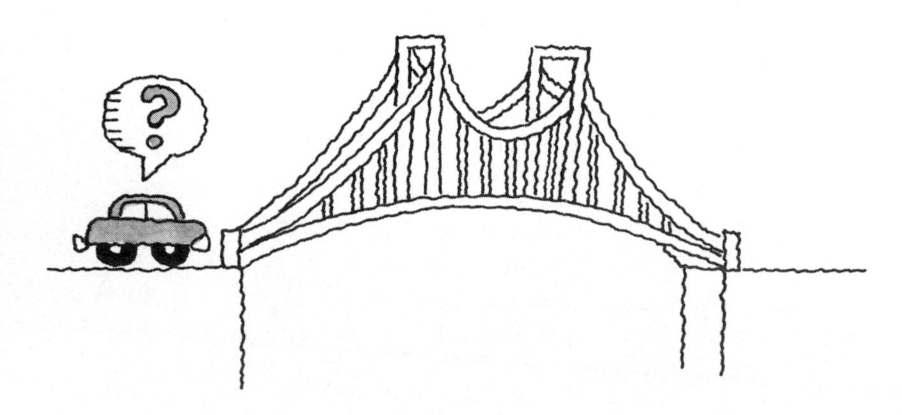

FIGURE 8–3. This bridge was built on a cool day such that a car did not have to go up and down to cross. Notice what happens on a hot day.

FIGURE 8–4. Rocker supports under a bridge.

(2) they are strong enough to withstand the stress produced by any such expansion.

Figure 8–6 illustrates a (perhaps) surprising property associated with the expansion of materials. Shown is a metal plate with a hole cut in it. As the metal is heated, one might think that since the metal is free to expand in all directions

FIGURE 8–5. Battling flame and ice, this piece of cookware survives. (Photo by Corning Glass Works.)

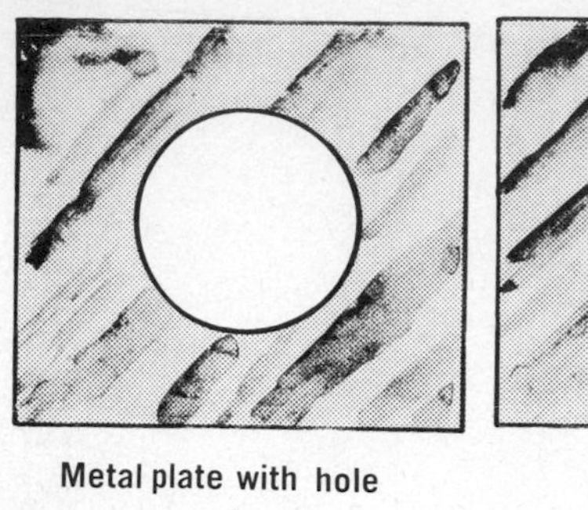

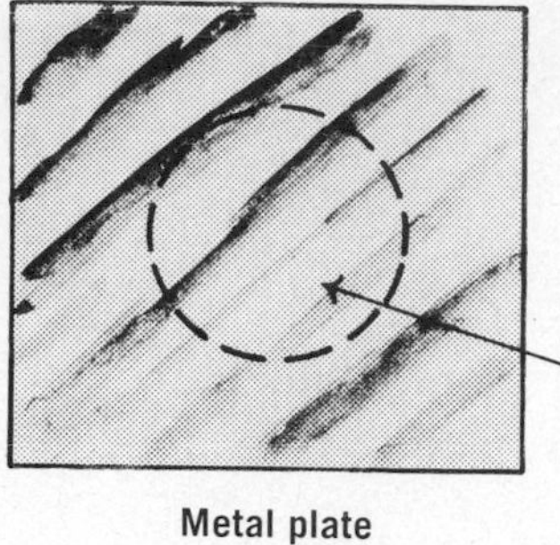

FIGURE 8–6. Upon heating a metal plate with a hole in it, does the hole expand or contract?

it would expand into the hole and make the hole smaller. The opposite is true, however: the hole expands in exactly the same way as does the metal. To convince you that this is reasonable, Figure 8–6*B* shows the same metal plate with the hole filled by the same kind of metal. If we now heat the entire plate, it increases in size uniformly. But what does the plate care whether there is metal or a hole in its center; it will expand in the same way in either case.

A thermostat is a common device that uses the expansion property of materials in order to do its job. The heart of a thermostat is a strip made by rigidly attaching two pieces of dissimilar metals together (Fig. 8–7*A*). The metals are selected such that one expands more than the other as the temperature rises. As a result, when the temperature of the strip rises, it bends (Fig. 8–7*B*). Figure 8–7*C* shows how the strip is put to work in a thermostat. At 65° Fahrenheit, the strip, made of brass and steel usually, holds an electrical contact arm away from the contact points on the device. But as the temperature rises, say to 70° F, the brass expands more than the steel and causes the contact points to touch, thus closing an electric circuit to start an air conditioner.

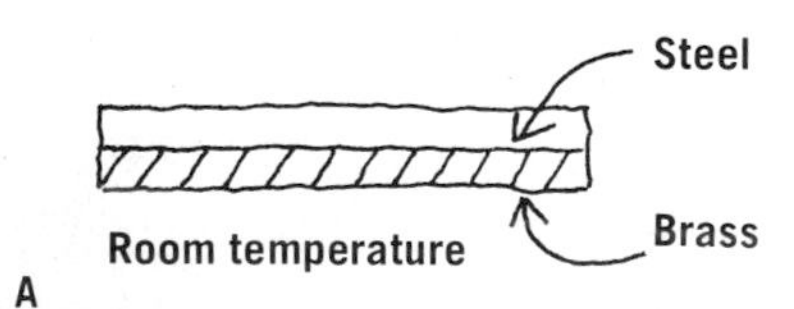

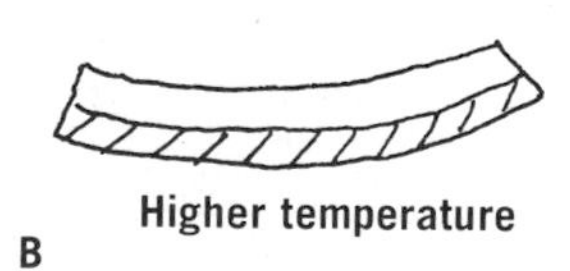

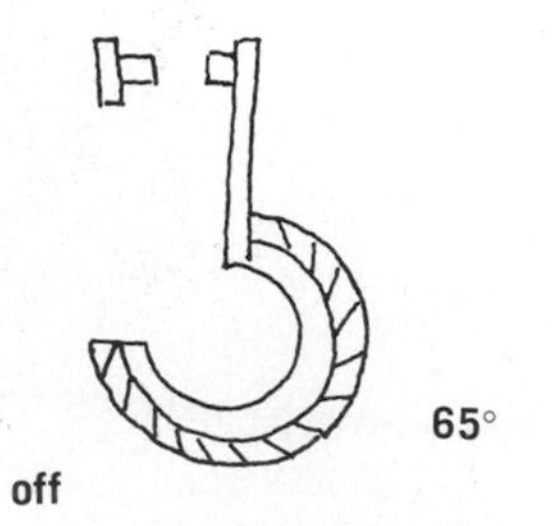

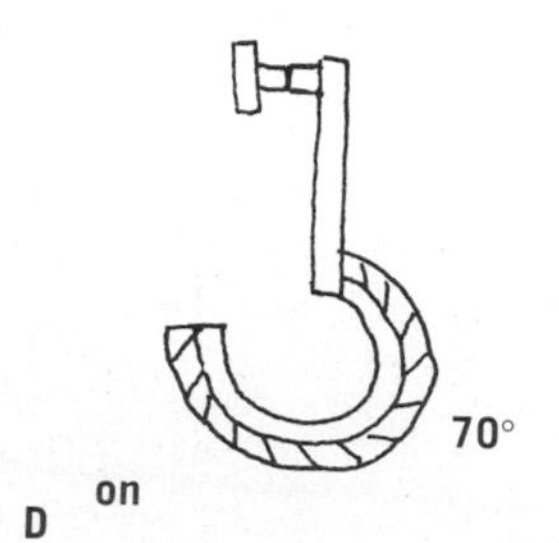

FIGURE 8–7. The principle of the thermostat.

Water is Unusual, Usually

Examples given so far have all centered around the expansion of materials as temperature increases. A notable exception to this rule is water. As its temperature is lowered from, say, 70° Celsius, water behaves as do other materials, by decreasing in volume, but at 4° C it begins to expand as the temperature drops further! At 0° C, water freezes, and in so doing, expands even more.

Fish rejoice at the foresight of water to behave in this way. If it were not so, lakes would freeze from the bottom up, and marine life would be presented with serious problems. (Ice skaters would also experience some difficulty. Skating on the bottom of a lake would be rough.) To see why lakes freeze as they do, let's examine what happens to a lake as its temperature drops. As a cold front passes through, the cold air lowers the temperature of the water. As an example, let's suppose the upper surface drops from 6° to 5° Celsius. As the surface water in the lake cools, it decreases in volume, becoming more dense than the water below it. This cool water sinks and is replaced by warmer water from below, which, in turn, is cooled and sinks. This mixing process continues until all the water has reached 4° C. But now the situation changes. A drop from 4° to 3° in air temperature produces a similar temperature change in the surface water. Now, however, the water at 3° is expanding and becoming less dense than the water below it. It therefore floats, and the mixing process does not take place. From 3° down to 0° C the colder water remains on the surface, effectively shielding deeper layers of water from further decreases in temperature. Finally, at 0° C the surface water solidifies. Get out your ice skates!

MEASURING HEAT

Long before the true nature of heat was known, methods had been devised for measuring it. In the two methods in common use, the units of heat are the calorie and the British Thermal Unit (BTU). *The calorie is defined as the amount of heat that will raise one gram of water one Celsius degree. The British Thermal Unit is that amount of heat which will raise one pound of water one Fahrenheit degree.*

The BTU has become a common part of American parlance through its application in air conditioning systems. The purpose of an air conditioner is to remove heat from a room and to reject it into the outside atmosphere. If an air conditioner is rated at 3000 BTUs, this simply means that the device will transfer 3000 BTUs every hour from inside the house to the outside. Except in such engineering usages—and then only in the United States and a few other countries—the calorie is the most widely used unit for measuring quantities of heat. We will restrict ourselves to the calorie in all that follows.

SPECIFIC HEAT

The next time you eat a pizza, take a moment to convince yourself that different substances have widely different abilities to store heat. When the pizza is set before you, piping hot and at the same temperature throughout, note that you can take a bite of the crust, but a bite of the cheese is likely to burn your tongue. Even though both are at the same temperature, the amount of heat each contains is different.

In order to talk about the ability of an object to store heat, a quantity called *specific heat* is used. The *specific heat of a substance is the amount of heat needed to raise the temperature of one gram of the substance by one Celsius degree.* If you will recall the definition of the calorie from the last section, you should agree that the specific heat of water is one. With the exception of ammonia, water has the highest specific heat of any substance with which you are likely to come into contact on a routine basis. As a comparison, the specific heat of aluminum is about 1/5 that of water. This means that while one calorie will raise one gram of water by one Celsius degree, only 1/5 calorie is needed to raise one gram of aluminum by the same amount. Another way to look at this is to consider a piece of aluminum foil and an equal mass of water at the same high temperature. Because of the water's higher specific heat, more heat energy had to be supplied to it than to the aluminum to get it to this high temperature. Notice this the next time you tear the aluminum foil off a TV dinner without burning your fingers but are unable to touch the water containing vegetables beneath the foil.

The high specific heat of water compared to that of land is responsible for the pattern of airflow at a beach. During the daylight hours the sun adds roughly equal amounts of heat to both the beach and the water, but the lower specific heat of sand causes the beach to reach a higher temperature than the water.[1] The air above the land reaches a higher temperature than that over the water, and as a result, the air rises as shown in Figure 8–8A. Cooler air from above the water is drawn in to replace this rising hot air, resulting in a breeze from ocean to land during the day. The hot air gradu-

[1]In addition, the fact that water mixes continuously adds to this effect.

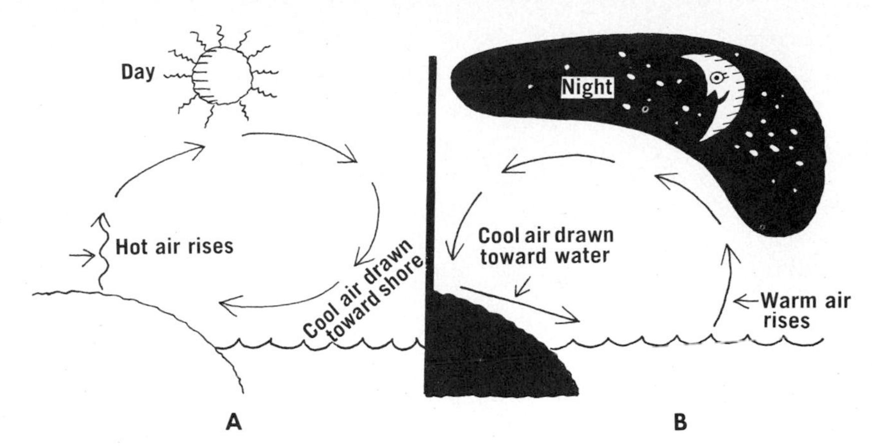

FIGURE 8–8. The breezes at a beach are partly the result of the fact that water and sand have different specific heats.

ally cools as it rises and thus sinks to set up a circulating pattern as shown in the figure. During the night, the land cools more quickly than the water, and thus the circulating pattern reverses itself because the hotter air is now over the water (Fig. 8–8*B*). You should not expect to observe this pattern every time you go to the beach because prevailing winds caused by other factors often obscure it.

The high specific heat of water is also responsible for moderating the temperatures of regions near large bodies of water. As the temperature of a body of water decreases during the winter seasons, it gives off its heat to the air, which carries it landward when prevailing winds are favorable. For example, the prevailing winds off the west coast of the United States are toward the land, and the heat liberated by the Pacific Ocean as it cools keeps coastal areas much warmer than they would be otherwise.

CHANGE OF STATE

Let us suppose than an overzealous refrigerator has cooled a one-gram ice cube down to −50° Celsius. Now if what we would like to have is one gram of *steam*, we can make the required transition with the addition of some heat. Let's examine the changes involved, one at a time, and as we do so, some surprising features will appear (Fig. 8–9).

(1) The first step is to heat up the ice to 0° C. No difficulty in this stage; we are merely heating a solid.[2]

(2) When the ice reaches 0° C, a surprising thing occurs: we continue to add heat to the ice, but its temperature no longer changes. In fact, the temperature stays right at 0° C until all the ice has melted. If we measure the amount of heat required to melt our one-gram ice cube, we would find that 80 calories are necessary. In fact, we would need 80 calories for every gram of ice we melt.

---

[2]The specific heat of ice, however, is 0.5. (Not 1 like liquid water.) Thus, it takes 25 calories to heat the one gram of ice 50 degrees.

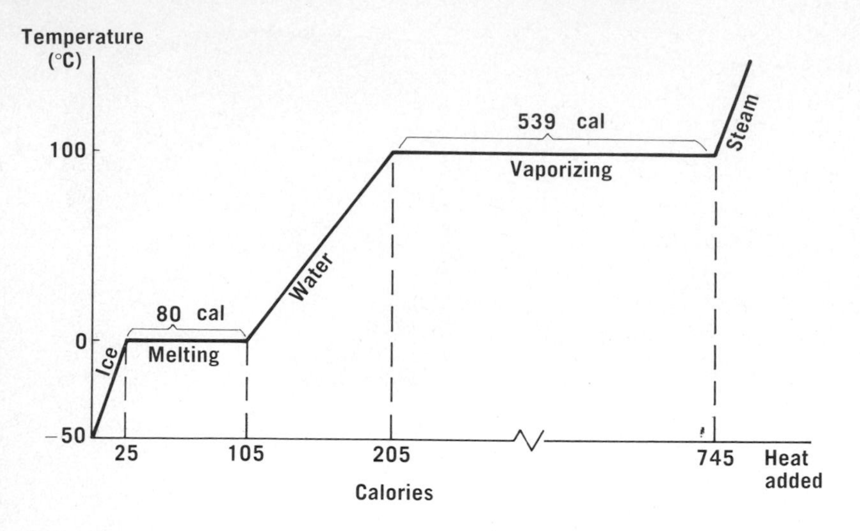

FIGURE 8–9. A graph showing the changes in temperature of one gram of water as heat is added to (or taken away from) it.

Most substances[3] behave in this fashion. A substance going from one state to another, as from solid to liquid, always absorbs heat without a temperature rise until the change has been completed. The amount of heat needed to produce the change varies from one material to the next. For example, only about six calories are necessary to change one gram of solid lead at its melting point (327° C) to one gram of liquid lead.

(3) As soon as the last speck of ice has melted, the temperature begins to rise again until the water reaches 100° C (its boiling point). Earlier we defined the calorie to be the amount of heat required to raise one gram of water one Celsius degree. We have now raised one gram of water from 0° to 100°, and in order to accomplish this, we had to add 100 calories.

(4) At 100° C, another change begins to occur: our water changes into steam. As in the melting process, this change occurs at a constant temperature. If we measure the amount of heat needed to completely vaporize our one gram of water, we would find that 539 calories are required. This is a large amount of heat needed to perform what seems to be so simple a task. This amount of heat, 539 calories, is sufficient to raise 539 grams of water by one degree, but we have used the 539 calories to change one gram from liquid to gas and have not changed its temperature from 100° C at all.

The amount of heat required to change the state of a material from a liquid to a gas also varies from one material to another. For example, helium requires only five calories to change one gram from a liquid to a gas, while copper requires about 1200 calories per gram to make the transition.

(5) Once the water has changed to steam, no new

[3]Butter, margarine, and glass are exceptions. They have no definite melting point, but instead *gradually* change from what we would call solid to what we call liquid.

wrinkles develop. If we continually add heat to the steam, its temperature will increase steadily.[4]

If we had started at the other end of the scale with one gram of steam at a high temperature and cooled it down, the steps would have reversed themselves. The temperature of the steam would decrease to 100° C, at which point it would begin to condense to liquid water. In order to convert it completely to one gram of water, we would have to remove 539 calories from each gram. Similarly, at the melting point (or freezing point in this case), 80 calories of heat would have to be removed to convert it into ice.

WHAT IS HEAT?

It is common knowledge that a hot object transfers heat to a colder one when the two are placed in contact. In the process, both come to some intermediate temperature. The question we raise now is the same as that raised by the first people who observed this phenomenon: What is it that is transferred? Prior to the eighteenth century, it was believed that objects possessed an invisible fluid, called caloric, which flowed freely from a hot object to a colder one, the process being not unlike water flowing downhill.

Count Rumford (1753–1814) was assigned the task of boring cannons[5] for the government of Bavaria. To prevent them from overheating, the cannons were filled with water as the cutting tools did their work. The water continually boiled away and had to be replaced. It was thought that as the metal fragments were cut into smaller and smaller bits they became less able to contain their caloric and thus were responsible for boiling the water. Rumford, however, noted that the boiling continued even when the tools were so dull that they did no cutting at all. On the basis of this observation, he proposed that the caloric idea was incorrect and that instead, heat was actually another form of energy. The rotation of the boring tool continually supplied energy to the metal of the cannon. This energy was transformed into heat. Later experiments performed by him, and subsequently by many others, proved his ideas to be correct.

Count Benjamin Thompson Rumford (1753–1814)

You must read Rumford's biography* to appreciate the fantastic life of this remarkable man. A few facts: born in Massachusetts, he spied for the British during the American Revolution. He was later thrown out of England under suspicion that he was spying for France. By experimenting with insulation, he discovered convection currents and then, hiring beggars, he established a factory to produce warm uniforms for the Bavarian army. He secretly introduced potatoes, which were then considered not fit to eat by Bavarians, into soup served in the army; and thereby established potatoes as a staple food in central Europe. This count of the Holy Roman Empire was a truly practical scientist (and perhaps somewhat of a scoundrel). Read his biography!

MOVEMENT OF HEAT

There are three processes by which heat can be transferred to a substance: conduction, convection, and radiation. We will examine these in turn.

---

[4]The specific heat of steam is—like ice—0.5.

[5]Cannons are made by boring a hole in a solid cylinder of metal. It might be boring, but it produces cannons.

*Brown, Sanborn C.: *Count Rumford, Physicist Extraordinary.* New York, Doubleday, 1962. This short interesting book is one of a number of biographies of Rumford.

Conduction

Insert one end of a key into the flame of a cigarette lighter and observe what happens. Unless you are a masochist, you will soon let go. The addition of heat, or thermal energy, to the end in the flame increases the vibration of the molecules there. As their to-and-fro swinging increases, so do their collisions with neighboring atoms. These collisions cause the neighbors to swing violently also, and bit by bit the increased vibrations creep toward your unsuspecting fingers. Finally, the swinging atoms near your fingers cause some atoms or molecules in your fingertips to swing also. Zap! No fingerprints.

In general, metals are the best conductors of heat, with copper and silver leading the list. Metals exhibit this property because they have a large number of free electrons. In addition to increasing the vibrations of the atoms inside the material, heat furnishes kinetic energy to these electrons. Collisions between electrons in the material is thus another means for conducting heat energy through a metal.

The ability of a given material to conduct heat depends on its area and its length. Figure 8–10 shows a bar placed between two objects at different temperatures. The rate at which heat energy is conducted through the bar will increase as the area of the bar increases and decrease as the length of the bar increases.

To illustrate the effect of length on the heat-conduction ability of a material, fill a paper cup with water and place it over a flame (Fig. 8-11). Because the thickness (the "length" involved) of the bottom of the cup is small, the heat is quickly transferred through it. As a result, the water can be brought to a boil before the temperature of the paper becomes high enough to burn.

Convection

In the preceding method, heat was transferred from one place to another by the vibration of the molecules inside the material (and, in the case of metals, by the transfer of kinetic energy between electrons). *In the convection process there*

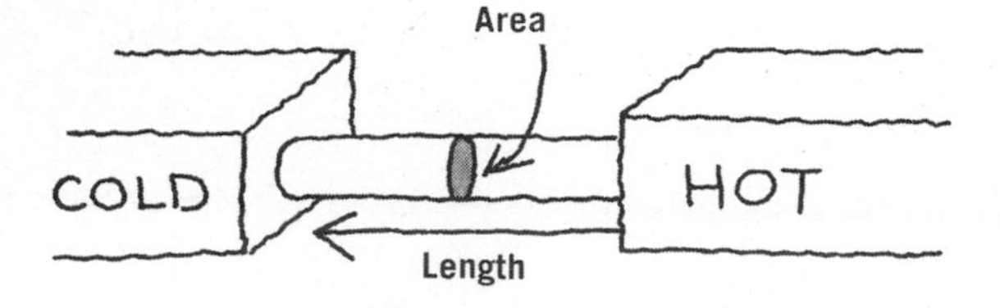

FIGURE 8–10. The rate of heat conduction depends upon area and length of the conductor.

PLATE I

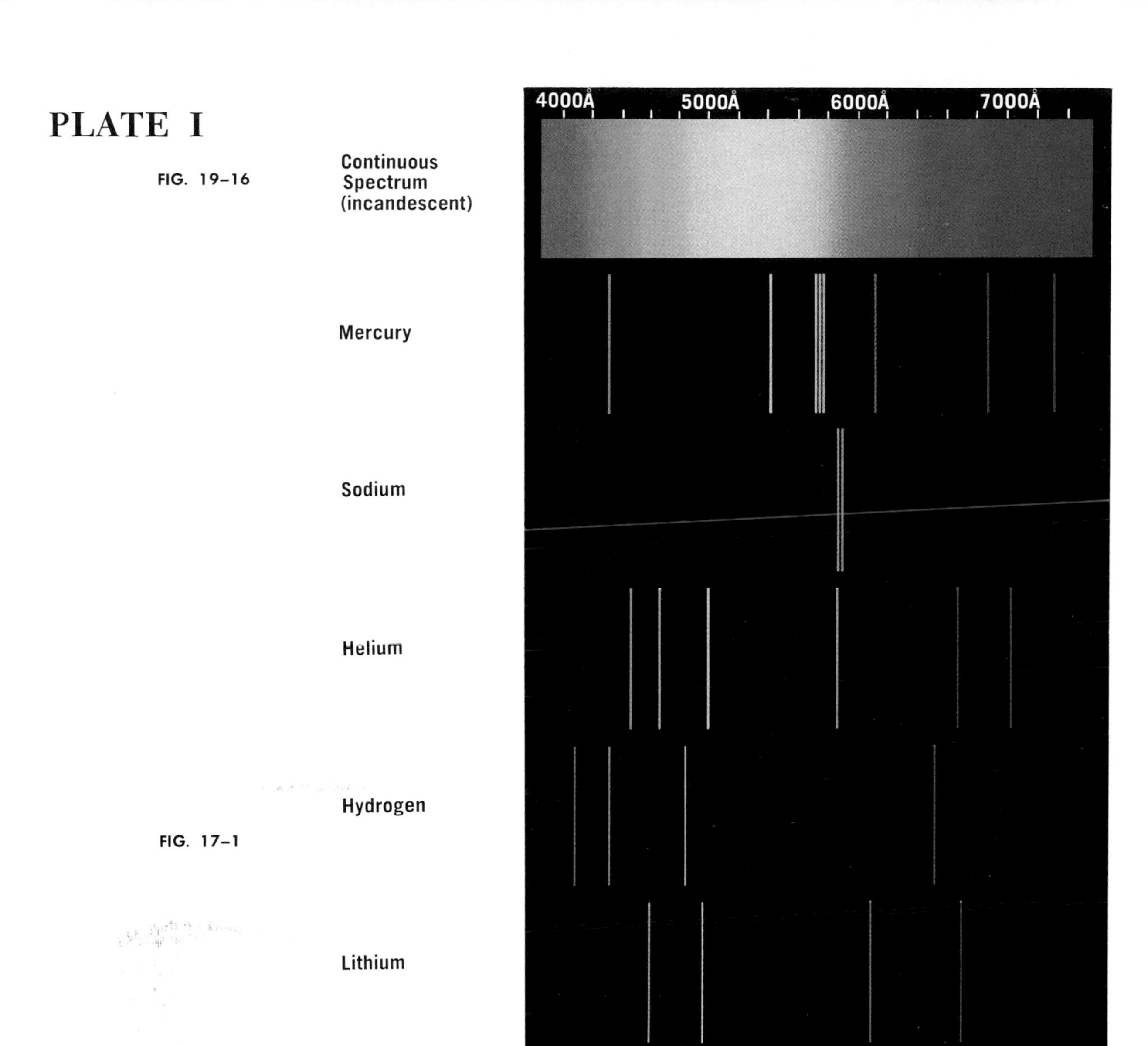

FIGURE 17–1. (See Figure 19–16.) The hydrogen spectrum contains only a few well-defined colors in the visible portion.

FIGURE 19–16. The continuous spectrum is shown at the top. Note that the line emission spectra from various elements differ from each other and that they consist of well-defined colors.

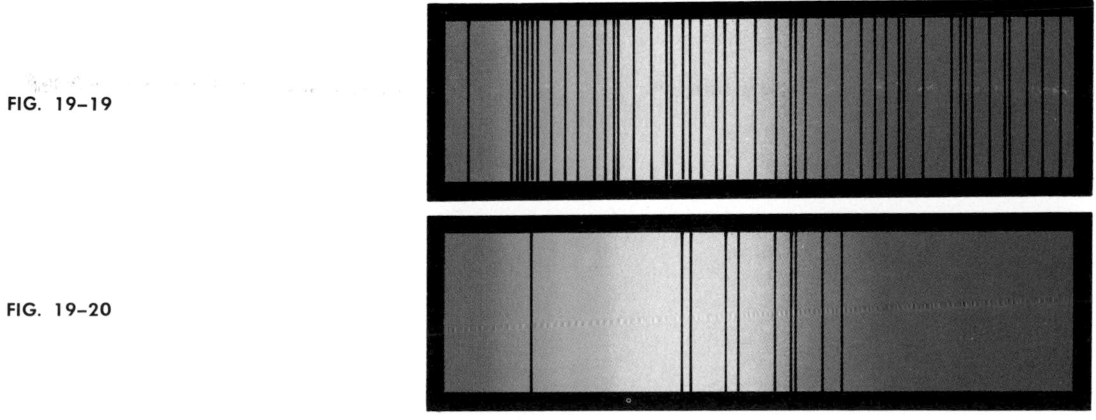

FIGURE 19–19. The solar spectrum is almost a continuous spectrum, but it has many absorption lines in it.

FIGURE 19–20. The spectrum of iron. Note the correspondence with the solar spectrum above it, indicating the presence of iron in the sun's atmosphere.

PLATE II

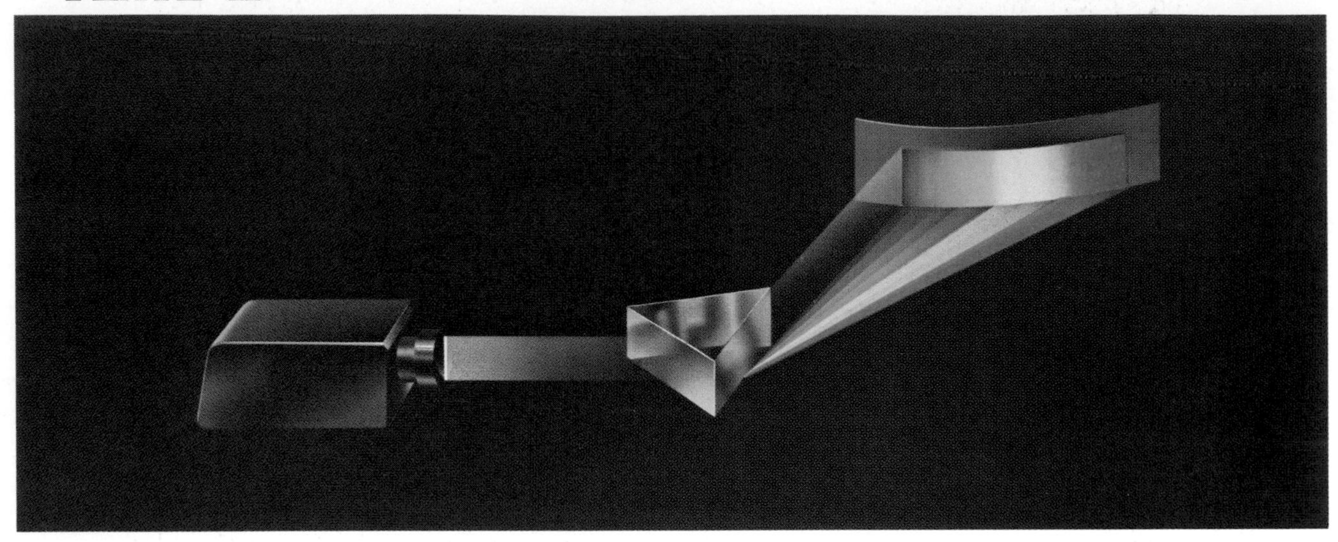

When a beam of white light passes through a transparent prism, it is bent. Short wavelengths of light (blue) are bent more than are long wavelengths (red). The result is a continuous spectrum, which demonstrates that white light contains all colors of the visible spectrum.

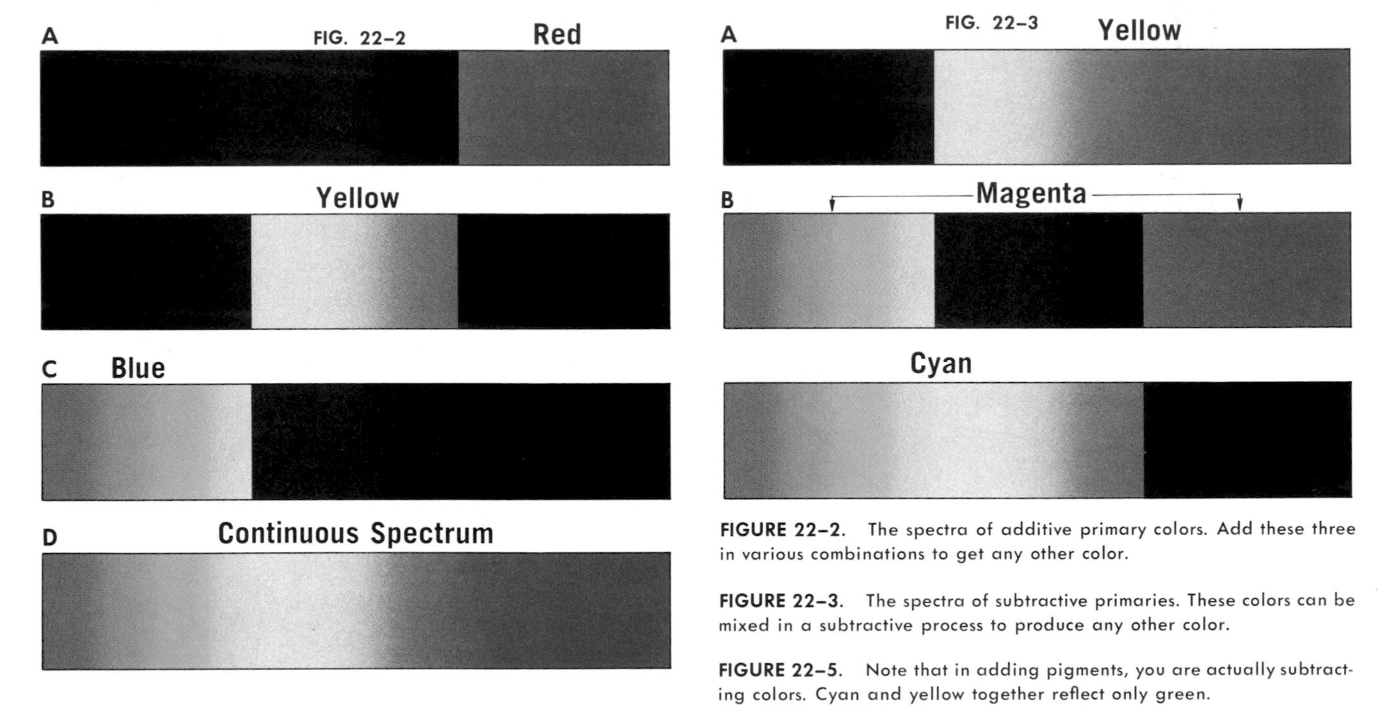

FIGURE 22–2. The spectra of additive primary colors. Add these three in various combinations to get any other color.

FIGURE 22–3. The spectra of subtractive primaries. These colors can be mixed in a subtractive process to produce any other color.

FIGURE 22–5. Note that in adding pigments, you are actually subtracting colors. Cyan and yellow together reflect only green.

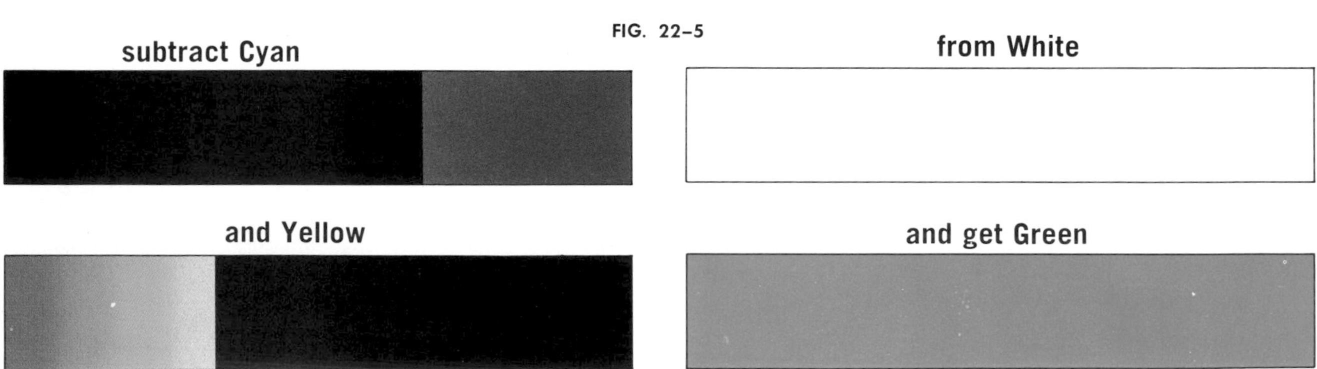

is an actual movement of the material from one location to another.

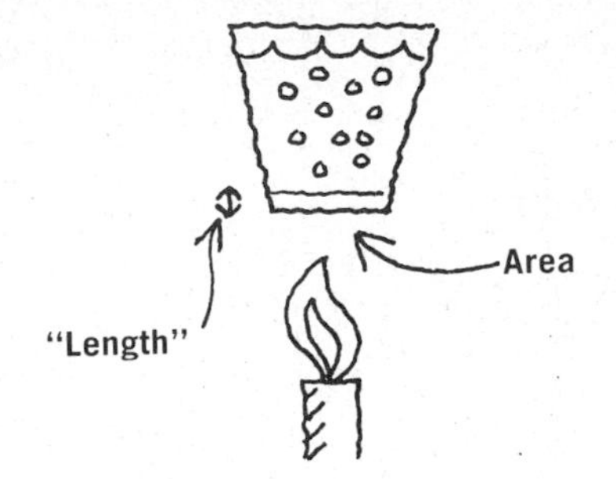

FIGURE 8–11. That is a paper cup. Will it burn?

At least once during every cold winter, the heating system fails in each dormitory in the nation. Consider Figure 8–12. To keep himself from freezing, the student pictured has swept the lint from beneath his bed and is burning it. As it burns, the air directly above the flame gets hot. The heated air expands, becomes less dense, rises, and is replaced by cooler air drawn in from other locations in the room. This air is in turn heated and also rises. As this heated air reaches the ceiling, it cools and begins to sink. The arrows in the figure illustrate the path that it will follow to eventually find its way back to the flame for reheating.

You should recall a similar situation discussed earlier, in which a circulating volume of air was responsible for breezes at a beach. Also, earlier in this chapter there was a discussion of why lakes freeze at the upper surface first. A review of that section will indicate that before the temperature dropped to 4°, there was a continual mixing of the water. This, too, is the convection process in operation.

FIGURE 8–12. The convection of heat.

Radiation

The final method of transmitting heat energy from one location to another can be illustrated by observing how our hands are warmed by a flame. If we hold a piece of metal in the fire and wait long enough, the conduction process will singe our flesh. If we place our hands over the flame, the rising convection currents warm us. But if we place our hands in front of the flame (Fig. 8–13*C*), the radiation process takes over. Air is a relatively poor conductor of heat, and thus conduction cannot supply much heat in this case; we are not in the direct path of a convection current, so this method also must be ruled out. The radiation process remains.

Radiation reaches us in the form of electromagnetic waves of a type called infrared waves. We will see in Chapter 16 that all warm objects emit these waves. The term "warm" here is a relative one. If you read the section in Chapter 7 dealing with the measurement of temperature, you learned that there is an absolute zero of temperature.

FIGURE 8–13. Methods of heat transmission.

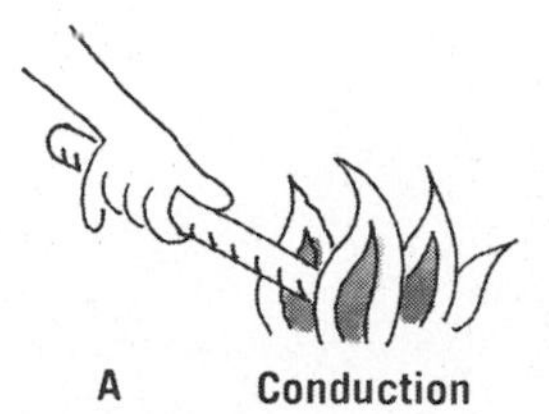
A Conduction

B Convection

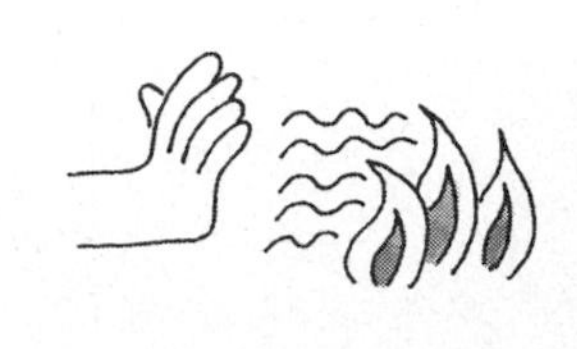
C Radiation

Compared to this, everything is warm, and hence, everything emits infrared waves. The higher the temperature of an object, the greater is its emission of such waves.

If your home is heated by a fireplace, the principal way that heat enters the room is by radiation. As Figure 8–14 shows, the chimney provides an outlet for convection currents to escape. The cooler air drawn into the room to replace the escaping air provides oxygen to feed the flames, but it also competes with radiation; the air cools the room as radiation from the flames heats it. If the home is not well constructed, entering drafts of cold air may offset the heating-by-radiation to the extent that the overall temperature of the room may not be increased appreciably by use of the fireplace.

Hindering Heat Transfer

The name of the game when heating or cooling a home is to prevent heat from escaping in the former case and to prevent it from entering in the latter. To bring this about, insulating materials are placed in the walls and above the ceilings. These materials, such as fiber glass, are poor conductors of heat. In addition, they are loosely packed and therefore have air trapped within them. Air is a poor conductor, and when trapped in small cavities, it cannot carry much heat by convection currents. Savings in fuel bills of as much as one third can be attained by proper insulation.

The Thermos bottle is a device used to prevent heat transfer (Fig. 8–15). The walls of the bottle are made of glass and are silvered on the inner surface. The highly reflecting silvered surfaces reflect radiation, hindering heat loss from your coffee by radiation and preventing heat from radiating

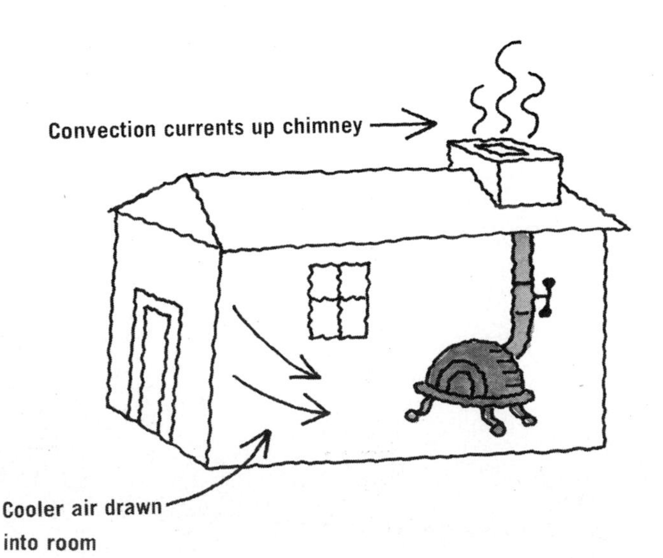

FIGURE 8–14. The radiation from a fireplace must add more heat to the room than escapes up the chimney by convection if the fireplace is to do its job.

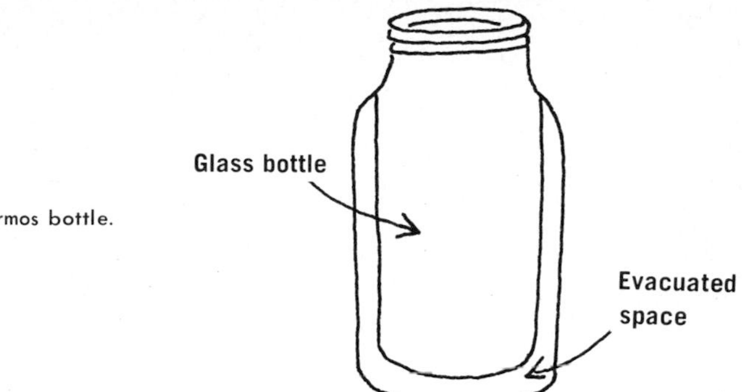

FIGURE 8–15. A thermos bottle.

in if you need a cool drink. Finally, to insure that as little heat loss as possible occurs due to conduction or convection, the space between the walls of the bottle is evacuated. A vacuum is certainly a poor conductor of heat; and if there is nothing there, convection currents cannot form.

THE EFFECT OF PRESSURE ON FREEZING

Everyone knows how to make a snowball. All you have to do is pick up some loosely packed snow, squeeze it together in your hands, and then—POW. The reason the ball remains packed the way it does is because of the effect of pressure on the freezing point of materials. It is found that *any liquid that expands when it freezes has its freezing point lowered by an increase of pressure.*[6] To see how this applies to the snowball, try the following.

take-along-do-it

Take two ice cubes and press them together firmly. When you release them, you will find that they are firmly stuck together.

At the points where the ice was in contact, the high pressure you exerted was enough to lower the freezing point of the ice. As a result, the ice melted, and the water was pushed aside to a point where the pressure was lower. At this location it promptly refroze, thus sealing the blocks together. Exactly the same thing happens in a snowball.

And Pressure's Effect on Boiling

The boiling point of a liquid is also influenced by pressure. In general, as the pressure on the liquid is increased,

[6]If the material contracts on freezing, as most materials do, pressure will raise its freezing point.

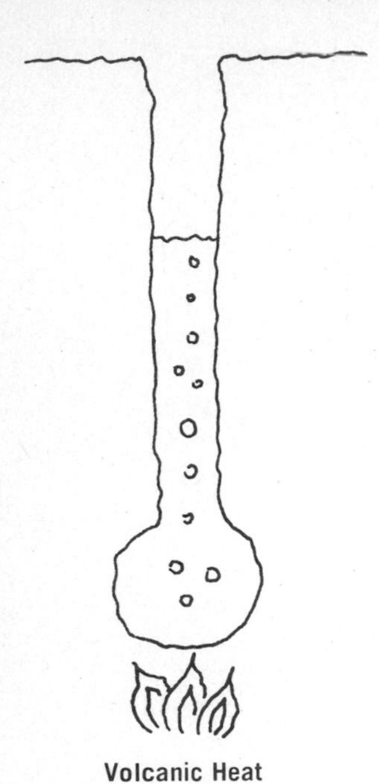

FIGURE 8–16. In a geyser, the water at the bottom rises above 100° C before boiling.

the boiling point also increases. This is reasonable to expect from the molecular interpretation of boiling given in the last chapter. An increase in pressure on a container makes it more difficult for molecules to escape the liquid, and hence the boiling point increases.

Geysers, such as "Old Faithful" at Yellowstone National Park, provide an example of this phenomenon in nature. The geyser, pictured as a narrow shaft in Figure 8–16, is filled with water that is heated from below by volcanic heat. The narrowness of the shaft prevents the heated water from rising by convection, and so it is, in effect, trapped. The pressure on the water at the bottom is due not only to the atmosphere but also to the weight of the column of water above it. This increase of pressure on the water raises its boiling point to about 130° C. When this temperature is reached at the bottom (where the source of heat is), it begins to boil, and the rising bubbles force the column of water above it out in a great eruption. As the water leaves, the pressure at the bottom is reduced, thus decreasing the boiling point and causing any remaining water to boil even more vigorously.

Just as an increase in pressure increases the boiling point, a reduction of pressure reduces the temperature at which a liquid boils. At a height of 12 miles above sea level the atmospheric pressure is much less than at sea level, and blood would boil there at 98.6° Fahrenheit—body temperature! Since you may not wish to verify this experimentally, let's look at another way the effect of pressure on boiling can be observed. Imagine that water is placed inside a container from which the air can be gradually pumped. Suppose the water is at room temperature when placed inside the container, and hence is far below its boiling point. As the air is pumped out, the pressure on the water is gradually reduced and so is the boiling point of the liquid. Finally, when the pressure is reduced to about 0.4 pounds/square inch (from about 15 pounds/square inch originally), the boiling point is reduced to room temperature and the water boils.

The boiling process is actually a case of rapid evaporation, which is a cooling process. Hence, as the water in the container boils, it also is being cooled. If you reduce the pressure a little more, the water will boil a little more; and its temperature will also decrease a little more. Finally the temperature of the liquid will reach the freezing point (even though it is boiling!). So if anyone ever asks you if you can boil water and freeze it at the same time—you can.

HEAT IN THE EXTREME—PLASMA

We have looked at a number of interactions of heat with material objects. We have observed that many mundane effects such as expansion or change of state occur, but now let's consider what happens when these same objects are exposed to large amounts of heat. Starting with almost any solid you choose, the addition of heat will eventually change it to a liquid and then to a gas. But is a gas the final state? What happens if we continue to add heat? The answer is that a gas is not the last stage. We have chosen to concentrate most of our attention on solids, liquids, and gases simply because they are the states of matter that we most often encounter as we go about our daily business. On our earth they are, by far, the most common forms in which we find substances. There is, however, a fourth state of matter, called *plasma*,[7] which can be produced in a variety of ways, one of which is the addition of heat to a substance.

In the atoms of a material we find electrons bound to and circling around a nucleus. This binding is rather tenuous, and under certain circumstances, electrons can be given enough energy to break free. When this occurs, we end up with many free electrons and many positively charged atoms (called ions) left behind. Matter existing in this condition is called plasma. The addition of large amounts of heat to a gas can increase the energy of its atoms to the point that when atoms collide, electrons can be knocked loose, thus creating a plasma. This occurs with great regularity throughout the universe. In fact, if you should take a grand tour of the universe, you would find far more plasma around than all of the other states of matter combined.

Plasma is "star stuff." Our own sun, for example, has a core which is tremendously hot—about 20,000,000°C. At this tremendous temperature the materials of the sun are in an almost perfect plasma state. The word "perfect" is used here to imply that not one or a few electrons have been removed from some fraction of the atoms but instead that all atoms have lost all their electrons. It is this separation of electrons from atoms that differentiates a plasma from a simple heated gas.

This condition is not the sole property of our sun but is the way of life for all the stars in the universe. But why do we care? Firstly, we care because as curious beings we want to know what makes up our universe and how it works. But there is a second, practical reason: there has been increased interest in the plasma state in the past few years because of the hope that we can utilize it to better our life on this planet. The interest has been prompted largely because of our dwindling supplies of fossil fuels: coal, oil, and natural gas. These have been sufficient to furnish us with our energy needs in the past, but now they are running out. In

---

[7]Do not confuse this with blood plasma. The two are vastly different!

search of energy alternatives, man has turned to plasma. In addition to being ionized in the plasma state, the atoms are hustling about at very high speeds. Under these conditions they can collide with sufficient energy to stick together and produce heavier elements. This process is called fusion, and we will examine it carefully in Chapter 25. In the fusion process, not only is a new heavier element formed but also there is a release of energy. This is the energy that is responsible for the high temperature of the sun and other stars. This fusion reaction and this release of energy can occur only when matter is in the plasma state. Our efforts to imitate the sun here on earth to produce energy for our needs has caused an intense interest in the study of the plasma state.

Perhaps what has been said so far leads you to believe that plasma is confined to the sun and stars—and perhaps to the scientific laboratory. Such is not the case, for plasma is about us now. The neon signs that proclaim the low cost of hamburgers or the excitement prevailing inside an establishment cast their light from the glow of a plasma. Here the plasma is created not by intense heat but by a voltage which is sufficiently high to partially ionize a gas. This plasma is imperfect in the sense that not all atoms are ionized, and those that are have lost only one electron. Mercury vapor lights, sodium lights, and fluorescent lights are other common examples. The lights in our polar skies (called the aurora borealis) stem from a glowing plasma produced when charged particles are dumped into the atmosphere. These phenomena and others will be considered when we study the production of light in Section V. You'll have to wait!

OBJECTIVES

After being heated by this chapter you should be able to:

1. Explain how a mercury thermometer operates, discussing the phenomenon of expansion-with-heat.
2. Perform a demonstration showing expansion upon heating.
3. Explain why uneven heating may result in breakage of a material.
4. Describe how water differs in expansion characteristics from most substances, and show how this results in water freezing at the top surface first.
5. Define calorie and BTU giving a numerical example of the use of each.
6. Describe the heat and temperature changes that occur as water changes state from liquid to solid and from solid to gas.

7. Explain the caloric theory of heat.
8. Relate Count Rumford's cannon-boring experience, and trace the reasoning that led from this observation to the concept of heat as energy.
9. Name and give examples of the three methods of heat transfer.
10. Explain how home insulation and thermos bottles perform their function of reducing heat transfer.
11. Describe the effects of increased pressure on the freezing and boiling points of water, giving an example in each case.
12. Define plasma (as used in physics), and name three places where it is found.

QUESTIONS — CHAPTER 8

1. Markings to indicate length are placed on a steel tape in a room which has a temperature of 65° Fahrenheit. A surveyor uses the tape on a day when the temperature is 100° F. If he measures the width of a lot to be 100 feet, is his measurement too long, too short, or accurate? Defend your answer.

2. A pendulum clock is designed to keep accurate time when the temperature of the room it is in is at 70° Fahrenheit. Will it run fast or slow when the temperature climbs to 100° F? Defend your answer.

3. One hundred grams of ice is at 0° Celsius. How many calories of heat are required to change all of this to steam?

4. Give examples to distinguish between temperature and heat.

5. How can heat conduction be explained by the caloric theory?

6. A piece of paper is wrapped around a rod made half of wood and half of copper. When held over a flame, the paper in contact with the wood burns, but the half in contact with the metal does not. Why?

7. If water is a poor conductor of heat, why can it be heated quickly when placed over a fire?

8. Why does a piece of metal feel colder than a piece of wood when they are at the same temperature?

9. If water did not display its unusual property of expanding as it cools below 4° Celsius, life would be much different on earth. List several examples to support this statement.

10. Updrafts of air are familiar to all pilots. What causes these currents of air?

11. State three ways in which addition or removal of heat changes an object.

12. What would happen if the glass of a thermometer expanded more upon heating than did the liquid inside?

13. Of what advantage is a pressure cooker? (Hint: Consider what happens to the boiling point of water as pressure rises.)

14. Why can you get a more severe burn from steam at 100° C than from water at 100° C?

15. The specific heat of copper is 0.09. How many calories are required to heat 5 grams of copper from 20° C to 30° C? How much would this same amount of heat raise the temperature of 5 grams of water?

16. The temperature is expected to drop below freezing; you have forgotten to put antifreeze in your car; you have no garage! Smile, you can prevent the water in your car from freezing by placing a large container of water under the radiator. Explain.

17. By what method (or methods) is heat lost through windows of a heated house? (Hint: Infrared waves do not pass through glass.) If the size of the window is doubled, how does the heat loss change?

18. What is meant by the term "ionize"?

SECTION THREE

WAVE MOTION - EMPHASIS ACOUSTICS

9

PRODUCTION AND DESCRIPTION OF A SOUND WAVE

"Simple pendulum, my eye—I'm out of control!"

The lilting beat of a popular song and the grinding clatter of a jackhammer on concrete have a common characteristic: Both create vibrations that set air into motion, and the air in turn causes our eardrums to vibrate. Our brains subsequently interpret the vibration as sound. In the next few chapters we will investigate some of the properties of these waves of sound.

THE PENDULUM AND FREQUENCY

A simple pendulum consists of a string and an attached weight which swings to and fro over the same path. Although this device is not very interesting to discuss, nor is it exactly in the mainstream of contemporary physics, still, it offers a good introduction to some of the essential ideas of wave motion. All wave motion, whether a sound wave, a radio wave, or any of a vast variety of others, has as its origin a vibrating object. With this in mind, a brief look at one of the simplest of vibrating objects, the pendulum, will allow us to move easily into more complicated and, hopefully, more interesting territory.

Heinrich Hertz (1857–1894)

Heinrich Rudolf Hertz was born in Hamburg, Germany, and studied physics in Berlin. His discoveries concerning electromagnetic waves led to the development of radio. Hertz is so revered in scientific circles even today that a few years ago the unit of frequency was officially given his name. His genius was lost to mankind when he died at the early age of 37.

The *frequency* of a pendulum is defined to be the number of cycles made in one minute (or other unit of time). If, for example, the weight swings back and forth 20 times in 10 seconds, the frequency is two cycles per second or 120 cycles per minute. The expression "cycles per second" is usually abbreviated to the single word "hertz" (pronounced *hurts* and symbolized Hz), so that the above frequency is two hertz. The concept of frequency is an important one, and we will be using the term often in the study of both sound and light, and occasionally in the study of electricity. It has been introduced in connection with a pendulum because a pendulum has a fairly low frequency—generally from about 50 to about 200 cycles per minute.[1] Thus it is easy to visualize the swinging pendulum and to understand the meaning of frequency as related to the pendulum. In our study of sound, however, we will encounter frequencies of up to about 20,000 cycles per *second* (20,000 hertz). And frequencies involved in the study of light are even higher.

THE TUNING FORK

The tuning fork is a device which is in many ways similar to the pendulum, though it bears little physical resem-

[1]But pendulums can be constructed with frequencies well outside this range.

FIGURE 9–1. A typical pendulum.

blance to it. It consists of two metal prongs, called tines, which, when struck, vibrate back and forth. The length and other physical characteristics are determined by the manufacturer, and each tuning fork vibrates with its own characteristic frequency when struck. Tuning forks in common use in laboratories for teaching physics may have frequencies from a couple of hundred hertz to a few thousand hertz.

A tuning fork produces a sound wave by disturbing the arrangement of the air molecules near it. As a tine moves to the right (Fig. 9–3*A*), it forces air molecules near it closer together (as indicated by the black lines in the figure). When air molecules are crowded close together, we say that the density of the air has been increased, and we refer to regions in which the density is higher than normal as *condensations*. In Figure 9–3*B*, the tine has snapped back to the left, leaving behind it a region into which air molecules can move. As the molecules swing back into this space, they move farther apart, and a region of lowered density, called a *rarefaction*, is produced.

These regions of condensation and rarefaction spread out from the tuning fork, and the resulting disturbance in the air is called a sound wave. In order to perceive what is happening in a different light, consider a long line of customers lined up to buy tickets for the hit Broadway production, "Physics in Ten Easy Lessons." An anxious customer gets in line at the end of the row, and, because of his enthusiasm, falls forward onto those in front of him. This causes a chain reaction of people falling which is similar to the condensation discussed above. When the anxious customer gets back up, those in front of him do so too, and this sends down the line a chain reaction of people getting back up. This compares to the rarefactions in a sound wave. If you can imagine all of this happening, you should have no trouble believing that our troublesome friend might continually fall

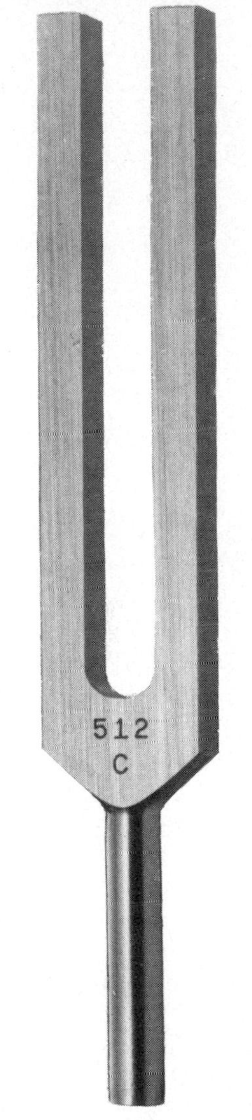

FIGURE 9–2. A tuning fork. (Photo from Riverbank Laboratories Inc.)

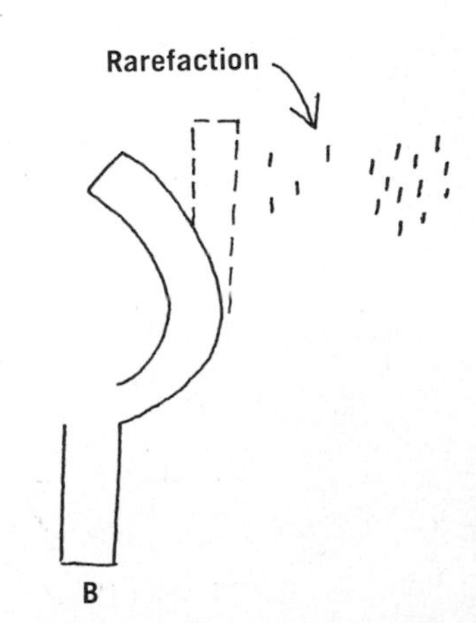

FIGURE 9–3. Tine of tuning fork vibrating to form a sound wave.

FIGURE 9–4. A wave in a theater line.

down and get back up at some constant rate, in which case a situation would eventually be set up as shown in Figure 9–4. Downed-people regions move toward the box office followed closely by upright people, and this effect would also be moving toward the box office. The people in line would call this a nuisance; with a tuning fork replacing the troublesome customer and molecules replacing the people in line, we call it a sound wave.

Figure 9–5A depicts the appearance of the air between a tuning fork and listener – if air molecules could be seen – at one instant of time. The graph in Figure 9–5*B* shows a simple way of representing the information shown in *A*. Plotted is a graph of density against distance from the tuning fork. The straight line on the graph represents the value of the air's density before the tuning fork was struck and sound emitted. The wavy line shows the density at some instant as sound passes through the air. At the points where condensations are indicated in *A*, the density is shown to be higher than normal in *B* and vice-versa for the rarefactions. The graphical method illustrated here can be used to represent almost any kind of wave disturbance. If we had chosen to discuss water waves rather than sound waves, the horizontal line would represent the undisturbed level of the water. The high points on the curve would represent crests of a water wave, and the low points would be troughs, or valleys.

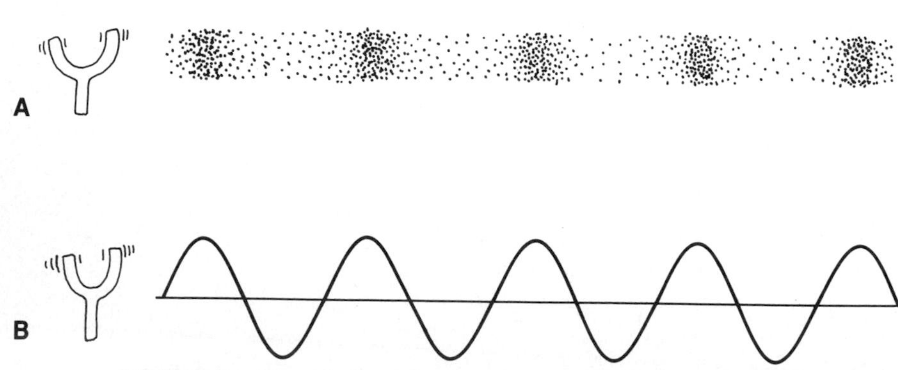

FIGURE 9–5. "Picture" of sound wave and its graphical representation.

TERMS COMMON TO ALL WAVES

Longitudinal and Transverse Waves

As a sound wave moves through space, the individual air molecules vibrate *back and forth* along the direction in which the wave is traveling. One may compare this to the back-and-forth motion of the people in the theater line. A wave motion of this nature is called a *longitudinal* wave. A longitudinal wave can be set up in a coiled spring by causing one end of the spring to move back and forth in a direction along the length of the spring.

A water wave behaves differently, however. A cork placed on water can be observed to bob *up and down* as the wave passes by. (Although there is some back-and-forth motion of the cork, the primary motion is up and down.) Individual water drops follow a similar path in the water; they have an up-and-down motion rather than the back-and-forth motion of a longitudinal wave. Waves of this kind, in which the moving particles vibrate at right angles to the direction in which the wave travels are referred to as *transverse* waves. We shall see in a later chapter that light waves are transverse waves.

It should be noted that the graphical representation of a wave is similar in appearance to an actual transverse wave in a rope (which can be set up by laying a rope on the floor and vibrating one end). The graph may also be used to represent the density changes within a sound wave, but it is important to see that it is not meant to be a picture of a sound wave. The vibrations taking place in a longitudinal wave differ fundamentally from those in a transverse wave.

Frequency

Sound waves, light waves, water waves, waves in a rope—all of these are vastly different, but by the fact that all are wave motions they have certain features in common. For one thing, all can be assigned a frequency.

The frequency of a tuning fork is the number of complete vibrations made by one of the tines in one second. This corresponds exactly to the definition as expressed for the pendulum. As the fork vibrates, it sends out a disturbance through the air that is called a sound wave only if its frequency is between 20 and 20,000 hertz, for it is only within these limits that the normal human ear is sensitive. A wave is sent out when the frequency of a vibrator is below 20 hertz, but we simply do not hear it.

Waves having frequencies above 20,000 hertz are called *ultrasonic* waves and can be heard by some animals, including dogs. It is of some interest to imagine how this fact

FIGURE 9–6. A sure sale.

could be used to sell dog food. A television set does not reproduce audio signals having frequencies above 20,000 hertz, but imagine for the moment that it does. If so, an imaginative Madison Avenue advertising executive could transmit a high-intensity ultrasonic signal into your living room. If intense enough, it could cause a dog in the room to be uncomfortable and to focus his attention on the source of his discomfort, the television set. Coinciding with this, of course, would be a dog food commercial. Misunderstanding his pet's interest in the commercial, the faithful owner would then rush out and buy several packages of the product. Free enterprise triumphs!

Amplitude

Figure 9–7 shows a drawing of a wave and also indicates a new term that needs to be added to our repertoire.

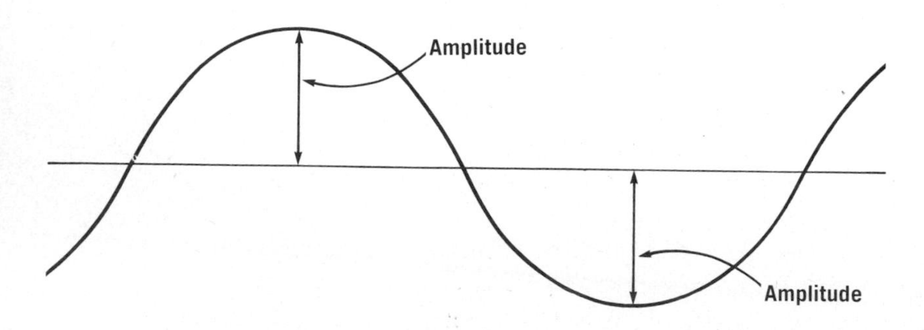

FIGURE 9–7. The amplitude of a wave.

The vertical arrows represent the amplitude of the wave. In terms of a sound wave, the *amplitude* tells us how much the density of the air has been raised or lowered. Note that in a simple wave, as indicated here, the amount by which the density is raised at a condensation is exactly equal to the amount it is lowered at a rarefaction. For a water wave, the amplitude refers to the amount the water level has been raised above normal at a crest or lowered at a trough.

Wavelength

The horizontal arrow in Figure 9–8 shows the distance between two successive points which are at exactly the same stage of their motion. This distance is called a *wavelength.* For a sound wave this could be the distance between two consecutive condensations or two consecutive rarefactions. For a water wave it could be the distance between consecutive crests or troughs.

Wave Speed

In the example presented earlier of the eager fellow causing a disturbance to move along a row of theatergoers, it should be noted that on the average, no one along the line ever moved any closer to the box office or receded from it as a result of the "wave" set up in the line. The motion of each person consisted only of oscillations about a fixed location. The same is true for the molecular motion that comprises a sound wave. If any one molecule in the path of a sound wave could be observed, it would be seen to undergo a to-and-fro vibration about its undisturbed position. The wave disturbance itself, however, moves from the source to the observer with a speed that depends on several factors, such as the temperature and the type of material through which the wave propagates (travels).

The speed of a car is relatively easy to measure. Its speedometer is calibrated to read in miles per hour, and as the units indicate, if you measure the distance an object travels and divide this by the time required to travel that distance, you have found the object's speed. Exactly the same procedure may be used to measure the speed of a sound wave.

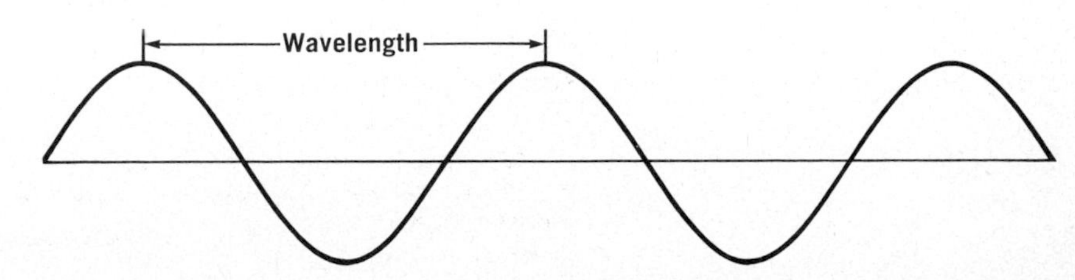

FIGURE 9–8. The wavelength of a water wave.

take-along-do-it

A wave pulse (a part of a wave) can be sent down a length of rope or a long spring (such as a "Slinky") by attaching one end to a wall and wiggling the other end up and down once. An observer, watching from the side, can measure the time elapsed during a pulse's trip from one end to the other. Dividing the distance traveled by the time elapsed yields the wave speed. To get reliable results you may need to take a number of readings and average the results. Why does this improve reliability?

As an expansion of this experiment you may want to determine the factors that influence the wave's speed. Test such variables as the amplitude and tension[2] in the rope or spring.

While this TADI indicates a method that can be used to determine the speed of a wave, indirect methods are more often used. The basis for these indirect methods revolves around a simple equation that is applicable to any type of wave motion. This equation can be developed by considering water waves moving under a fishing dock. Suppose fishing is slow and the fisherman starts wondering about waves. He estimates that the wavelength of the waves he sees is three feet and notes that they hit the dock with a frequency of one wave per second. What, then, is the speed of the waves? If each wave is three feet long and one hits the dock every second, the waves must be moving at three feet per second.

Ripples in a bathtub might typically be 10 centimeters in length and hit you four times per second. Their speed—40 centimeters per second. Generalizing these examples to an equation we see that

$$\text{Speed} = \text{Wavelength} \times \text{Frequency}$$

If the wavelength is measured in meters and the frequency in hertz, the speed will be in meters per second.[3]

Now back to sound waves. The frequency of a sound wave is relatively easy to determine. If its source is a tuning fork, the kindly manufacturer usually has stamped it directly on the base. This frequency, when multiplied by the wavelength, gives the speed of the wave. As will be seen later, the wavelength of a sound wave is easily determined, and thus by a straightforward procedure the speed can be calculated accurately.

We are using very few equations in this text, but we emphasize this one because of its simplicity and its universal applicability to all types of waves. You should not only memorize it but should try to get the "feeling" that it makes sense. Let's consider one more example of its use. A particu-

[2]The tension in the spring is the amount of force that is being used to stretch it.

[3]Note that the wavelength unit can be thought of as meters/cycle. Thus $\frac{\text{meters}}{\cancel{\text{cycle}}} \times \frac{\cancel{\text{cycles}}}{\text{second}}$ yields, by algebraically canceling "cycles," $\frac{\text{meters}}{\text{seconds}}$.

lar tuning fork produces a 256-hertz note—this is a typical value since it corresponds to middle C on a piano—and sets up a sound wave with a wavelength measured to be 4.3 feet (1.3 meters). Substituting into the wave equation we find

$$\text{Speed} = 4.3 \text{ feet} \times 256 \text{ hertz} = 1100 \text{ feet/second}$$

or, metrically,

$$\text{Speed} = 1.3 \text{ meters} \times 256 \text{ hertz} = 335 \text{ meters/second}$$

At normal temperatures this represents a reasonably accurate value for the speed of sound in air: 1100 feet per second, or 335 meters per second, or 750 miles per hour.

Thus two methods for determining a wave's speed have been discussed: (1) measurement of distance and time, and (2) measurement of wavelength and frequency. Results from either method will give the same answer and also will show that the speed of sound is very low when compared to the speed of light waves. Similar calculations to those above, if performed for a light wave, would produce a value for the speed of light of 186,000 miles per second or 3×10^8 meters per second.[4] To illustrate the enormity of this value, by the time you have read the phrase "186,000 miles per second," a beam of light would have enough time to make eight trips around the earth.

A baseball fan who has been unfortunate enough to find that his ticket leads to a seat in center field has seen a vivid demonstration of the great difference between the speeds of sound and light. The visual information that a player has hit the ball arrives almost instantaneously, but the sound of bat hitting ball arrives later. If the seat is 500 feet from home plate, the delay would be about ½ second.

This difference in speed also allows a person to estimate the distance between himself and a thunderstorm. The flash of light from a lightning bolt arrives almost at once. Upon arrival of the flash, the number of times that the observer is able to count to five gives an estimate of the number of miles between him and the storm (because sound requires about five seconds to travel one mile). Many people feel that if they are able to count at all after a flash of lightning, that is all that is important—they made it!

Almost all of the old cheaply made western movies contain a scene in which one cowboy removes his white hat, places his ear to the train track, and says, "The train's a-comin'." Our hero has made the observation that sound travels better through a solid (the rail) than it does through a gas (the air). This enables the cowboy to hear distant sounds that would otherwise be too faint. This example indicates that sound travels farther through some materials than it

[4]If you started the book with this chapter and are not familiar with powers-of-ten notation, turn to Appendix I. You'll need an understanding of it as you proceed.

does through others before becoming too faint to be heard. Additionally, it is found that the *speed* of sound also depends upon the nature of the medium.[5] Sound travels fastest through a solid, next fastest through a liquid, and slowest through a gas. To be more precise, sound travels about four times faster through water than through air, and on the average, about 15 times faster through a solid than through air. (It does not travel through a vacuum at all since sound requires a medium for transmission.) The reason for this variation is that there are differences in the stiffness, or elasticity, of the different materials. The elasticity of a substance describes how fast it returns to its original shape after it has been distorted. In a highly elastic material, the forces between individual molecules cause them to act as though they were connected by springs. This makes it easier for a disturbance of one molecule to be transferred to its neighbors. In our earlier example of the theatergoers in line, if each person had been holding on to the person in front of him, the disturbance sent down the line would have traveled faster. Solids are more elastic than liquids, which in turn are more elastic than gases, thus accounting for the observed differences in speed.

A final factor affecting the speed of sound is the temperature of the medium. As the temperature of a substance increases, so do the speeds of the molecules. As a result, the atoms collide more often, and thus a disturbance of one atom is transmitted more quickly to a neighboring atom. This is especially important in the case of gases, because the molecules of a gas exert forces on one another – and thereby transmit vibrations – only when they collide. It is found that a difference of one degree Fahrenheit causes a variation of about one foot per second in the speed of sound in air.

ENERGY[6]

After you strike a tuning fork it emits a tone for only a short while. If you hit it harder, you expect it to ring longer; but if someday you hit one and it continues to sound, on and on, you'll surely look for some outside cause for this strange behavior.

When we hit a tuning fork, we give it something; and as it vibrates, it loses that something. We call that something *energy*. The amount of energy it gets depends on how hard we hit it. As it vibrates, it causes the air near it to vibrate and thereby gives some of its energy to the air in the form of

[5]A medium is simply something to travel in. If a person claims to be a medium, he carries messages – but in a different manner.

[6]If you started this book at the beginning, we apologize for repeating here some ideas with which you are familiar. But a review shouldn't hurt.

sound. There are also frictional forces inside the fork, which cause some of the initial energy to be turned into heat. If we take very careful measurements, we find that a tuning fork gets slightly warmer after being struck several times. (Likewise, bending a piece of wire quickly back and forth causes it to heat up. If you've never experienced this, try it.)

To expand this idea of energy, suppose that after we strike the fork, we hold the "handle" of the fork against a table top. We note two things: the sound is louder, and the fork stops vibrating sooner than it would have if it had not been touching anything. The increased loudness of the sound occurs because the vibrating fork causes the table top to vibrate also. The large surface area of the flat table is able to set a lot more air into motion than the fork when acting alone. In this case some of the energy of the tuning fork must be used to cause the table to vibrate. As a result, its energy is used up more quickly and the fork is not able to vibrate for as long a time. The extra energy of sound in the air results in a louder sound.

Energy is a very important concept in physics—one of the most important. This is because we can define many different kinds of energy, such as electrical, gravitational, chemical, nuclear, heat, sound, and light; and although energy can be changed from one form into another, it cannot be created or destroyed. The total amount of energy in existence remains constant; we say that energy is "conserved." In the preceding examples, the tuning fork could vibrate only until all of its energy was changed to heat and sound.

The law of conservation of energy works even on a small scale. Consider the two tines of the tuning fork as they vibrate in and out. If you could watch a tuning fork in slow motion, you would see the tines come inward, stop, move outward, stop, come inward, . . . Yet we say that they have the same amount of energy at each instant (except for the slight amount given to the air with each swing of the tines). If this is true, the energy must be of a different type at different parts of the fork's cycle.

When the tines are moving inward (or outward), they possess energy because of their motion. This energy is the same type that a fast-moving baseball acquires from its contact with a fast-moving bat. Energy in this form is called *ki-*

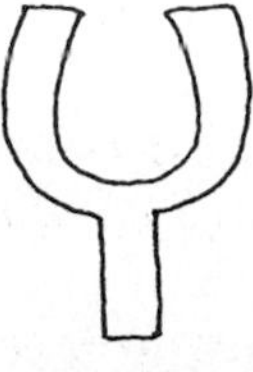

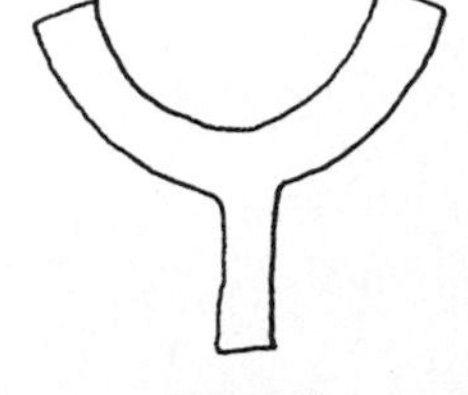

FIGURE 9–9. A tuning fork during its cycle.

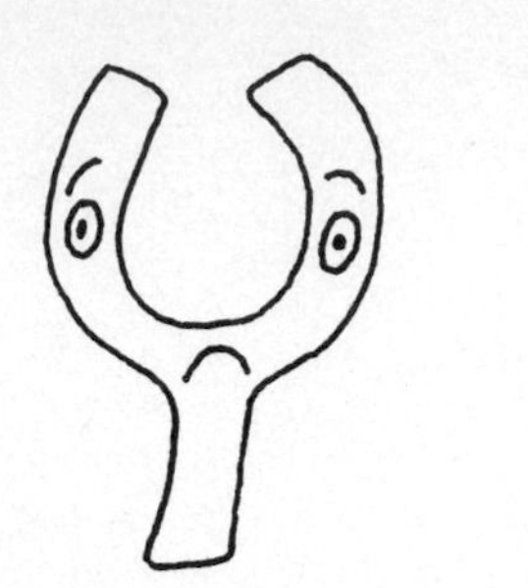

FIGURE 9–10. A tuning fork at one end of its swing.

netic energy. When the tines are at rest at either end of their swing, this kinetic energy is gone. In its place the tines have an energy which is caused by the elasticity of the fork. The metal of the fork at these times is under stress, and that is the reason the tines immediately start into motion again. This energy is similar to the energy possessed by a wound-up watch spring or an archery bow pulled back ready to shoot an arrow, and is called *potential* energy. There are other forms of potential energy, each one resulting from an object's position or state of tension.

Thus an object can have energy due to its motion—kinetic energy—and it can have energy because of its position or state of tension—potential energy. The vibrating fork involves a back-and-forth change from one of these energy-types to the other. But the important point is that the energy never disappears. It changes back and forth between kinetic and potential energy while at the same time losing a little bit with each swing as it gives energy to the air in the form of sound and to itself in the form of heat. But the sum total of all these types of energy remains constant even as one form is changed to another.

OBJECTIVES

After completing Chapter 9 you should be able to:

1. Define and give representative numerical examples of the frequency of a pendulum.
2. Describe how vibration of a tuning fork (or other surface) results in a sound wave in the air.
3. Represent a simple sound wave graphically if given the positions of condensations and rarefactions.
4. Given two different graphical representations of sound waves, state which represents the louder sound and which the higher frequency sound.
5. Describe both longitudinal and transverse waves and give an example of each.
6. State the lower and upper frequency limits of audible sound.
7. Given two of the three quantities—speed, frequency, and wavelength—calculate the third.
8. Describe a practical method for measuring the speed of sound when a great amount of accuracy is not needed.
9. Perform an experiment to measure the speed of a pulse in a long spring.
10. Describe at least two common observations which illustrate that the speed of sound is much less than that of light.
11. Explain why a tuning fork stops vibrating sooner when

its handle is held against a flat surface, and explain why the sound is louder in such a case.

12. Describe the energy changes taking place in a tuning fork as it vibrates and finally ceases its vibration.

QUESTIONS — CHAPTER 9

1. A bee, when held captive, moves its wings at a frequency of 190 times a second. What is the wavelength of the sound wave it gives off if the speed of sound in air is 1100 feet per second? What is the answer in metric units? (Speed of sound = 335 meters/second.)

2. Some 1776 musicians played as a single band in the 1973 inaugural parade. Discuss problems involved in getting the long band to march in unison.

3. The starter for a track meet stands near the runners when he fires the starting pistol. If he stood at the finish line instead, how could a smart runner always get a jump on his opponents? Estimate the jump he would get for a 100-yard-dash. Would it matter where the judge who is looking for false starts stands?

4. A meteor explodes at a height of 10 miles. How much time is required for the sound to reach the earth? The light?

5. Does sound travel faster at the foot or the top of a mountain? Why?

6. What is the difference between a longitudinal and a transverse wave? Which kind can be set up in a Slinky spring? In a rope?

7. If a bell is ringing inside a bell jar, we cease to hear it when the air is pumped out of the jar; but we can still see it. What difference does this indicate in the properties of sound waves and light waves?

8. What is meant by the statement "Energy is conserved"? How, then, can moving things slow down?

9. Explain how energy changes its form as a pendulum vibrates.

10. When Lincoln delivered the Gettysburg Address, his vocal cords caused the air molecules near him to vibrate. These molecules caused others to pick up the vibration, and on and on. Discuss the possibility of building a device sensitive enough to reconstruct the address in Lincoln's original voice.

11. The distance between a condensation of a sound wave and the next rarefaction is observed to be two feet. What is the wavelength of this wave?

12. What is the frequency of the second hand of a clock? The hour hand?

10

ACOUSTICAL PHENOMENA

Thus far we have described how sound waves are produced, what they are, and how they travel through matter. A complete description of sound, however, requires us to examine the fundamental differences between the signals emanating from different sources. We must also discover what occurs when sound waves overlap. The insight developed in this chapter will enable us to better understand why we hear what we hear.

THE INTENSITY OF SOUND

When you ask someone how intense a given sound is, he answers according to how loud it sounds. A sound that seems to be very loud will be described as very intense and vice-versa. This, however, is a subjective judgement on the part of the listener. In order to provide a more accurate way of specifying the intensity of a sound, the *decibel scale* was developed.

The decibel scale is constructed so that the lowest perceptible sound for the normal ear is assigned a zero decibel level. A sound of 10 decibels carries 10 times as much energy as the zero decibel sound. A reasonable guess is that a 20 decibel sound will carry 20 times as much energy as the zero decibel sound, right? Wrong! Actually a 20 decibel sound carries 100 times as much energy as the zero decibel sound or 10 times as much energy as a 10 decibel sound. In short, on the decibel scale an increase of 10 decibels means that the intensity increases by 10 times. For example, a 50 decibel sound is 10 times more intense than a 40 decibel sound.

Why was such a scale developed? The reason lies in the unusual ability of the human ear to respond to a vast range of intensities. If a given sound is barely audible, one that is 10^{12} times as intense will be on the verge of being painful to the ear. It is inconvenient to talk about such a wide range of numbers, and the decibel scale has the advantage of compressing such numbers into a more manageable range. For example, the barely audible sound is at the zero decibel level, and the one 10^{12} times as intense has a 120 decibel sound level. (You might prove this for yourself by reasoning as in the last paragraph.) Table 10–1 indicates the decibel level of several common sounds.

TABLE 10–1. REPRESENTATIVE NOISE LEVELS*

Noise Level (Decibels)	Intensity of Sound	Source of Sound
180	10^{18}	Rocket engine
150	10^{15}	Jet plane
125	3×10^{12}	Rock music
120	10^{12}	Threshold of pain
100	10^{10}	Riveter
70	10^{7}	Busy street traffic
60	10^{6}	Ordinary conversation
10	10	Rustle of leaves
0	1	Threshold of hearing

*Intensities are relative to the barely audible sound, which is given an intensity of one.

It was noted earlier that noise levels of about 120 decibels and above vibrate the eardrum with such ferocity that they cause the sensation of pain. If such sounds are experienced for long periods of time, deafness can result. In primitive cultures, where people have not been exposed to the constant clamor of noise that people in more advanced cultures accept as routine, hearing ability is much greater. It has been demonstrated that hearing ability for the average 70 year old in these "deprived" societies is about the same as for the average college student in New York City. Not the least of the contributing factors is the very loud music to which many people expose themselves. A sad fact about this hearing loss is that it is irreversible.

In addition to hearing loss, the constant din of day-to-day existence can lead to other problems, both psychological and physiological, such as increased irritability and hypertension. Tests have been performed in which animals were exposed to loud noises for various periods of time. It was found that when pregnant animals were exposed to noises of from about 70 to 100 decibels for six minutes of every hour, about half of the pregnancies resulted in miscarriage. Such findings cannot be assumed to apply to humans, but they are nonetheless frightening. The problem of noise pollution has been recognized by the federal government, and laws have been enacted to regulate noise exposure in manufacturing plants.

REFRACTION OF SOUND

When a sound wave passes from one medium to another of different density (or to different densities within the same medium), the path of the sound wave is bent away from its normal straight line. This bending is called *refraction.* If you have ever sat beside a quiet lake on a calm summer night, you probably noticed that you could hear distant voices and sounds quite distinctly. Figure 10–1 illustrates why this occurs. The dashed lines represent some waves sent out by the caller under normal conditions. As the waves move up into the warmer air, however, they travel faster, and therefore the upper part of each wave advances beyond the point where it would be without this speed increase. Notice how this results in the sound waves being bent down to the shore where the beachcomber is sleeping. Normally only sound waves that started in his direction would reach the listener, but under the conditions of warmer air layers over cold air, a part of the wave which started toward the stars is bent back down. This bending, or refraction, makes it possible to hear sounds at great distances when conditions are right.

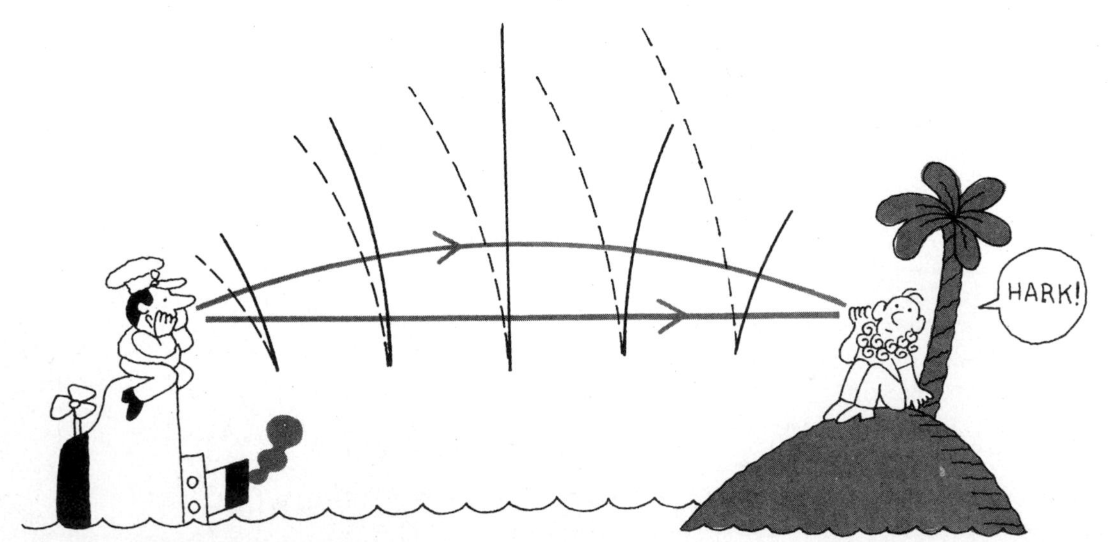

FIGURE 10–1. The dashed lines show the positions at which wave crests would fall if the air were of uniform density. The solid lines represent the wave crests as they actually travel. Note that in the warmer air high above the water they get ahead and thus refract back down.

take-along-do-it

Blow up a balloon, and hold it between a watch and your ear. You should be able to hear the watch's ticks better when the balloon is in place than when it is removed.

The results of this TADI may come as a surprise. It would seem that the balloon would block the sound waves, but in fact it increases the intensity of sound reaching your ear. When you blow into a balloon you put some carbon dioxide gas from your breath into it. Sound waves travel more slowly in carbon dioxide than in air, and this results in refraction of the waves (Fig. 10–2). Note that the part of the wave reaching the balloon first is slowed down, allowing other parts of the wave crest to catch up. Then as the wave emerges from the other side of the balloon, the top and bottom parts of the wave emerge first, speeding up and getting ahead of the center section. The result is that the sound is focused toward your ear. We will see that this is very similar to the way that a glass lens focuses light: you have, in fact, created a "sound lens."

REFLECTION OF SOUND

When a sound wave encounters the surface of a material different from the one in which it is traveling, a portion of it is reflected. If the surface is extremely rigid and smooth, a very large portion is reflected.

Domes and curved walls and ceilings act for sound waves in the same way that mirrors do for light. This can create architectural problems in the design of large concert halls. The following examples show that problems other than those of an architectural nature can also be created.

A confessional in a cathedral in Sicily was situated near a curved ceiling which acted as a reflector. This caused the whispers of penitents to be carried to a distant part of the

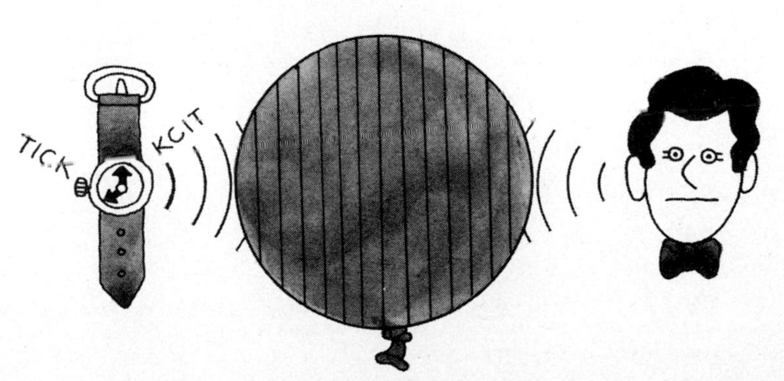

FIGURE 10–2. Refraction of sound by the carbon dioxide in a balloon.

structure, where they could be heard distinctly. The story goes that a young married man discovered this and amused himself and his friends by listening to secrets intended for the attending priest only. The entertainment aspect of his eavesdropping is said to have ceased on the day his wife was the penitent and both he and his friends were regaled by some stories concerning infidelity on her part which were previously unknown to him.

A dungeon in Syracuse (Greece, not New York) gained the name "The Ear of Dionysius" because of its peculiar construction. The curvature of the walls was such that whispers of the inmates were carried over a long distance to their imprisoner, the tyrant Dionysius. Perhaps some solace can be taken in the thought that modern-day conveniences such as wire taps and bugged rooms have their roots firmly fixed in history.

Echoes are produced when a reflecting surface is sufficiently far away that we can distinguish the reflected sound from the direct one. Knowing the speed of sound enables us to estimate the distance between us and a reflecting surface. Suppose we yell the sentence, "Physics is fun" while in the Alps (Fig. 10–3). If the speed of sound is 1100 feet per second and an echo returns in one second, then a quick calculation shows that the reflecting mountainside must be 550 feet away. Note that in the calculation one must remember that the sound has to travel to the mountain and back.

Echoes produced in a mountainous region may be a beautiful accompaniment to a frustrated baritone, but when present in a lecture or concert hall they can cause severe difficulties. To combat this problem, the furnishings of such buildings are selected to absorb a large fraction of the sound

FIGURE 10–3. An echo in the mountains of Germany.

energy. This can be achieved by padding seats in the auditorium and covering walls with drapery. The distinct echoes in an unfurnished building are caused by the absence of furniture and other absorbing materials.

SUPERPOSITION

Often two or more waves attempt to travel through the same region of space. The combined effect of these waves can be determined by the principle of superposition, which states that the resultant wave is found by adding together the individual waves point by point. Figure 10–4*A* and *B* represents two different waves. Adding these two together point by point gives a resultant wave (Fig. 10–4*C*), which has an amplitude exactly equal to the sum of the amplitudes of the component waves. Sound waves coming together such that condensation meets condensation and rarefaction meets rarefaction are said to be *in phase.* They are also said to be undergoing *constructive interference,* because the resultant wave is bigger than either of the two waves which cause it.

Figure 10–5*A* and *B* shows two waves in which the condensation of the first one coincides with the rarefaction of the second. These waves are said to be undergoing *destructive interference.* The combined effect is shown in *C* and is seen to be a state of complete cancellation. The condensation in wave *A* has tried to force air molecules close together while the rarefaction in *B* has tried to pull the molecules far-

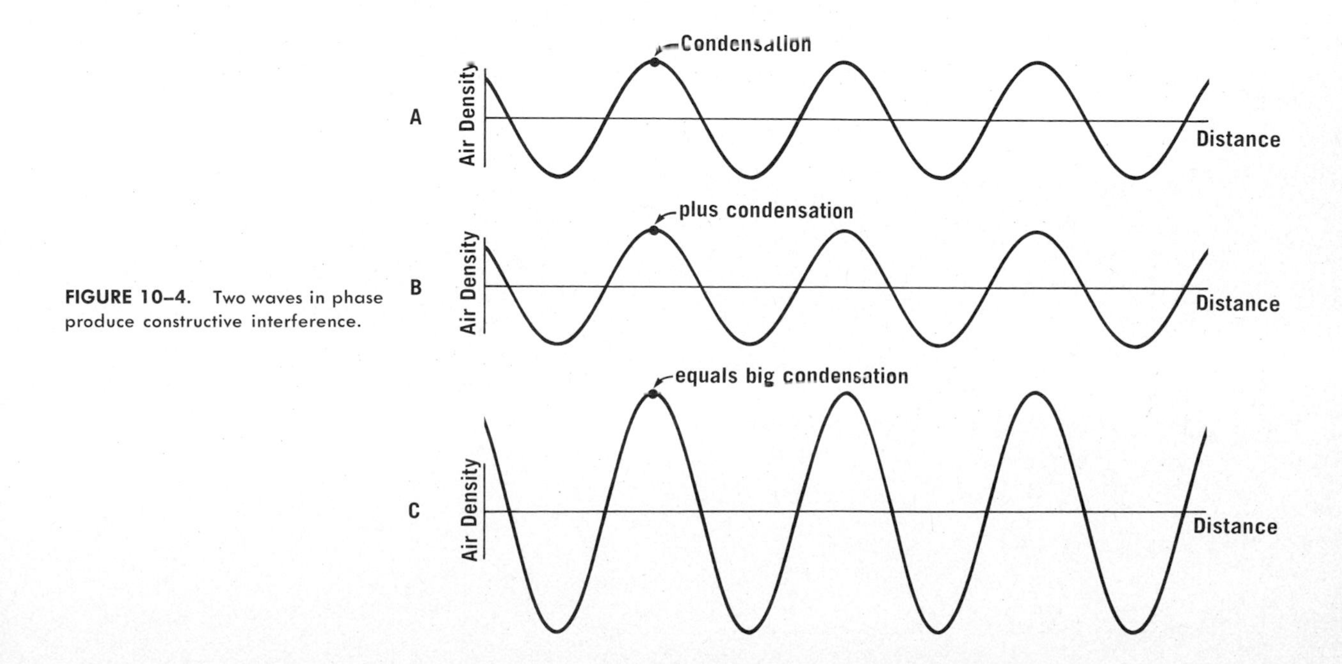

FIGURE 10–4. Two waves in phase produce constructive interference.

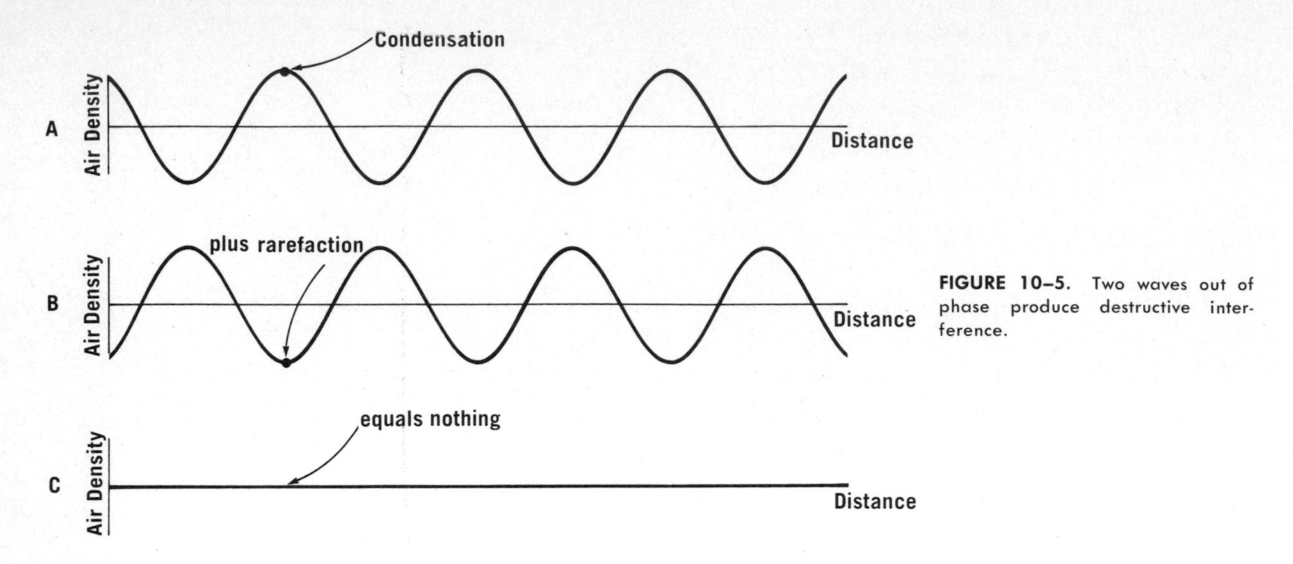

FIGURE 10–5. Two waves out of phase produce destructive interference.

ther apart. The net result is a condition in which the air remains in the state it would have been in if no waves were present.

Superposition in Action: Interference

As Figure 10–6 shows, if the tip of a vibrating object is dipped into a still pool of water, waves radiate outward from it in a circular pattern. When two such objects are dipped into the water, waves spread out from each and overlap. When the waves meet so that crest meets crest and trough meets trough, constructive interference occurs. At other points, crests are meeting troughs, and destructive interference is taking place. The resultant crossing of the waves is shown in Figure 10–7. Figure 10–8 is a photograph of actual waves in such a situation. Note that along some lines radiating out from the vibrators, no waves are shown. These are

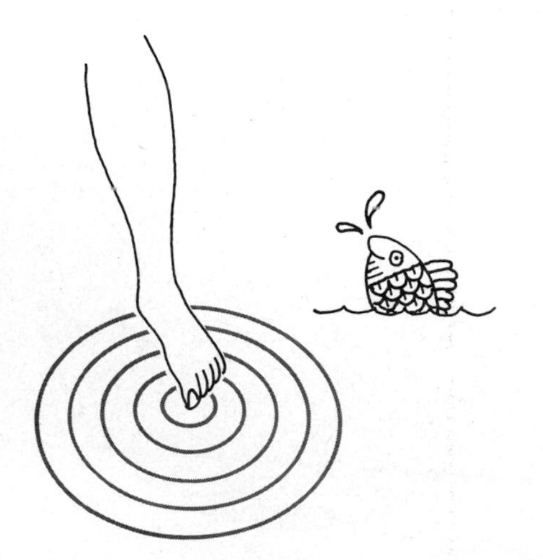

FIGURE 10–6. A vibrating object sends out a circular wave pattern in water.

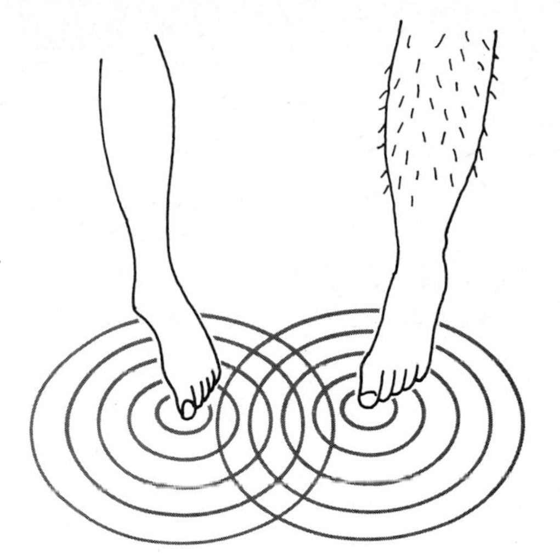

FIGURE 10–7. Two vibrating objects produce an interference pattern.

areas of destructive interference. Strong waves occur in other areas, however. A cork placed at point X would bob up and down in the waves passing under it, while a cork at point Y would feel no waves.

Now assume that the two objects are not water vibrators but vibrators in the air with a frequency greater than 20 hertz. In this case, the interference pattern will be formed in the air and will be an acoustical phenomenon. The resulting pattern will be the same as shown in Figure 10–9. If one should walk from point A to point F in the figure, one would find points such as B and D where the sound is louder than it would be if only one vibrator were emitting a signal. At other points, such as C and E, no sound would be heard.

Let us suppose that a high-level international conference has been called between our nation and some adversary. Since we desire our side to be successful, imagine the result if we should replace the vibrating objects with two high-frequency speakers hidden in a wall and beaming a weak signal toward the participants seated at a table running

FIGURE 10–8. Water waves emerging from two vibrators and interfering. (By permission of The Ealing Corporation.)

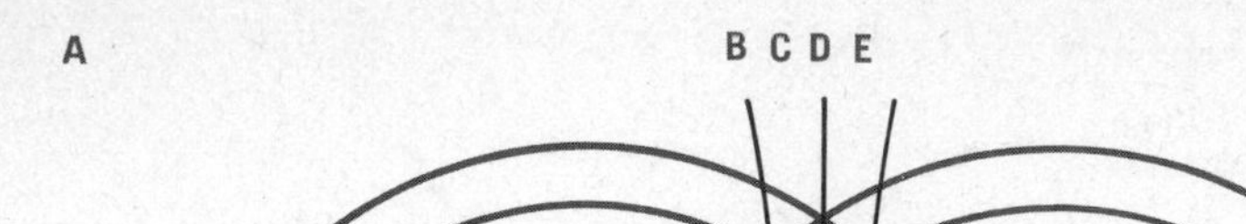

FIGURE 10–9. Interference pattern in air.

from A to F. Those at points B and D (the bad guys) would be in the constructive part of the interference pattern and would be distracted by the annoying sound, while those seated at points C and E (our team) carry on with business as usual.

This interference phenomenon can actually be experienced if you hold a tuning fork near one ear. Each tine acts as a source of sound, and an interference pattern exists near the fork. Move your ear (and your head, too) around slowly and listen for differences in loudness.

take-along-do-it

One can see the superposition principle at work in a long spring such as a "Slinky." Have two people hold opposite ends of the spring stretched along a long table (or use a smooth floor). Then, at the same instant, have both people hit the spring sharply so as to send a pulse down it. Observe what happens when the two pulses meet. The superposition lasts only for the short time that the two pulses overlap, and you must look carefully to see the effect. Figure 10–10 is a drawing of two pulses passing through each other, in which the total displacement of the spring is shown to be the sum of the individual displacements.

Superposition in Action: Beats

If two sound-emitting objects have slightly different frequencies, another interesting interference effect can be heard. The interference pattern is no longer stable. At times the waves will interfere constructively and a loud sound will be heard, but moments later the sound will diminish in intensity owing to destructive interference. Figure 10–11*A* shows two waves of slightly different frequencies traveling along together (toward the right). As point X on the waves strikes a listener's ear, no sound is heard because the two waves cancel one another (Fig. 10–11*B*). This also occurs at some later time, as point Z strikes the ear. Point Y illustrates an increase in amplitude where the two waves are exactly in phase. Thus as the resultant wave (shown in *B*) strikes the listener's ear, he hears a trembling effect known as *beats*. The transition between loudness and silence will occur at a

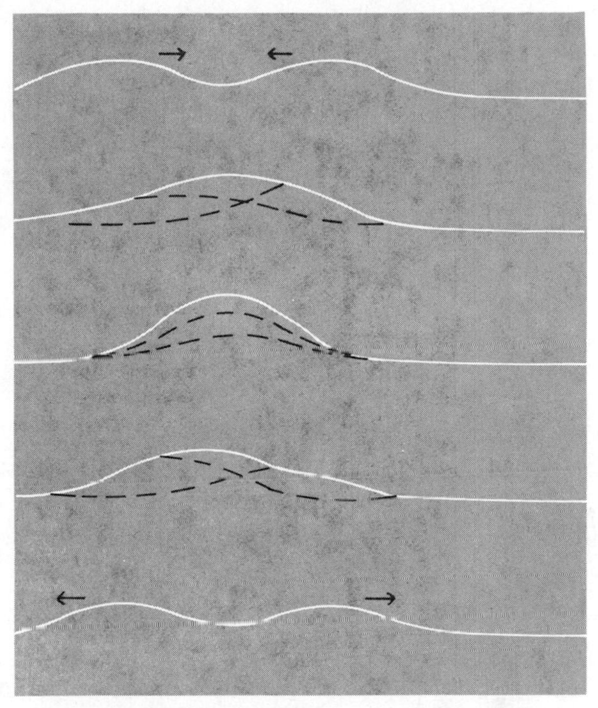

FIGURE 10–10. This is a motion picture sequence of two pulses in a string crossing one another. In the top frame they are shown moving toward one another. Note that when they cross in the center, the string takes the shape of the total of the two.

rate equal to the difference in frequencies between the two sources. If one source has a frequency of 504 hertz and the other a frequency of 500 hertz, a beat frequency of 4 hertz will be heard. A piano tuner uses beats to bring an instrument into tune. The note A on a piano corresponds to a frequency of 440 hertz. The tuner sounds the note on the piano and simultaneously listens to a 440-hertz tuning fork. If the piano is out of tune, beats will be heard and the piano

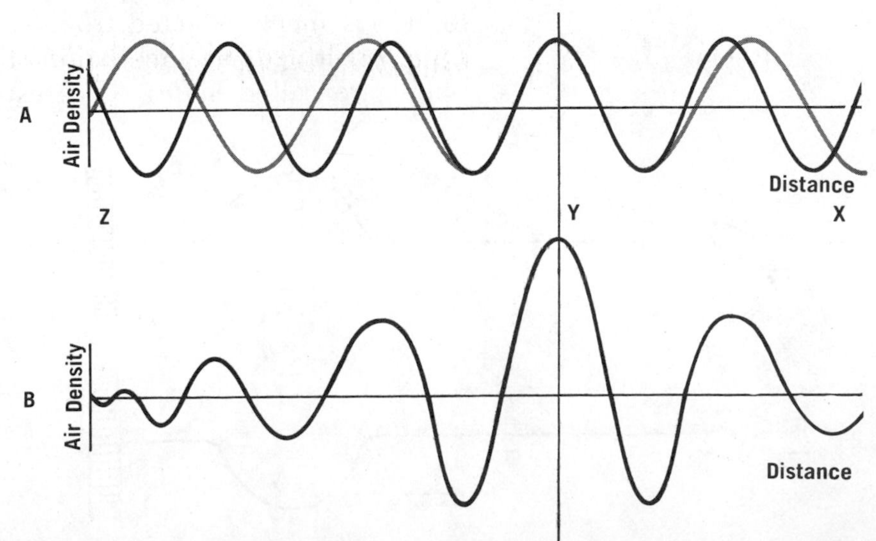

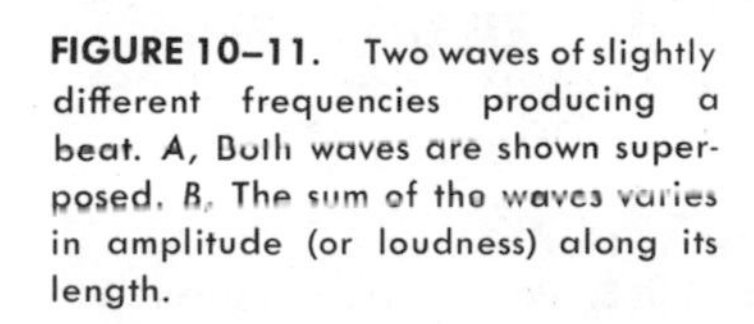

FIGURE 10–11. Two waves of slightly different frequencies producing a beat. A, Both waves are shown superposed. B, The sum of the waves varies in amplitude (or loudness) along its length.

wires can be tightened or loosened until the beats disappear.

take-along-do-it

If two guitars are available, try plucking them both at the same time and at almost the same tone. With careful adjustment so that the frequencies are almost the same, beats may be heard.

STANDING WAVES

Many musical instruments depend on vibrating strings to produce their musical tones. In order to discuss these instruments, some additional interference effects need to be considered. Until now we have not looked at what happens when a wave is reflected back upon itself.

take-along-do-it

Tie one end of a rope to a wall and give its free end a sharp upward jerk. A pulse travels down the rope and is reflected back when it hits the wall. If the pulse went down on top of the rope, does it return on top or does it flip over and come back on the bottom?

If you tried the TADI you saw the pulse flipping over as shown in Figure 10–12. We say that the pulse (or wave) undergoes a 180 degree phase change when it strikes the wall. While you have the rope attached to the wall, try shaking the free end up and down at various frequencies. In this case you are causing a *wave* to be set up in the rope; the wave travels down the rope, and all points on it undergo a phase change when they strike the wall. If you vibrate the rope at just the right frequency, you can get ongoing crests to always meet reflected troughs at the same point on the rope, resulting in a cancellation of motion at that point. Such points are called *nodes*. The resulting pattern is such that

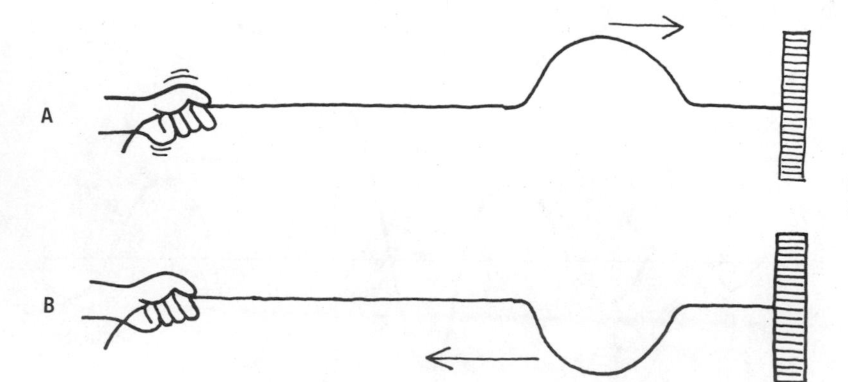

FIGURE 10–12. Pulse in a rope. *A*, Before hitting wall; *B*, After reflection from the wall.

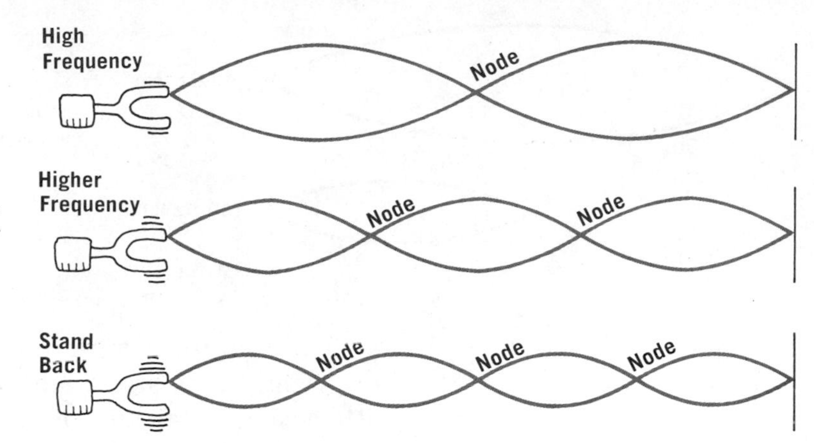

FIGURE 10–13. The faster the string vibrates, the more nodes are set up and the shorter the wavelength.

some points vibrate with large amplitude while the nodes do not vibrate at all. As shown in Figure 10–13, the faster the string is vibrated, the more waves are set up. (Remember the relationship between frequency and wavelength.) The pattern set up on the rope is such that the wave does not move along the rope at all. It just stands there; hence its name, a *standing wave*.

PITCH

In a musical instrument such as a guitar, the strings are stretched between two fixed supports and the vibrations are usually initiated by plucking them. Since a vibrating string is not capable of setting very much air into motion, the strings may be mounted on a sounding board. The larger surface area of the board is able to move much more air than the string alone. (See the energy discussion of Chapter 9). The vibrations of the wooden sounding board also set up vibrations in a cavity inside the instrument to intensify the sound.

We can describe the vibration of the string in terms of the frequency of its vibration. However, in musical terminology the description is usually in terms of its pitch. The pitch we hear is determined primarily by the frequency of the sound wave, a high frequency producing the sensation of a high pitch. The frequency or pitch produced by a vibrating string depends on the string's length. If the string in Figure 10–14A emits a 40-hertz sound when vibrating as shown, a string half as long would produce one of 80 hertz, a string one third as long would produce a 120 hertz frequency, and so on. The violin or guitar player elevates the pitch of any string by moving his finger to shorten the length of the vibrating portion. In the harp and piano, the low and high notes are produced by long and short strings respectively.

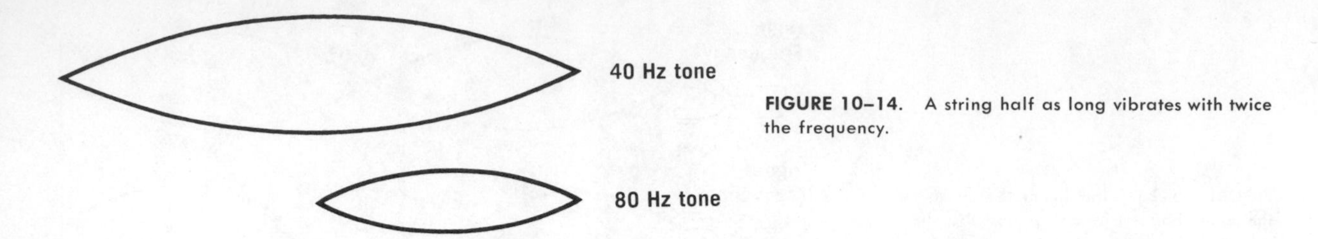

FIGURE 10–14. A string half as long vibrates with twice the frequency.

The frequency emitted by a string also depends upon how tightly it is stretched, and this provides the musician with a simple means of tuning his instrument. A tighter string will vibrate with a higher frequency and thus sound a higher pitch. A final factor in determining the frequency with which a given string vibrates is its weight. A heavy string gives off a lower pitch than a light one. In the piano and guitar, the bass notes are produced by a string which has a metal wire wrapped around it.

QUALITY

A string vibrating with the pattern shown in Figure 10–15*A* is said to be vibrating in its fundamental mode. The vibrations in *B*, *C*, and *D* are called the first, second, and third overtones, respectively. From the drawing we can see that the wavelength decreases as one moves toward higher overtones. When wavelength changes, frequency must change also; if the fundamental mode has a frequency of 220 hertz, the first overtone's frequency is 440, the second overtone's frequency is 660, and so on.

When a string, as on a guitar, is plucked, the frequency that we hear is basically that of the fundamental, but the string does not vibrate with a simple pure frequency. Instead, the string vibrates such that it emits not only the fundamental but also several overtones as well. This means that

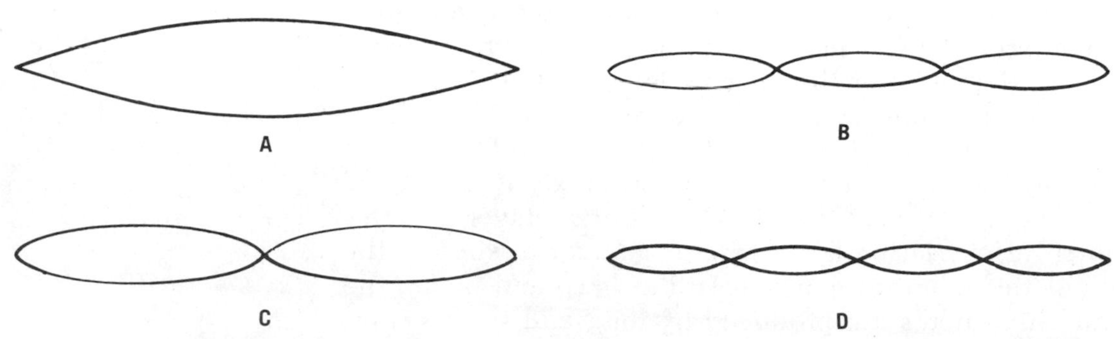

FIGURE 10–15. A fundamental vibration and the first, second, and third overtones.

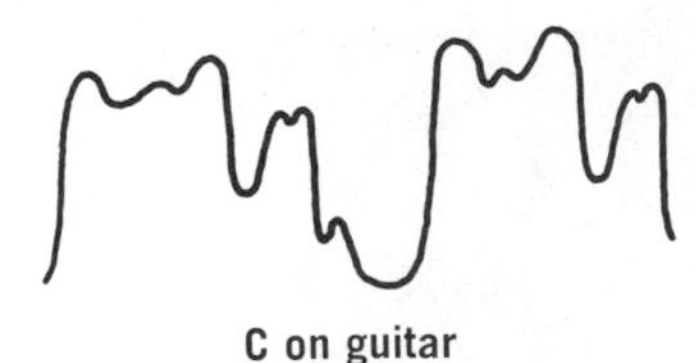

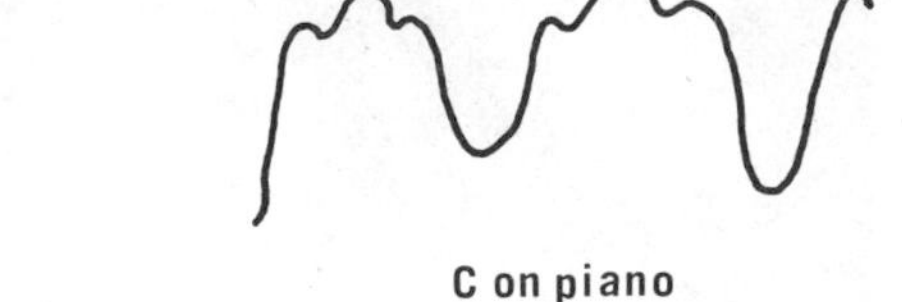

FIGURE 10–16. The waveforms of a note on a guitar and a piano.

if we try to cause a string to vibrate at a frequency of 220 hertz, the resultant sound will really be 220 hertz plus a little 440 plus a little 660, and so on. All of these frequencies add together according to the rules of the superposition principle to produce a rather complicated sound signal (Fig. 10–16*A*). If we now attempt to produce a 220 hertz note on a piano, the resultant signal will again be the 220 hertz note plus, perhaps, 440 and 660 hertz overtones that are slightly more intense than in the guitar. As a result, the waves added together will produce a sound pattern that would look like that of Figure 10–16*B*. Because the wave emitted by each instrument is different, even when playing the same tone, we are able to distinguish a piano from a guitar. In the study of music, the mingling of overtones to produce the characteristic sound of any instrument is referred to as the *quality* of the sound. We say that the note C on a guitar differs in quality from the same C on a piano.

We have discussed quality with reference to stringed instruments, but all musical instruments produce overtones which mix in with the fundamental. As a result, the number of overtones and their relative intensities give all instruments their own characteristic sound. A concert would be much less enjoyable if this were not so.

RESONANCE

Every object that is free to vibrate has a particular set of frequencies at which it "prefers" to vibrate. For example, a guitar string prefers to vibrate at its fundamental or one of its overtones. When we pluck or push an object in such a way that our pushes match one of the object's preferred frequencies, we say that *resonance* exists. Under resonance conditions, the amplitude of vibrations of the object can become extremely large. Opera singers have demonstrated this vividly by breaking crystal goblets with their powerful voices. In this case, resonance occurs when the wavelength of the sound wave emitted by the singer is the same length as the distance around the rim of the glass. When this condition is met, the glass of Figure 10–18*A* can go into a resonant vibration as shown in Figure 10–18*B*. If the singer is

FIGURE 10–17. A glass shattered by the amplified sound of a human voice. (Photograph courtesy of Memorex Corp.)

able to sustain the note, the amplitude of the vibration will increase to the point that the glass shatters.

In order to achieve resonance in an object, the incoming sound must have a frequency which is the same as the natural vibrating frequency of the object. If the sound is of a different frequency, the object will not build up its vibration. An analogous situation is observed if you watch a child just learning to dribble a basketball. He has not yet learned to synchronize his pushes (the input) to the ball's bounce (the

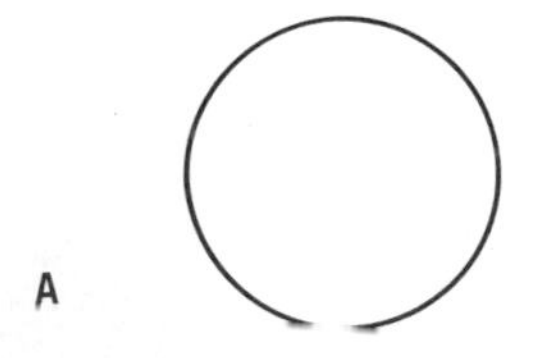

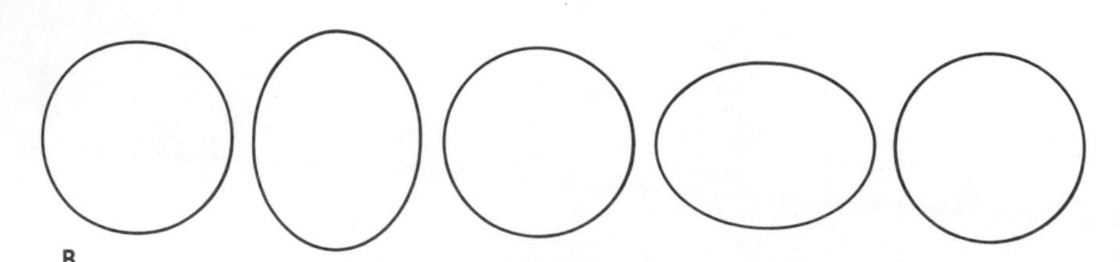

FIGURE 10–18. *A*, A drinking glass as seen from above. *B*, The same glass resonating.

vibration). He often pushes down at the wrong time or fails to push at the right time, and he is unable to get the ball to continue to bounce. An experienced dribbler, on the other hand, pushes "in tune with" the ball's bounces.

The possibility of resonance vibrations in a bridge requires soldiers to break step when crossing, for if they marched across in step and the frequency of their step happened to match the natural frequency with which the bridge would like to vibrate, the bridge could build up vibration until it came crashing down. Folklore has it that it is dangerous to cross a bridge while leading a cat on a leash lest the resonant frequency of the bridge correspond to the pitterpat of its feet. One should remember that friction at the bridge supports tends to absorb some of the vibrational energy of the bridge, and this factor is sufficient to overcome any effects the cat might produce. However, if the bridge is not very strong and if you have a big cat. . . .

Resonance in Air Columns

Many instruments produce their musical sounds by the vibrations of enclosed columns of air. For example, in an organ pipe, air is forced through a small opening and strikes a thin reed. The vibration of the reed sets into motion a column of air confined to a tube. The length of the pipe determines the frequency of sound emitted.

take-along-do-it

Hold the mouth of a bottle near your mouth and blow across it. If you blow at the correct angle, the air inside the bottle will be set in vibration and you will hear a hum. Pour some water into the bottle and try the same thing. The hum should be of a higher pitch.

Have a friend hold the mouth of a bottle next to his ear while you blow across an identical bottle to produce a good loud note. Your friend will hear a weaker sound from his bottle, although of the same note.

Just as a solid such as a tuning fork or a crystal goblet has a natural frequency of vibration, so does the air in a cavity. This frequency depends upon the length of the air column, as indicated in the first part of the TADI. When you blew across the bottle, you set the air into vibration, and it vibrated at this natural frequency. When you shortened the air column by adding water, the wavelength of the sound produced was decreased, and the frequency thus increased.

In the second part of the TADI, the two bottles were identical and therefore had the same resonance frequency. The sound waves you produced by blowing across your bottle caused vibration of the air in the second bottle, and because the frequency of the vibration you caused corresponded exactly to the natural frequency of that bottle's column, its vibration was built up enough so that the bottle became a separate source of sound.

A slightly different arrangement provides a simple method for measuring the wavelength of a sound wave. If a tuning fork is sounded above a tube for which the length can be varied as shown in Figure 10–19, resonances (or loud sounds) occur for certain lengths of the tube. These resonances are caused by a reflected sound condensation returning to the mouth of the tube just as another condensation is about to be transmitted down the tube. The first such resonant point will occur when the length of the tube is about $1/4$ wavelength. A second point occurs when the tube length is $3/4$ wavelength. Thus one can determine the tube length for each of these cases, subtract them, and know the wavelength for that particular frequency.

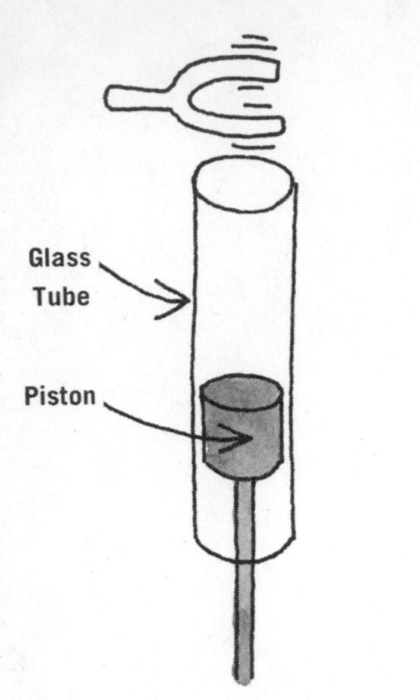

FIGURE 10–19. Determining resonant lengths of a tube.

This simple measurement provides an easy method for measuring the speed of sound, because if the wavelength is known, one need only note the frequency (printed on the tuning fork), and multiply the two to obtain the speed. But you know all this from Chapter 9.

OBJECTIVES

If you have finished studying Chapter 10, you should be able to do the following:

1. Distinguish between intensity and loudness, frequency and pitch.
2. Give an example of sound refraction in air, describing an experiment which demonstrates the effect.
3. If given the time it takes for an echo to return to its source, calculate the distance to the reflecting object.
4. Define constructive and destructive interference and state the necessary conditions for each.
5. Describe the appearance of two-source interference of water waves.
6. Describe two-source sound interference.
7. Carry out a demonstration of interference using a Slinky spring.
8. Explain what causes beats and what conditions are necessary for their production.
9. Explain how beats can be used to tune a musical instrument.
10. Use a long rubber hose (or a rope) to set up standing waves, and demonstrate the relationship between frequency and wavelength.
11. Describe how the length of a string on a musical instrument is related to the pitch of the sound emitted.
12. Describe the effect on the pitch of the sound emitted as

a string of a stringed instrument is varied in (a) length, (b) tension, and (c) weight.

13. Define "overtones" and give quantitative examples of fundamental tones and their overtones.
14. Define "quality of sound" and explain how sounds of the same pitch can differ in quality.
15. Give an example of resonance produced by a sound wave.
16. Describe resonance in an air column and explain how this can be used to measure the speed of sound.

QUESTIONS—CHAPTER 10

1. Why is an echo weaker than the original sound?

2. The sound of a hiccup is returned to the ear from the face of a cliff six seconds after it is hiccupped. How far away is the cliff?

3. At certain speeds—but not at all speeds—the front of a car is noticed to vibrate wildly. What might cause this?

4. The wave pulses shown in Figure 10–20 are moving toward one another on a string. What happens when they superpose?

5. How could you determine whether two pulses like those in Question 4 pass through one another or reflect from one another? (Try it and see if you can tell.)

6. Why does the sound of a tuning fork get louder when its handle is pressed against a table?

7. Why do you sound so "full voiced" when singing in a shower?

8. How many times as much energy does a 70 decibel sound carry as a zero decibel sound? A 30 decibel sound?

9. Of the following sounds, which is more likely to have a decibel level of 60: a rock concert; the turning of a page in this text; normal conversation; a cheering crowd at a football game; background noise in a church.

10. Estimate the decibel levels of each of the sounds listed in Question 9.

11. Why does wrapping a second wire around the main wire help to produce low notes on a piano or guitar?

12. Suppose the only guitar string you have available is lighter than a broken one you wish to replace. How would you compensate for this so that the string produces the proper tones?

13. You sound a 4000 Hz tuning fork at the same time that the highest C on a piano keyboard is hit, and you hear two beats per second. You loosen the string very slightly and hear four beats per second. What is now the frequency at which the string is vibrating?

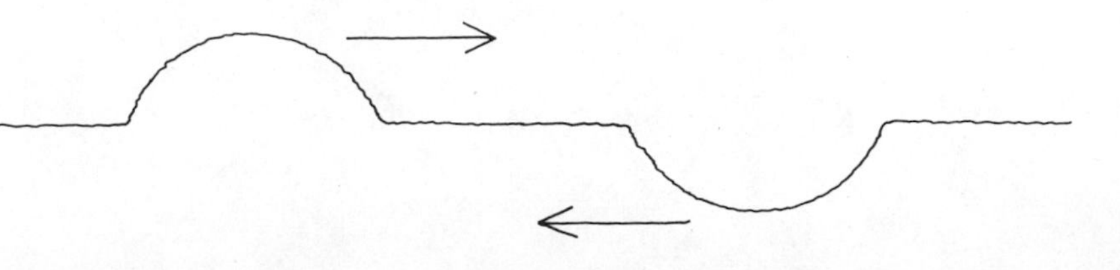

FIGURE 10–20. Two pulses in a string moving toward one another.

Reprinted with the permission of Sidney Harris.

11

WHAT DOES IT SOUND LIKE?

In the last two chapters you have been presented with many facts about sound waves. Many important questions are still unanswered. Is a high fidelity system really high? Does a sonic boom? Is the human ear only a passing fancy? For the answer to these and other questions, read on.

HIGH FIDELITY AND STEREO

A useful summary of many of the ideas presented in the preceding two chapters can be achieved by examining a few of the characteristics of a high fidelity or a stereo system. Figure 11–1 shows a typical arrangement of the components of a high fidelity system. An electrical signal originates in the cartridge of the record player. It is then strengthened, or amplified, and fed to the speakers. An examination of each of these stages will serve as a review of many of the properties of sound waves.

A record is produced by cutting a continuous spiral groove in a vinyl plastic disc. An artist sings into a microphone, which converts the sound into an electrical signal.

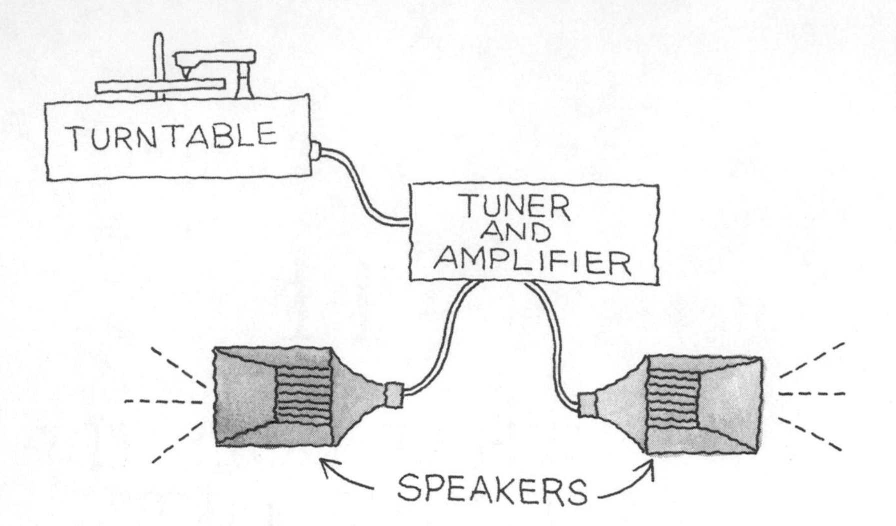

FIGURE 11–1. A high fidelity system.

This signal is then made to cause a needle to vibrate from side to side in the record groove, carving out small notches along the side. The higher the pitch of the sound being recorded, the closer together are the notches along the sides of the groove; the louder the sound, the deeper the notches. Because most sounds consist of a number of overtones along with the fundamental frequency, notches must be carved on notches. For example, all the details of the piano's note shown in the representation of its wave (Fig. 11–3) must be recorded on the record.[1]

[1]A discussion of recording on magnetic tape will be included in a later chapter.

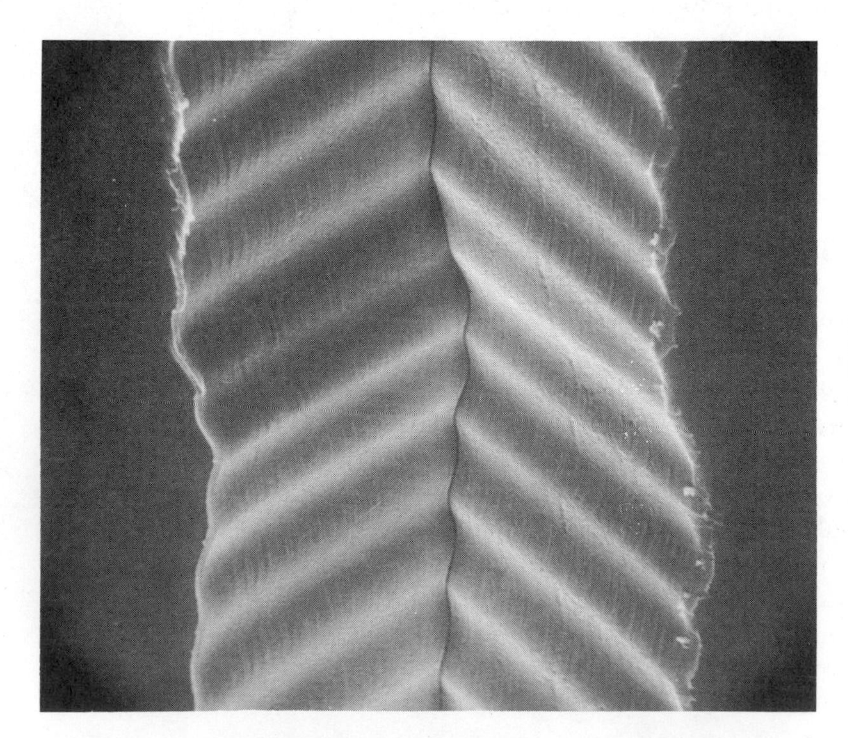

FIGURE 11–2. Close-up view of a groove in a record, showing the notches along the sides that cause the needle to vibrate. (By permission of RCA.)

The purpose of the turntable in a hi-fi system is to convert the information in these grooves back to an electrical signal. This is accomplished by allowing the needle in the set to follow the groove and reproduce the side-to-side vibrations caused by the notches. This side-to-side vibration of the needle generates an electrical signal in the cartridge. This signal is subsequently amplified and used to drive a speaker.

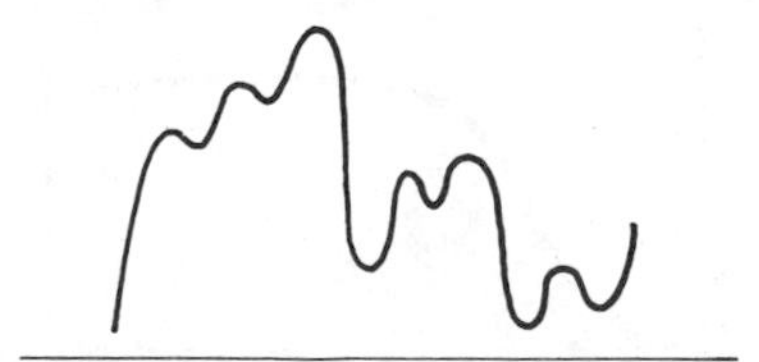

FIGURE 11–3. Wave-form of a piano note.

In order to attempt to bring the realism of a live production into your living room, the stereo recording system was developed. In its simplest concept, two separate recordings are made, using two separate microphones. Each microphone is directed toward a different part of the stage, because we hear with *two* ears and are thereby sensitive, at a live performance, to the origins of sounds from the various parts of the orchestra. To produce the stereo recording, a set of ridges is cut into the right side of the record groove from one microphone and into the left side from the other. The needle on your set then picks up separate signals from each side. They are amplified separately and fed to two different speakers.

Until a few years ago, stereo was accurately advertised as "the next best thing to being there." The newest innovation to challenge this claim is the quadraphonic or four-channel system. This system derives four separate signals from four separate recordings (on magnetic tape), and these are played simultaneously by four separate speakers. The listener is thus surrounded by sound. The obvious goal of all such improvements is to increase the feeling of "presence" for the listener.

Frequency Response

Because high fidelity means accurate reproduction, one of the requirements of a good system is that it not add anything to or subtract anything from the original music. Since the human ear is sensitive to all sounds having frequencies between 20 and 20,000 hertz, signals within this range must be faithfully reproduced. This characteristic is specified in terms of the frequency response curve. Such a curve for a good hi-fi system is compared to the curve for an inexpensive set in Figure 11–4. Notice that the cheaper set does not reproduce very low bass or very high treble tones well. The requirement for good frequency response is satisfied by a good amplifier and good speakers. "Good" is a relative term, however. What sounds good to one person will fall short for someone else, and the fantastic differences in cost between different stereo sets correspond—somewhat, at least—to a vast difference in quality. Thus having a "tin ear" can save you some money.

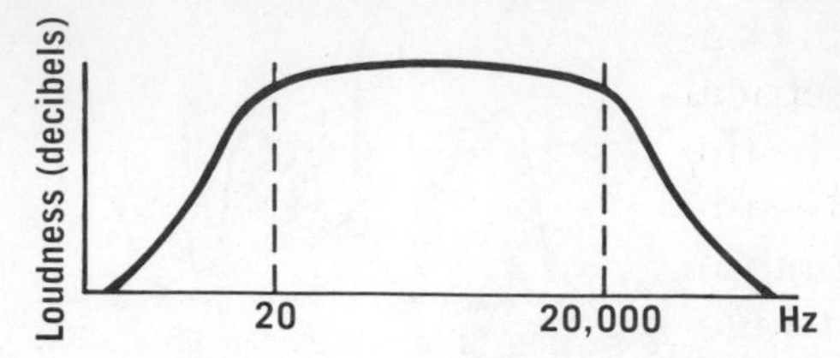

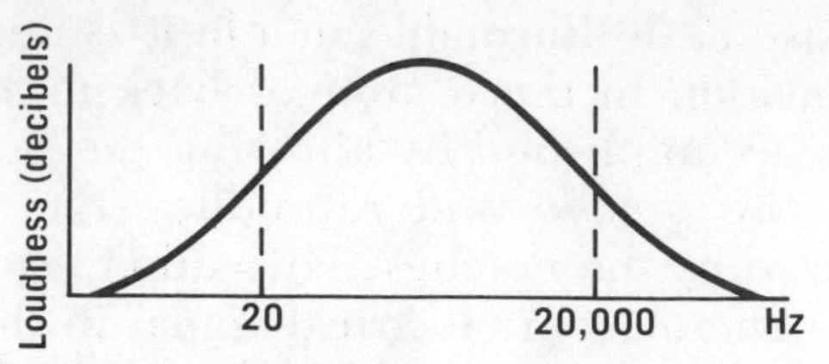

FIGURE 11–4. Frequency responses of two hi-fi systems. The one at left represents the better system.

Even in a set which has good frequency response, the environment in which the system is played and the condition of the records (or tapes) will alter the final sound. For this reason, several knobs are provided on the tuner to correct for observable variations from the desired sound. The bass and treble controls are the most important of these. At low volumes, the human ear does not respond as well to bass as it does to treble tones. A balance can be achieved by increasing the bass response while slightly decreasing the treble. High frequencies are absorbed more efficiently by furnishings, such as draperies and padded furniture. To compensate for this effect, a slight increase in treble is required. Older records often hiss when played, and a slight decrease in treble reduces this high-frequency annoyance. Other controls on the tuner allow the listener to choose between AM and FM radio. A discussion of the meaning of these settings will be given in a later chapter.

The speakers in the system are perhaps the most important pieces of equipment in the combination. Since a speaker tends to reproduce bass notes less efficiently than it does the higher frequencies, the cabinets are constructed as sounding boards or resonant cavities to enhance the lower frequencies. Many speakers require large amounts of electrical power to drive them. When used in the comfort of your dormitory or living room, this extra power is usually unnecessary. A system supplying about 10 watts[2] of power is sufficient to reproduce adequately most percussive sounds such as drums or cymbals, and power in excess of this is not necessary unless you wish to entertain in a coliseum.

THE DOPPLER EFFECT

A most unpleasant sound is the wail of an approaching police siren. If a police car has ever caught up with and passed you, perhaps you observed a distinct change in the pitch of the note as it passed. However, it is quite likely that you were too busy wiping sweat from your palms to notice.

Christian Johann Doppler (1803-1853)

This Austrian physicist was almost an American physicist. When he was 31, he made plans to emigrate to the United States because he could not find a college teaching position in Europe. A job offer was made just before he left.

Doppler tested his mathematical explanation of the effect bearing his name by placing trumpeters on a flatcar and having it pulled repeatedly past listeners chosen because they were able to accurately estimate pitch. He predicted a similar effect for light waves but was unable to demonstrate it because of the great speed of light.

[2]The watt is the measure of power. See Chapter 13.

When the siren was approaching, it seemed to have a high pitch, and when receding, the pitch was lowered. This phenomenon is called the *Doppler effect.*

To illustrate the cause of this apparent change in frequency, consider the following situation. A student has just completed his first history examination and upon leaving the classroom begins to scream with a constant pitch. If he stands stationary, the sound waves emanate from him as shown in Figure 11–5. Not satisfied with the reaction he is getting from his instructor, he begins to race madly down the corridor in search of a high-rise building from which to leap, squealing all the while. If we could observe the sound-wave pattern now, we would find the distance between consecutive condensations to be less in front of him than behind him. This occurs because during the short time interval between the emission of two consecutive wave crests the student moves forward and thus the crests in front of him are closer together. The pattern would look like that shown in Figure 11–6. A listener in front of the student finds more condensations reaching him per second than when the student was at rest and concludes that he is now emitting a higher frequency scream. A listener behind him concludes just the opposite, because fewer condensations per second are reaching him.

FIGURE 11–5. Student experiencing tough exam emits sound waves as shown.

Applications of the Doppler effect more often involve radar waves than sound waves. Radar waves will be examined in a later chapter, but elements of their use can be understood here. A police cruiser directs a beam of radar waves down a straight stretch of roadway. An approaching car will reflect these waves back to a receiver at the police unit. If the reflecting car is stationary, the wave sent back will have the same frequency as the initial wave. If, however, the reflector is in motion, the wave sent back will have a higher frequency. Electronic devices within the police car translate this change of frequency into a speed for the car.

FIGURE 11–6. A student seeking a high-rise building from which to leap causes a different wave pattern.

Moments later you may find that you have just been observed on Doppler effect radar.

The Doppler effect also applies to light waves. If we examine the light directed toward us from a distant galaxy, we observe that the frequencies of all the colors have been shifted away from what we expect them to be. These visible frequencies have been shifted toward the frequency on the red end of the spectrum, which is the low-frequency end. One interpretation of this observation is that distant galaxies are moving away from us. On this observation rests the now-accepted theory that the universe is expanding. There are various scientific theories concerning the origin of this expanding universe. Each must explain this Doppler "red shift."

Sonic Boom

Sonic boom may be described as the Doppler effect carried to the extreme. When a source of sound (such as the disturbed student or an airplane) moves, it squeezes together the waves in front of it. As its speed becomes greater, the squeeze becomes greater, as shown in Figure 11–7. Finally, when the plane has reached the speed of sound, waves emitted in the forward direction cannot get ahead of the plane. In fact, all of the condensations being produced are now "stacked up" in front of the plane. Crossing this superposed wave is referred to as "breaking the sound barrier," because the high density of the condensation causes a barrier through which the plane must pass. Water skiers crossing the towing boat's bow wave experience a similar problem.

There is a common misconception that a sonic boom is produced only when a plane is *crossing* the wave barrier. Not true! Figure 11–8 represents waves sent out at various points along the path of a plane traveling at supersonic speed. The wave represented by the largest circle was initially produced when the plane was at point A. The second

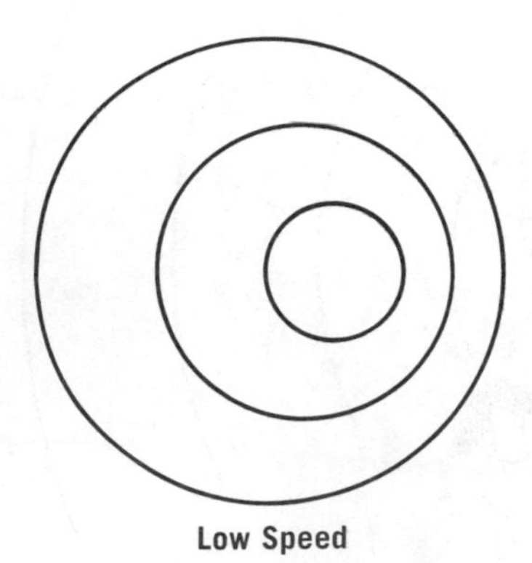

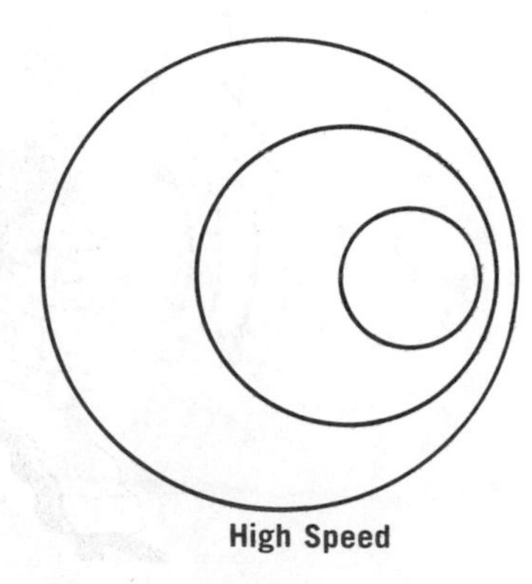

FIGURE 11–7. The higher the speed, the greater the squeeze.

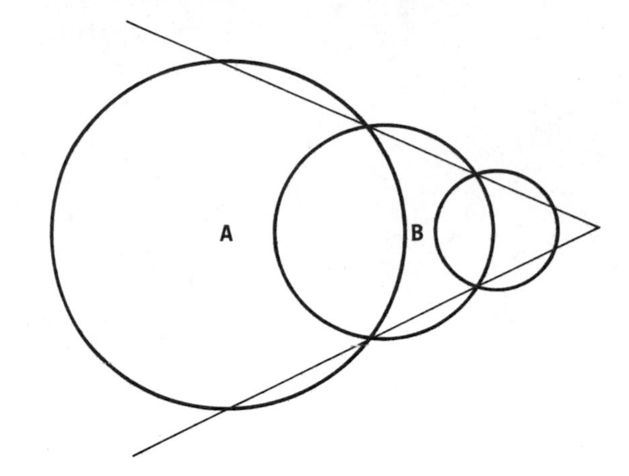

FIGURE 11–8. Wave pattern for a plane at supersonic speed.

wave started at B, and so on. Note that along a "V" starting at the tip of the plane, the crests of the waves overlap and form a straight wave front. From the previous discussion of superposition we can see that this wave front, being formed from the superposition of all the condensations produced along the plane's path, would have a very high density. This wave is "dragged along" by the plane (Fig. 11–9). It is when it sweeps across an observer on the ground that it is heard. Thus in the figure the person to the left hears it before the one on the right.

Returning to our water wave analogy, a boat traveling faster than water waves (which are pretty slow) produces a bow wave which forms a "V" from the front of the boat. When this wave sweeps down the shore, it submerges sun worshipers lying too close to the water. The two situations are very similar. A major difference, however, is that while the boat's wave travels only along the surface of the water, the plane's shock wave spreads not in a "V" but in a cone—to the sides as well as up and down.

THE EAR

The subject of how we hear falls more into the field of physiology than physics, and thus only a brief sketch is

FIGURE 11–9. A shock wave behind an airplane.

presented here. Let's return to the line of theatergoers still waving back and forth (see Chap. 9). What is happening at the box office? The fellow buying his ticket gets hit repeatedly in the back and pushed into the bars at the window. The effect is that the box office is being alternately pushed and released. If it is not very rigid, it will vibrate (or sway) along with the disturbance.

Figure 11–10 shows the structure of the human ear. Sound waves travel down the ear canal to the eardrum, which acts like the box office and vibrates at the frequency of the disturbance. Behind the eardrums are the three small bones of the middle ear, called the hammer, the anvil, and the stirrup because of their shapes. These bones serve as conductors to transmit the vibration to the inner ear, which contains the cochlea, a snail-shaped tube about one inch in length. Running along its inside is the basilar membrane. This membrane varies in thickness and tension along its length, and different portions of it resonate at different frequencies. (Recall that the natural frequency of a string depends upon its thickness and the tension on it.) Along the basilar membrane are 30,000 nerve endings, which sense the vibration of the membrane and in turn transmit impulses to the brain. The brain interprets the impulses as sounds of varying pitch depending upon the location along the basilar membrane of the impulse-transmitting nerve.

Because of his earlier study of resonance in an air tube, surely the reader has immediately said, "But the ear canal should act like an organ pipe does. It should resonate." Correct! The resonance frequency is in the range of 3000 to 4000 hertz. It is for this reason that we are more sensitive to sounds in that range.

The ear has its own built-in protection against loud sounds. The muscles connecting the three bones to the

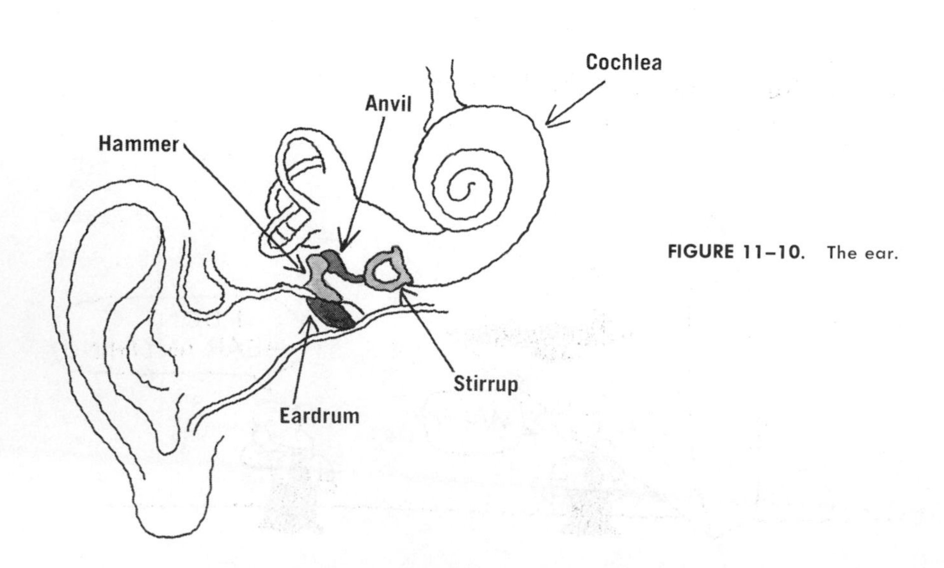

FIGURE 11–10. The ear.

Reprinted with permission of Saturday Review/Sidney Harris.

walls of the middle ear serve as a volume control by changing the tension on the bones as sound builds up and thus hindering their ability to transmit vibrations. In addition, the eardrum becomes stiffer. These two occurrences cause the ear to be less sensitive to sound. There is a time delay between the onset of loud sound and the ear's protective reaction, however, so that a sudden very loud sound can still cause damage to the ear.

OBJECTIVES

A study of Chapter 11 should enable you to:

1. Define "high fidelity."
2. List the steps by which sound signals are recorded on a phonograph record and then reproduced.
3. Describe the difference between a good and a poor high-fidelity set as to frequency response.
4. Explain why the nature of the room in which a hi-fi is played would affect the proper bass-treble setting.
5. Explain how and why the pitch heard from a moving object varies as the object's speed changes relative to the listener.
6. Describe at least two examples of the Doppler effect.

7. List the major parts of the ear and describe the function of each.
8. Explain the advantages of stereo over monaural sound reproduction.
9. Explain the cause of a sonic boom, and correct the common misconception that a sonic boom is heard only when a plane is in the act of breaking the sound barrier.
10. Draw a diagram of the ear, labeling at least six parts and describing their functions.

QUESTIONS CHAPTER 11

1. Would the Doppler effect occur if the source of sound were stationary and the listener in motion? Why? What would the listener have to do in order to hear a higher frequency? A lower frequency?

2. Give an example to illustrate that the Doppler effect can occur for water waves.

3. Why might a hi-fi system capable of reproducing tones with a frequency greater than 20,000 hertz be of value?

4. Why does listening to stereo produce a more pleasing response than does the older monaural system?

5. What is the equivalent of a sonic boom for a boat on open water?

6. The frequencies reproduced on a telephone are between 500 and 4000 hertz. Discuss the limitations of the telephone in transmitting music.

7. The electrical signals along a telephone line travel *much* faster than sound waves. Describe a cross-country telephone conversation if they did travel at the speed of sound.

8. In some classrooms and lecture halls it is easier to understand the speaker when the room is full than when there are only a few people in attendance. Explain why this occurs. (Hint: It is especially noticeable in a room with unpadded seats.)

9. Why are sonic booms less intense when made by high-flying planes?

10. How would our sense of hearing differ if our ear canals were twice their normal length?

SECTION FOUR

Touching the dome of a Van de Graaff generator gives this student a charge.

12

CHARGES AT REST

The earliest known study of electricity was conducted by the Greeks about 500 BC. Their results were modest, but from these roots have sprung the enormous electrical distribution systems so much a part of the civilized world. A trite expression applies: the longest journey begins with but a single step. Thus the modern electronic computer began with a Greek thinker studying what we now call static electricity.

THE BEGINNINGS

It all began when someone noticed that a waxlike substance called amber, after having been rubbed with wool, would attract small objects. Since then we have learned that this phenomenon is not restricted to amber and wool, but that a similar effect can be observed (to some degree) when almost any two nonmetallic substances are rubbed together.

You can carry out experiments of a similar nature on your own. Walk across a rug and touch something metallic, such as the knob of a television set. A spark jumps from your hand to the knob. Rub a cat's hair in a direction opposite to

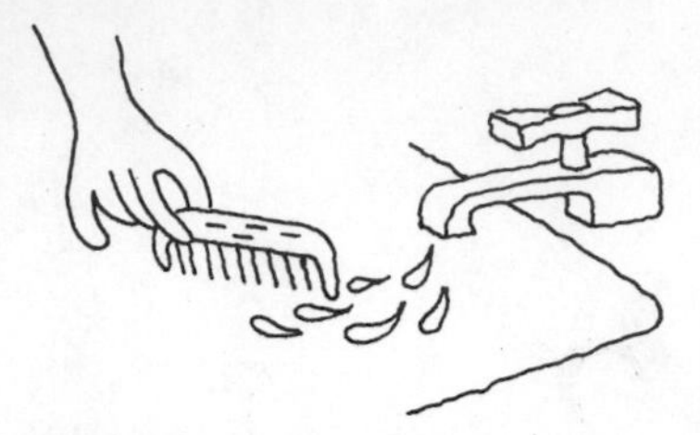

FIGURE 12–1. The charged comb attracts running water.

the way it grows. If you are in a darkened room, you can observe sparks crackling about your hand. Run a comb through your hair, and use it to attract small bits of paper. Or better yet, hold the comb near a small trickle of water and watch the water run "uphill" (Fig. 12–1). If you haven't found anything you want to do yet, the next suggestion is guaranteed to make you popular. In the interest of science, scoot across a plastic seat cover while wearing wool pants or skirt and kiss someone. Sparks of nonbiological origin are sure to fly.

take-along-do-it

This TADI is yet another example similar to those above, but it also demonstrates another property of charges. Rub nylon hose with a plastic drycleaner bag. The nylon hose will expand like a balloon (Fig. 12-2). The reason for the expansion will soon become obvious.

All of the effects listed above are electrical in origin. Before we can adequately explain what is occurring in these situations we must review briefly our model of the atom. In the planetary model, we picture electrons as swirling in orbits, like planets, around a central core called the nucleus. A nucleus is not homogeneous but is, instead, composed of building blocks called protons and neutrons. The neutron will be of little interest to us in the next few chapters because, as its name indicates, it is neutral—it carries no electric charge. The proton and electron do have the characteristics of charged objects, however, with the proton possessing a positive charge and the electron a negative charge. The terms *positive* and *negative* have no fundamental meaning: they are used only to indicate that there is some difference between the kinds of charge possessed by these two tiny particles. The strength of the positive charge carried by the proton is exactly the same as the strength of the negative charge carried by the electron. All the effects mentioned above can be explained by the action of these charges. In fact, these charges are responsible for an almost unbelievable variety of phenomena, ranging from the transmission of signals along nerve fibers to the spectacular discharge of lightning.

CHARGING BY FRICTION

Normally, if we select any object around us and test it in some way, we find that it has no net charge (that is, it is neither positively nor negatively charged but is neutral). If we could carry our investigation into the interior of the object, many charges would be found. Every object is composed of atoms, and these contain both protons and electrons: in fact,

FIGURE 12–2. Charged nylon hose.

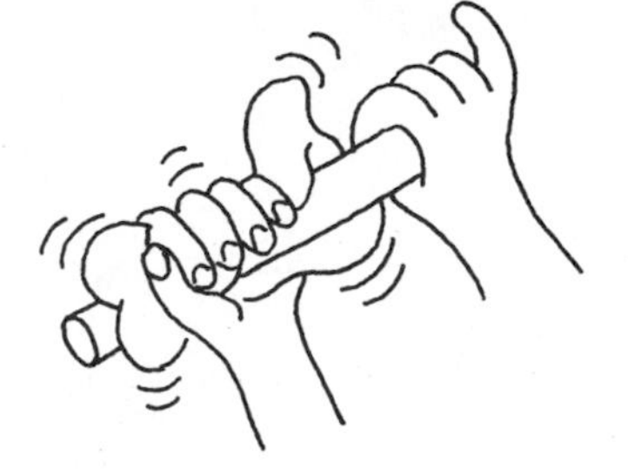

FIGURE 12–3. Rubbing a glass rod with silk separates the charges as indicated.

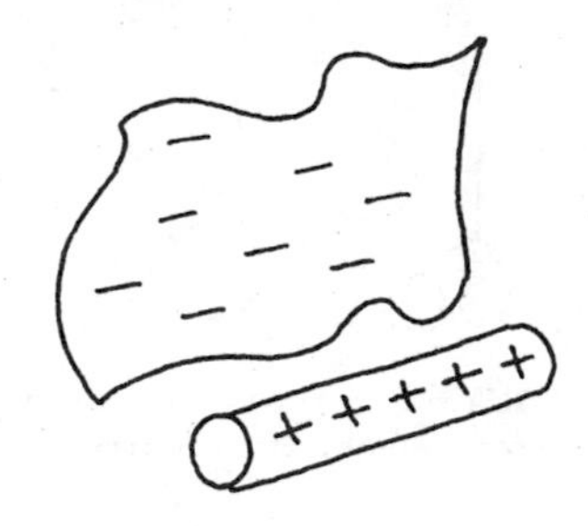

we would find billions of billions of billions of protons and electrons inside the substance. But for every positively charged proton we find, we would also find an electron with its negative charge. These two would compensate each other, and so would all other such pairs. The only way that we can give an object a net charge is to upset this delicate balance in some way.

Studies of solid objects reveal that their atoms are essentially immobile. They are fixed at precise locations about which they can vibrate, but they cannot wander about at liberty. It is a difficult task to free a proton from the atom (as we will see when we study nuclear physics), and as a result, the protons are immobile like the atom. Electrons, however, are bound only loosely to the atom, and inside a solid many of them become detached and are able to wander freely throughout the material. Consequently, when charges move, it is always the electrons that are shuffled around. Some materials have a greater tendency to hold on to their electrons than do others, and so, when two substances are rubbed together, the one that has the weaker hold on its electrons loses some of them to the other.[1] As an example, glass retains its electrons less well than does silk. As a result, when glass is rubbed with silk, electrons are transferred to the latter, leaving it with more electrons than it had previously. Since it still has the same amount of positive charge before and after the rubbing, the result is a net negative charge on the silk. The glass, having fewer electrons than before (but the same number of protons), is left with a net positive charge. The transfer is shown in Figure 12–3.

A rubber rod, on the other hand, acts like the ancient Greeks' amber and gains a net negative charge when rubbed with wool. Figure 12–4 shows this transfer of charge. The earliest studies of electricity were carried out using objects charged in this fashion, and, not wishing to break a good tradition, we'll continue with glass and rubber rods. Keep in mind, however, that the principles discussed apply to *any* charged objects.

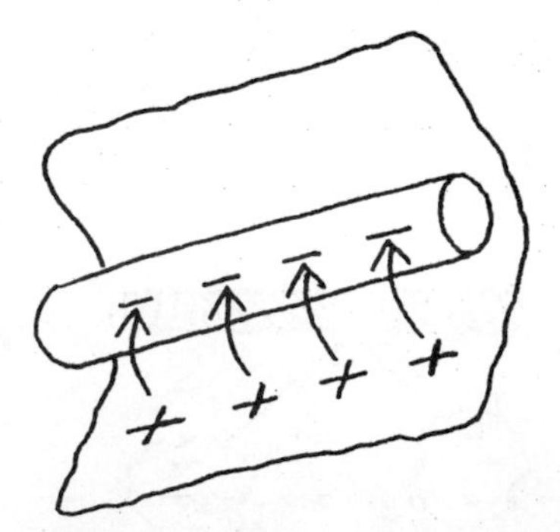

FIGURE 12–4. Upon being rubbed together, a rubber rod gains electrons from a piece of wool.

[1]The only effect of the rubbing is to bring various portions of the two materials into intimate contact. A few electrons would even be transferred without rubbing. The effect is still called "charging by friction," however.

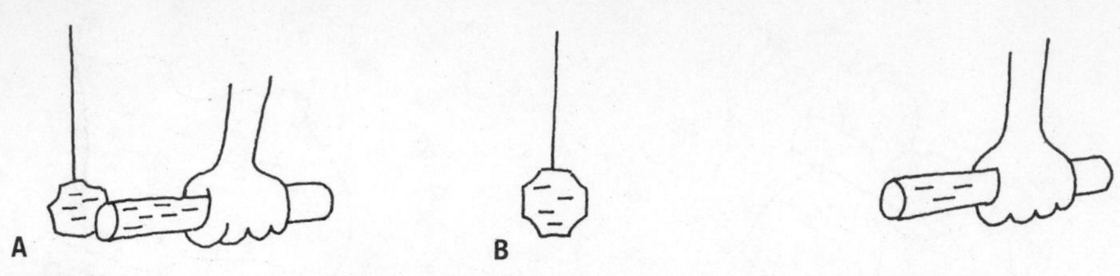

FIGURE 12–5. When the charged rod is touched to the aluminum foil ball (*A*), some charges are transferred so that the foil becomes charged (*B*).

The Conservation of Charge

Rub a rubber rod with wool, and the rod ends up with a negative charge and the wool a positive charge. But all that has really happened is a transfer of electrons from wool to rubber. Nothing has been created or destroyed as the materials gained their charge. This serves as an example of a well-established law called the *conservation of charge.* This law says nothing more than that *electric charge does not appear or disappear* out of thin air. We will encounter some interesting cases later when charges do appear and disappear, but they always go or come in pairs. You can bet that if a negative charge ever goes out of existence, a positive charge goes with it.

FORCES BETWEEN CHARGES

A small piece of aluminum foil suspended by a light thread can be used to demonstrate some of the important characteristics of charged objects. In Figure 12–5*A*, a rubber rod that has been charged (negatively) touches the foil. At the point of contact many of the electrons on the rod move onto the foil. When the contact is broken (Fig. 12–5*B*), the foil is left with an excess number of electrons and hence has a negative charge. When two pieces of suspended foil which have each been charged in this manner are brought close together, they stand apart (Fig. 12–6). An identical experiment can be performed with a charged glass rod. When it is placed in contact with the foil, electrons go from the foil onto the glass (Fig. 12–7*A*). The foil in this case is left with a net positive charge when the glass rod is removed. Two pieces of foil charged in this way also stand apart, as Figure 12–8 shows. Finally, bring a negatively charged foil and a positively charged foil close together, and they will be attracted toward one another, as in Figure 12–10.

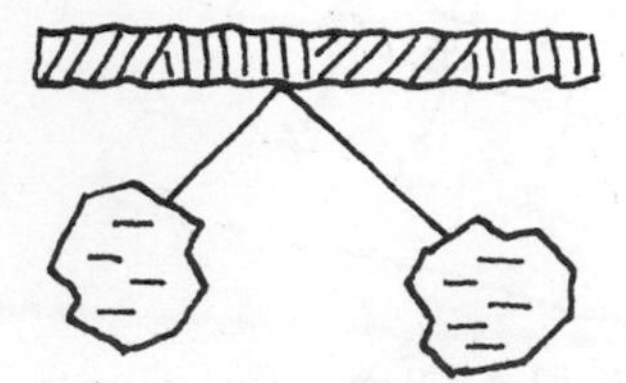

FIGURE 12–6. Two balls charged negatively repel one another.

The several observations above can be stated as a general observation: *like charges (those of the same sign) repel each other; unlike charges (of opposite sign) attract each other.* It might be worthwhile here to repeat the TADI

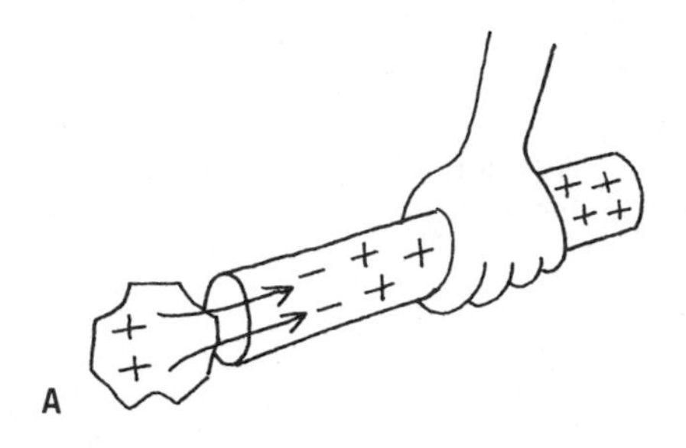

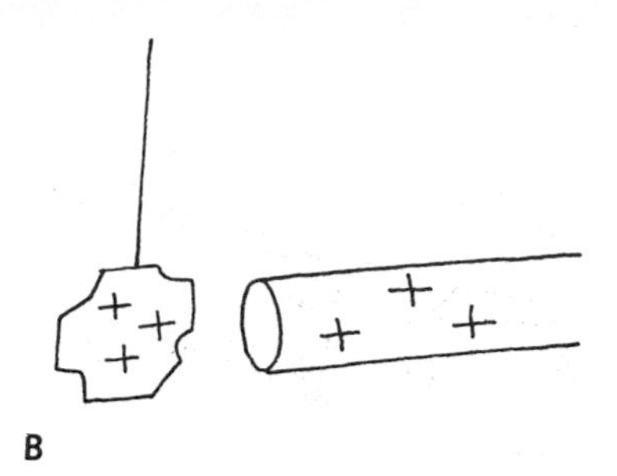

FIGURE 12–7. *A*, The positive rod, when touched to the foil ball, takes some electrons from the foil, leaving it positive (*B*).

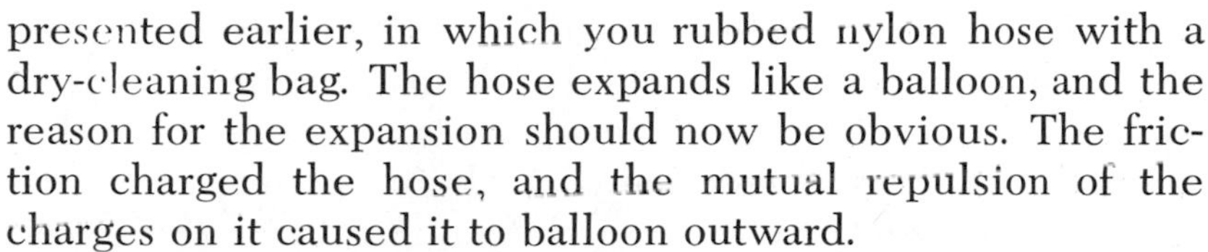

presented earlier, in which you rubbed nylon hose with a dry-cleaning bag. The hose expands like a balloon, and the reason for the expansion should now be obvious. The friction charged the hose, and the mutual repulsion of the charges on it caused it to balloon outward.

The strength of the electrical force of attraction or repulsion between charged objects was investigated by Charles Coulomb in 1789. He found that the force depended on the amount of charge on each object: double the charge on one object, and the force doubles; double the charge on both and the charge quadruples. Distance was also found to be a factor: if the separation between two charged objects doubles, the force between them is reduced to 1/4 of its original value; if the separation is tripled, the force decreases to 1/9 of the original value. In equation form:

$$F = k \frac{Q_1 Q_2}{d^2}$$

Q_1 and Q_2 are the charges on the objects,[2] d is the distance between them, and k is a constant.

With Coulomb's law established, we can explain such phenomena as bits of uncharged foil being attracted by a charged comb. When the comb is drawn through hair, it receives a negative charge because of the friction. When the comb is brought near a piece of foil (Fig. 12–11), it causes a rearrangement of charges within the foil, as shown. Negative charges are repelled by those on the comb and move as far away from them as they can. When they move to the opposite side of the foil, they leave the side near the comb with a net positive charge. The force of attraction between these positive charges and the comb is strong enough to cause the foil to move toward the comb. (A repulsive force also exists between the comb and the *negative* side of the foil, but since the negative side is farther from the comb than is the positive side, this force is weaker, and the attractive force wins the tug of war.) Upon contact, several negative charges from the comb move onto the foil, causing both comb and paper to have a net negative charge. The piece of foil then flies away from the comb.

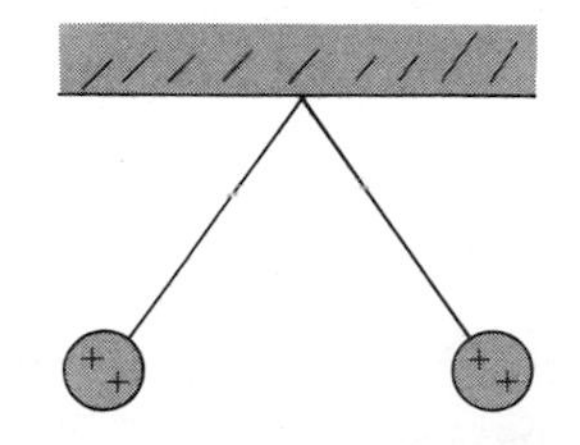

FIGURE 12–8. Two positively charged balls also repel each other.

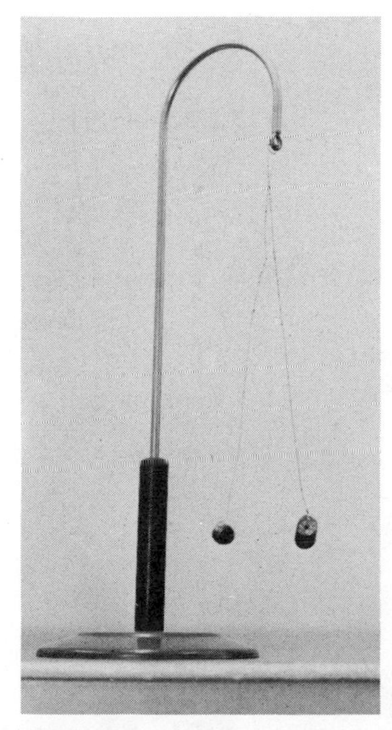

FIGURE 12–9. Two small objects repel each other, indicating that they both have a charge of the same sign.

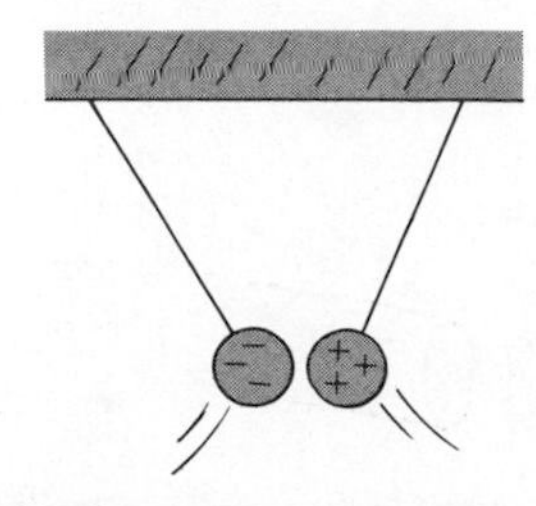

FIGURE 12–10. Oppositely charged balls attract each other.

[2]Charges are usually measured in units called coulombs. It requires 6.25×10^{18} electrons to produce a charge of one coulomb.

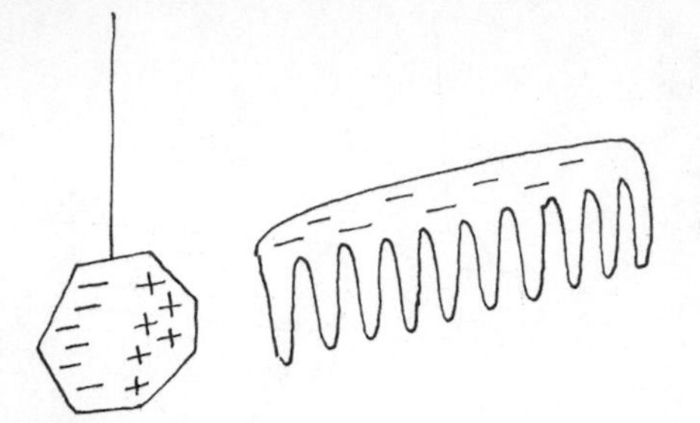

FIGURE 12–11. Because of the motion of charges within the foil, it is attracted to the comb.

When a suspended piece of foil is touched by a negatively charged rubber rod, the foil gains a negative charge. Similarly, a positively charged glass rod touched to a foil gives the foil a positive charge. In both of these cases the foil ends up with a charge of the same sign as the object which touched it. As another example of how charges move about, let's consider a method by which foil can be given a positive charge by a negatively charged rod or vice-versa. The steps are indicated in Figure 12–12. A negative rod is brought near a suspended piece of foil, causing a rearrangement of charges in the foil as the electrons in it attempt to get as far away from the rod as possible. In Figure 12–12*B*, an observer touches the foil with a moist finger. A moistened finger is a fair conductor, so the electrons now see a pathway by which they can move even farther away from the charged rod. They pass through the finger and onto the observer's body. Thus, when the finger is removed (Fig. 12–12*C*), the foil is left with a net positive charge.

Discharging into the Air

An object that receives a charge will not keep it forever. Air, particularly moist air, is a conductor of electricity, although a poor one. This provides a pathway for the charge to leak from an object and into the air. This leakage of charge goes on at a faster rate if there are sharp points on the object. Figure 12–13 shows why points have the effect they do. Figure 12–13*A* shows a piece of metal with a flat surface on one end and a rather sharp point on the other. The metal has

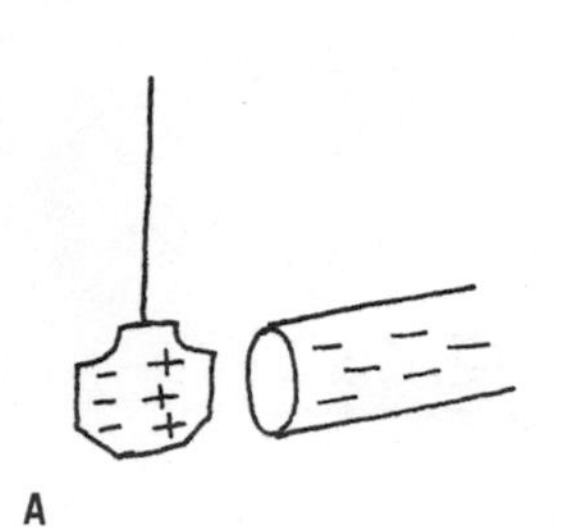

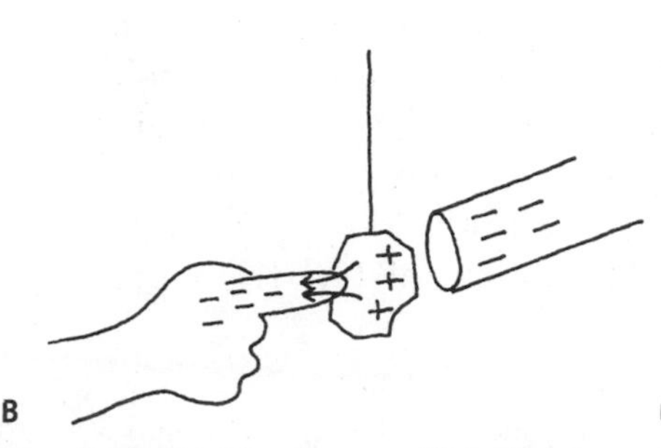

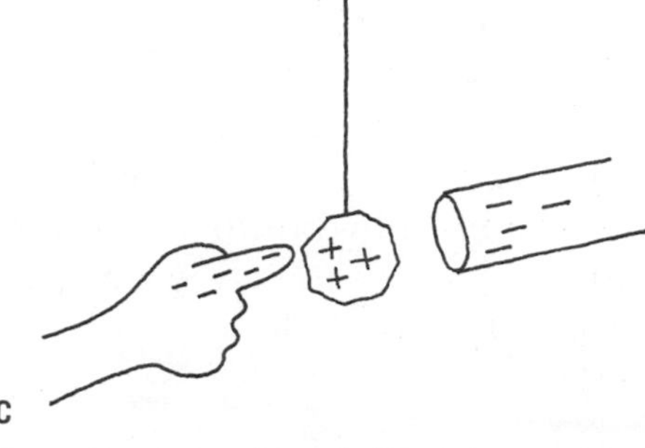

FIGURE 12–12. The steps involved in charging something without actually touching it to a charged object.

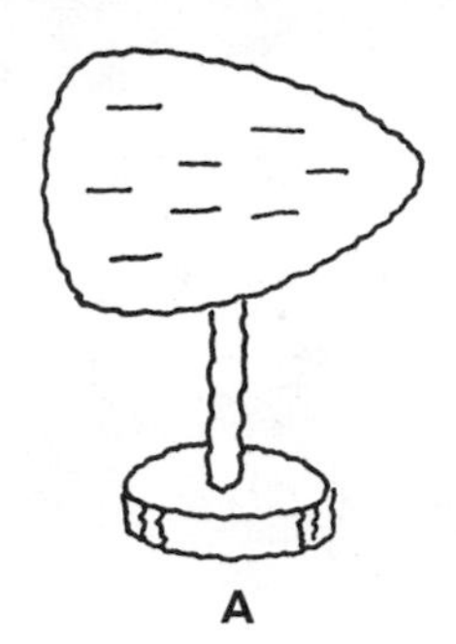

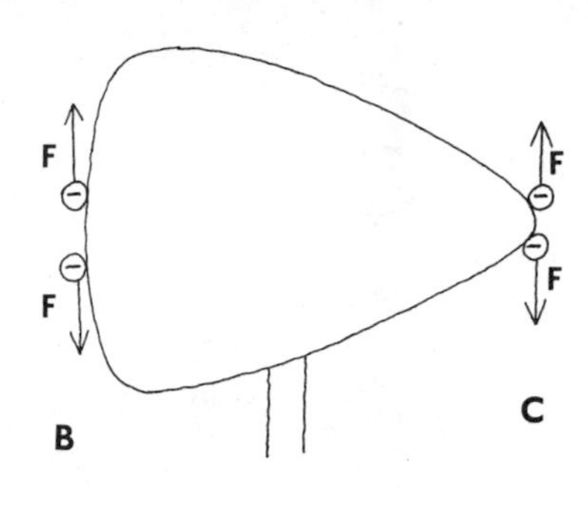

FIGURE 12–13. On relatively flat surfaces (*B*), repulsive forces cause electrons to move away along the surface, but near points (C), the same force is directed more away *from* than along it.

been given a net negative charge, and a close-up view of two of these negative charges located on the flat part of the surface are shown in Figure 12–13*B*. Because they are of like sign, these charges repel one another; the electrical forces, F, pushing them apart are represented by arrows in the figure. Since these forces are directed along the surface, their primary effect is to cause the two charges to move away from one another along the surface, until the shoves from other electrons stop their motion. Figure 12–13*C* shows what happens on the pointed end. The forces are still repulsive, but as the arrows indicate, the shoves are more away from the metal surface than they are along it. Each electron pushes on all the other electrons near it, which can produce a considerable force on each electron if many are gathered quite close together. In fact, the net push away from the surface can become so great that electrons are literally sprayed into the air.

LIGHTNING AND LIGHTNING RODS

The common phenomenon of lightning provides a large-scale example of many of the topics that we have covered. Lightning occurs when charges jump from one cloud to another or between a cloud and the earth. The method by which clouds get their charge in the first place has never been explained to the satisfaction of all scientists, but it is believed that the process is at least somewhat as follows.

As rain falls, updraughts of air pass the falling raindrops. These air currents seem to cause large drops to split into two drops, one large and one small. The larger one carries a net positive charge after the split and the smaller one a net negative charge. The larger drops continue to fall, but the smaller ones are swept upward by the rising air. The result is a separation of charges between layers of clouds. The difference between this charge separation process and the friction process described earlier is that much larger amounts of charge are separated by the process described here.

Two clouds which have received opposite charges are

FIGURE 12–14. If clouds have opposite charges, lightning may jump between them.

shown in Figure 12–14. The negative charges in one cloud are strongly attracted by the positive charges in the other, and if the attraction becomes great enough, the charges leap from one cloud to the other. This jagged leap of charge is lightning.

In Figure 12–15 a positively charged cloud is shown hovering over a house. Electrons from the earth are attracted by the cloud and are spread over the house, giving it a net negative charge. Protection against a destructive discharge between the two can be effected by lightning rods, which are pieces of metal with sharp points on the end, attached to the house in such a way that they extend higher than any other portion of the building. Wires leading from the rods are connected to metal posts driven several feet into the earth. Because charges tend to accumulate at points and be sprayed off (or onto) them, any discharge occurring will be between cloud and lightning rod. The connecting wires carry the flow of charge away from the house and to the ground.

Benjamin Franklin (1706-1790)

No stranger in American history books, this fifteenth child in a family of 17 often does not receive due credit for his technological work. He was more an inventor than a scientist, inventing such things as an improved stove and bifocal glasses, but he also made noteworthy scientific contributions to our knowledge of electricity, including the invention of the lightning rod.

As a final example of the consequences of electrical forces, consider the following experiments. A piece of metal is shaped into the form of a sphere so that it resembles a solid copper bowling ball. Imagine that we place several excess negative charges near the center of the ball (Fig. 12–16*A*). The negative charges exert forces of repulsion on each other, which causes them to move as far apart as possible. This means that they will quickly move to the surface of the sphere (Fig. 12–16*B*). Regardless of the shape of the metal, if a conductor has an excess charge, the charge will always be located on the outside surface. It is this phenomenon which makes an automobile a safe place during a lightning storm. If lightning strikes the car, it will charge up the car during the instant that it is passing through to the ground. But because the passengers are surrounded by the metal of the car, the charges reside on the outside surfaces of the car and do not affect the inhabitants. (Fright is not classified as "affecting" them.)

FIGURE 12–15.

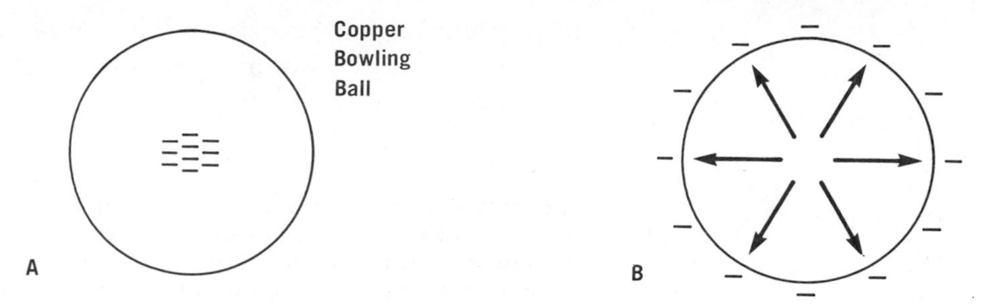

FIGURE 12–16. Repulsive forces cause charges to distribute themselves on the surface of a metal sphere.

THE ELECTROSCOPE

When a rubber rod is stroked lightly with wool, the rod becomes slightly negatively charged; a more vigorous rubbing gives it a greater negative charge. In an effort to determine the amount of charge on an object such as the rod, an instrument called the electroscope was invented. It has been largely replaced today by more sensitive electronic instruments, but because of its simplicity and its place in the history of electricity, we will consider it here. As pictured in Figure 12–17, the electroscope consists of a metal ball and a rod to which is attached a small strip of gold or aluminum foil. The rod is supported by a rubber insulator, which acts

FIGURE 12–17. A charged electroscope.

to prevent the charges from leaking onto the case. The case protects the foil from air currents and has windows in it so that the foil can be observed.

take-along-do-it

A simple electroscope made from a small jar sealed with a cork is shown in Figure 12-18. Passing through the cork is a metal wire that is bent on the lower end to support a thin strip of aluminum foil.[3] (Make sure the insulation is removed from the wire.) You can use it to test the effects discussed in the remainder of this section.

FIGURE 12–18. A homemade electroscope.

In Figure 12–19*A*, a negatively charged rod is touched to the ball on top of the electroscope. Some of the electrons leave the rubber rod and distribute themselves over the metal knob, rod, and foil. Repulsion between the charges on the metal rod and those on the foil causes the light foil to be pushed away (Fig. 12–19*B*). If the rubber rod had been given a much greater charge, more electrons would have left it, and the foil would have been pushed away much more. A scale shown in the figure may be used to compare the charge on different rods.

In many instances, it would be undesirable to perform an experiment in the manner described in the preceding paragraph. When the rubber rod was touched to the electroscope, some of the charge left the rod and was distributed over the metal parts of the electroscope. If the rubber rod is now removed, it will have less charge on it than it did before. In this, and in *any* experiment, it is an undesirable feature of a measuring instrument if it changes the condition

[3]Ordinary kitchen aluminum foil is much too heavy to be used for this purpose. Your physics instructor should be able to supply the type you need from the departmental stockpile. Or use foil from a gum wrapper after removing the paper backing.

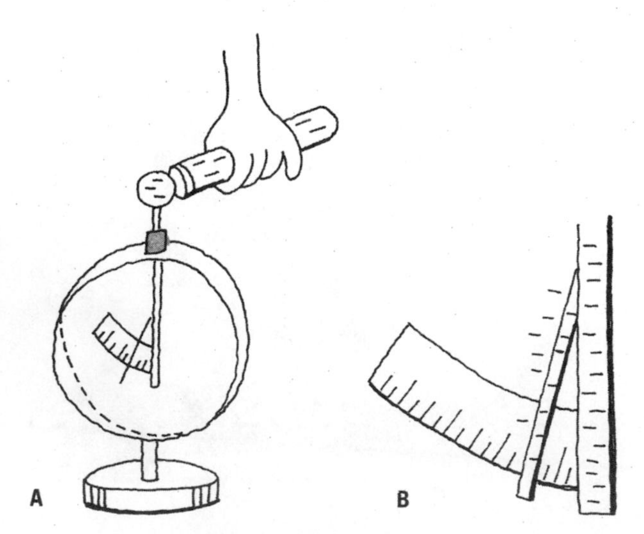

FIGURE 12–19. An electroscope equipped to measure (rather than merely to detect) charge.

of whatever it is measuring. The electroscope, to be an effective instrument, needs to measure the amount of charge on an object like the rubber rod without removing any charge from it. This can be accomplished by bringing the negatively charged rod near the ball but not in contact with it (Fig. 12–20). The negative charge on the rod repels the electrons in the knob so that they distribute themselves over the foil and metal stem. As before, the amount that the foil is pushed away serves as an indication of the charge on the rubber rod, but in this case no charge was removed from the rod in the measurement process. (Repeat these observations using your homemade electroscope.)

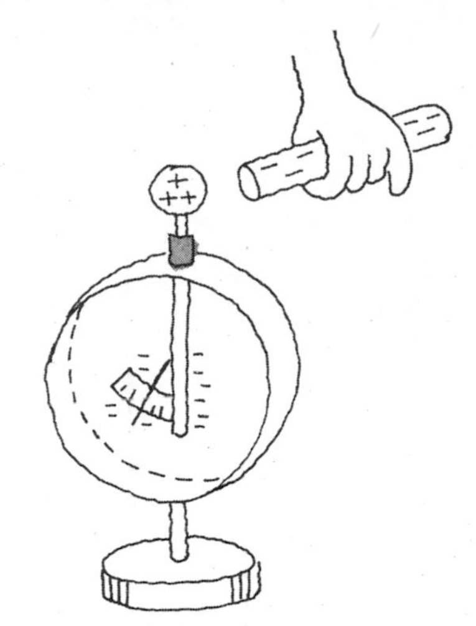

FIGURE 12–20. The electroscope deflects even without being touched by the charged rod.

THE PHOTOELECTRIC EFFECT

In Figure 12–21*A*, a piece of metallic zinc is connected to the ball on the top of an electroscope. The electroscope and zinc have been charged negatively by touching the zinc with a charged rubber rod, and unless the air surrounding the electroscope is very humid, the scope will remain charged—and the leaf extended—for quite a long period of time (an hour or more). However, if ultraviolet light falls on the zinc, the electroscope loses its charge in a matter of seconds. The photoelectric effect (see Chapter 20) is responsible for the behavior of the foil. Photons (bundles of energy) in the incident beam of light eject electrons from the zinc, and as the metal loses electrons, the leaf returns to its vertical position. To demonstrate that it is electrons which are

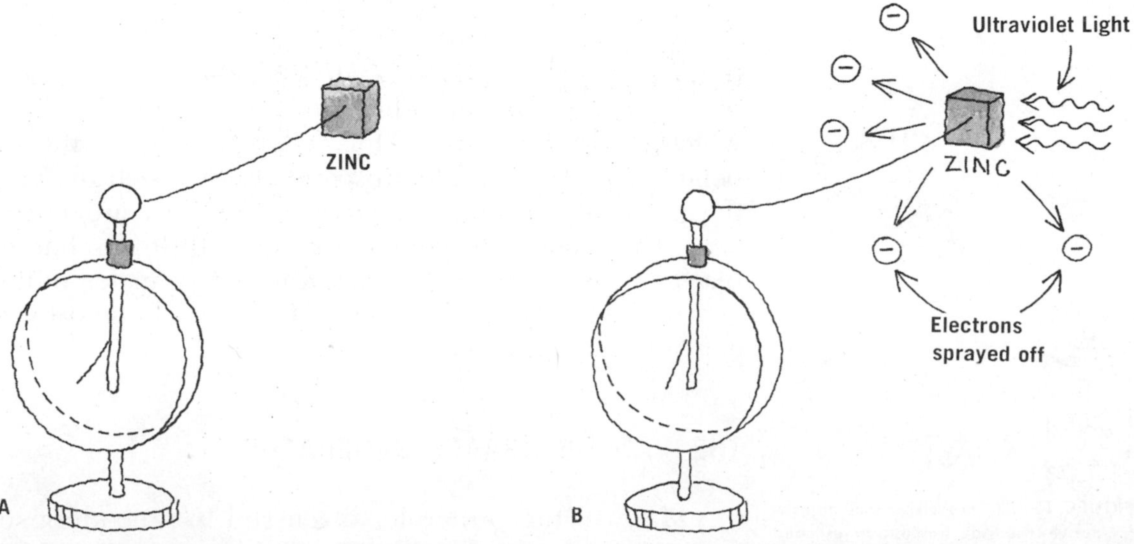

FIGURE 12–21. Shining ultraviolet light on the zinc causes it to become positively charged.

emitted by the zinc and that protons cannot be emitted, one may give the zinc a positive charge and shine ultraviolet light on it. Result—no discharge. In fact, electrons are ejected from the zinc regardless of its original state of charge. As an example, consider the zinc to be connected to the knob of an uncharged electroscope. If ultraviolet light falls on the zinc, the leaf will slowly stand out, indicating the presence of charge. In this case, electrons are being ejected from the zinc, leaving it and the leaf with a net positive charge. In the next chapter, we will discuss several practical applications of the photoelectric effect.

CONDUCTORS AND INSULATORS

take-along-do-it

Charge your homemade electroscope by touching it with some object you have charged by friction (Fig. 12-22). Hold a pencil by one end and touch the other end to the top of your scope. Do the leaves fall? Try it again, using other objects such as a piece of paper or a glass bottle. Do the leaves fall for any of the objects? If your trials did not include a metallic object such as a scissors or a belt buckle, try one.

You probably found that none of the plastic, rubber, or paper objects would carry away the charge on the electroscope but that the metal objects did. The explanation of the difference between these two types of objects is fairly simple. Some materials have a large number of electrons which are able to move freely. All metals fall into this category and are called *conductors*. If your electroscope was negatively charged, when you touched it with the metal object some of the electrons went from the electroscope through the metal and onto your hand. If your electroscope was positively charged, electrons flowed from your body through the metal and onto the electroscope, thus discharging it.

Insulators, on the other hand, have very few electrons which are free to move. Thus, when you touch the wooden pencil to a negative electroscope, a few electrons pass onto the wood at the point of contact, but they are not carried along the wood into your body. Very little discharge takes place, therefore. (The fact that a few *do* transfer to the stick explains why the electroscope can be discharged by repeatedly touching it with an insulator.)

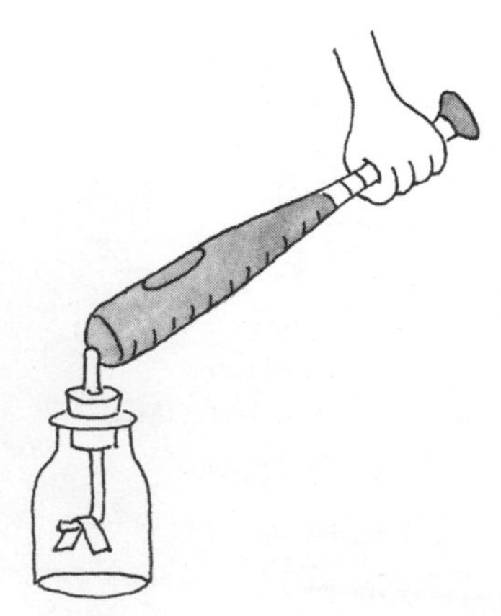

FIGURE 12–22. Touch various objects to your charged electroscope to test conductivity.

THE VAN DE GRAAFF GENERATOR

Many of the principles discussed in preceding sections are utilized in the Van de Graaff generator (Fig. 12–23). A motor in the base drives a pulley, causing the rubber belt to turn, as indicated by the arrows. Friction between the belt and the pulley gives the belt a negative charge. This charge is carried upward by the belt until metallic points at the top "comb" it off in much the same fashion that a lightning rod extracts charge from a cloud. The charges move from the points to the outside surface of the metallic dome. In this fashion, it is possible to give the dome a quite large negative charge. (By selectively choosing the materials of which the belt and pulley are made, we could have given the dome a positive charge instead.) When sufficient charge is accumulated, it can leap off, creating miniature lightning bolts.

In Figure 12–24 a person has placed his hand on the metallic dome and is standing on a piece of insulating material. He now becomes a part of the Van de Graaff generator, acting much like a part of the dome. Negative charges are distributed over him and remain there, unable to escape through the insulator to the earth. An interesting demonstration of the repulsion between charges occurs if the person has straight, fine hair. The charges distribute themselves uniformly on the hair, and because of their repulsive effects, cause the hair to stand out.

The generator serves as an excellent teaching tool to demonstrate many of the phenomena discussed heretofore. Over and above this use, however, it has played a major role in the study of nuclear physics. We will see in Chapter 24 that an important pursuit of the nuclear physicist is to speed up charged particles and to allow them to collide with atoms. Figure 12–25 illustrates how this could be done. A tube is connected to the positively charged Van de Graaff as shown, and a proton is released at position P. The repulsion between the charges on the dome and the positively charged proton causes the latter to move rapidly along the dashed path. The tube has had the air pumped out of it so that the proton will not collide with air molecules that might otherwise be in its path. At the end of the tube it is allowed to strike a target made of the atoms to be studied. The "nuclear reactions" produced in such collisions will be discussed in detail in Chapter 24.

Robert J. Van de Graaff (1901-1967)

Van de Graaff was born in Tuscaloosa, Alabama, and graduated from the University of Alabama in 1922. He received a Rhodes scholarship to study at Oxford University, where he received his Ph.D. in 1928. While a professor at the Massachusetts Institute of Technology, in 1931, he invented the high-voltage generator which bears his name.

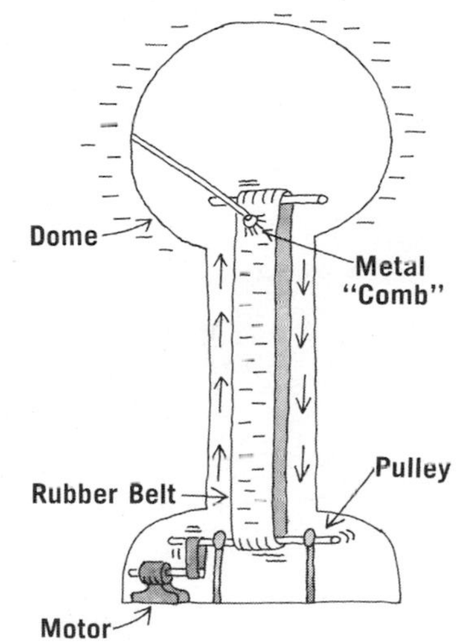

FIGURE 12–23. The Van de Graaff generator.

FIGURE 12–24. The charge goes on to the person also. Note that his hair is on end.

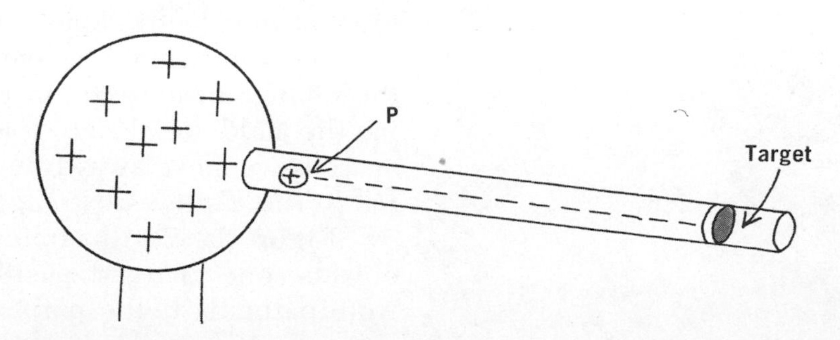

FIGURE 12–25. The proton zaps down the evacuated tube toward the target.

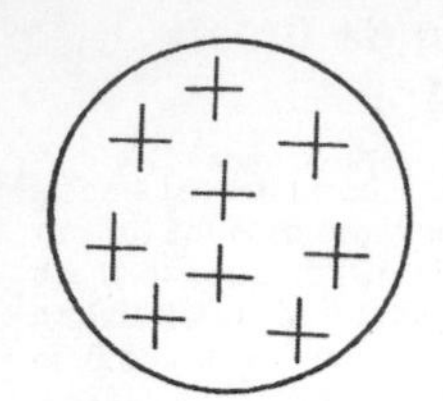
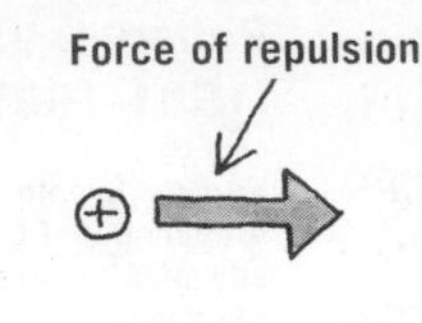

FIGURE 12–26. The small positive charge feels a repulsive force from the charged sphere.

THE ELECTRIC FIELD

We have seen that charged objects exert forces of attraction or repulsion on each other even if there is no physical contact between them. In such circumstances, it is often helpful to introduce the concept of fields, which helps us to visualize the situation. Using this approach, we say that any charged object has an electric field existing in the space around it. We will see in this section that a direction and a strength can be assigned to an electric field and that a knowledge of these two characteristics of a field will enable us to predict the movement of other charged objects when they enter the field.

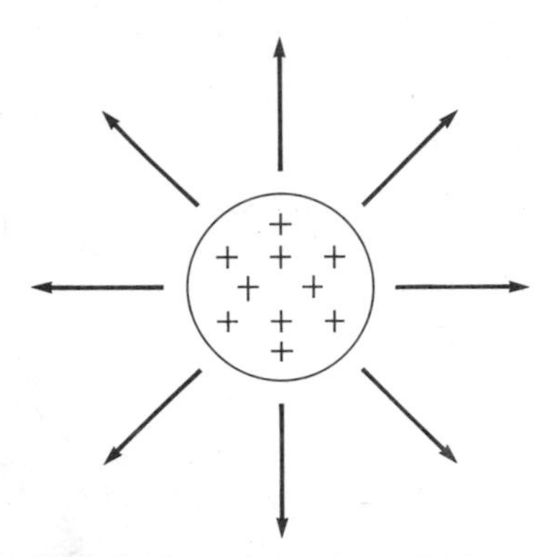

FIGURE 12–27. The electric field near a positively charged sphere.

Figure 12–26 shows a spherical object, which could be the dome of a Van de Graaff generator, that has been given a positive electric charge. Another positively charged object brought nearby will experience a force of repulsion. We say that the object is in the electric field produced by the charges on the sphere. In order to define a direction for an electric field, we imagine a small positively charged object to be brought into the field. In the circumstances of the figure, the small positive charge feels a push outward, away from the sphere, and so we say the direction of the field is outward. We represent this by drawing lines to represent the field (Fig. 12–27). Thus, *the direction in which a small positive test charge would move is defined to be the direction of the electric field at that point.*

But the lines which represent the field are drawn to convey more information than simply the direction of the field. When the small charge is very near the sphere, it will be pushed away with a very great force. This force of repulsion gradually diminishes as the test charge moves farther and farther away. In terms of the electric field concept, we say that the electric field is quite strong near the sphere and decreases in strength as we move away from it. To indicate this change in strength, in Figure 12–27 the lines representing the field are closely spaced near the sphere and farther apart as we move away from the sphere. Thus, *the density of the lines tells us the strength of the field.*

Figure 12–28 illustrates the electric field between two objects, one charged positively and the other negatively. Note particularly the points where lines are closely spaced;

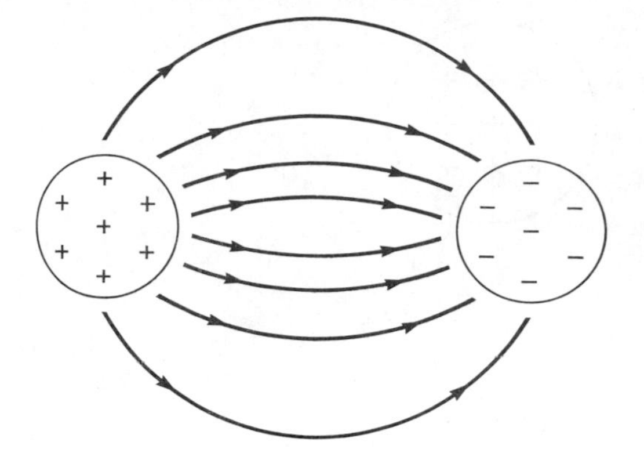

FIGURE 12–28. The electric field between two oppositely charged objects.

in these regions, a charged object would experience the greatest force. The arrows on the lines indicate the direction of the electrical force felt by a positive charge at each point in the field. A negative charge feels a force in the opposite direction.

Figure 12–29 illustrates the electric field around an irregularly shaped object which has a negative charge. It is evident that a positive charge would be attracted toward the object, and therefore, the electric field is shown to point inward. The spacing of the lines indicates that near the more pointed end of the object, the electric field is considerably stronger than near the flatter portions.

ELECTRICITY AND ENERGY

Since the concept of energy is one which runs throughout all of physics, it obviously will find a role to play in electricity. In Figure 12–30*A*, a small positive charge is located far away from a positively charged sphere. If the small charge is pushed closer to the sphere (Fig. 12–30*B*), the pusher will encounter a force back on him because of electrical repulsion between the objects. Clearly an expenditure of energy by the pusher is required to place the charge near the sphere; therefore we must look for the location of this

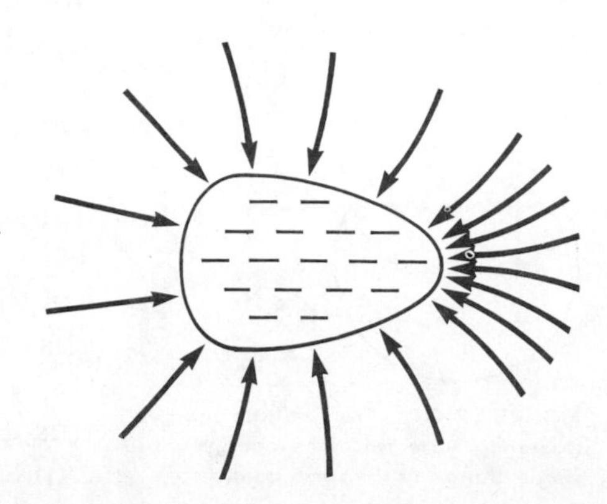

FIGURE 12–29. The electric field near an irregularly shaped object is stronger at points where the object is more sharply pointed.

FIGURE 12–30. The small charge has more energy in *B* than in *A* because energy was necessary to move it to the closer position.

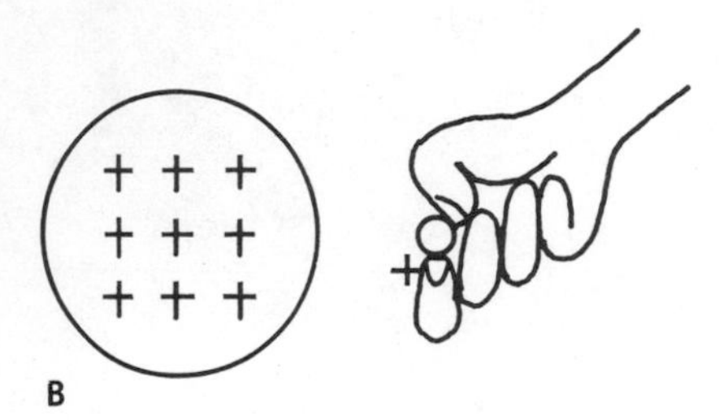

energy when the small charge is near the sphere. The answer is that the charge possesses this energy because of its location; we call this *electrical potential energy.* If released when it is near the sphere, the charge takes off, moving in a direction away from the sphere. The electrical potential energy gradually changes into kinetic energy as the charge moves away.

Suppose that instead of one small positive charge we put two charges at the position shown in Figure 12–31. We will ignore any forces between these two so that we can concentrate on the much greater effects produced by the sphere. In this case, the total energy possessed by the two charges is twice as great as that of a single charge. Similarly, three small charges at the same point possess three times as much electrical potential energy. It turns out to be convenient in the study of electricity, instead of considering the total energy of charges, to speak of the *energy per charge.*

An example is in order. Figure 12–32 is similar to Figure 12–31 except that two positions are indicated near the sphere. Suppose that a small positive charge has 8 chunks of energy at point HI and 4 chunks at point LO. Two charges would then have 16 and 8 chunks of energy at these locations; three positive charges, 24 and 12 chunks. Note that if we divide the amount of energy in each case by the amount of charge, we always obtain the same numbers. Two charges at point HI have twice as much energy as one, but they also have twice as much charge; so the energy per charge remains the same.

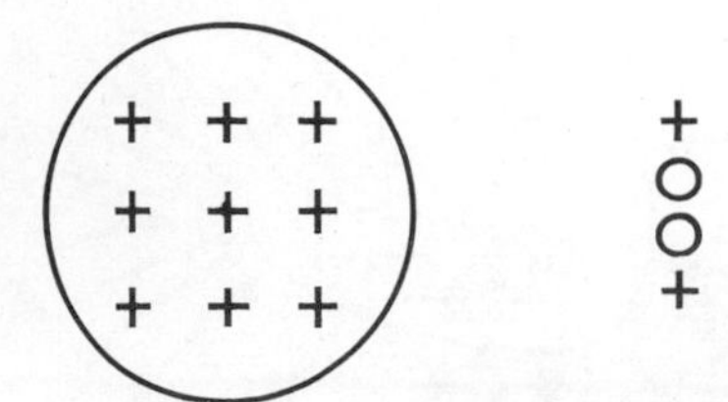

FIGURE 12–31. The two little charges (together) have twice the energy of a single charge at the same point.

The energy per charge concept is important enough that it is given a name: *voltage* (because of the unit in which it is expressed—volts). The significance of voltage lies in the fact

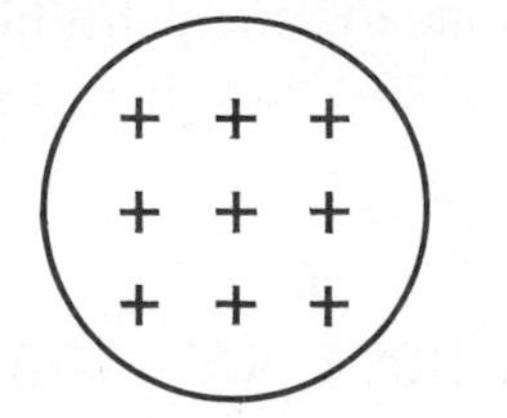

FIGURE 12–32. Point HI is a point of higher electrical potential energy per charge than point LO.

that a definite value for it can be assigned to a location irrespective of whether or not a charge exists in that location. As we will see in the next chapter, a 12 volt battery is a device in which one location, the positive terminal, is maintained at a voltage 12 volts higher than another location, the negative terminal. Such voltage differences are important because charges will move between points differing in voltage. Use Figure 12–32 to convince yourself that a positive charge will move from a point of high voltage to a point of low voltage, whereas a negative charge will move from a point of low voltage to a point of high voltage. In both cases, electrical potential energy is changed to kinetic energy (or other forms of energy) during the movement of the charge.

ELECTRICAL FORCES AND THE ATOM

Our planetary model of the atom depicts electrons as orbiting about the nucleus. We have not yet asked why the electrons stay in their orbits instead of flying off into space. The answer should be obvious now: the nucleus is composed of protons, which, because of their positive charge, exert a force of attraction on the electrons. The electrical force exerted on an electron at various points in its path is shown in Figure 12–33. If this force should vanish, the electron would go sailing off in a straight line. The situation in the atom is analogous to circumstances in our solar system. The planets are kept in their endless orbits about the sun by a similar force of attraction, which in this case is a gravitational force.

When we investigate the atom more closely in Chapter 16, we will find that normally the electron circles in an orbit near the nucleus called the ground state orbit. In order to move the electron to higher orbits—excited states—energy must be given to the electron. This transfer of energy is usually accomplished when an incident missile, generally either another electron or a photon, strikes the electron. The energy discussion of the preceding section should make this statement seem reasonable. In its lowest orbit, the electron has potential energy, but in an excited state it has more energy because it is farther away from the positive charge of the nucleus. The conservation of energy law says that the only way an electron can move from an orbit with a low po-

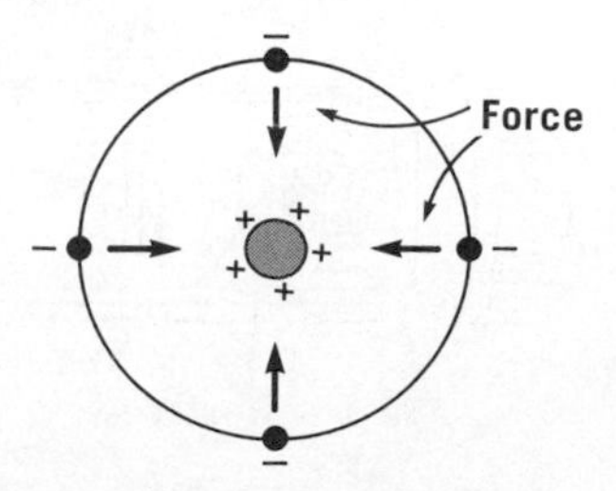

FIGURE 12–33. The force between the positive nucleus and negative electrons is constantly attractive.

tential energy to a higher orbit with a higher potential energy is to gain the excess energy from some outside source.

THE CATHODE RAY TUBE

The cathode ray tube is an evacuated container which contains the basic parts shown in Figure 12–34. By connecting the tube to various pieces of electrical equipment, it can be made to display visual information on a screen at its front. In this section we will examine how it works, and in later sections we will see how it is used in one of its most basic applications: the picture tube of a television set.

Refer to Figure 12–34. In the neck of the tube, an electric current flowing through the heater causes its temperature to rise and to heat a metal cylinder, called a cathode. The cathode gets hot enough so that electrons are "boiled" out of it. Located a short distance down the neck of the tube is an anode, which is shown here to be a wire screen. Other electrical circuits, not pictured, maintain a positive charge on the anode, and this positive charge attracts the electrons that were freed from the cathode. Since the anode is a screen with many holes in it, most of the electrons pass on through without striking it, and if left alone, they would then travel in a straight line until they strike the face of the tube. The tube face is coated with a material that fluoresces when electrons strike it. (The coating on the screen of a black-and-white television set emits white light when struck by electrons.) An observer watching the tube face would see a bright spot of light at the point where the electrons strike. To produce a little more excitement, two sets of plates are installed in the neck to make the electron beam, and hence the spot, move around; these are the horizontal and vertical deflection plates.

In order to see how the deflection plates work, let's con-

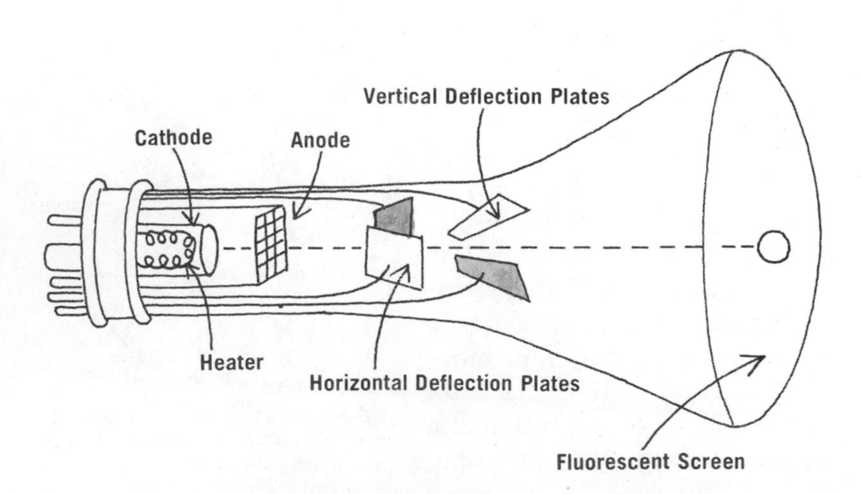

FIGURE 12–34. Cathode ray tube.

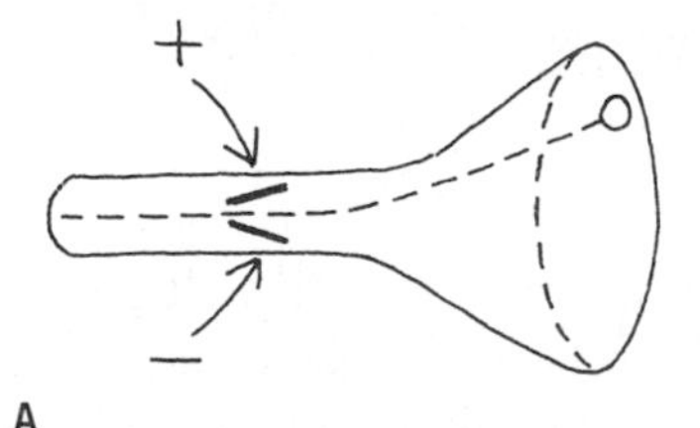

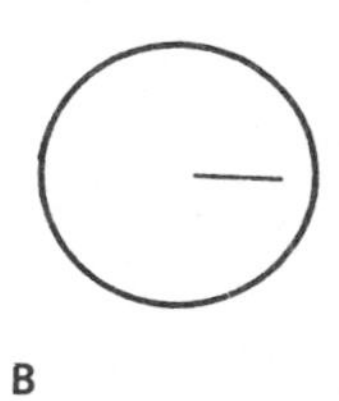

FIGURE 12–35. Top view (A) and front view (B) of a cathode ray tube whose beam is being deflected by the horizontal deflection plates.

sider them separately. First, consider the horizontal deflection plates. Initially let's assume that the electron beam travels directly down the center of the tube and forms a spot at the center of the screen. External electrical circuits can change the amount of charge present on the horizontal deflection plates; so let's assume that positive charge is gradually placed on one plate and the same amount of negative charge is gradually placed on the other. This creates an increasing electric field between the two plates, which will cause the electron beam to be bent more and more away from its straight path. A top view of the tube (Fig. 12–35*A*) shows the path taken by the beam at some instant of time. The spot of light will now be off center, a little to the right of its center position. Figure 12–35*B* shows what would be seen. The tube face is slightly phosphorescent and as a result will glow briefly after the electron beam moves from one point to another. Slowly increasing the charge on each of the horizontal plates causes the electron beam to move gradually from the center toward the right of the screen. Because of the phosphorescence, however, we see a line extending across the screen instead of the simple movement of the dot.

The vertical deflection plates act in exactly the same way as the horizontal plates, except that changing the charge on them causes a *vertical* line to be drawn on the tube face. In practice, the horizontal and vertical plates are used simultaneously. To see how it can display visual information, let's examine how we could observe the vibration of a tuning fork on the screen. For this purpose, the charge on the horizontal plates is caused to change such that the bright dot sweeps across the face of the tube at a constant rate. The tuning fork is then sounded into a microphone, which changes the sound signal to an electrical signal that is applied to the vertical plates. The combined effects of the horizontal and vertical plates cause the beam to sweep across the face of the tube horizontally and up and down at the same time, with the vertical motion corresponding to the tuning fork signal. The pattern swept out is shown in Figure 12–36 and represents the wave form of the tuning fork. A device used in the manner just described is called an oscilloscope. These instruments are widely used by electronic technicians as test equipment to observe electrical wave forms.

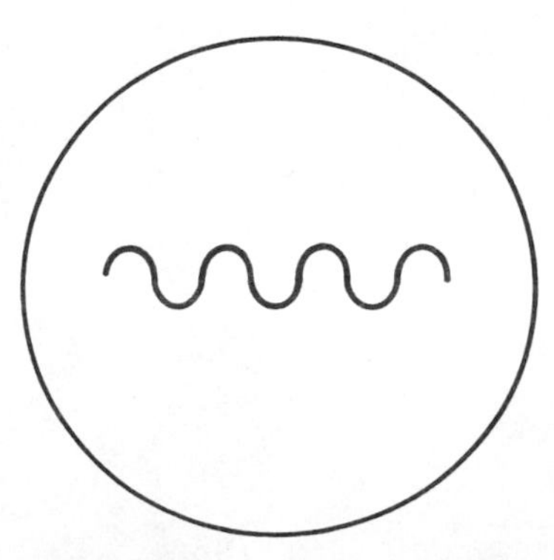

FIGURE 12–36. As the beam sweeps across, the vertical deflection plates move it up and down, forming a wave on the screen.

OBJECTIVES

A study of Chapter 12 should enable you to:

1. Briefly describe the historical beginnings of our knowledge of electrical charges.
2. Perform an experiment demonstrating static charge effects.
3. Explain the atomic origin of electrical effects and how this origin results in only negative charges being mobile in a conductor.
4. Describe, on the atomic level, what happens when objects are charged by rubbing.
5. State, explain, and give one example of the law of conservation of charge.
6. State the attraction-repulsion rule of static electric charges and describe demonstrations which illustrate this rule.
7. Explain why a charged object attracts a *neutral* object.
8. Draw a figure illustrating the charge distribution on an irregularly shaped conductor and explain why the charges distribute themselves as they do.
9. State and explain two beneficial effects of lightning rods.
10. Relate a widely accepted theory of the cause of charge separation in clouds.
11. Explain why an automobile is a relatively safe haven in a lightning storm.
12. Construct an electroscope and explain its principle of operation, including an explanation of the effect observed when a charged object is brought near it but not placed in contact with it.
13. Describe the use of an electroscope to demonstrate the photoelectric effect.
14. Perform an experiment with an electroscope to test the conductivity of various objects.
15. Explain the differences between conductors and insulators.
16. Draw a diagram showing the essential parts of a Van de Graaff generator and explain its operation, including its use as a particle-accelerator.
17. Draw the electric field lines around each of the following: (a) a single positive charge, (b) a single negative charge, and (c) a pair of oppositely charged objects.
18. Define voltage and explain the concept by using an example.
19. Use the ideas of forces between charges to explain the orbiting of electrons and relate this to the orbiting of planets in the solar system.
20. Draw a diagram of a cathode ray tube, describing the function of the following: (a) the cathode, (b) the

anode, (c) the deflection plates, and (d) the fluorescent screen.

QUESTIONS – CHAPTER 12

1. If a metal object receives a positive charge, what happens to its mass? Does it increase, decrease, or stay the same? What happens to the mass if you give the object a negative charge?

2. Assume someone proposes a theory that says people are bound to the earth by electrical forces rather than by gravity. How could you prove him wrong? In fact, could you get close enough to him to tell him he is wrong?

3. A rod with a positive charge is brought near, but not touching, an electroscope. Its leaves stand apart. Why? What kind of charge is on the leaves?

4. A rod with a positive charge is brought near, but not touching, an electroscope. We now touch the knob of the electroscope. When we remove our finger and the positively charged object, do the leaves move apart? Why? What kind of charge is on the leaves?

5. Two objects are charged: one positive, the other negative. What happens when they touch?

6. Assume you have a solid metal ball and a hollow ball of the same size and material. Which one can you place the most charge on? Why?

7. A charged comb often will attract small bits of dry paper which fly away when they touch the comb. Why?

8. Why is it more difficult to charge an electroscope on a humid day than on a dry day?

9. Why is a lightning rod dangerous to a house if it is not connected to wires leading to the earth?

10. Are the inhabitants of a steel-framed building safer than those in a wood frame house during an electrical storm or vice-versa? Explain.

11. Discuss the differences between insulators and conductors.

12. Sketch the electric field between two objects with a positive charge. Why do you space the lines as you do?

13. Sketch the electric field between a positive and a negative object. Why do you space the lines as you do?

14. An object has a negative charge. Does a small positive charge have more potential energy when near it or when far away? What about voltage? Repeat for a small negative charge.

15. Why does a charged object attract a neutral object? Does it matter whether or not the neutral object is a conductor or an insulator? Explain.

16. Write the equation relating the charge on two objects, the distance between them, and the force between them. Beside this write the equation (from Chapter 3) relating mass, distance, and gravitational force. Note the similarities. Which of the two forces can be repulsive as well as attractive?

17. Isaac Asimov claims that Ben Franklin's invention of the lightning

rod was a major turning point in the history of science and its relation to society. Read Asimov's short essay concerning this in *The Stars in Their Courses* (Doubleday, 1971) and write a report.

18. Why should a ground wire be connected to a television antenna?

19. Why are trucks that transport flammable materials required to have metal chains dangling from the truck and making contact with the road?

20. What effect, if any, would you expect the temperature of a material to have on the ease with which electrons can be ejected from it in the photoelectric effect?

21. It was stated in the section on the photoelectric effect that an electroscope could be charged by shining light on a piece of zinc attached to the knob of the electroscope. Why will it be impossible to get a very large deflection of the leaves in this way? (Hint: After some electrons have been emitted from the zinc, why is it difficult for more to leave?)

Photo courtesy of LaDa.

13

THE SHOCKING STORY OF CHARGES IN MOTION

For centuries man has known that objects can be "electrified" by friction. The study of charged rubber rods, glass rods, and so forth, has given experimentalists a good deal of insight into the nature of electricity. Beyond this, in our everyday life, such *static* electricity is a curiosity or, at times, an annoying distraction. (As an example of the latter, repeat the observation of the last chapter in which you donned wool clothes, scooted across a plastic seat cover, and kissed a partner.) In this text we have studied charged objects in order to build a foundation for other more important applications.

It was only when we learned how to make these electrical charges flow in a controlled way that a new day dawned for electricity. Charges in motion have become an inseparable part of our lives. They power our telephones, radios, television sets, air conditioners, and refrigerators; ignite the gasoline in our cars' engines; light our lights; and perform countless other invaluable tasks.

We'll start with a look at a common source of moving electrical charge, the battery, and then we'll move on to a discussion of some of the effects of moving charges.

THE BATTERY

There are many different kinds of batteries in use today. The most common type is usually referred to as a "flashlight battery," although its uses far exceed those implied by its name. These batteries are produced in many shapes and sizes, but they all work in basically the same way. Figure 13–1 is a diagram of the interior of such a battery. The case of the battery is made of zinc and holds ammonium chloride, a pastelike chemical. Inserted into the center of the ammonium chloride is a carbon rod.

When these materials are assembled in this fashion, two chemical reactions take place; one occurs at the zinc case, the other at the carbon rod. We will consider each of these in turn. At the case, zinc atoms leave the surface to enter the ammonium chloride solution, where they combine with chlorine. (The chlorine atoms are present because a small percentage of the ammonium chloride breaks up, leaving some free chlorine atoms in the solution.) The fate of the zinc is less important to us here than are the events that take place when it leaves the surface of the case. As each zinc atom is removed from the surface, it leaves behind two electrons. As more and more zinc atoms leave, more and more electrons accumulate, leaving the case with a net negative charge.

The ammonium chloride molecules that broke up supplied chlorine at the case. This chlorine became involved in a chemical reaction with the zinc, which, as we saw, left the case *negative*. The remainder of the molecule eventually

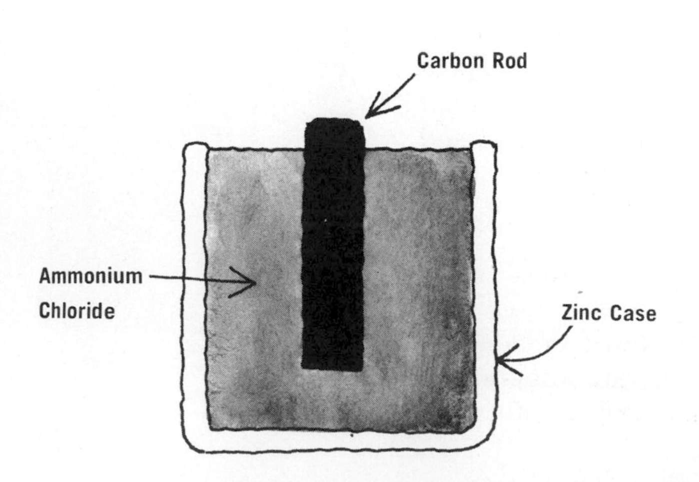

FIGURE 13–1. The principal parts of a flashlight battery.

makes its way to the carbon rod. When this remnant originally broke free of the chlorine, the chlorine took two extra electrons with it. The remainder of the molecule then shows up at the carbon rod, short two electrons. This deficiency is corrected by the carbon rod, which supplies the needed electrons, and in the process, the carbon rod ends up with a net *positive* charge.

These chemical reactions do not continue for very long. The zinc case finally achieves such a strong negative charge that the zinc atoms can no longer get away, because as each zinc atom separates from the case, it always leaves behind two electrons. Thus it leaves with a positive charge, and the electrical attraction between the negative case and positive *ions* (electrically charged atoms) finally becomes so great that no more zinc can escape. A similar saturation occurs at the carbon rod. The chemical reactions can continue only if some electrons are removed from the case and if some positive charge is removed from the carbon rod.

The Storage Battery in An Automobile

The method by which an automobile storage battery functions is in many ways similar to that of the flashlight battery and is represented in Figure 13–2. Shown are two plates immersed in sulfuric acid; one of the plates is covered by spongy lead and the other by lead dioxide. The sulfuric acid molecules break up into positively charged hydrogen ions and sulfate ions which are doubly charged (SO_4^{--}). At the lead plate, the sulfate ions react chemically with the lead to form lead sulfate, with electrons left over to leave this plate with a net negative charge. At the lead dioxide plate, another chemical reaction occurs simultaneously. This reaction is between the lead dioxide, the sulfuric acid, and the hydrogen. The products of the reaction are lead sulfate,

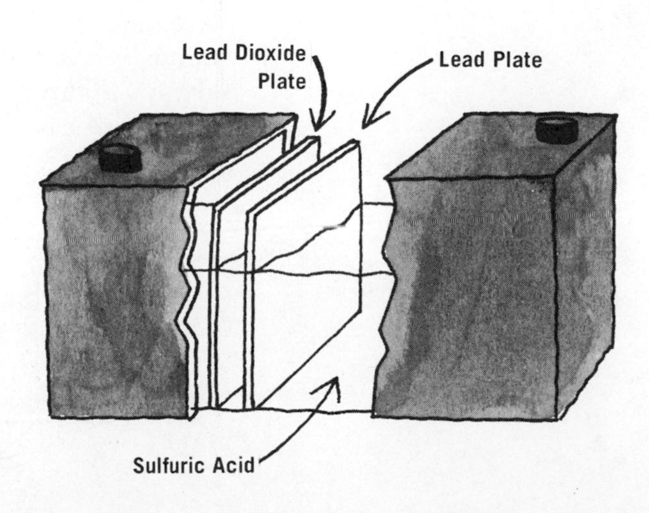

FIGURE 13–2. The automobile storage battery.

which stays on the plate, and water, which goes into solution with the sulfuric acid. Extra electrons must be supplied by the lead dioxide plate to trigger the reaction, and thus, the plate is left with a net positive charge.

As is true in all cases, the law of conservation of energy applies to batteries. As the battery is used to start the car and to supply power to its radio and lights, the chemical energy stored in the battery is changed to electrical energy. If your car does not start easily, you may find that the energy stored in the battery is insufficient—the battery runs down. This happens when the chemical reaction in the battery can no longer proceed. At both plates in the battery, lead sulfate is formed, and after a short time it completely covers both plates. When this occurs, the plates are chemically alike, and the chemical reactions can no longer take place. If, however, you are fortunate enough to get the clunker started just before the battery dies, the generator of the car takes over and charges the battery again, using energy from the burning of gasoline. In the recharging, the chemical reaction in the battery is reversed, and the lead and lead dioxide plates are restored to their original form; the battery is recharged. Because of this ability to recharge, car batteries are correctly called storage cells; they store *energy*.

ELECTRIC CURRENT

Figure 13–3*A* depicts a flashlight battery in which the case has received a negative charge and the carbon rod a positive charge. When one end of a metal wire is touched to the case and the other end to the rod (Fig. 13–3*B*), electrons from the case can move through the wire to the positive rod. As electrons leave the case, its negative charge begins to decrease, and upon arrival at the rod, each electron cancels a positive charge there. This simultaneous decrease in the charge at the case and rod allows the chemical reactions inside the battery to start again, and thus the supply of negative and positive charges is replenished almost instantaneously. As long as the chemical reactions occur, there is always a supply of electrons available to move through the wire from case to rod.

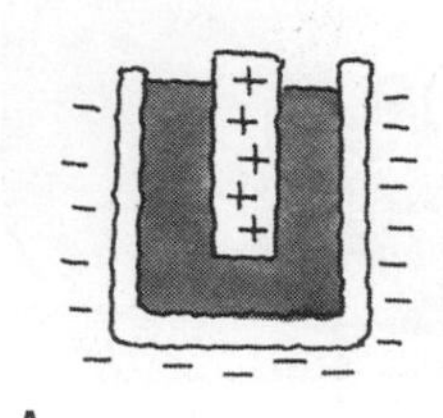

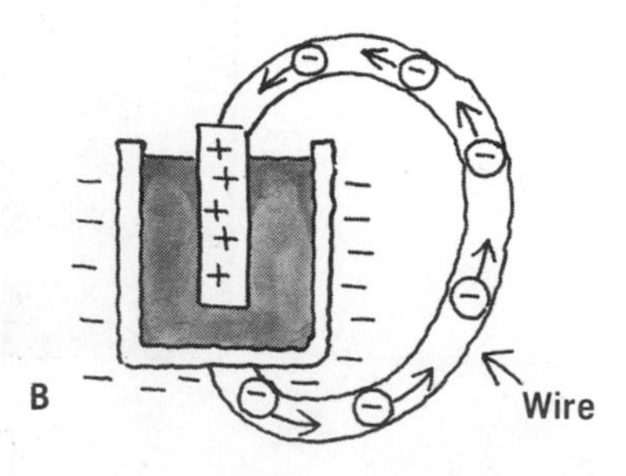

FIGURE 13–3. *A*, Flashlight battery with its charge. *B*, Wire connecting the positive end to the negative end.

FIGURE 13-4. For every ship-deserter who enters the lifeboat on the right, one falls off on the left.

To appreciate some of the details of the movement of electrons in a wire, consider the analogy illustrated by Figure 13-4. Shown are passengers abandoning a sinking ship. The last lifeboat is, unfortunately, filled, but the passengers continue to enter. As one enters from the right, an unfortunate soul must be pushed out on the left. The one going overboard is not the same as the one entering, but if more and more people enter on the right this new boarder will also, eventually, be pushed out. This may take a long, long while, depending on the rate at which people enter the boat. The same procedure applies to the electrons moving through a wire connected to a battery. An electron enters the wire at the negative case at the same time that one leaves the wire at the positive rod, but they are different electrons. An individual electron actually moves through the rod at a very slow speed, which may be as low as a fraction of a centimeter per second.

This flow of electrons through a wire is called an *electric current* and is analogous to drops of water flowing through a pipe. It is the battery that causes the motion of the electrons in much the same way that a pump forces water to move through a pipe. It should be noted that a battery causes electrons to move in one direction only; that is, from the negative case to the positive rod. A current that always travels in the same direction is called a *direct current*; in a subsequent chapter, we will find that there are devices which cause electrons to flow first in one direction, then reverse and flow the opposite way through a wire. A current which alternates in direction of flow is appropriately called an *alternating current*. The current that flows from a wall outlet is of this type.

take-along-do-it

The use of chemical reactions to produce electric current, as in a battery, can be illustrated with a lemon or orange, a piece of zinc, and a piece of copper. The zinc will be the most difficult item to locate, but the case of a discarded flashlight battery is a likely source. The lemon should be rolled

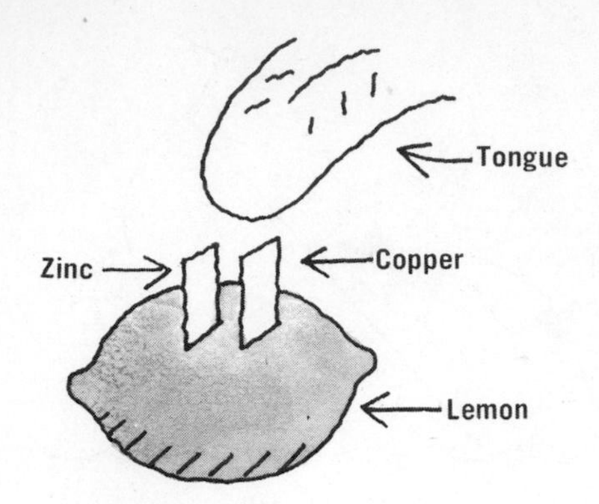

FIGURE 13–5. Touch your tongue to both metals at the same time.

until it becomes mushy. Then push the zinc and copper strips into the lemon, being sure that the strips do not touch inside the fruit.

Chemical reactions at the metal strips produce charge separation just as the battery does. To cause a current to flow and at the same time to detect it, touch your tongue between the strips as shown in Figure 13-5. The moisture on your tongue provides a path for electron flow just as a wire would. The tingling sensation on your tongue is a result of the current.

Andre-Marie Ampere (1775-1836)

Ampere grew up in France during the days of the French Revolution and the Reign of Terror and received little, if any, formal schooling. His father, a retired merchant, and the French "encyclopedie" served as his teachers. His unhappy life included the following: when he was 18 his father was guillotined; his first wife died after five years of marriage; and his second marriage, two years later, failed.

He is known today primarily for establishing the relationship between electricity and magnetism, and is honored by having his name serve as the worldwide unit of electric current.

Imagine that you can shrink to a size small enough to be able to enter a wire and watch the electrons flow past a point as in Figure 13–6. You observe a stream of electrons flowing past you, headed from the negative terminal of the battery to the positive terminal. *The amount of charge flowing past a given point in the wire in one second is a measure of the current in the wire.* Electric currents are measured in *amperes* (usually shortened to "amps"). If you observe that 6.25×10^{18} electrons move past you each second, we say that the current is one amp; half that number—half an amp; twice that number—two amps.

CURRENT AND VOLTAGE

An electron always moves from the negative terminal of a battery, through the wire, to the positive terminal. We

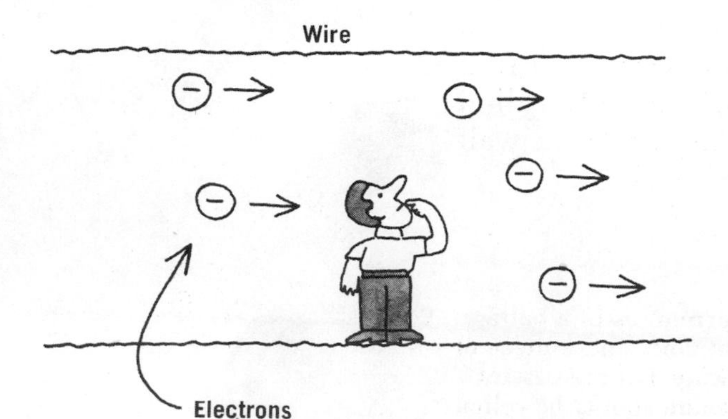

FIGURE 13–6. You are inside a wire, watching electrons move by.

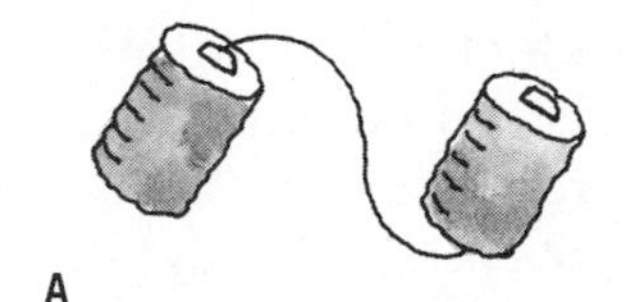

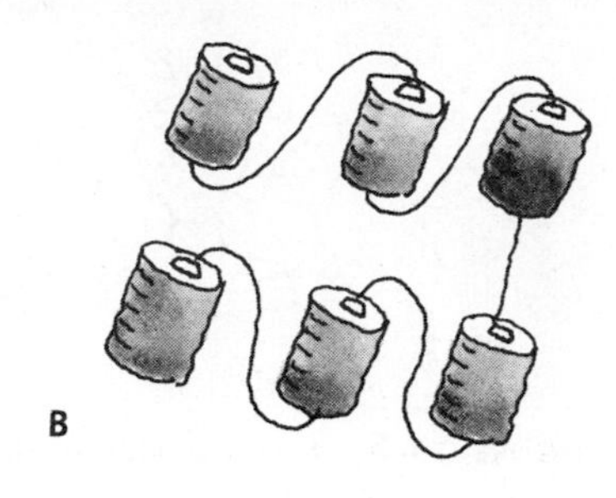

FIGURE 13–7. *A*, 2 × 1.5 = 3 volts. In *B*, 6 × 1.5 results in a total of 9 volts.

recall from the last chapter that an object with a negative charge feels a force away from a point of low voltage and toward a point of high voltage. An electron is at a position of low voltage when it is at the negative case and at a higher voltage when it reaches the carbon rod. We can express this observation by saying that there is a difference in voltage between these two points: for a flashlight battery, this difference is 1.5 volts. Regardless of how large we make the zinc case or the carbon rod, and regardless of how much ammonium chloride we dump in, the voltage difference can never be made greater than 1.5 volts. If we increase the physical size of the battery, it will last longer, but the chemical reactions set a maximum of 1.5 volts on the voltage the flashlight battery will produce.

This small voltage is not great enough to allow many of our electrical devices to operate properly, but more voltage can be obtained by connecting together two batteries as shown in Figure 13–7*A*. Two batteries connected in this fashion will produce three volts. (This is typical for flashlights.) The connection in Figure 13–7*B* produces nine volts. Note that the batteries are connected so that the positive side of one is joined to the negative side of the next. Batteries connected in a continuous line like this are said to be in *series*.

It is a common mistake when replacing batteries in a device to connect them as shown in Figure 13–8, with the positive terminals in contact and the batteries opposing one another. An electron attempting to flow through the wire joining the two negative terminals receives equal pushes from both batteries. One of the batteries tries to make current flow clockwise in the figure, the other counterclockwise: the result is no current at all.

RESISTANCE

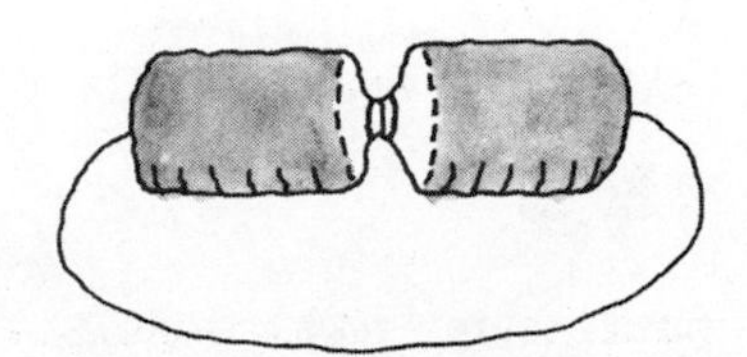

FIGURE 13–8. Two batteries connected in opposition: no current flows.

In the discussion of electric currents, we investigated what would happen if a piece of wire were connected from the negative end of a battery to the positive end. This ex-

FIGURE 13–9. A simple electric circuit.

ample serves quite well to define electric current, but simply connecting a wire in this fashion has no practical use. In practice, we connect other devices along with the wire to form an electric circuit. Figure 13–9 shows one of the simplest useful circuits.

Electrons leaving the negative side of the battery possess potential energy, which is converted to heat and light inside the light bulb. The light bulb is able to perform this conversion because of its *resistance.* Every material, even the best conductor, offers some resistance to the flow of electrons through it. The cause of this resistance can be explained in terms of the atomic theory of matter. This theory states that all materials are composed of atoms, which in a solid such as wire are fixed in place: they can vibrate about their locations but do not wander about inside the material. If an electron leaves the negative terminal of the battery to move through the wire, it is moving from a point of high potential energy to a point of low potential energy. As it moves, its speed gradually increases, and it gains kinetic energy. Periodically, however, it collides with the atoms inside the wire, and when one of these collisions occurs, the electron is slowed down. The kinetic energy lost by the electron has to go somewhere, and it is the struck atom that picks up this energy. The extra energy causes the atom to vibrate more (with larger amplitude) than before. After the collision, the push of the negative charges on the battery speeds up the electron again, only for it to collide with another atom. *It is this hindrance to the motion of electrons that we call the resistance of the material.*

As a current flows in a wire, the collisions between electrons and atoms increase the amplitude of vibration of the atoms. Thus, inside a material that has resistance, the electrical energy of the charge is being transformed into heat. If the material gets hot enough, it will glow, first with a dull red light, and then as it gets hotter, with a white light.

The Light Bulb

Figure 13–10 illustrates how the resistance of a material can be put to a useful purpose in a light bulb. Two wires support a fine coil of tungsten wire (the filament), which is responsible for the emission of light. A current will flow through the bulb when connections from a battery are made at points A and B on the base of the bulb. When a current of about one amp passes through the filament of a 100-watt light bulb, the filament reaches a temperature of about 4500° F. At this temperature, it becomes white-hot and emits light.

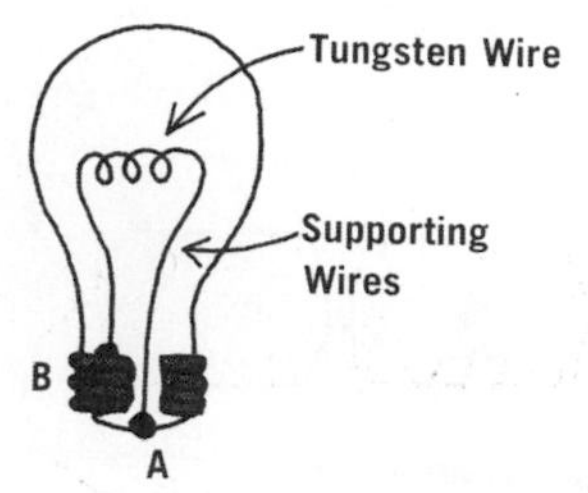

FIGURE 13–10. The fine tungsten wire glows white when it gets hot enough.

We will suggest a number of experiments which require light bulbs, flashlight batteries, and some connecting wires. Since these items are inexpensive, we suggest a visit to a hardware store to purchase three flashlight bulbs, two batteries, and a few feet of wire. (The bulbs should be designed to operate at a low voltage—about three to six volts.) Ordinary flashlight batteries are okay, but the kind having screw terminals on the top are more convenient to use. Ordinary lamp cord is fine for the wire; just be sure you strip the insulation off the ends of the wire when you make a connection, as to a battery. These items are essential, and so is a convenient way to hold your light bulb. You can buy inexpensive sockets for this purpose, but a better way is illustrated in Figure 13-11. Three uncoated metal nails are driven into a cork to hold the light bulb in the position shown. The equipment needs are summarized in Figure 13-12.

When you have all the materials assembled, try connecting the circuit shown in Figure 13-13A. Note how brightly the bulb glows, and then try connecting the circuits shown in B and C. Is there any regularity to the brightness of the bulbs for the three cases?

Resistance Causes Heat

The heat generated when current flows through a resistive material is used in many common devices. A cut-away view of the spiral heating element of an electric range is shown in Figure 13–14. The material through which the current passes is surrounded by an insulating substance in order to prevent the current from flowing through the cook when he or she touches the pan. A material that is a good conductor of heat surrounds the insulator.

In Figure 13–15, a fan blows air past heating coils. In this case, the warmed air is used to dry hair, but on a greater scale this same principle is used to dry clothes and to heat buildings. A final example of a household appliance that uses the heating effect of electric currents is the steam iron shown in Figure 13–16. A heating coil warms the bottom of the iron and simultaneously turns water into steam, which issues from jets in the bottom.

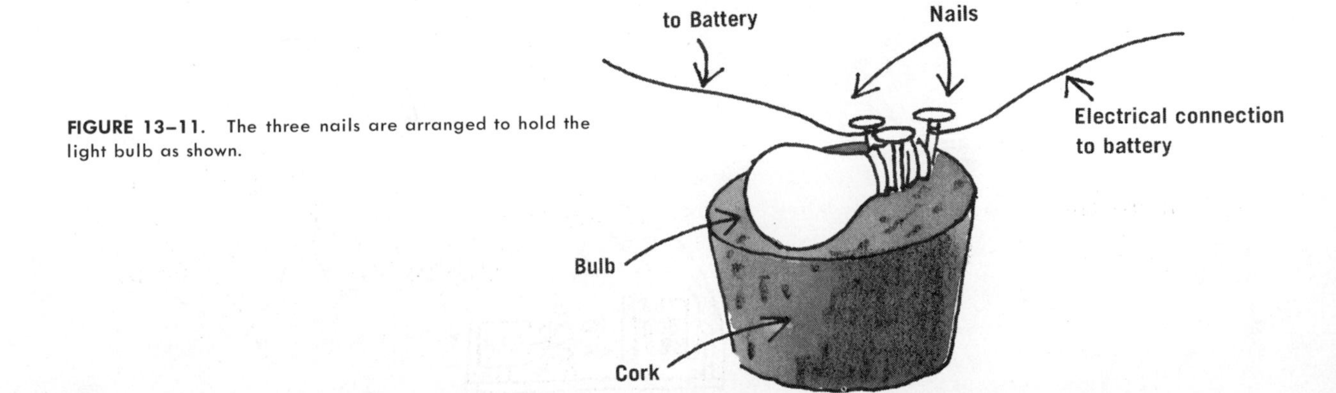

FIGURE 13–11. The three nails are arranged to hold the light bulb as shown.

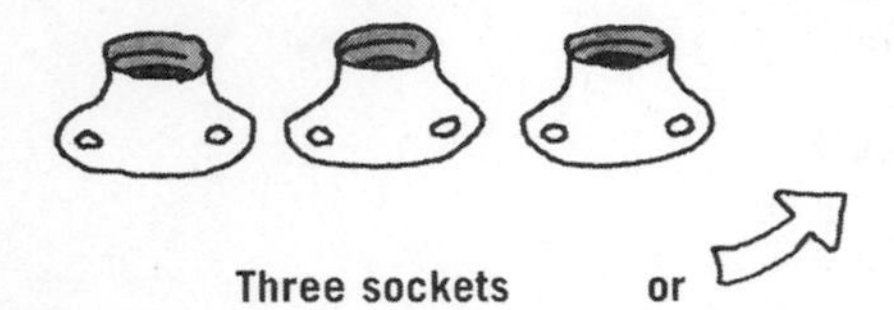

FIGURE 13–12. Equipment for this (and other) TADIs.

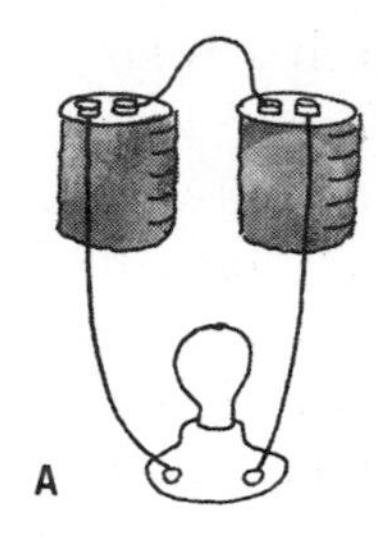

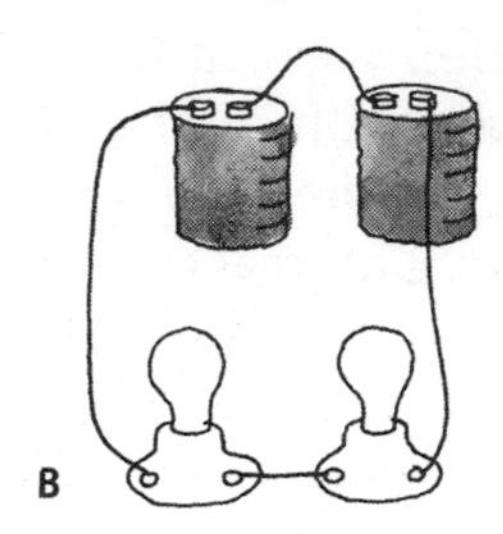

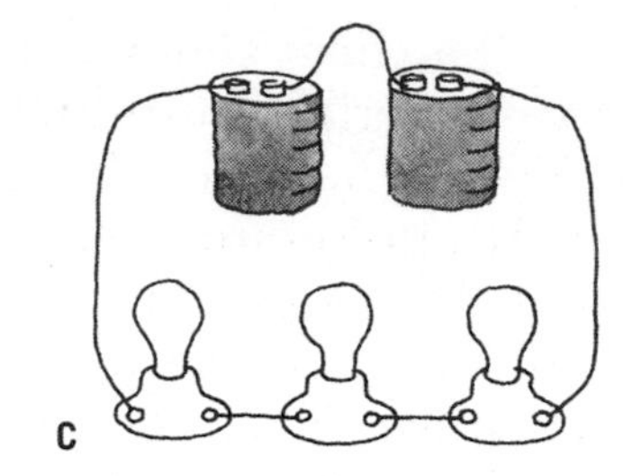

FIGURE 13–13. Three circuits to set up. In which case is the light(s) brightest? Dimmest?

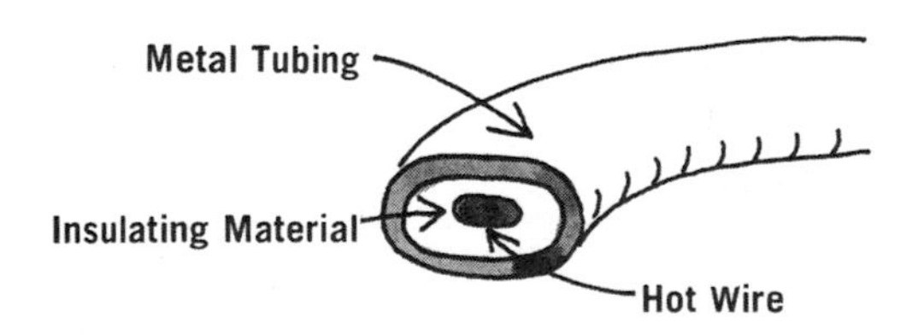

FIGURE 13–14. A cross-sectional view of an electric stove element.

FIGURE 13–15. A hot-head.

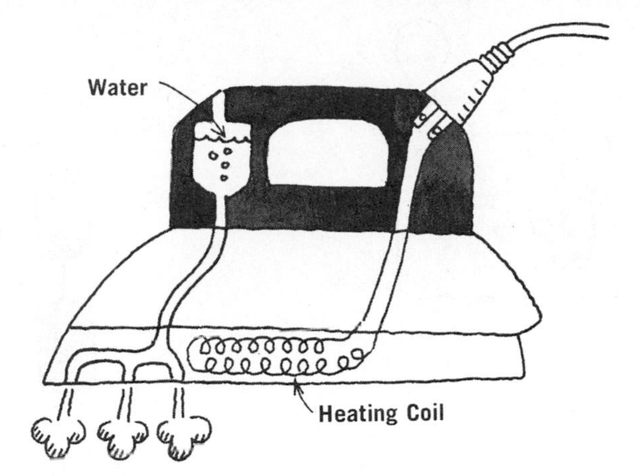

FIGURE 13–16. The coil heats not only the ironing surface but also the water.

Resistance and Current

take-along-do-it

Arrange a circuit as shown in Figure 13-17A and use it to test the resistance of a number of objects. First connect wires A and B directly together and notice how brightly the bulb glows. Now disconnect the two wires and randomly place a variety of objects, metallic and nonmetallic, between A and B (Fig. 13-17B). In each case, observe how brightly the bulb glows.

In the preceding TADI you should have noticed that the brightness of the bulb changed as different objects were inserted into the circuit, but in no instance did it glow more brightly than when the wires were connected together. Any decrease in brightness was caused by a decrease in the current going through the bulb. In general, the more resistance that you place in a circuit, the more difficult it becomes for current to flow, and subsequently the current flow is decreased. As an analogy, consider the water flowing through the pipes in a house. If the pipes are new, the water flows freely, but with age, a scale may form inside the pipes. We can say that the resistance of the pipes increases, and if the scale becomes thick enough, the water flow could be reduced to a trickle.

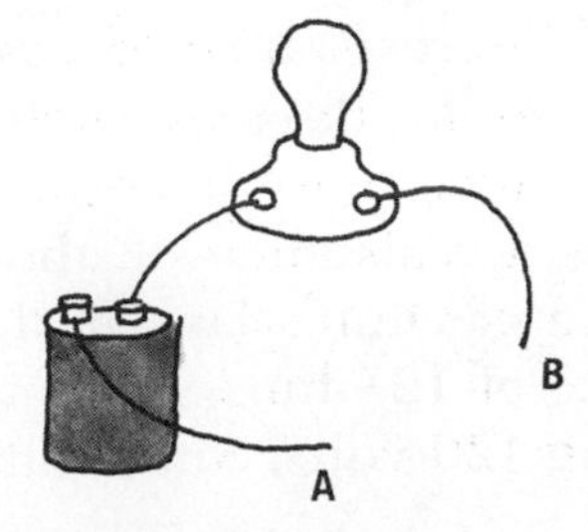

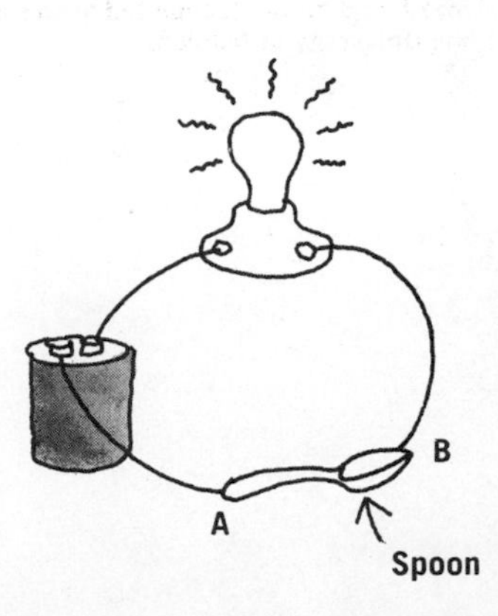

FIGURE 13–17. A, A simple resistance tester. In B, we see it in action.

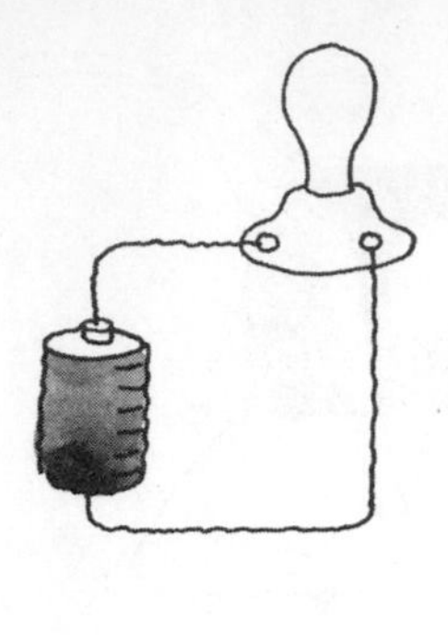

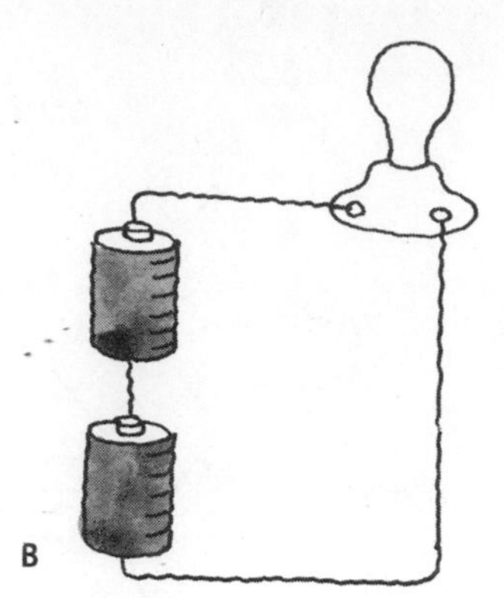

FIGURE 13–18. The second battery (shown in *B*) causes the bulb to burn more brightly.

take-along-do-it

As another example of the ways in which one can change the current in a circuit, try the set-ups shown in Figure 13-18. Note the brightness of the bulbs when the circuit is that of A, and then note that the brightness is increased when the second battery is added, as in B.

Georg Simon Ohm (1787-1854)

Ohm made his only first-rate contribution to science for an interesting reason: he had a position as a high school teacher and wanted to teach at a university. He thus sought to make some research discovery and chose the field of electricity. Because of his lack of wealth, he made his own equipment, and in using his home-made wires (his father was a mechanic) he discovered the relationship between current, voltage, and resistance. However, he did not get the university appointment and in fact was later forced (because of criticism of his teaching) to resign his high school job. After he lived for six years in poverty, the importance of "Ohm's law" was recognized, and he was appointed professor at the University of Munich.

This TADI has indicated to us another factor affecting the current in a circuit: voltage. Since it is the voltage of the battery which supplies the push to cause the current in the first place, an increase in voltage should produce an increase in current. In 1826, Georg Ohm (pronounced *ōm*) related the several observations of this section in the form of a law, appropriately called Ohm's law, which is expressed as follows:

$$\text{Current} = \frac{\text{Voltage}}{\text{Resistance}}$$

From this relationship, we can easily see that increasing the voltage applied to a circuit increases the current in it, and increasing the resistance of a circuit will produce a decrease in the current.

Ohm's law also provides a convenient method for illustrating how the resistance of a device is measured. If a one-volt battery produces a one-amp current, Ohm's law indicates that the resistance must be one unit; the unit is called an ohm. The resistance of a 100-watt light bulb is about 150 ohms, whereas devices such as an iron or toaster, which has to produce more heat—and therefore requires more current—has a resistance of about 10 ohms. As an application of the use of Ohm's law, find the current in a toaster with a resistance of 12 ohms when it is connected to a wall outlet supplying 120 volts. Substitution gives:

$$\text{Current} = \frac{120 \text{ volts}}{12 \text{ ohms}} = 10 \text{ amps}$$

Now *you* determine the resistance of a light bulb that draws 1/2 amp when connected to a 120-volt source.[1]

The Carbon Microphone. The carbon microphone is an example of a device in which varying resistances and currents can be put to a useful purpose. The object is to cause an electric current to fluctuate in exactly the same way that the density in a sound wave fluctuates. Figure 13–19 is a diagram of a microphone in the mouthpiece of a telephone. A flexible steel diaphragm is placed in contact with carbon granules inside a container. These carbon granules act as the primary resistance in a circuit containing a source of current (shown here as a battery) and a transformer, whose action will be described in a later chapter. When connected as shown, a current flows. When a sound wave strikes the diaphragm, the diaphragm flexes in and out, sometimes causing the carbon to be compressed more compactly than normal, and sometimes causing the granules to be farther apart than normal. When the granules are pressed together tightly, the resistance of the circuit is low, and electric current can flow easily. As the diaphragm springs out, the granules become loosely packed, with more space between them, and it is more difficult for current to pass. These changes in the current correspond to the changes in the sound wave: a high molecular density, or condensation, at the mouthpiece produces a high current; a low density, or rarefaction, a low current. These variations in current are ultimately fed into the telephone transmission lines by the transformer. A speaker in the listener's earpiece converts these electrical signals back to a sound wave.

The carbon microphone has a very poor frequency response. It reproduces frequencies below 4000 hertz adequately and is therefore suitable for transmission of speech, since frequencies emitted in normal conversation are usually below this value. Its capabilities fall off rapidly at higher frequencies, rendering it useless for high-fidelity purposes, which require reproduction of all frequencies between 20 and 20,000 hertz.

---

[1]Answer: 240 ohms.

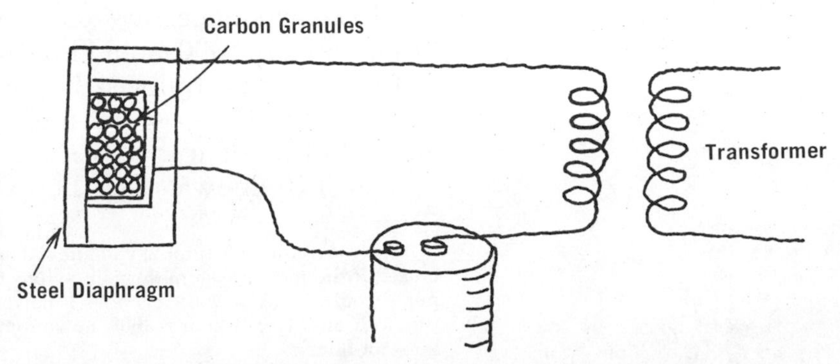

FIGURE 13–19. The carbon granules are compressed and released by the sound wave, which causes their resistance to change. Thus the current in the wire varies with the same frequency as the sound wave.

"The air pollution is virtually gone since they switched to electric saucers."

Reprinted with the permission of Audubon/Sidney Harris.

COST OF ELECTRIC ENERGY

We have seen a number of examples in which electrical energy from a battery or other source is converted into other forms of energy, primarily heat and light. The original electric energy must be purchased. This is done in the case of the battery when we buy it, but the electrical energy supplied to our homes is paid for by monthly bills.

James Watt (1736-1819)

Watt grew up as a sickly youth in Scotland. He went to England, became an instrument-maker, and returned to Scotland to work. His major accomplishment was a vast improvement in the steam engine. In measuring the power of his steam engine, he established the size of the "horsepower" as it is used today. His name lives in the metric unit of power, the watt.

The rate at which a device converts energy to some other form is called power and is expressed in watts. For example, a 200-watt light bulb converts electrical energy into heat and light at twice the rate of a 100-watt light bulb. Appliances that produce large amounts of heat, such as clothes dryers, ovens, water heaters, or electric furnaces, require large amounts of electric energy and therefore are by far the most expensive devices to operate in our homes. What we actually pay for is the amount of energy we use; this is measured in kilowatt-hours (kWh).[2] As an example of how the cost is calculated, consider the following problem: how much does it cost to burn a 100-watt light bulb for 24 hours if electricity costs five cents per kilowatt-hour?

It is first necessary to convert 100 watts to kilowatts. "Kilo" means 1000, so a kilowatt is the same as 1000 watts. A 100-watt light bulb is then a 0.10 kilowatt bulb. To find the amount of energy we must pay for, we multiply by the amount of hours it is used:

$$0.10 \text{ kilowatt} \times 24 \text{ hours} = 2.4 \text{ kilowatt-hours}$$

---

[2]If you studied Chapter 2, you know that energy is the product of force and distance. In the metric system this is newtons times meters. A watt is a newton-meter per second. Thus 1 watt-second = 1 newton-meter (usually called a "joule" in physics), and 1 watt-hour = 3600 newton-meters. How many newton-meters is a kilowatt-hour?

If energy is purchased at five cents per kilowatt-hour, the cost to us is

$$2.4\ \text{kWh} \times \$.05/\text{kWh} = \$.12$$

Twelve cents will keep the light bulb going for a day. This is a small amount, but as devices become larger and more complex, the costs go up rapidly. Demands on our energy supplies have made it necessary to be aware of the energy requirements of our electrical instruments. This is true not only because they are becoming more expensive to operate but also because with the dwindling of the coal and oil resources which ultimately supply us with electrical energy, increased awareness of conservation becomes necessary. Electrical appliances have the necessary information on them to allow a calculation of their power (or "wattage") requirements, and hence operating costs can be calculated before purchasing them. The wattage is often stated directly on the device, as on a light bulb. In other cases, the amount of current used by the device and the voltage at which it operates are given, and this is sufficient to allow us to calculate the cost.

To see why current and voltage are sufficient to calculate power requirements, consider the meaning of each term. Current tells us how much charge passes through a device each second. Voltage tells us how much energy each of these charges gives up to the device. So the product of current and voltage gives the total amount of energy supplied to the device each second, or its power

$$\text{Power} = \text{Current} \times \text{Voltage}$$

As an example of the use of this equation, consider a television set which draws three amps of current when plugged into a 110-volt wall outlet.

$$\text{Power} = 3\ \text{amps} \times 110\ \text{volts} = 330\ \text{watts}$$

The steps in calculating the cost proceed from here as in the previous example. Carry them out to show that you can watch a complete World Series lasting 21 hours for about 35 cents. (Again, assume that the cost of a kilowatt-hour is five cents.)

CIRCUITS AND SWITCHES

Series Circuits

In a series circuit, connections are made such that there is only one path that electrons can take in their journey from the negative to the positive side of a battery. A typical series

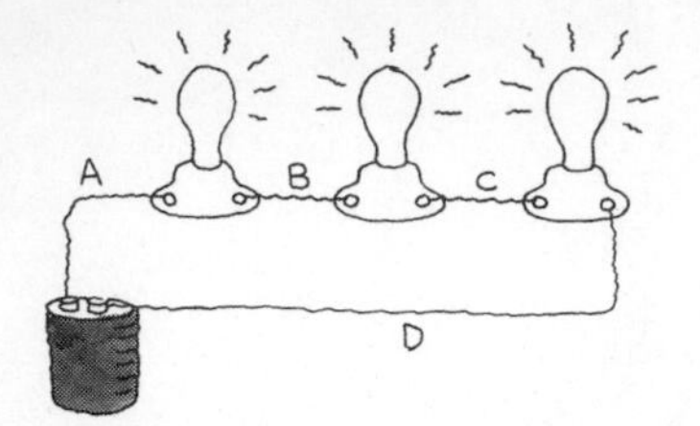

FIGURE 13–20. A series circuit. An electron must pass through all three lamps to get from the negative to the positive side of the battery.

circuit consisting of three light bulbs connected to a battery is shown in Figure 13–20.

A characteristic of all series circuits is that the *current is the same at all points*. For example, if we find a current of 0.10 amp at point A in the circuit, we would also find 0.10 amp flowing at points B, C, and D. If the current should turn out to be different at various points in the circuit, it would imply that electrons were either appearing or disappearing as if by magic in the circuit. This is not the case; electrons are supplied at the negative terminal of the battery by a chemical reaction, and since charge is never created or destroyed, they eventually make their way to the positive terminal.

The charges flowing in the circuit are given a certain amount of energy by the battery. This energy is distributed among the light bulbs, with the bulb having the highest resistance getting most of it. Since the light bulbs must share the energy of the charges passing through them, they will glow less brightly when connected in series than when each one is connected individually to the same battery.

A disadvantage of such a circuit is that when one of the bulbs burns out, all of them go out. A light bulb burns out when the filament breaks. When this breakage occurs, there is no longer a path for the electrons to travel, and so the current stops. Some Christmas tree light sets are connected in this way, and the agonizing experience of determining which is the culprit that is burned out is a familiar one. Frustrating experiences such as this illustrate how inconvenient it would be to have all appliances in an entire house connected in series.

take-along-do-it

To experiment with series circuits, connect the circuit shown in Figure 13-21A. Note how brightly the bulb glows, and then connect, successively, the circuits in B and C. You should notice the bulbs growing dimmer in each successive case. While you have the bulbs connected as in part C, disconnect one of them to indicate what happens in the circuit when a bulb burns out.

Parallel Circuits

Another common circuit is shown in Figure 13–22. In this arrangement, three bulbs are connected such that there are three different paths that electrons can take in their travels from the negative to the positive side of the battery. *Any circuit in which there is more than one path for current is called a parallel circuit.* When some amount of current leaves the battery, it divides at the junction, but it does not necessarily divide equally. An old adage says that people,

FIGURE 13–21. Three circuits for you to set up. In which is the bulb (or bulbs) brightest? Dimmest?

particularly politicians, follow the path of least resistance; the same is at least partially true for electrons. Most of the current goes by the path which has the lowest resistance (in this case, electrical resistance).

Regardless of the amount of current flowing through any one bulb, each electron passing through it has all the energy that it had when it left the battery. This energy is given over to the light bulb. It is a characteristic, then, that each device – in our case each light bulb – is supplied by the same voltage as if it were alone in the circuit. Moreover, if one of the bulbs burns out, there are still two paths left for current to take, and thus the other two light bulbs would continue to operate as if nothing had happened. This makes such a circuit ideal for home usage.

Figure 13–23*A* shows a typical home circuit. The two wires are connected to the utility lines of the power company such that 110 volts is maintained between them. The current supplied is alternating current, but all the ideas developed thus far in our discussion remain valid. In Figure 13–23*B*, a light bulb has been connected between the wires. Let's suppose that the current is one amp. In *C*, a radio has been added. The light bulb and the radio are in parallel: either can be off, on, or burned out without affecting the other. When the radio is turned on, let's suppose that three amps flow through it. Three amps for the set plus one for the bulb gives four amps being drawn through the lead-in wires. In *D*, we add another device, a heater requiring 15 amps to operate. With the addition of this device, a total of 19 amps is passing through the home wiring, a dangerously high level for these wires to carry. They have some resistance, which causes them to heat up, and the danger of overheating to the point of producing a fire is evident. In order to insure that the current drawn cannot go beyond safe limits, a 20-amp fuse is inserted in the circuit (Fig. 13–23*E*). Fuses are protective devices which allow currents to pass through them

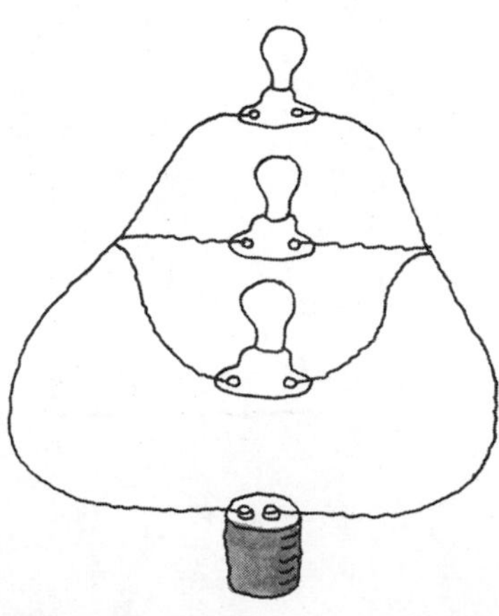

FIGURE 13–22. In a parallel circuit, an electron has a choice of paths to travel from (–) to (+).

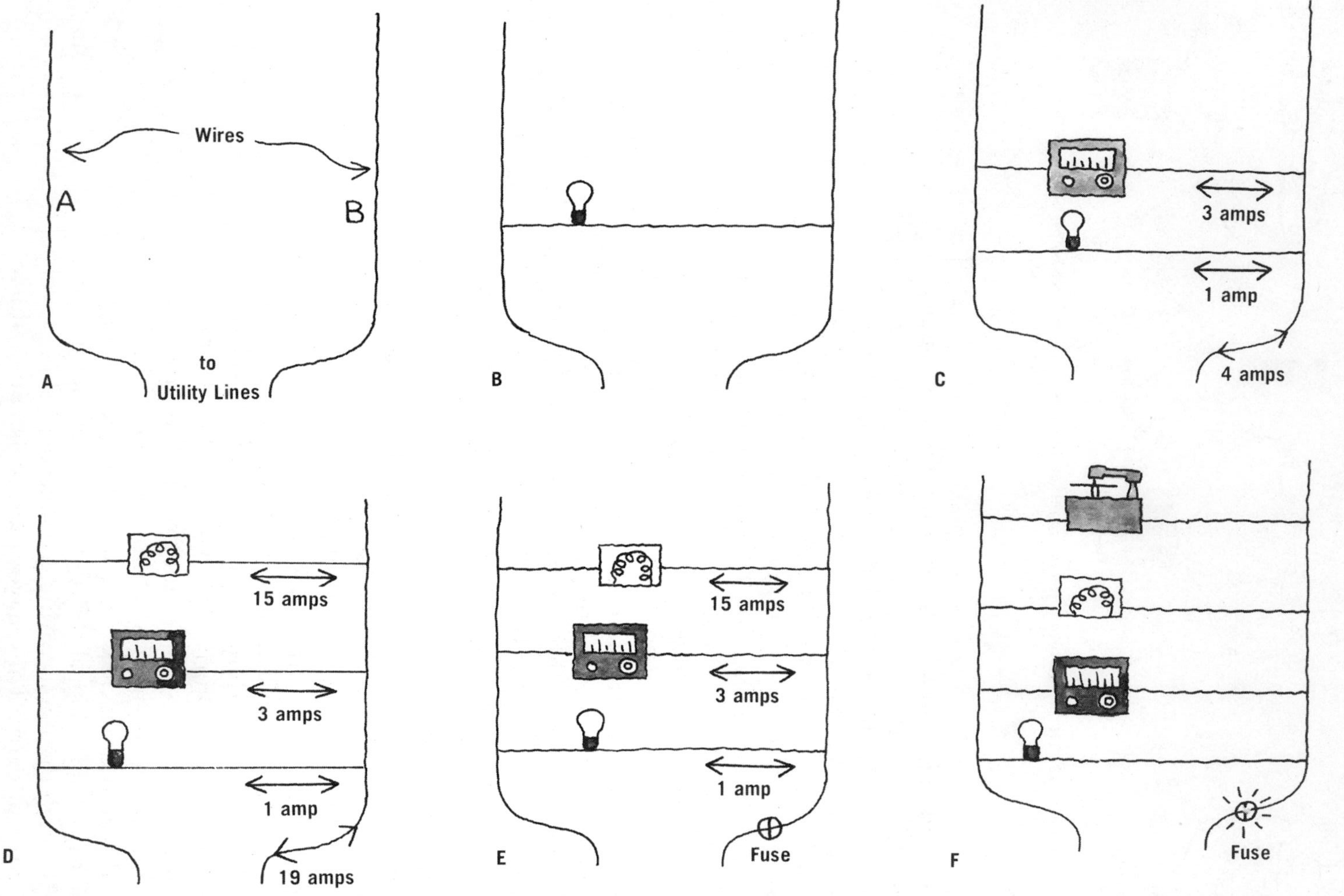

FIGURE 13–22. The basic uses of the circuit in a home. Note that the addition of each device increases the total current until the 20 amp fuse is blown.

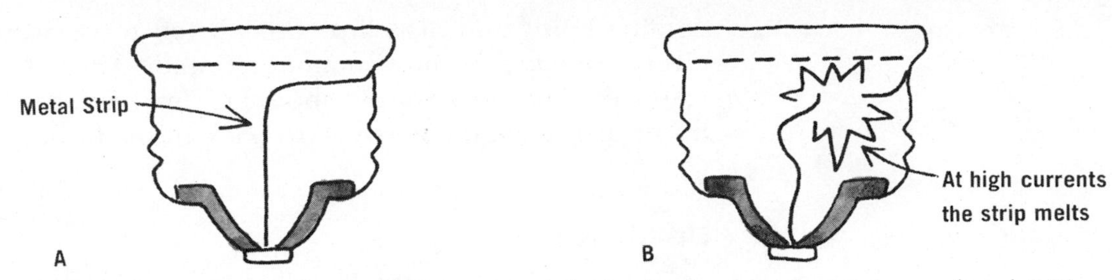

FIGURE 13–24. A, A fuse. The metal strip is made such that it melts when a certain current passes through it (B).

only up to a certain level, 20 amps in this case. Beyond this point, they "blow," and all current stops. In *F*, we add the straw that breaks the camel's back. We turn on a stereo requiring two amps to operate, thereby raising the total current drawn to 21 amps. The 20-amp fuse blows, and the current to all the devices stops.

A cut-away view of a fuse is shown in Figure 13–24. The current passing through the fuse flows through a small metallic strip that has a low melting point. The current heats the strip, and if the current rises high enough, it causes the strip to melt. By suitable designing of the size of the strip, a wide variety of fuses can be made that will tolerate vastly different amounts of current before they blow.

In most new homes, a circuit breaker is used instead of a fuse. (See Figure 13–25.) In this device, current passes through a bimetallic strip, which bends as the current heats it (see Chapter 8 for a discussion of bimetallic strips). If it bends far enough to the left, the strip settles into a groove in the spring-loaded metal strip. When this occurs, the metal strip drops downward sufficiently to open the circuit at the

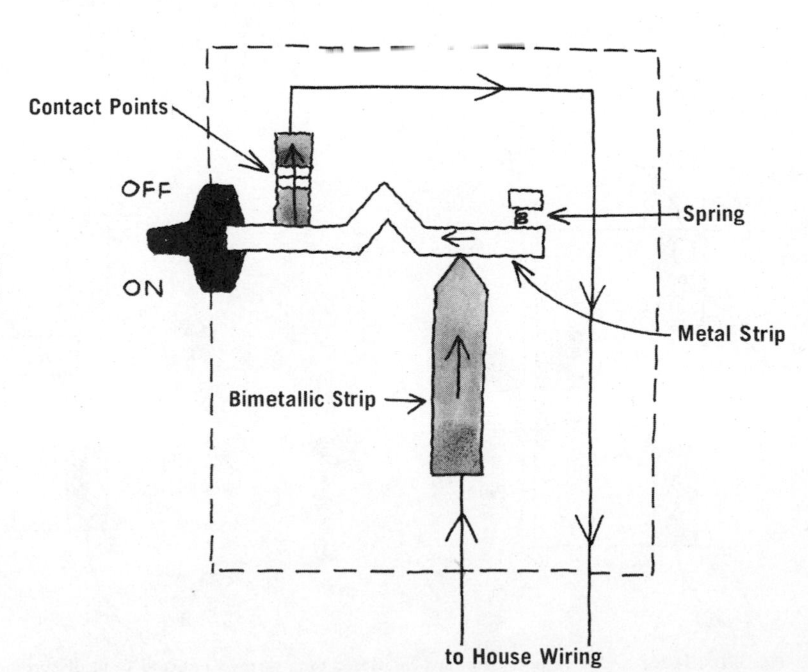

FIGURE 13–25. A circuit breaker can be used instead of a fuse. It needn't be replaced when overloaded. (Arrows indicate path of current.)

contact point and also flips the switch to indicate that the circuit breaker is now nonoperational. After the overload (the stereo in our example above) is removed, the switch can be flipped back on, and you're back in business.

Electric Switches

In order for electrons to move through a circuit, there must be a continuous conducting path for them to follow. The progression of electrons through a wire is somewhat similar to traffic flow along a busy highway. If someone opens a drawbridge, the traffic stops. The switch in an electric circuit produces much the same effect when it is opened, except in this case, electrons stop on both sides of the "drawbridge."

A typical home wall switch is illustrated in Figure 13–26. When in the position shown in *A*, the current path is continuous through the switch, and current flows along the path indicated by the arrows. Moving the lever down, as in *B*, breaks the contacts at A and B and the current no longer has a path along which to flow.

A mercury switch, often used in thermostats, is shown in Figure 13–27. In the position shown in *A*, current flows along the path indicated by the arrow. Mercury is a good conductor of electricity, and when the metal contacts are immersed in it, a continuous path for current flow is produced. When the switch is rotated to the position in Figure 13–27*B*, the liquid mercury moves away from the contacts, and a continuous path is no longer formed.

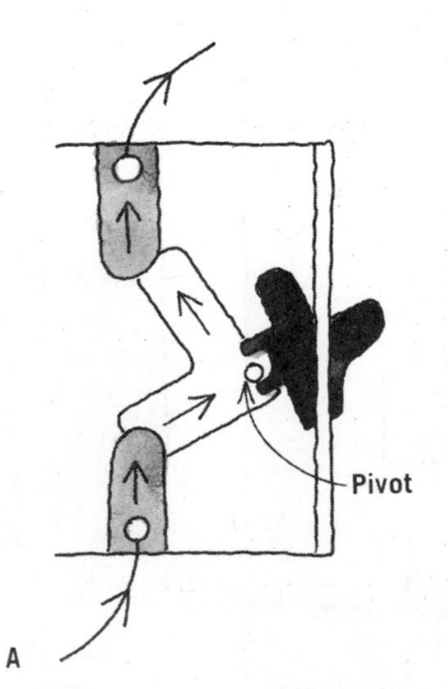

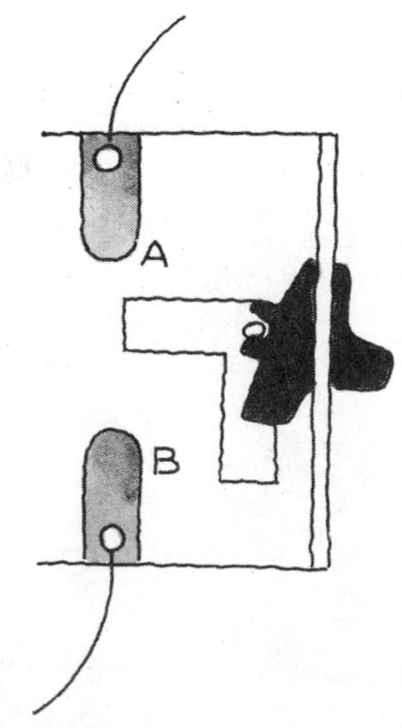

FIGURE 13–26. *A*, A regular wall switch, closed (on). Arrows indicate path of current. In *B*, the switch is open, and no current flows.

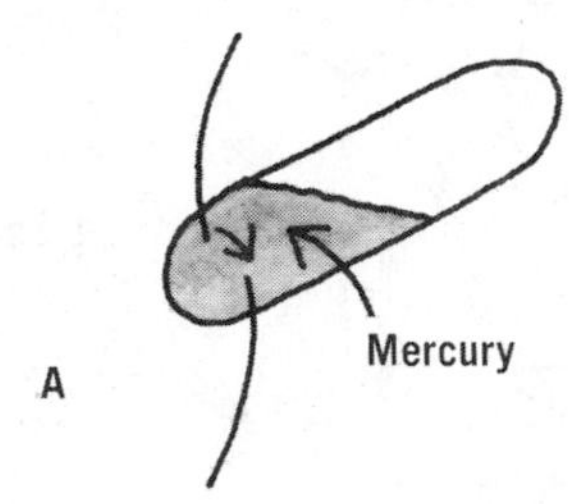

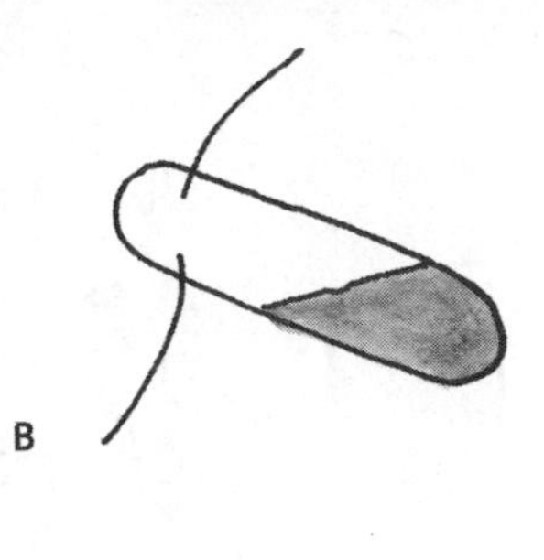

FIGURE 13–27. The heart of a mercury switch. *A*, Closed; *B*, Open.

Photoelectric Circuits

The previous section dealt with common electrical switches. When closed they allow current to flow; when open they prevent it from flowing. Another kind of switch is shown in Figure 13–28. This one employs the photoelectric effect. The positive side of the battery is connected to a wire, and the negative terminal is connected to a cesium-coated metal[3] plate, which emits electrons by the photoelectric effect when light shines on it. This wire-and-plate assembly is contained within a glass cover from which the air has been removed. Such an arrangement is usually called a phototube.

With the circuit connected as in Figure 13–28*A*, the phototube acts as an open switch. There is no pathway for electrons to get from the negative to the positive side of the battery, because there is no way for them to escape from the negative plate and jump across the gap. When light shines on the cesium surface, however, electrons are kicked out by the light. As Figure 13–28*B* shows, electrons are now able to get across the gap, and the phototube now acts as a closed switch. As long as light shines on the tube, a current exists

[3]Cesium is used here rather than zinc, which was used in the example in Chapter 12, because zinc requires ultraviolet light for photoelectric emission, while cesium works with the less energetic photons of visible light.

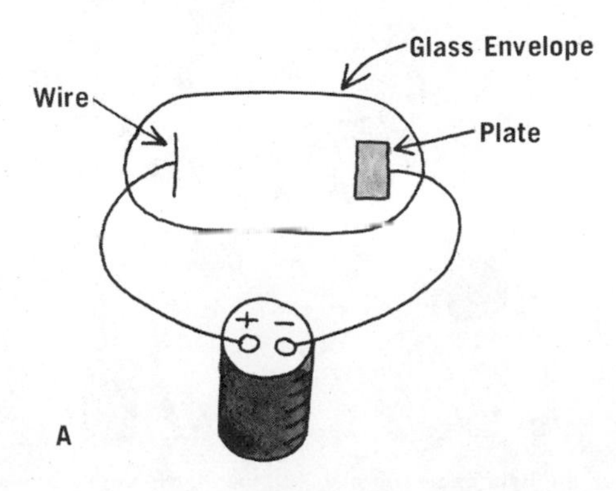

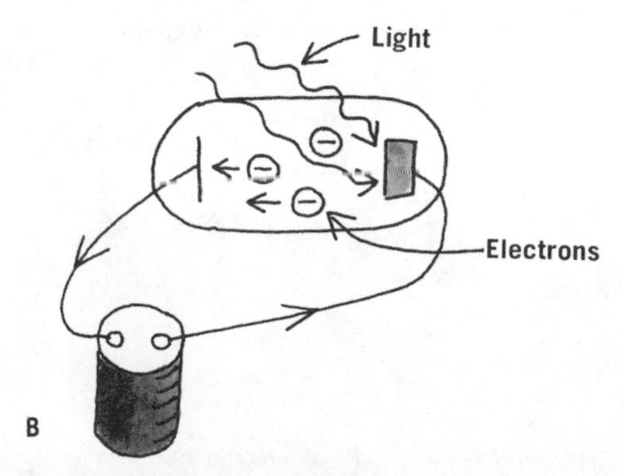

FIGURE 13–28. When light shines on the plate of the phototube, current flows.

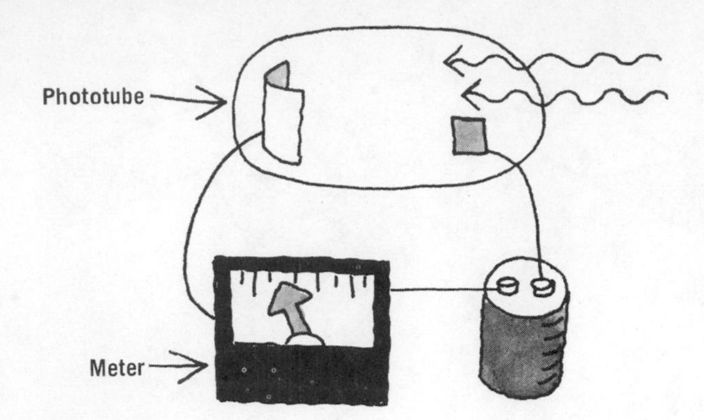

FIGURE 13–29. A light meter is actually a phototube and an electric current meter.

in the circuit. We will now investigate three ways in which the photoelectric effect is used.

Light Meters. An obvious example of the use of the photoelectric effect is the light meter. As its name indicates, its principal use is to measure the intensity of light falling on it. A common application is in photography, because the photographer must know the illumination of an object in order to set his camera for the proper exposure. In operation, light reflected from the object to be photographed strikes a photoelectric surface, causing it to emit electrons, which then pass through a very sensitive current meter. The strength of the current depends on the strength of the light. The circuit for a light meter is shown in Figure 13–29.

Burglar Alarms. A second example of the use of the photoelectric effect is the burglar alarm. These devices sometimes use ultraviolet rather than visible light in order to make the presence of the beam less obvious. In operation, a beam of light passes from the source to a photosensitive surface. The electrons flying from the surface to the collector plate complete the circuit and act as an electric current. The current passes through an electromagnet, which attracts a metal rod (Fig. 13–30A). As long as the rod is in the posi-

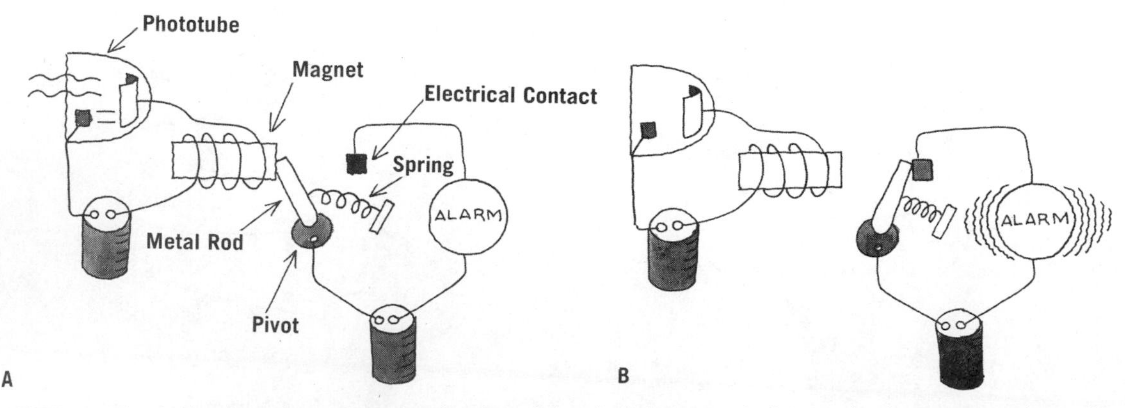

FIGURE 13–30. *A*, The phototube used as a burglar alarm. When the light beam is broken (*B*), the electromagnet releases the switch, allowing current to pass through the alarm.

tion shown in *A*, the alarm is deactivated. If an intruder breaks the light beam, the current to the electromagnet switches off, and the spring pulls the iron rod to the right (Fig. 13–30*B*). In this position, a completed circuit allows current to pass to the alarm system. A catch is included which prevents the rod from returning to the left when the burglar no longer blocks the light path.

Sound on Movie Film. In addition to visual information, a motion-picture film also carries a sound track. Microphones on stage convert sound waves into electrical signals, which cause a shutter in the movie camera to open and close, exposing a portion of the movie film along its edge. The optical pattern of light and dark lines shown in Figure 13–31 can be used to reproduce sound. In order to do this, a lens in the movie projector directs a beam of light through the sound track toward a photoelectric cell (Fig. 13–32). The variation in shading on the sound track varies the amount of light reaching the phototube. The changing amounts of light falling on the plate of the cell change the number of electrons ejected from it and vary the current in a circuit connected to it. This changing current electrically simulates the original sound wave and reproduces it in a speaker.

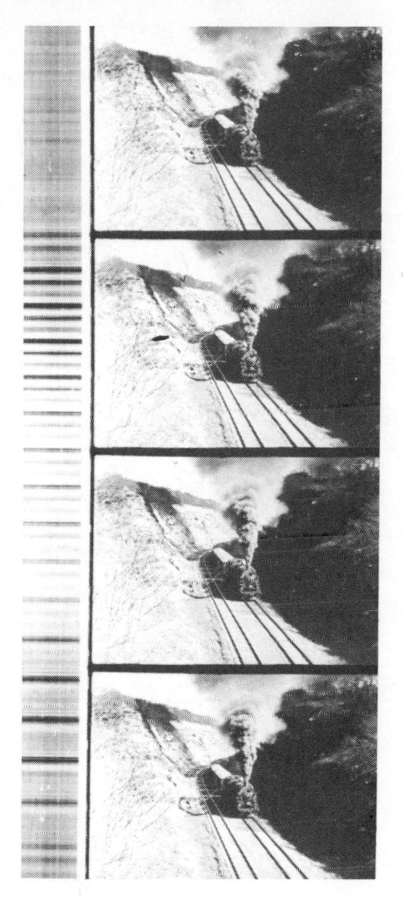

FIGURE 13–31. The sound signal for this piece of movie film is carried by the irregularly shaded strip running down the side.

TELEVISION

We end this chapter with a discussion of a device that has practically replaced the dog as man's best friend: the television set. The underlying physical principles of its operation include a multitude of phenomena from the fields of electricity and optics.

Television employs two tricks of vision, one of which is common to the movies, the other to print reproductions of photographs. Before moving to the electrical aspects of the problem, let's examine these tricks. A motion picture uses the deception that a series of still pictures, when scanned rapidly enough by the eye, is perceived as continuous motion. A theater movie projector flashes 24 different still pic-

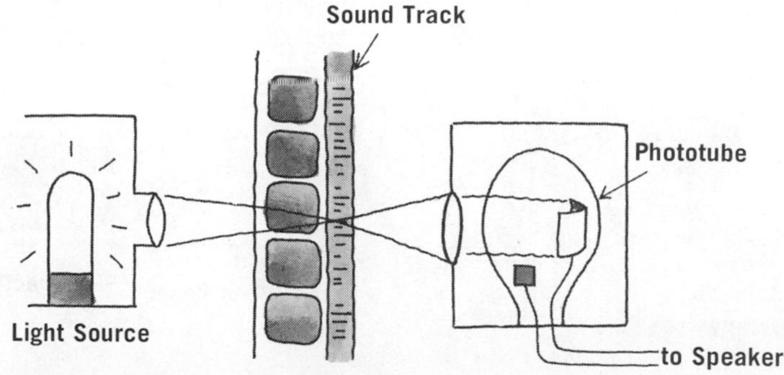

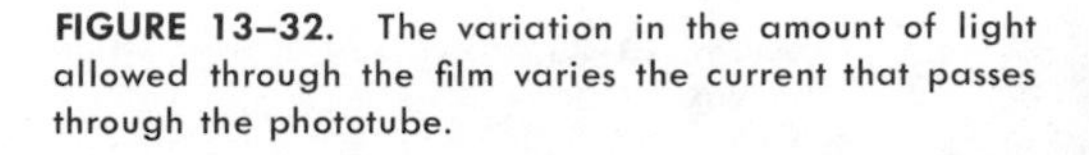

FIGURE 13–32. The variation in the amount of light allowed through the film varies the current that passes through the phototube.

tures before your eye each second. Each picture is only slightly different from the one that preceded it, and the persistence of vision of the eye blends them together as a continuous action. In television, 30 different pictures are flashed before your eye each second. The second deception employed is best understood by careful examination of a photograph in a newspaper. Close scrutiny will reveal that the picture is not continuous at all but instead is an assembly of a huge number of tiny dots. At large distances, the eye is not able to distinguish the individual dots but instead blends them together into a continuous scene.

The television station sends its signal through the air as electromagnetic waves, just as a radio station does (see Chapter 16). Two carrier waves are used by television—one for the sound, one for the picture. We will focus our attention on the latter.

The function of the television camera at a studio is to break down an individual picture into a series of tiny dots as in the newspaper photograph and to produce 30 complete pictures each second, one dot at a time, on the carrier wave. The set in the home must receive these strings of dots and reassemble them into a picture on the face of the screen of a cathode ray tube. In order to see how this is accomplished, let's start at the television camera.

The heart of the television camera is a tube (Fig. 13–33). Light from the scene to be televised is focused by lenses onto a surface covered by thousands of light-sensitive dots. To simplify our discussion, let's assume we have only 16 such dots and that they are arranged as shown in Figure 13–34. In this figure, the lenses of the camera have focused a picture of a cross on the dots (the dots are shown as squares). These squares lose electrons by the photoelectric process when light hits them. At points where the screen is dark, few electrons are given off; where it is light, many are emitted. We now have a series of squares which have given off varying amounts of electrons. The number of electrons emitted by each square must now be "read." This reading is done in the same way that you read a book: your eye follows

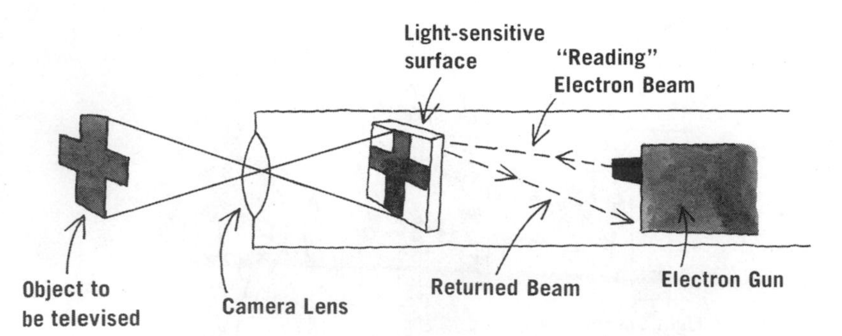

FIGURE 13–33. The heart of a television camera.

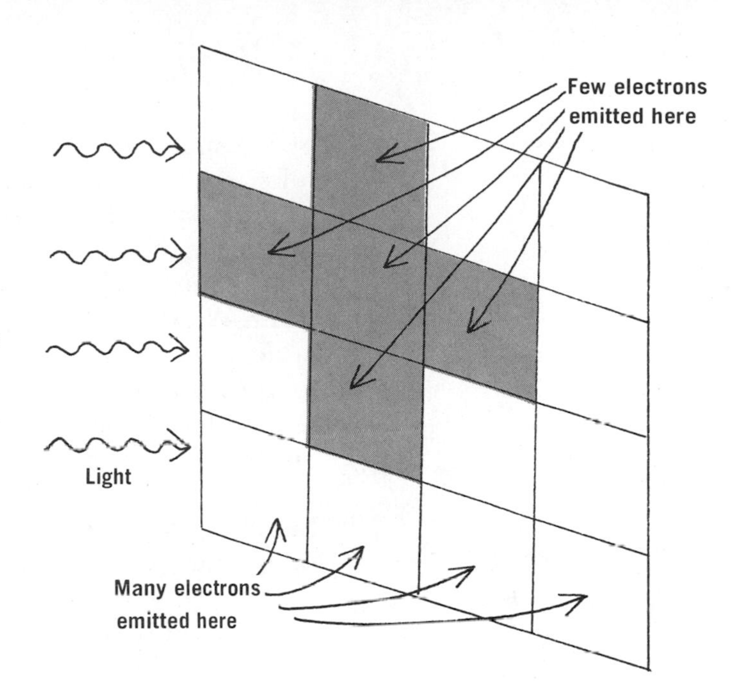

FIGURE 13–34. Where no light strikes the light-sensitive surface, few electrons are released.

each word across a line, then skips to the next line, and reads each word in it. The same is done in the tube of the television camera, except the reading is done by a beam of electrons rather than with eyes. In the tube is an electron gun. Its function is to direct a beam of electrons at each of the squares in turn, to determine how many electrons are missing from each square, and to send out this information via the carrier wave. The beam moves from one dot to the next by the deflection process described in the section on cathode ray tubes in the preceding chapter. (The method actually employed, however, is one of magnetic rather than electric deflection.) Each square, or dot, read corresponds to a dot of the final picture, as in the newspaper photograph. When the beam strikes a square from which many electrons have been removed, some of the electrons in the beam are used to replace them, making the light-sensitive dots electrically neutral again. Those electrons not used in this way are returned to the base of the tube (Fig. 13–33). A very light portion of the screen returns few electrons; a dark portion returns many. These variations in the number of returning electrons are carried to the home by the electromagnetic wave.

The Picture Tube

The cathode ray tube, or picture tube, in the home receiver is synchronized with the message sent to it by the carrier wave. This message has markers inserted, indicating the start of a new line or the top of a new picture. Thus, if a bright spot is formed at the upper left-hand dot on the camera tube, the television set in the home is ordered to also form a bright spot at this corresponding location. An electron gun located in the neck of the picture tube shown in Figure 13–35 sends bursts of electrons toward a fluorescent screen at its front. When the camera at the studio picks up a light area at a particular location on its screen, it directs the picture tube to turn on its electron gun when it points toward the corresponding location on the picture-tube screen. A dark area on the studio camera screen tells the electron gun in the picture tube to turn off. Magnetic fields deflect the electron bursts from the gun in the picture tube from point to point on the tube face in synchronization with the camera tube. As a result, a varying pattern of light and dark information is transmitted which blends into a complete picture.

Color cameras and television sets operate in the same basic way except that the camera at the studio contains three tubes instead of one. A series of separating filters and mirrors (Fig. 13–36) breaks down the light from the scene into the three primary colors, red, green, and blue. Information about the strength of each of these color signals is then transmitted to the home receiver. The color picture tube has three electron guns (Fig. 13–37*A*), each of which shoots a beam whose strength depends on the red, blue, or green signal reaching it. These electron beams are directed toward fluorescent dots on the tube face, as in *B*. The signal from the "blue" electron gun strikes a dot which gives off blue light when electrons hit it; the red and green guns produce their respective colors. These dots are sufficiently close together that the eye cannot distinguish them individually. Instead, it blends them together to produce a color that depends on the amount of light coming from each dot. If each dot emits the same amount of light, the colors blend to produce a white spot at that location. The sensation of any color can be produced by a proper combination of these three primary colors (see Chapter 22).

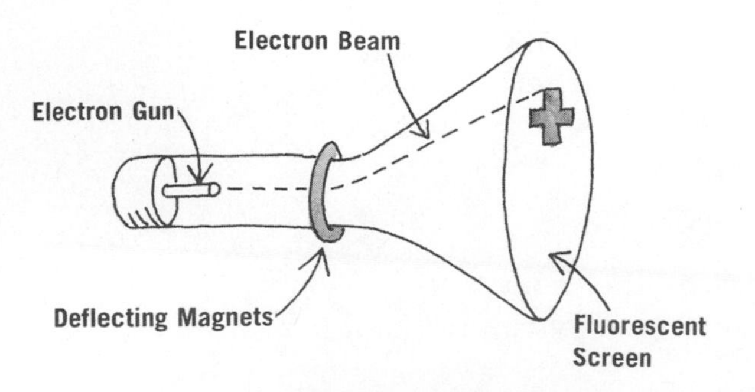

FIGURE 13–35. The picture tube in a television set uses an electron beam to cause the screen to fluoresce at the right places.

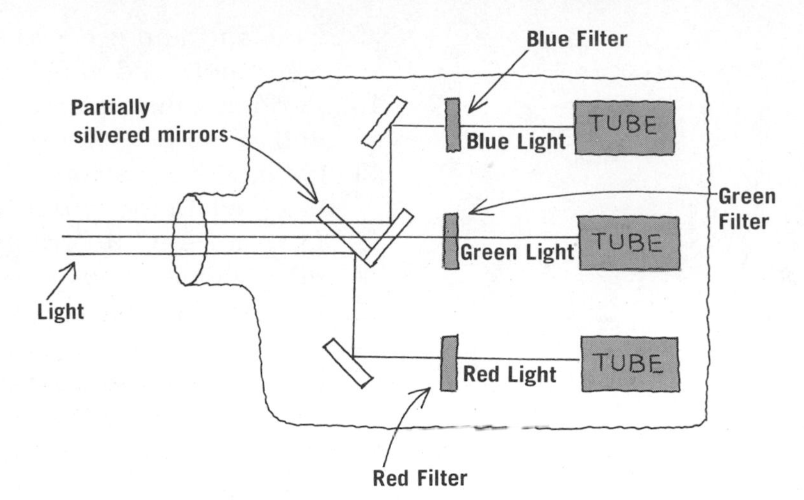

FIGURE 13–36. In a color television camera, color filters separate the incoming light into three parts, and an image is recorded for each. Each "tube" here is similar to the one depicted in Figure 13–33.

OBJECTIVES

Concerning current, you should currently be able to:

1. Describe the internal operation of the flashlight battery and the automobile storage battery.
2. Define the following terms: electric current, direct current, and alternating current.
3. Use a piece of fruit and two metal strips to produce an electric current.
4. Distinguish between current and voltage, and show by diagrams how flashlight batteries can be connected to produce three volts, 4.5 volts, etc.
5. State the relationship between current, voltage, and resistance.
6. Describe the operation of a light bulb, using a diagram of a bulb to show its internal circuit.
7. Use a flashlight bulb, batteries, and wires to set up a circuit.
8. Give two examples of the use of electrical resistance to produce heat.

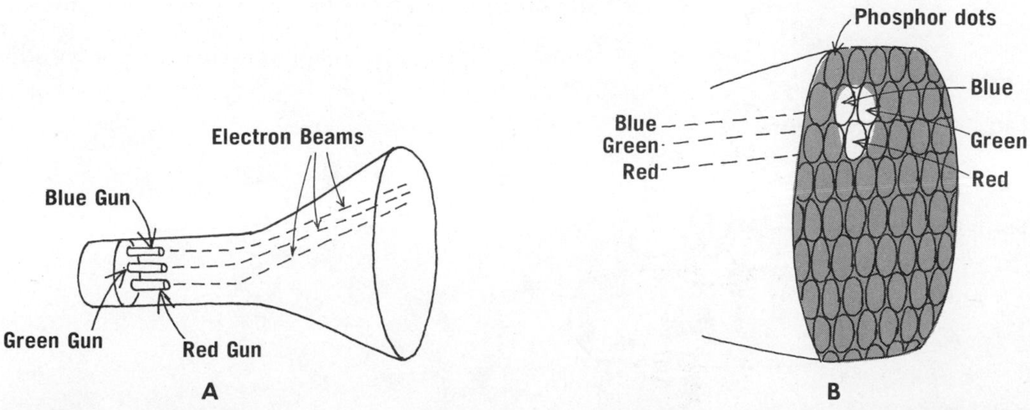

FIGURE 13–37. Color television picture tube has three electron guns (*A*), and three sets of colored dots on its screen (*B*).

9. Construct and use a simple resistance-tester with batteries, a bulb, and wire.
10. Perform a demonstration to show the effect of increased voltage on the brightness of a flashlight bulb.
11. Given the resistance of a circuit and the voltage across it, calculate the current.
12. Diagram and describe the functioning of the carbon microphone.
13. Given the power rating of an electrical appliance (or the current and voltage involved), the time of use, and the cost per kilowatt hour of electrical energy, calculate the cost of operating the device.
14. Using a battery and three light bulbs, connect the bulbs in series.
15. Predict the relative brightness of connecting one, two, or three light bulbs in series across a given voltage.
16. Diagram a typical household circuit and explain the operation of a fuse in such a circuit.
17. Given a diagram of a circuit breaker, explain how it operates.
18. Diagram and describe the common wall switch and the mercury switch.
19. Explain how the basic photoelectric circuit functions and describe its use in three applications.
20. Describe the operation of the television camera and picture tube.
21. Explain how the color television camera and picture tube differ from the black-and-white equipment.

QUESTIONS–CHAPTER 13

1. You need a 27-volt battery but all you have is a box of several 1.5-volt batteries. Sketch a circuit that would enable you to get your 27 volts from them.

2. Figure 13–38 illustrates two resistances–one large and one small–connected to a battery. In which resistor is the current largest? Why? Which resistor has the most voltage across it? Why?

3. Historically, the direction of electric current was defined to be the direction in which a positive charge moves. Is this an important error?

4. How could we prove that our homes are not wired in series?

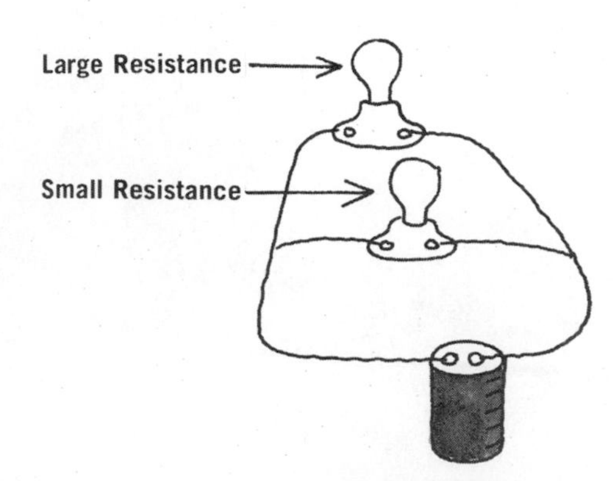

FIGURE 13–38.

5. If charges flow very slowly through a metal, why does it not require several hours for a light to come on when you throw a switch?

6. Why is it unsafe to replace a fuse with a penny?

7. How many electrons pass a point in a circuit each second if the current is 5.5 amps?

8. If electric power costs 6 cents per kilowatt hour, how much is your electricity bill for a month during which you use 3500 kilowatt hours?

9. If electric power cost 10 cents per kilowatt hour, how many kilowatt hours did you use in a month in which your bill is $30?

10. A 110-volt heater draws 15 amps. How much does it cost to run this for one week if electric power costs 5 cents per kilowatt hour?

11. Given three light bulbs and a battery, sketch as many different electric circuits as you can.

12. Discuss the changes in energy that occur as a charge moves through a circuit.

13. A clever burglar dresses in clothing covered by glowing light bulbs. Why?

14. Use the atomic theory of matter to explain why the resistance of a material should increase as its temperature increases.

15. A 12-volt battery is connected to a $\frac{1}{2}$-ohm starter motor in a car. What current is furnished?

16. A light bulb draws $\frac{1}{4}$ amp when connected to 110 volts. What is the resistance of the bulb?

17. A battery supplies five amps when connected to a resistance of six ohms. What is the voltage of the battery?

18. Manufacturers supply either the wattage requirements or the current and voltage requirements on a sticker attached to electrical devices. Use this information, along with an estimate of the time it is on, to calculate the cost of operating for one month any electrical device you own. Assume electricity costs of ten cents per kilowatt hour or, better yet, call the utility company to find out the rate for your area.

14

MAGNETISM

The first magnetic materials discovered were in the form of mineral deposits found in Magnesia in Asia Minor. The mineral was called lodestone or magnetite. The term "lodestone" is derived from the Saxon word *laedan*, meaning to lead. And lead they did. The earliest use to which these materials were put was as a direction finder. The Chinese, in about the first century AD, discovered that when a magnet is suspended by a string, it will align itself in a north-south direction, and the use of magnets as compasses soon followed.

In this chapter we will examine magnetic materials and attempt to discover what they do and why they do it. We will find that the study of magnetism cannot be divorced from that of electrical currents. For example, it is easily observed that a compass will be affected when placed near a wire carrying a current. As we will see, such observations provide us with a useful understanding of magnetic effects. We will break up our study in this chapter into two parts: the first deals with phenomena associated with magnetized materials, and the second relates to the magnetic effects of electric currents.

MAGNETIC PHENOMENA

The study of magnetism begins much as our study of electricity began. There we charged an object by friction and attempted to explain its behavior. We can magnetize a material such as iron or steel in a fashion quite similar to the friction process, by stroking it with a magnet, making sure our strokes are always in the same direction (Fig. 14–1). So there is a similarity between magnetism and electricity, although it is rather tenuous, in that rubbing or stroking a substance with another of the proper kind produces each effect. There are other similarities as we will see.

If we place a magnet in a container filled with small iron filings, the filings will cling most strongly to the ends of the magnet (Fig. 14–2). Those areas where the filings cling are called the poles of the magnet.

FIGURE 14–1. Stroking an iron bar with a magnet magnetizes the bar.

Vital equipment in most of the experiments that we will suggest in this chapter include a magnet and a compass. These are inexpensive and can be purchased at a novelty or hardware store. It's always better to see what happens than it is to read about it, so it's worth your time to purchase them.

A paper clip attached to a rubber band provides a convenient method for observing the strength and location of your magnet's poles. With the clip and band resting on a ruler, as in Figure 14-3, gradually pull the band until the clip breaks free from the magnet. Do this at several locations along the magnet, such as A, B, and C, noting how much the rubber band has to be stretched at each point before the clip comes free. You will find a large force (or stretch) is needed near the ends—the location of the poles—and little or none near the center. You can also use this method to compare the relative strength of different magnets. How?

When suspended by a thread, a magnet aligns itself so that one of the poles, called the north-seeking or simply the north pole, points north. It is found that if two north poles are brought near to one another they repel each other. Two south poles also repel each other, but a north pole and a south pole are found to be attracted to each other (Fig. 14–4). We can summarize these observations by saying that *like poles repel and unlike poles attract each other.* Again, the similarity between magnetic effects and electrical effects is striking. Due to the close correspondence between these two phenomena, it is tempting to regard magnetic effects as being caused by tiny north and south poles, somewhat like

FIGURE 14–2. Iron filings collect about the poles of a magnet.

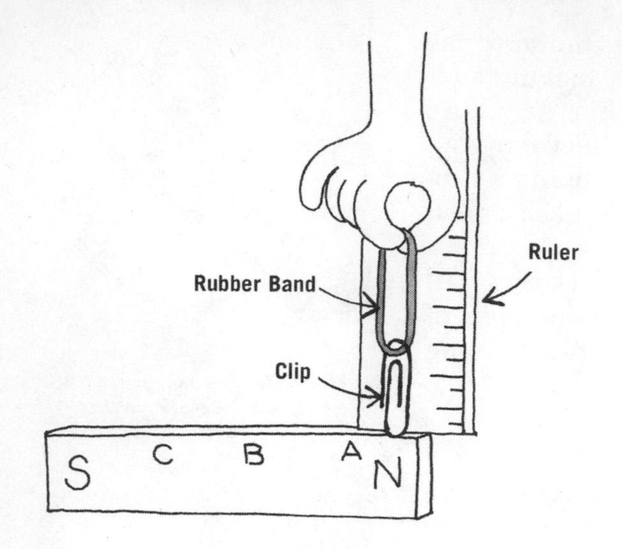

FIGURE 14–3. A homemade magnetic-strength tester.

the tiny positive and negative electric charges. If true, then perhaps what is happening when we magnetize a material is that the stroking causes a separation of these tiny poles, with the north species accumulating at one end and the south species at the other.

If such tiny magnetic charges do exist, it would seem that a simple way to observe their action would be to break a magnet into smaller and smaller pieces until finally we isolate a single little magnetic charge. However, if we break a magnet in half, we do not end up with a north pole in one hand and a south pole in the other. Instead, we simply end up with two magnets, as shown in Figure 14–5. We could continue our breaking-in-half process indefinitely, but all we would end up with is a stack of magnets like the one we started with—albeit shorter.

Numerous attempts have been made to isolate magnetic poles from one another, but all have proved unsuccessful. We will later find that magnetism is explained on completely different grounds than have been suggested so far. It required a discovery relating magnetic effects to electric currents to satisfy our curiosity and to allow formulation of a workable theory of magnetism, as we will soon see.

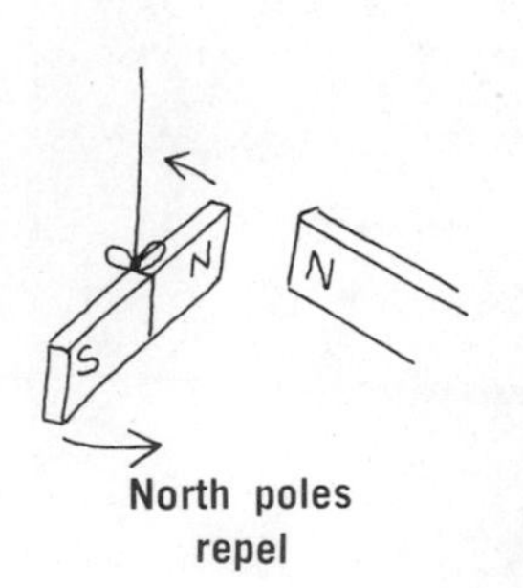

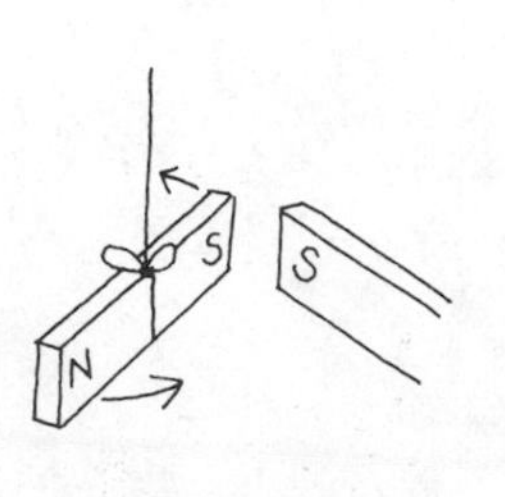

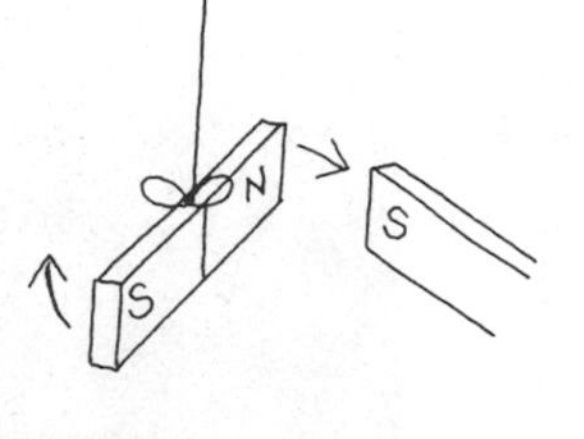

FIGURE 14–4. Like poles repel and unlike poles attract.

MAGNETIC FIELDS

The concept of electric fields was introduced earlier to provide a visual description of electrical forces; in this section we will treat forces which are magnetic in origin in the same way. The magnetic fields which we will discuss will be represented by lines from which we can read the field direction and strength at any location.

The method used to trace the lines of a magnetic field requires a small compass. When the compass is brought near the end of a magnet (Fig. 14–6A), it points as shown. The small north pole of the compass is repelled and points away, while the south pole is attracted and points toward the stronger pole of the magnet. Moving slightly farther away (Fig. 14–6B), we find the compass taking a different alignment. The north end of the compass is repelled by the north end of the magnet at this position, and the south pole of the magnet attracts the north end of the compass, but only slightly because of the large separation between the two. The net result is a slightly skewed direction for the compass needle. We could continue this process for many more points and many more paths, eventually tracing out several lines of the magnetic field, like those shown in Figure 14–7. We assign a direction to a magnetic field at any point as being the direction in which the north end of a compass would point at that location. The arrows in the figure indicate the direction of the field.

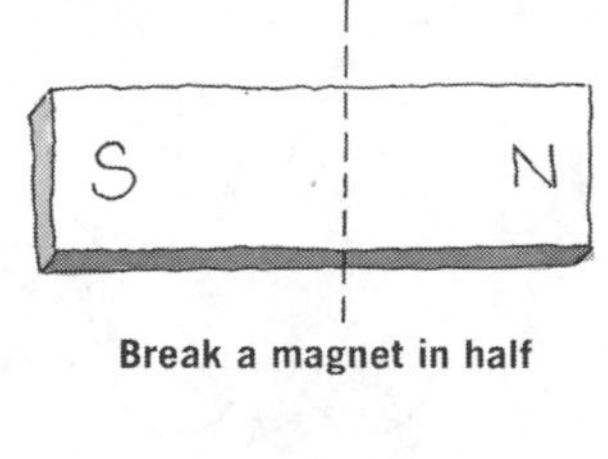

Break a magnet in half

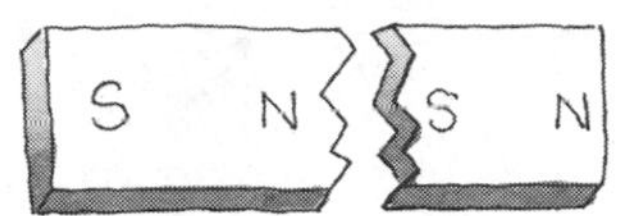

and you get two magnets

FIGURE 14–5. You get baby magnets from big magnets.

You can trace out a magnetic field line for yourself by using the method just described. Place a magnet on a sheet of paper resting on a flat surface, and bring a compass near it. Place a dot at the tip of the compass with a pencil (Fig. 14-8A). Then scoot the compass until its tail is at the dot, and mark another dot at the tip of the compass, as in B. Continue this procedure until you trace out a complete line. The direction of the field line at any location is in the direction the compass points at that location.

You can trace out a magnetic field with iron filings also. Any machine shop will supply the filings (soak them in a soap solution to remove grit and oil). Scatter them lightly over the surface of a paper covering the magnet, and then lightly tap the paper to jar the filings into alignment. Figure 14-9 is a photo of filings on a piece of paper which covers a bar magnet.

Figure 14–10 indicates why the iron filings act as they do. The figure shows two nails being picked up by a mag-

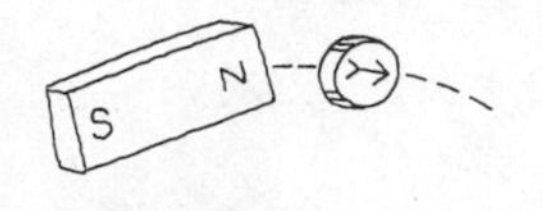

FIGURE 14–6. Determining one line of the magnetic field.

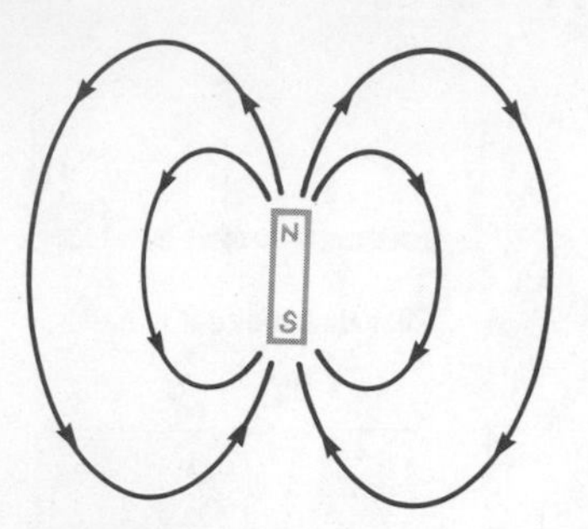

FIGURE 14–7. The magnetic field of a bar magnet.

net. While in contact with the magnet, the first nail takes on magnetic properties and attracts the second nail. The small bits of iron on the paper act in exactly the same fashion. In the field of the magnet they each become magnetized, just as the nail did, and cling together to trace out the field pattern.

Figure 14–9 indicates that the field lines gather more closely at the ends of the magnet. We say that the magnetic field is strongest at these locations. As in the case of electric fields, the strength of the field is indicated in our drawings by the density of field lines: the closer together the lines are, the stronger is the field. The magnetic field of a horseshoe magnet is sketched in Figure 14–11. The points between and near the poles are the strongest parts of the magnetic field, as the close spacing of the lines indicates.

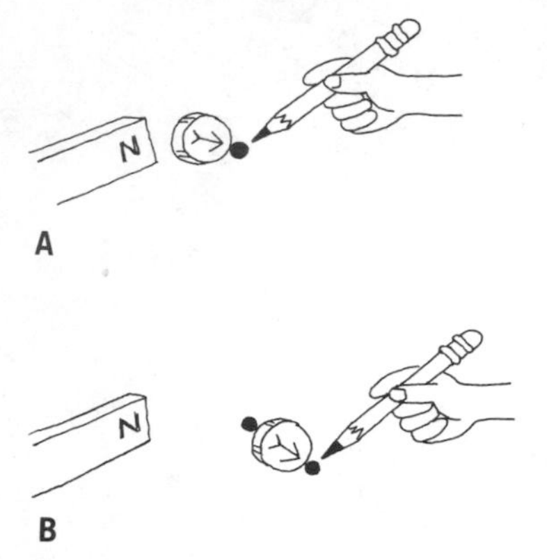

FIGURE 14–8. You can trace out the magnetic field.

DIRECTION OF ELECTRIC CURRENT

We are now prepared to investigate the relationship between electric currents and magnetic fields. In order to do so, however, we will find it necessary to define a direction for electric current, and the method that we choose may be somewhat surprising to you. In the preceding chapters dealing with electricity, we pointed out that the current in circuits is produced by the motion of electrons through the wires. It seems, as a result, that the logical approach would

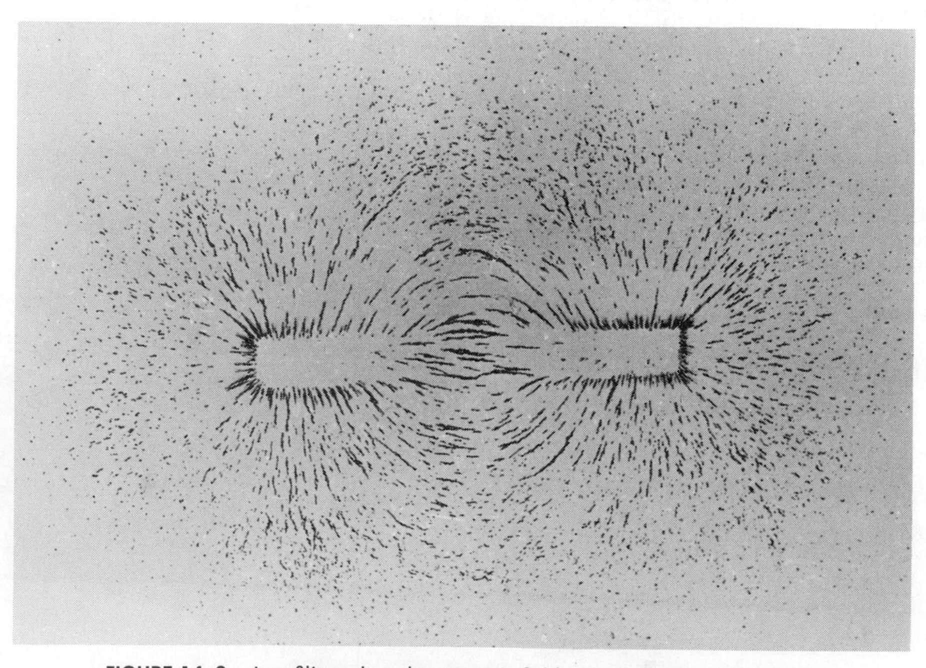

FIGURE 14–9. Iron filings show the magnetic field pattern about a bar magnet.

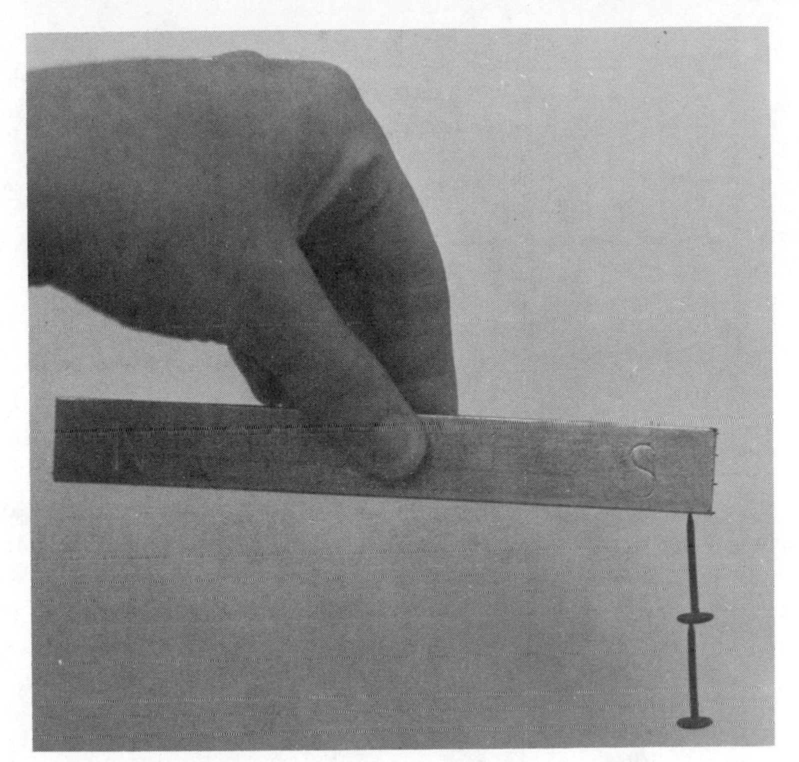

FIGURE 14–10. The upper nail is magnetized by the magnetic field of the bar magnet. It thus supports the lower one.

be to define the direction of current to be the direction in which electrons move. Unfortunately, for historical reasons, *the direction of current is defined as the direction in which positive charges would move if they were free.*

Don't make a big deal out of this point. Simply determine the direction in which an electron would move in a circuit and then say that the current is in the other direction. It really makes no difference; doesn't a negative charge moving from left to right produce exactly the same end result as would a positive charge moving from right to left? Think about it!

MAGNETIC EFFECTS OF ELECTRIC CURRENTS

It is said that the first indication that electric currents could produce magnetic fields was discovered accidentally. Hans Oersted in 1820 was lecturing to a class on the topic, "The lack of a connection between electricity and magnetism." In order to prove his point, he brought a compass near a wire through which an electric current flowed. Lo and behold, the compass needle deflected wildly. Such is the fate of many demonstrations during physics lectures. The conclusion that can be drawn from this experiment is that electric currents produce magnetic fields. A method for

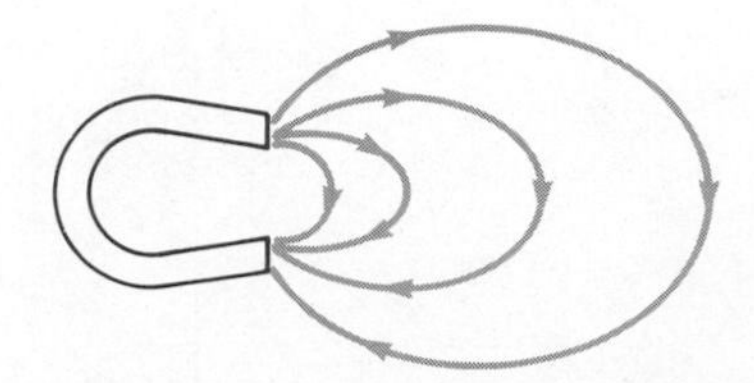

FIGURE 14–11. The field of a horseshoe magnet.

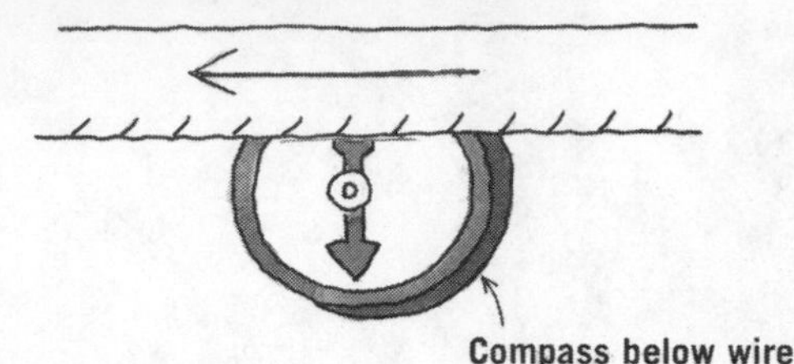

FIGURE 14–12. This demonstration shows that there is a magnetic field associated with an electric current.

studying these fields is given in Figure 14–12, which shows a current-carrying wire placed above a compass. When the current flow is from right to left in the wire, the magnet aligns itself at a right angle to the wire, with its north pole pointing as shown. If one traces out the magnetic field lines around a current-carrying wire, they are found to be circular (Fig. 14–13). As an aid in remembering the direction of the magnetic field around a wire, the so-called right-hand rule was developed. To see how this works, grasp the wire in your right hand, as shown in Figure 14–14. With your thumb pointing in the direction of the current, your fingers curl around the wire in the direction of the magnetic field. Figure 14–15 shows a wire with current flowing down. Use the right-hand rule to verify that the direction of the magnetic lines is as indicated.

If the wire is formed into a circle, the strength of the magnetic field inside the loop is increased, because the contributions from each little section of the circle add up. Application of the right-hand rule to the circular turn in Figure 14–16 indicates that the magnetic field at the center of the coil is as shown. Figure 14–17*A* is a side view of a coil of wire carrying a current; Figure 14–17*B* shows the magnetic field set up by a small bar magnet. The striking similarity between the two fields gives us our first clue as to the origin of all magnetic effects. We will return to this concept in a later section.

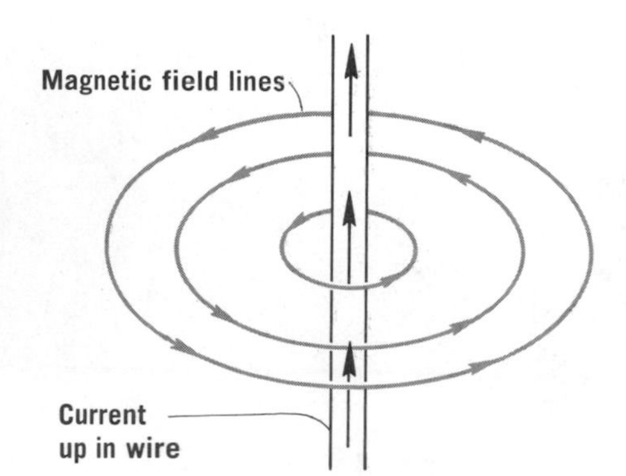

FIGURE 14–13. The magnetic field circles a current-carrying wire.

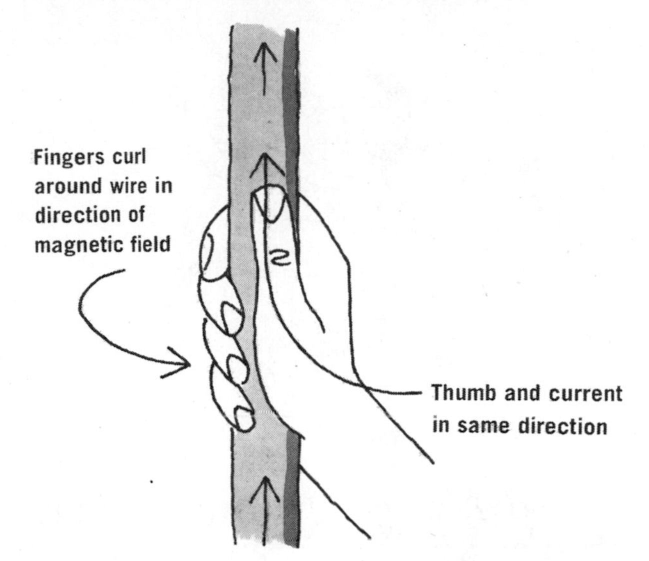

FIGURE 14–14. A "handy" way to remember the field's direction.

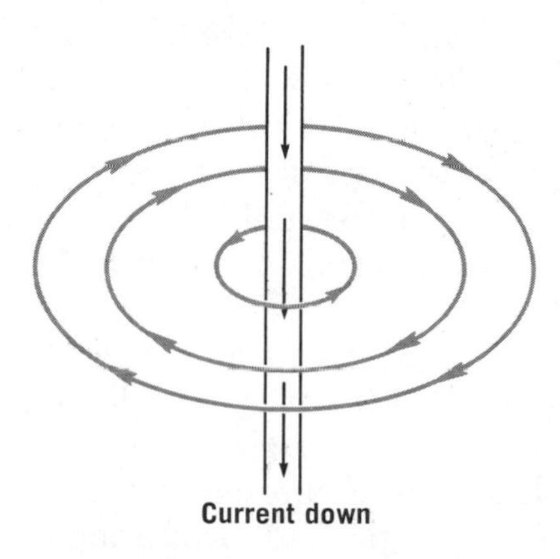

FIGURE 14–15. When the current reverses, so does the field.

FIGURE 14–16. Here, the same rule shows the direction of the field through a current loop.

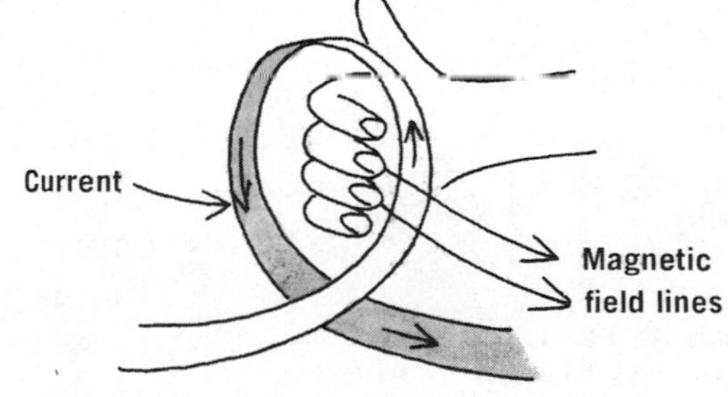

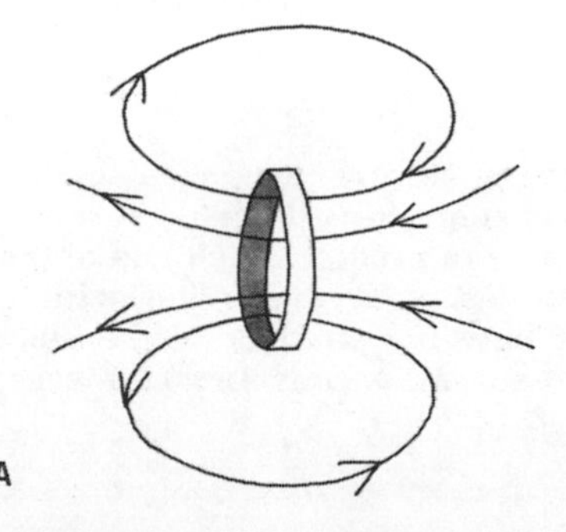

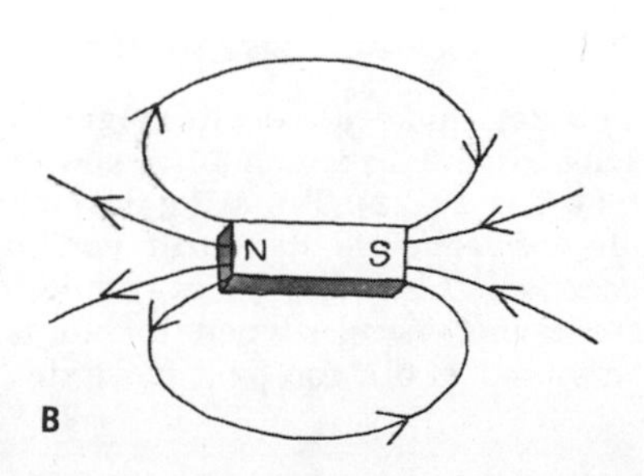

FIGURE 14–17. Note the similarity between the magnetic fields caused by a current loop *(A)* and a bar magnet *(B)*.

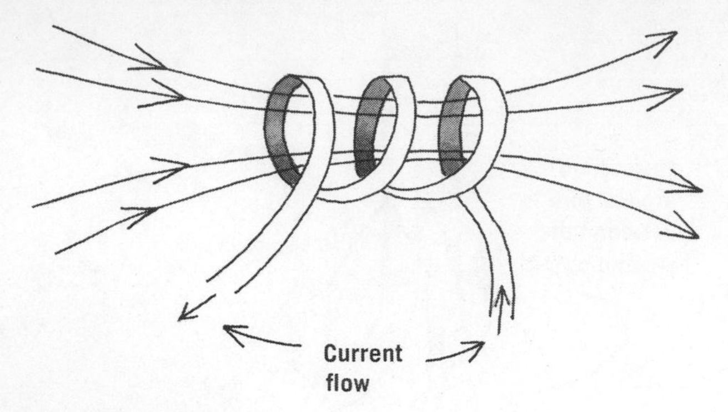

FIGURE 14–18. Each loop adds to the magnetic field through the center.

FIGURE 14–19. A slightly different handy rule for determining the direction of the field through a coil.

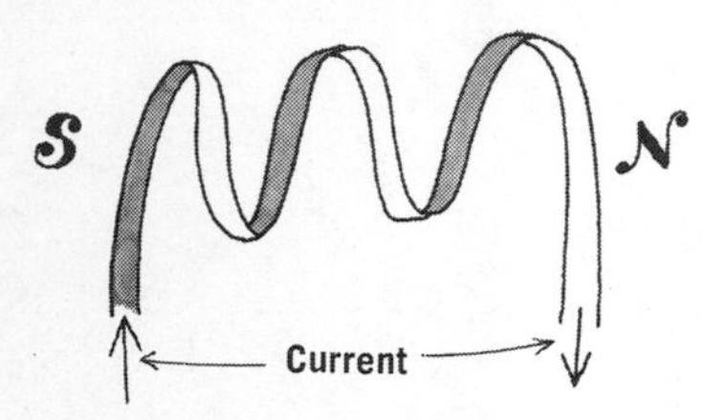

FIGURE 14–20. In what direction is the field through the loop?

FIGURE 14–21. An iron core adds its own field to increase the strength of the field at each end.

THE ELECTROMAGNET

The electromagnet employs the philosophy that if one coil of wire will produce a magnetic field, why not use several coils and produce a stronger one. Figure 14–18 illustrates these several turns and the magnetic field produced. The right-hand rule developed in the preceding section can be used to verify that the direction of the field is as indicated in the figure; however, a second right-hand rule is more easily applied.

If you liked our first right-hand rule, you are sure to love this one. Figure 14–19 shows how to use right-hand rule number two. When the fingers of the right hand are curled around the coils in the direction of the current, the thumb points in the direction of the magnetic field through the coil. Figure 14–20 gives you a chance to use this rule to prove to yourself that the north pole of the electromagnet is on the end shown.

The strength of an electromagnet can be increased by inserting a piece of iron into the coils (Fig. 14–21). Just as iron is easily magnetized by another magnet, it also becomes magnetized when in the magnetic field of the coils. The magnetic effect of the iron core adds to that of the coils, producing quite a strong magnet. Iron used as a core also has the beneficial property of losing its magnetism quickly when not in a magnetic field. As a result, if the current in the coil is turned off, the entire device—including the core—ceases to be magnetic.

You can make an electromagnet by wrapping several turns of insulated wire around an iron bolt (or some nails) and connecting the wires to a battery (Fig. 14-22). Try using the right-hand rule to predict which end of the electromagnet is its north end, and then test your prediction with a compass. Also, use your compass to see how the strength of the electromagnet changes when the bolt is slipped out of the coils. In which case, in or out, is the compass most affected?

The Doorbell

An example of the use of an electromagnet is the electric doorbell. The bell shown in Figure 14–23*A* uses a battery as its source of current, but an ordinary bell uses the electrical system of the home. When the button is pushed, current flows through the coils of the electromagnet, down the clapper to the screw, and back to the battery, thus completing the circuit. As the clapper is attracted toward the magnet, it strikes the bell (Fig. 14–23*B*). But the contact between the clapper and the screw is now broken (at point A). This opens the circuit, causing the current to stop; the electromagnet no longer attracts the clapper, and the spring causes it to return to its original position. Once in its original position, current flows again, starting the cycle over. All of this happens in a fraction of a second, so as long as the door-to-door salesman keeps his finger on the button, the cycle cycles and the bell keeps ringing.

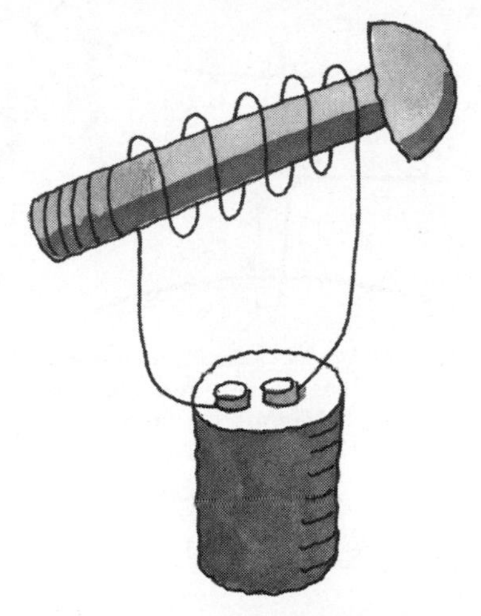

FIGURE 14–22. Use a battery, wires, and a bolt (or nail) to make an electromagnet.

WHY SUBSTANCES ARE MAGNETIC

The initial attempt to explain why a magnetized material acts as it does used the idea of tiny magnetic poles. This hypothesis held that a north pole of a magnet was a location at which several tiny north poles had collected, somewhat

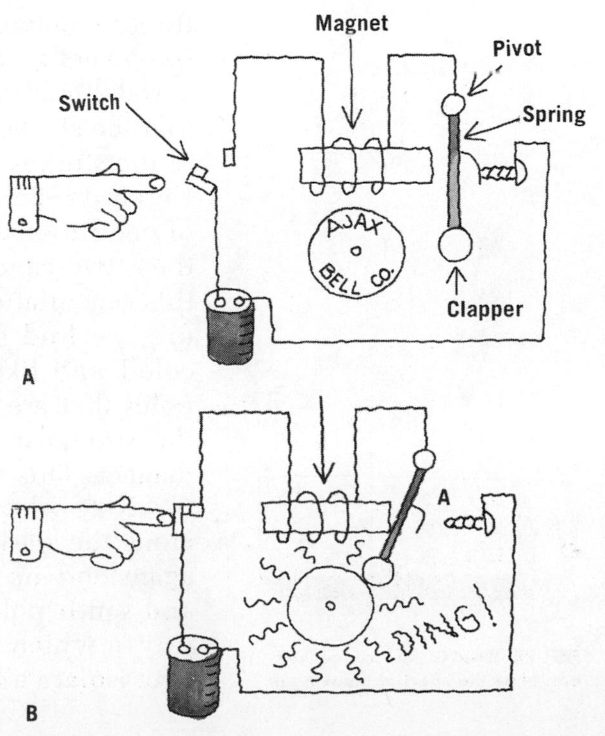

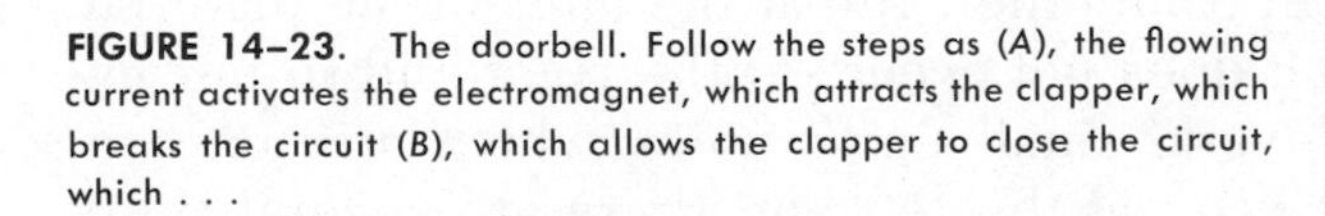

FIGURE 14–23. The doorbell. Follow the steps as (A), the flowing current activates the electromagnet, which attracts the clapper, which breaks the circuit (B), which allows the clapper to close the circuit, which . . .

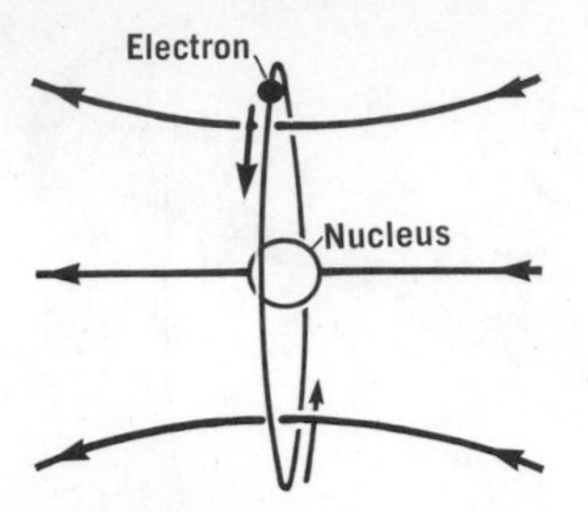

FIGURE 14–24. The electron revolving around the nucleus, forms a current loop, and therefore the atom acts as a magnet.

like charges on the dome of a Van de Graaff generator. At the south pole, the tiny counterparts of these north poles were similarly collected. Attempts to isolate these poles, however, were unsuccessful, and other avenues of investigation had to be opened. Figure 14–17, which demonstrates the similarity of the magnetic field of a current-carrying loop of wire to that of a bar magnet, gives us an important clue to the ultimate origin of magnetism. On an atomic scale, such current loops exist. As the electron circles in its orbit about a nucleus, it acts as an electric current and therefore sets up a magnetic field. In Figure 14–24, the field set up by a single electron orbiting a nucleus is shown. Use the right-hand rule to verify that the magnetic field is represented correctly. (Remember, if the electron goes in one direction, the current is in the opposite direction.)

In addition to the magnetic fields set up by its orbital motion, the electron also sets up another field because it is spinning on its axis in much the same fashion that the earth spins on its axis. A spinning electron is a charge in motion, and a charge in motion constitutes a current, thereby producing a magnetic field. In fact, the magnetic field produced by the spin of the electron is the predominant factor in producing the external fields of such common magnetic materials as iron.

FIGURE 14–25. In an unmagnetized material, the magnetic fields are randomly oriented. The magnetism of the small arrows is indicated by the arrow below the piece of material.

In a piece of unmagnetized material there are many such atoms, each with its own tiny magnetic field, but they are oriented at random directions with one another. If we represent each atom as a tiny bar magnet, we can picture the state of affairs inside an unmagnetized piece of material as in Figure 14–25. Because of the randomness of direction, the tiny magnetic fields all tend to cancel, and the material has no net magnetism. Earlier we said that we can magnetize certain materials by stroking them with a magnet such that the strokes always are in the same direction. The effect of the strokes is to cause the tiny bar magnets to align (see Figure 14–26). At all points inside the material, a north end of one tiny magnet rests near the south end of another, and these two cancel each other. But at the ends of the material, this cancellation does not occur. At the right end in our figure, we find several small north poles which are not canceled, and likewise, at the left end there are several south poles that are not canceled. We have previously noted that the strongest magnetic effects occur at the extremes of a magnet. This is the origin of the poles of a magnet.

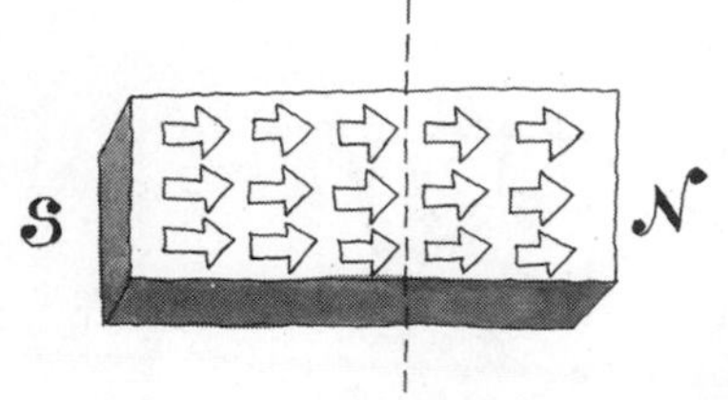

FIGURE 14–26. A magnetized iron bar. Note the effect of breaking it.

Note what would happen if we broke our magnet in half along the dashed line indicated in the diagram. We would again end up with a situation in which uncanceled north and south poles would be at each end. There is, then, no way in which to break a magnet into small enough pieces to ever isolate a free north pole and a free south pole.

A material such as iron does not retain its magnetism for long, because shortly after it has been magnetized as shown in Figure 14–26, it quickly returns to the state represented by Figure 14–25. Several materials, such as steel, will keep their atoms frozen into the position indicated by Figure 14–26 for long periods of time. Magnets made of such materials are usually called permanent magnets. Even permanent magnets can be made to lose their magnetism, however. One method of doing this is to heat them. Heating causes the atoms within the magnet to vibrate faster. As they do this, they are knocked from their alignment with one another and the magnetism is destroyed. An even simpler method is to hit the magnet with a hammer. Here again the atoms are jarred from their alignment, and permanent magnetism is lost.

At first glance it would seem that the basic mechanism we have used to explain the origin of magnetic effects would cause all materials to be strongly magnetic. Certainly all materials have electrons spinning within their atoms, but in most atoms the electrons spin in different directions, and the magnetic field created by one electron is canceled by that of another in the same atom. It is only in certain materials that this cancellation does not occur and that external magnetic effects are exhibited.

THE MAGNETIC FIELD OF THE EARTH

In the magnetic field of the earth, the north pole of a compass aligns itself toward a point near the geographic north pole of the earth. As a result, if you recall that there is an attraction between north and south magnetic poles, you should agree that the magnetic pole in the north of the earth really has to be a south pole. The magnetic field of the earth is shown in Figure 14–27, and is quite similar to the kind of field that would be set up by a large bar magnet deep inside the interior of the earth. Note again that the magnetic north pole of the earth is really the south pole of this internal magnet.

The actual source of the earth's magnetism is unknown. There are large deposits of iron ore inside the earth, but they certainly are not responsible for the earth's field, because the high temperatures inside the earth would destroy any magnetism of these deposits. At present, the best guess is that the magnetism is caused by electric currents circling in the liquid interior of the earth. These electric currents could be caused by the movement of charged atoms and electrons, which are set into motion as the molten core of the earth is churned by convection currents. Some evidence indi-

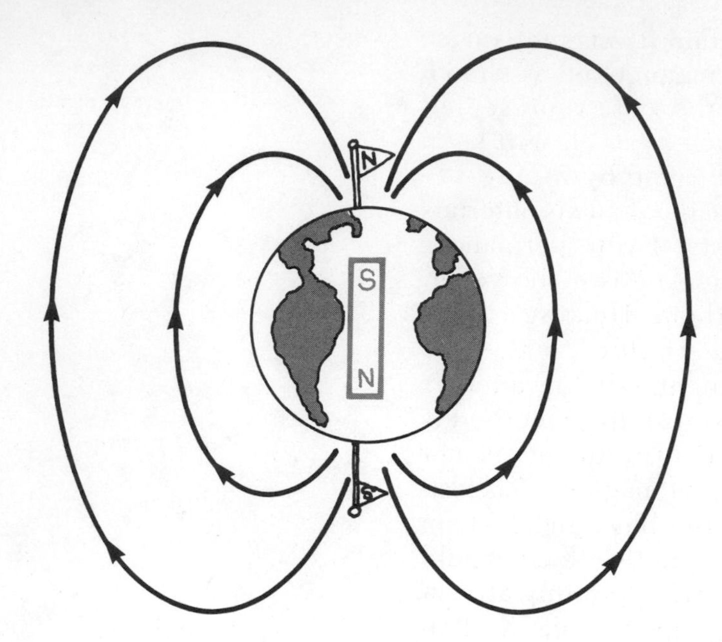

FIGURE 14–27. Note that a south magnetic pole is at the earth's North Pole.

cates that the strength of the earth's field is dependent upon the rate at which the planet rotates. Space probes indicate that planets which rotate faster than ours (such as Jupiter) have a stronger field, while slower rates of rotation (as on Venus) lead to weaker fields. Even though a complete answer as to how (or even if) these factors, either singly or in combination, produce the resultant magnetic field of the earth cannot be given, we can nonetheless examine the properties of this field.

If a compass needle is suspended from its center by a string, it will swing horizontally only when near the equator. In the northern hemisphere the north end of our compass would point more and more down toward the ground as it is carried north, and in the southern hemisphere the north end would point skyward (Fig. 14–28). In 1832, just north of Hudson Bay, a point was located at which the compass needle became vertical. This is the location of the north magnetic pole of the earth. The location of the north magnetic pole differs from the geographic north pole by about 1300 miles. A similar procedure locates the south magnetic pole of the earth about 1200 miles away from the geographic south pole.

take-along-do-it

A compass needle that will allow you to determine the variation from the horizontal of the magnetic field in your locale can be constructed as in Figure 14-29A. A small needle is suspended so that it is balanced and rests horizontally in a north-south direction. It is then magnetized (by stroking with a permanent magnet) and returned to its original position. It will now swing to a position somewhat like that shown in B.

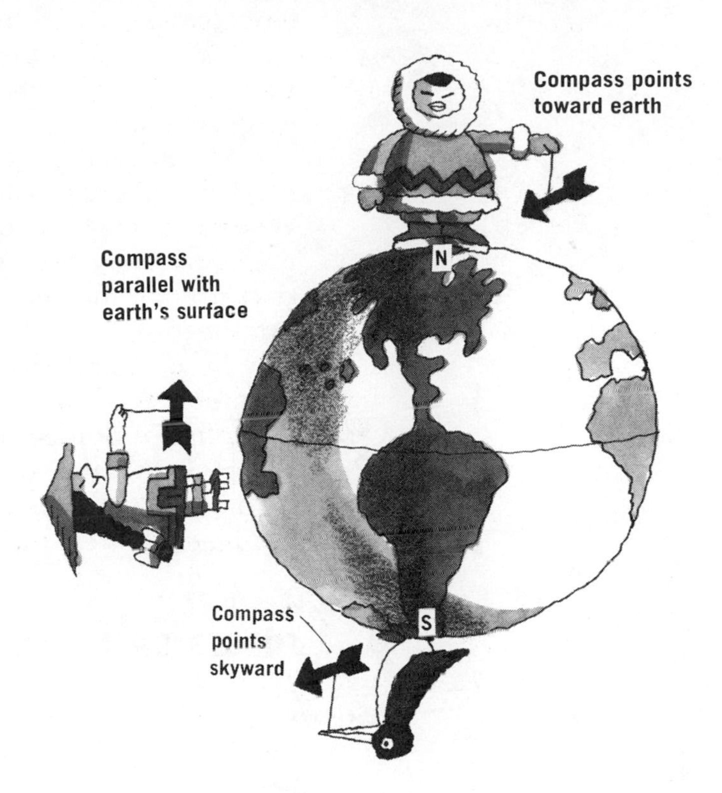

FIGURE 14–28. Free-swinging compasses other than at the equator do not line up horizontally.

Using the Earth's Field to Magnetize

The strength of the earth's magnetic field is extremely small when compared to that of even a tiny bar magnet. A typical magnet purchased at a novelty store produces a magnetic field as much as a thousand times stronger than the earth's field. Even so, the field of the earth is strong enough to magnetize objects placed in it. A piece of soft iron can be magnetized by placing it in a north-south direction and either heating it or hitting it with a hammer. The violence done to the iron enables the tiny atomic magnets to orient themselves more easily along the lines of the earth's field than if simply left alone. In the absence of heating or striking, the piece of iron will acquire some magnetism, but it takes longer. Often a long-standing structure such as an iron

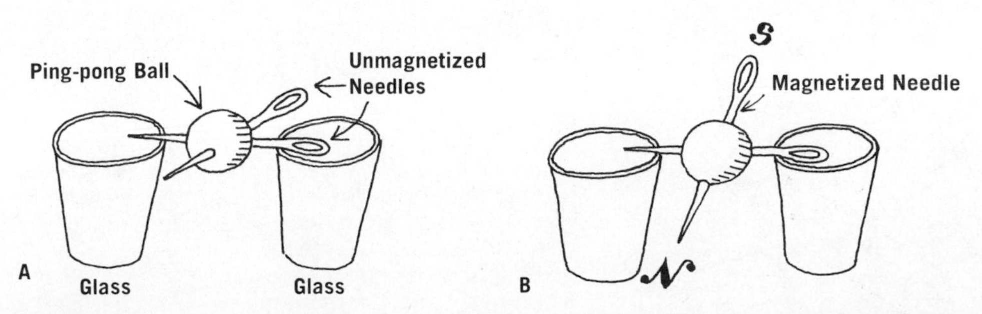

FIGURE 14–29. Balance a needle as shown in A. Then magnetize it and align it north-south. It will point as in B.

fence, when tested by a compass, will be found to be magnetic. This is also the reason why some naturally occurring deposits of ore, such as the lodestone discussed earlier, are magnetized.

Magnetization of material by the earth's magnetic field has been going on since the earth has had a magnetic field. When a rock is formed which contains iron, the iron is slightly magnetized by the earth's field. Such magnetization occurs regularly on the ocean floor when volcanic action spews out iron-bearing lava from deep in the earth. This lava solidifies and in so doing records for us the direction of the earth's magnetic field at that time. By examining the ocean floor for both age of rocks (by nuclear methods to be discussed in Chapter 26) and direction of the magnetic field, it was learned that the poles of the earth have actually reversed a number of times during the earth's history.

FORCES ON MOVING CHARGES

When two positively charged objects are brought close together, there is a force of electric repulsion between the two (Fig. 14–30*A*). However, Figure 14–30*B* indicates that a magnet brought near the suspended charged object produces no effect at all. There is no interaction between magnets and charges as long as there is no motion, but when the charges begin to move, we find that a magnetic force acts on them.

Figure 14–31 shows a positive charge moving upward on the page in a region in which a magnetic field exists (possibly due to a nearby magnet). The X's are used to indicate that the direction of the magnetic field is *into* the paper. In this case there is a magnetic force on the particle toward the left. Figure 14–31*B* indicates a memory device for determining the direction of any one of the three quantities – velocity, magnetic field, and force – if the directions of the other two are known. We have developed two right-hand rules for other purposes, and now, by popular demand, we

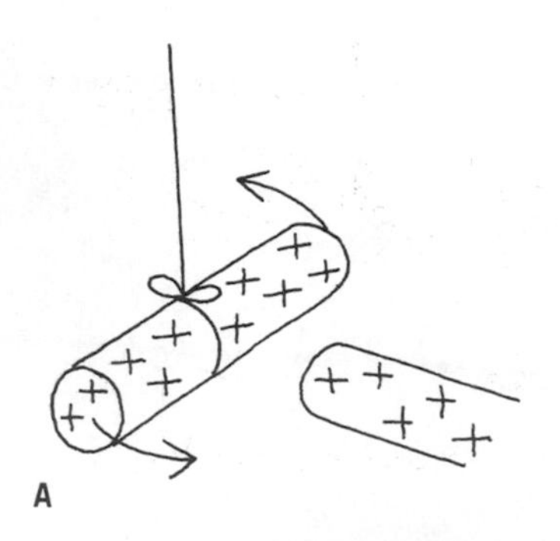

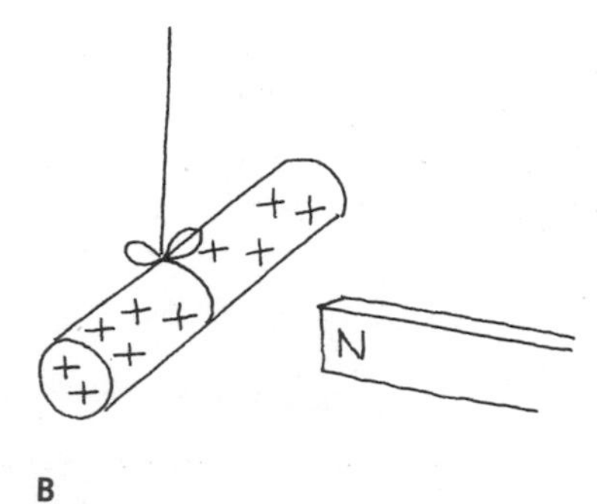

FIGURE 14–30. Electric charges at rest show interaction (*A*), but a magnet exerts no force on a stationary charged object (*B*).

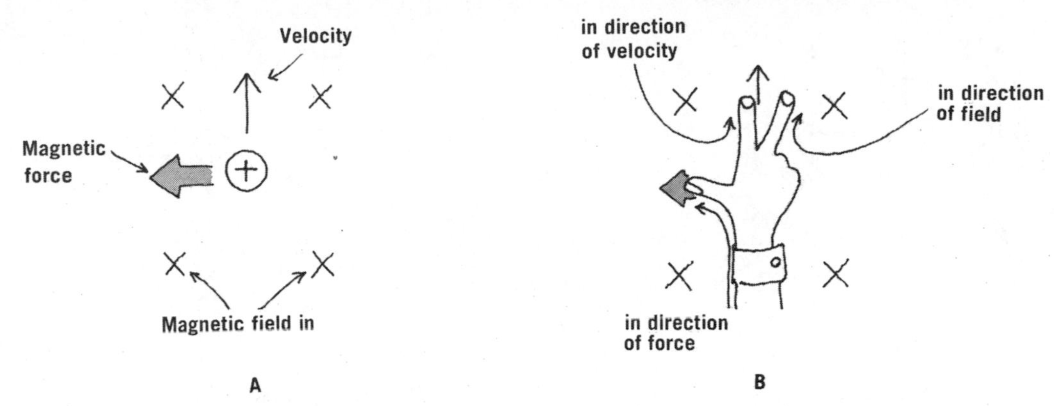

FIGURE 14–31. A positive charge moving through a magnetic field experiences a force as shown in *A*. *B*, a way to remember the directions involved.

develop a third. As shown, if you point the index finger in the direction of motion of the positively charged particle and the middle finger in the direction of the magnetic field, the thumb points in the direction of the force on the moving charge. Figure 14–32 shows four different situations that you can use to test your skill with this new right-hand rule.

You should make particular note of the fact that the right-hand rule applies only to situations in which the moving charge is positive. If it is negative, the same procedure is used, but you have to use your left hand. Figure 14–33 presents the same four situations as in the preceding illustrations, but the moving charge is negative. Again, test your skills as a contortionist.

In the last chapter we discussed the picture tube of a television set. At that time we simply stated that the elec-

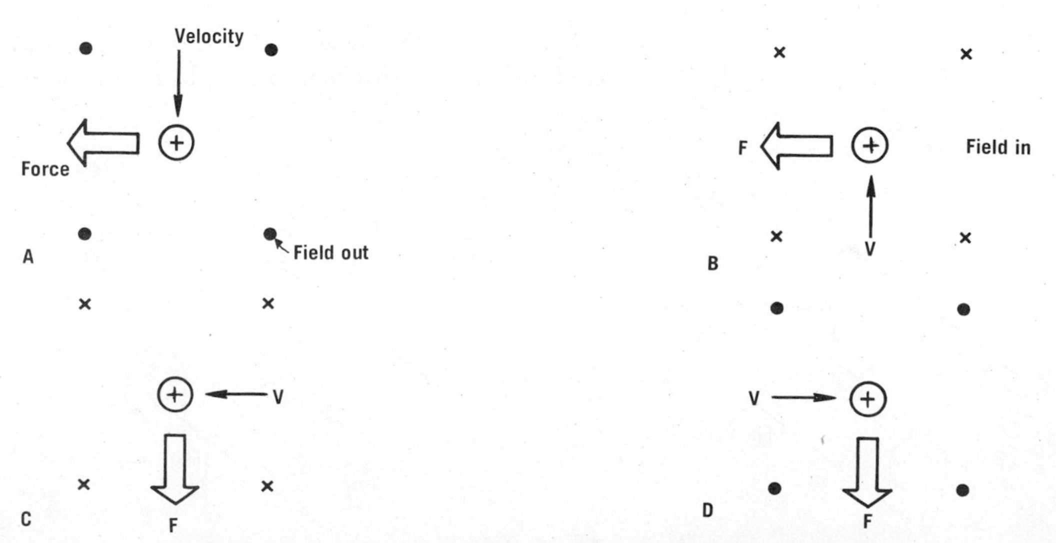

FIGURE 14–32. Check each of these by applying the latest right-hand rule.

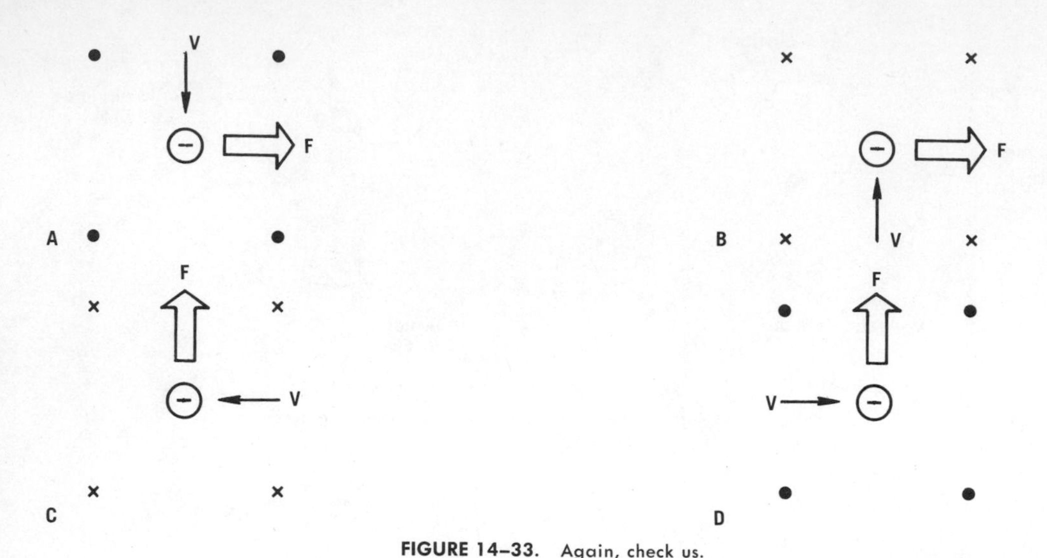

FIGURE 14–33. Again, check us.

trons emitted from an electron gun were bent by magnetic fields. This statement should now make sense. The electrons emitted by the gun are charged particles, and when they pass through a magnetic field, the magnetic force changes their direction. By varying this field, the stream of electrons from the gun is directed to scan across the face of the tube.

The rule for determining the direction of force on a moving charged particle has been given, but one special case should be noted: *if the particle happens to move parallel to the magnetic field lines, it will feel no force at all.* Figure 14–34 indicates two situations in which a particle can move oblivious to the field.

Magnetic Bottles

Figure 14–35 shows a positive charge moving in a magnetic field that is uniform and that extends over a large area.

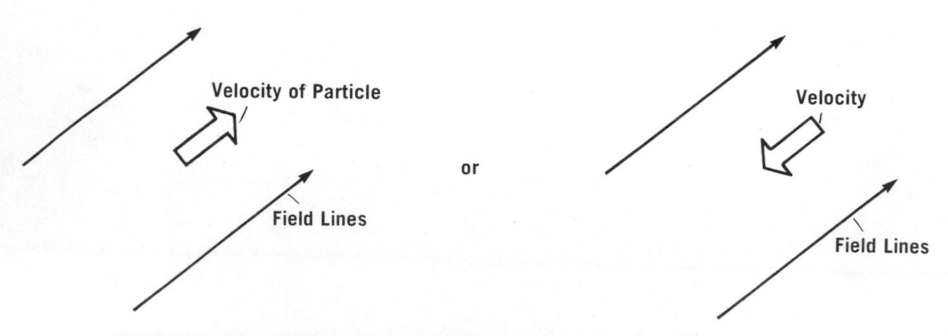

FIGURE 14–34. When the charged particle moves *along* the field, no force is felt.

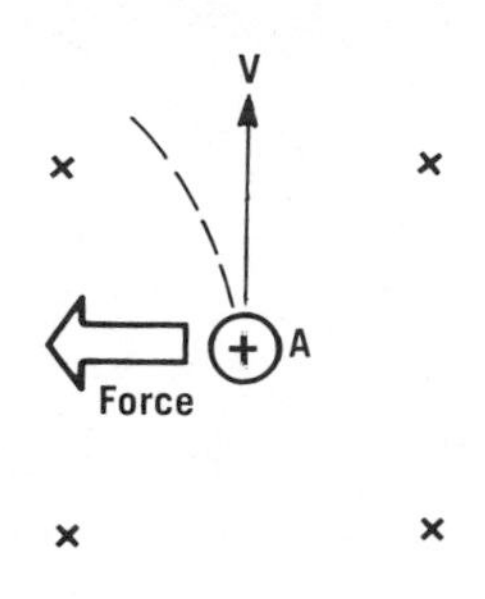

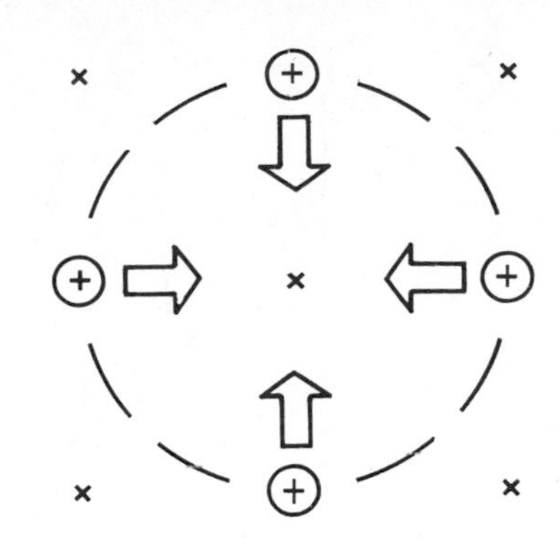

FIGURE 14–35. The resulting motion of a charged particle moving perpendicular to a uniform magnetic field is circular.

At position A, the force on it is to the left, causing the particle to change direction and to follow the dashed path. The total path followed is indicated in Figure 14–35*B* and is seen to be circular. The arrows in *B* indicate the direction of the force on the particle at various points in its path. If the particle is not moving exactly perpendicular to the magnetic field, the circular path will not be closed but, instead, will be helical (Fig. 14–36).

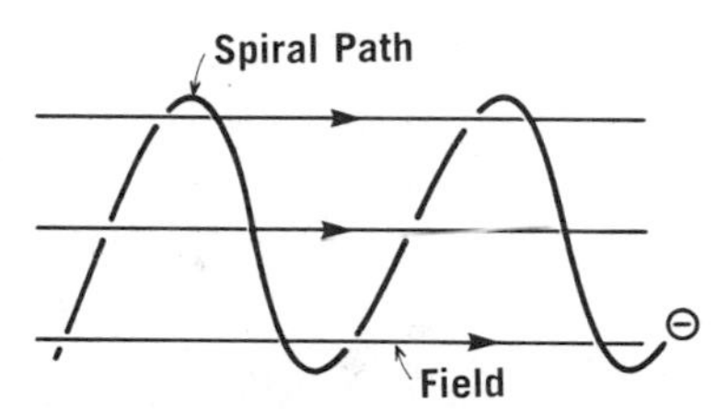

FIGURE 14–36. In this case, an electron spirals around as it moves down along the field.

"Magnetic bottles" can be made which can be used to contain charged particles.[1] The trapping process employs the following principle: if the strength of the magnetic field increases, the radius of the circular path in which the charge moves gets smaller. Figure 14–37 shows a situation in which the magnetic field gets stronger as we move toward the right in the diagram, as indicated by the closer proximity of the magnetic field lines. Because of this increasing field, the particle gradually moves in a tighter and tighter spiral. When the radius of the spiral gets small enough, the particle can actually reverse direction and move back toward the left in the diagram. If the field gets stronger off to the left of the page, the particle will undergo a similar reversal at that point also. The result—a particle continually spiraling back

[1]Such methods for trapping particles occur in nature in the Van Allen radiation belts, which will be discussed in Chapter 27. They also are being tested in an attempt to contain particles in order to make use of a new source of energy. This process will be discussed in Chapter 25.

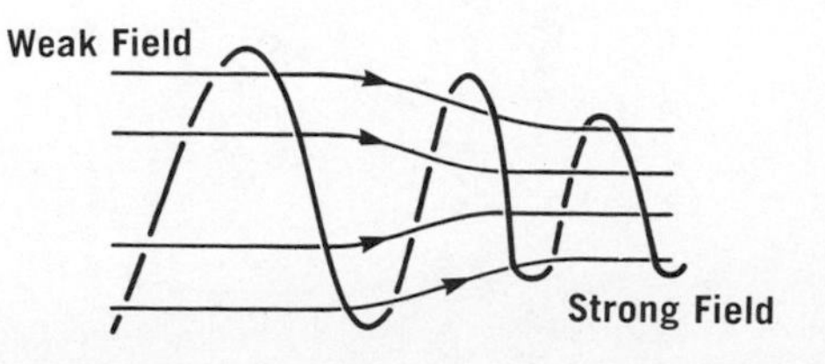

FIGURE 14–37. As the field changes strength, the charged particle changes its motion.

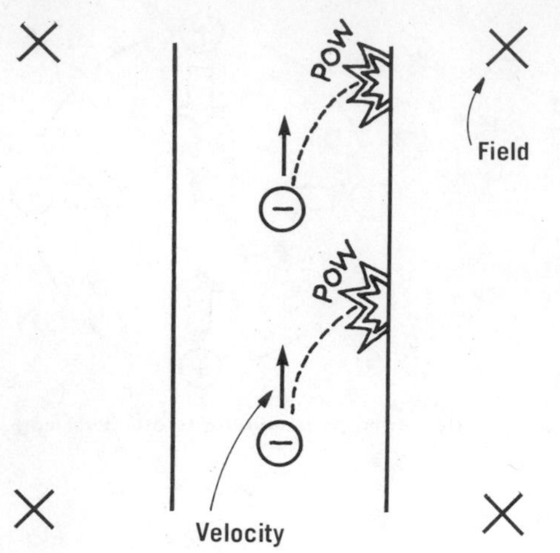

FIGURE 14–38. The electrons in the wire feel a force from the magnetic field, but the result is that the entire wire feels the force.

and forth, trapped in the magnetic field, never colliding with any solid walls.

FORCE ON CURRENT-CARRYING CONDUCTORS

Charges moving in a magnetic field experience a force whether they are in air or in a material. In Figure 14–38, a section of wire carries electrons upward on the page. The X's indicate the presence of a magnetic field that is directed into the page. Remembering that these are (negative) electrons, we must use the *left*-hand rule to determine the force on them. We find that there is a force on them toward the right. They will therefore attempt to follow the dashed paths. Collisions with the "side" of the wire, however, prevent them from moving in this direction for long. One electron colliding with the edge of the wire gives the wire a slight nudge to the right. The force exerted on the wire is extremely small when one considers the push given by a single electron, but in a typical electric current, countless numbers of electrons are moving. Each is deflected, and each gives its own tiny push. The result is that the wire receives a push strong enough to cause it to actually move toward the right: the larger the current, the greater the push.

Note that if we had used the right-hand rule on the *positive* current, we would have found that the force on the wire is also toward the right. In what follows we will ignore the fact that it is actually electrons causing the effect and will use current direction and the right-hand rule.

Figure 14–39 shows several wires carrying currents. Again, use your skill with the right-hand rule to demonstrate that the forces acting on the wires are as shown. Note case *D*, in which current is along the direction of the magnetic

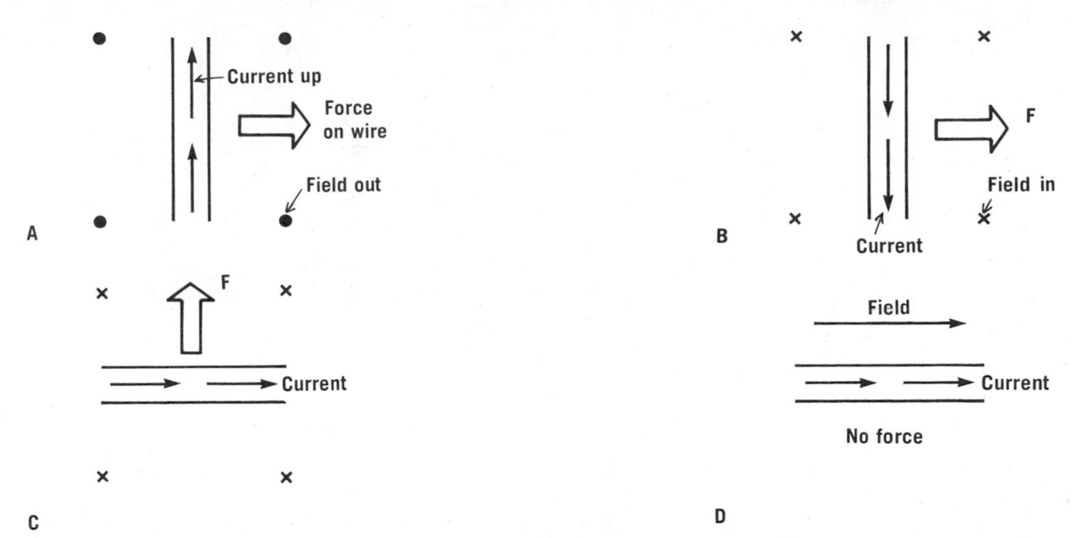

FIGURE 14–39. Each of these situations is predicted by the right-hand rule.

field lines. In this case there is no force exerted on the electrons, and hence, there is no force on the wire.

THE AMMETER

The ammeter (contraction of "ampere meter"), a basic instrument used to measure electric current, is based on a simple fact: an electric current passing through a wire in a magnetic field can produce a large enough force on the wire to move it. Figure 14–40 shows a coil of wire (carrying current in the direction indicated by the arrows) located between the poles of a permanent magnet. Application of the right-hand rule indicates that the force acting on segment AB of the coil is directed outward, away from the paper, and that the force on side CD is directed into the page. No

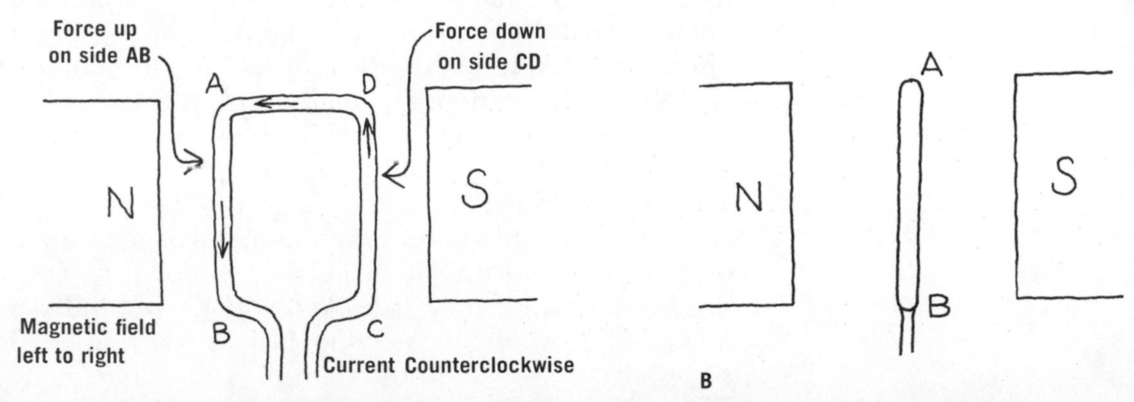

FIGURE 14–40. *A*, The force on side AB is out of the paper; on CD, into the paper. The coil tends to rotate to the position shown in *B*.

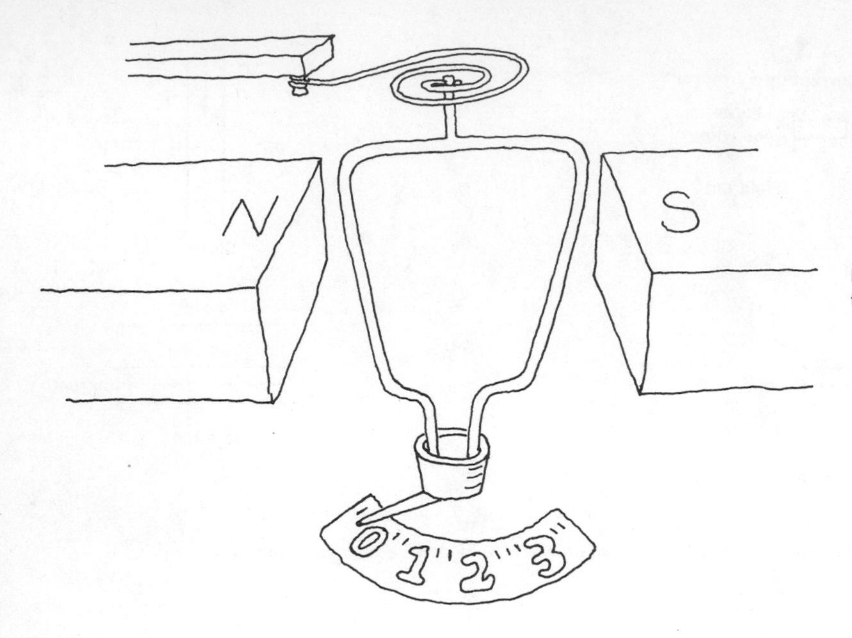

FIGURE 14-41. A spring keeps the coil from rotating to the position of Figure 14-40*B*.

forces act on sides AD and BC, which carry current along the direction of the magnetic field lines. These sides will be neglected in our discussion because they play no role in the operation of the meter except to carry current. There will be forces on these sides when the coil rotates out of the plane of the paper, but these forces do not tend to produce rotational motion, which is important here.

The forces on sides AB and CD cause a twisting motion of the coil, which would make it rotate if it were suspended so that it could turn freely. In fact, if it were free to rotate, it would turn until it moved to the position shown in Figure 14-40*B*. At this position the forces would no longer tend to rotate it.

In order to hinder its motion, a small spring is attached to the coil (Fig. 14-41). As the coil rotates, the spring tightens to stop its motion. A stronger current flowing in the coil would cause greater turning forces to be exerted on it, and the spring would have to tighten considerably before it hampered the motion enough to stop the rotation. The amount that the coil rotates before the spring tightens enough to stop it is an indication of the strength of the current—a small rotation = a small current; a large rotation = a large current. A pointer attached to the coil and a scale whose numbers indicate the strength of the current are shown in Figure 14-42.

Figure 14-43 shows how the meter is used in an actual circuit. In *A*, an unknown current is flowing. To measure this current, the circuit is broken at the point indicated. Any other point would have done just as well since this is a series circuit, in which the current is the same at all points. The meter is then inserted as shown in *B*.

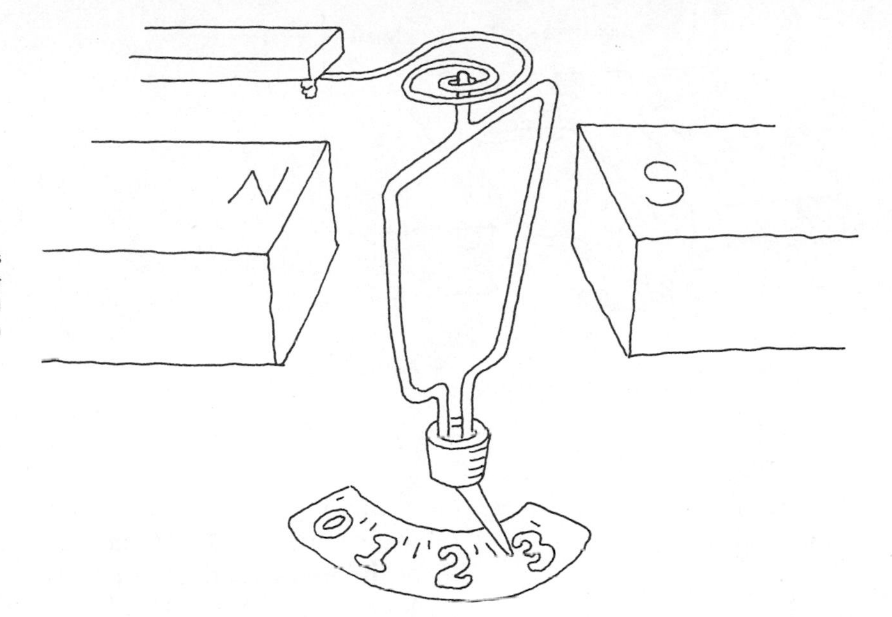

FIGURE 14–42. When current flows through the coil (coming in and going out by wires not shown in the figure), the coil is forced to turn, indicating the strength of the current.

THE ELECTRIC MOTOR

The principle of operation of the electric motor is the same as that of the ammeter: a current-carrying wire in a magnetic field experiences a force. The basic parts of a simple motor designed to operate on direct current are shown in Figure 14–44. A coil of wire is mounted so that it is free to rotate, and as the coil rotates, it causes a shaft attached to it to turn. Connected to the coil are two semicircular pieces of metal, and rubbing against these half rings are electrical contacts called brushes.

In order to see how the motor works, we will consider the steps indicated in Figure 14–45. In *A*, the battery is the source of current, which flows as indicated by the arrows. The current leaves the battery, moves through one of the contacts to one of the rings, goes around the coil, and then leaves through the other ring and contact to return to the battery. As the charges move through the coil, they move through a magnetic field and thus experience forces. Apply the right-hand rule to verify that the directions of the forces are as given in the figure. These forces acting on the coil

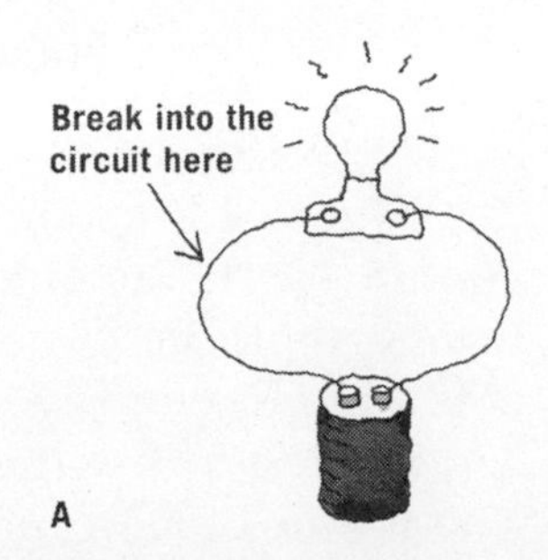

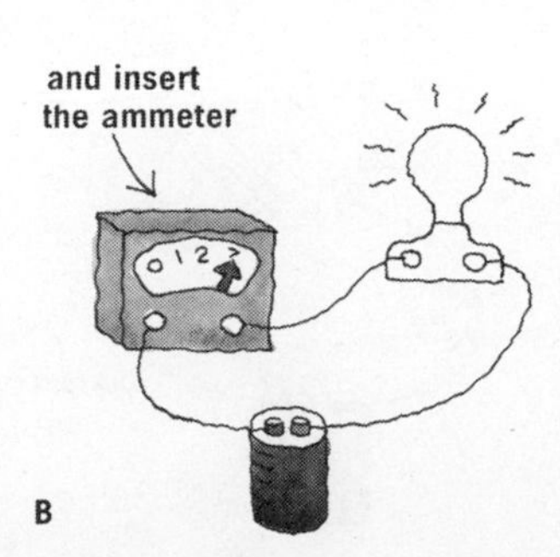

FIGURE 14–43. The ammeter is connected into the circuit so that all of the current passes through it.

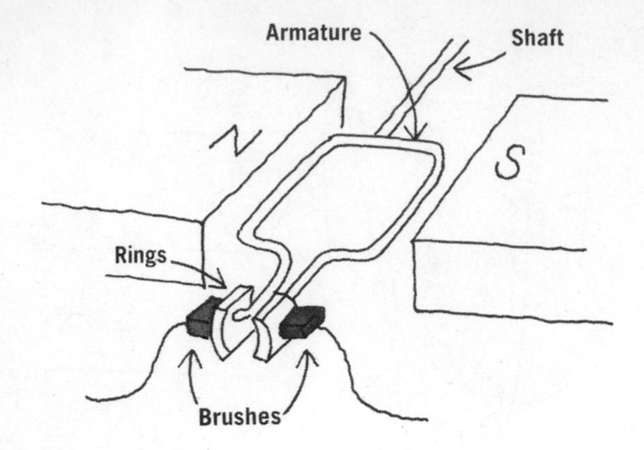

FIGURE 14–44. The essentials of an electric motor.

cause it to turn until it reaches the position in *B*. At this position, the contacts momentarily become disconnected from the rings, and as a result, no current passes through the coil, and no forces exist. The angular momentum of the coil, however, is sufficient to cause it to continue to turn to the position shown in *C*. Sides CD and BA have now switched places. The current now goes from B to A to D to C.

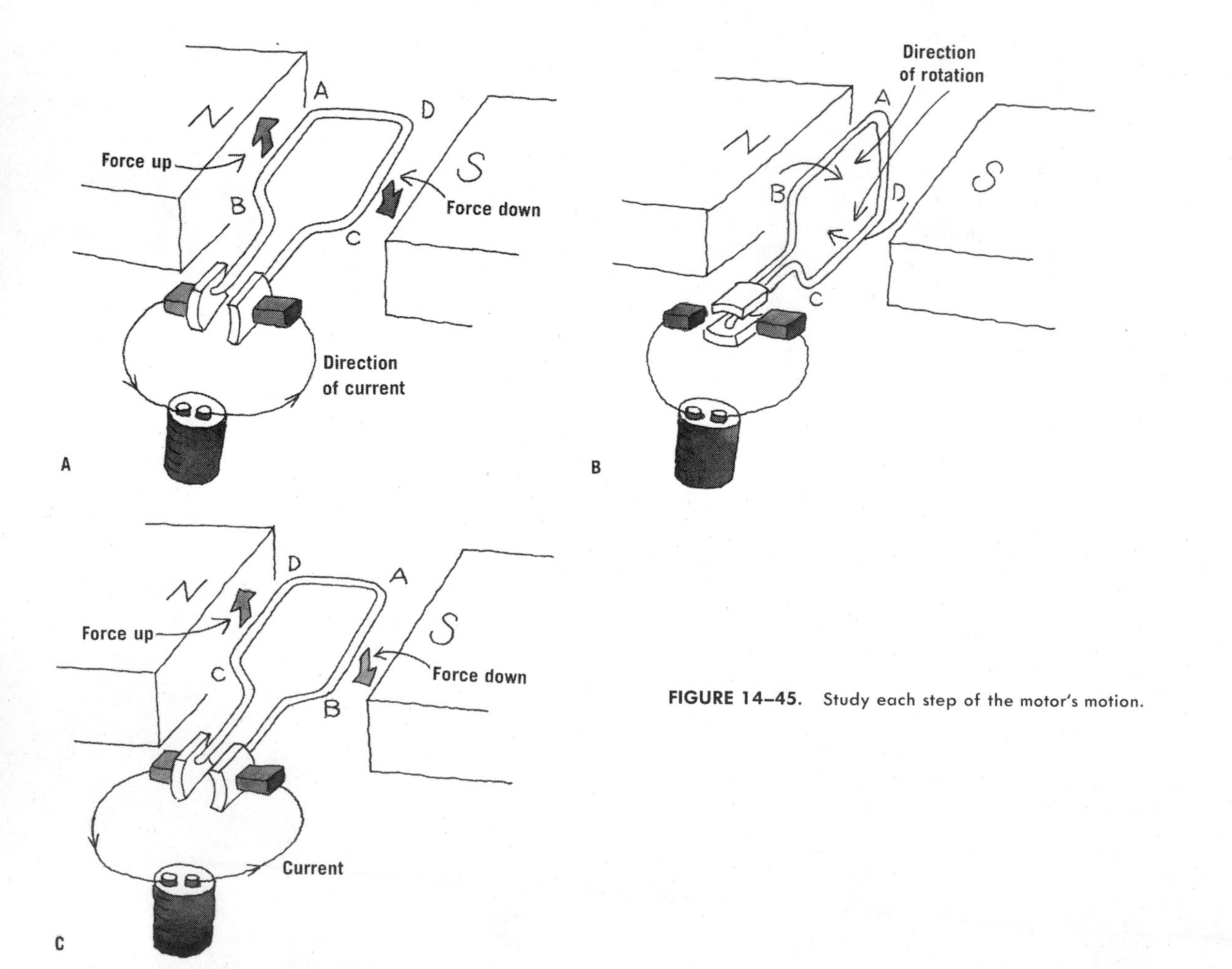

FIGURE 14–45. Study each step of the motor's motion.

The forces on the wires are as shown in the figure (again, use the right-hand rule to verify the directions). Thus the coil continues to rotate, always in the same direction.

The useful applications of the electric motor are literally uncountable. Most applications require a smoother operation than our simple motor would provide (it would "jerk" twice with each turn). To obtain this smoother action, a greater number of magnets are used, and more coils are wound, so that there is always a strong force acting on the coils.

OBJECTIVES

You've studied this chapter, so you should be able to:

1. Explain how an iron bar can be magnetized by using a permanent magnet.
2. Construct a simple device to measure the relative strengths of magnets.
3. State the general attraction-repulsion rule of magnetic poles.
4. Define the term "magnetic field" and explain the meaning of the arrows used to represent a magnetic field.
5. Map the magnetic field of a permanent magnet by the use of (a) a small compass, and (b) iron filings.
6. Given a wire carrying an electric current in a known direction, describe the magnetic field near the wire.
7. Use the right-hand rule to relate the direction of the current through a coil to the direction of the magnetic field through the coil.
8. Construct an electromagnet with a battery, wires, and a bolt or nail.
9. Describe, by means of a diagram, how a doorbell functions.
10. Use the concept of electron orbits to explain permanent magnetism.
11. Sketch the magnetic field of the earth.
12. Perform a demonstration which shows that the direction of the magnetic field is not generally parallel to the surface of the earth.
13. Use the magnetic field of the earth to magnetize an iron rod.
14. Given a drawing of an electric charge moving in a magnetic field, predict the direction of the force on the charge.
15. Explain how magnetic fields can be used to trap charges in a "magnetic bottle."
16. Given a current-carrying wire in a magnetic field, predict the direction of force on the wire.
17. Use a diagram to explain the operation of an ammeter,

and describe how such a meter is hooked into a circuit.

18. Diagram the essential parts of an electric motor and explain its operation.

QUESTIONS— CHAPTER 14

1. A wire carries a current from south to north. If a compass is placed above the wire, in what direction does its north pole point? Repeat for a compass placed beneath the wire.

2. A beam of electrons moves from the electron gun of a television tube toward a screen. How would a bar magnet have to be placed in order to deflect the electrons upward? Sketch.

3. List several similarities and differences of electric and magnetic forces.

4. A magnet attracts a piece of iron. The iron can then attract another piece of iron. Explain, on the basis of alignment of the atomic elementary magnets, what happens in each piece of iron.

5. Two parallel wires carry electric currents. Do they affect each other? Why?

6. Devise a hypothetical method for levitating a flying carpet by using a huge current flowing through a wire in the carpet, and the magnetic field of the earth. (Extra credit will be given if you build a working model.) Would your carpet work equally well near the equator as near the magnetic poles?

7. You are an astronaut stranded on a planet with no test equipment or minerals around. The planet doesn't even have a magnetic field. You have two pieces of iron in your possession; one is magnetized, one is not. How could you determine which is magnetized?

8. There are irregular variations in the strength and direction of the earth's field at points on the surface having the same latitude. What could cause these variations?

9. A beam of particles shoots through your dormitory room. If you would like to know whether they are charged positively or negatively, how could a magnetic field resolve your problem?

10. Why does the magnetic field of a wire become stronger when you form the wire into a circular turn?

11. List several factors that influence the strength of the magnetic field of an electromagnet.

12. Will a nail be attracted to either pole of a magnet? Explain what is happening inside the nail.

13. An ammeter is inserted in a circuit in series with a light bulb in order to measure the current through the bulb. Why must ammeters be low-resistance instruments?

14. What would happen to a current-carrying coil if you suspended it by a string so that it could rotate freely in all directions?

15. A voltmeter is a device which measures the voltage difference between the points to which its terminals are connected. The internal construction of this instrument is basically the same as that of an am-

meter except that its resistance must be much higher. Why? (Hint: Before you can answer this question, you need to determine how you would connect a voltmeter to measure the voltage across, say, a light bulb. How would you do it?)

16. The north-seeking pole of a magnet is attracted toward the magnetic north pole of the earth. Yet like poles repel. What is the way out of this dilemma?

17. Suppose that the ammeter in Figure 14–43*B* were hooked into the circuit on the other side of the light bulb. Would it read less, more, or the same in that location? Why?

15

INDUCED VOLTAGES

These generators in the powerhouse of Watts Bar Dam have a capacity of 150,000 kW. (Courtesy of Tennessee Valley Authority.)

We have seen that an electric current causes a magnetic field. We now look at the other side of the coin: can a magnetic field produce a current? You probably already know the answer—perhaps not so much from experience with magnetic fields, but from experience with books. If magnetic fields could not produce a current, the question would not have been used to begin a chapter!

KEEP THE MAGNET MOVING

Michael Faraday (1791-1867)

Faraday was one of ten children of a poor blacksmith in England. As a boy he attended a lecture given by Sir Humphrey Davy, a noted scientist. He took notes during the lecture and sent them to Davy, who was so impressed that he later offered Faraday a position as his assistant at the Royal Institution in England. Faraday belonged to a now-extinct religious sect which did not approve of worldly goods, and for this reason he refused the many awards offered him during his lifetime, including knighthood and the presidency of the Royal Society. One of his particular pleasures was giving scientific lectures to children every year at Christmas.

The critical experiment to answer our opening question was performed by Michael Faraday in 1831. An updated version of his experiment follows. As illustrated in Figure 15–1, a bar magnet is held near a coil of wire connected to an ammeter. (The ammeter used here has its zero reading at the center so that it indicates current in either direction.) When the magnet is held stationary in the position shown, the ammeter reads zero—there is no current. But as the magnet is brought closer to the coil, the meter deflects. If the magnet is moved away from the coil, a deflection again occurs, but this time the ammeter indicates that the current is in the opposite direction. Only when the magnet is in *motion* does the meter indicate a current in the coil. In addition, the faster the magnet is moved, the greater is the deflection of the needle, and hence, the greater is the current. A final factor influencing the amount of current produced is the number of loops of wire in the coil. If the coil has several turns, the magnet affects each turn individually; twice as many loops results in twice as much current.

In all previous examples, when a current was found to be present in a wire, it was set up by a voltage source such as a battery; in the present case, there is a current with no apparent source. In these circumstances we say that the effect of the magnetic field has been to *induce* a voltage in the coil and hence a current.

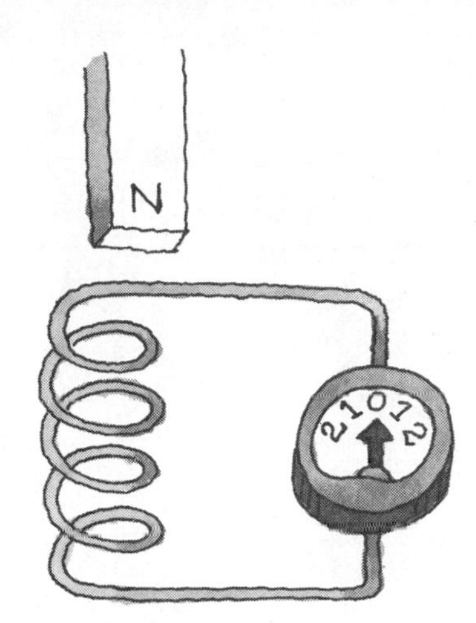

FIGURE 15–1. When the magnet is being pushed into or pulled out of the coil, current flows.

DIRECTION OF INDUCED CURRENT

We can use our picture of the magnetic force lines of a bar magnet to predict the direction of the induced current in a circuit. Remember that the spacing between the field lines indicates the strength of the field; the closer together the lines are drawn, the stronger is the field. In Figure 15–2*A*, a magnet is placed so that its north pole points toward a single turn of wire. The magnet is not close to the coil, and so the magnetic field at the coil is not very strong. We can visualize this concept by saying that only a few lines of the magnetic field pass through the loop of wire. Figure 15–2*B* shows the loop as we look through it toward the north pole of the magnet. The dots indicate that only a few magnetic field lines (two in the figure) pass through the loop and that these are pointing out of the paper. In Figure 15–3*A*, the magnet is being moved toward the coil. As it moves, the magnetic field at the coil is getting stronger, and hence more and more field lines pass through it. We represent this fact in Figure 15–3*B* by showing more lines of the field passing through the coil (four in the diagram).

In order to determine the direction of the induced current, let us assume the following: *in a coil of wire, a current is set up in a direction such that the magnetic field produced by the current will try to prevent any change in the number of field lines passing through the loop.* In other words: (1) If the original field is *increasing*, the field set up by the current induced in the coil will be in the opposite direction; (2) If the original field is *decreasing*, the field set up by the induced current will be in the same direction as the original field. These statements are rather long and involved but will be simple to use after a few examples are given.

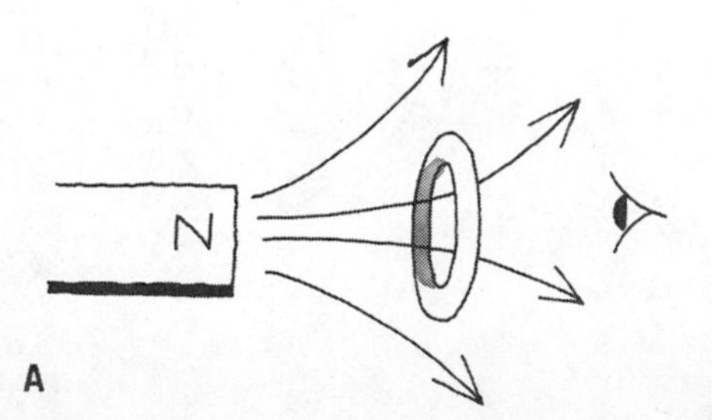

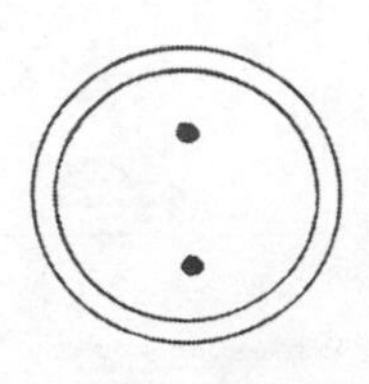

FIGURE 15–2. *A*, Side view of two field lines passing through a loop of wire. *B*, View looking toward the magnet.

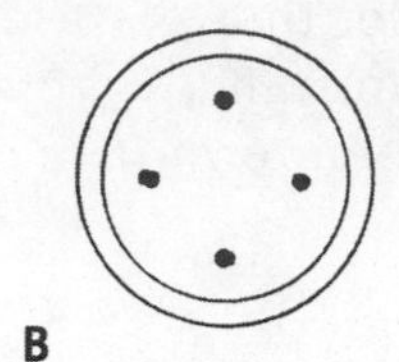

FIGURE 15–3. Now the magnet is moving closer to the loop. Note that four lines pass through the circle of wire.

In Figures 15–2 and 15–3, the magnet was moved such that the number of field lines passing through the coil changed from two to four. The coil "prefers" that the number stay at two rather than increase, and thus it sets up a magnetic field in the direction opposite to the original field in an effort to cancel the additional lines. This field is shown in Figure 15–4. Application of the right-hand rule shows that a current will have to flow clockwise, looking toward the magnet, in order to set up a field in an opposing direction. Once the magnet stops moving, there is no longer a change in the number of lines passing through the coil, and the current stops.

Clockwise Current

FIGURE 15–4. The current induced in the wire (shown as arrows) causes a field (X's) that opposes the increased field.

Now let's see what happens when the north pole is moved from a position near the coil to one farther away. When the magnet is close, several lines pass through the coil (four in Figure 15–5*A*), but this number decreases to a smaller value (two in Figure 15–5*B*) as the magnet moves away. The coil sets up a field in the same direction as the lines from the magnet in an effort to keep the number from diminishing. The field that is set up and the direction in which the current must flow in order to produce the field are shown in Figure 15–5*C*.

Figure 15–6 repeats the preceding examples, but in this case the south pole of a magnet is brought closer to the coil.

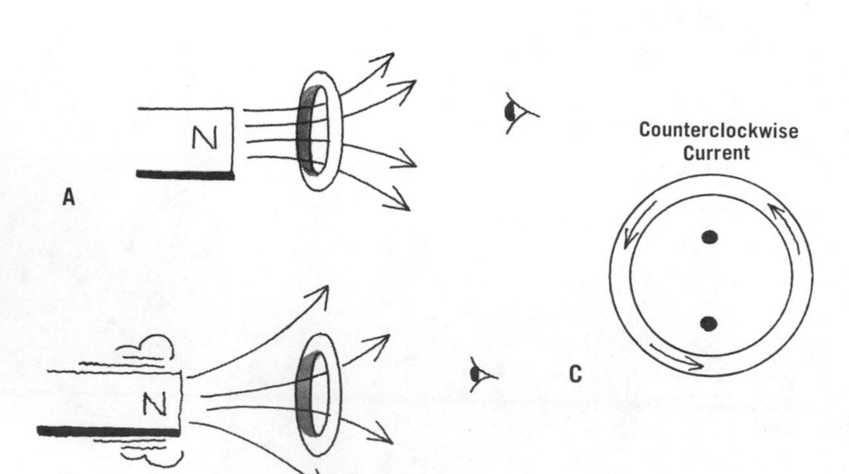

FIGURE 15–5. *A*, The magnet is close; in *B* it is moving away. This causes an induced current and its field (C).

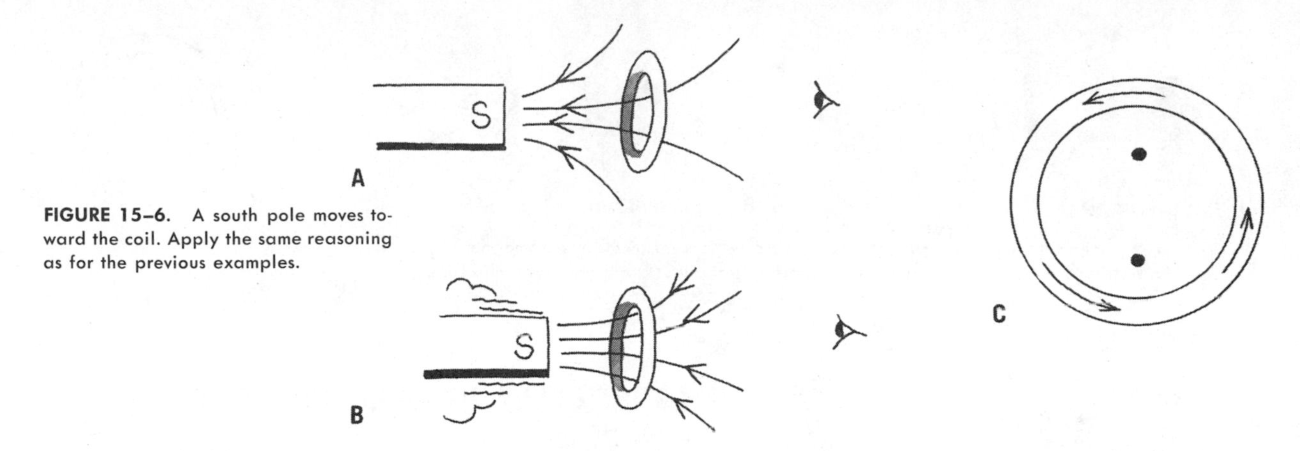

FIGURE 15–6. A south pole moves toward the coil. Apply the same reasoning as for the previous examples.

Use the rules discussed above to demonstrate that the induced current will be as shown in *C*. In Figure 15–7 the south pole is moved away from the coil, and again Figure 15–7*C* indicates the direction of the current.

Induction of Current without Moving a Magnet

As a final example, an electromagnet near a coil of wire is shown in Figure 15–8. Initially no current flows in the wires of the electromagnet, so no field is set up. When the current is turned on, the end nearest the coil becomes a north pole. (Review the previous rule for determining the polarity of an electromagnet, and verify that this end is a north pole.) As the number of field lines increases from zero to some number in the direction shown in *B*, the coil sets up a current to oppose this change. The resultant current is as shown in *C*.

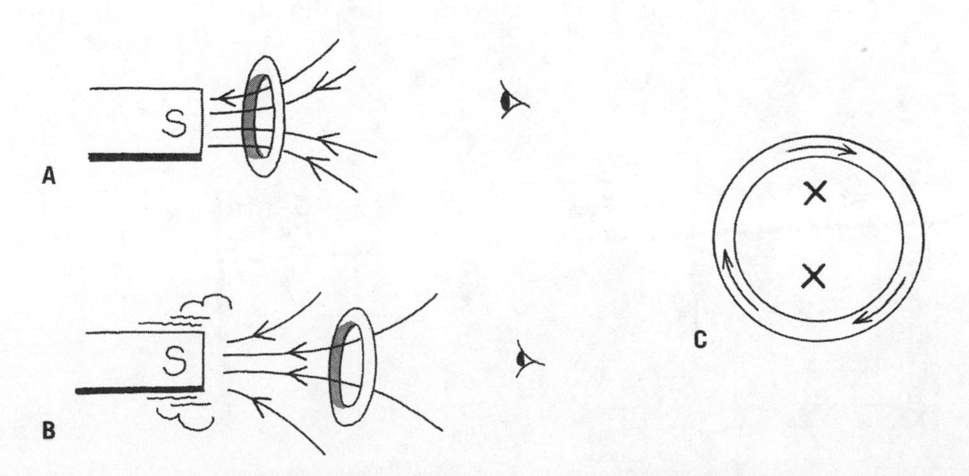

FIGURE 15–7. Now the south pole moves away. Again, check us.

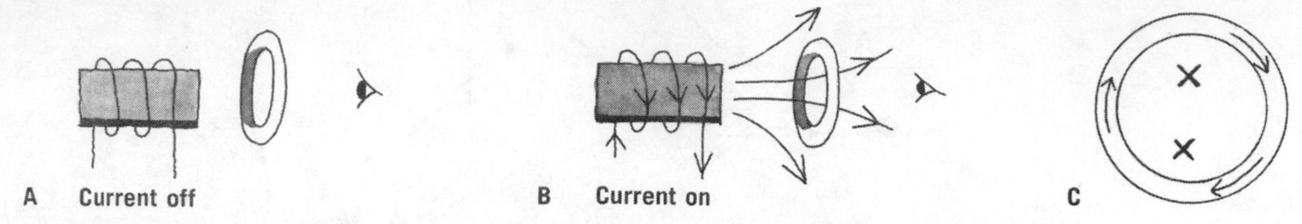

FIGURE 15–8. *A*, There is no motion here, but when the current in the electromagnet causes an increase in magnetic field through the loop (*B*), a current is induced in the loop to oppose that field (C).

FLOW OF MAGNETICALLY INDUCED CURRENT

It must be discouraging for a student to start a new chapter such as this one and be presented with a new phenomenon or a new principle which seems to bear no relation to previous principles. You probably hope that there is a connection between the phenomena of this chapter and previous ones. Good! This desire is at the heart of scientific thought. The scientist seeks to relate the various phenomena he sees. What is more, our old friend the law of conservation of energy will play a part in making the connection.

Magnetically induced currents are not essentially different from those discussed in Chapter 14. The fact that a magnetic field will exert a force on a moving charge (seen in the preceding chapter) enables us to explain the essentials of this phenomenon. In Figure 15–9, a conductor has been bent into a "U" shape, and resting on it is a piece of metal that is free to roll. The arrows indicate that there is a magnetic field directed into the paper and passing through the region enclosed by the metal. In Figure 15–9*B*, we focus our attention on the movable metal rod, which is shown to be rolling toward the left. In particular, we concentrate on one electron in the metal. As the rod moves to the left, the electron has a velocity, and it is in a magnetic field; hence there is a force acting on it. Application of the left-hand rule (electrons!) indicates that the force is upward on the page, as

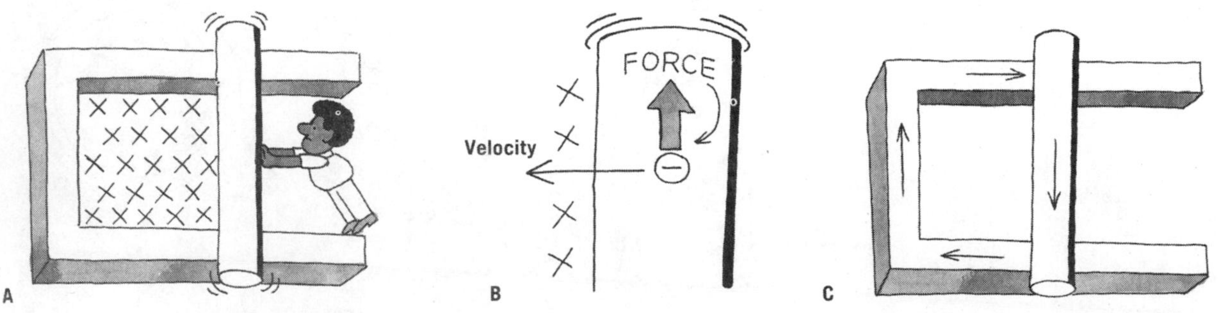

FIGURE 15–9. *A*, Rod is being rolled across the magnetic field. *B*, An electron in the rod feels an upward force and *(C)* causes a current. (Note that the positive current in C is in the opposite direction of the force on the electron in *B*.)

shown. This electron and all the others inside the rolling rod feel the force acting on them. The net result is a motion of electrons counterclockwise (and a current clockwise) around the circuit composed of the U-shaped wire and the rolling rod (Fig. 15–9*C*).

As an alternative approach to this problem, consider Figure 15–10. In *A*, the metal rod is in a position such that several magnetic field lines thread through the enclosed area (shown as six lines in the diagram). As the rod rolls to the left, it will move to a position as in *B*, where fewer lines thread through the area. In an effort to maintain the original number of lines, a current will be set up to create a magnetic field *into* the paper. Such a current must flow clockwise. This argument utilizes the reasoning developed in the previous section. It is slightly different from our other examples, however, in that there was a motion of a piece of metal in this one, whereas in all the others the metal wires were stationary while the magnetic field changed. As far as the electrons in the wire are concerned, it makes no difference; they feel a force acting on them in either case. Either of these approaches can be used to determine the direction of a current set up by a change in the number of lines of force threading through a conductor. We will use them both in the future.

Why the Coil Cares How Many Lines Pass Through It

We have used the idea that the current set up in a coil is in a direction such that the magnetic field caused by it tends to keep the total magnetic field from changing. For example, when the movable wire was rolled inward so that it decreased the number of lines through the loop, a current was set up in a direction such that it appeared that the loop was trying to maintain the original number of lines through it. But loops can't count lines! Is there any physical significance to this apparent "desire" to keep the number of lines constant? Yes, there is, and it depends upon the conservation of energy.

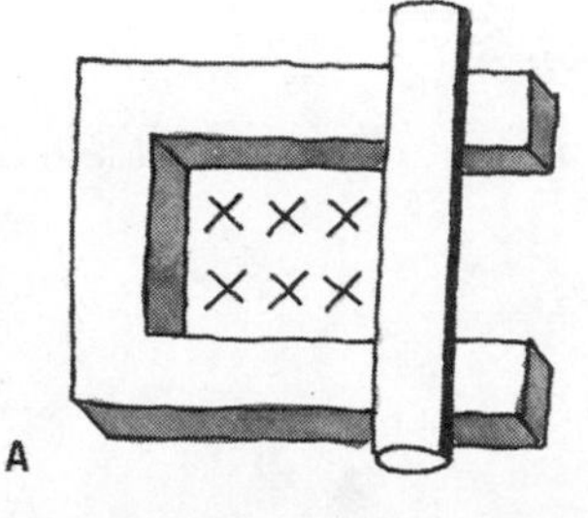

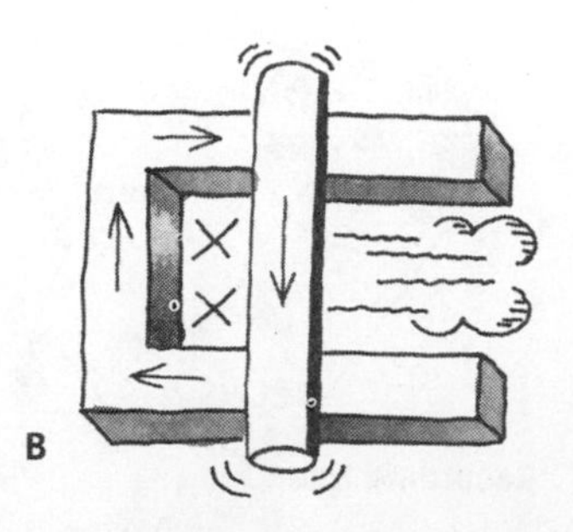

FIGURE 15–10. *A*, The rod moves in such a way that the number of magnetic field lines through the closed loop is decreased. *B*, A current is induced in the loop to oppose this decrease.

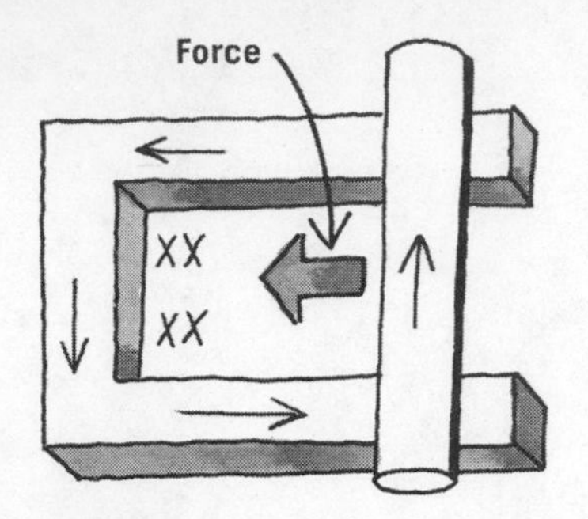

FIGURE 15–11. Suppose you pushed the rod inward and the current went in the direction indicated. (It doesn't, actually.)

Suppose that you started rolling the movable wire inward, establishing a current as shown in Figure 15–11 (instead of the way it *actually* goes). Now apply the right-hand rule to determine the direction of the force on this current-carrying wire. You will find that the force is *inward,* which would make the wire move faster. The increased speed should cause more current, which should make the wire move faster, which should cause more current, which.... Such an event is impossible because it would mean that if you barely nudged the wire, it would *take off.* Its kinetic energy would be increasing, but no other energy would be decreasing. The law of conservation of energy will not allow this. If a current is induced, it must be in the other direction; its associated magnetic field must exert forces to oppose the motion.

TAPE RECORDERS

In 1960, improved electronic recording techniques eliminated the faint hissing sound that had made the tape recorder a poor second to the phonograph player, and as a result the tape player has now made decisive inroads into the stereo enthusiast's world. There are three types of tape recorders: reel-to-reel, cassette, and cartridge. Our examples will be confined to the reel-to-reel instrument, but the basic principles are the same for all.

Figure 15–12 depicts the crucial parts of a tape recorder. Tape moves from the supply reel past a recording head and a playback head to the take-up reel. The tape itself is a plastic ribbon which is coated with iron oxide or chromium oxide. We will examine the function of both of the heads, starting with the recording head.

The steps of recording are illustrated in Figure 15–13. A sound wave is translated by a microphone into an electrical current. This signal is amplified and allowed to pass through a wire coiled around a doughnut-shaped piece of iron, which functions as the recording head. The iron ring and the wire constitute an electromagnet, which contains the

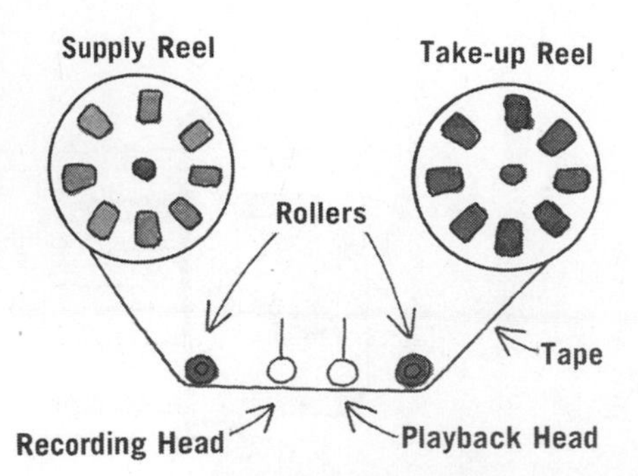

FIGURE 15–12. The basic components of a tape player.

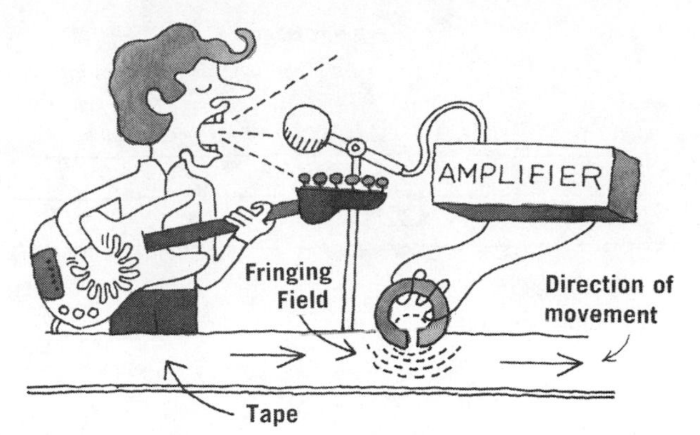

FIGURE 15–13. The changing electrical signal varies the magnetic field, which magnetizes the moving tape.

lines of the magnetic field completely inside the iron except at the point where a slot is cut in the ring. At this location the magnetic field fringes (or spreads out), and it will magnetize the small pieces of iron oxide imbedded in the tape. As the tape moves past the slot, it becomes magnetized in a pattern which reproduces both the frequency and loudness of the sound signal entering the microphone. Thus the recording process uses the fact that a current passing through an electromagnet produces a magnetic field.

To reconstruct the sound signal, the tape is allowed to pass through a recorder again, this time with the playback head in operation. Figure 15–14 shows that this head is very similar to the recording head in that it consists of a similar doughnut-shaped piece of iron with a wire coil wound on it. When the tape moves past this head, the varying magnetic fields on the tape produce changing field lines through the wire coil. These changing lines cause a voltage and current to be set up in the coil which corresponds to the voltage and current in the recording head that produced the tape originally. This changing electric current can be amplified and used to drive a speaker. The playback process is thus an example of induction of a current by a moving magnet.

FIGURE 15–14. As the tape moves by, its varying magnetization causes variations in the electric current induced.

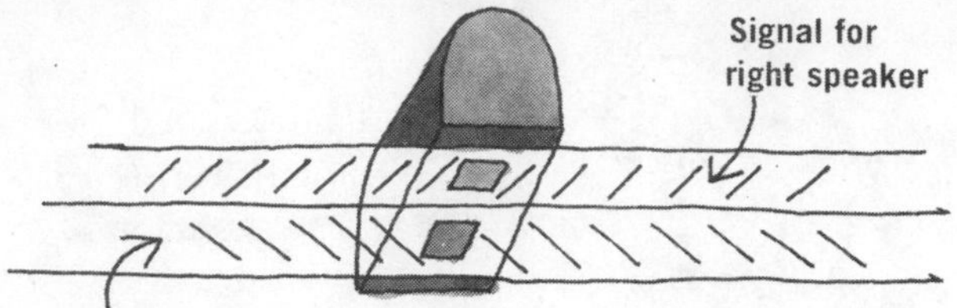

Playback Head

Signal for right speaker

Signal for left speaker

FIGURE 15–15. For stereo, two separate tracks are recorded.

The discussion of stereo recording on records in Chapter 11 indicated that two signals were inscribed in the record's groove. This procedure is simplified on tape, because the separate signals can be recorded on two different tracks of the tape, one above the other (Fig. 15–15). The signals from both tracks can be sent separately to different speakers as in disc recordings.

THE ELECTRIC GENERATOR

The electric generator is nothing more than an electric motor run backward. By this we mean that the motor uses a current to produce rotary motion, while the generator uses rotary motion to produce a current. For the moment we will put aside questions concerning the causes of motion in a generator and simply assume that a coil inside a magnetic field can be caused to rotate. Figure 15–16 shows a coil that is free to rotate between the poles of a magnet. The coil is connected to rings which rub (or brush) against stationary contacts, called brushes, which allow for connections to some external circuit that could be an entire city (but in our case is a single light bulb). When the coil is horizontal, as in *A*, no magnetic field lines pass through it, but a small rotation in the direction indicated by the arrows causes several lines to pass through, directed from left to right in the diagram. The rapid change in lines passing through the coil induces a large voltage in it. Application of the right-hand rule indicates that the voltage produces a current from A to B, which passes through the light bulb in a left-to-right direction.

As the coil moves from the horizontal to the vertical (Fig. 15–16*B*), the current decreases. When the coil is nearly vertical, a small rotation of it does not greatly change the number of magnetic lines passing through it, and the current drops, becoming zero when the coil is exactly vertical. In *C*, the coil has again rotated to the horizontal position, and again the current becomes large. Application of the right-hand rule shows that the current in the coil is now from B to A and will flow through the light bulb from right to left. A continued rotation of the coil again takes it to the vertical position, as in *D*, and again the current drops to zero.

When we examine the direction of current in the light bulb, we find that it follows these steps: (1) It has a large value from left to right; (2) It decreases and goes to zero; (3) It increases to a large value from right to left; (4) It decreases and goes to zero. A current of this nature that changes direction regularly is called alternating current and is the type supplied to your home by the utility company. Remember that a battery supplies current which is always in the same direction. The basic principles of operation of most electrical equipment are the same for both types of current, however. The resistance of a light bulb, for example, still causes the electrical energy possessed by the moving charges to be converted into heat and light energy. The alternating current produced by generators in this country varies at 60 hertz. This means that the current goes through a complete cycle 60 times per second.

The problem of causing the coil to rotate is solved in a variety of ways; a common method is to direct falling water against the blades of a device called a turbine (Fig. 15–17). The water striking the blades produces rotational motion of

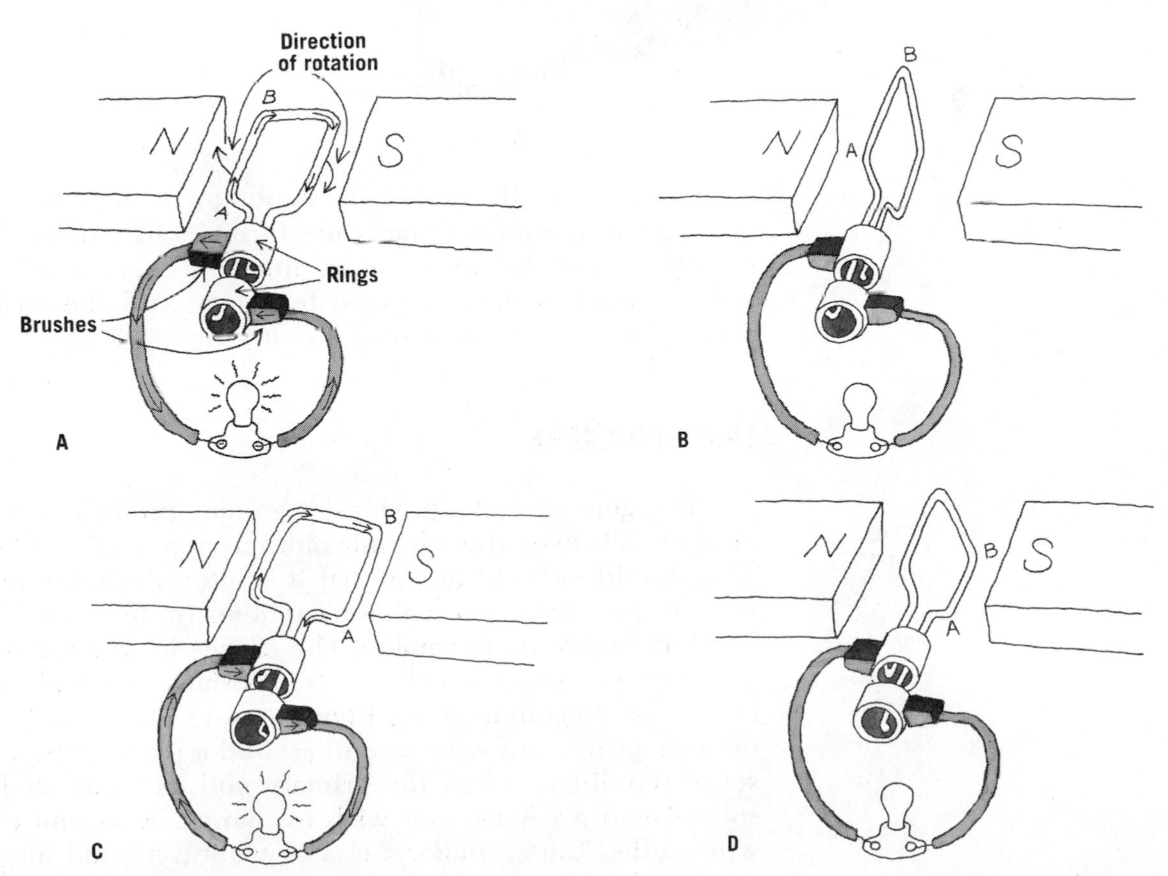

FIGURE 15–16. The electric generator. Note that the current stops in *B*, reverses its direction in *C*, and stops in *D*. The cycle then returns to *A*.

FIGURE 15–17. A turbine being replaced at TVA's Wheeler Dam. (Courtesy of Tennessee Valley Authority.)

a shaft on which the coil of the generator is mounted. Coal, oil, and nuclear power plants use turbines in a similar manner. However, in these cases, burning coal or oil heats water, which is then changed to steam, and the steam is directed toward the blades of the turbine.

TRANSFORMERS

The generator at a power station may produce a voltage of 2200 volts even though your radio requires only 110 volts. This should indicate to you that it is often desirable to convert voltages from one value to another (unless you like to watch radios go up in smoke). The device in which this conversion takes place is called a transformer. A simple transformer is diagrammed in Figure 15–18. It consists of a number of turns of wire wound around a piece of iron. One set of windings, called the primary coil, is connected to a source whose voltage you wish to change. A second coil of wire, called the secondary, is also wound around the iron, and it will deliver the changed voltage to the point where it is to be used.

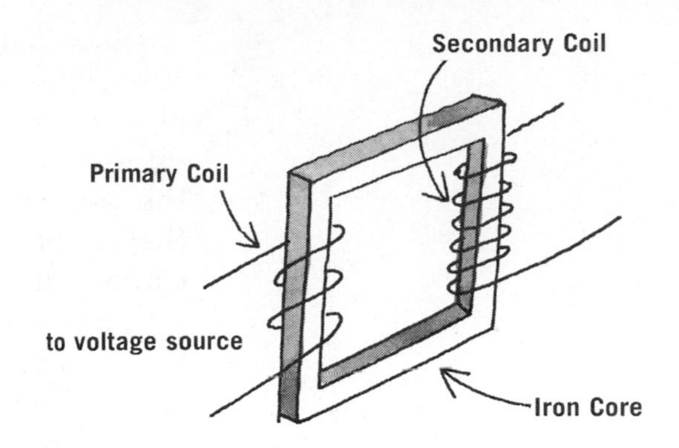

FIGURE 15-18. The differing number of turns in the primary and secondary coils is the key to the transformer's operation.

As an example of transformer operation, assume that we have an alternating voltage source, such as a generator, which furnishes 10 volts. An alternating current in the primary produces a changing magnetic field. As shown in Figure 15-19, these changing lines of the magnetic field thread through the turns of the secondary coil. The changing number of lines induces a voltage in the secondary, which can be either more or less than the voltage induced in the primary, depending upon the number of turns of wire in the primary and secondary. If there are twice as many turns in the secondary as in the primary, the voltage is twice as great at the secondary. If the secondary has fewer turns than the primary, the voltage is less at the secondary.

A casual glance at one of the statements in the preceding paragraph should cause some raised eyebrows. If the number of turns in the secondary coil is greater than in the primary, you get an increase in voltage. Does this mean that

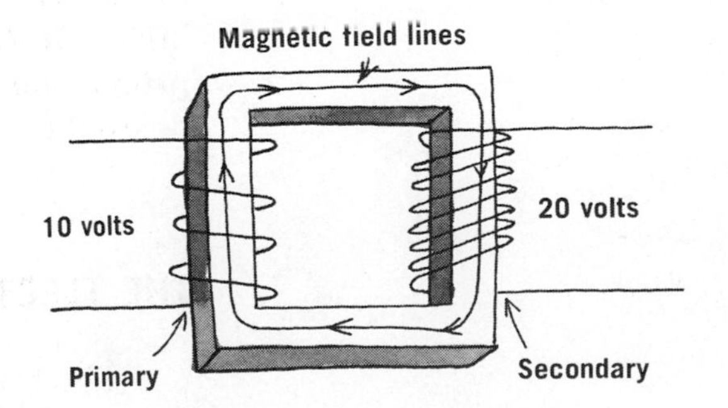

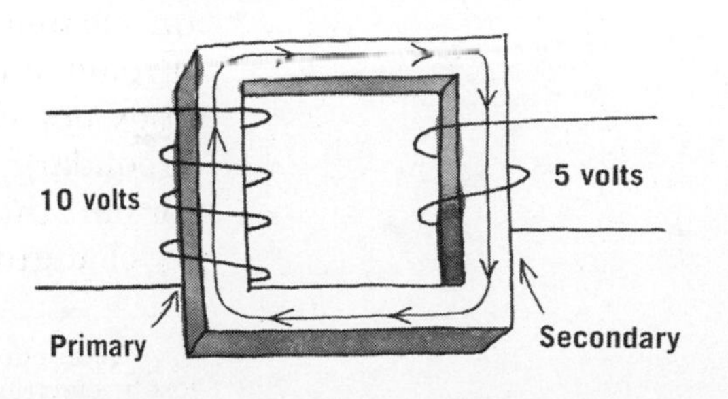

FIGURE 15-19. Note that in each case the ratio of voltages equals the ratio of turns (A, 20/10 = 8/4; B, 5/10 = 2/4).

we are getting something for nothing? It really does not. When the voltage is increased, the maximum current attainable is decreased by the same amount. The rate at which energy is supplied to the primary is equal to the current in the primary times the voltage applied to the primary. (See the chapter dealing with electric currents.) The secondary cannot supply more energy than is supplied to it, or all our principles of energy conservation would become inoperative. This implies that if the voltage is doubled in a transformer, the current must be reduced by half and vice-versa.

As an example of the use of transformers, let's follow the path of current and voltage from a power station to your home. Let's assume that the generator at the power station furnishes 2200 volts. In order to transmit enough energy to supply a city at this voltage, a very high current would be required. Such a high current could overheat the power lines, resulting in a great deal of energy loss. To overcome this problem, a large transformer at the power station is wound with many more turns on the secondary than on the primary. Suppose that there are 100 times more turns on the secondary than the primary. As a result, the voltage is changed from 2200 volts to 220,000 volts, and the current decreases to 1/100 as much as previously. (The current is responsible for heating the power lines; thus a reduction in current produces a tremendous saving in energy.)[1]

At substations near the home, this high voltage is reduced to a value similar to that originally produced at the generator. A transformer with 100 times more turns on the primary than on the secondary would reduce the voltage to 1/100 of its former value—from 220,000 back to 2200 volts. This voltage is still too high for home use, so a final transformer is needed. This last transformer is located near the home. It has fewer turns on the secondary than on the primary, and as a result it changes the voltage to levels desired for home appliances.

THE ELECTRICAL SYSTEM OF AUTOMOBILES

The transformers discussed in the preceding section are designed to operate on alternating voltages. They depend on changes in the number of magnetic field lines passing through a coil. If the number of lines through the secondary does not change, no current or voltage is produced in the secondary, and since there are no moving parts in a transformer, the only way that the magnetic field can change is by changing the current producing it. For this reason trans-

---

[1]If the current is reduced to 1/100, the energy lost is reduced to 1/10,000. Energy lost by electrical resistance equals the current squared times the resistance.

FIGURE 15–20. The transformer changes the high voltage from the wires on top to 220 volts across the lower wires.

formers are not suitable for use with direct currents except for a situation such as the following. In an automobile engine, spark plugs produce an intense spark in order to ignite a gasoline-air mixture in the cylinders. In order for the plug (Fig. 15–21) to work, a voltage must be created which is high enough to cause a spark to jump across the gap shown. The method for creating this high voltage is indicated in Figure 15–22. The primary coil of the transformer (the transformer itself is called a coil in an automobile) is connected to the 12-volt battery of the car. Because batteries provide a direct current, there are no changing magnetic field lines passing through the secondary, and as a result no voltage or current is present in the secondary. The secondary is connected to a spark plug through a device called the distributor.[2] When a spark plug is supposed to fire, a switch breaks the circuit between the battery and the primary. This interrupts the current in the primary, and because of the cessation of the current, the magnetic field quickly drops to zero. This rapidly changing magnetic field through the coils of the secondary produces a voltage and current in it. The transformer is wound with several hundred times as many turns in the secondary as in the primary, and the resulting voltage in the secondary is quite high—high

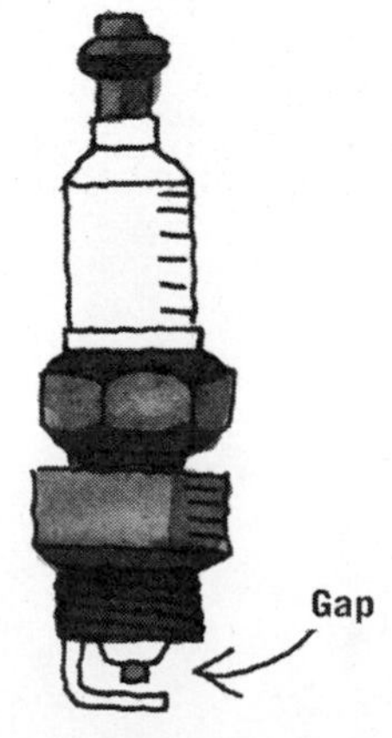

FIGURE 15–21. A spark plug. The electrical connection between the top terminal and the center pin at bottom is not shown.

[2]The distributor is a mechanical device that connects, in turn, each of the spark plugs in the engine to the secondary coil at the instant the spark plug is supposed to fire.

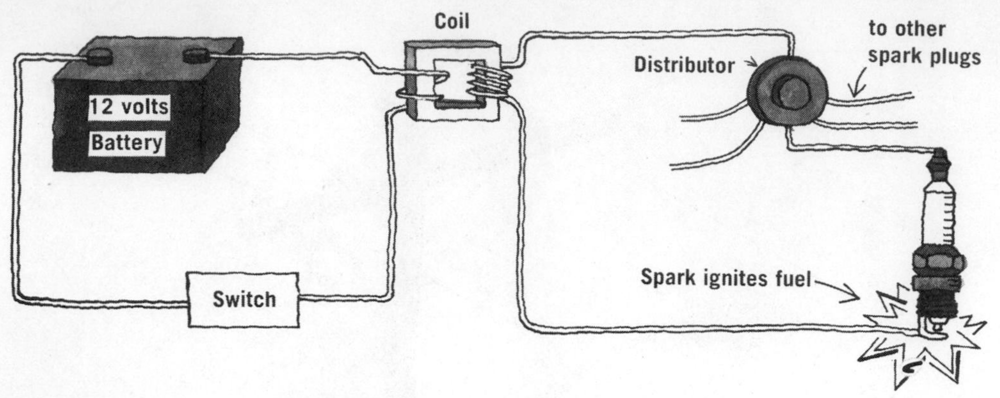

FIGURE 15–22. The ignition system of a car.

enough to cause a spark to jump when it is applied across the air gap of the spark plug. It is this spark which ignites the fuel in the cylinder.

A FINAL RIGHT-HAND RULE

In the last two chapters you have been besieged with right-hand rules used to determine directions of magnetic fields and currents. Usually after so many of these rules, a feeling of frustration begins to set in. And so, in an effort to relieve your tension, we propose this final right-hand rule. Curl the fingers of your right hand around the tops of the pages of this text such that they encircle all the pages of the last two chapters. With your thumb pointing toward the binding, slowly but firmly pull downward. When you have finished, the loose pages in your hand will not add to your knowledge of physics, but they may produce a degree of satisfaction.

OBJECTIVES

The authors hope that Chapter 15 will prepare you to:

1. Describe an experiment demonstrating induced current.
2. Given a situation of a north or south magnetic pole moving toward or away from a coil of wire, predict the direction of the current induced.
3. Describe a way to induce current in a coil without actually moving either a magnet or the coil.
4. Explain by using (a) the conservation of energy, and (b) the right-hand rule, why current is induced by a changing magnetic field.
5. Use knowledge of the phenomena of induced voltage to show how an audio tape player operates.

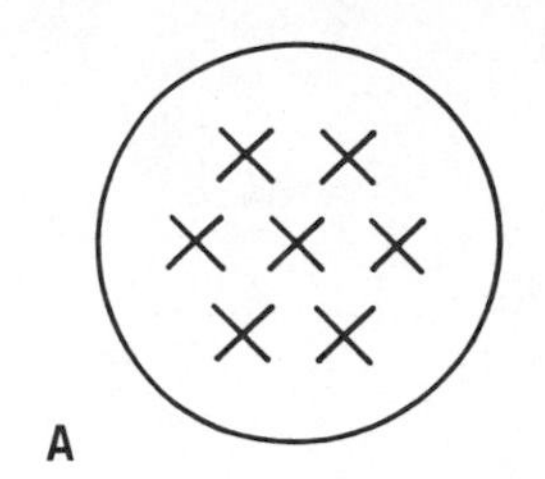

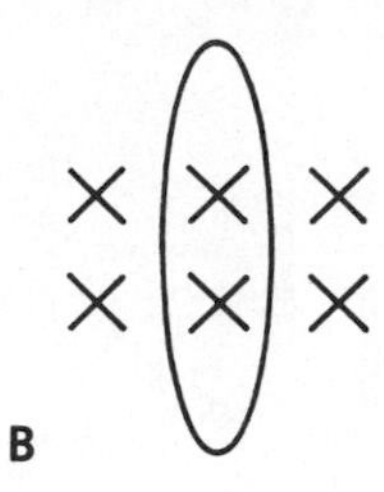

FIGURE 15–23. Mashing a loop of wire.

6. Explain how stereo is recorded on tape.
7. Use drawings to describe the operation of an electric generator, including statements of the current direction during each part of the generator's cycle.
8. Explain the functioning of a transformer, and given the number of windings on both primary and secondary, predict the ratio of voltages in the coils.
9. Discuss the use of transformers in transmitting power from the generating station to the home.
10. Explain how a transformer is used in the ignition system of a car.

QUESTIONS – CHAPTER 15

1. In Figure 15–23A, a loop of wire is in a magnetic field directed into the paper. If the loop is grasped at the sides and mashed into the shape shown in *B*, in what direction does a current flow while the change is occurring?

2. A coil of wire is located in a magnetic field which points out of the paper. How would the magnetic field have to change in order to produce a clockwise current in the coil?

3. A transformer has five turns on the primary. If ten volts are applied to the primary, how many turns must be in the secondary in order to produce 60 volts? Six volts?

4. Why are primary and secondary coils of a transformer wrapped on an iron core that passes through both coils?

5. Compare and contrast the electric motor and the electric generator.

6. Will dropping a magnet down a long copper tube produce a current in the tube? Explain.

7. A satellite circling the earth has a coil of wire in it. An astronaut notes that there is a current in the coil although no battery is connected to it and there are no magnets on the spacecraft. What is causing the current?

8. The authors have devised a plan to solve the problems of generating electricity for centuries to come. We propose to use a motor to turn a generator which will produce electricity to turn our motor and simultaneously furnish electricity to a city. (We may use a transformer to increase voltage.) Do you see anything wrong with our idea? If not, stock certificates are for sale.

9. Would a generator work if the coil were held stationary and the magnet rotated? Explain.

SECTION FIVE

WAVE MOTION—EMPHASIS

16

ELECTROMAGNETIC WAVES

James Clerk Maxwell, the man who in 1864 developed the theoretical framework which has made possible not only our better understanding of light waves, radio waves, and x-rays but also the many devices of our everyday life which depend upon them, said of his work:

> I have a new theory, which until I am convinced to the contrary, I hold to be great guns.

We will now examine the area of physics to which Maxwell was referring.

PARTICLE OR WAVE?

It seems innocent enough to speak glibly about waves of light, but the idea of calling light a wave was not one that was met with quick acceptance by the physicists of 300 years ago. These early physicists were familiar with different types of waves, such as water waves and sound waves. All the varieties of waves which they knew about had a common feature: they all required a substance (or medium) through which to travel. A sound wave cannot be transmitted without gas, liquid, or solid—it will not pass through a vacuum; a water wave also obviously has a medium for its propagation. It was believed that all waves must have a medium through which to pass or else there could be no wave, but by the seventeenth century it was an accepted fact that the vastness of space was an almost perfect vacuum. If this were so, physicists reasoned, "how can it be that we can see the sun? There is no medium between the earth and the sun in which these light waves could travel." Thus some adjustment of the theory of light was necessary. The early physicists could have decided that somehow light waves require no medium for transmission, but this would mean that wave theory would have to be changed. It would somehow have to allow for waves moving through a vacuum, which would mean that the waves would be waving where there was nothing to be waved. Such a change seemed unreasonable and unwarranted.

Another possible approach would be to consider light not as a wave but as a stream of particles, a theory favored by Isaac Newton (1642–1727). He was able to explain successfully all of the known phenomena of light by considering it to be a stream of particles that are emitted by an object and which enter the eye to produce the sensation of light. Because of Newton's tremendous success in other areas of physics, some of which were discussed earlier, his ideas held sway for a century.

Thomas Young (1773-1829)

The young Young could read when he was two years old and had read the Bible through twice by the age of four! He received his degree in medicine when he was 23 and opened practice in London. He was the first to describe how the lens of the eye changes shape to focus on objects at different distances, and to explain astigmatism. It was from his interest in vision that he progressed to a study of light, wherein he made the discoveries for which he is known today. His contributions also include such widely diverse areas as a study of the surface tension of liquids and the deciphering of ancient Egyptian hieroglyphics.

In 1803, however, an experiment was performed that showed conclusively that light could not be a particle but had to be a wave. This experiment by Thomas Young, an English physician, showed that two beams of light could be made to interfere. We have seen that sound waves interfere: when they overlap under certain conditions, areas of loud sound and other areas of no sound are created. This phenomenon was easily explained by the superposition of sound waves. We will delay the details of Young's experiment until a later chapter, but what is important to us now is the fact that interference could not be explained if light consisted of a stream of particles. There is no imaginable way that one particle can cancel out another.

So light must be a wave! What about transmission through a vacuum, then? If outer space is a vacuum, and waves require a medium for transmission, and Young's experiment showed that light acts like a wave, we seem to be at a dead end (or at least in a vicious circle). But, fear not, a way out was found.

The Ether

The approach that was taken to solve the problem of light waves is ingenious in its simplicity. Since there had to

FIGURE 16–1. Physicists have been known to debate issues with some vigor.

be a medium for transmission of light, physicists concluded that such a medium exists and that they simply had not yet found it. This substance would have to be so subtle that it would fill all space, it would fill the human body, it would elude all chemical tests, and when all the air had been pumped out of a bottle by a vacuum pump, this substance would still remain. So convinced did they become of its existence that they gave it a name—"the luminiferous ether."[1] An object which was emitting light would somehow set the ether into motion. The waves that were formed in the ether would propagate outward at a speed later found to be 186,000 miles per second and would finally strike the eye, causing us to see the object which emitted them.

This idea was accepted for nearly 100 years, until serious attempts were undertaken to detect the ether. Finally, in 1887, a famous experiment by Albert A. Michelson and Edward W. Morley proved conclusively that ether was difficult to detect because it simply *does not exist.*

So physics was back to its vicious circle. Only one conclusion remained: light must be a special cat. It must be a wave which does not require a medium for its propagation. Whatever it is that makes up a light wave must be able to exist in a vacuum.

So here we stand. Light *does* act like a wave, and in the next few chapters we will investigate its wave nature and describe the phenomena of light-as-a-wave. But when you find us writing that light "acts like" a wave instead of writing that light "is" a wave, remember the history of the controversy (and prepare for trouble in Chapter 20).

---

[1]Do not confuse this "ether" with the gas of the same name which is used in medicine to make you forget another kind of painful trouble.

PRODUCTION OF AN ELECTROMAGNETIC WAVE

The name "electromagnetic wave" destroys the suspense by telling us the composition of the wave. "Electro" indicates that an electric field is involved, and "magnetic" indicates that a magnetic field is also present. The question to be considered in this section is: how do we produce these waves composed of electric and magnetic fields?

To investigate this question we will perform an imaginary experiment, using two metal rods each about three feet long. In any piece of metal there are a large number of electrons that are free to move around. There are an equal number of protons, but they are not as mobile as the electrons, and they tend to stay in fixed locations. Connected to these metal rods is a device which makes the electrons move up and down in the rods (Fig. 16–2).

This movement of electrons is an electric current. In an earlier chapter we saw that an electric current produces a magnetic field around the wire. Thus, as the charges are jostled up and down in the rods they produce a magnetic field. When the electrons are moving down, a magnetic field is set up around the rods as shown in Figure 16–3*A*. When the electrons reverse direction and move up in the rods, the magnetic field also reverses (Fig. 16–3*B*). After a short time, a magnetic field pattern is formed around the wire as shown in Figure 16–4. At some locations the magnetic field points upward on the page, while a short distance back toward the rods the magnetic field is pointing down on the page. The entire pattern is rapidly spreading out from the rods just as a water wave would spread out from an object vibrating up and down in water. To more firmly appreciate the correspondence between water waves and this vibrating magnetic field, imagine the points where the magnetic field is pointing up on the page to be like the crests of a wave and

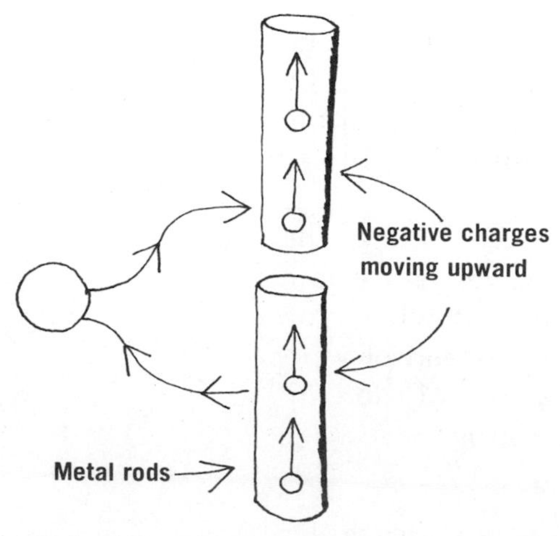

FIGURE 16–2. Two metal rods are connected to a device that causes electrons to move up and down through the rods.

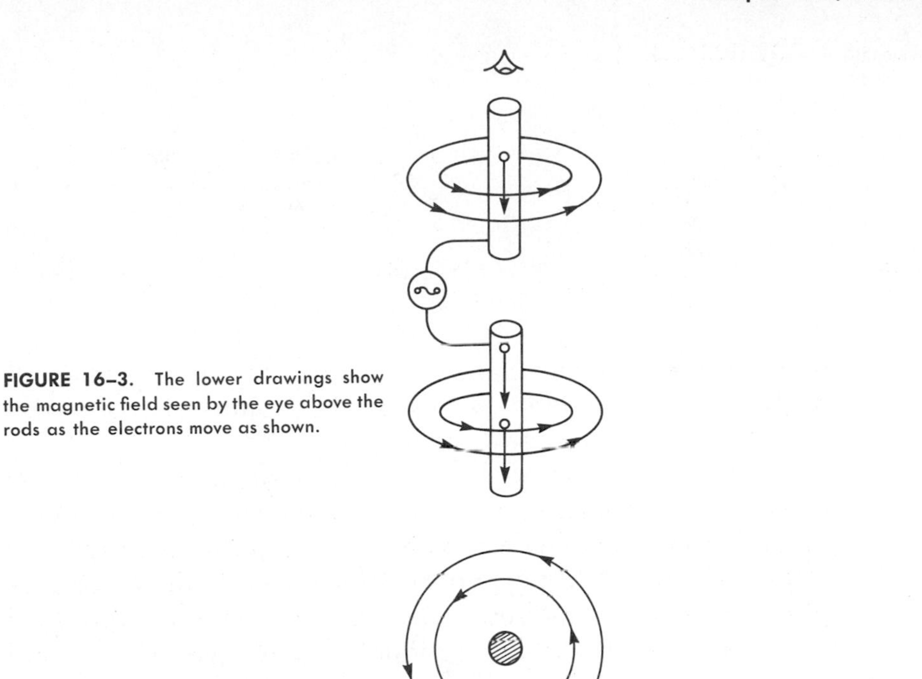

FIGURE 16–3. The lower drawings show the magnetic field seen by the eye above the rods as the electrons move as shown.

points where the field is down to be like the troughs of the wave. This allows us to represent the magnetic part of the wave by the curve shown in Figure 16–5 (which should be familiar to you from your study of sound waves). All the old terms which we have used to represent a wave must be called back into play.

The wavelength, as we have seen, is the distance between any two successive points on the wave which are in the same condition. In Figure 16–5 the wavelength is shown to be the distance between two points in which the magnetic field has its maximum value in a given direction. The speed with which the wave spreads out from its source is the wave speed, which is found to be 186,000 miles per second. The concept of wave frequency can also be incorporated into our discussion. We form a wave on a string by moving one end up and down at a constant rate, and the number of up-and-down motions per second is the frequency of the vibration and, therefore, the frequency of the wave. In the case of the "magnetic wave," the up-and-down motion of the electrons corresponds to the up-and-

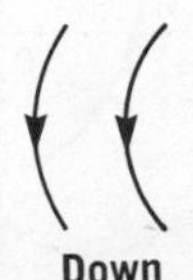

FIGURE 16–4. Magnetic field set up by oscillator (as seen from above the rod). At some points field is up; at others it points down.

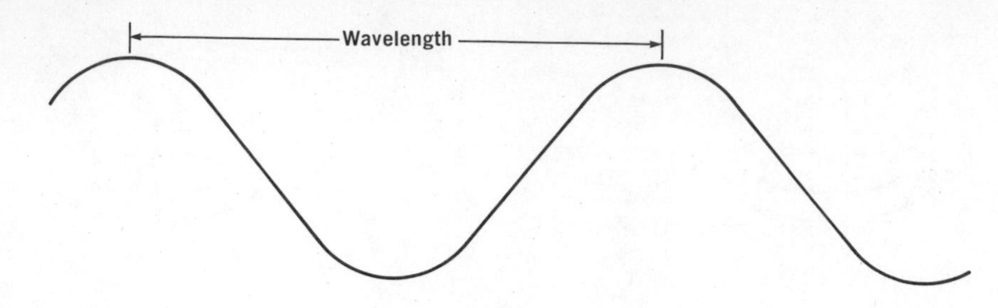

FIGURE 16–5. The definition of wavelength is the same as in the subject of sound.

down motion of our hand. Typically, the frequencies of useful electromagnetic waves range from somewhat less than 10^6 to approximately 10^{18} hertz.

We have discussed how a "magnetic wave" could be produced, but remember the name "electromagnetic wave"; in addition to the magnetic field portion of the wave there is also an electric field portion. We must have both, or we simply do not have an electromagnetic wave. We can produce an electric field without producing a magnetic field and vice-versa, but we cannot produce an electromagnetic wave without both.

So we now take up the electric field part of the wave. As the electrons surge up and down in the metal rods, there will be times when a large number of them will be found at the top. Before this oscillation of charges started, there were as many protons in the rods as there were electrons, which means that a large number of positive charges are left behind in the bottom rod (Fig. 16–6*A*). Also shown in this figure is the electric field now present around the rods. A short time later the negative charges will have moved to the bottom rod, and the configuration of the electric field will be as shown in Figure 16–6*B*. The arguments concerning the way in which a "magnetic wave" spreads out from the rod can now be adapted to the "electric wave." The final result is that in the space around the metal rod, a magnetic wave and an electric wave co-exist. To make matters simpler, let us now cease to refer to a magnetic wave and an electric wave since they are not really separate waves at all. Instead

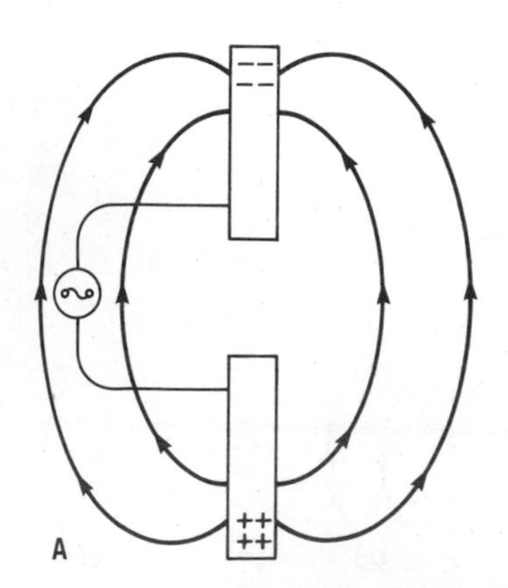

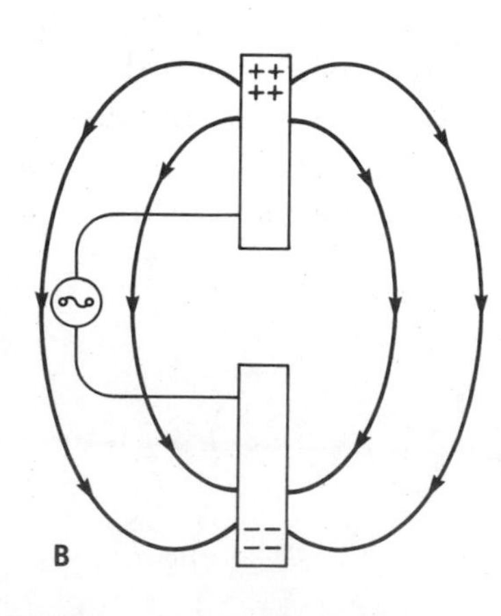

FIGURE 16–6. The electric field produced by two different charge distributions in the rods.

they form a unit, a wave which we call an electromagnetic wave.

At first thought this may seem to be an example divorced from reality, but it is taking place every minute of every day all over the world. A radio station antenna is nothing more than our metal rods, in which charges are caused to oscillate up and down at a frequency assigned to the station by the Federal Communications Commission and international agreements.

SPEED OF ELECTROMAGNETIC WAVES

In 1856, James Clerk Maxwell was doing a theoretical study of electromagnetic waves. Much was already known about electric and magnetic forces, and Maxwell was calculating the details of what should happen when charges are caused to oscillate in rods (as we discussed in the preceding section). Among the many predictions that resulted from his calculations was the prediction that these electromagnetic waves should travel at the speed of 186,000 miles per second. "But hark!" Maxwell thought, "that is the same as the speed of light!" Was a coincidence of this magnitude possible? Of all the possible speeds, could it be mere chance that electromagnetic waves travel at the same speed as light? Surely not! The conclusion must be that there is a relationship between light and electromagnetic waves. Indeed, we have found that light waves are only a particular type of a general category of waves called electromagnetic waves.

James Clerk Maxwell (1831-1879)

Maxwell, born in Edinburgh, Scotland, was a graduate of Cambridge University and later returned as a professor of experimental physics. While at Cambridge he founded the Cavendish laboratory, which became widely known in later years as a major center for research in nuclear physics. Much of his work was directed toward a mathematical study of gases, but he is best remembered for his research in electromagnetic theory.

THE ELECTROMAGNETIC SPECTRUM

The full range of useful electromagnetic waves is broken up into groups according to the frequency of the waves. We call this classification "the electromagnetic spectrum" (Fig. 16–7). Waves of low frequency (and therefore long wavelength) are called radio waves. The frequency of these waves is normally in the thousands of kilohertz (one kilo-

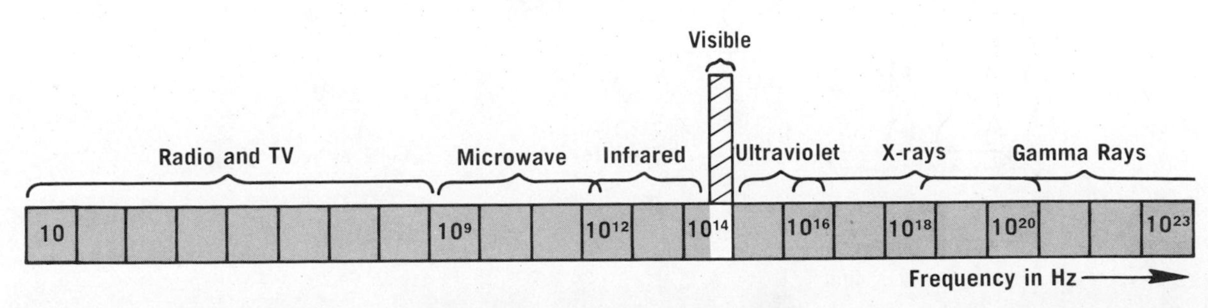

FIGURE 16–7. The electromagnetic spectrum.

hertz = 1000 hertz) range. Just above this band in the spectrum is the television band, extending into the range of hundreds of megahertz (*mega* = million). Moving up in frequency in the spectrum, one encounters the microwave and then the infrared region. Following the infrared is a very small region having frequencies ranging from 4.0×10^{14} hertz to 7.5×10^{14} hertz, and it is only this small band that is visible to the human eye. The lower frequencies in this visible range are perceived by the eye as red light, and as we move up to higher frequencies we see orange, yellow, green, blue, and—at the highest visible frequencies—violet.

Beyond the visible are the ultraviolet, the x-ray, and the gamma ray regions. The latter two are important to the medical profession. Gamma rays have sufficient energy to penetrate the body and can cause severe damage. Under controlled conditions, however, these rays can be used to kill cancerous cells. In Chapter 26 we will investigate several biological applications of gamma rays.

This power of penetration manifests itself in a useful fashion in the case of x-rays also.[2] If a beam of x-rays is focused on a hand, the penetrating power of the waves is sufficiently great to traverse the hand and expose a photographic film on which the hand rests (Fig. 16–8). Because it is easier for the waves to penetrate fatty tissue than the bones of the hand, examination of the exposed film shows the bones standing out in vivid detail (Fig. 16–9).

Infrared Radiation

The human eye is not sensitive to infrared radiation, but we do have a means of detecting it. When you put your hand near something warm, the sensation you feel is due to the infrared radiation emitted by that object. For this reason, infrared radiation is often called "heat radiation." Heat lamps,

[2]In fact, the lower frequency gamma rays are indistinguishable from x-rays. They differ only as to their source and will be discussed in Chapter 24.

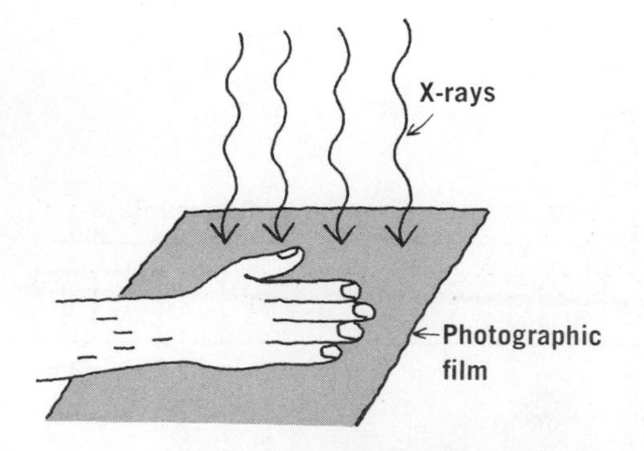

FIGURE 16–8. Hand exposed to x-ray beam.

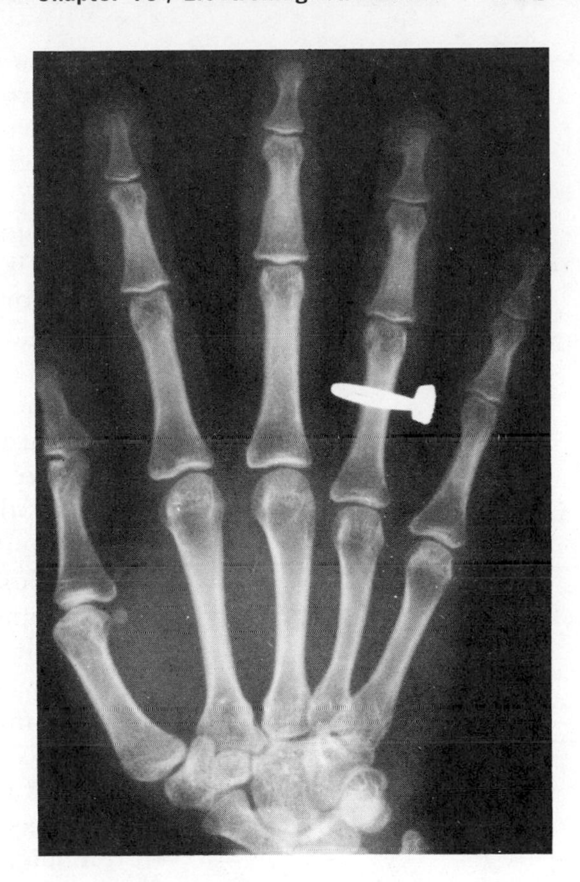

FIGURE 16–9. A hand exposed to an x-ray beam (with x-ray film on the other side of the hand). (From Poznanski, A. K., Garn, S. M., Nagy, J. M., and Gall, J. C., Jr.: Metacarpophalangeal pattern profiles in the evaluation of skeletal malformation. Radiology, *104*:1-11, 1972.)

which are used to warm tired aching muscles and are sometimes put in bathroom ceilings to warm and dry the bather, emit most of their radiation in the infrared region of the spectrum. Incandescent lights (regular light bulbs) also emit much infrared radiation, which can be felt, but we use them primarily for the visible light they emit. Fluorescent lamps, on the other hand, emit much less infrared light. Since one normally uses lights only for the visible radiation produced and not for the heat, fluorescent lamps are much more efficient in terms of energy consumed than are regular incandescent lamps.

Some types of photographic film are sensitive to infrared radiation. Cameras equipped with such film can take advantage of the fact that infrared rays penetrate fog and haze better than visible light. When cameras with film sensitive to the infrared are used in airplanes or satellites for photographing large land areas, they give more information than can be obtained using only visible light, even when fog and haze are not present. This is because there are sharp differences between the amount of infrared radiation emitted by the ground, by water, and by crops. In fact, although a field of corn and a field of soybeans may look very much the same from a high-flying plane, the amount of infrared radiation emitted by each crop is different, so that infrared photography is very useful in large-scale crop surveys. And there is an additional benefit: a diseased crop differs from a healthy crop in the amount of infrared emitted. Thus diseased areas of very large fields (including forests) can be pinpointed using this technique.

Warm objects emit more infrared radiation than do cooler objects. Thus infrared photography from airplanes and satellites (even at night) shows such features as population centers, factories, military troop concentrations, and ships. That there are political and military implications is obvious.

Telescopes have been developed which immediately change an image produced by infrared radiation into a visible image. A person equipped with such a telescope can locate another person in the dark simply by looking for a source of infrared radiation. The person as seen through such a telescope would not correspond to our normal image of the human body, however. Since the exposed areas of the body emit more infrared rays than do the cooler clothes which cover most of the body, the uncovered areas would show up much more vividly. Thus the head and hands of a fully dressed person would appear much brighter than the rest of his body.

Just as some animals are capable of detecting sound frequencies beyond the capability of the human ear, some forest creatures are able to detect parts of the electromagnetic spectrum other than that which we call visible. For example, some snakes are capable of seeing into the infrared region. A warm object would render itself visible to a snake even in what we would call total darkness. Facts such as this should be borne in mind by all who take their warm bodies into snake-infested dark places for whatever reason.

RADIO TRANSMISSION

One of the parts of the electromagnetic spectrum has been referred to as consisting of radio waves. The name obviously refers to their use as carriers of information from a radio transmitting station to your radio receiver. There are two fundamentally different kinds of radio stations: one is commonly called an AM station and the other, FM. These are the initials for amplitude modulation and frequency modulation. The meaning of these terms will become apparent shortly.

All radio stations are assigned a carrier frequency by the federal government. This assigned frequency is the basic frequency of the electromagnetic wave which that station is allowed to broadcast. If the station is an AM station, its carrier frequency will be between about 500 and 1600 kilohertz (thus the numbers from 5 through 16 on your radio dial). FM stations operate at higher frequencies, ranging from 88 to 108 megahertz.

AM Radio

Consider for the moment an AM station with an assigned frequency of 1000 kilohertz. Some means must be used to enable this electromagnetic wave to carry a message. (The term "message" in this context refers to any intelligible signal, usually musical or vocal.) One method might be for the station to turn its transmitter on and off in a dot and dash fashion and train all of its listeners to read Morse code. The difficulty here is, of course, that such a system would not reveal the character of the human voice (and music by Morse code is hard to imagine). A solution to the problem rests with the principle of amplitude modulation.

The term "modulation" simply means altering or changing. To visualize the modulation process of our AM station, let the carrier wave be represented as in Figure 16–10*A*. Suppose an innovative programmer decides to broadcast a program called "The 100 hertz Tuning Fork Hour." During the program he broadcasts the sound of a tuning fork producing a single continuous note. The note is picked up by a microphone at the studio and converted to an electrical signal represented by the wave form shown in Figure 16–10*B*. Electrical equipment at the station (the description of which is beyond the scope of this book) superimposes this signal from the tuning fork onto the carrier wave (the basic 1000 kilohertz wave) in such a way that the amplitude of the carrier wave is caused to vary at the frequency of the tuning

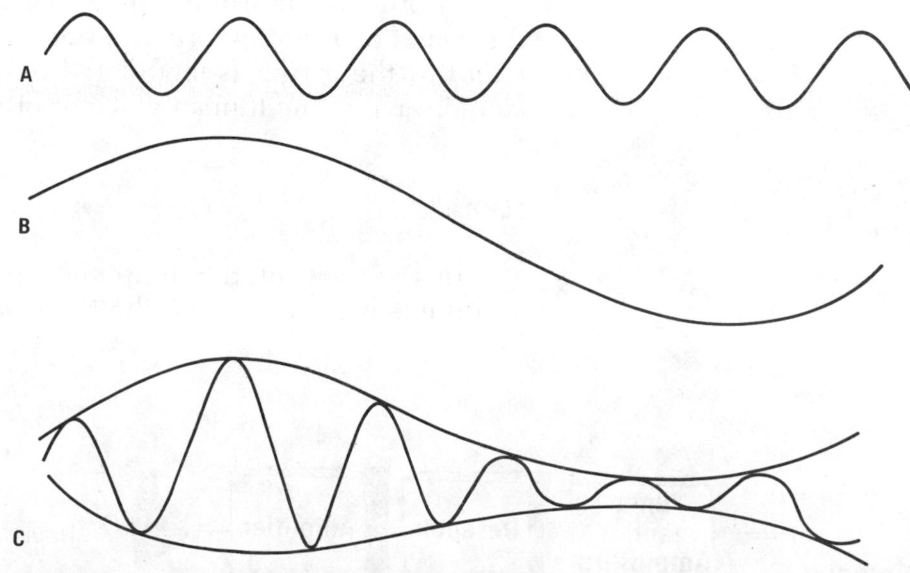

FIGURE 16–10. The signal *B* is used to modulate the amplitude of wave *A* to produce the AM wave (C).

fork. The shape of the modulated electromagnetic wave that actually leaves the transmitter is shown in Figure 16–10*C*.

The modulated carrier wave travels through space; an unnoticed presence until it strikes the antenna of a radio receiver. This antenna can be either a piece of metal extending into the air (as is usually the case on a car radio) or a coil of wire, as is built into the home radio. In the first type, the changing electric field in the electromagnetic wave causes electrons inside the metal to surge up and down. This constitutes an electric current that corresponds exactly to the current that produced the wave originally, at the transmitting antenna. The coil of wire also has a current induced in it which emulates that at the transmitter antenna, but in this case the current is produced by the changing magnetic field of the wave which threads through the loop. (Review the section on induced voltage, Chapter 15, to see how this works.)

Impinging on your receiving antenna are not one but several such electromagnetic waves – one from every radio station within range. The tuner of your receiver picks out the one, from many, that you desire to hear (Fig. 16–11). As you turn the station selector, you match your tuner's frequency to the carrier frequency of the desired station. The set then responds to this particular frequency only and rejects all others. The current from the antenna is amplified and goes to a section of the receiver called the detector, which removes the carrier wave. All that now remains is the original audio signal that was placed on the carrier at the radio station (our 100 hertz tuning fork note). This signal is amplified and sent to the loudspeaker, where it causes a paper cone to vibrate. These vibrations reproduce the original sound wave.

In order to transmit something more interesting than the sound of a tuning fork, the same process takes place except that the carrier is modulated by the more complicated sound wave of the human voice or of music.

FM Radio

In the case of the frequency modulated station, the amplitude is left alone and the *frequency* of the carrier wave

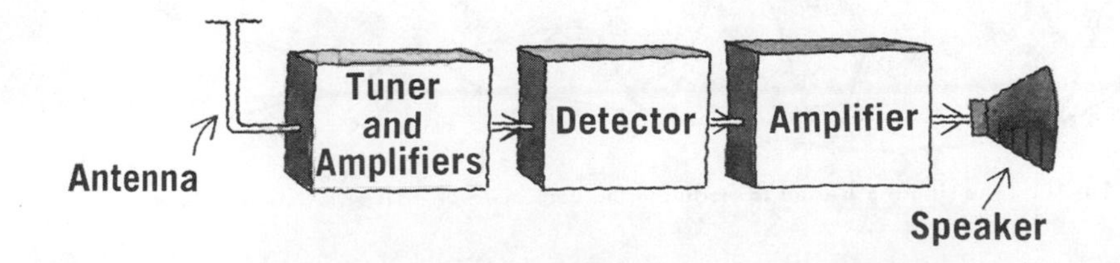

FIGURE 16–11. The elements of a radio.

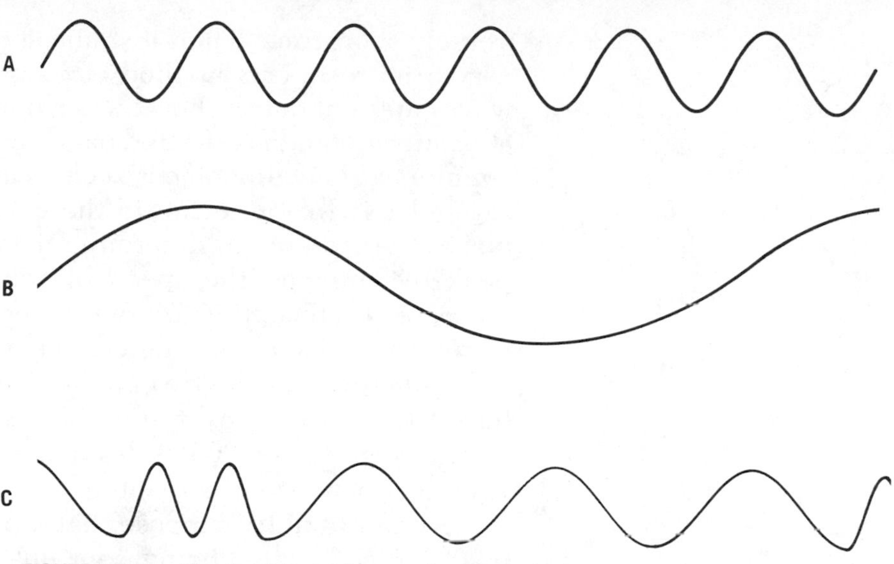

FIGURE 16–12. The signal *B* modulates the frequency of wave *A* to produce the FM wave (*C*).

is altered. Figure 16–12*A* represents the unmodulated carrier wave of an FM station; *B* again shows our tuning fork signal, and *C* shows the carrier wave after its frequency has been modulated. The wider-than-normal portions indicate that the carrier has a lower frequency than usual at those points, and the closely spaced portions indicate points where its frequency is greater than usual. (Remember the wavelength-frequency relationship.) The function of the radio receiver is, as before, to pick out the tuning fork signal and discard the carrier. The major difference between the AM and FM receivers is in the type of detector used; otherwise, all the steps remain essentially the same.

The enjoyment of recorded music was enhanced with the advent of stereo recordings. We saw in Chapter 11 that the added dimension gave listeners the feeling of being "on the scene." In 1962, a similar boost was given when FM stereo stations began to broadcast. By means of what is called a "subcarrier," FM stations can broadcast two signals. The information carried by these separate signals is fed to different speakers, as in normal stereo reproduction, by a receiver in the home.

RADAR

The term "radar" is derived from the phrase "radio detection and ranging." Radar has received primary application as a long-range searcher for enemy ships and aircraft and, on a more peaceful level, as a means for tracking weather disturbances and traffic violators. The principle of operation bears many similarities to radio broadcasting. As in radio broadcasting, an antenna is used to transmit an electromagnetic wave (at microwave frequencies in this case). In radar, however, the transmitter is turned on for only an extremely short time. Then it shuts off to come on again after a very brief rest. This method of transmission means that the waves are sent out in short bursts, or pulses. Each pulse radiates outward and is reflected back toward the antenna if it encounters a reflecting object such as an airplane. The returning pulse is like an echo (in the case of sound waves), and just as we are able to determine the distance to a mountain peak by knowing the speed of sound and the time from emission to reception of the reflected sound signal (the echo), so are we able to determine distance with radar. The only difference in the two methods is that they use two different kinds of waves, the radar wave traveling at a much higher speed—186,000 miles per second as compared to 1100 feet per second for sound.

As an example, suppose that a pulse returns as an echo 0.0002 second after being sent out. During this period of time the pulse has traveled a total distance of 37.2 miles. (Check our arithmetic). The pulse had to leave the antenna, strike the reflector, and return over the same path: therefore the reflecting object is actually 18.6 miles away. The radar operator would not be required to perform these calculations; he can read the distance directly from the face of the viewing screen (thus avoiding problems like those of Figure 16–13). Other systems present in a sophisticated radar set can also indicate the height of an airplane and the direction in which it is traveling.

In the case of police radar, electronic systems make use of the Doppler effect to indicate the speed of the car under observation. If the car is traveling toward the radar set, the reflected wave has a slightly shorter wavelength and thus a higher frequency. (A review of the discussion of the Doppler effect in Chapter 11 may be in order here.) The equipment in the police cruiser translates this slight frequency difference into a reading of the car's speed.

PRODUCTION OF A LIGHT WAVE

We have examined the question of how electromagnetic waves are produced. The discussion of the production of these waves began by suggesting the use of two metal rods, each about three feet long. This length was not chosen at random. In the part of the electromagnetic spectrum that we refer to as radio waves, a calculation would show that the wavelength of such waves is approximately 12 feet—twice the total length of our two rods. Some radio waves are longer, some are shorter, but 12 feet is a representative length for them. It is found that in general, electromagnetic waves are radiated most efficiently by an antenna that is equal to one half the wavelength of the wave; this situation is similar to the resonance phenomenon discussed for sound

FIGURE 16–13. A place for a computer.

waves (Chap. 10). In that case we found that any object free to vibrate has a particular frequency that it prefers; we called this its resonance frequency. We can attempt to make an antenna broadcast a wave of any frequency, but as in the case of sound resonance, there is one that it *prefers*. The resonance frequency for our antenna is determined by its length.

Physicists have discovered that light is an electromagnetic wave and that its wavelength is about one millionth of a meter. This gives us, therefore, an indication of the size of the "antenna" that would broadcast light waves. In the next chapter our search for the source of light will take us inside the atom. We find there that the situation will become much more complex than this chapter would lead one to believe.

OBJECTIVES

Having completed this chapter, you should be able to:

1. Outline the history of the wave-particle controversy, from Newton through the electromagnetic theory.
2. Explain why interference demands a wave explanation.
3. Explain what "ether" is and why it was hypothesized.
4. Describe, using an example, how physical theories develop.
5. Draw the electric field around a rod in which the positive charges are concentrated at one end and the negative charges at the other.
6. Describe and draw a representation of the magnetic wave as it exists around rods in which electrons are oscillating back and forth.
7. Define wavelength and frequency for electromagnetic waves.
8. Explain how James Clerk Maxwell first realized that light waves might be electromagnetic waves.
9. Describe what is meant by the electromagnetic spectrum and list at least five portions of it in order from low frequency to high frequency.
10. Give an example of a beneficial use of gamma rays.
11. Define infrared radiation (by locating it in the electromagnetic spectrum) and describe practical uses of the ability to detect this radiation.
12. Explain why an infrared photograph of a person looks different from a photograph taken with visible light.
13. Explain what is meant by AM and FM and explain the difference between the two as to how a signal is transmitted.
14. List the steps involved in a radio receiver in converting the incoming electromagnetic wave to a sound wave.
15. Explain how an FM stereo signal is broadcast.
16. Explain how radar operates as a distance-finder.
17. Explain how police radar works to determine a car's speed.
18. Describe the relationship between the length of a transmitting radio antenna and the wavelength of the transmitted radio wave.

19. Given any two of the following three: wavelength, frequency, and speed of an electromagnetic wave, calculate the third.

QUESTIONS—CHAPTER 16

1. Sound waves are described as longitudinal waves. How would you describe a light wave? Why?
2. List as many similarities and differences as you can between sound waves and light waves.
3. What does a radio wave do to the charges in the antenna that provides a signal for your car radio?
4. Write two verses and a chorus to a rock song entitled "The Electromagnetic Wave," using as many of the characteristics of such a wave as you can. The instructor guarantees five points extra credit for this exercise if your song makes the top forty.
5. Was the idea of ether a stupid idea? Defend your answer.
6. Compare the phenomenon of color for electromagnetic waves to pitch for sound waves.
7. What is the frequency of a wave that has a wavelength of four feet?
8. Relate the history of the wave-particle controversy to the five steps of the scientific method found in Chapter 0.
9. A radar pulse returns to the receiver after a total travel time of 0.0003 second. How far away is the object that reflected the wave?
10. What is the difference between AM and FM radio?
11. What is the wavelength of the carrier wave broadcast by an FM station with a frequency of 90 megahertz?

17

ATOMS AND LIGHT

"Will the formula offer a clue to the structure of the atom, or will it merely equal zero? Tune in tomorrow..."

Reprinted with the permission of Electronic Age/ Sidney Harris.

The discussion of the production of electromagnetic waves has led us to the idea that when charges are caused to oscillate, waves are created and transmitted into space. While the theory as presented is capable of explaining the production of radio waves, we will find that much still needs to be explained concerning the production of other varieties of electromagnetic waves, particularly light waves. It is easy to imagine that an atom might emit a light wave if we could somehow set one of its electrons into oscillation, causing it to act like a tiny antenna. The reality of the situation, however, is much more complex than this simple picture.

THE BOHR ATOM

Almost everyone has seen what happens when white light is passed through a prism: the light is fanned out into a continuous spectrum of colors.[1] But if the light is from a heated gas, a different result is found; instead of a rainbow only a few distinct colors are seen. For example, for hydrogen there are four prominent colors: red, blue-green, blue, and violet (Fig. 17–1, Plate I). Every element, when heated in the gaseous state, emits a characteristic set of colors. The question of why *all* colors are not emitted by the heated gas was a puzzle which was finally solved (partly, at least) by a refinement of the planetary model of the atom.

[1]To refresh your memory, see Figure 22–1.

In the planetary model we pictured the electrons as circling about the nucleus like planets about the sun. At first glance it seems that this picture leaves much to be desired if one tries to use it to explain the emission of light. From what was known about electricity and magnetism around 1800, there seemed to be no way by which an atom could emit only certain colors and no others. There was, however, an even greater problem. It was known that if an electric charge was accelerated it would emit light. As the electron moves in its circular path, it is accelerated by the attraction toward the nucleus, but if it were to continuously emit light as it accelerated, it would lose energy and very quickly spiral into the nucleus. In effect, the atom would collapse. Since this does not seem to be the case, an alternate explanation is necessary. Enter Niels Bohr.

In 1913, Bohr made two radical assumptions that helped to explain the nature of the atom. He selected the hydrogen atom to study because it is the simplest of all the elements, consisting of a single positive charge with an electron revolving around it. However, he assumed that there were certain orbits in which the electron could revolve without giving off energy and that the electron will never revolve in any other orbit except one of these. (These "nonradiating" orbits get us out of the predicament of having the electron continuously radiating energy and thereby "running down.")

The idea that there exist only certain "allowed" orbits for the electron is a strange one. When you lift something from the ground, you can hold it at any height you want: there are not certain allowed heights with forbidden territory in between. But for the electron there are! The situation is analogous (slightly) to your situation when you buy a belt. You can get a belt ranging from 20 inches up to 50 inches around, as long as you want one with a whole number of inches. You can't get a 32½ inch belt, or a 26.5843 inch belt.[2] Likewise, there are only certain levels at which the electron can orbit the nucleus of an atom.

When you go from the first floor of a building to the second floor, an energy change takes place. The chemical energy in your muscles is converted into gravitational potential energy—energy of position—as you climb the stairs. Similarly, in order for an electron to move from an inner orbit of an atom to an outer orbit, it must gain potential energy—electrical potential energy in this case. This is because just as there is an attraction between your body and the earth, there is an attraction between the negative electron and the positive nucleus. If by some process the elec-

Niels Bohr (1885-1962)

"The father of atomic energy" was the first-born of a Jewish mother and a Christian father in Copenhagen. During college he and his brother, Harold, became famous throughout Scandinavia as soccer players. After college, he studied in England under J. J. Thomson (the discoverer of the electron) and then Ernest Rutherford (the discoverer of the nucleus—see Chapter 24). When he was 37, he became the youngest Nobel prize winner up to that time. In 1943, he fled the Nazis in Denmark, using a small fishing boat to get to Sweden and riding in the bomb bay of a plane to England. After the war, he became a great force in the effort to harness nuclear energy for peaceful uses.

---

[2]You can, however, get a 81.28 cm belt since this is 32 inches. Try asking for one in metric units the next time you buy one.

tron does gain the energy to get to a higher orbit, we say that the atom is excited or is in an excited state. An atom in an excited state is somewhat like a stretched spring or a flexed tuning fork – it has potential energy.

De-excitation of the Atom – Bohr's Other Assumption

Our atom will not stay in an excited state for long. Just as a person might fall from the second floor back to the first, the electron will return to its lowest orbit as quickly as possible. Typically the electron stays in an excited state for one hundred millionth of a second (10^{-8} sec) before returning to the lower orbit.

A difficulty with our picture now arises. In order to get to the higher orbit, the electron had to gain some energy. It cannot get back down to the lower orbit unless it can get rid of this excess energy. The means by which it disposes of energy is by the emission of a small bundle of electromagnetic waves. This short burst of electromagnetic radiation is sometimes interpreted as light by our eye and is usually called a photon of energy, or simply a photon. In emitting its excess energy in this way, the electron returns to a lower orbit (Fig. 17–2).

We see that light does indeed come from the atom, but it is released in short bursts of light energy rather than in a long continuous wave. The energy carried by the photon is related to the frequency of the light by the relation

$$\text{Energy} = \text{Frequency} \times h$$

in which h is a number (called Planck's constant) which must always be included in the equation when energy or frequency is being calculated. We will not perform any calculations with the equation, but it should be noted that a high-energy photon has a higher frequency than a low-

Max Planck (1858-1947)

Max Karl Ernst Ludwig Planck was born in Germany. His first son was killed in World War I, and his second was executed by the Nazis for conspiring to assassinate Hitler. During an air raid on Berlin he lost his home and valuable library. His work in theoretical physics laid the basis for quantum theory, which revolutionized physics in this century. In 1918, he won the Nobel prize for physics, and his name has been immortalized in the name of the constant which relates photon energy to the frequency of radiation: Planck's constant.

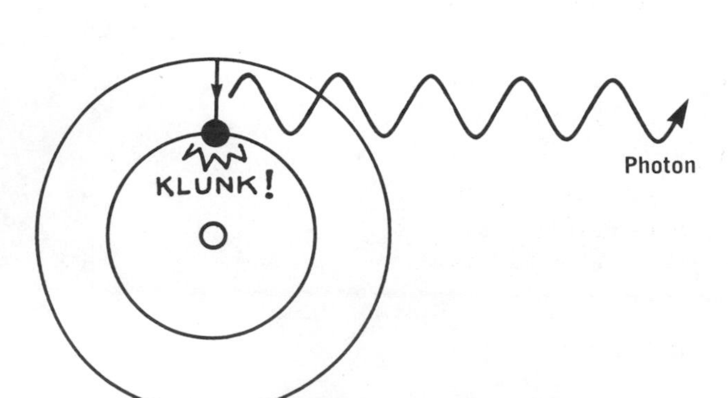

FIGURE 17–2. A photon is emitted when an electron falls to a lower energy orbit.

energy photon. For example, a photon carrying 20 units of energy[3] would have twice the frequency of a photon carrying 10 units of energy. Remembering that the frequency of light determines its color, we see that the two photons thus represent light of two different colors.

Not only is visible light produced in this way but so also are infrared and ultraviolet radiation. If the energy carried by a photon is low, then the frequency is also low, and if it is low enough the wave is not in the visible part of the spectrum at all but is in the infrared. A photon carrying away a large amount of energy from the atom would have a high frequency and could be in the ultraviolet part of the spectrum. (See Figure 17–3.)

The assumptions made by Bohr were radical. Because of their importance we'll restate them:

(1) An electron circles the nucleus in certain definite orbits from which it radiates no energy.

(2) Electromagnetic radiation is emitted in single bursts when an electron goes from a high-energy orbit to an orbit of lower energy.

The validity of these assumptions could only be determined by their success. From them, Bohr was able to derive equations with which the frequencies that should be emitted by the hydrogen atom can be calculated. In short, the theory was a tremendous success, and as radical as these assumptions are, their success leads us to conclude that they must have some validity. They have limitations which were uncovered later, but in general we still find Bohr's ideas useful today. However, his model picturing perfectly defined orbits is largely not accepted. The modern viewpoint is that the electrons form something like a fuzzy ball or cloud around the nucleus, and as a result, we speak of energy levels rather than of orbits. Nonetheless, for our purposes, we will adopt Bohr's orbit model in the following discussion, and we will use the terms "orbit" and "energy level" interchangeably.

---

[3]The energy of photons, protons, electrons, and so forth are usually expressed in electron volts. One electron volt is the amount of kinetic energy an electron would receive if it were allowed to move freely from the negative to the positive terminal of a one-volt battery. For convenience we will adopt arbitrary units for our discussion.

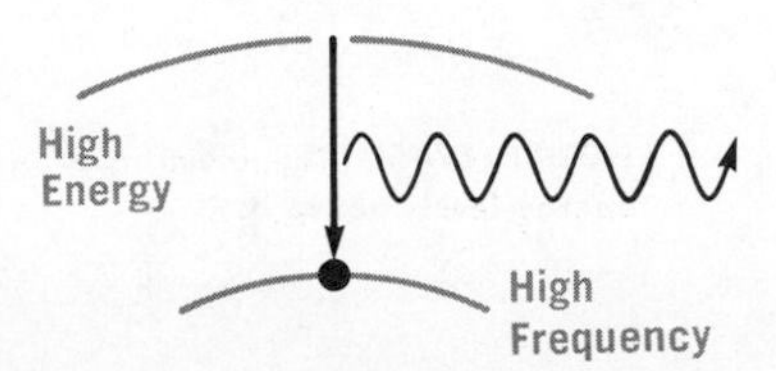

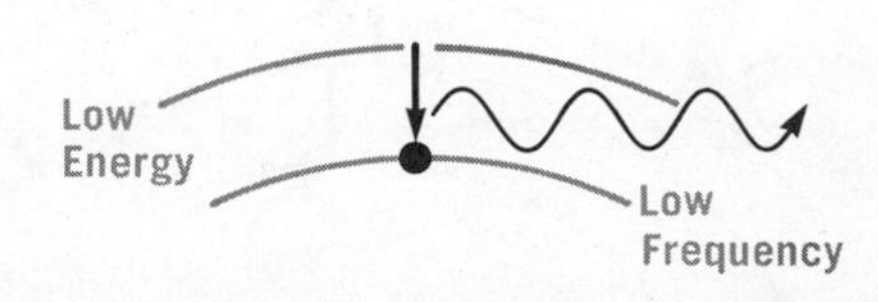

FIGURE 17–3. A photon of great energy has a high frequency, and one of less energy has a lower frequency.

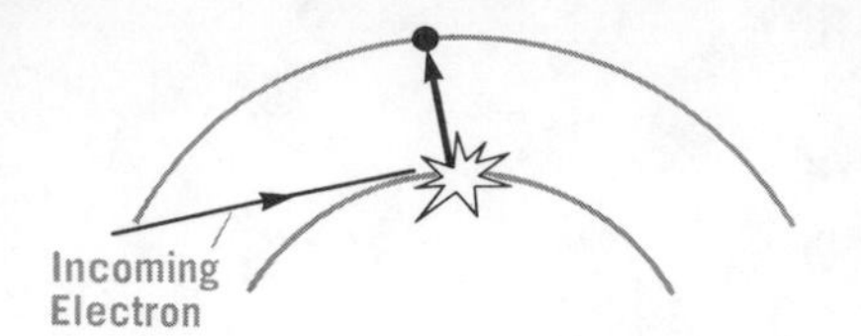

FIGURE 17–4. An electron may be excited to a higher orbit by collision with an incoming electron.

Exciting an Atom

There are many processes in nature that can cause an electron to jump to a higher orbit. For example, let's suppose that a high-speed electron comes flying toward the atom as if it had been shot out of an electron cannon. The electron bullet has kinetic energy because of its speed, and a collision with an orbiting electron can kick the orbiting electron into an orbit of greater energy (Fig. 17–4). In this process, kinetic energy of the incoming particle has been given to the atom as potential energy.

To illustrate the concept of energy levels (or orbits) more clearly, let us examine a number of examples. Consider a hypothetical atom as shown in Figure 17–5, with the electron in its lowest state, and above it are two other excited levels to which it may be raised. (An actual atom has many possible excited levels, but for convenience we will consider only these two.) The energy difference between the lowest state and the first excited state is shown to be 10 units of energy, and the difference in energy between the first excited state and the second is 8 units of energy. Note that the spacing between levels gets closer as you move toward higher excited states.

Now suppose that an electron bullet comes in to strike our orbiting electron. If the incoming missile has an energy of 5 units, what happens? Surprisingly, nothing. The bullet simply bounces off and moves on its merry way (Fig. 17–6). The reason for this behavior lies in the fact that the first excited level is 10 units above the lowest state, and since no intermediate levels exist, the orbiting electron must be given at least enough energy to surmount this 10-unit barrier. No incoming missile with energy less than 10 units can produce any effect. If the bullet has exactly 10

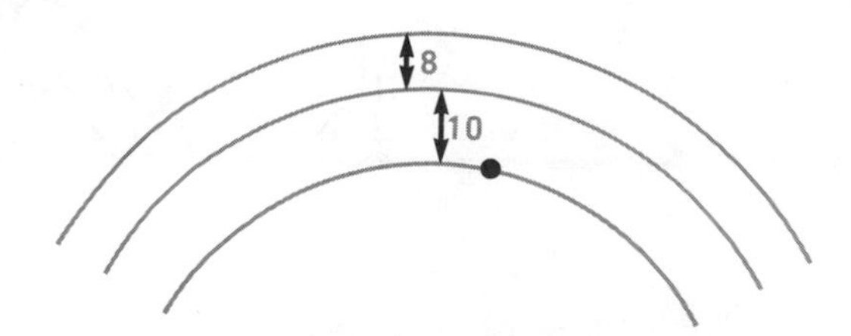

FIGURE 17–5. The electron is in the lowest level and has two possible energy levels above it.

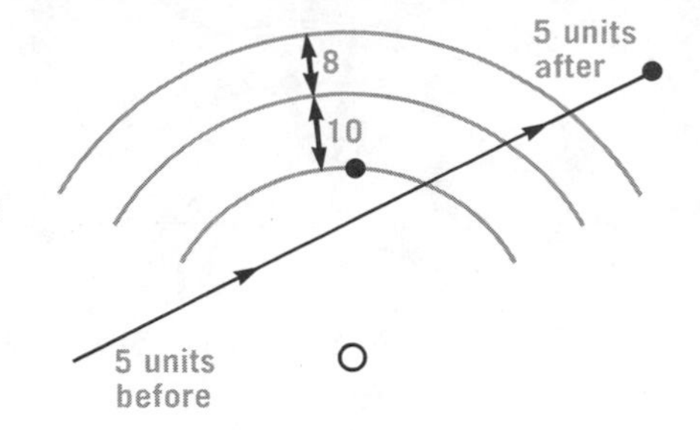

FIGURE 17–6. An incoming electron with five units of energy cannot excite the orbiting electron.

units of energy, it can kick the electron to the first excited state and in turn, having lost all its energy, be reduced to rest (Fig. 17–7).

If the missile has 15 units of energy, it can kick the target up to the first excited state. Because this action requires 10 units of energy—no more, no less—the missile has 5 units left over and goes on its way at a much slower speed than before the collision (Fig. 17–8).

With an incoming energy of 18 or more units, two possibilities exist in the collision: (1) Missile gives target 10 units, target goes to first excited state, and missile moves away with the remaining energy, or (2) Missile gives target 18 units, target goes directly to second excited state (bypassing the first), and missile moves away with the remaining energy (Fig. 17–9).

EMISSION OF PHOTONS BY ATOMS

Using the same two excited states of the atom, let's examine the way in which photons of light are emitted and in particular the way that different frequencies can be emitted by this atom. (Since frequency and color are interchangeable concepts, an equivalent question would be: how can different colors of light be given off by this atom?) Figure 17–10 shows the various possibilities and indicates that three different frequencies (or colors) can be emitted. When the orbiting electron is hit by the incoming electron, the possibilities are as follows:

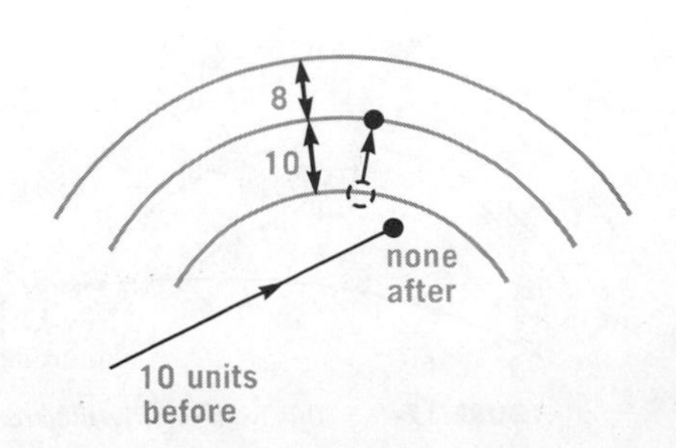

FIGURE 17–7. A 10-energy unit electron gives up all its energy to bump the orbiting electron one level.

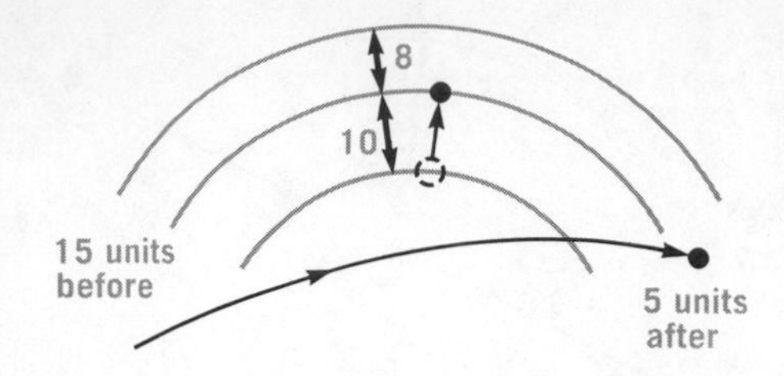

FIGURE 17-8. The 15-energy-unit electron has energy left over after doing its job.

(1) Electron goes to first excited state upon being hit. It then emits a 10-unit photon and returns to its lowest orbit—one color emitted.

(2) Electron goes directly to second excited state. It then falls back to the lowest orbit and emits an 18-unit photon—second color emitted.

(3) Electron goes to second excited state and instead of returning directly to the lowest orbit, drops to the first excited state, emitting an eight-unit photon—third color emitted.

Superficially, it seems that a fourth color might be given off. In case (3) above, with the electron stopping momentarily in the first excited state, why doesn't it emit another photon when it finally reaches the lowest orbit? It does, but this photon has 10 units of energy, which is exactly the same as the photon in case (1). Since

$$\text{Energy} = h \times \text{Frequency}$$

the two photons having the same energy also have the same frequency (and therefore the same color). Try this procedure for an atom with three excited states (Fig. 17-11) to determine the number of colors possible.

THE ELECTROMAGNETIC SPECTRUM (AGAIN)

One surprising feature seems to occur based on the example of the previous section. When counting the number

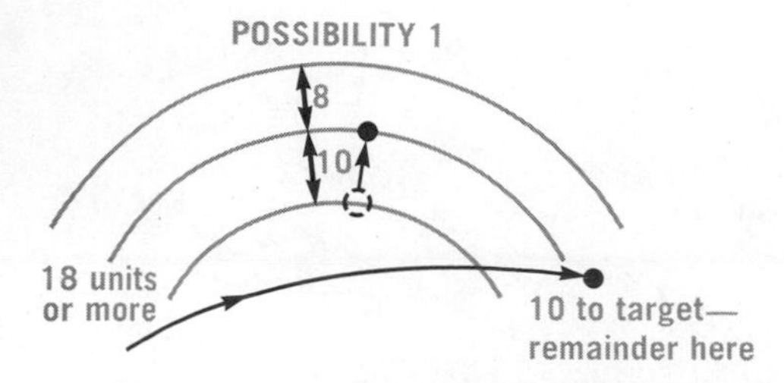

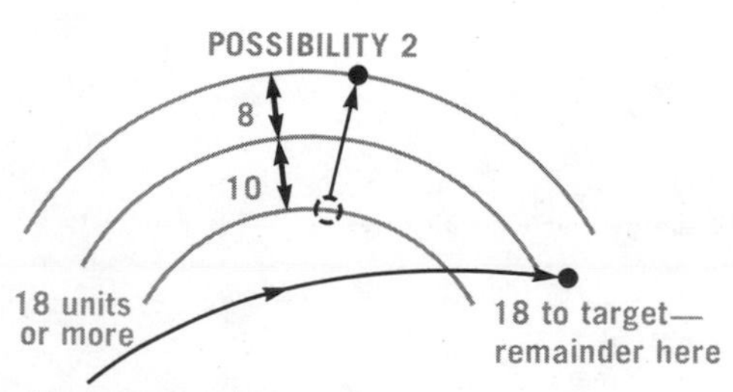

FIGURE 17-9. This high-energy electron has a choice of two possible actions on the orbiting electron.

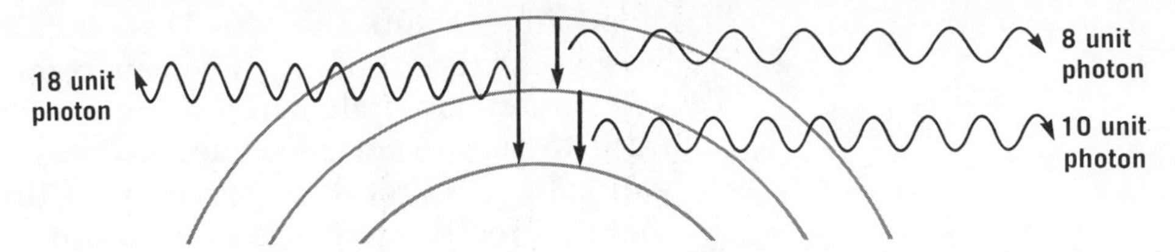

FIGURE 17–10. Different possible transitions leading to emission of light in an atom having two excited states.

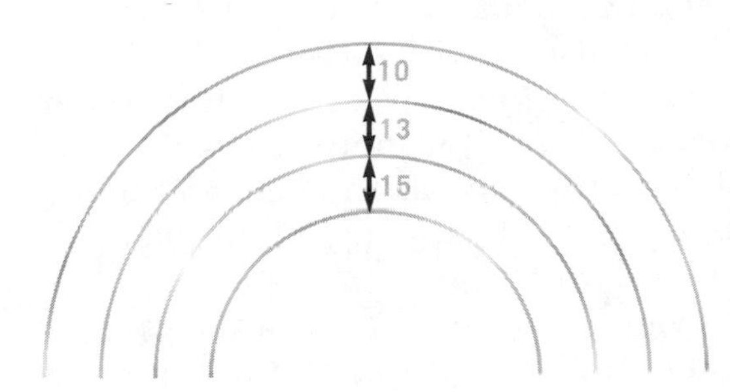

FIGURE 17–11. A three-level do-it-yourselfer.

Reprinted with the permission of American Scientist/Sidney Harris.

of possible colors that our "two-state" atom can emit, we found only three colors. (If you analyzed the three-state atom, you should have calculated six colors for it.) This seems contrary to our experience: an ordinary light bulb gives off *white* light. Careful examination of this white light shows that it is really a mixture of all the possible colors from red to violet, with every hue appearing. All these colors are produced because the source of light in a light bulb is a solid, a tungsten filament, and the simple picture we have drawn does not apply in the case of a solid. In solids, atoms are packed closely together such that neighboring atoms can greatly influence one another. This alters the spacings between the energy levels and also provides a means by which an electron from one atom can jump to a level in another atom. In fact, so many possibilities for jumps exist that any frequency in the visible spectrum has a fair chance of being produced, and indeed, that is what happens. All frequencies (or colors) together produce white light.

Our example as previously outlined does work well for a gas. If neon gas is heated until it emits light, the gas atoms are so far apart that they do not influence one another, and a relatively small number of possible jumps between energy levels is possible. In the case of neon, these jumps result in colors that are primarily at the red end of the visible spectrum. This is the reason for the familiar red light of neon signs. If mercury—the silvery liquid found in many thermometers—is heated to a vapor, it produces only a few vivid colors, the most intense of which are in the blue and violet part of the spectrum. This is the reason your skin takes on a bluish tinge when you are under a mercury vapor street light.

PARTICLE OR WAVE?—REVISITED

At the beginning of Chapter 16 we touched on a few of the controversies surrounding the question of whether light is a wave or a particle. Reference was made at that time to the experiment of Thomas Young, which proved that light waves can interfere as sound waves do. Because particles cannot interfere with one another in a way which might make them cancel each other and disappear, we concluded that light must act like a wave. We now have seen that an atom gives off light not as a continuous wave, but instead in little bursts or "chunks" of energy called photons. Each photon given off by an atom has a certain frequency associated with it—it acts like a wave of a particular frequency and wavelength.

It would be satisfying if this section could end with the preceding paragraph, but unfortunately, that is not the full story. Experiments performed early in this century demonstrated beyond a shadow of a doubt that under certain circumstances light acts as a particle and *not* as a wave. One of these experiments—the photoelectric effect—will be discussed in Chapter 20. And so our photon must have characteristics other than those associated with a wave; sometimes it must act like a particle. This split personality, or duality, is one with which we must live. Perhaps we were naive to think that light must be either only particle-like or only wavelike. Light is light, and it is the job of science to describe it as fully as possible. Our model of light simply has to be complicated enough to allow for this dual nature.

OBJECTIVES

Upon completion of Chapter 17 you should be able to:

1. Describe the Bohr model of the atom.
2. List and explain Bohr's two assumptions concerning radiation from the atom.
3. Compare the solar system and the Bohr model of the atom, pointing out similarities and differences.
4. State the relation between frequency and the energy of a photon.
5. Explain why energy is needed to raise an atom to an excited state, and apply the principle of the conservation of energy to explain what eventually happens to the energy.
6. Use an example with arbitrary energy units to show how an incident electron with various amounts of energy can result in different amounts of excitation of the atom.
7. Explain why an ordinary light bulb emits white light rather than particular colors.
8. Explain what is meant by the dual nature of light.

QUESTIONS—CHAPTER 17

1. Which has more energy, a photon of ultraviolet "light" or a photon of yellow light?
2. If an electron in an atom "sees" three excited levels above it, as in Figure 17–11, show the jumps that would cause photons of 10, 13, 15, 23, 28, and 38 units of energy to be emitted.
3. Does the light produced by a neon sign constitute a continuous spectrum or only a few colors? Defend your answer.
4. Compare and contrast light and sound in terms of (a) origin, (b) speed, and (c) a medium for transmission.
5. Why was it necessary for Bohr to assume the existence of nonradiating orbits?
6. Suppose the orbiting of satellites obeyed Bohr's first assumption

(which applies to electron orbits). How would satellites behave differently from the way they do now?

7. It has been said that there is no such animal as a photon of white light. Is this true? Defend your answer.

8. Is light a particle or a wave? Discuss.

9. Suppose an electron is in the lowest level of an atom having levels as shown in Figure 17–5. What will happen to the electron as a result of a collision with a photon with 8 units of energy? 12 units?

10. Why is an electron orbiting in one of the Bohr orbits being accelerated? (Hint: review Chapter 1.)

Photo by John A. Wright.

18 LIGHT IN ACTION

The light produced as an atom excites and de-excites leaves its source and eventually may reach our eyes. Along the way it may bounce off a mirror or be bent in other ways. An understanding of some of the things that can happen to light is necessary in order for us to understand why we see what we see and to explain many of the simple optical devices which we encounter daily.

MEASURING THE SPEED OF LIGHT

The speed of sound can be measured quite easily. Its relatively slow speed lends itself to measurement with simple tools such as a meter stick and a clock. Light travels so fast, however, that it seems to go from source to observer *instantaneously*. One of the early attempts to determine whether or not any passage of time was involved with light propagation was proposed by Galileo Galilei in about 1600. Figure 18–1 illustrates the approach. Two observers with lanterns positioned themselves on hills a few miles apart. One opened his shuttered lantern. When the second observer saw the signal, he opened his lantern. The plan was simple: the first man should be able to determine whether or not the light took time to get to his partner and back. As

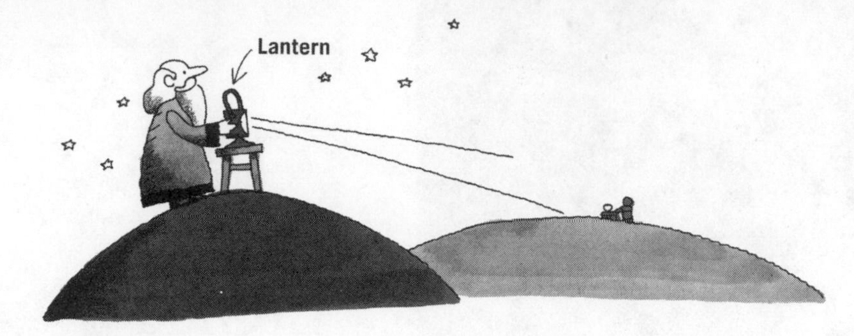

FIGURE 18–1. Galileo's method for measuring the speed of light.

you might suspect, the results showed that light either propagated instantaneously or that it traveled at a *very* high speed. We now know that light travels at 186,000 miles per second. This means that the hills in the experiment would have to be 93,000 miles apart in order for one second to elapse in the light's round trip.[1] It is little wonder that with only a few miles separating the hills, the only thing the observers were able to measure was their own reaction times. Considering such technical difficulties as finding two hills 93,000 miles apart, it is not surprising that the first accurate measurement of light's speed relied on astronomical observations.

An Astronomical Method

The planet Jupiter has 13 moons[2] which we can observe circling Jupiter just as our moon circles Earth. The periods of these moons undergo some puzzling variations, which were first explained by the Danish astronomer, Olaus Roemer. Using one of Jupiter's moons, Io, we will explain Roemer's reasoning.

Io circles about Jupiter and is observed to disappear behind the planet every 42.5 hours. (We say that its period of revolution is 42.5 hours.) In Figure 18–2, P and Q represent two positions of the earth as it moves in its orbit about the sun. Roemer noted that a disappearance (or eclipse) of Io occurred at a particular time of night when the earth was at point P. With this information and a knowledge of the period of Io, he was able to predict in advance at what time a similar disappearance should begin three months later, when the earth had moved to point Q.

To illustrate, suppose that on June 13, 1676, Io was seen to disappear at exactly 9:10 PM. A calculation would show that another eclipse should begin on September 13, 1676, at

[1]Since the diameter of the earth is only about 8000 miles, such hills would be hard to find.

[2]Four of these are visible through small telescopes and were seen first by Galileo. A fourteenth was tentatively discovered in September, 1975.

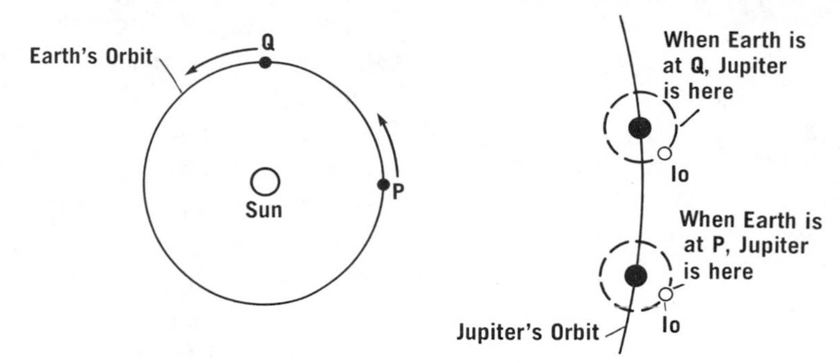

FIGURE 18–2. Light takes longer to get to earth from Jupiter when earth is at point Q than when it is at point P. This allows a measurement of the speed of light.

exactly 11:10 PM. (This would be the fifty-second eclipse since the one on June 13.) The September 13 eclipse was not seen until 11:18:20, however! What took so long? The answer is that the eclipse occurred on time but that the extra 500 seconds were required for the light to reach earth at its greater distance from Jupiter. If the earth had still been at its June 13 location, the eclipse would have been right on schedule—at 11:10. Half the distance across the earth's orbit is about 93 million miles (150 million kilometers). In order to cover this distance in 500 seconds the speed of light would have to be 186,000 miles/sec (300,000 km/sec).[3]

An Earthbound Method

Earthbound measurements have been performed many times by many people, but one of the most noteworthy experimenters was Albert Michelson, who measured the velocity of light on numerous occasions, starting in 1877. We will discuss a method he used in 1926, which is illustrated in Figure 18–3. A beam of light from source S is directed

[3]The first measurement—because of inaccuracies in timing and in the earth-to-sun distance—actually yielded 136,000 miles/sec. The important point, though, is that it was the first measurement which yielded even a rough idea of the speed of light.

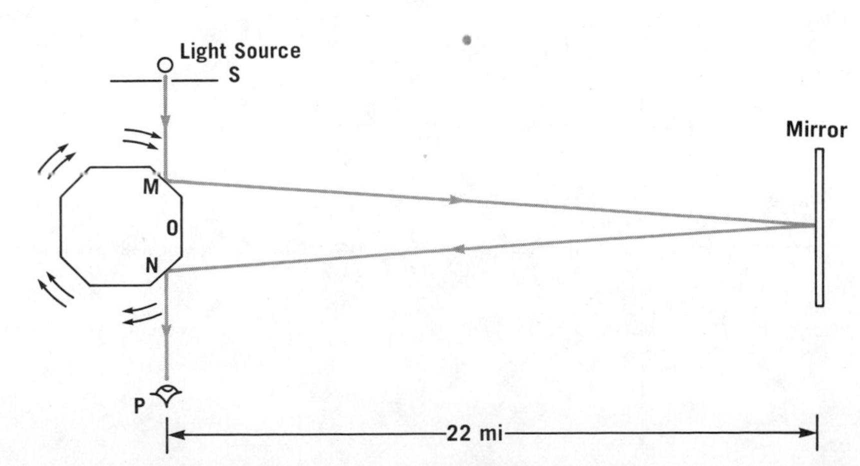

FIGURE 18–3. Michelson's method. Light goes from S to M to mirror to O to P. (O has by that time moved to N.)

toward an eight-sided mirror and is reflected from side M in such a way that it travels 22 miles to a mirror which reflects it back. Suppose that the eight-sided mirror is stationary. The light will then reflect from side N into the observer's eye. But now the experimenter rotates the mirror. Light hits one of the sides at position M, goes to the distant mirror, and reflects back. But when the light gets to position N, that side of the mirror will have moved on, and the light will not reflect from it to the observer's eye. If the mirror's spinning speed is adjusted correctly, however, side O will have moved up to take N's place just when the light beam gets there. In that case the light will again be reflected to the observer.

Albert A. Michelson (1852-1931)

Born in Prussia, Michelson and his family emigrated to the United States when he was two years old. When he was 17, he traveled alone from San Francisco to Washington, D.C., to seek an appointment to Annapolis. President Grant was so impressed by him that he was admitted even though no vacancy existed. Although he is best known for his work with Edward Morley on the experimental search for the ether, his work in optics included such achievements as the first accurate measurement of a star's diameter. In 1907, he became the first American scientist to win a Nobel prize.

The distance traveled by light in Michelson's experiment was about 44 miles for its to-and-fro trip. During this trip the mirror rotated $1/8$ turn, which allowed Michelson to calculate the time for one round trip (since the rate of rotation was known). Thus the speed of light was calculated. Michelson's results (he repeated the experiment many times) and other measurements have resulted in our present best value of about 186,364 miles per second.

So we know the speed of light. But light is only one part of a vast variety of waves known as the electromagnetic spectrum. Speeds have been measured for all the many varieties in the spectrum, and it has been found that *all of these waves travel with this same speed in a vacuum.* That sure makes it easier to remember!

RAYS OF LIGHT

When an object emits light, the waves spread out uniformly in all directions. We will discuss the motion of these waves in terms of *rays. A ray is a line drawn in the direction in which waves are traveling.* A sunbeam passing through a darkened room traces out the path of a ray. Figure 18–4

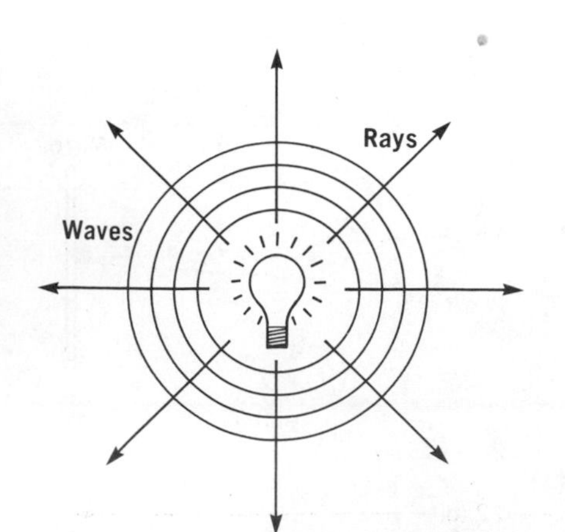

FIGURE 18–4. Waves and rays from a light source.

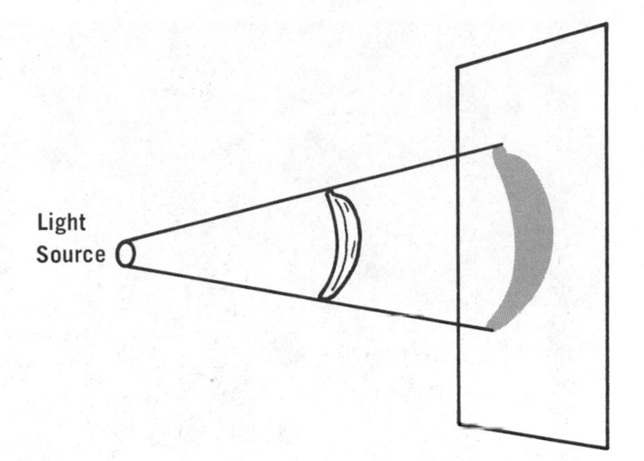

FIGURE 18–5. A shadow cast by using a very small light source.

shows a small object emitting light waves, with the circles indicating crests of the waves. The same kind of picture would be drawn for water waves spreading out from a vibrating object in water. In water, however, the waves travel along the surface only, whereas light waves spread out from the source in all directions. Several rays are indicated in the figure. Observation indicates that these *rays of light always are straight lines as long as the light is not passing from one material into another*—for example, from air into water.

The fact that light travels in straight lines is responsible for the formation of shadows. The very small object in Figure 18–5 is sending out light in all directions. The banana placed in its way will not allow light to pass through, and so a shadow is formed on the wall. In practice, the light emitter is often not such a small object, in which case, the edge of the shadow will be fuzzy. In Figure 18–6 the bug at point A is in total darkness behind the ball. Although light from the center and bottom of the light source cannot hit the bug at B, light from the top can hit it. So bug B is in a partial shadow, a gray area.

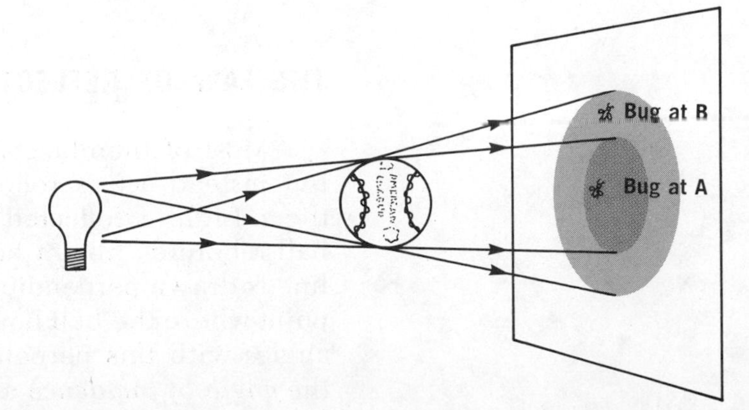

FIGURE 18–6. A larger light source results in fuzzy shadow-edges.

Photo by Paul Lambert.

take-along-do-it

To illustrate the different shadows formed by a large and a small light source, try this. Hold a table lamp (with a regular light bulb in it) about one foot above a table. Put your finger halfway between the light and the table and observe the shadow. Now, if a "high-intensity" lamp (the type with a small clear bulb) is available, put it in place of the table lamp and notice your finger's shadow. A substitute for the high-intensity lamp is the table lamp with a piece of cardboard close in front of it. Punch a hole about the size of a pencil in the cardboard and let this hole serve as your small source of light.

THE LAW OF REFLECTION

Most of the objects around us are not emitters of light but instead act as reflectors of the light which falls upon them. Light is reflected in much the same way that a tennis ball rebounds from a hard court. In Figure 18–7, a dashed line is drawn perpendicular to the surface of the court at the point where the ball bounces. The figure shows that the two angles with this perpendicular are equal. Angle i is called the *angle of incidence* and is formed with the perpendicular

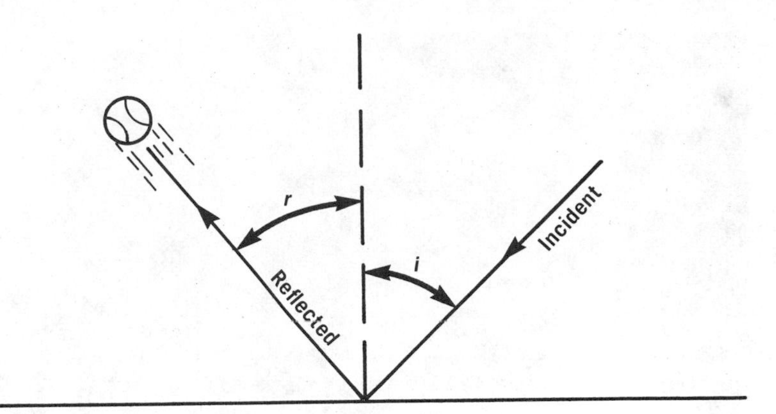

FIGURE 18–7. Angle r (nearly) equals angle i when a tennis ball bounces.

by the incoming tennis ball; r is the *angle of reflection* between the perpendicular and the path of the reflected tennis ball. A careful measurement in an actual experiment would show angles i and r to be *almost* equal.

To adapt this situation to the case of light rays, replace the court with a reflecting surface such as a mirror and let the path of the tennis ball be replaced by a ray of light. Careful measurements for this situation would indicate that angles i and r are *exactly* equal. This observation allows us to state the law of reflection: *The angle of incidence is equal to the angle of reflection.*

Our visual information about the world around us is provided by rays of light reflected to our eyes according to the law of reflection. When an object struck by a ray of light has a rough surface, the multitude of little bumps on it scatter the rays in all directions. Light scattered in this manner is said to be *diffusely reflected.* Figure 18–8 shows that at all points the rays are obeying the law of reflection. Even the most perfectly polished object diffuses light to some extent. Microscopic irregularities and/or dust particles present on its surface are sufficient to diffusely scatter the light and allow it to reach our eyes no matter from what angle we view it.

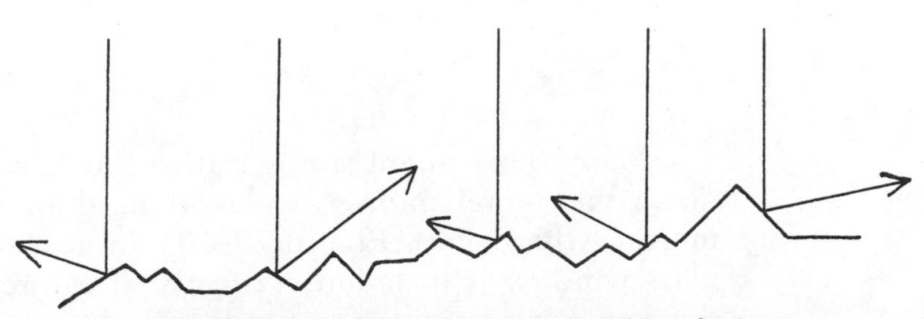

FIGURE 18–8. A rough surface results in diffuse reflection.

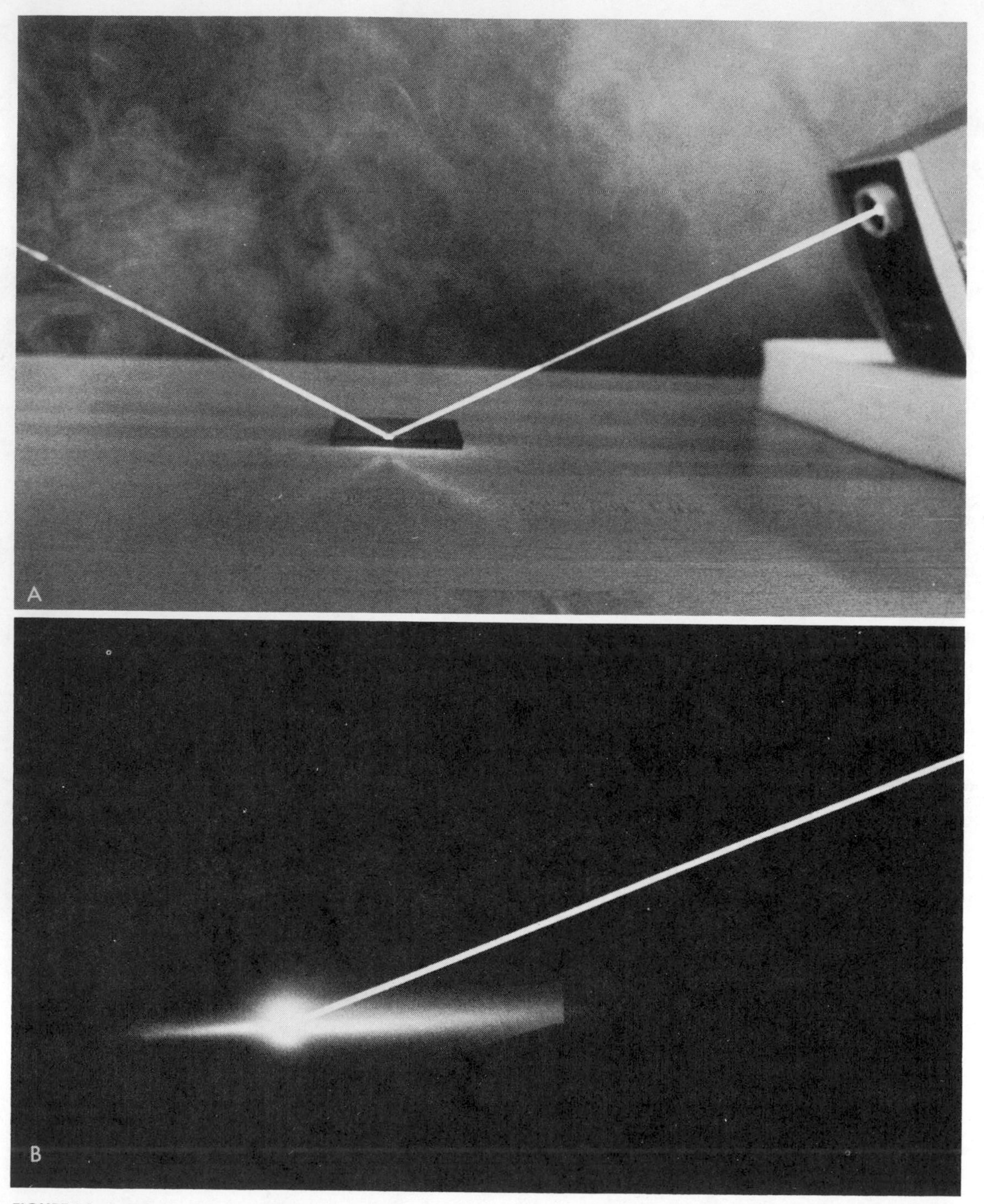

FIGURE 18–9. *A*, Beam of laser light strikes a mirror and bounces away, obeying the law of reflection. (Note the smoke, which was blown through the air to make the beam visible.) *B*, When the laser beam strikes a piece of paper, it is diffusely reflected. (Courtesy of Jim Shepherd.)

Many department stores utilize the law of reflection to make the merchandise displayed in their store windows more easily visible. Figure 18–10A indicates that if a plate glass window is installed vertically, it can reflect images of the sun, street lights, automobile lights, and so forth, directly into the eyes of the prospective customer, interfering with his view. If, however, the glass is installed at an angle with the vertical (Fig. 18–10*B*), annoying reflections are bounced away from the eyes of the beholder.

Mirror Images

In Figure 18–11 a small candle has been placed in front of a flat mirror. In order to see how the mirror forms an image of this object, we must follow the path of at least two different rays of light that leave the same point of the object. One of these, labeled A in the figure, leaves the object to strike the mirror at a 90° angle. Recalling the law of reflection, we can see that it is reflected straight back on itself. A second ray of light, B, is seen to leave the same point on the object, then to strike the mirror at some arbitrary angle, and

FIGURE 18–10. A worthwhile use of the law of reflection for controlling unwanted light.

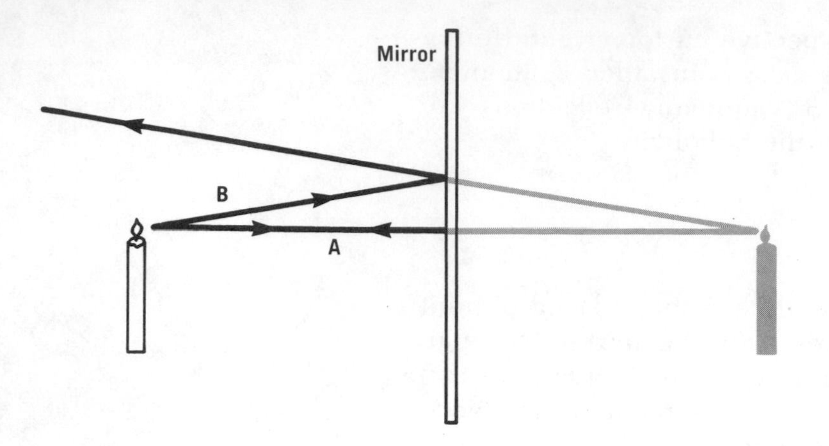

FIGURE 18–11. Formation of a virtual image.

to reflect—obeying, of course, the law of reflection. Awaiting these two rays of light (and numerous others) is the eye of an observer. In order to form an image, the eye traces the two rays back along their paths to the point at which they cross. The shaded lines indicate that this point is behind the mirror, and it is at this location that the eye sees an image of the candle. Measurements taken from the figure would show that the distance between the image and the mirror is the same as between the object and the mirror. This is always the case for a flat mirror. Note also that the image formed is the same size as the object; flat mirrors never magnify.

Images formed in this manner are called *virtual images*, because the light only appears to come from the image; which is in contrast to the type of image formed when rays of light are actually brought to a focus, as is done by a movie projector forming an image on a screen. This kind of image is called a *real image*. The light entering your eye from the image on the screen has actually come from the screen.

The method by which the eye forms an image is important and should be well understood. Whenever light enters our eyes, we interpret the location of the light's source according to the direction from which the light comes. For the case of the flat mirror, when the light reached our eyes after reflection, it appeared to have come from a point behind the mirror, and it is at that point that we see an image. Whether the mirror is flat or curved, whether it is a mirror or a lens or any other optical device, the eye acts in the same way.

As a second example, Figure 18–12 indicates the way in which the eye forms an inverted image of an object in a horizontal mirror. Rays of light leave the candle and are reflected by the glass into the viewer's eye. The eye follows these reflected rays back (along the dashed line) to the point from which they apparently originated. This image is also a virtual image.

FIGURE 18–12. Another virtual image.

REFRACTION OF LIGHT

As long as light is traveling within a uniform material, it follows a straight line path. If the light goes from one material into another, such as from glass into water, or if some property of a material (such as its density) varies, the ray of light may undergo a change in direction. This bending of a ray of light is referred to as *refraction.*

To understand the refraction process, consider a long line of men all running at the same speed and carrying a snake as shown in Figure 18–13A. All goes well as long as

FIGURE 18–13. *A,* The snake-carriers are approaching the mud area. What will happen? *B,* The snake-carriers keep the snake straight when all are on concrete, but when some hit the mud, they slow down, and the snake must bend.

the men are all running on concrete, but the situation alters when one end of the chain steps into a muddy field (Fig. 18–13*B*). Those in the mud are no longer able to run at the same speed as those still on the concrete. They slow down immediately, and by the time the other end of the line of runners has advanced into the field, the direction of travel of the line will be altered as shown.

The value 186,000 miles/sec for the speed of light is a value only for its speed in a vacuum (although the speed of light in air is only slightly less). For any other medium, such as glass, the speed is slower. For example, the speed of light in glass is about 125,000 miles/sec, and in water it is about 140,000 miles/sec. A beam of light from a flashlight will then act much like our chain of runners (Fig. 18–14). The lower part of the beam encounters the water first and is slowed down just as were the men entering the muddy field. The ultimate effect is a bending, or refraction, of individual rays of light. This bending always occurs at the *surface* between the two media. As long as the light is in air, its path is straight, and as long as it is in water, it also follows a straight line.

Examination of a single ray of light during refraction will enable us to formulate the laws governing the refraction process. In Figure 18–15, a ray of light in air strikes the surface of water and is bent. In order to analyze this situation, we draw a dashed line perpendicular to the surface at the point where the light ray hits it. Compare the angle of incidence (A = air) to the angle of refraction (W = water). After observing a number of beams of light entering materials in which the light slows down, we can form a general law: *As light passes into a material in which it goes more slowly, it bends toward the perpendicular.* In other words, the angle with the perpendicular is *smaller* in the material through which light travels more slowly.

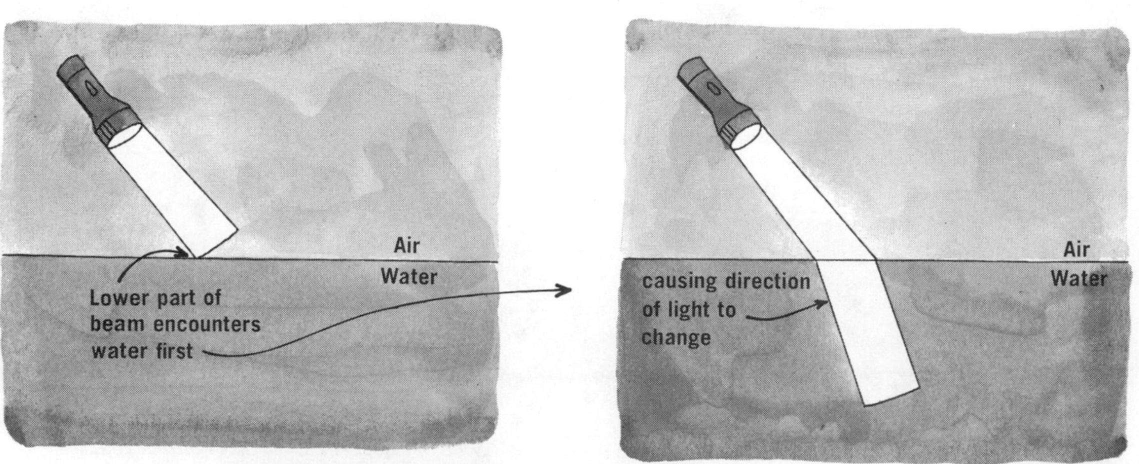

FIGURE 18–14. The bending of a light beam. Compare to Figure 18–13.

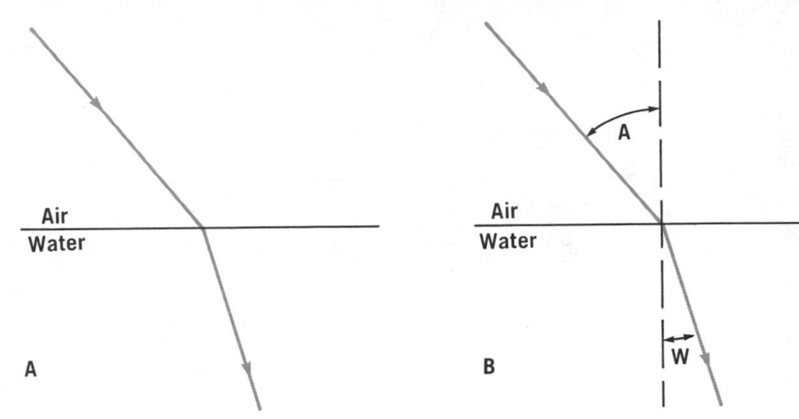

FIGURE 18–15. Note that the angle with the perpendicular is less in the water than in the air.

Now, consider a case in which the source of light is in the water and sends a beam through the surface and into the air (Fig. 18–16*A*). If this figure looks a lot like the preceding one, it is no accident. The fact is that if light is sent backward toward a surface by the same path it entered, it will bend such that it still continues along the same path. So a general drawing is shown in Figure 18–16*B*. No arrowhead is placed on the ray of light to indicate direction. The direction doesn't matter! Thus we can further generalize our general law: *When light passes from one material into another, it bends in such a way that the angle with the perpendicular is less in the material where the light goes slower.* This is a statement of the *law of refraction.*

There is one special case: When the ray of light hits the surface at a zero angle with the perpendicular—hits it "head on"—it does not bend at all, but passes straight through. (This does not really violate the law, however; the angle is zero in both materials.)

TOTAL INTERNAL REFLECTION

Figure 18–17 depicts a light source beneath the surface of a pool of water. Shown are four rays of light leaving the

FIGURE 18–16. Compare *A* to Figure 18-15. Since the direction of light makes no difference, *B* here shows no arrows on the ray of light.

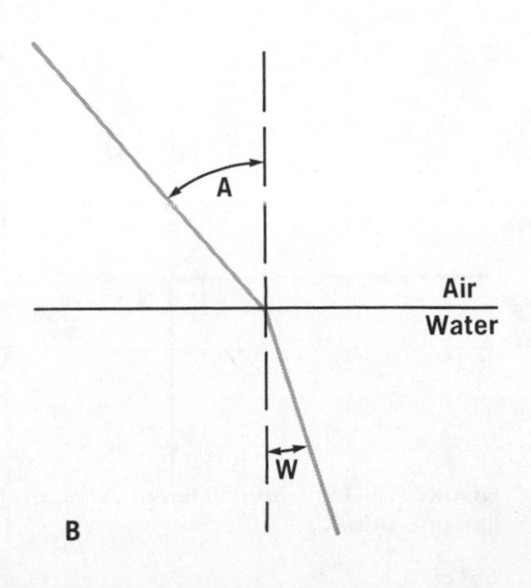

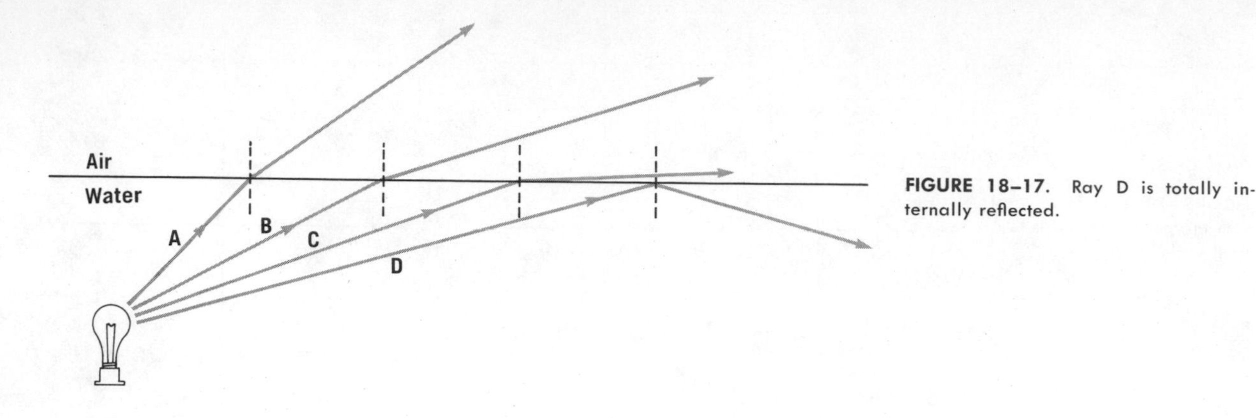

FIGURE 18–17. Ray D is totally internally reflected.

source. Ray A strikes the surface and bends away from the perpendicular, as it must. Ray B strikes the surface, and it also is bent away from the perpendicular. Since its angle with the perpendicular in the water is greater than that of ray A, B's angle in the air must also be greater than A's, and so it is bent away from the perpendicular to a much greater degree. Finally, ray C strikes the surface at an angle of incidence such that when it passes out of the water, it skims right along the surface. Any ray of light, such as D, striking out farther than C, simply cannot get out of the water at all. It is reflected from the surface just as though the surface were a mirror. This ray is said to undergo *total internal reflection.*

Total internal reflection can occur only when light is traveling from a material in which it travels slowly to one in which it travels faster. This principle finds its greatest degree of practical application in the case of glass. Figure 18–18 shows light incident on the face of a prism having one 90° angle. The light strikes side A and passes straight into the glass without refraction; it then strikes side B, where it undergoes total internal reflection and is thus reflected toward side C, where it can leave the glass. A combination of two such prisms would produce a change in direction of the light as shown in Figure 18–19. The combination of prisms in *A* could be placed in a tube and used as a periscope. If you prefer to see where you have been rather than where you are going, a periscope like that in *B* would be more to your liking. (Note that in this case the image will be inverted.)

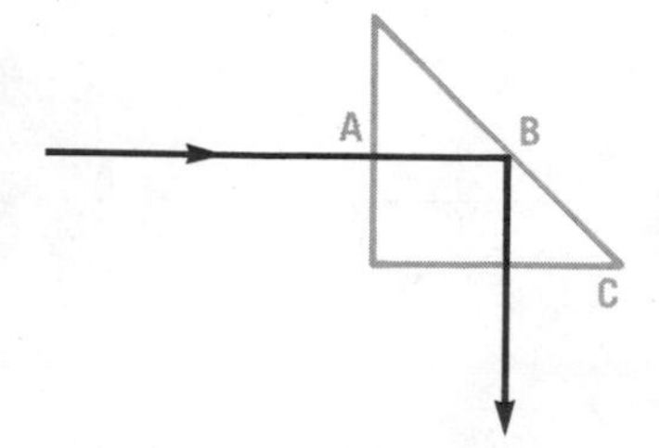

FIGURE 18–18. Total internal reflection in a prism.

An interesting use of the phenomena of total internal reflection is in fiber optics. Figure 18–20 shows a curved rod made of transparent plastic. Light entering one end travels in a straight line until it encounters the side of the rod, but unless the rod is bent at a sharp angle, the light will strike such a glancing blow on the side that it will be totally internally reflected. It will then continue to reflect down the

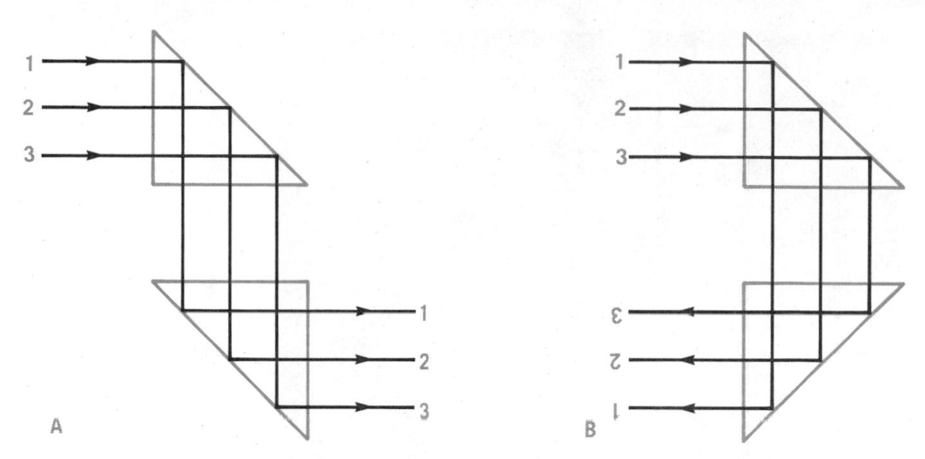

FIGURE 18–19. Two uses of a pair of prisms. Note the inversion in the second case.

tube until it strikes the other end, which is struck at a small enough angle that the light is transmitted. In practice, fine plastic threads, which are flexible, are used rather than large rods. Several applications follow:

(1) Some ornamental lamps make use of flexible plastic fibers to produce interesting effects, as shown in Figure 18–21. The plastic threads carry light from a bulb in the center to their ends, producing many little light sources that move in a breeze.

(2) Some cars have a light on the dashboard to indicate whether or not the headlights are on. This could be done with a small light bulb on the dash, but a simpler and more dependable method is to run a group of small fibers from the headlight to the dash. When the headlight is on, light comes through the fibers and emerges to appear as a light visible to the driver—nearly foolproof!

(3) An interesting medical use of fiber optics is in the photography of ulcers, inside a person's stomach, without surgery. A tube with a great number of small fibers inside is inserted via the throat into the patient's stomach. Some of the fibers are used to carry light down into the stomach. A

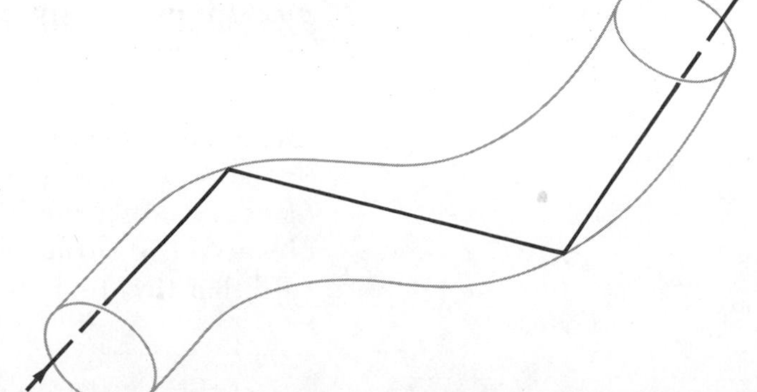

FIGURE 18–20. Total internal reflection is the basic principle of fiber optics.

FIGURE 18–21. Fiber optic decorative lighting. (Courtesy of Poly-Optics, Inc.)

tiny lens on the end of the pipe then focuses the illuminated stomach lining onto the remaining fibers. The fibers receiving light from the lens transmit the light along the tube, so the same image appears on the other end of the tube. All that remains is to magnify that image, and the physician can see the lining of the patient's stomach.

(4) This one is probably imaginary. A tube somewhat like the medical tube described in (3) could be held by the CIA director above a top-secret message. If the fibers of the tube were scrambled instead of running straight through, the image on the other end would be jumbled. So the director takes a photo of this scrambled image and sends it to his spy in the field. The agent has his own scrambling fiber optics gadget – identical to the boss's – and he uses it in reverse to reproduce the message.[4]

EXAMPLES OF REFRACTION

In these times of student awareness, there exists a need for greater relevancy in the classroom. With this as a backdrop, let us now investigate the question burning on the lips of every concerned student – why does a fish appear to be closer to the surface of water than it really is? Never let it be said that this text avoids controversial issues of our day!

[4]The CIA director may contact either author for details.

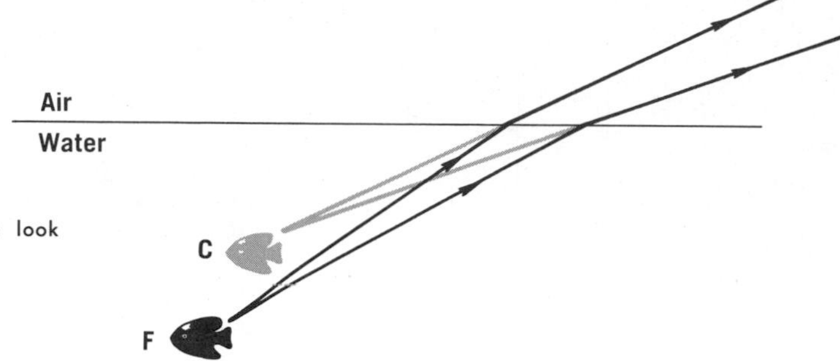

FIGURE 18–22. Refraction of light causes a fish to look closer to the surface than he really is.

In Figure 18–22 are shown two rays of light leaving a fish. The two rays pass through the surface, bending according to the dictates of the law of refraction. These two rays enter the eyes of the observer, and according to their habits, the eyes follow the rays back to the point from which they appear to have originated (along the shaded lines) and see an image of the fish at point C.

take-along-do-it

You probably don't have a fish in your bathtub to test the preceding phenomenon, so we suggest the following: Ask someone to put some beads (or sand, or gravel, or pennies, or anything that will not float) into a cup up to a depth unknown to you. Then have him fill the cup with water. Now look down into the water and place your finger on the outside of the cup at the point where the top of the beads appears to be. Unless you cheated, the true level of the beads is below the position of your finger.

To further illustrate this effect, pour the water out of the cup, leaving only the beads. Now move your head until the beads are just out of view as shown in Figure 18-23A. Have someone slowly add water to the container. The water causes the light rays to bend as they pass from the water into the air, and slowly the beads come into view (Fig. 18-23B).

When the container is filled with water, stick a pencil into the water and observe the apparent bending of the pencil.

We have seen that a fish in water appears to be closer than it actually is to the surface. This lifting effect also explains why the immersed pencil in the TADI appears to be bent: each point of the pencil that is underwater is apparently raised toward the surface, just as was the fish.

FIGURE 18–23. Refraction.

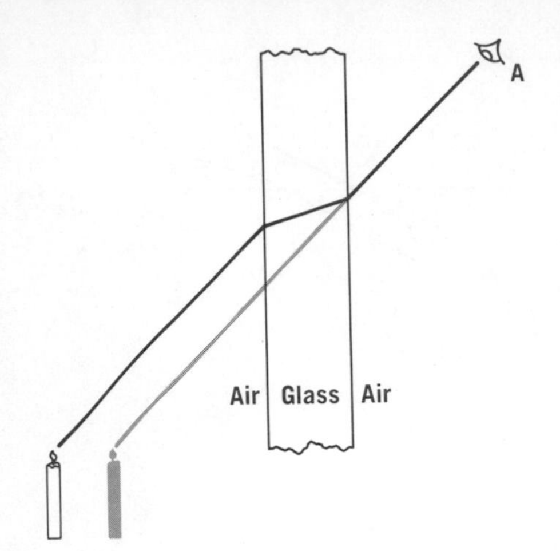

FIGURE 18–24. Note that each refraction follows our general rule of refraction and that the outgoing ray is parallel to the incoming one.

Passage of Light through Windows and Prisms. When light strikes the surface of flat glass as illustrated in Figure 18–24, it is bent toward the perpendicular inside the glass. While in the glass, it travels in a straight path until it encounters the second surface. Here the light bends away from the perpendicular. An observer at A would still see the source in the same general direction as he would in the absence of the glass, but it would appear to be moved slightly.

Suppose that instead of a sheet of glass with parallel sides, we have one shaped like a wedge (Fig. 18–25). Applications of the law of refraction at each of the faces of this wedge would indicate that the ray of light will be bent as shown. In this case, the eye would trace the refracted ray back to its apparent direction of origin, and the source will appear to be displaced in the direction indicated. Such a wedge of glass is called a prism and may be any of a number of shapes, depending upon the angle at its corners. The prism has another important use besides that of simply bending rays of light. It was mentioned earlier that rays of light are refracted because light travels with different speeds in different materials. It is also found that light waves of different frequencies travel with slightly different speeds inside the same material. The frequency of visible light that we perceive as violet in color travels the slowest, and the wave having a frequency corresponding to that of

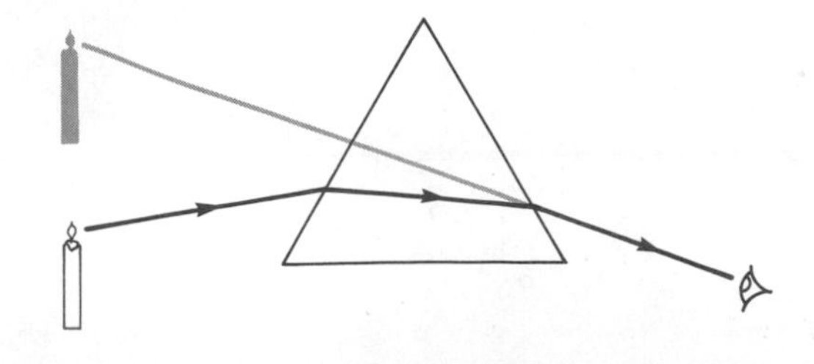

FIGURE 18–25. Refraction in a prism may result in a great deal of displacement of the object.

FIGURE 18–26. Violet carriers bend the snake more than red carriers because they are slowed down more by the mud.

red light travels the fastest. To see the effect of this principle on light passing through a material, consider the analogy illustrated in Figure 18–26. In *A*, a team of runners carrying a snake steps off concrete into a muddy field. This team has a V on their sweatshirts, indicating that they are members of the violet team. The characteristic of this team is that it runs well on concrete but moves very slowly in mud. As a result, when the runners step into the field, they slow down a great deal, and the team is bent, or refracted, considerably from its original direction. The team in *B*, the red team, is characterized by their ability to run on concrete at exactly the same speed as the violet team, but they are able to travel faster in mud than the violets. As a result, they are refracted less when they pass into the mud.

To apply this analogy to light, remember that white light is a mixture of all the possible colors existing in the spectrum of visible light, and consider the red team to correspond to red light and the violet team to violet light. Thus, when white light goes from air into glass, the violet light is bent the most and red the least, while intermediate colors such as orange, yellow, green, and blue are bent at angles between those of red and violet. This difference in bending will separate the colors. As an aid to remembering the main spectral colors in the order in which they appear, remember the boy's name ROY G. BV (the last name is pronounced *Biv*). Each letter in his name is one of the principal colors in the spectrum (red, orange, yellow, green, blue, and violet).

White light is incident on a prism as shown in Figure 18–27. At the first surface, the colors are separated as indicated. The separation is enhanced at the second surface to finally fan out the colors so that they appear on a screen in the order shown.

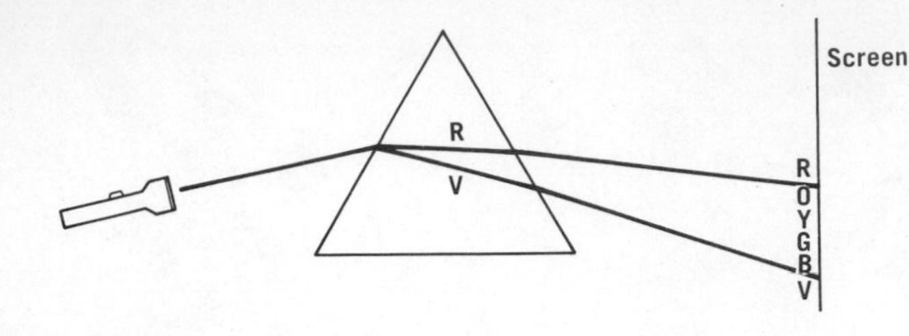

FIGURE 18–27. Because different colors refract by different amounts, a prism can separate white light into a spectrum.

take-along-do-it

A simple prism can be constructed by filling a transparent plastic pie-slice carrier (the type you might use to carry a piece of pie to eat during lectures) with water. Use a mirror to deflect sunlight through your water prism, or use a flashlight in a dark room. Which color is bent most by your prism?

LENSES

Imagine that in an idle moment you should decide to stack matching prisms together as shown in Figure 18–28. Two rays of light from an object far to the left would be refracted by each of the prisms and brought together at point P. Deciding that you may have hit on a wonderful idea, you grind this shape out of a single piece of glass and finally round off the edges to form it into a circular, symmetrical shape, as shown in Figure 18–29. Your final application for a patent would have to be denied, however, because the Chinese thought of this idea in about 2000 BC. Your discovery is nothing more than a particular kind of lens called a *converging lens.*

The use of the term "converging" to describe this lens should be obvious if one considers its effect on a group of parallel rays of light (Fig. 18–30). Rays of light that travel parallel to each other as shown in the figure can be obtained by placing the light source at a great distance from the lens. In passing through the lens, they are refracted from their original path and are converged to a focus as indicated.

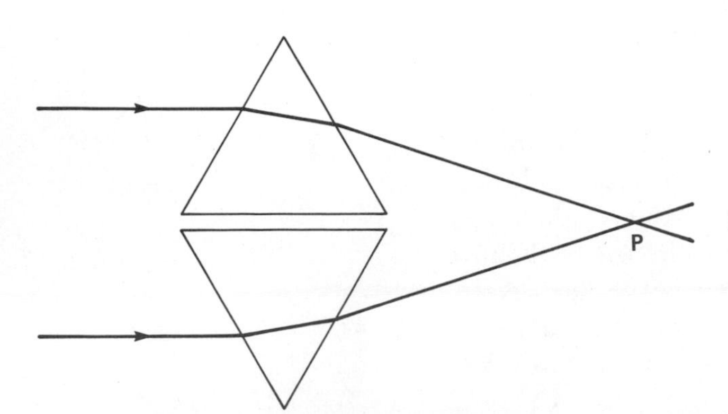

FIGURE 18–28. Two prisms doing their thing.

Another occupation for your idle moment could have been to place your prisms in the configuration indicated in Figure 18–31. The individual prisms now alter the paths of two parallel rays as shown. A piece of glass ground in this fashion and again smoothed into a circular, symmetrical shape is called a *diverging lens* (Fig. 18–32). The term "diverging" is used to indicate that parallel rays are diverged away from their original path upon passing through the lens.

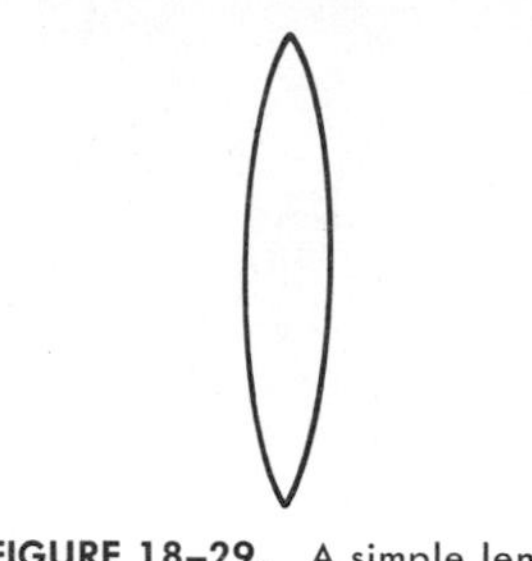

FIGURE 18–29. A simple lens.

It is a simple matter to distinguish between a converging and a diverging lens; a converging lens is always thicker at the center than at its outer rim, and the opposite is true for a diverging lens. Eyeglasses use lenses ground as shown in Figure 18–33. The reason is very pragmatic: the inside surface of the lens must curve away from the eye so that the wearer's eyelashes do not hit it.

Focal Length and Focal Point

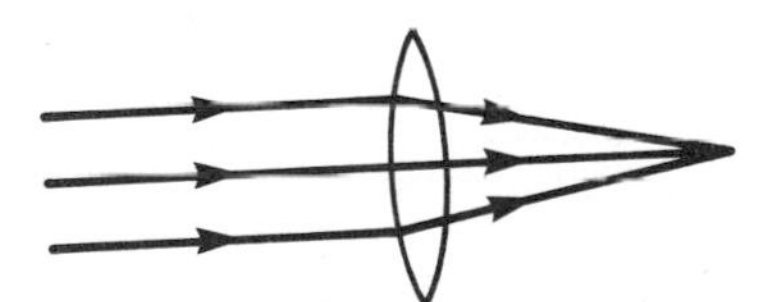

FIGURE 18–30. A converging lens brings parallel light to a single focus point.

Parallel rays of light striking a converging lens as in Figure 18–30 are converged to a focus. This point of convergence is called the *focal point* of the lens, and the distance from the center of the lens to the focal point is called the *focal length*. The focal length is a distinguishing characteristic of a particular lens. If you tell someone that you received a convex lens for your birthday, you really have not told him very much. If, on the other hand, you say that you received a converging lens of focal length 10 centimeters, he will be very impressed. He knows that if he places a light bulb at a great distance and then uses your lens to form an image, the image will be formed 10 centimeters away from the lens. Optical instruments that use lenses require them to be of specific focal lengths to perform their function. For example, a converging lens of just any focal length may not improve your vision when used in eyeglasses. The appropriate focal length must be determined by an eye examination.

As indicated in the discussion of refraction, light rays

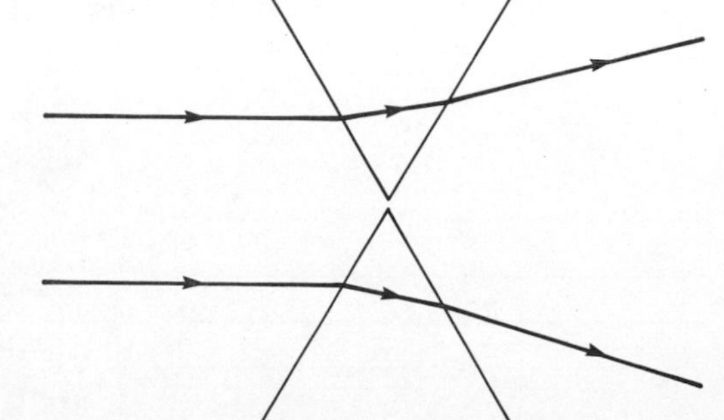

FIGURE 18–31. Two more prisms.

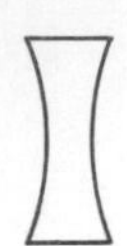

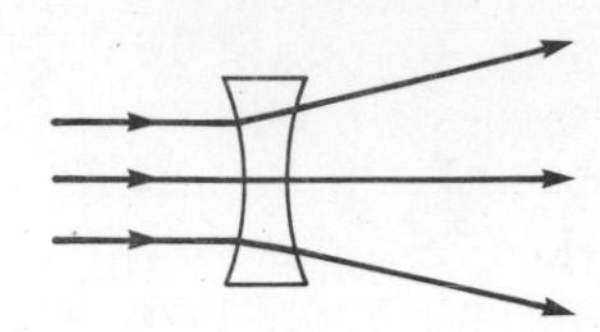

FIGURE 18–32. A diverging lens and its effect on parallel light.

follow the same path if their direction is reversed. This means that if a source of light were placed at the focal point, its rays, after passing through the lens, would be refracted in parallel lines (Fig. 18–34). An application of this will be shown in the next chapter.

Diverging lenses also have focal points. Figure 18–35 shows two parallel rays of light passing through a concave lens, which causes them to be diverged. If these diverged rays are extended back along the dashed lines to the point where they intersect, that point is defined as the focal point, and the distance from the center of the lens to this point is called the focal length.

Vision Correction with Lenses

FIGURE 18–33. A differently shaped converging lens.

The most common application of lenses is the correction of imperfect eyesight. In a normal eye, the light from an object passes through the lens of the eye and is refracted by this lens and brought to a focus on the retina, located at the back of the eye. The stimulation of tiny nerve endings in the retina is transmitted by the optic nerve to the brain. If the image is not perfectly focused on the retina, the image will appear blurred.

In the defect commonly called nearsightedness (myopia), the eye is too deep or the eye lens too strong, so that the image is brought to a focus in front of the retina (Fig. 18–36*A*). The distinguishing feature of this imperfection is that distant objects are not seen clearly. To correct this problem, an ophthalmologist or an optometrist fits the eye with a diverging lens, which prevents the light from coming to a focus until it reaches the retina (Fig. 18–36*B*).

A second common defect of the eye is called farsightedness (hyperopia). Here, distant objects are seen clearly,

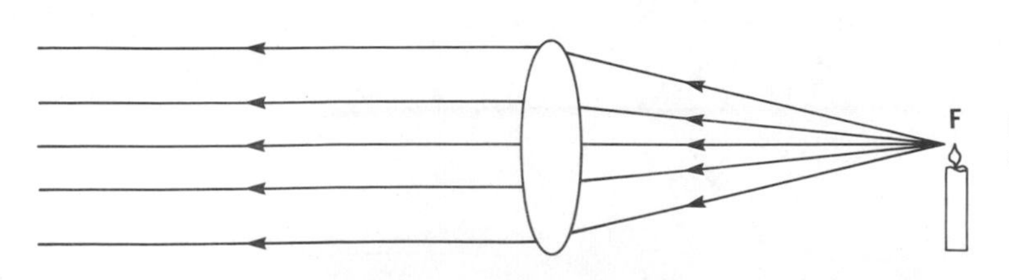

FIGURE 18–34. Rays from the focal point of a converging lens are made parallel by the lens.

but near objects are indistinct. The lens of the eye is unable to bring diverging rays of light to a focus on the retina and instead "tries to" form its image behind it. Figure 18–37 shows that a converging lens placed before the eye causes the light rays to converge, or come to a focus, on the retina, thus correcting the imperfection.

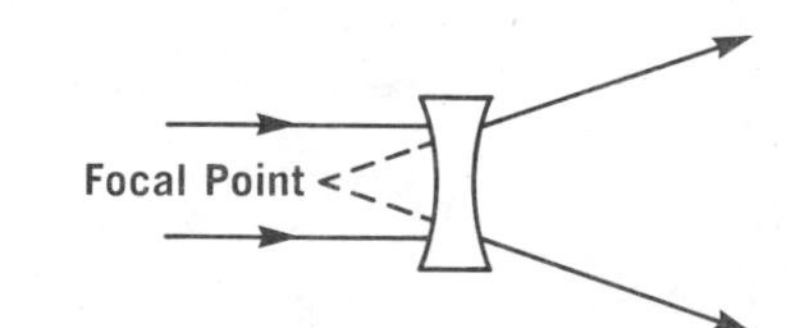

FIGURE 18–35. Parallel light, after passing through a diverging lens, appears to have come from the lens's focal point.

take-along-do-it

To study the formation of images by lenses, obtain a pair of eyeglasses having converging lenses. (Since converging lenses are used to correct farsightedness, and because this eye defect is more common in the elderly than in young people, your best chance of obtaining such glasses is by a visit to Grandma.) In a semidarkened room, light a small candle. Using only one of the lenses, move the glasses, as shown in Figure 18-38, until an image is formed on the sheet of paper. Is the image erect (that is, does the flame burn upward as you observe the image), or is it inverted? Is the image larger than the candle or smaller?

Now move the glasses closer to the candle, and, holding them in this position, move the paper until another image is formed. Describe it. If you have trouble with this part of the experiment, don't get alarmed. Sometimes lenses are ground to correct another visual defect called astigmatism. This correction may make the image so blurred that you cannot distinguish very much about its features. The remainder of this TADI will work in any case.

Try moving the glasses far away from the candle (10 to 15 feet) and note the position of the paper when an image is formed. The candle in this case is sufficiently far away so that the rays of light are almost parallel when they pass through the lens. This means that the distance from the lens to the image is equal to the focal length of the lens.

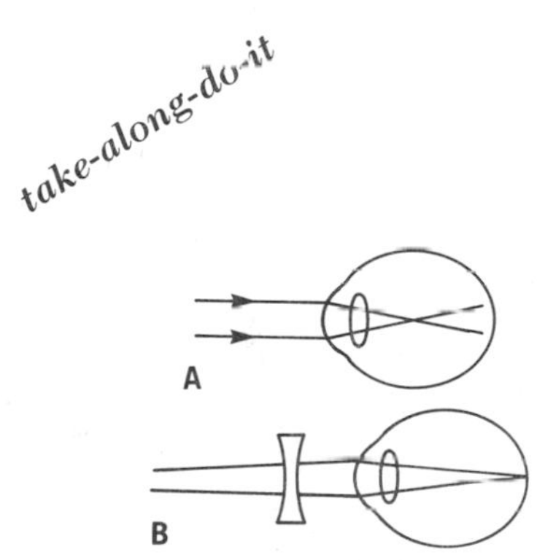

FIGURE 18–36. A nearsighted eye and correction by use of a diverging lens.

Lens Defects

The similarity between a lens and a combination of prisms should indicate to you that the image formed by a lens may not be perfect. Figure 18–39 indicates the problem. White light passing through the lens is fanned out into a spectrum, as it was when passing through a prism. The violet light is bent the most and red light least. Since every

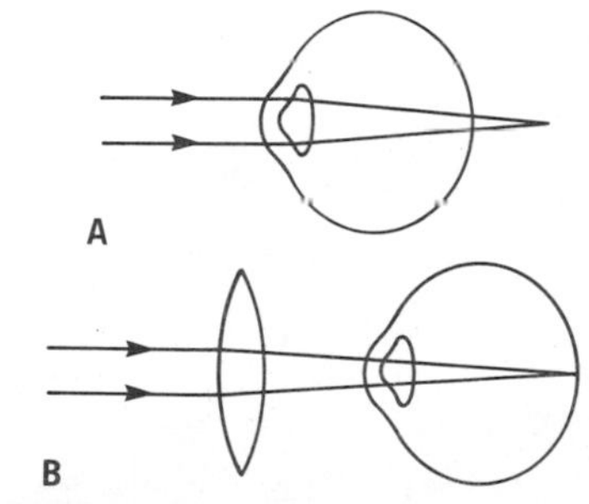

FIGURE 18–37. A farsighted eye and its correction.

FIGURE 18–38. Using a converging lens, you should be able to form an image of a bright object on a screen. You will have to position the three objects at much greater distances than indicated here, however.

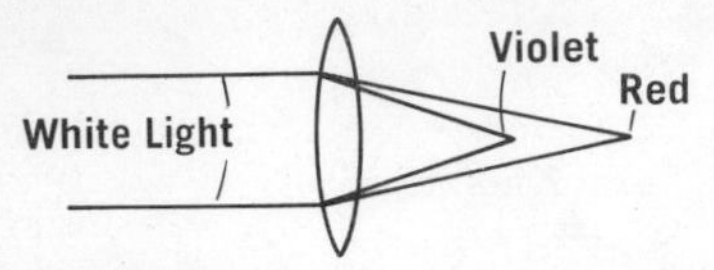

FIGURE 18–39. Chromatic aberration.

color is bent by a different amount, the various colors come to a focus at different points, thus forming a different image for every color. This lens defect is called *chromatic aberration.*[5] It is possible to avoid this effect in all optical instruments by using combinations of lenses such that the aberration produced by one lens is canceled by a second lens.

A second common lens defect is called *spherical aberration,* which occurs because the spherical shape to which lenses are normally ground is not the ideal shape for them. It is used simply because it is easy to grind a lens to this shape. As a result, even for light of a single color, the rays which pass through the outer edges of the lens do not come to a focus at the same point as those passing through closer to the center (Fig. 18–40). Blurring of the image can be minimized by placing a diaphragm in front of the lens that will block off the rays passing through the outer edges.

MIRAGES

Few of the old western movies would have been complete without a scene in which the old prospector, his faithful donkey dead, struggles across a hot desert to warn the white-hatted hero of impending danger. Ravaged by thirst, he sees before him a sparkling lake with shadows of trees glimmering in the cool water. He rushes forward, only to plunge into a sand bank. On a less dramatic scale, many automobile travelers have seen pools of water glistening before them on a long stretch of hot roadway. Both of these examples have a common characteristic—the ground and consequently the air near the ground are hot. Since hot air is less dense than cooler air, we have a situation in which a layer of air near the earth is less dense than layers of cooler air above it. This means that the layers of air as we move upward from the ground have a varying density—a primary requirement if refraction is to occur in air, because light travels faster in less dense air.

In Figure 18–41, a ray of light from the top of the tree is shown heading toward the ground. As it nears the ground,

[5]It is interesting to note that Isaac Newton developed the reflecting telescope in an effort to avoid the chromatic aberration present in refracting telescopes.

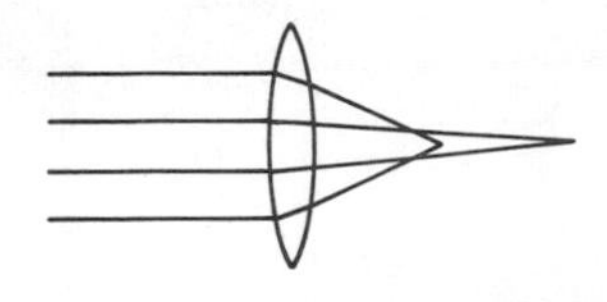

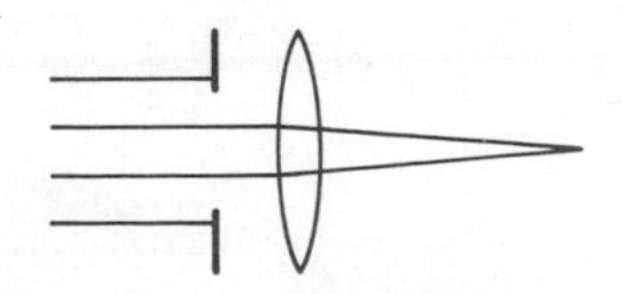

FIGURE 18–40. Spherical aberration and its correction by blockage of the outer rays. How does this affect the brightness of the image?

FIGURE 18–41. That water looks good!

however, it passes into the less dense air and is gradually refracted to follow the path shown. When this ray of light enters the eye of our prospector, he perceives it to be coming along the dashed line. Simultaneously, another ray follows the straight line from the tree to his eye. The ultimate result is that he sees an inverted tree, as well as an erect one. Leafing through his memory bank of past experiences, he recalls that the only other occasion in which he has observed a similar situation was when a tree had been reflected in a pool of water. In addition, light from the blue sky is refracted from the desert near the area where the refracted tree appears. It is no wonder he thinks he is seeing a pool of water.

On the open sea, a situation often occurs which is diametrically opposite to the mirage. It is not uncommon for layers of air near the ocean surface to be cooler than those higher in the air. (This occurrence leads to the amplification of sounds on open water as discussed in Chapter 10.) The refracting effect of these layers of air with varying densities can direct rays of light from a ship back toward the surface of the sea (Fig. 18–42), enabling the observer in ship A to see ship B even though it may be over the horizon.

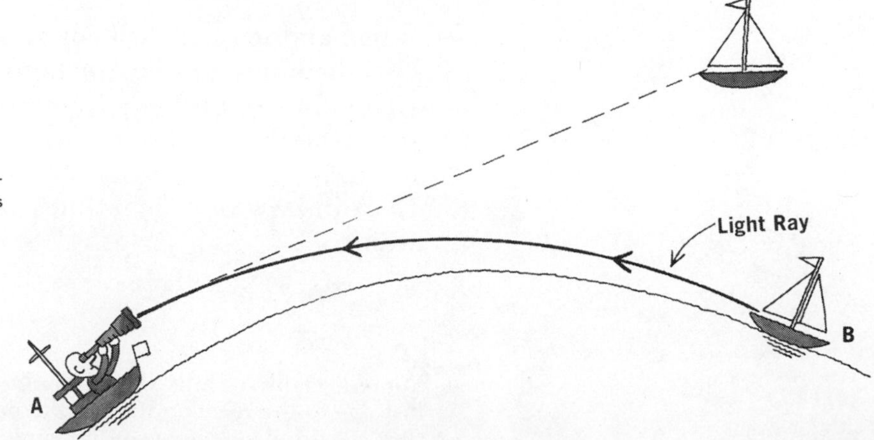

FIGURE 18–42. Light bent by different densities of air at different heights above the water causes this effect.

OBJECTIVES

After Chapter 18 you should be able to:

1. Describe Galileo's attempt to determine whether light travels instantaneously and show numerically why he got the results he did.
2. Describe Roemer's method of measuring the speed of light, giving a numerical example.
3. Describe a successful earthbound method of measuring the speed of light, showing the set-up in a diagram.
4. Perform a demonstration to show the difference in the shadows produced by a large light source and a small one.
5. State and explain the law of reflection of light.
6. Describe diffuse reflection, using a diagram.
7. Draw a diagram and use it to explain image formation in a flat mirror.
8. State rules regarding the size and location of an image formed in a flat mirror.
9. Explain the difference between real and virtual images, and give an example of each.
10. Use a diagram to explain total internal reflection.
11. Show how total internal reflection allows one to use a prism to reflect light.
12. Relate the refraction of light to the speed of light in different materials.
13. State and explain, with a diagram, the law of refraction of light.
14. Perform a demonstration to illustrate the bending of light as it goes from water into air.
15. Use knowledge of the difference in speeds of red and violet light in glass to explain how a prism separates white light into colors.
16. Distinguish between a converging and a diverging lens, and draw a representation of each, showing the paths taken by (previously) parallel light after passing through each lens.
17. Name and explain two common lens defects.
18. Use diagrams to explain both nearsightedness and farsightedness and the correction of each with lenses.
19. Describe the conditions necessary for the formation of a mirage.
20. Explain how one sometimes sees "over the horizon" on water.

QUESTIONS– CHAPTER 18

1. Suppose Galileo's hills were 10 miles apart. Assuming that *no* time is lost due to the reaction time of the person 10 miles away, how much time would elapse between opening the first lantern and seeing the second?

2. The earth is about 93 million miles from the sun. At 186,000 miles per second, how long does it take light to reach us from the sun? If the sun went out five minutes ago, would we know it yet?

3. A solar eclipse occurs when the moon gets between the earth and the sun. Use a diagram to show why some areas of the earth see a total eclipse, other areas a partial eclipse, and most of the earth *no* eclipse.

4. Figure 18–43 is a bird's-eye view of a person looking into a mirror. Use the law of reflection to draw two rays from his right ear bouncing off the mirror. Do the same for his left ear. Do you see why he sees himself "backwards" in the mirror?

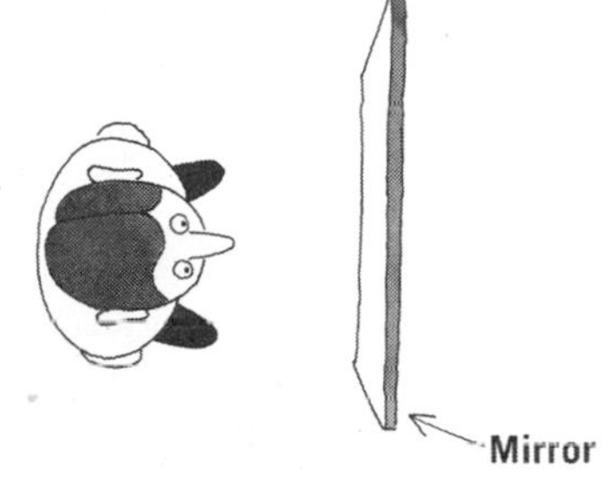

FIGURE 18–43. Draw his image.

5. Look at your image in a mirror. Try to imagine the appearance of a person who is the same distance from you as is your image. Is your image the same size as that person?

6. How long must a mirror be in order that a person 5′6″ tall can see her whole image (standing) in the mirror? (Be careful—the answer is not 5′6″.)

7. Suppose you are told only that two colors of light (X and Y) are sent through a prism and that X is bent more by the prism than is Y. Which color travels slowest in the glass of the prism?

8. In practice, air currents prevent a mirage from being a crisp, clear image—only a blur is seen. What psychological factors come into play to convince a thirsty desert-wanderer that an oasis is ahead?

9. Figure 18–44 shows a single beam of light striking the lower edge of a prism at such an angle that it is totally internally reflected. Draw two rays starting parallel to the beam of light and show that a prism used in this way causes an inverted image.

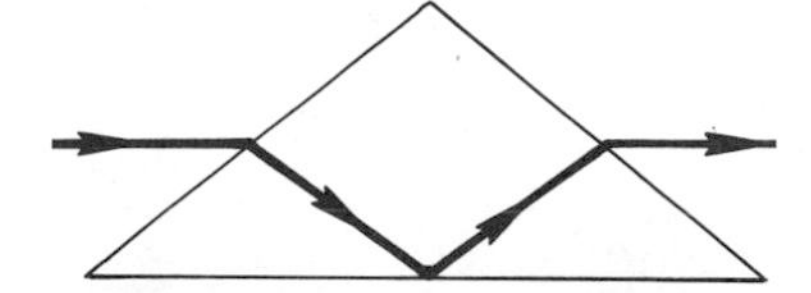

FIGURE 18–44. Draw in two more rays starting parallel to this one.

10. Why is the word "AMBULANCE" written backwards in Figure 18–45?

FIGURE 18–45. Why is the name backwards?

19

OPTICAL INSTRUMENTS

The study of a variety of optical instruments is deemed by some to be an essential part of the study of physics. Although an understanding of reflection and refraction is necessary to understand these devices, their development falls more into the field of technology than that of pure science. It is interesting to note that as science has progressed, we have been able to make better optical instruments and that these instruments have then been used to further the advance of science. This advance allowed development of better instruments, and these instruments. . . . So the two are closely connected. Let's look at some of the results of technology.

THE SIMPLE MAGNIFIER

A simple magnifier is the most basic of all optical instruments, in that it consists of nothing more than a converging lens. Figure 19–1 shows how it works. The object to be examined is placed so that it will be closer to the lens than is the focal point. Two rays of light, A and B, are shown leaving the top of the arrow. Ray A, passing through the top of the lens, is refracted and then passes to the eye. Ray B passes straight through the center of the lens and is not deviated from its path. The eye traces these two rays back along the dashed paths to the point from which they appear to come; this is the point A′. Similar rays are shown leaving the bottom of the arrow at C and are traced backward to C′. The image formed at A′C′ is a virtual image; it is also erect and enlarged. (If it were not for the latter, we would have to think of a new name for the simple magnifier.) The descriptive term *erect* means that if the original arrow points upward, so does the image. We will encounter instances later in which the image is inverted with respect to the object.

The simple magnifier can also be used to set fires. Rays from the sun are parallel when they reach the earth and can be refracted by a lens to come to a focus at the focal point. This focusing of the rays can produce a small spot of intense heat sufficient to set a fire.

The two examples discussed above indicate a property common to all converging lenses: If the object is inside the focal point, a virtual image is formed; if the object is outside the focal point, a real image is formed. As a result, a converging lens can be used as a simple magnifier only when the object to be examined is inside the focal point, and it can be used to "set fire" (or to form a real image) only when the light source is outside the focal point.

take-along-do-it

Obtain a converging lens. Hold it near a letter on this page so that you can use it as a simple magnifier, and examine the image to see that it corresponds to Figure 19-1. Is the letter's image larger than the real letter you are looking at? (It had better be!) Does it seem farther away from you than the real letter? Is it right side up?

Magnification is defined to be the ratio of the size of the image to the size of the object. A simple method used to determine magnification is shown in Figure 19–2. The lens is held above a sheet of lined paper, and by examining the comparison between the line spacing as seen through the lens and on the actual paper, we see that the magnification is two. An instrument used for magnification purposes will usually express its capabilities as 2x, 3x, and so on.

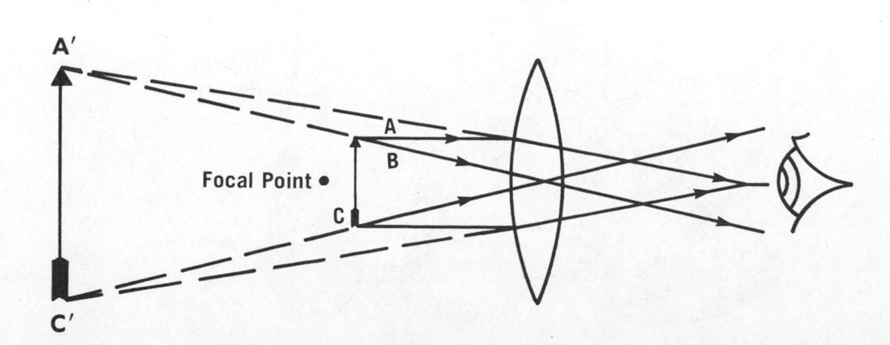

FIGURE 19–1. The optics of a simple magnifier.

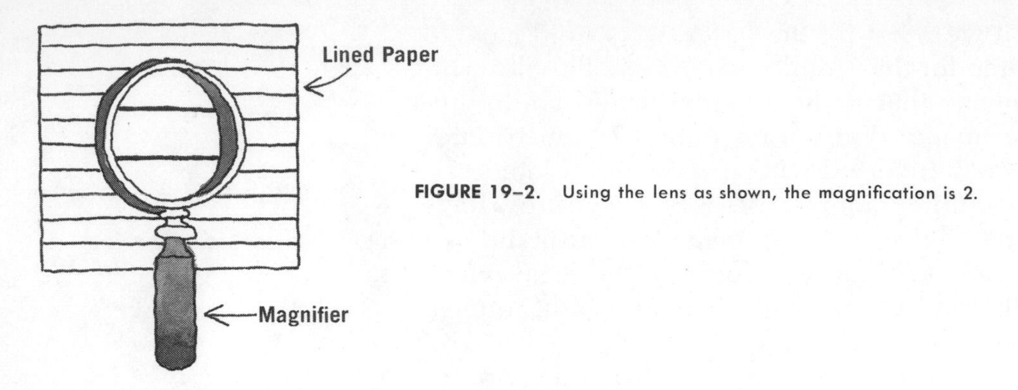

FIGURE 19–2. Using the lens as shown, the magnification is 2.

THE MICROSCOPE

If one converging lens can magnify an object, why not use two lenses and magnify it even more? The answer is that you can do so, and the device is called a microscope. The mirror shown in Figure 19–3 reflects light through the object to be viewed. Light from the object passes first through a small lens (called the objective lens) as shown, and a real, enlarged, intermediate image is formed at A′B′. (Why must the objective lens have a short focal length? Review the last section if you are unsure of the answer.) The microscope provides an example of a situation in

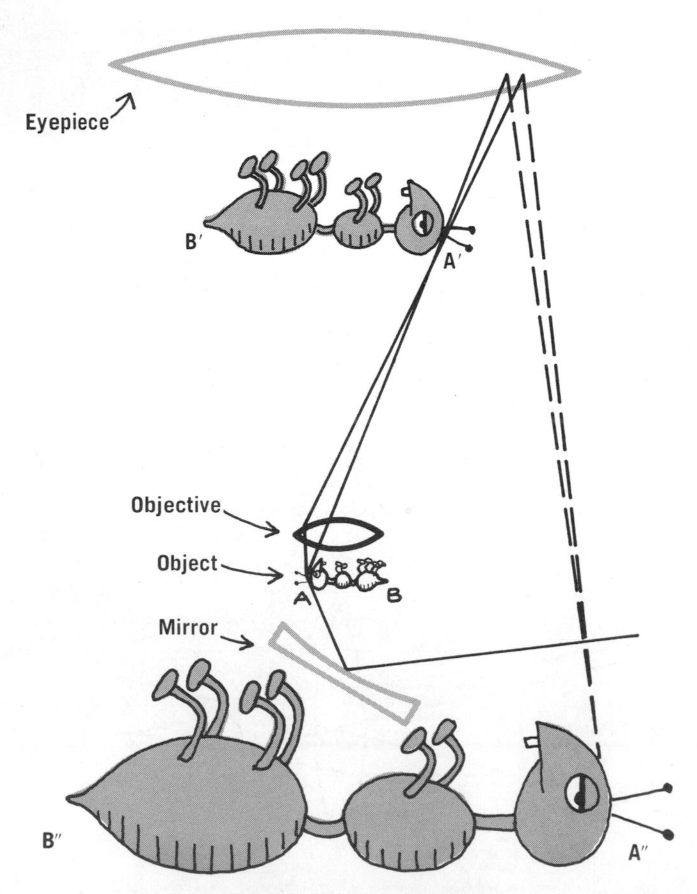

FIGURE 19–3. The optics of a microscope. A″B″ is the final image.

which the intermediate image is inverted with respect to the object. The purpose of the eyepiece lens is to magnify the image A′B′; this lens thus performs the function of a simple magnifier. (Why must the image formed by the objective be formed inside the focal point of the eyepiece?) The final image formed is shown as A″B″ and is considerably magnified.

Man's vision has penetrated to the unknown depths of the incredibly tiny with the microscope. Improved precision in grinding lenses has increased the capabilities of these instruments. Perhaps, you ask, if we were patient enough and careful enough, could we construct a microscope that would enable us to see the atom? The answer is "No!" as long as we use light to illuminate our object, because the object being viewed under a microscope must be at least as large as a wavelength of light in order to be rendered visible. An atom is many times smaller than this value, and so its mysteries have to be probed in other ways.

take-along-do-it

The dependence of the "seeing" ability of a wave on its wavelength can be illustrated by water waves set up in a bathtub. Vibrate your hand in the water until waves having a wavelength of about six inches are progressing across the surface. (The authors recommend that since you will be interested in the scientific aspects of this investigation, someone else should be in the tub with you to set up the waves.) As the waves progress toward you, place a small object, such as a toothpick, in their path. Note that the wave is not much disturbed but instead continues along its path, oblivious of the small object. Now insert a larger object, about the size of this textbook, in the path of the wave and notice that the wave is considerably "disturbed."

In the first case, the toothpick was smaller than the wavelength of the wave, and as a result, the wave did not "see" it. In the second case, the size of the object was about the same size as the wavelength of the wave and it was "seen." Light waves behave in the same general way, and even though the wavelength of light is incredibly small, it is many times larger than the size of an atom. So, make all the improvements you like in a microscope, but as long as it depends upon light to view an object, you will never be able to see an atom with it.

TELESCOPES

There are two fundamentally different kinds of telescopes, but both have the same purpose—to view objects that are at a great distance. The classifications are: (1) the *refracting* telescope, which uses a lens combination, and (2) the *reflecting* telescope, which uses a curved mirror.

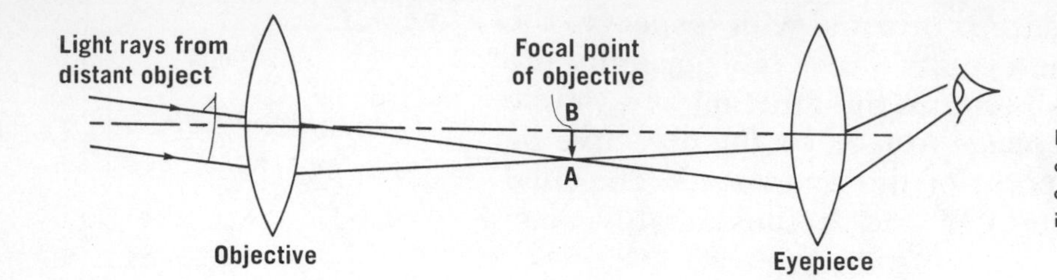

FIGURE 19–4. AB is the image formed of a distant object by the telescope objective. The eyepiece magnifies this image.

The Refractor. The refracting telescope is very similar to the compound microscope. Light is collected by the first lens, as shown in Figure 19–4. Because distant objects may be very faint, it is necessary to collect as much light from them as possible, and for this reason, a lens with a large diameter is usually chosen. The incoming rays of light are nearly parallel, so the image AB formed by the objective is located at its focal point. It is smaller than the object and is inverted. (Is it real or virtual?) This image is then magnified by the second lens, the eyepiece, forming a final image, which is also inverted with respect to the original object. This inversion of the image is not important for an astronomical telescope since it makes very little difference whether we observe a star right side up or upside down, but it is important for telescopes used in earthbound observations, such as scanning the girls' dormitory from a distant building. This inversion is often corrected by insertion of two additional lenses between the first lens and the eyepiece, which function to reinvert the image but do not produce additional magnification.

The Reflector. The principal features of the reflecting telescope are indicated in Figure 19–5. Light passes into the barrel of the telescope and is collected by a large curved mirror at the bottom. This mirror is shaped in such a way that the light reflected from it would form an image at point A in the diagram. However, before the light reaches A, a small flat mirror, M, reflects it out the side of the tube so that the image can be seen without having to stick one's head in the path of the incoming light.

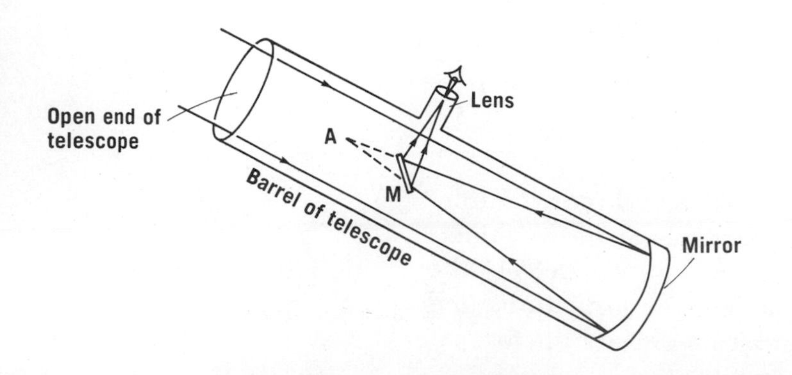

FIGURE 19–5. A reflecting telescope.

THE CAMERA

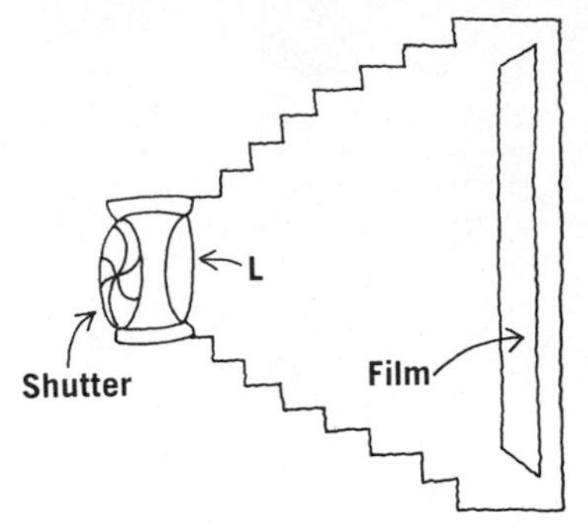

FIGURE 19–6. This camera may look like an old-timer, but the principles apply to every camera.

If you quickly review the optical instruments described so far, you will note that they have a common characteristic—they are exceedingly simple in principle. A few lenses placed in the proper position can produce phenomenal results. Just as simple as all the others is the camera. A lens, a light-tight box, some film, and you're ready to capture forever the picture of some buxom (or muscular) creature, which can be stapled into this text as a centerfold to while away the endless hours of lectures.

A cut-away view of a camera is shown in Figure 19–6. A lens, L, is attached to a light-tight enclosure, such that its distance from the back of the box can be varied. In operation, the camera is focused on the object by changing the distance from lens to film so that the image formed on the film will be clear. A shutter placed before the lens is then opened for a fraction of a second, allowing light to enter the box and expose a piece of film located at the back of the camera. The film is usually made of a compound called silver bromide suspended in a gelatinous material. When light strikes the silver bromide, some of the silver breaks away from the bromine to which it is attached. A lot of light causes a lot of silver to be dislodged; little light, little dislodged silver. When the film is placed in suitable chemicals, additional silver accumulates at locations where silver was dislodged by the light. The silver bromide that was present on the film originally can now be washed away. The film has dark regions of silver still clinging to it at locations where it had been exposed to light. This is the negative, which is dark where the picture should be light and vice-versa.

Because a negative is not aesthetically satisfying, the roles of dark and light on the negative must be reversed. This reversal is accomplished by the procedure shown in Figure 19–7. Light is passed through the negative and allowed to fall on light-sensitive developing paper. The dark regions on the negative will not allow light to penetrate, and

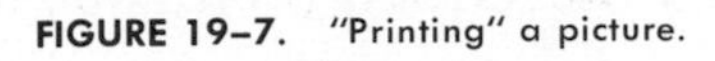

FIGURE 19–7. "Printing" a picture.

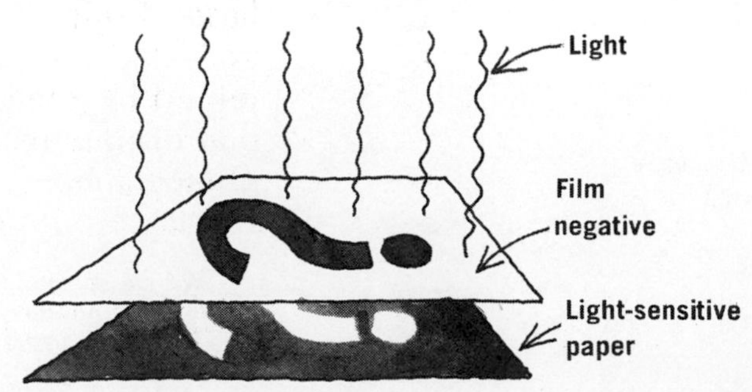

these regions on the developing paper will not be exposed. The opposite is true for areas that are transparent. The development of the light-sensitive paper proceeds in an almost identical fashion to the steps previously described. The image on the paper is referred to as a "positive" because it corresponds to the true re-creation of light coming from the original source.

Understand Your Camera

If you have ever read any of the literature that comes with a new camera, you may have encountered some strange terms, the most common of which are "f-stop" and "shutter speed." Because shutter speed is the easiest to understand, let's consider it first. When open, the shutter allows light to enter the camera, but if it is open for too long, the subject may move, resulting in a blurred photograph. This is particularly disturbing when one attempts to photograph an object in motion, such as a basketball player driving for the basket. The shutter-speed adjustments allow the photographer to set the length of time that the shutter is open and to take "stop-action" photographs. Typical shutter speeds are $1/30$, $1/60$, $1/125$, and $1/250$ second. A stationary object is normally shot with a $1/60$ second shutter-opening time.

The f-stop is a number that describes the width of the shutter opening when it is tripped and is defined as the focal length of the lens divided by the diameter of the shutter opening. A typical camera[1] might have f-stops of 2.8, 4, 5.6, 8, 11, 16, and 22. A setting of f/2.8[2] for this camera would allow the maximum amount of light to enter. Using the definition of f-stop, can you figure out why this is true? A hazy or cloudy day would require a setting in this range. As you move toward higher numbers, the amount that the shutter opens becomes smaller and smaller. (The higher numbers represent a *smaller* diameter opening.) An f-stop of 22 would be used only for a brightly illuminated subject.

The careful reader may note an apparent redundancy in the construction of the camera. Isn't roughly the same amount of light allowed to enter a camera when one uses a small f-stop and a fast shutter speed as when one uses a large f-stop and a slow shutter speed? The answer is yes – and the correct combination to choose is best determined by practice. It is the "feel" for the proper settings that distinguishes the skilled professional from the amateur photographer.

---

[1]We are speaking here of adjustable cameras. Pocket cameras usually have a single fixed f-number.

[2]This is the usual way of listing f-stop – f/2.8, f/4, and so on.

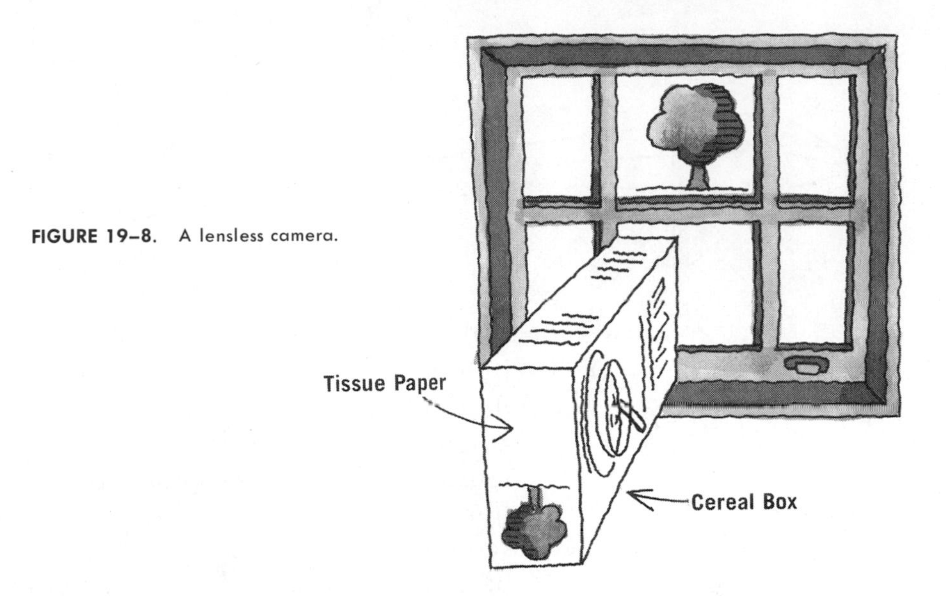

FIGURE 19–8. A lensless camera.

take-along-do-it

If you still have that converging lens you used as a magnifier, you may now use it to demonstrate the physics of a camera. Hold the lens near the wall of a room, opposite a well-lit window. With careful adjustment of the distance to the wall, you should be able to get the lens to form an image of the window on the wall. Such adjustment corresponds to focusing of a camera. Note that a slightly different distance is needed to focus the window frame than is needed to put the outside scene into focus. Is the image erect or inverted?

Now punch a hole in a piece of paper by sticking a pencil through it. Place the paper in front of the lens so that you reduce the amount of light entering the lens. (This corresponds to increasing the f-stop setting on a camera.) Note the effect on the image.

If a lens is not available, you can make a crude pinhole camera as follows: Using a shoe box or cereal box, punch a small hole (just the thickness of a pencil point) in the center of one end. Cut off the other end and replace it with a piece of tissue paper (Fig. 19-8). Now hold the box so that the hole faces a brightly lit window. You should be able to see an image on the tissue paper. No lens is involved here. Can you explain why an image is formed? If not, examine Figure 19-9. Gradually make the hole larger and observe the effect on the image.

THE SLIDE PROJECTOR AND THE MOVIE PROJECTOR

Figure 19–10 shows a slide projector. An intense source of light sends a beam through lens L_1, which directs the

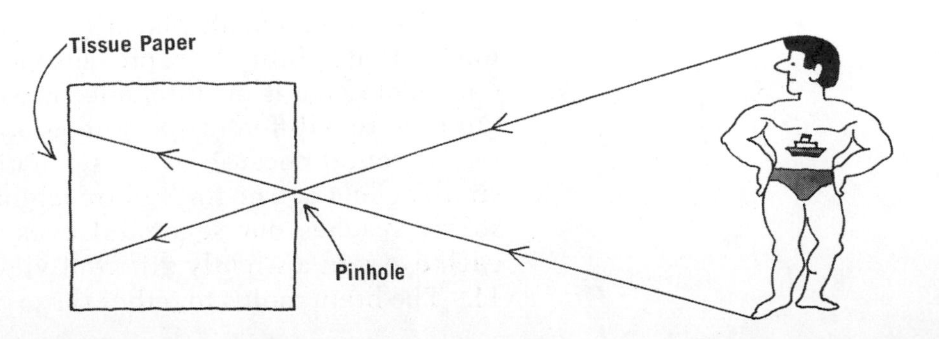

FIGURE 19–9. The image is formed because all the light that enters the box passes through the pinhole and proceeds straight on to the "film."

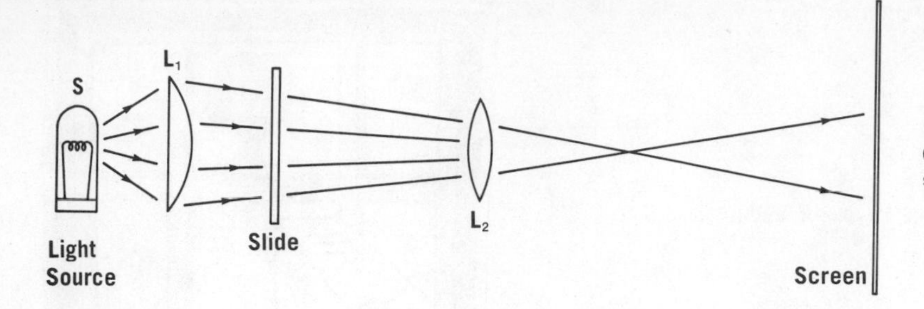

FIGURE 19–10. A slide (or movie) projector. (Distances are not nearly to scale—the screen should be across the room from the projector.)

light toward the slide located as shown. Certain portions of the slide are dark, others are light, and therefore varying amounts of light are allowed to penetrate and reach the next lens, L_2, which is the projecting lens. When properly positioned, the projecting lens brings the light to a focus on a screen. Can you explain, from the diagram, why a slide must be put in upside down if an upright image is to be formed on the screen?

As described and pictured, the slide projector is the essence of simplicity; the only thing wrong is that it will not work. Sources of light emit infrared waves (or "heat waves") in addition to visible light, and if these heat waves are allowed to reach the slide, its useful lifetime will be reduced to about two seconds. A thick piece of heat-absorbing glass is placed at some point between the light and the slide to counteract the effect of the heat waves. The only function of this glass is to absorb the infrared waves and not to affect the visible light to any appreciable degree.

If you add a couple of mechanical contrivances to the slide projector, it is converted into a movie projector. The slide is replaced by a reel of film, and mechanical carriages place successive pictures or frames of the film in the light path at a rate of about 24 per second. Thus the image on the screen is actually a series of still pictures, and the scene jerks from one picture to the next. The image of each frame persists on the retina until the next frame appears, however, and this persistence of vision leads the viewer to believe he is watching continuous motion.

A STEREO FOR THE EYE

The stereo record player (see Chapter 11) owes its popularity to its ability to represent more accurately the music one would hear if in attendance at a concert. Music reaching our ears from different speakers produces a directional effect for the sound because of the separation between our ears. A similar effect is true for light reaching our eyes. Light from a source reaches our separated eyes by different paths, and each eye sees a slightly different view of the object (Fig. 19–11). The brain melds together these two images to produce a

FIGURE 19–11. Because of their separation, the right eye sees a slightly different view of a scene than does the left eye. *A* shows what the left eye would see, and *B* is what the right eye would see. The brain melds together these two images to form a three-dimensional picture.

single three-dimensional picture. A person who has only one eye is deprived of this ability and, consequently, loses his depth perception.

take-along-do-it

You can do this by yourself, but it works better with a partner. Have your partner hold up one finger in front of him. Now you close one eye and slowly move one finger downward toward his finger until they meet tip-to-tip. Feel uncoordinated? It's simple with two eyes, because you then have depth perception.

A popular device in the parlor of all happy homes around the turn of the century was the stereoscope, and latter-day versions (for example, the Viewmaster) are still quite prominent. This instrument is used to view a set of two photographs made by a camera with twin lenses spaced 2½ inches apart—the average separation of the human eyes. These photographs, at positions L and R in Figure 19–12, are then viewed as shown. Light from L is passed through a lens, L_1, and an image of it is formed at point C. Light from R is allowed to pass through a different lens, L_2, and its image is also formed at C. The final scene as interpreted by the viewer has the same depth as he would have observed if he had been at the site of the camera. A new technique to be described in Chapter 21 shows promise of someday bringing the third dimension to movie and television screens. This imaging technique (referred to as holography) makes use of certain unique features of laser light. Until further developments allow for a breakthrough in this field, however, our entertainment will have to remain flat. Interpret the last sentence in whatever way you choose.

THE EYE

The most nearly perfect of all optical instruments is the eye. It is self-focusing, has a built-in light meter, is very compact and portable, is self-cleaning, is not harmful to children, and its magnificent design enables us to view in living color every hue present in the visible spectrum. As we progress through an understanding of the way it works, you should compare the various parts of the eye to those of a camera.

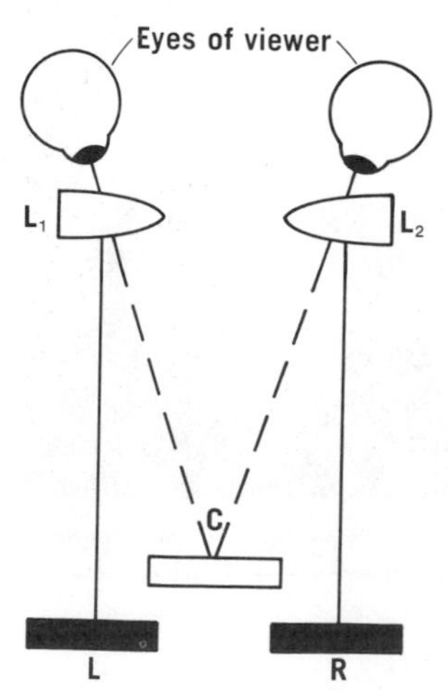

FIGURE 19–12. The principle of the stereoscope.

The outer, white segment of the eye is called the sclerotic coat. The most important portion of this coat is a small convex area at the front of the eye called the cornea. (See Figure 19–13.) Directly behind the cornea is a liquid called the aqueous humor, which has approximately the same optical properties as water. Behind the aqueous humor is a colored shutter called the iris. It is the iris's color that we refer

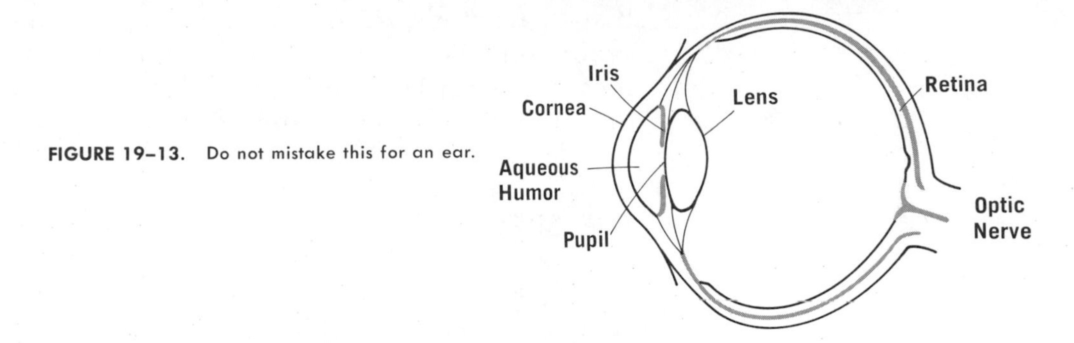

FIGURE 19–13. Do not mistake this for an ear.

to when we speak of blue eyes or brown eyes. Light entering the eye is first refracted by the cornea, and if you recall the properties of a surface of this shape, you should note that the rays of light passing through the cornea are converged toward a focus. These rays are further converged by the lens-like shape of the aqueous humor and are allowed to enter the interior of the eye through a small hole in the iris called the pupil. Using the eye's built-in light meter, if the amount of light coming from the scene viewed is low, the pupil opens (dilates) to allow more light to enter. If too much light enters the eye, the pupil closes.

The shape of the cornea, the aqueous humor, and of the eye as a whole, are fixed. This means that all light refracted by them is bent in the same fashion and would always be converged toward the same position. Thus, we would be able to see objects only if they were at a certain definite distance from us. To overcome this problem, a lens-shaped gristle called the crystalline lens is located behind the pupil. Muscles attached to this lens can cause it to change its shape and thereby bring into focus on the retina the light from objects at varying distances; this is the eye's self-focusing feature. The retina is a delicate layer of nerve fibers at the back of the eye, which acts as a screen on which the image is formed. These fibers are offshoots of the optic nerve, which enters the eye at the rear. Stimulation of the nerve endings by light is subsequently interpreted by the brain.

Although the eye is nearly perfect, certain defects are present in many cases. Two of them, nearsightedness and farsightedness, have been discussed in Chapter 18. Farsightedness is a deficiency of the eye common in the elderly. As we grow old, the crystalline lens loses its ability to converge light enough so that it is properly focused on the retina. As children, we are able to focus on objects as close as ten inches from the eye, but as we grow older, we find our nearest point of focus getting farther away as the lens gradually loses its flexibility. Finally, we wake up to the realization that a book must be held at arm's length to be read. If you recall the properties of lenses, you will remember that a pair of glasses containing a converging lens is now in order (to make up for the lack of curvature of the crystalline lens). A discussion of color blindness, another common defect, will be deferred until Chapter 22.

THE SPECTROSCOPE

White light passing through a prism is fanned out into a complete spectrum. The word "complete" is used here to mean that all of the colors are present. This is in contrast to the situation discussed in Chapter 17, where it was noted that heated gases do not emit a complete or continuous spectrum but instead emit only a very few colors or frequencies. A particular heated gas will emit only colors characteristic of it. Sodium, when heated, emits only yellow light. Mercury gas emits several frequencies, the most prominent of which are a green, an orange, and a violet. These colors act as "fingerprints" for the gas. If we can measure these frequencies accurately, we can determine exactly the gas that emits them. The effect of a prism on hydrogen gas is indicated in Figure 19–14. A beam of light from heated hydrogen gass falls on the prism from the left. As one views the screen from a point such as A in the figure, one finds that the colors have been dispersed. Looking from top to bottom, one sees a red line, then a green one, and finally a violet line.

An instrument used to observe these spectral colors is the spectroscope, which is diagrammed in Figure 19–15. In order for it to work properly, the rays of light which strike the prism must be parallel. Parallelism is ensured by lens L. Light from the heated source passes first through a narrow slit, S, which is located at the focal point of lens L, causing the light that passes through L to be formed into a beam of parallel rays. The light then strikes the prism and is dispersed. As an aid to viewing the colors, lens L_1 is used to bring the slit to a focus. Because the colors observed are always in the shape of the slit (a line), this kind of spectrum is often called a line spectrum. Figure 19–16 (Plate 1) shows a continuous spectrum and line spectra of a number of common elements in gaseous form.

Absorption Spectra

The importance of the spectroscope as an analytic tool is sufficiently great that we will give additional examples of ways in which it is used. In the previous section we observed that a heated gas emits a bright line spectrum that can be used to identify the gas. Figure 19–17 indicates anoth-

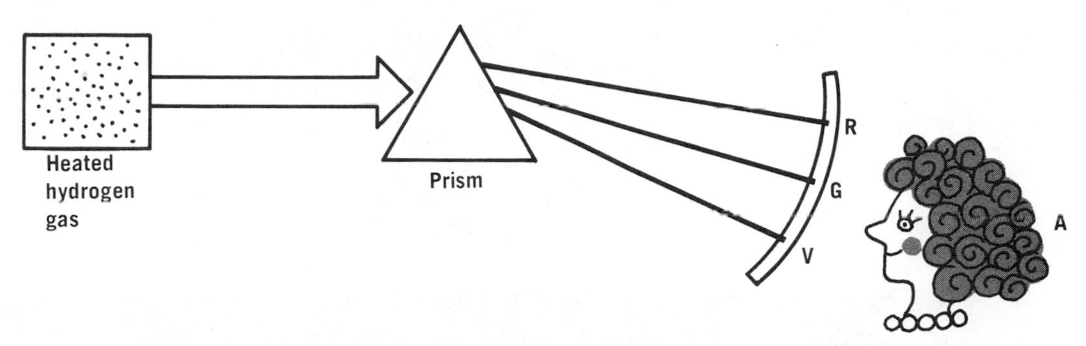

FIGURE 19–14. The viewer sees only certain characteristic colors from the hydrogen gas.

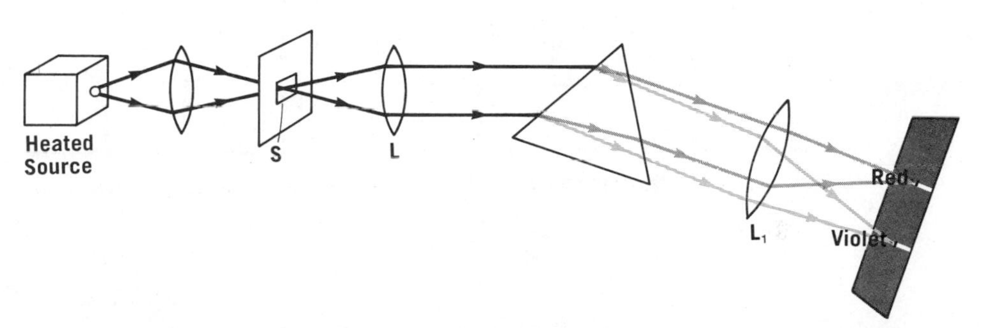

FIGURE 19–15. The spectroscope is simply a refinement of the idea shown in Figure 19–14.

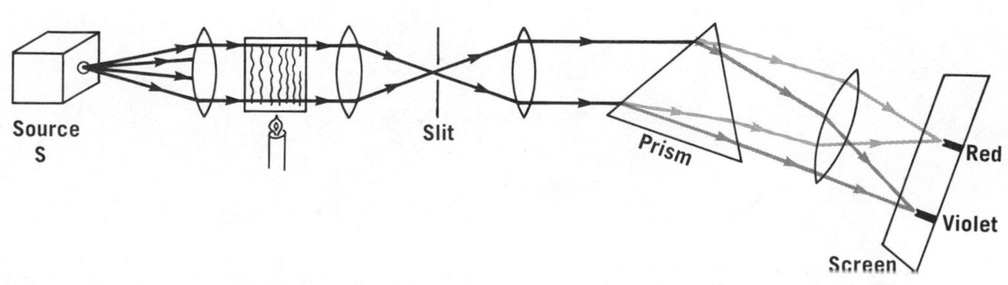

FIGURE 19–17. Note that the gas is now absorbing, rather than emitting light. In this figure the lines leading from the prism to the screen show paths along which no light travels.

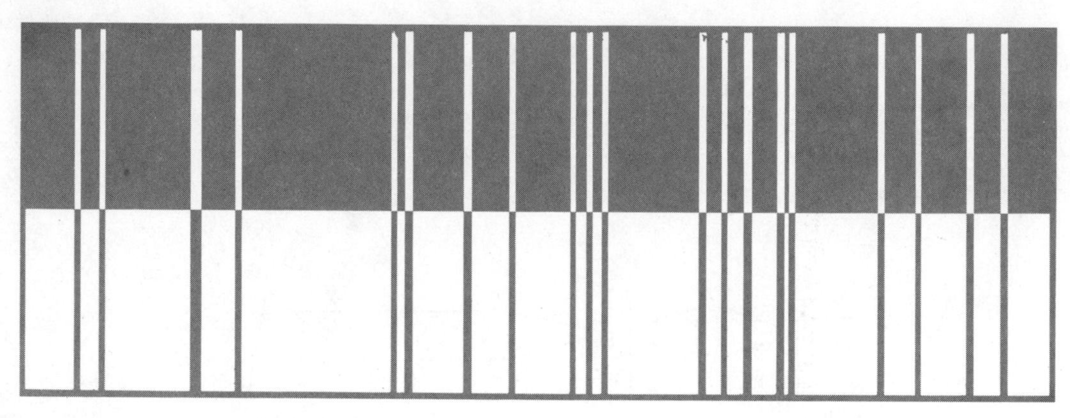

FIGURE 19–18. Bright-line spectrum, with absorption spectrum below.

er method that can be used for the same purpose. The source, S, emits a continuous spectrum, and this light is allowed to pass through vapor from an unidentified substance. For illustrative purposes we will assume that the substance is hydrogen and that it is not at a temperature high enough to emit light. After the light has passed through the vapor, it can be analyzed with the spectroscope. The continuous spectrum will now be found to have dark lines passing through it, as indicated in Figure 19–18. A spectrum of this kind, from which certain frequencies are missing, is called an *absorption spectrum.* The dark lines indicate that certain colors are being absorbed out of the incident beam. In Figure 19–18, the bright line spectrum of an element is compared to its absorption spectrum below. It is obvious that the frequencies emitted by the intensely heated element are identical to those absorbed by the vapor.

When an electron drops from an excited state to the ground state, a photon is emitted by the atom; in the absorption process, the reverse occurs. A photon is absorbed from the beam, causing the electron to move to an excited state. The electron will soon return to the ground state and emit another photon, but this "new" photon can be sent out in *any* direction, and thus very few photons of this frequency pass through the spectroscope. As a result, certain colors are not found when the spectrum is observed.

The Solar Spectrum

In 1817, Joseph von Fraunhofer studied the spectrum reaching earth from the sun. He found it to be a continuous spectrum from which a number of lines had been removed by the absorption process discussed in the previous section. Figure 19–19 (Plate 1) depicts the solar spectrum. In order to explain its appearance, one must assume that the sun initially emits a continuous spectrum. Before reaching the surface of the earth this light must pass through two relatively cool layers of gas—the solar atmosphere and atmosphere of the earth. In both instances certain frequencies are removed by absorption.

The study of the absorption lines from the sun enables us to identify the chemical elements that are present there. In Figure 19–20 (Plate 1), the solar spectrum is compared to an absorption spectrum of iron. We note that there are matching absorption lines, indicating the presence of iron in the solar atmosphere. This illustrates a method for identification of an element, but it actually does not tell us whether the element is present in the solar atmosphere or in our own since the light must pass through both. We can make use of the

Doppler effect to determine which atmosphere produces the absorption lines.

In our discussion of the Doppler effect as applied to sound waves, we found that a sound has a slightly different frequency depending upon whether its source is approaching us or receding from us. This principle also applies to light waves. The sun rotates, just as does the earth, and thus, if we look at the sides of the sun, one side rotates toward us, the other away. This motion causes a shift in the light frequencies from the edges. It should be remembered that sound waves are shifted toward lower frequencies when the sound source is receding. The light frequencies from the edge of the sun that is turning away from us are similarly decreased. The light from the side rotating toward us has its frequency slightly increased. Only those lines which undergo this Doppler shift originate in the sun's atmosphere, as shown in Figure 19–21. Lines that undergo no shift must originate in our atmosphere, which is stationary with respect to the observer.

It is interesting to note that when the solar spectrum was first being studied, some lines were found that did not correspond to any known element. A new element had been discovered on the sun! The Greek word for sun is *helios*, and the element was thus named helium. It is the next-to-lightest element and has since been found on the earth.

We are able to examine the light from stars other than our sun in this fashion, and we have never detected the presence of any element not present on the earth. Only about two thirds of earthly elements have been seen on the sun. The others are probably there but in such small amounts that they do not absorb enough light to be detected.

The Doppler effect can be used to determine whether a distant celestial object is approaching or moving away from us. When the light from other galaxies is examined, we always note a red shift. This means that some identifiable absorption lines are moved slightly toward the red end of the spectrum – toward lower frequencies – and thus, we conclude that other galaxies are moving away from us. We saw in Chapter 11 that this is the basis for the idea of an expanding universe.

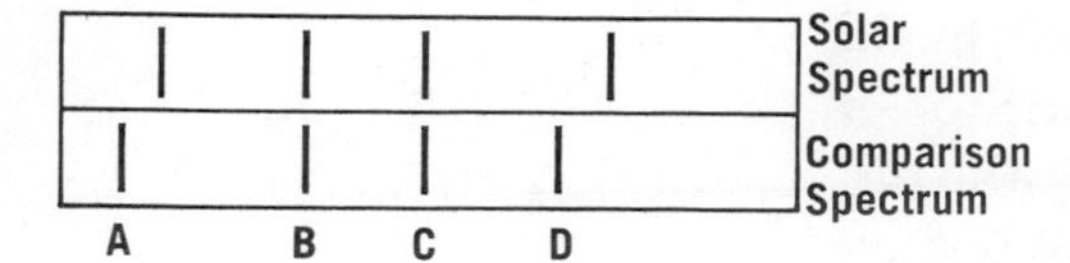

FIGURE 19–21. Lines A and D in the solar spectrum have undergone a Doppler shift, indicating that they originated in the sun's atmosphere. B and C are caused by the earth's atmosphere.

OBJECTIVES

After study of Chapter 19, here's what you should be able to do:

1. Draw a diagram to show how a simple magnifier works.
2. Explain, with diagrams, how microscopes, refracting telescopes, and reflecting telescopes work.
3. Explain with the aid of a demonstration how the nature of light itself places a lower limit on the size of objects which can be seen.
4. Explain why an astronomical telescope is unsatisfactory for observing distant objects on earth.
5. Diagram a simple camera, and explain what is meant by f-stop and shutter speed.
6. Outline the steps involved in obtaining a final photo from a negative.
7. Perform a demonstration with a lens to show the functioning of a camera, and demonstrate the effect of different f-stops.
8. Construct a pinhole camera and explain its operation.
9. Explain how we see in "three dimensions."
10. Perform a demonstration to show that we depend upon having two eyes for perception of distance.
11. Compare the various parts of the camera to corresponding parts of the eye.
12. Describe the operation of the spectroscope, and list three of its uses.
13. Distinguish between a continuous spectrum, a line spectrum, and an absorption spectrum, and explain how each may be obtained.
14. Describe how the spectroscope may be used to explore the sun's atmosphere, and how one distinguishes elements in the earth's atmosphere from those near the sun.
15. Explain how the spectroscope played a part in the evidence for an expanding universe.

QUESTIONS—CHAPTER 19

1. Explain why a fish in a spherical goldfish bowl appears larger than it really is.
2. What difficulty would astronauts encounter in finding their landing site on the moon if astronomers were not aware of all the properties of the telescope?
3. Estimate the shutter speed and f-stop necessary to photograph a center driving for the goal in a well-lighted basketball gymnasium. Repeat for a race car driving the final lap on a cloudy day. Why will the latter probably be blurred anyway?
4. Compare and contrast the eye and a camera. What corresponds to the iris, the retina, and the cornea in a camera?

5. By referring to the line spectrum of mercury gas, describe the appearance of its absorption spectrum.

6. What would happen to an absorption line in the spectrum from the sun if the sun suddenly started to move rapidly away from us?

7. Large telescopes are usually reflecting rather than refracting. Can you think of a practical reason for this?

8. The optic nerve and the brain invert the image formed on the retina. Why do we not see everything upside down?

9. In observing the absorption spectrum from a star, what advantage would an orbiting space station have over an observatory on earth?

10. Draw a diagram to show why a slide must be inserted upside down in a projector in order to get an upright image.

PLATE III

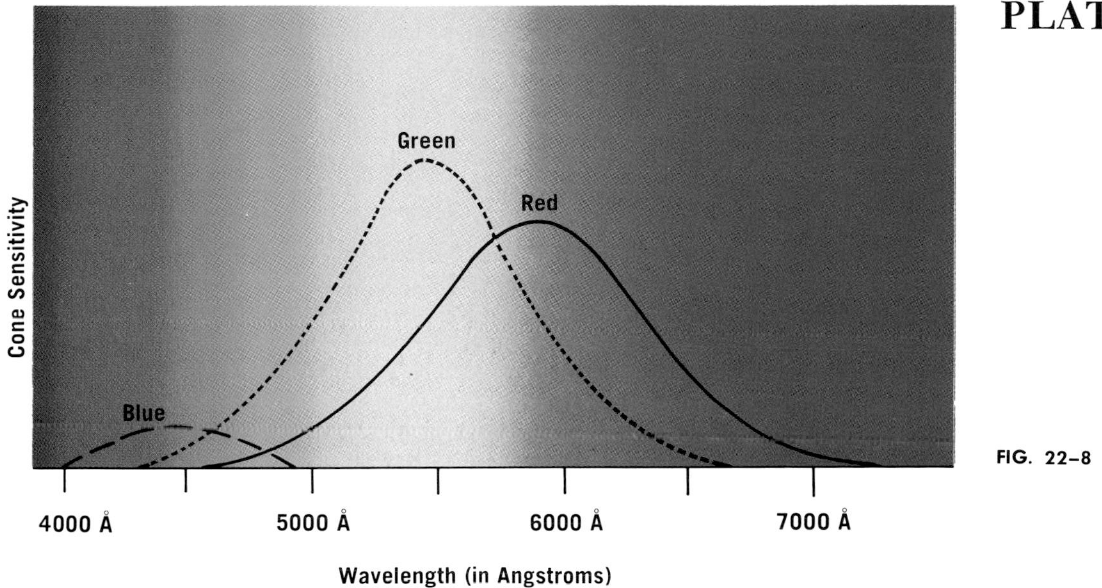

FIG. 22–8

FIGURE 22–8. Graph of the color sensitivity of each of the three kinds of cones of the eye. The height of each line shows the sensitivity of that kind of cone for each wavelength of light. Note that the blue-sensitive cones respond to a very narrow range of colors.

FIGURE 22–9. If you are unable to see the hidden number in this bunch of dots, you may be red-green colorblind. Or the light in your room may be out.

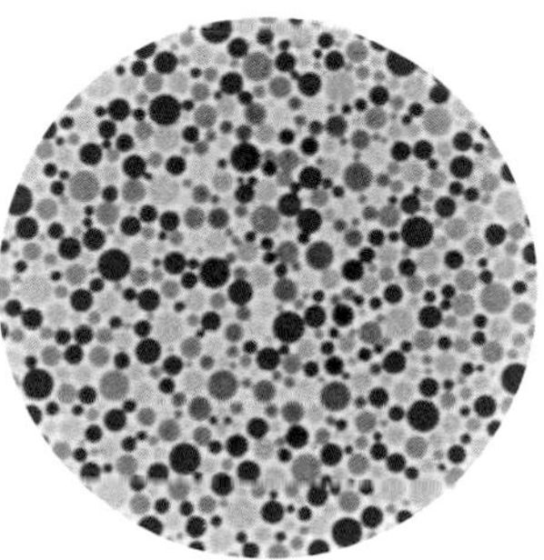

FIG. 22–9

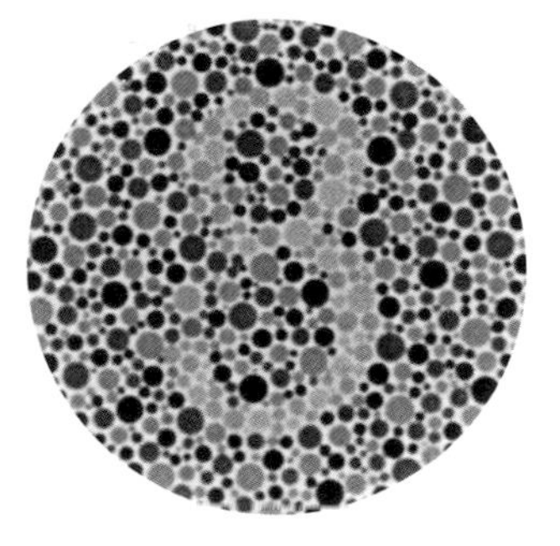

FIGURE 22–10. Fatigue test: Stare at lower right star for 30 seconds, then look at a white wall.

FIG. 22–10

PLATE IV

WHAT IS PHYSICS?

It's sunlight reflected off waves in a brook. (Photo by John Wright.)

It's force and motion. (EKU Photo.)

It's the violent discharge of lightning.

It's time and temperature. (Photo by John Wright.)

It's the sound of five drummers drumming. (EKU Photo.)

IT'S ABOUT YOU.

Photo by John A. Wright.

20

LIGHT DOES STRANGE THINGS

In this chapter we will discuss some properties of light which may be termed unusual, including such seemingly impractical events as the bending of light around corners and the cancellation of light waves by other light waves. Yet phenomena such as these are responsible for the beautiful colors in a peacock's tail. Not so impractical—for peacocks, anyway.

BENDING OF LIGHT WITHOUT REFRACTION

Figure 20–1 illustrates a common property of sound waves. The pedestrian on Madison Avenue in New York City screams in pain because he has caught his foot in a bear trap (a common environmental hazard among big-city dwellers). The fellow around the side of the building can hear the scream of pain, thus demonstrating that sound waves bend around corners. (A TADI immediately comes to mind to illustrate this phenomenon, but we will not suggest it. We doubt that you have a bear trap.)

FIGURE 20–1. Refraction of sound.

Sound is a wave, and light also acts like a wave. Why is it, then, that we cannot see around corners? Why is it that light does not bend around corners as does a sound wave? We emphasized in Chapter 18 that light travels in a straight line, and we used this fact to explain the formation of shadows. But, as we will see, light actually *does* bend around corners. The difference between the bending of sound and the bending of light is a difference in degree only.

We are able to speak to someone through an open door in a room even though the person to whom we speak may be out of our view, because the sound waves bend around the corners of the doorway to reach the listener. A typical sound wave that we might emit to call our friend would have a wavelength of about one meter, which is about the same size as the opening in the door. This gives us an important clue as to how we might be able to observe the bending of a ray of light. Perhaps what we need is an opening for light to pass through which is about the same size as a wavelength of light. The following observations indicate that this is indeed the case.

take-along-do-it

Hold two fingers in front of your eye and observe a distant street light between them. Slowly close your fingers until the light is finally pinched off. Just before the light disappears, it will spread out in a streak across your fingers. A second method of observing the bending of light is to look at a distant light through a stretched handkerchief or an opened umbrella. In this case, the light passing between threads of the cloth spreads out to form streaks.

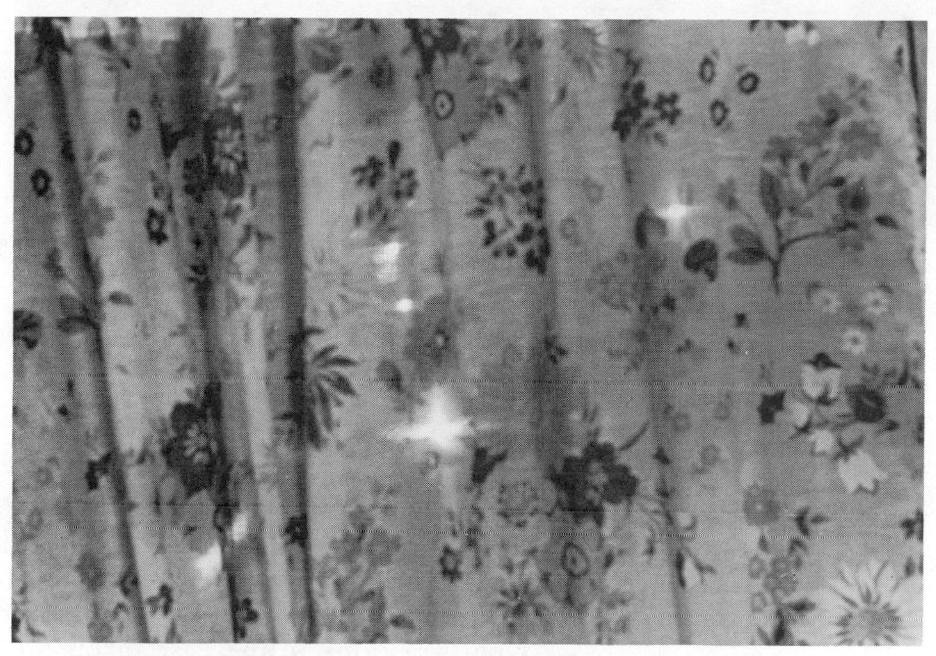

FIGURE 20–2. Interference caused by street lights shining through a curtain.

The observation that light spreads out as it passes through a slit will enable us to investigate a variety of interference phenomena, as we did for sound waves.

INTERFERENCE OF LIGHT

In our discussion of sound we described how sound waves emitted simultaneously from side-by-side speakers interfere, producing areas of loud sound and areas of diminished sound. Our first inclination in attempting to see whether light will act in the same way might be to place two small light bulbs side by side and to look for light and dark areas on a screen. You can try it if you like, but the light and dark pattern will not appear. The reason is that the light from the two sources is emitted in a much more haphazard fashion than is the sound from two separate loudspeakers. This is precisely because the light sources are separate; the loudspeakers were driven by a single sound amplifier.

For constructive interference to occur, two different waves must overlap such that crests meet crests and troughs meet troughs. If for some reason one of the waves undergoes some random change, this perfect overlap might be destroyed. Light from an ordinary light source undergoes such random changes about every 10^{-8} second. So if at some instant of time, constructive interference was being produced on a screen by two separate light sources, 10^{-8} second later one or both of the sources might change, and constructive interference could be changed to destructive interference or

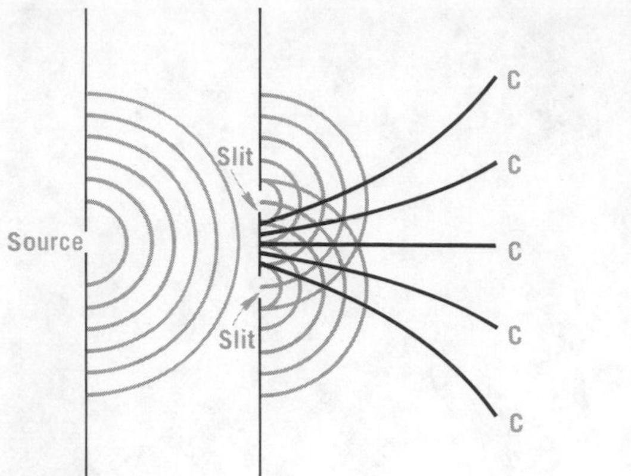

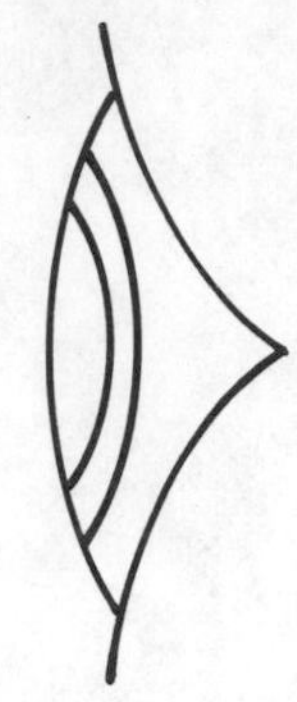

FIGURE 20–3. Light comes from the single source and then spreads out upon passing through each of the two slights. Interference is observed by the eye on the right.

to some intermediate state. The result is a uniform illumination of the screen, with no observable interference effects present. This situation can be overcome by breaking into two parts one wave from a light source (which, after all, is what was done in the case of the sound signal). A random change in the light emitted by the source will occur in the two separate beams at the same time, and interference patterns can be produced.

In 1803, Thomas Young produced an experiment that demonstrated the interference of light. To illustrate his experiment, we scratch two fine lines on a painted piece of glass and allow light from a small source to pass through the openings. The light spreads out from each slit and interferes as shown in Figure 20–3. This figure should bring to mind a similar drawing used in discussing the interference of sound waves. Along certain lines (labeled C in the figure), the waves of light emerging from the two slits interfere constructively, while along other lines, midway between these, destructive interference is occurring. This means that if we place our eye in the position shown, we will see alternating bright and dark lines, as in Figure 20–4.

This demonstration of interference gave the wave model of light a hefty boost. It was inconceivable that particles of light coming through these slits could cancel each other in a way that would explain the regions of darkness. Today we still use the phenomenon of interference to distinguish wavelike behavior in any observation; if no interference can be observed, we do not accept a wave explanation for the effect.

An Alternative Explanation of Interference

In order to understand interference, it is often helpful to redraw Figure 20–3 in a slightly different way. This alternative approach is shown in Figure 20–5. Two waves are shown leaving the two slits and striking a screen. Both of

FIGURE 20–4. This photo shows what the eye sees when looking through two narrow vertical slits at a light source.

these waves start in phase; that is, when the top wave has a crest, so does the lower one. After leaving the slits, the waves travel an equal distance and come together on the screen at point P. They start in phase and they travel the same distance; therefore, they will arrive in phase. Arriving in phase is a requirement for constructive interference, and so at this central point, we will see a bright area.

In Figure 20–6 the two light waves again start in phase, but the upper wave has had to travel one wavelength farther to reach its destination at point Q on the screen. The upper wave has fallen behind the lower one by one wavelength, but still crest meets crest and trough meets trough, so a second bright area is seen. The point R midway between P and Q is shown in Figure 20–7. At this position, the upper wave has fallen one half wavelength behind the lower wave. This means that the trough from the bottom wave overlaps the crest from the upper wave, and thus destructive interference occurs and a dark region is observed. At points between P and R and between R and Q the path difference is such that

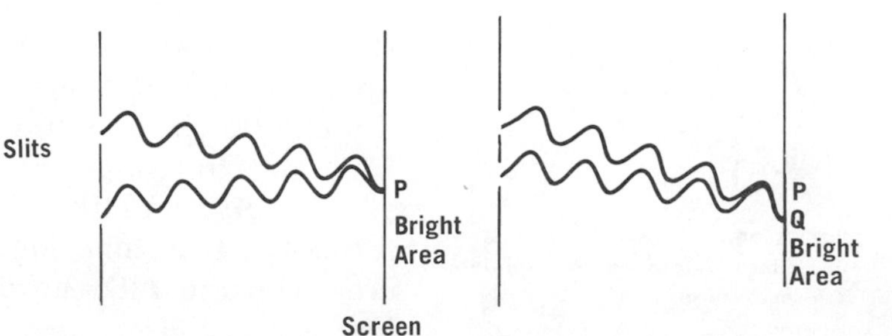

FIGURE 20–5. We now concentrate on waves moving from each source toward point P.

FIGURE 20–6. The waves moving toward Q also interfere constructively.

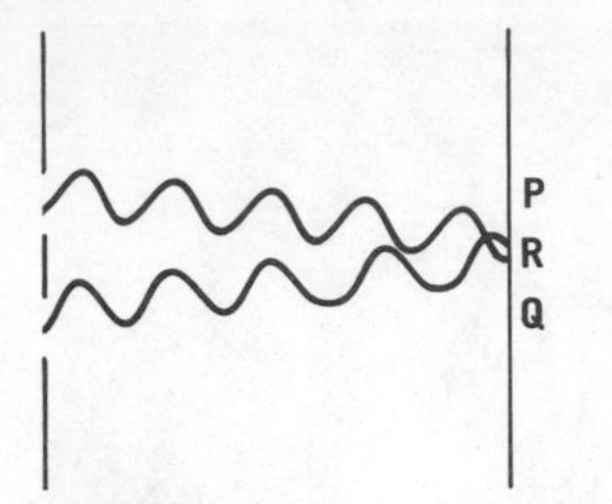

FIGURE 20–7. Destructive interference results at R.

neither constructive nor destructive interference occurs. These areas are neither very bright nor very dark but are instead dimly illuminated. The pattern fades from bright to dark and then back to bright and then. . . .

THE DIFFRACTION GRATING

Our discussion so far has concerned interference effects produced by two slits, but the arguments used can be extended to any number. A common device that employs a large number of slits is called a *diffraction grating*. Diffraction gratings are usually made by engraving fine, closely spaced, parallel grooves on a piece of glass, such that there are typically about 15,000 grooves per inch. The grating allows light to pass through unobstructed at points between the grooves but to be reflected out of the way when it strikes a groove (Fig. 20–8). Our original two slits have been increased to 15,000, but the theory remains the same.

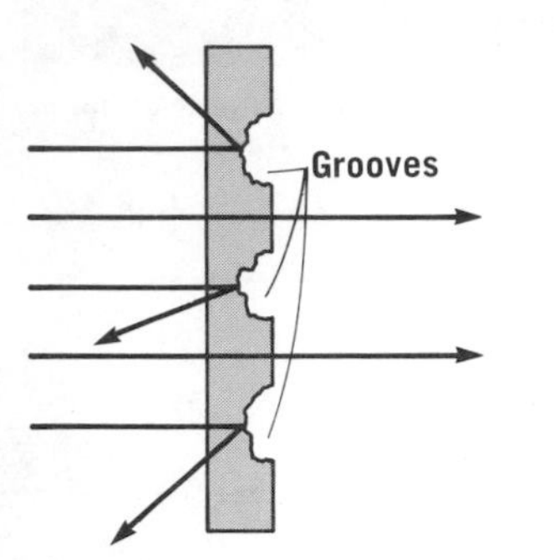

FIGURE 20–8. A greatly magnified side-view of part of a diffraction grating.

Figure 20–9 shows light passing through three slits of a diffraction grating. To investigate the pattern produced when the light strikes a screen, we will consider only one direction at a time. The figure shows three rays of light—one from each slit—which are moving toward the same point on a screen. Light from the bottom slit, A, and from the middle slit, B, move toward the screen, but we have chosen the direction such that light from B will arrive exactly one wavelength behind that from A (because of the extra distance it has to travel). As a result, they arrive in phase, and constructive interference will occur between them. The next slit, C, also sends light toward the same point on the screen, but this light is one wavelength behind that from B and two wavelengths behind that from A, which means that constructive interference occurs between all three waves. In fact, if this argument is carried out for the light from all the slits in the grating, we find that constructive interference always occurs, and as a result, a bright area is seen at the point on the screen where the rays meet. For the light which is directed toward a point such that the wave from B is one half wavelength behind the wave from A, and C's is one half wavelength behind B's, destructive interference will occur and darkness results.

Particular note should be taken of the fact that for constructive interference to occur, the light coming through an opening must fall one wavelength behind the wavelength directly below it. White light is composed of all the colors from red to violet. Each of these colors has a different wavelength, and therefore a red bright area will not be formed at the same location on a screen as a green bright area. Thus, just like a prism, a grating can be used to fan out

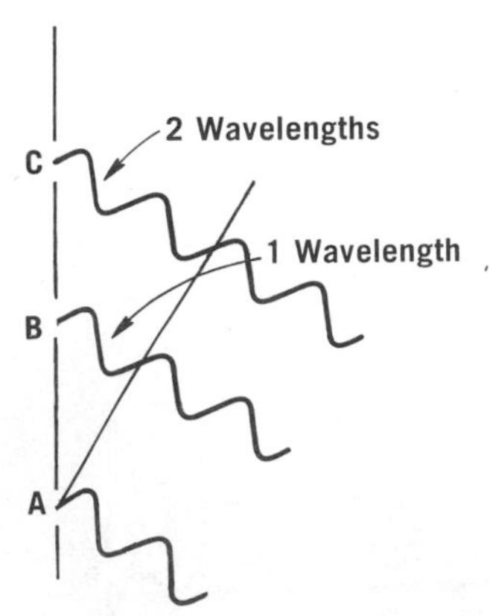

FIGURE 20–9. Waves from three slits of a diffraction grating. They will interfere constructively.

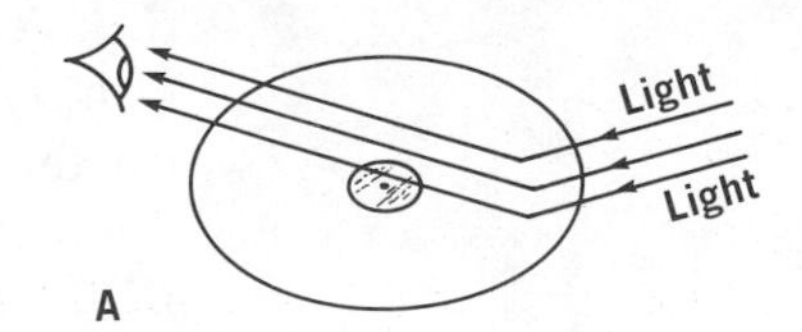

FIGURE 20–10. An easily observed interference effect.

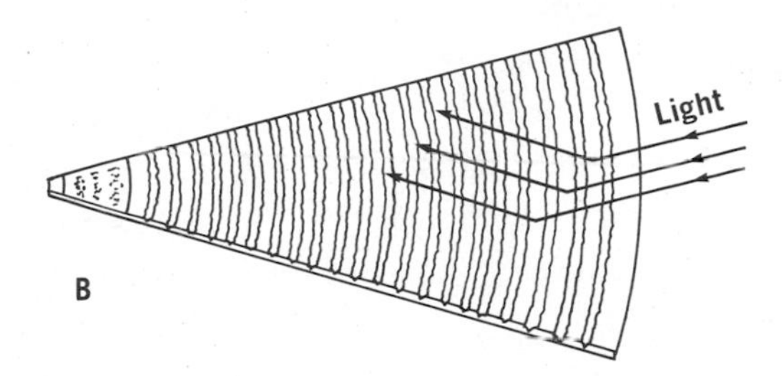

a beam of white light into its spectrum. In fact, diffraction gratings are more often used to study spectra in devices such as the spectroscope than are prisms.

The diffraction grating effect is easily observed with everyday equipment. For example, hold a long-playing record so that light is reflected from it at a very glancing angle (Fig. 20–10A). The spaces between the many grooves in the record act as individual reflectors of light, and as light reflects from each, it interferes and causes you to see various colors in the reflected light. Although the phenomenon is now one of reflection rather than transmission, the theory is essentially the same. In like manner, the fine feathers of a peacock's tail act as tiny, regularly spaced reflectors, and interference is responsible for some of the vivid colors of the fanned tail.

Interference by Reflection

A common example of light interference is seen when you observe the colors produced by spilled oil on a wet pavement. Figure 20–11 represents a greatly magnified puddle of oil. White light (from the sun) falls on the oil's surface. Some of the light is reflected at A, while some passes through this surface and is reflected at the lower surface, B. When these two waves recombine at A, they may be in phase or they may be out of phase, depending upon their wavelengths and the thickness of the puddle. The color seen at a given point is determined by the color that is undergoing destructive interference at that point. For example, if yellow light is interfering destructively, the color seen is that which is produced when yellow light is subtracted from white light. The subject of mixing and subtracting colors will be

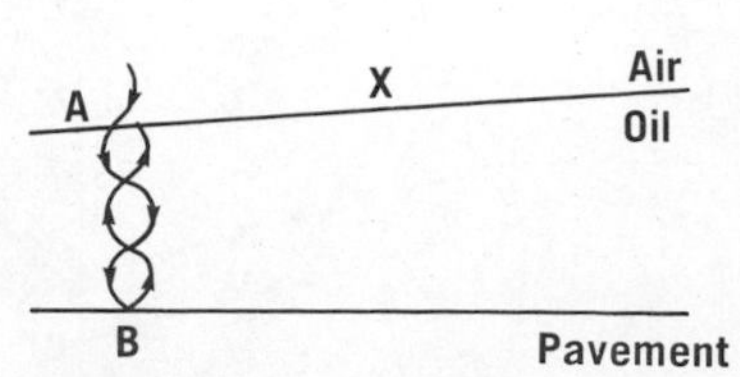

FIGURE 20–11. Destructive interference occurs for the wavelength shown at point A on the oil puddle.

delayed until Chapter 22, and so for the moment we will simply state that the color blue is seen. At some other point on the film, such as X, the thickness of the puddle may be such that a different color interferes destructively and the film takes on another color at this point.

The beautiful colors seen in soap bubbles are likewise the result of interference of light as it reflects from the two surfaces of the thin soap film. If you blow a soap bubble and watch it for a few seconds, you will see that the colors change as the thickness of various parts of the bubble changes.

Most optical instruments of high quality utilize destructive interference to cut down on stray reflected light. As an indication of the trouble that stray light can cause, consider the situation of Figure 20–12*A*. A ray of light is shown passing through a lens that brings it to a focus at point A on the screen. At the back surface of the lens, part of the ray is reflected, however. This part bounces back to the front surface, where part of it is reflected again and finally reaches the screen at point B. Because the light is coming to a focus at two different points, A and B, the image will be slightly blurred. This is certainly an undesirable feature if the screen is the film in a camera. In order to produce a sharp image, the front surface of the lens is coated with a thin film (Fig. 20–12*B*). The thickness of the film is controlled so that the stray light will undergo destructive interference after reflecting from the front and back surfaces of the coating. Stray sunlight causes most of the problem, and since all wavelengths are present in this light, a coating of a single thickness cannot completely cancel all of it. Because yellow light is predominant in sunlight, the coating is controlled to cancel this color. White light that has had yellow removed from it appears blue, and lenses that have been coated in the manner described have a characteristic bluish tinge.

High quality optical instruments also require lenses that have been ground smoothly enough to remove any waviness that might produce distortions. Interference can be used to test the precision of the workmanship. The lens to be tested is placed against a perfectly flat piece of glass (Fig. 20–13), and light of a single color is shined on the upper face of the lens. At position A in the figure, some of the light is re-

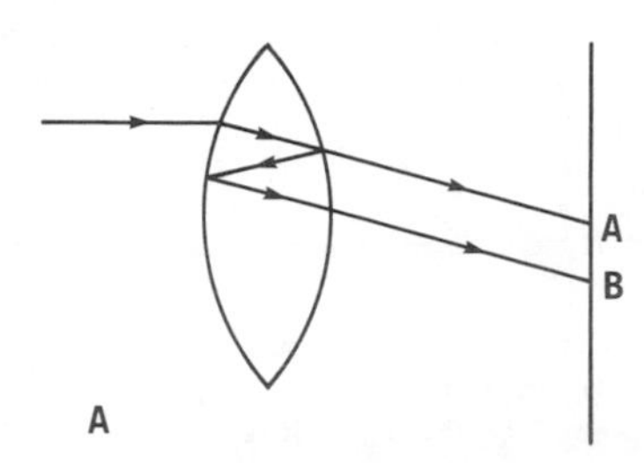

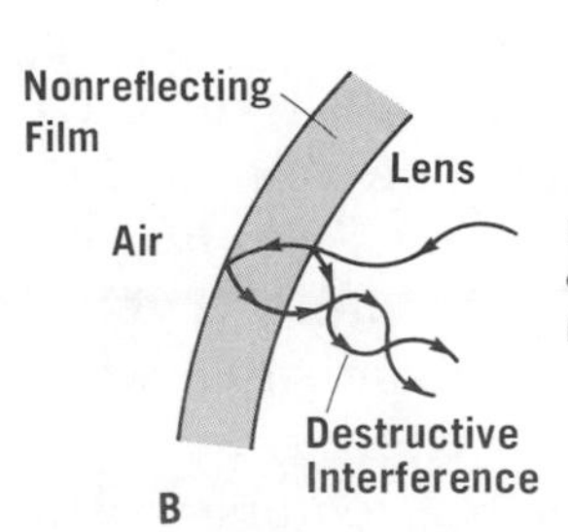

FIGURE 20–12. The reason for and the principle of the coated lens. Note that in *B* the reflected waves are out of phase.

FIGURE 20–13. The use of interference to check lenses.

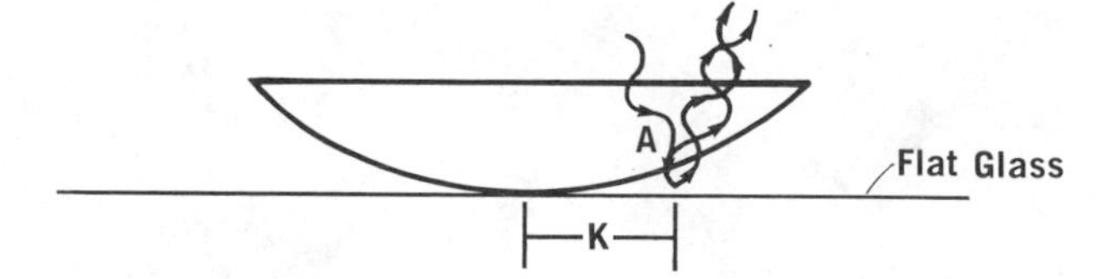

flected from the spherical face of the lens, while part of the light passes on through and is reflected from the flat glass plate. Assume that when these reflected waves recombine they produce destructive interference. If the lens is perfectly spherical, we will see darkness at all points which are at the distance K from the point of contact of the lens with the flat plate. This results in a circular ring of darkness about the point of contact. There will be several points at which such destructive interference can occur, and also, of course, several points where constructive interference occurs. The resultant pattern is shown in Figure 20–14*A*. If the lens is not perfectly ground, its shape will not be spherical, and the rings will appear as shown in Figure 20–14*B*.

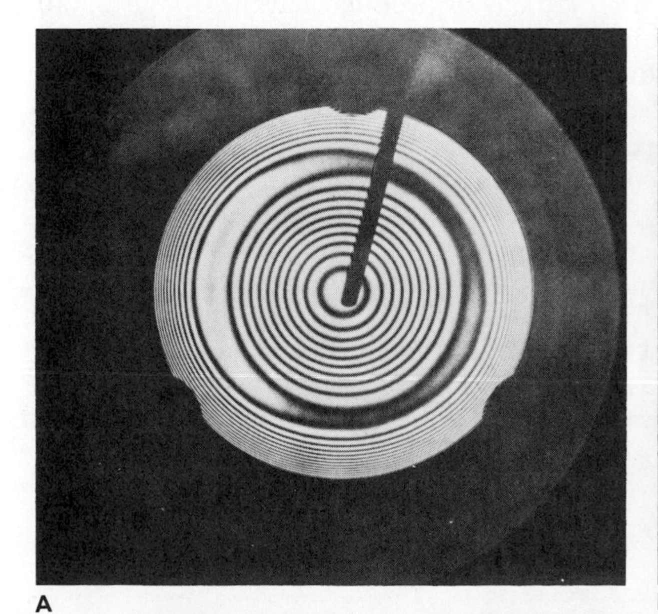

FIGURE 20–14. *A*, The circular rings of the lens indicate that the lens is a good one (at least in the area of the rings). *B*, The lens does not have circular rings, demonstrating that it is not a good spherical lens. (From Physical Science Study Committee: College Physics. Lexington, Mass., D.C. Heath & Co., 1968.)

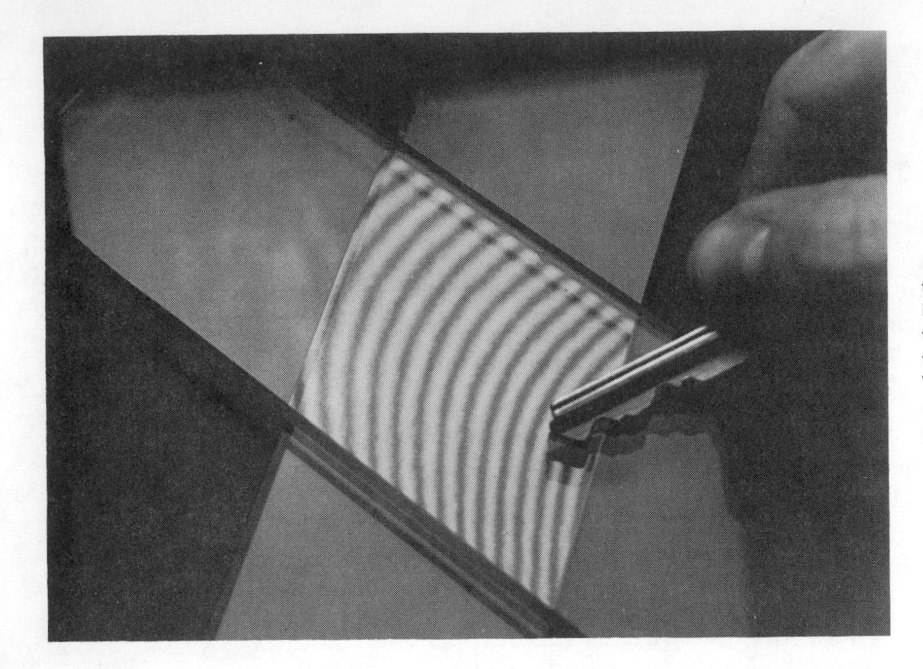

FIGURE 20–15. Interference between two crossed plates of glass. The interference rings are curved because the pressure from the key pushing on the top plate has bent it very slightly. Visit a biology lab and you can find microscope slides sufficiently stuck together to show this effect.

Campaign for More Interference

The trouble with the interference of light is that it can occur on a small scale only. The areas of destructive interference where light cancels out light in the double-slit experiment are just too small to be of any large-scale use. For example, imagine being able to darken several acres of land during the daytime. Drive-in theaters could operate all day! At present, we use what are called "lights" to produce light. What we need are "darks" to produce darkness. We could put up "dark poles" along Lovers Lane and let the beloved region do 24-hour duty. But alas, light's short wavelength will limit us to small-scale darkness forever.

POLARIZATION

The light emitted from a single atom is shown in symbolic form in Figure 20–16. Light is an electromagnetic wave composed of a changing electric and magnetic field. For simplicity, assume that Figure 20–16 illustrates only the electric field part of the wave, but remember that anything

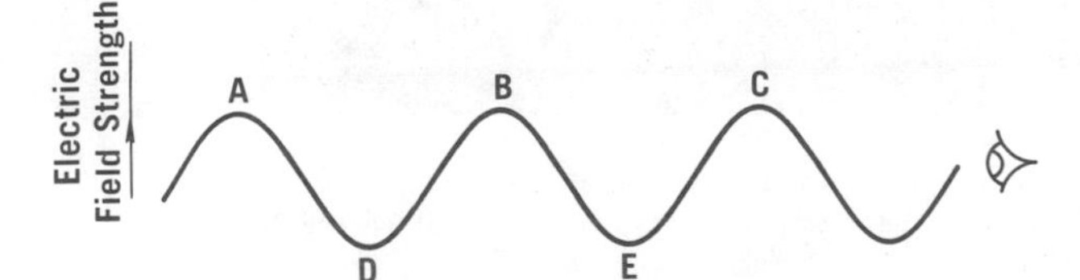

FIGURE 20–16. A simple wave moving toward the eye at right.

we say about the electric field also applies to the magnetic field. At points such as A, B, and C, the electric field points up, and at places such as D and E, it points down. If the eye could actually "see" these electric fields, the end-on view of the wave of Figure 20–16 would look like Figure 20–17. A wave such as this, in which all the electric field lines point along a single direction, is called a *polarized wave.*

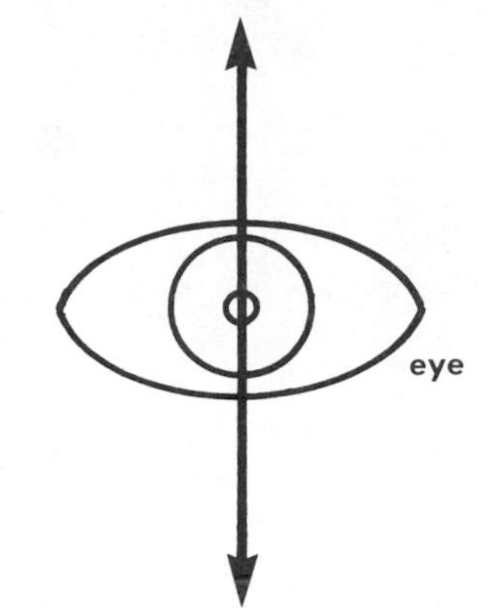

FIGURE 20–17. An end-on representation of a polarized wave.

The atoms in a source of light are oriented at random with respect to each other. This means that the light waves which they send out are randomly directed also. An end-on view of several waves emitted by a source of light would look like Figure 20–18, because the wave sent out by one atom can be oriented at any angle with respect to the wave sent out by another atom. Such a mixture of waves is typical of all light sources, from the sun down to the smallest of light bulbs. Such light, in which all directions of vibration occur, is not polarized.

There are several ways by which unpolarized light can be polarized. One method is by allowing the light to pass through certain substances which allow only vibrations along a single direction to penetrate. Figure 20–19*A* illustrates the way that these materials affect light; shown is a man with several sticks in his arms, being chased by a dog through a picket fence. Figure 20–19*B* shows what has happened to the sticks after the man goes through the fence.

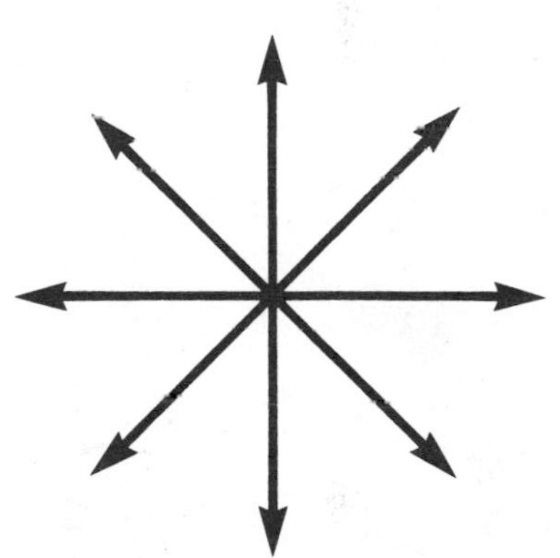

FIGURE 20–18. An unpolarized wave represented end-on.

In the substances which change regular light to polarized light, the molecules are all lined up like the pickets in the fence, and the electric field waves in ordinary light are like the sticks in the man's arms. Only the sticks that are

FIGURE 20–19. Polarizing the unpolarized.

able to pass between the pickets make it through—the rest are broken. Only the electric fields pointing along the direction in which the molecules line up make it through; the others are absorbed by the material.

take-along-do-it

Obtain two pairs of polarizing sunglasses (Polaroid is a common brand of such glasses). Put on one pair of the glasses and close one eye. Hold up a lens of the other pair in front of your open eye (Fig. 20-20) so that light must pass through a lens of each pair before entering your eye. Now rotate the second pair of glasses around as you look through it. As you rotate the lens, the light reaching your eye will be almost completely cut off at some orientation of the two lenses and will pass freely at others.

To understand this TADI, consider what happens when light passes through the first lens. It emerges as polarized light. Suppose that this light is polarized in an up-and-down (vertical) direction. If the second lens is oriented such that it also allows vertically polarized light to pass, the light goes on through. If, however, the second lens is in a position where only horizontally polarized light gets through, it blocks the up-and-down light, and darkness results. At intermediate angles, part of the light gets through and part is absorbed.

FIGURE 20–20. Like this.

Anyone who has ever driven at night is aware of the dangers of being temporarily blinded by the bright lights of approaching cars. It was suggested at one time that the problem could be solved by a method similar to that described in the previous TADI. According to the plan, the headlights of all cars would be covered with a material that would polarize the light leaving the headlights. The windshields of all cars would also be fitted with a similar polarizing material rotated so that only a small fraction of the light from oncoming cars could penetrate. The idea had to be rejected because the method would cut down on light not only from oncoming cars but also from all outside sources. Thus, everything would be darker than before, and night driving would become more, rather than less, hazardous. The manufacturers of dimmer switches breathed a sigh of relief.

When light is reflected from a surface, it becomes partly polarized. If the reflecting surface is horizontal, the resulting polarization is also horizontal (Fig. 20–21). The value of polarizing lenses for eyeglasses lies in the fact that this horizontal light will not pass through the lenses when they are oriented to transmit vertically polarized light only.

take-along-do-it

On a sunny day, rotate your polarizing glasses in front of your eye and observe light reflection from a window. Note the change in the amount of "glare" (light reflected directly from the surface) for various orientations

FIGURE 20–21. Light becomes polarized (to some degree) by reflection.

of the glasses. For an additional surprising effect, rotate your glasses while observing various areas of the sky. You will find that skylight reaching you from directions at right angles to a line from you to the sun is highly polarized.

Figure 20–22 indicates the reason why skylight is polarized. Shown is an incident unpolarized beam from the sun on the verge of striking an air molecule. For simplicity, we will confine our attention to the horizontal and vertical components of vibrations of the incident beam. When this beam strikes the air molecule, it sets the charges in the molecule into vibration. These vibrating charges act like the vibrating charges in an antenna, except that these charges are vibrating in a very complicated pattern. The horizontal part of the electric field in the incident wave causes the charges to vibrate horizontally, while the vertical part of the electric field simultaneously causes the charges to vibrate vertically. During the horizontal motion of the charges, they send out a horizontally polarized wave, which is sent down to the observer. During the vertical motion of the charges, a vertically polarized wave is sent out, but it is transmitted in directions parallel to the earth. Proper orientation of our sunglasses will cut out the horizontally polarized light which reaches us.

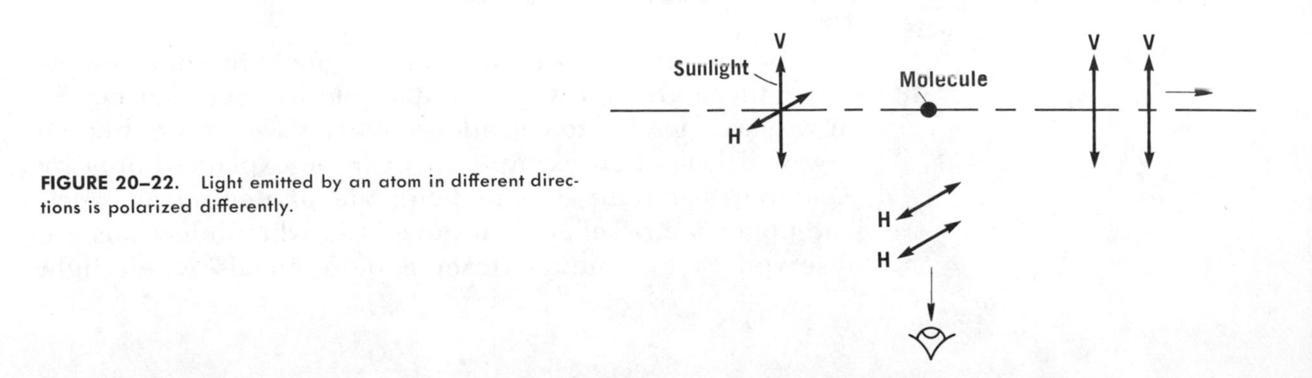

FIGURE 20–22. Light emitted by an atom in different directions is polarized differently.

FIGURE 20–23. *A*, Photo taken without the use of a polarizer in front of the camera. *B*, Note that the polarizer reduces the amount of reflected light so that the boy is more visible.

THE PHOTOELECTRIC EFFECT

On a number of occasions in this book, the question as to whether light is a wave or a particle has been raised. So far we have needed to consider only its wave nature, but we now will look at an example that can be explained only by assuming that light is a particle: the *photoelectric effect.* The photoelectric effect is a process in which electrons are observed to be emitted from certain metals when light shines on them (Fig. 20–24). (A method for observing this effect was discussed in Chapter 13.)

In a metal, some electrons are reasonably free to move around, because they are detached from the atoms of the material, but they cannot get out of the metal. It is as though the metal surface were a fence that the electrons, with their small amount of energy, cannot leap. If, by some means, an electron does get enough energy, it can escape the metal; light shining on the surface is one way of giving the electron this extra energy boost. Initial careful studies of this process revealed some results that could not be explained by the simple wave nature of light.

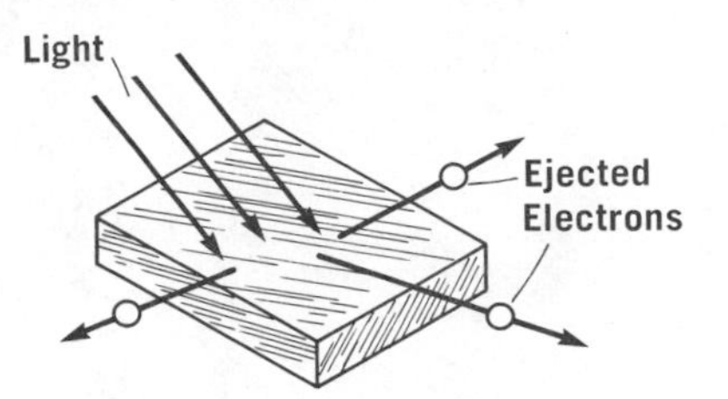

FIGURE 20–24. The photoelectric effect.

A typical set of observations for one particular metal is listed below.

(1) The electrons are bounced out by high-frequency light (such as ultraviolet or violet) but not by low-frequency light (such as red).
(2) Dim violet light causes the effect, but even the brightest red light has no effect.
(3) The electrons begin to leave the surface *as soon as* the light strikes it—there is no time lag at all.
(4) The electrons come out at a higher speed when ultraviolet hits the metal than when violet hits it.

At the time of these observations (late in the nineteenth century), it had become generally accepted that light was a wave. Let's examine the above observations point by point, using the wave theory. Inadequacies arise quickly.

(1) Both high-frequency light and low-frequency light carry energy; therefore it should make no difference what frequency we use—electrons should emerge for both types.
(2) Presumably, dim light would carry less energy than bright light. As a result, bright red should eject electrons much sooner than dim violet. Instead, bright red ejects none, whereas dim violet ejects electrons with ease.
(3) Calculations could be performed to determine the amount of energy carried by the light waves. Also, the amount of energy needed to get electrons out of the metal is known. These two data indicate that the light waves could give enough energy to the electrons to allow them to jump out only after shining on the surface for several minutes. But in actual fact, there is no time lag observed.
(4) Again, it would be expected that the speed of the ejected electron would depend on the amount of energy that the electron received from the wave. The brightness of the light, and not its frequency, should determine the speed.

Explanation by the Photon Theory

The wave theory of light fails at each step. In order to explain the observation, we must turn to our picture of light as being made up of small bundles of energy called photons. Each of the photons emitted by an atom is assumed to behave much like a particle. The amount of energy carried by an individual photon depends directly on its frequency, according to the equation

$$\text{Energy} = \text{Frequency} \times \text{h}$$

We have observed before that the h in the above equation is a number that must be used to convert a frequency value to an energy value. It should be noted that the higher the frequency of the light, the higher is the energy of the photon, and thus, photons of ultraviolet light carry more energy than do photons of red light. When a photon strikes an electron inside the metal, it instantly gives all of its energy in one lump sum to the electron and then disappears. A collision of this kind is similar to one that another particle might have with the electron.

With this picture in mind, let's examine all the experimental observations once again.

(1) Photons of violet light have more energy than photons of red light. Apparently, the red-light photons do not have enough energy to get an electron out of the metal.
(2) The brightness of a beam of light depends on the number of photons that it contains. Bright light has many photons; dim light has fewer photons. Bright red light, then, has many photons, but *none* of them has enough energy to get an electron out. Dim violet light has fewer photons, but each of its photons carries enough energy to free an electron.
(3) Time is not a problem. As soon as a photon carrying enough energy hits an electron – boom! There is no need for a gradual accumulation of energy by an electron. Either a single photon does it or else it isn't going to get done.
(4) Ultraviolet photons have a higher frequency than do violet photons and hence carry more energy. Only a certain amount of this energy is needed to free the electron. Therefore, more energy is left over when the ultraviolet photon has struck than when the violet has struck, and the excess energy makes the electron fly out at a higher speed.

Thus, at the end of the last century, when the wave theory of light seemed to be well established, the photoelectric effect stood as an unexplainable phenomenon. This was one of those instances in science in which an accepted theory was found to be inadequate. The photon theory of light, however, easily explains the photoelectric effect, and so it takes its place in our effort to understand nature. So, what is light? Is it a wave or is it a particle? Interference phenomena can be explained only if we assume a wave nature for light. The photoelectric effect, on the other hand, can be explained only by assuming the existence of particle-like photons. In short, light has a split personality, or dual nature; sometimes it acts like a wave, sometimes like a particle. Luckily for us, it never acts like both in the same experiment.

FLUORESCENCE

The photoelectric effect has indicated to us that light at times acts like a particle. This means that when light strikes an electron in an atom, it should be able to excite the atom just as a particle would. In Figure 20–25, an ultraviolet photon is shown striking an atom. The photon loses all of its energy and disappears. Part of this energy is absorbed by the electron, causing it to move to a higher orbit. Any remaining energy is given to the entire atom, causing it to move faster if it is an atom in a gas, and to vibrate faster if it is an atom in a solid.

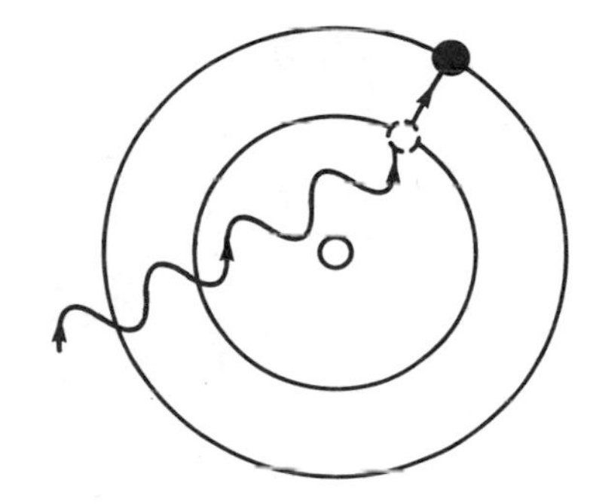

FIGURE 20–25. A photon of ultraviolet light exciting an atom.

When the excited electron returns to the ground state, it emits a photon. This photon has less energy than the original ultraviolet photon which struck the atom (because some of the original energy was used to speed up the entire atom). Because it has less energy, it also has a lower frequency, and as a result may produce light within the visible part of the spectrum. The common fluorescent light makes use of this principle. Figure 20–26 shows the construction of one end of a typical fluorescent bulb. The filament at the end of the tube is heated to a temperature high enough to cause

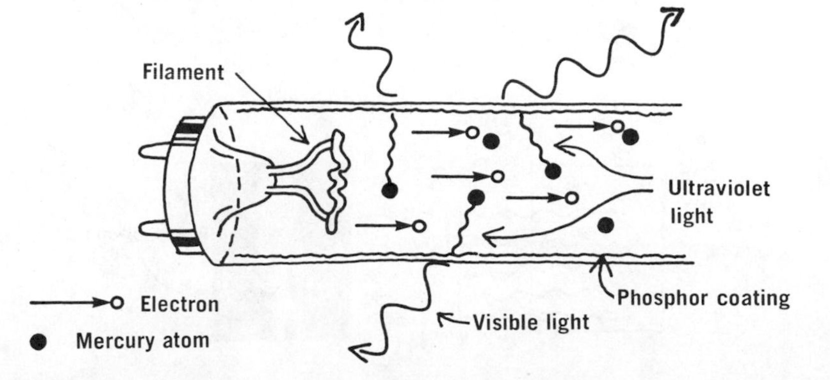

FIGURE 20–26. The process occurring in a fluorescent light.

electrons to be "boiled" out of it. Electrical connections to the tube cause these electrons to sweep down the tube, and while traveling down the tube, they collide with atoms of mercury vapor which are present. Many mercury atoms are raised to excited states in these collisions and emit ultraviolet light as they de-excite. The atoms of the phosphor coating on the inner surface of the tube absorb the ultraviolet photons, become excited, and then de-excite, emitting visible light.

Different types of phosphors on the tube emit light of different colors. There are "cool white" fluorescent lights, which emit nearly all visible colors and are very white in appearance. "Warm white" fluorescent lights have a phosphor that emits more red light and thereby has a "warm" glow to it. Note that the fluorescent tubes above the meat counter in a grocery store are usually "warm white" to make the meat look redder.

The fluoroscope, used in medicine, employs the same principle as the fluorescent light, except the radiation used is in the x-ray region rather than the ultraviolet. (Remember that x-ray photons are of even higher energy than ultraviolet photons.) The x-rays are allowed to penetrate an object such as the hand in Figure 20–27 and then to strike a screen containing a phosphor. At certain points, x-rays will penetrate the hand quite easily and cause the phosphor screen to glow, while at other points, bones may intervene and absorb much of the radiation. No radiation strikes the screen in the areas behind the bones, and as a result, a sharp outline of the bones appears on the screen.

There is a very common fluorescing substance, which you may have in your medicine cabinet, that can produce wild effects: Murine, the liquid used to relieve tired eyes. Put some on your face[1] and stand under a "black light" of the type used to illuminate fluorescent posters. Your face will glow! (The so-called black light is simply a lamp which emits ultraviolet light along with some visible violet-blue light.)

[1]This new wonder-paint will not harm the eyes if "accidentally" applied to the eyelids!

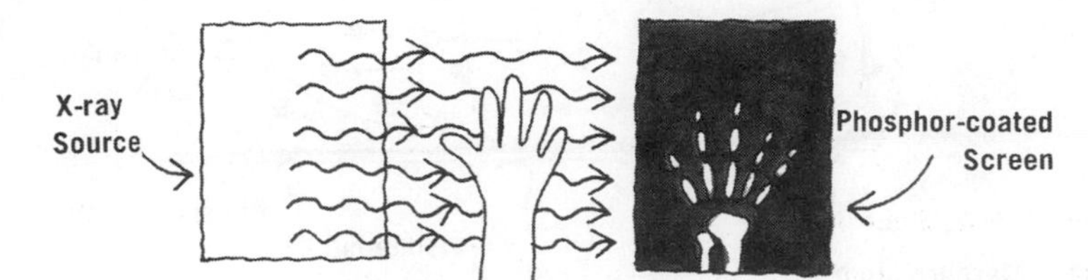

FIGURE 20–27. Fluorescence used in the x-ray fluoroscope.

PHOSPHORESCENCE

When electromagnetic radiation strikes a material and causes it to glow, we call this phenomenon "fluorescence" if the material glows only when the radiation is falling on it. However, some materials continue to glow long after all illumination has been removed. Materials of this nature are said to be *phosphorescent.*

The first steps in the phosphorescing process are the same as those in fluorescence. A high-energy photon, for example ultraviolet light from the sun, strikes an atom of the material and raises an electron to an excited state. But in this case the electron does not immediately return to the ground state; it is effectively "hung." Eventually it will de-excite, but the jump occurs at a later time, from a few seconds to several hours later. When it does drop back to the ground state, a photon is emitted in the visible part of the spectrum. This slow return of electrons to an unexcited state causes a phosphorescent material to emit light long after being placed in the dark. Paints made from these substances are often used to make the hands of clocks visible during the night or to outline doors and stairways in large buildings for use in case of power failures.

CONCLUSION

You should by now see that it would be a waste of time to try to draw a single simple conclusion about the nature and properties of light. Light is fascinating. Many of its phenomena are explained fairly easily: in many ways light is analogous to sound. But it has strange ways, too. It appears to travel in straight lines and to cast shadows. But when we look at light after it has passed through a narrow opening, we find that it bends around corners. We find that it has difficulty in passing through a "picket fence" of molecules, as occurs in polarization. But most incredible of all is its dual nature: sometimes it acts like a wave and sometimes like a particle.

OBJECTIVES

Chapter 20 should enable you to:

1. Perform a demonstration which shows the bending of light around corners.
2. Explain why two separate lamps cannot give an interference pattern, and describe how a two-source interference pattern can be obtained for light.
3. Use drawings of waves emerging from two slits to explain the two-slit interference pattern.

4. Describe the diffraction grating and explain how it produces an interference pattern.
5. Explain why a diffraction grating breaks white light into colors.
6. Explain why colors appear in an oil-spill on wet pavement, and describe a way in which this phenomenon can be used to examine the surfaces of lenses and to reduce stray reflections in high-quality lenses.
7. Explain why regular light as emitted by a lamp is not polarized, and use an analogy to explain how polarization occurs.
8. Perform a demonstration with polarizing sunglasses to show how the amount of light passing through two polarizers can be controlled and to observe polarization by reflection.
9. List four observations concerning the photoelectric effect, and show why each demands a photon model of light for its explanation.
10. Explain fluorescence, making reference to the principle of the conservation of energy, and give two examples of everyday observations of fluorescence.
11. Define and explain phosphorescence.

QUESTIONS—CHAPTER 20

1. Can a sound wave be polarized?
2. Interference patterns can be produced by the two light waves shown in Figure 20–28. Explain.
3. A soap film is suspended vertically in a ring as shown in Figure 20–29. As the soap slowly settles downward, the colors of the film change. Why?
4. To reduce the glare from light reflected off water, should polarizing glasses cut off vertically or horizontally polarized light? Explain.
5. Explain why visible light will not eject electrons from some metals but ultraviolet light will.
6. Summarize carefully the reasons why the wave theory of light will not explain the photoelectric effect.
7. Explain why the phosphorescent hands of a clock will cease to glow if never exposed to light.
8. Explain why "black-light" posters glow under ultraviolet light.

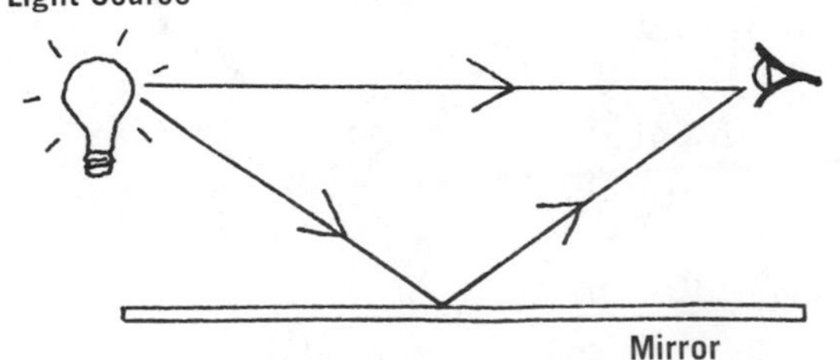

FIGURE 20–28.

9. In order for you to observe diffraction from a record, we specified use of a long-playing record. Why?

10. Show, by a sketch, that the waves emitted by a radio antenna are polarized.

11. Can fluorescence be caused by a photon of infrared light striking an atom?

12. What effect, if any, would you expect the temperature of a material to have on the ease with which electrons can be ejected from it in the photoelectric effect?

13. Why does the interference through the curtain in Figure 20–2 show two sets of streaks at 90° from each other?

14. Is light a wave or a particle? Discuss.

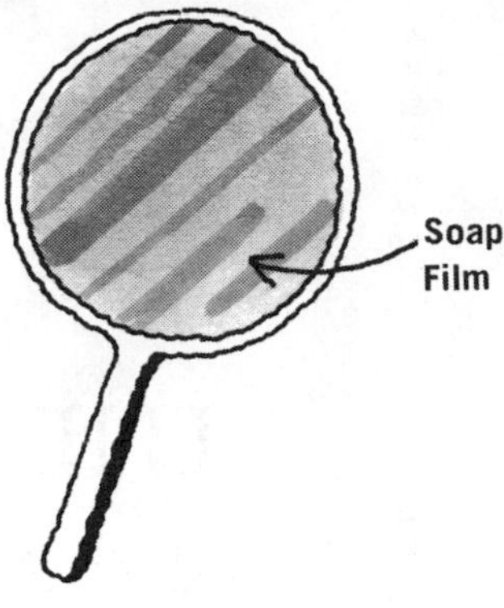

FIGURE 20–29.

21

THE FANTASTIC LASER

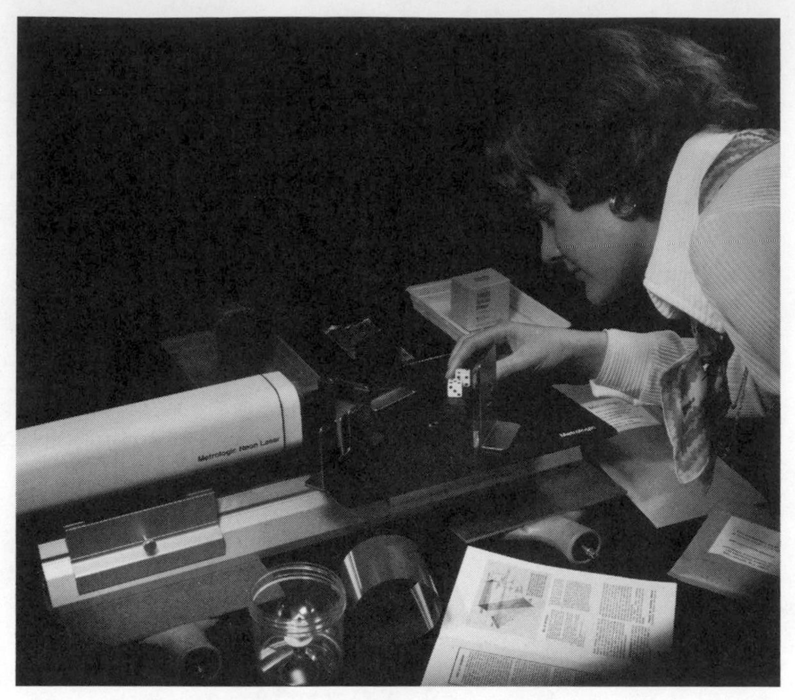

By permission of Metrologic Instruments, Inc.

If a visitor from another planet were to try to judge western civilization's ultimate desires based upon our movies, television shows, and commercials, he might come to some interesting conclusions. Our advertisements might make him think that our major problems are dirty clothes, body odor, and hair control, while the goal of humanity in many of our movies seems to be development of the ultimate weapon of destruction. Science fiction has, in the past, been amazingly accurate in its forecasts of the future, and in science fiction stories this ultimate weapon is often a death ray which dissolves the bad guys in a cloud of smoke. Fortunately, such instruments are not with us yet. The closest thing we have now is the laser, an interesting device whose use thus far has been mostly peaceful. (But don't give up hope!)

DEVELOPMENT OF THE LASER

The theory behind the laser was developed in 1958 by Arthur L. Schalow and Charles H. Townes,[1] and the first

[1]Townes was awarded the Nobel Prize in physics in 1964 for his work leading up to the development of the laser.

working laser was constructed by T. H. Maiman in 1960. It burst upon the scene like a fire storm. The promises it held for new developments were trumpeted to the world by the news media. Many of these promises have not been fulfilled—no death ray yet—but there are hundreds of thousands of lasers doing practical work for man right now, and the end of new developments in laser use is not in sight.

The word "laser" is an acronym for Light Amplification by Stimulated Emission of Radiation. We'll start at the end—"stimulated emission of radiation"—and work backward.

Stimulated Emission of Radiation

In Chapter 17 we examined the means by which an atom emits light. First, some source of energy raises an electron from its lowest orbit to a higher orbit (an excited state) as indicated by Figure 21–1. Then the electron quickly jumps back down to the lowest orbit again and in doing so gives off a burst of light energy called a photon. The time spent by the atom in the excited state is normally extremely short (10^{-8} second), but in some cases an electron may encounter difficulties in returning to the lowest orbit. Certain excited states of atoms act "sticky"; electrons get "stuck" in them for as long as 10^{-3} or 10^{-2} second. (One hundredth of a second may not seem long to us, but it's a million times as long as an electron usually stays in the excited state.)

There are a number of ways by which a stuck electron can free itself, but the way that is important to the laser is a process known as stimulated emission. Consider the atom in Figure 21–2*A*. The electron in an excited state is experiencing some difficulty in returning to a lower orbit. If it could jump back down, it would produce a photon of red light in doing so. In Figure 21–2*B* a very special photon happens upon the scene: its wavelength is exactly the same as that which the stuck electron would emit if it could return to the lower orbit. This incoming photon acts like a good Samaritan. It jars (or stimulates) the stuck electron and frees

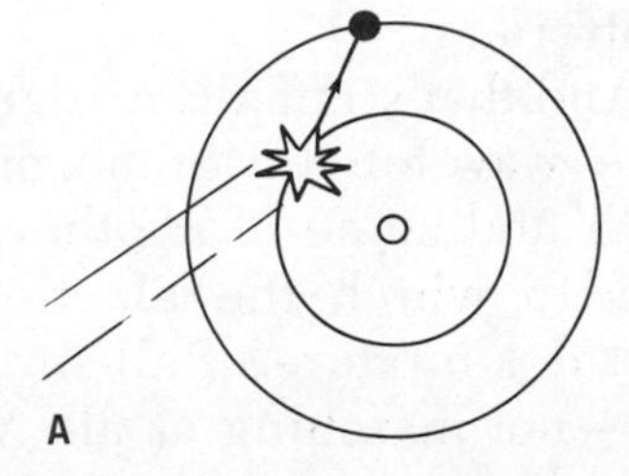

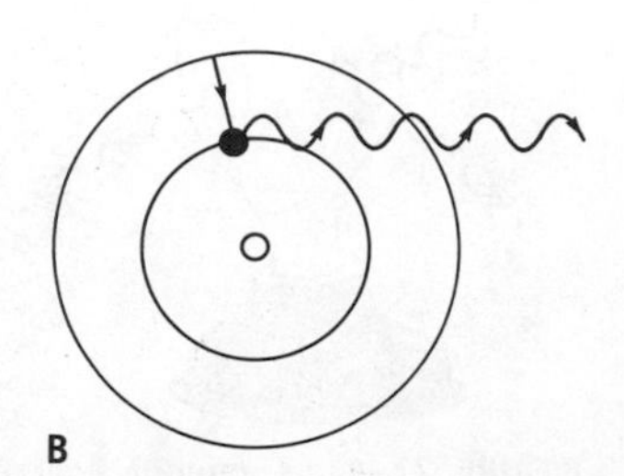

FIGURE 21–1. *A*, A source of energy increases the atom's energy, causing the atom to give off a photon of light (*B*).

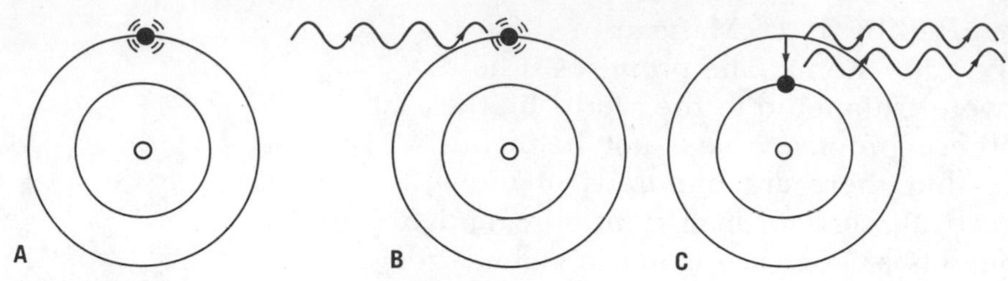

FIGURE 21–2. A passing photon stimulates the excited atom to emit its energy.

it, allowing the freed electron to jump down and emit its photon of red light. The two photons now go marching off together (Fig. 21–2*C*). The term "marching" is not chosen casually. The two photons go off together such that their crests and troughs are exactly in step (Fig. 21–3). Thus the stimulated emission causes two important results: (1) the two photons move in exactly the same direction, and (2) they are in phase – in step.

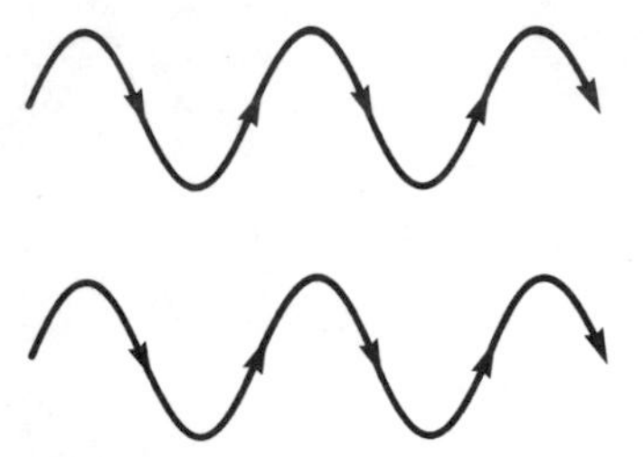

FIGURE 21–3. Two photons in phase.

If these photons encounter another excited electron waiting for help, they stimulate it to emit a photon, and off the three go – in one direction and in step. If a number of such stuck electrons are around, they too are stimulated. An avalanche effect is produced that is at the heart of the operation of the laser.

Regular Light – Out of Step

To better appreciate the difference between the laser's stimulated emission of radiation (light) and regular light, let's back up and consider how regular light is emitted. When we turn on a light, we cause an electric current to excite atoms in the bulb's filament or in a fluorescent tube's gas. This excitation occurs randomly: first one atom is excited, then another, then another, in no particular order or sequence. And this is exactly the way that the electrons return to the lower orbit and emit energy as light photons. We can imagine the result to resemble Figure 21–4, with the photons coming out not only in all directions but also with no particular relationship to one another. They are not in step. It's as if you dribbled sand into a bucket of water – little waves are created everywhere, with no particular order or pattern.

FIGURE 21–4. A regular light emits its photons randomly.

And that's not all. A regular light source does not emit a single wavelength. Some of the photons are of one wavelength and some of another (because of the many different orbits to which the electrons were excited.) The overall result is a mixture of all sorts of photons going in all directions – not marching at all. What a mess! But that's life, or rather, light.

FIGURE 21–5. In a fountain, little waves are created everywhere with no particular order or pattern. In this respect, these water waves are similar to light waves emitted from an ordinary light bulb. (Photo by Donald F. Wallbaum.)

THE RUBY LASER

The first laser to be developed used a material which you may normally think of as belonging in a jewelry box: ruby. The ruby used for the laser is basically the same as the red material used in jewelry, but for the laser the ruby is cut into the shape of a rod (perhaps a few inches in length and about 1/4 inch in diameter). The two ends are polished and given a silvery mirror coating. One end is fully silvered in order to make it reflect as much light as possible, and the other is only partially silvered (with a very thin coating) so that some light will pass through and some will be reflected.

A spiral flash lamp is placed around the ruby rod (Fig. 21–6). When this lamp is turned on it emits a quick intense burst of light, similar to the light emitted by the flash bulb of a camera. Light from this lamp passes through the ruby, and some of it is absorbed by atoms of the ruby, causing electrons to go to an excited state, where they get stuck. Figure 21–7 indicates a number of atoms in the excited state.

So there the atoms expectantly sit, until about 0.01 second later, when some of them fall to their lowest orbit and

Flash Lamp
Ruby

FIGURE 21–6. The heart of the ruby laser.

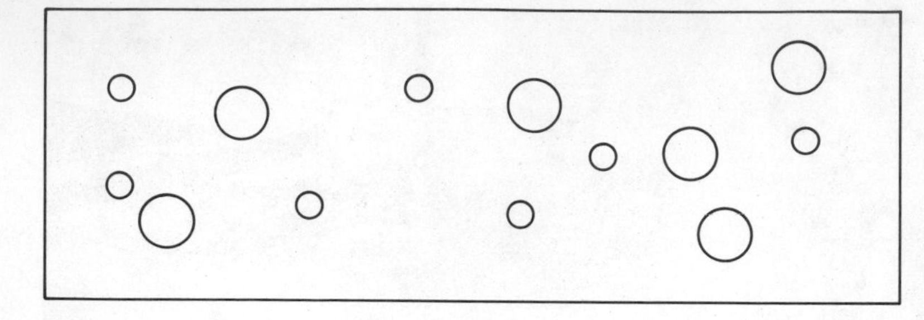

FIGURE 21–7. The large circles represent excited atoms in the ruby; the small ones represent unexcited atoms.

emit photons of light, which for a ruby laser are photons of red light. If these photons go out the side of the ruby rod, nothing of any significance happens. But finally a photon will be given off that travels directly down the length of the rod, as the photon from atom A in Figure 21–8. When this occurs, the action starts—by stimulated emission. The photon soon encounters another excited atom and stimulates it to emit a photon. This photon goes marching off with the first one as if they were two soldiers looking for a parade. More atoms are encountered, and they too are stimulated to emit more photons. The amount of light—that is, the number of photons—increases as shown in Figure 21–9. The *L*ight is being strengthened, or *A*mplified.

The soldier photons finally strike the left end of the ruby rod (in our drawing), where they are reflected back. As they move back toward the right, they stimulate more photon emission. The strength of the beam is growing rapidly as more photons are added. (It should be noted that without those "sticky" excited states of the atoms, all the excited atoms would have emitted their light by this time and would not be sitting there waiting for stimulation.)

The initial soldier photon became a pair of soldier photons; then a squad was formed; then a company; then a brigade. Finally more brigades are marching together than have ever been assembled on earth since time began. All in

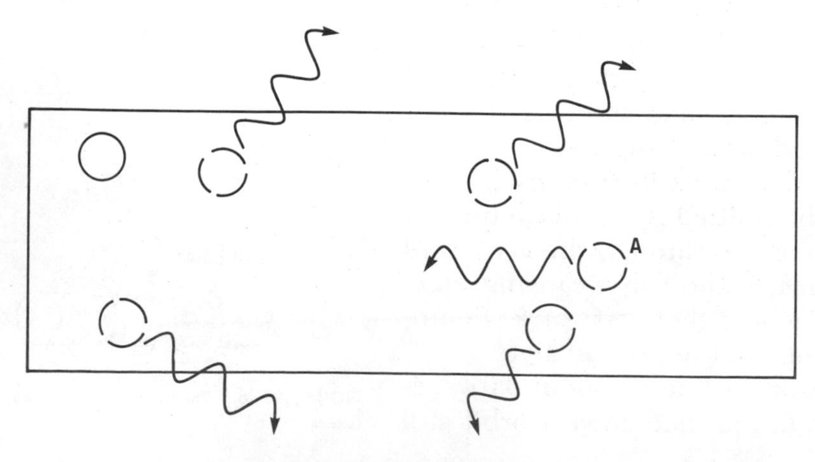

FIGURE 21–8. The large broken circles represent atoms which have emitted photons. Note that atom A's photon is going right down the tube.

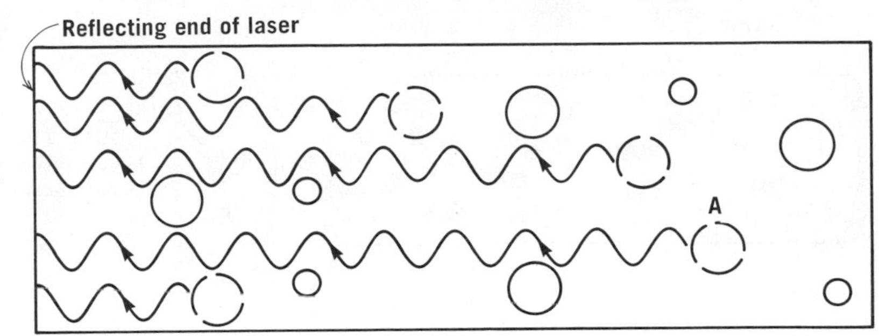

FIGURE 21–9. More and more atoms are stimulated to emit their photons in phase.

the same direction, all in step—enough to make the bravest general shed a tear.

When our army of photons hits the right end of the ruby, some of them get through and the rest are reflected (because of the partial silvering on this end). Those that are reflected continue their back-and-forth march through the ruby, stimulating as they go. Each time they get to the partially silvered end, some photons escape (Fig. 21–10). These make up the laser beam.

Clearly, this process must eventually stop. After a very short period of time all the excited atoms will have gone back to their lowest orbits and the beam will have died out. The action can be started again, however, with another flash of light from the flash tube. The ruby laser, therefore, does not give a continuous beam of light; its light comes in pulses, one for each flash of the flash tube.

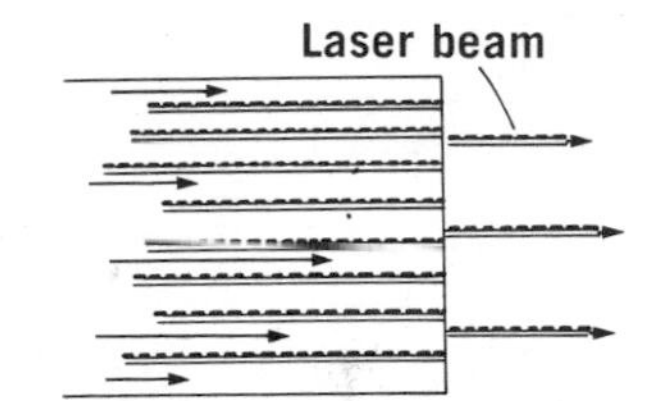

FIGURE 21–10. The partially silvered end of the ruby emits a small fraction of the light striking it.

We know that the amount of energy produced by the laser cannot exceed the amount supplied. An intense beam of light is emitted, but energy had to be added to the atoms to excite them in the first place. The principle of conservation of energy must apply in this case (as in all cases), and much of the light energy from the flash lamp escapes and does not excite atoms. In addition, many of the photons of red light from the excited atoms are lost through the sides of the tube and do not take part in the lasing action. So although you get energy out of the laser in the form of light, you receive much less than you supply. Such a process would not be worthwhile were it not for the unusual properties of laser light: its "single-direction-ness" and its "in-step-ness." Before looking at applications which take advantage of these properties, let's examine another type of laser.

THE GAS LASER

The ruby laser was the first laser developed. The most common lasers in use today, however, are gas lasers, and the most common type of these uses a mixture of helium and neon gases to produce its laser action. The principal aspects of its operation are quite similar to the ruby laser, however.

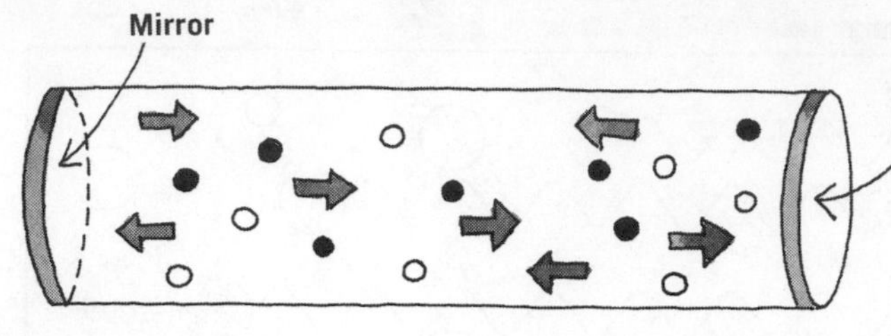

FIGURE 21-11. The helium-neon laser. Black circles are helium atoms, white circles are neon atoms, and arrows represent electrons.

The mixture of helium and neon is confined to a glass tube about one foot in length. It is sealed at the ends by two mirrors, one completely silvered and one partially silvered—so far, the same as before (Fig. 21-11). Not shown in the diagram is a piece of electrical equipment that causes electrons to sweep back and forth inside the tube. You may recall (from Chapter 20) that electrons sweeping through a fluorescent tube collide with the atoms of gas in the tube and excite them. The same reaction takes place in the laser; the helium and neon atoms are energized to excited states. There is a difference here, though: one of the excited states of helium is a "super-sticky" state. Electrons seldom jump from this state to a lower state and emit photons. Instead, the most likely way for the excited helium atom to lose its excess energy in the present circumstances is in a collision with a neon atom. Such a collision results in the neon atom's being put into an excited state. Figure 21-12 illustrates these steps.

So we have neon atoms excited in two different ways: by collisions with surging electrons and by collisions with excited helium atoms. The result is a great many excited neon atoms in the tube, which is just what is needed for laser action. Now, just as with the ruby laser, if an emitted photon happens to start down the length of the tube, it will encounter one of the many excited atoms and initiate the stimulated emission. The mirrors at the end reflect the photons back and forth, stimulation continues, and out of the partially silvered end comes the laser beam.

An important feature of the helium-neon laser is that it will work continuously. The electrons can be made to continue sweeping up and down the tube, which means that neon atoms will continually be raised to an excited state and

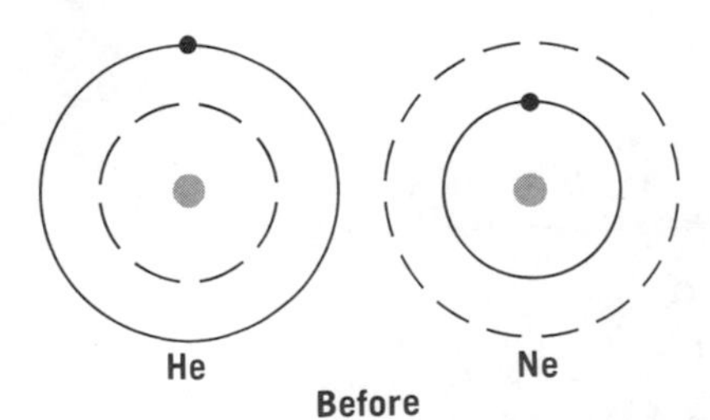

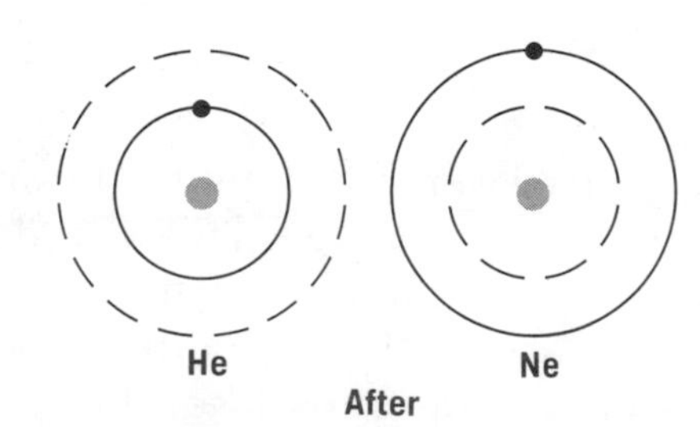

FIGURE 21-12. Before and after a collision between an excited helium atom and unexcited neon atom.

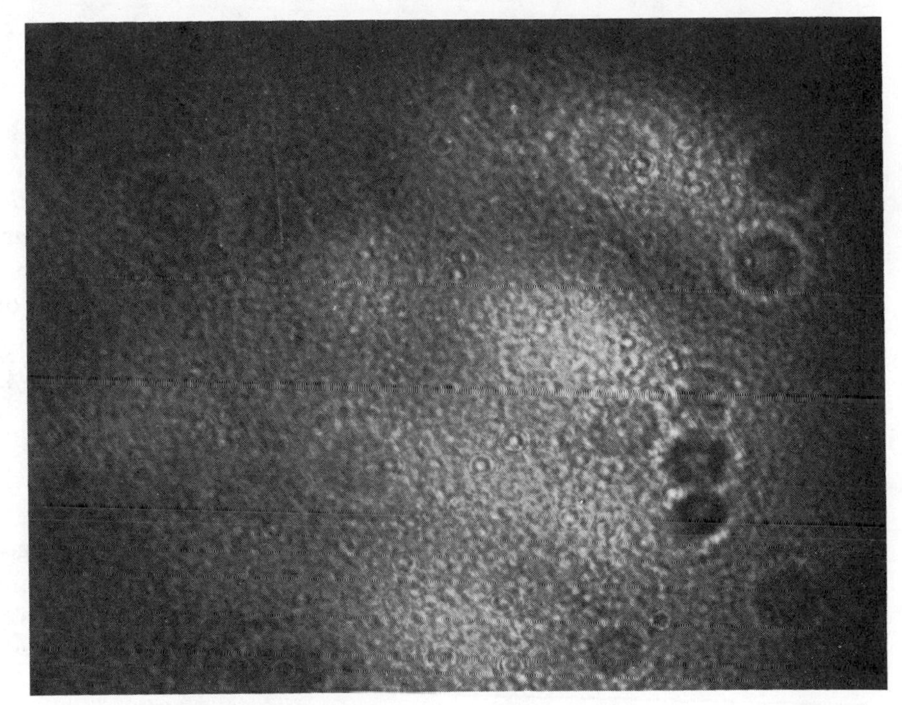

FIGURE 21–13. A hologram. (Photo by Jim Shepherd.)

will be ready and waiting for stimulated emission to occur. Thus a helium-neon laser emits a continuous beam rather than the pulsed beam of the ruby laser. Now that we have a laser that can produce a steady beam, we can examine some of its uses.

THE HOLOGRAM

Figure 21–13 is a photograph of a hologram, which is a picture made in a special way with laser light. Exciting, isn't it? Not really—in fact, it is just a bunch of swirling lines that look like fingerprints made by an FBI rookie. But if you allow laser light to pass through it and look back along the direction from which the beam comes, an amazing thing happens. Hanging suspended in midair will be a three-dimensional image.

FIGURE 21–14. Hold your fingers up like this.

take-along-do-it

Hold two fingers in the air in front of your face with one about six inches behind the other (Fig. 21-14). As you move your head from side to side the finger in back will disappear behind the other and then reappear. Or holding the two fingers still, close one eye and then the other. You will see that you get a different view of your fingers from each eye. This illustrates what is meant by a three-dimensional view.

An ordinary photograph of your two fingers would always show them in the same position with respect to each other. If one is hidden behind the other in the photograph it will always be hidden regardless of how you move your head in viewing the photo. But if you photograph your fingers by holography, the entire three-dimensional scene is recorded in the hologram. This means that if you look into the laser light passing through the hologram, you see the two fingers. And if you move your head from side to side, you observe the appearance and disappearance of one finger behind the other. If you look with one eye and then with the other, you see two different views, just as in real life. Figure 21–15 shows a number of regular photos taken through the same hologram but from different angles.

The applications of holography promise to be many and varied. For example, it is hoped that someday your television set will be replaced by one using the hologram photographic process. Instead of being flat, the picture you view will be in three dimensions. Normally when watching television, we distinguish between the foreground and the

FIGURE 21–15. These photographs taken through a hologram at different angles clearly show the three-dimensional aspects of the image. (By permission of Metrologic Instruments, Inc.)

background primarily by comparing sizes. But with 3DTV, the background will actually appear farther away. Any bump on a person's body, for example the nose, will actually appear to protrude on the television screen.

How a Hologram is Made

The hologram is merely an interference pattern formed by two overlapping beams of light. Figure 21–16 shows an arrangement that can be used to produce the pattern. Light coming from the laser is first split into two parts by a half-silvered mirror, B. One beam of light goes through the half-silvered mirror and strikes lens L_1, which causes the narrow beam of the laser to spread out and strike the subject. The half-silvered mirror bounces the other half of the laser beam toward lens L_2, which spreads it out. This beam is reflected by the two mirrors, M_1 and M_2, and finally strikes a sheet of photographic film placed at F. Also striking the film is the light reflected from the subject. The two beams overlap to form an extremely complicated interference pattern on the film. The pattern is so intricate, in fact, that no vibration of the parts can be allowed while the hologram is being made. A movement or vibration as small as one ten-millionth of an inch can ruin the interference pattern and hence the photograph, which when developed, is the hologram.

In order to obtain the interference pattern, the phase of the waves in the two light beams must be related in some way and must maintain this relationship throughout the exposure of the film. This is possible only if the light is from a laser—in step, in phase. Ordinary light waves from a source such as the sun or a light bulb are undergoing random

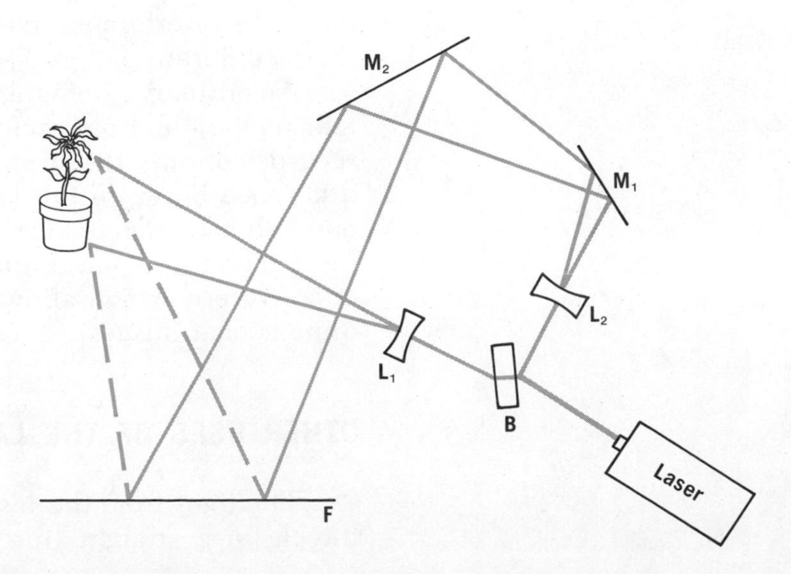

FIGURE 21–16. To make a hologram, light from the laser is split into two beams by the half-silvered mirror B. Lenses L_1 and L_2 spread the beams, which eventually combine at the film F.

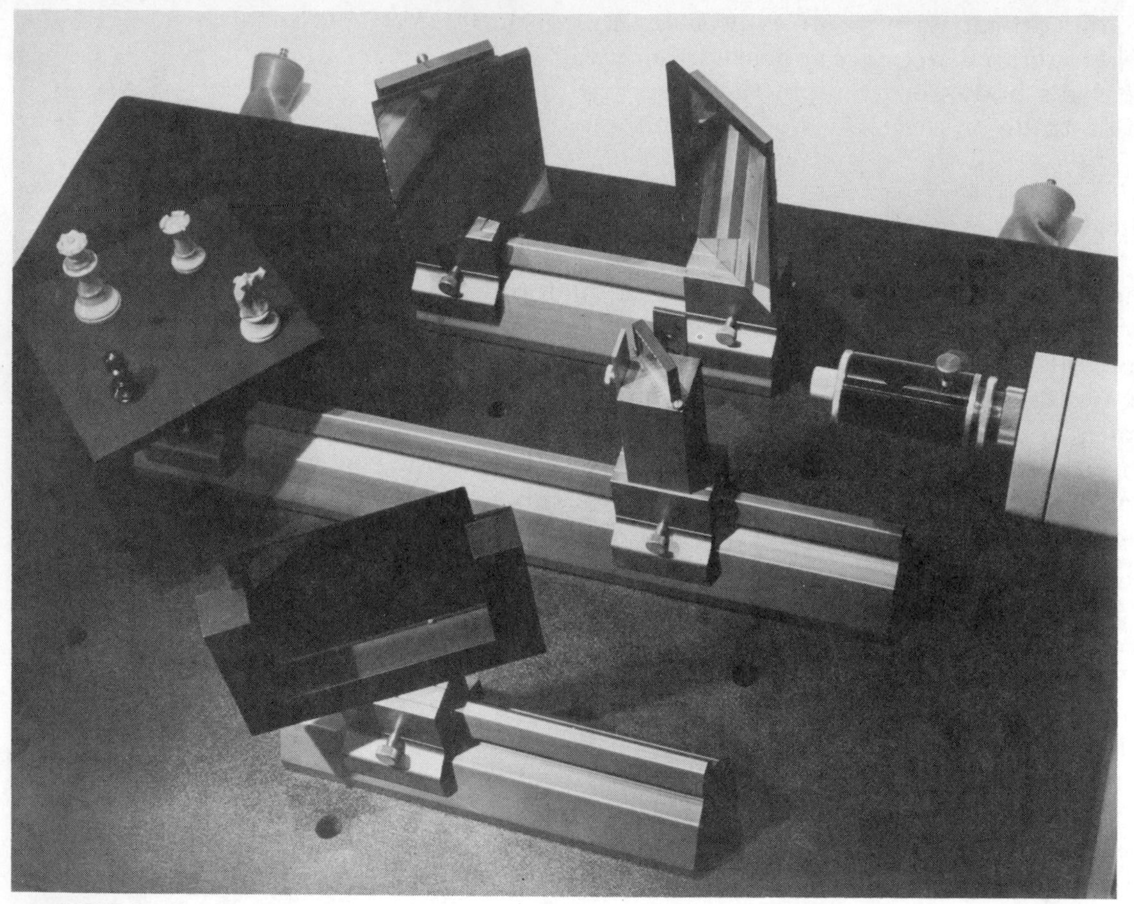

FIGURE 21–17. A set-up to make a hologram of some chess pieces. (By permission of Metrologic Instruments, Inc.)

changes with respect to one another which would wash out any interference pattern that might result on the film. In addition, ordinary light consists of a great number of different wavelengths of light. (Even light which has passed through a colored filter has many different wavelengths in it.) Laser light, however, is but a single wavelength. Thus there is only one interference pattern produced on the film rather than a different pattern for each wavelength.

An ordinary photograph records the amount of light that falls on the film from each point on the object. The hologram records not only the strength of the light but also the phase difference between the two halves of the laser beam after one half has reflected from the object. It is this last piece of information, the phase relationship, that causes the interference pattern which allows the hologram to produce three-dimensional images.

OTHER USES OF THE LASER

The light from the laser leaves the end of the tube and travels in a straight line with very little spreading. This

Reprinted with the permission of American Scientist/Sidney Harris.

property of the laser beam brought about the laser's first important application: use as an alignment tool. If one wants to align equipment very accurately, one need only shine the beam along the desired direction and place the equipment along the beam. This ability to mark out a very straight line is useful in many applications, from aligning airplane wings in construction to the procedure shown in Figure 21–18.

FIGURE 21–18. A mini-revolution sired by the laser.

The laser can also be used to measure distance. Astronauts, during each of the moon landings, placed reflectors on the surface of the moon. Light from an earthbound laser was reflected from them, and the earth-moon distance was determined by the same method used in radar determination of distance—pulses were sent and their travel time was measured. In this way we can determine the distance from earth to the moon at any time to an accuracy of a few feet.

The power of the laser is another of its most outstanding characteristics.[2] Figure 21–19 shows a laser cutting a piece of metal. The laser performs such feats by actually melting the metal. Lasers are now made that are powerful enough to melt holes through diamonds. Why would anyone want to have a hole in a diamond? Very fine wires are made by pulling thick wires through a very small hole. If the hole is drilled in relatively soft material, eventually the hole, or rather the material, wears out. Diamond, of course, is extremely hard, and with the laser to drill holes in it, "very hard holes" become possible.

[2]We will see in a later chapter that research is now underway in which the power of lasers is used to produce the tremendous heat necessary to start a thermonuclear reaction. Such controlled reactions would solve man's energy problems forever.

FIGURE 21–19. A laser cutting titanium. (By permission of Coherent Radiation.)

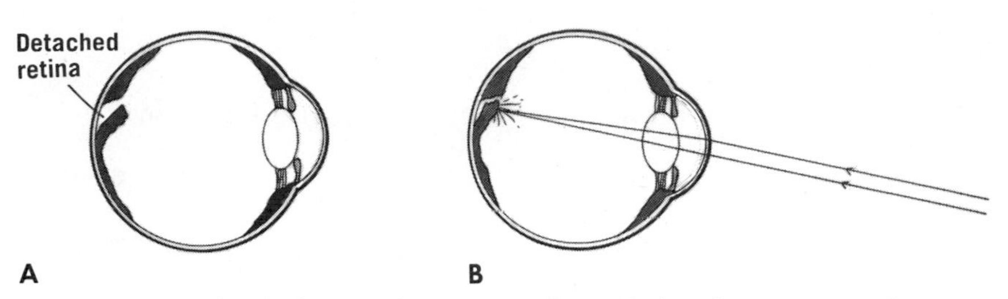

FIGURE 21–20. The use of the laser to correct a detached retina (*A*) is common today. *B*, The laser beam is scarring the retina, which will detach no further after it has healed.

Application of the laser in medicine has shown much progress. Probably the most prevalent use is in curing a common disorder of the eye in which part of the retina becomes detached (Fig. 21–20). The laser beam (from an argon gas laser, normally) is shined into the eye and is focused by the eye lens onto the retina. The intense beam burns a very small spot on the retina. As the burn heals, it bonds the retina to the back of the eye, thus preventing further detachment of the retina.

The laser replaces the surgical knife in correcting defects of the eye; the analogy with a knife does not stop there. Just as a surgeon's knife in the hands of an amateur is a dangerous weapon, so must care be taken with the use of lasers. Although the common classroom helium-neon laser is not powerful enough to burn your skin, if it is shined into your eye the lens of your eye will focus it onto the retina and damage will result. Anyone using a laser must be careful of eye damage—the beam should never be allowed to shine directly into the eye.

Perhaps of all avenues of exploitation of the laser none has received as much attention as its potential in communication. Light is an electromagnetic wave just as is a radio wave and a microwave. Both of the latter are used for communication, so why not the laser? With increasing demands being made on our present communications systems, it is no wonder that other methods are being investigated. There would be problems associated with constructing a laser communications system, however, because laser light is blocked by clouds, rain, and mountains, just as is regular light. In order to avoid these obstacles, laser beams would have to be sent from place to place through tubes. When a change in direction is required, suitably placed mirrors could do the trick. Another possibility is the use of light pipes, which could carry laser beams in a manner analogous to wires carrying electrical signals. Both of these alternatives require a more expensive system than the regular telephone line or microwave system. However, the great number of simultaneous messages which the laser is capable of carrying make it an attractive alternative to consider.

OBJECTIVES

After studying Chapter 21 you should be able to:

1. Explain the origin of the word "laser."
2. Explain the function of "sticky" excited states of electrons in the operation of the laser, and define stimulated emission.
3. Explain the difference between regular light and laser light in terms of emission from the atom and as regards the phase relations of photons.
4. Describe the operating principles of the ruby laser.
5. Relate energy conservation to the case of the laser.
6. Explain the function of helium in a helium-neon laser.
7. Describe how a hologram is made, how it is viewed, and the difference between the image seen in a hologram and in a regular photo.
8. List and describe at least four applications of the laser.

QUESTIONS—CHAPTER 21

1. If the spiral flash tube surrounding a ruby laser gave off only red light when it was discharged, could the beam from the laser be green? Explain. If the flash is green could the laser light be red? Explain. (Hint: consider photon energies.)
2. Would a laser operate if one end were made of a transparent material that reflected no light? Why or why not?
3. Why does the ruby laser emit only one visible color?
4. What would be the advantages and disadvantages of three-dimensional television?
5. Why can even a small vibration of the hologram set-up ruin the image?
6. Why can a hologram not be made with regular light?
7. There have been moves in some state legislatures to ban the use of lasers in the classroom. Discuss some reasons for such "laserphobia."
8. The forerunner of the laser involved microwave radiation. What does "maser" mean?
9. What is the difference between regular light and laser light as to their production? As to the wavelength(s) of the light?
10. Use an encyclopedia (or other textbook) to find a laser application not mentioned in this text.
11. Figures 18–8*A* and *B* were made by blowing smoke into the air in front of the laser and using a fairly long time exposure. Why was the smoke necessary?
12. If "sticky" states for electrons did not exist, could lasers be made? Explain.

Photo courtesy of National Broadcasting Company, Inc.

22 COLOR

Color is something that most of us take for granted because it seems such a natural everyday phenomenon. It is one of the very first things we mention when we describe an unfamiliar object. Yet to most creatures on earth, it is not familiar at all—most animals are completely colorblind. (Imagine seeing the world as it appears on black-and-white television.)

In this chapter we will see why our colorful world looks as it does. We will go from the world about us, where we will study the way in which colors are produced, to the interior of the eye, where we will learn how it responds to these colors. And all of this will be brought to you in living color!

THE COLOR SPECTRUM

When a beam of white light passes through a prism, the light spreads out into the spectrum of colors shown in Figure 22–1. In this continuous spectrum the colors blend into one another as you look from red to violet. The point at which red becomes orange is somewhat arbitrary—one person would locate it at one point, someone else at another.

FIGURE 22–1. Producing the spectrum by shining white light through a prism.

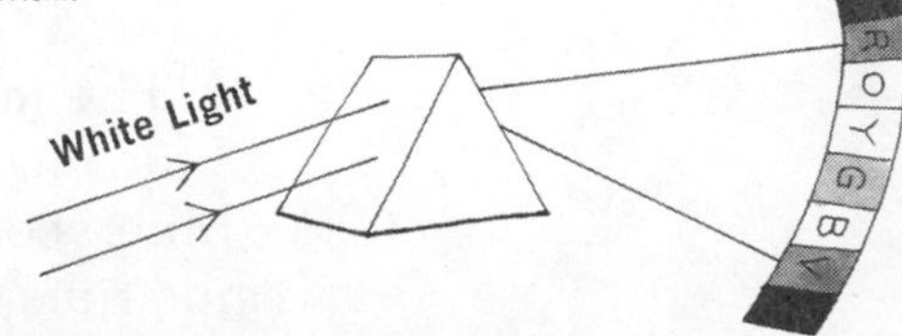

The colors of the spectrum are normally listed as red, orange, yellow, green, blue, and violet, but such a selection is subjective and not a property of the light itself. Some people may not see the orange as a separate color worthy of mention, while others may list a separate blue-green color, and still others may see the color indigo between blue and violet. In this text we will speak of the spectrum as consisting of red, orange, yellow, green, blue, and violet, but keep in mind that these are not separate well-defined types of light.

take-along-do-it

Closely examine the screen of a color television tube. To do this you have to get up close and look at the screen itself rather than at the picture. A magnifying lens will help. You will see that the screen is actually made up of a repeating pattern of three colored dots (or three colored streaks in some newer sets). The colors are red, green, and blue. Note that when you look at the part of the screen which appears white from a distance, all three dots are glowing. From the distance you see only the total effect of the three colors combined. The effect—white.

The TADI illustrates an interesting fact about the way we perceive color: when a number of different colors strike our retina at (about) the same spot, we do not perceive the separate colors but instead see a combination of the colors, which may be something entirely different. In the TADI, the combination of three colors looks white to the person watching television.

Adding Colors

Now let us consider how colors other than red, green, and blue are obtained on a color television set. If we take the light from the red dot and send it through a prism, we obtain the spectrum shown in Figure 22–2A (Plate 2). Note that the spectrum contains not only red but also orange and some yellow. Here again we experience the odd fact that the combination of these colors appears red to us; we cannot perceive the presence of the other colors at all. In the same manner, Figure 22–2*B* is the spectrum of the green dot, and Figure 22–2*C* is the spectrum of the blue dot. If all three dots are glowing and we are too far away from the screen to distinguish the individual dots, the light entering our eye contains all the colors of the spectrum, as in Figure 22–2*D*, which was obtained from white light passing through a prism.

The method used in color television is called *color addition* and is seen in cases other than television. If we shine a blue spotlight on a screen in a darkened room and at the same time shine a red and a green one on the same spot, the spot will appear white. Each of the three lights is *adding* its colors to the others, and we see the *total* of the three.

How, then, is yellow obtained on television (or with spotlights)? Suppose only the green and the red dots are glowing. The light entering our eyes will be composed of all the colors contained in the red spot and in the green spot, which includes red, orange, yellow, and green. These colors are shown in Figure 22–3*A* (Plate 2,) and their combination is seen by us as yellow.

Violet is an even more interesting case. To obtain violet, the blue and red dots are lit up. The combination contains the colors from each end of the spectrum, and omits the center colors (Fig. 22–3*B*). Such a combination is called "magenta," and although it is not really the same as violet, it is close enough to be an effective substitute on your television set. Because of their ability to reproduce the sensation of any color when added in varying amounts, red, green, and blue are called the *additive primaries*.

take-along-do-it

A color twirler can be used to demonstrate the addition of colors to obtain white light. It is made by coloring segments of a disc as shown in Figure 22-4A. The disc is twirled by strings (Fig. 22-4B). When rotating, an individual color is focused at a certain position on the retina and then is quickly replaced by another, and another. An image or color formed on the retina persists for a short time, and this persistence of vision allows all the colors to be brought together on the retina to produce white light. This causes the disc to appear white when twirled.

Make another disc, containing only red, orange, blue, and violet. Twirl it, and the results will be the same as when the primary colors red and blue are combined: the disc will appear violet.

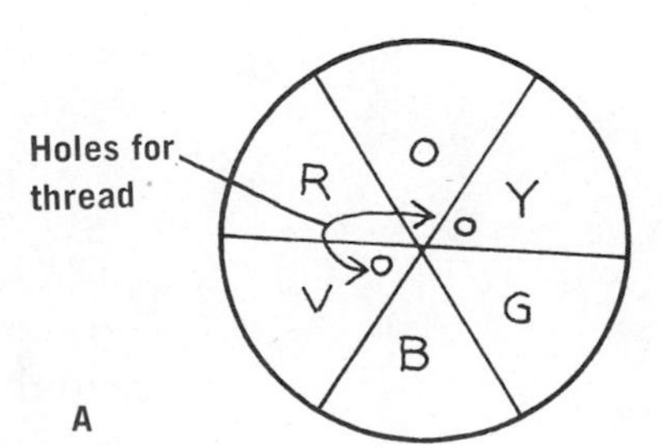

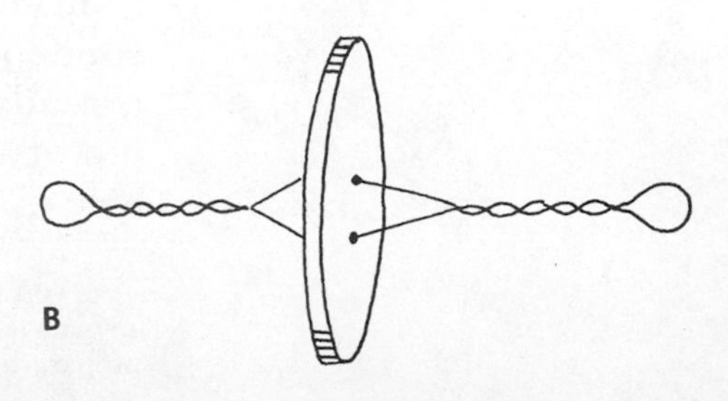

FIGURE 22–4. *A*, How to color your twirler; *B*, How to twirl it by alternately pulling and releasing the string.

Subtracting Colors

If you know an artist, ask him to name the primary colors. He is not likely to list blue, green, and red. The problem is not that he doesn't know what he is talking about but that he is not referring to the additive process. To illustrate:

take-along-do-it

Put a few drops each of red, green, and blue food coloring (or paint or ink) into a glass and mix them. White, right? Wrong—you get a muddy color!

This TADI did not produce a white mixture because the mixing of pigments is a *subtractive* rather than an additive process. To explain this distinction, we need to consider why a yellow-bellied sapsucker's yellow belly looks yellow. When white light is shined onto a piece of white paper, all colors are reflected back from the paper in about the same proportion. But if white light falls on a yellow surface, the light reflected yields the spectrum shown in Figure 22–3A. Recall from the last section that this combination of green, yellow, orange, and red appears yellow to the eye. White light hits the yellow surface, but the pigment in the surface *subtracts* out the blue and violet while reflecting the others, which the eye perceives as yellow. Likewise a green leaf appears green because it subtracts from the incident white light the colors violet, blue, orange, and red; the combination of remaining colors appears green to the eye.

It is important to understand the difference between the subtractive process being discussed here and the additive process discussed earlier. In the additive process we considered the light coming from a television set and the light reflected from a white screen when lights of various colors were shown on it. If we shine red light onto the screen and at the same time shine green light onto it, the screen reflects the *total* light that hit it—we are *adding* red and green. The pigment in a paint does not work that way, however. It emits no light of its own but simply reflects some of the light which strikes it. Suppose we shine white light on a piece of red cloth, analyze the reflected light, and find it to consist of red and orange light. The cloth *subtracted* the other colors from the white light which hit it. To further illustrate this concept, suppose we shine blue light on the cloth. The cloth is ready to reflect red and orange, but blue light contains no red or orange, so the cloth reflects no light at all and appears black.[1]

---

[1]You may try this if you have a blue light bulb. Some colored light bulbs allow a small amount of all colors through, however. So unless your bulb is deep blue, prepare to see a cloth that is dark but not quite black.

The Subtractive Primaries. We have seen that three colors—red, green, and blue—can be combined in an additive process to yield any other color. We called these colors the additive primaries. The subtractive primaries, the three colors which when mixed in various combinations as pigments will produce all colors, are yellow, magenta, and cyan (Fig. 22–3). As has been stated, magenta includes the two ends of the spectrum and looks very much like violet. Cyan includes all of the spectrum except red and orange (and a little yellow) and is sometimes referred to as blue-green.

Consider what happens when cyan pigment and yellow pigment are mixed and white light is shone on the mixture. The cyan pigment subtracts out red and orange light from the white light, while the yellow pigment subtracts blue and violet. So altogether the following colors are removed from the white light: violet, blue, orange, and red. (See Figure 22–5, Plate 2.) Green and yellow remain, which appears green. So a subtractive combination of cyan and yellow results in green.

In like manner you should be able to predict the color produced by mixing of magenta and yellow pigments[2] and by mixing of magenta and cyan.[3] Note that the color produced in each case includes only the colors which are reflected by *both* parts of the mixture, because if a color is not reflected by both, one of the pigments will subtract that color.

Disclaimer

At this point we must add a disclaimer. If you get some paints and try some of the suggested examples, you may have a difficult time producing exactly the predicted results. The reason is that in our examples we chose particular hues of each color and spoke of the results using these hues. We stated, for example, that the spectrum of red includes red and orange, but you may select for your experiment a red paint that reflects some yellow and violet along with the red and orange, or it could conceivably be more restrictive than our example and reflect *only* the color red. Each of these red paints will appear slightly different to you, but each would be considered red.

Before you can make an accurate prediction of the result of mixing two paints, you need to know exactly the spectral colors reflected by each paint. Knowing this, and understanding the subtractive color process, you can make such predictions. Writing a book is easier than mixing paints; we simply say that we use a red paint that reflects red and

---

[2]Answer: Orange and red, seen as red.
[3]Answer: Blue and violet, seen as blue.

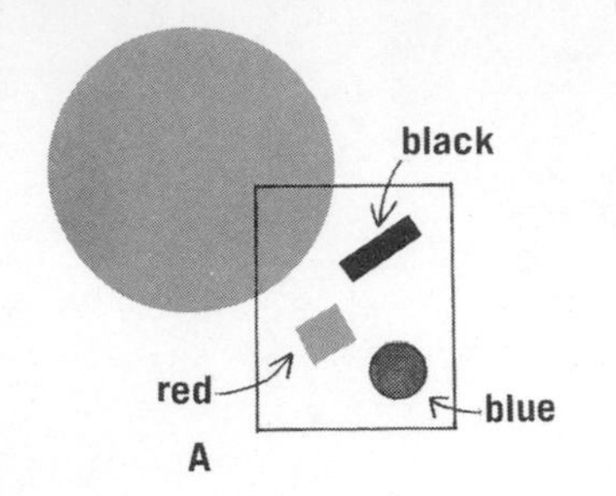

FIGURE 22–6. *A*, The filter near the paper. *B*, The filter covers the paper.

orange light. The paint mixer must use the paint that is available, regardless of its spectral characteristics.

A Summarizing Experiment

take-along-do-it

For this TADI you have a choice of equipment. The best choice is a piece of transparent red plastic or glass. You should be able to see clearly through the material, but it should have a full red color. If it is just slightly tinted you may need to use several layers. If this is unavailable, find a glass, jar, or plastic dish with a flat clear bottom that does not seriously distort an object viewed through it. Put three or four drops of red food coloring into the container and add water to about 1/4 inch. You should be able to see clearly through the water just as you would through the red material above.

Finding the equipment is the hard part. The procedure now is simply to look through your red filter (whether solid or liquid) at something written with red ink on white paper. Then look at some black and some blue writing. If you have a good red filter, the red writing should disappear, while the blue writing will look just like the black. Don't take our word for this; try it!

The light that hits the paper with red, blue, and black marks on it in the TADI is white. The white paper reflects all colors; the red writing reflects red light (and possibly orange); the blue writing reflects blue light (and possibly violet and green); and the black writing reflects no light. (See Figure 22–7.) The figure indicates light of the appropriate color reflecting from each part of the marked paper (including the fact that no light reflects from the black mark).

After reflecting from the paper and the marks, the light heads for the red filter, which works by a subtractive process, separating out other colors and allowing only red light to pass through it. Notice what happens when light from the paper and from each of the marks hits this filter. From the white light reflected to the filter, all colors are subtracted except red, so the paper looks red. Light from the red writing is red, and it therefore passes on through the filter, causing the red writing and the paper to look the same; both appear red. Note in the drawing that after the beams come through the red filter, you cannot distinguish the one originating at the red mark from those originating at

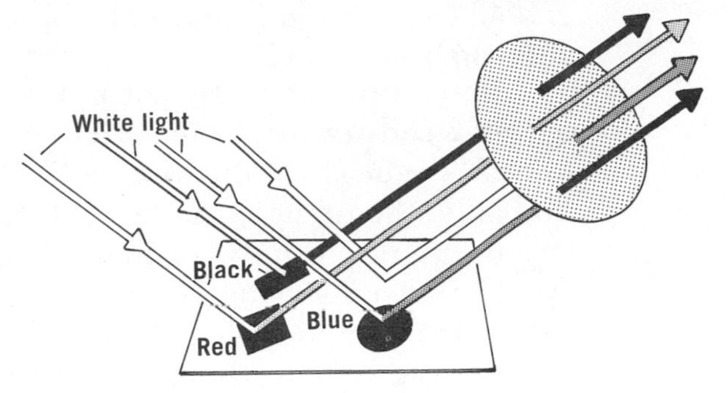

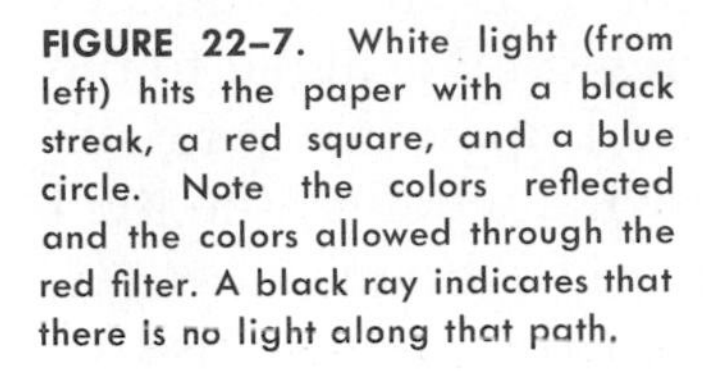
FIGURE 22–7. White light (from left) hits the paper with a black streak, a red square, and a blue circle. Note the colors reflected and the colors allowed through the red filter. A black ray indicates that there is no light along that path.

the paper. It is therefore impossible to detect the red writing. (It is perhaps not quite correct to say that you cannot *see* it, for light from it enters your eye. You are, however, unable to *distinguish* it from its background, the paper.) The light from the blue writing, on the other hand, cannot pass through the filter, so no light comes to your eye from the place occupied by the blue writing. The writing thus appears black, just as does the truly black writing.

COLOR VISION

We saw in Chapter 18 how the eye lens focuses light onto the retina to form images. Sensors on the retina then change the light into electrical signals, which are transmitted to the brain. The light sensors on the retina are of two types, called *cones* and *rods.* The rods are sensitive enough to respond to dim light and to slight changes in light intensity. They cannot distinguish between wavelengths of light, however, and are therefore completely "colorblind." It is the cones which are responsible for color vision.

There is not, at present, a completely satisfactory theory of color vision. The most successful one was first proposed by Thomas Young in 1801. (This is the same man who performed the important experiment showing interference of light, discussed in Chapter 20.) The following is a modification of his theory.

The cones of the eye are of three different kinds, each sensitive to one of the three additive primary colors. The cones sensitive to red light respond (that is, send signals to the brain) when struck by red, orange, yellow, or (to a lesser extent) green light (Fig. 22–8, Plate 3). The signal to the brain from the red-sensitive cones is the same no matter which of these colors strikes the cone. Likewise, the green-sensitive cones respond primarily to green and yellow and, to a lesser extent, to blue and orange. The blue-sensitive cones are sensitive to blue and violet.

Now suppose that light enters the eye and stimulates *only* the red-sensitive cones. The cones send a signal to the brain: "Light is stimulating me!" Based upon this information only, the brain has no way of knowing whether the light is red, orange, yellow, or even green. The brain, however, "sees" that no signals are coming from the green-sensitive cones. It concludes that since only the red-sensitive cones are sending signals, the light must be red light.

If, on the other hand, pure yellow light (say of wavelength 5.8×10^{-5} centimeters) enters the eye, it stimulates both the red- and green-sensitive cones. The brain interprets such stimulation as yellow. It is interesting that light of this single wavelength appears yellow to us but that we see the same yellow if a combination of green and red wavelengths (say 5.4×10^{-5} cm and 6.5×10^{-5} cm) strike the retina. Such a combination would appear yellow to us even though there is actually no light of a yellow wavelength present. The reason is that the red and green wavelengths stimulate both the red-sensitive and the green-sensitive cones just as does light of a single yellow color.

Colorblindness

Most animals are completely colorblind. Except for primates and a few other species, notably bees, the human being is the only animal which sees the world with more color than a 1950 television set. Complete lack of color vision is very rare in humans, but partial colorblindness is common. About one man in 12 and one woman in 120 has what is known as red-green blindness. This means that their eyes respond in the same way to red and to green and send the same message to the brain when either color is seen.

Although red-green blindness is a common eye defect, these two colors can be as different as stop and go at street corners. The colorblind person gets extra help in the fact that red is almost always put at the top of traffic lights. In areas where the old-fashioned lights with red at the bottom (on two sides) are still in use, however, watch out!

Figure 22–9 (Plate 3) shows a common method of testing for colorblindness. If the figure looks like a bunch of dots to you, and you can't see the hidden number, you may be red-green blind. This test is not conclusive, however. A reliable test must be administered under controlled conditions by a qualified person.

Color Fatigue

Stare at the lower right star of the flag of Figure 22–10 (Plate 3) for about 30 seconds. Then stare at a white wall.

You should then "see" the flag in its correct colors. The reason for this odd behavior of the eye is that when color sensors of the retina are stimulated continuously by a certain color, they become fatigued. The orange-yellow color in the field of the stars caused fatigue of the red-sensitive and green-sensitive cones where the image of the field appeared on the retina. When you looked at a white wall, these cones did not respond as much as did the blue-sensitive cones (which were not fatigued), so those points on the retina seemed to you to be receiving blue light from the wall. You should be able to explain by similar reasoning how the red and white stripes appear.

THE RAINBOW

A rainbow can be seen anytime you are between the sun and a rain shower. A ray of light passing over your head strikes a raindrop and is refracted as shown in Figure 22–11. The drop of water acts much like a prism in that it fans out the white light into a spectrum as the light enters. The violet light is bent the most, red the least. These red and violet rays (and all the others) continue until they hit the other side of the drop. At this surface much of the light is reflected, and when these colors pass out of the drop they are separated as shown.

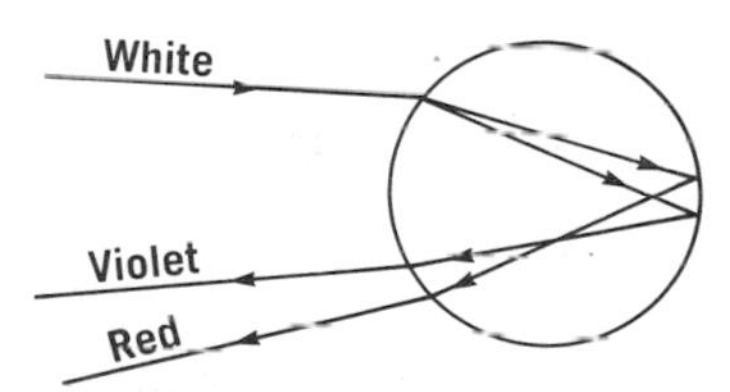

FIGURE 22–11. The white light is broken into a spectrum by the drop of water.

take-along-do-it

This experiment must be done with a bright sun low in the sky, within a couple of hours of sunrise or sunset. Fill a clear glass[4] with water and place it in the sunlight. Now hold some paper in the position shown in Figure 22-12. Note the little spectrum produced. Are the colors produced in the order shown in the figure, with the red farthest from the sun's rays?

Your inclination may be to look directly into the light coming off the glass. Don't do it! Even though some of the sun's light is lost at each reflection and refraction, it is still bright enough to damage your eye.

Consider the man shown in Figure 22–13. If he observes a raindrop high in the sky, the red light from it reaches his eye, whereas the violet light passes over his head. The observer would interpret the color of this drop to be red. A drop lower in the sky would direct violet light toward his eye and would be interpreted as being that color. (The red light from that drop would strike the ground and not be seen.) Colors between the red and violet extremes would be sent to the eye by drops located between these

---

[4]Because raindrops are nearly spherical rather than cylindrical, this demonstration will work better if you can obtain a spherical glass container such as a flower vase or a wine bottle.

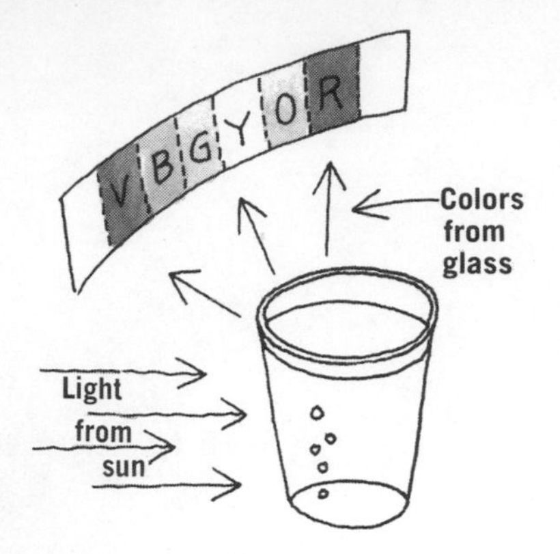

FIGURE 22–12. White light entering from left is broken into a spectrum by the glass of water. DO NOT LOOK INTO THE REFLECTION! LET IT SHINE ON THE PAPER.

two, and thus the rainbow would be red on the outside and violet on the inside. If the observer takes a step in any direction, the light that reaches him comes from other drops and a "different" rainbow is seen.

take-along-do-it

A rainbow can be formed by spraying water into the air from a hose in bright sunlight. If the droplets are falling close to you it is possible to see two rainbows; a different one is seen by each eye. Close one eye and only one bow is seen.

An observer high in the sky as in Figure 22–14 sees light coming from drops below him as well as from above. Numerous drops at the sides also send light to his eyes and so the rainbow forms a great circle (with no end for the pot of gold). For us at ground level, of course, the earth cuts off the lower part of the circle, and we see a half-circle rainbow.

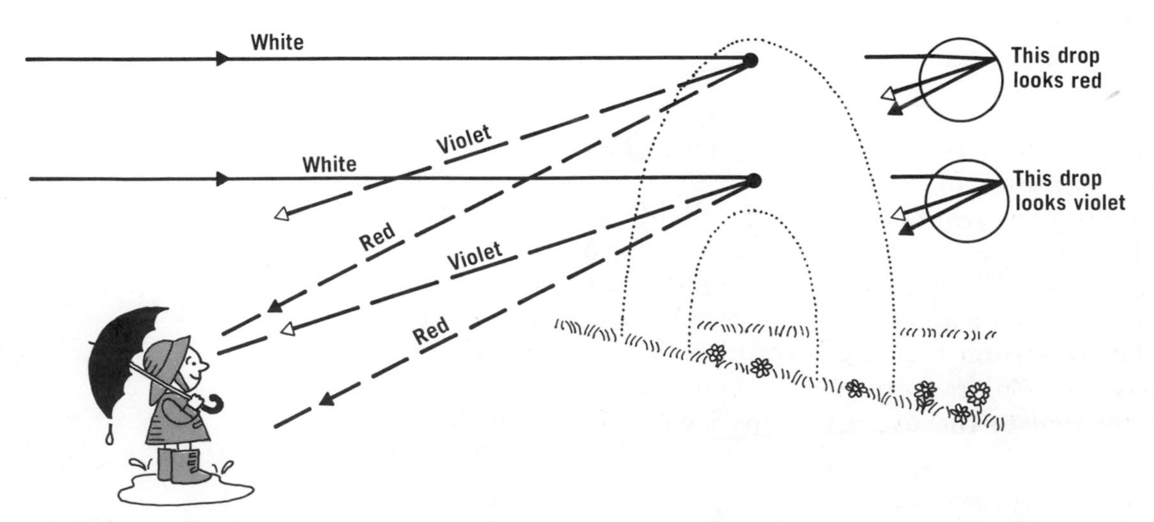

FIGURE 22–13. The observer sees red light coming from higher in the sky than violet light. The other spectral colors appear between the red and the violet.

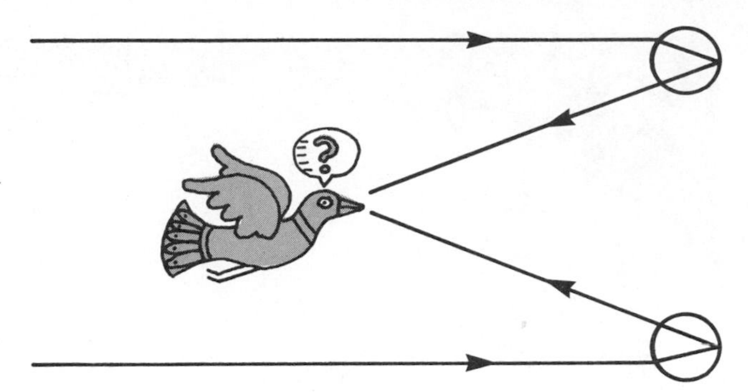

FIGURE 22–14. A circular rainbow seen by a high-flying observer.

WHY THE SKY IS BLUE

When light passes through a gas such as the earth's atmosphere, the electrons in the atoms of the gas absorb energy from the beam and begin to oscillate. These vibrating charges then act as antennae and radiate light having the same frequency as did the light that set them into motion. Just as a tuning fork has certain frequencies at which it will resonate, the atoms of the gas will respond only to certain frequencies of light by absorbing and re-emitting energy in this way. The atoms "resonate."

Because the re-emitted light might go in any direction from a given molecule, only a small portion of it ends up going in the same direction as the original beam. This has the effect of scattering light of certain wavelengths out of the original beam. The molecules of the earth's atmosphere are more effective in scattering light of a short wavelength. Violet is scattered most, then blue, green, yellow, orange, and finally red. Consider Figure 22–15. Suppose you are standing in the yard of a prison. You naturally are looking up at the sky. Most of the light from the sun comes straight through the atmosphere to you by a direct path, and thus you can see the sun overhead. Consider, however, a ray of sunlight which was originally going toward freedom on the other side of the wall. Some of this beam is scattered, primarily the shorter wavelengths. So when you look out over the wall at the sky, most of the light coming to your eye is from the blue-violet end of the spectrum. As is reasonable to conclude from our discussion of color, a combination of violet, blue, less green, and even less yellow appears blue to the eye. So the sky is blue.

FIGURE 22–15. Because light from the blue end of the spectrum scatters most, the sky is blue.

WHY THE SETTING SUN IS RED

Figure 22–16 represents the path of sunlight through the atmosphere over northern Africa. The sun is directly above a yachtsman off the coast of Morocco. Note that the

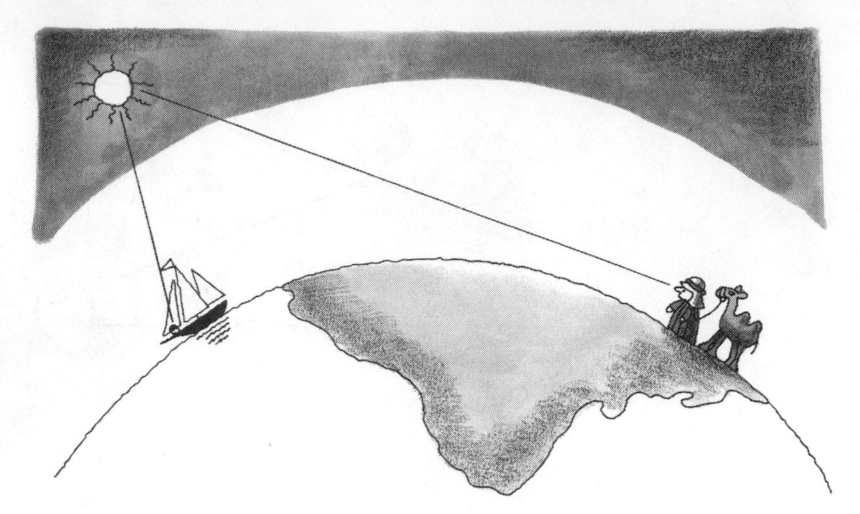

FIGURE 22–16. The sun appears white (or yellow) to a man in the boat. The traveler at sunset sees it as red because its long path through the air has caused blue to be scattered out.

sunlight's path through the atmosphere is shortest at that point. As we turn our attention to the east, we see that the sunlight there hits the atmosphere at a smaller angle and thus has a longer trip through the atmosphere. The light entering an oil tycoon's eye in Saudi Arabia (where it is evening) has passed through a lot of air. (Note that as the light enters the atmosphere, no refraction is shown in the figure. Refraction actually does occur, but we need not consider it for the present discussion.) Now remember that as the sun's light passes through more and more air, more and more short-wavelength light is scattered out of the beam. The Moroccan yachtsman sees the noonday sun as yellowish-white because only a small portion of the blue light has been scattered from the beam, but the oilman sees the sunlight after its long trip through the air and after much of the blue-violet has been scattered out. That which remains after violet, blue, much green, and some yellow are removed appears red.

This selective scattering effect can be noticed most dramatically when observing a rainbow as the sun sets. Beginning with violet, the colors gradually disappear as the sun gets lower, until only a faint red bow remains.

CONCLUSION

This chapter concludes our study of light. Throughout this unit we have referred to similarities between the phenomena of sound and of light. These similarities result, of course, from the fact that both are the result of wave motion. We hope that some of the unity of nature has shone through.

OBJECTIVES

Your study of color should enable you to:

1. List the colors of the natural spectrum in order from red to violet.
2. Explain color addition, listing the spectral colors present in the additive primaries, and predict what color will be perceived when various combinations of the primaries are mixed.
3. Construct a color disc and use it to demonstrate color addition.
4. Explain the subtractive color process, list the spectral colors present in the subtractive primaries, and predict what color will result when any two subtractive primaries are mixed.
5. Given the spectral characteristics of a surface and of a light, predict the appearance of the surface when illuminated by the light.
6. Demonstrate color subtraction by using color filters.
7. Explain the theory of color vision, and using this theory, explain color fatigue.
8. Draw a diagram showing how white light is separated into its colors by a raindrop to produce a rainbow.
9. Use a glass of water to demonstrate the formation of a rainbow.
10. Explain how scattering of light by the atmosphere results in blue skies and red sunsets.

QUESTIONS—CHAPTER 22

1. Is white a color? Is black?
2. Explain why two observers do not see the same rainbow.
3. Artists divide colors into categories called cold and warm. Is there reason for such a division from the standpoint of physics?
4. Which could be ignited quickest by a simple magnifier, black or white paper? Why?
5. In what direction can you see a rainbow early on a rainy morning?
6. In addition to the method discussed in the text for the formation of rainbows, another process forms what is called a secondary rainbow. Figure 22–17 illustrates how light is refracted within the drop which produces this bow. Does this bow have red or violet on the outside?
7. Why would the secondary rainbow of Question 6 be fainter than the primary bow?
8. Why does the sky appear black to an observer on the moon?
9. A 3-D effect can be achieved in movies by showing two superimposed pictures on the screen, one blue and one red, and watching the movie with a different color filter in front of each eye. Explain how this process works. (Hint: remember the results of the TADI on p. 440.)

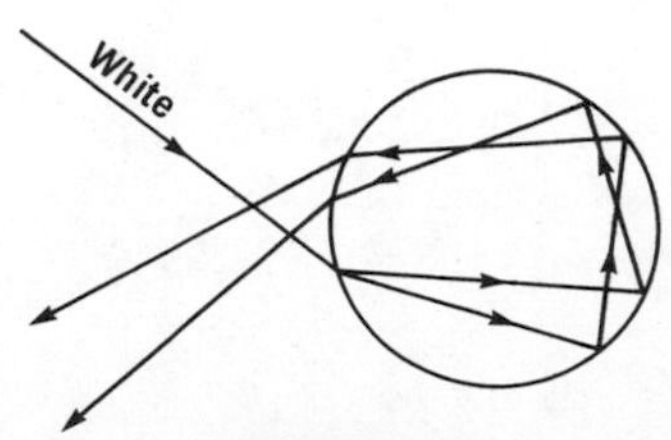

FIGURE 22–17.

10. After staring at a bright blue-green dot for quite a long time and then quickly looking at a white wall, what do you see? Explain.

11. Assuming "ideal" colors, as discussed in this chapter, what color results from mixing yellow and cyan paints?

12. Again assuming "ideal" colors, what color appears when red and blue spotlights are shined on a white wall? On a blue wall?

13. The text points out that refraction of the light entering the earth's atmosphere is neglected in Figure 22–16. Redraw the light rays in that figure to include this effect.

SECTION SIX

$E=MC^2$ and all that

$Fg = Gmm'/r^2$

$x = A \sin(\omega t + \theta_0)$

$m = m_0/\sqrt{1 - v^2/c^2}$

$F = ma$

$E_k = \frac{1}{2}mv^2$

$x = x_0 + v_0t + \frac{1}{2}at^2$

$W = \int F \cdot ds$

Photo by Hayes.

23
RELATIVITY

The mention of Albert Einstein's name immediately causes people to think of the theory of relativity. When questioned about the postulates or the predictions of this theory, these same people often throw up their hands in awe. Their idea that the theory is mathematically difficult is valid, yet a discussion of some of its predictions is not beyond the scope of this text.

Many of the predictions will seem to violate our common-sense understanding of how the world works. As you read this chapter and encounter these brain-breaking ideas, imagine yourself back in the sixteenth and seventeenth centuries. People of that era thought of the earth as the center of the universe. To them it was obvious that the sun and the planets move around the earth. But Copernicus, Galileo, Kepler, Newton, and others showed that the sun is really the center of such motion and succeeded in describing the motion of the planets in detail. This description of the solar system, which is now something that "every schoolboy knows," was an alien idea at that time and was just as mind-boggling to people then as the ideas of Einstein are to us when we first encounter them.

HOW SPECIAL IS A GENERAL?

The theory of relativity can be separated into two parts: the first is called special relativity, the second, general relativity. Special relativity applies only to objects which are at rest or moving with constant velocities. The astounding predictions made by Einstein (based on this theory) have been proven experimentally so conclusively that they must be accepted.

General relativity, as its name suggests, is a more general theory and includes situations in which an object is accelerating. Its postulates have been less well justified, and its predictions are much harder to test experimentally; in a sense, the jury is still out. We will confine our discussion to special relativity.

OLD-FASHIONED RELATIVITY

The man on the street, if asked, would probably tell you that relativity was discovered by Einstein. This is not the case, however. Actually, the idea of relative motion dates back to the days before Newton, but it was Newton who placed it on a firm foundation. Einstein's contribution was to point out that a very short-sighted view of the problem was being taken and to incorporate some fundamental assumptions that altered it considerably.

As a simple introduction to relativistic thought, imagine yourself traveling at a constant speed in a railroad car such that you are completely sealed off from the outside world. There are no telephone poles whizzing by to give you a feeling of motion, and no engine noises reach you; in fact, no outside stimuli of any kind are present to inform you that you are moving. In such surroundings, with little else to do, you decide to perform a physics experiment. You take a ball from your pocket and drop it. Common sense tells you what is going to happen: it falls straight down to the floor. But if an outside observer by the tracks could see the experiment, he would observe the ball to travel the path shown in Figure 23–1. You, in your closed environment, see exactly the same thing happening as you would see if you were not in motion at all.

This simple experiment could be greatly expanded to include very complex investigations, but the results would always be the same. You would see objects obeying the same laws of physics while you are moving with constant

FIGURE 23–1. An observer by the tracks sees the ball move along the dashed line as the train car moves from X to Y.

velocity as you would if you were at rest. This observation is one example of the principle of Newtonian relativity:

> All laws of mechanics which are found to be valid from the point of view of one observer are equally valid from the point of view of an observer moving at constant speed relative to the first. Thus, based on the laws of mechanics, there is no way to determine which observer is actually moving and which is not.

As examples of this principle, consider the following situations. Have you ever been sitting in your car in a parking space, daydreaming, when a car next to you slowly begins to pull out of its space? Often your reaction will be that it is not the other car moving forward at all but your car rolling backward! The embarrassment registered when you slam on your brakes, only to discover the truth, is a common experience. But is it really correct to say that there is a "truth" here? Were you really at rest and was the other car moving? Perhaps it all depends on your point of view.

As a second example, imagine yourself in an airplane soaring high above the earth. As you look down, you see the landscape passing beneath you, and if you let your mind wander a little, it's awfully easy to imagine that you are right, and the world is wrong. In other words, it's easy to feel that you, in the plane, are sitting still and that the earth is moving backward below you. An observer on the ground will, of course, violently disagree. To him it is obvious that you and the plane are in motion and that he is at rest.

Our ideas of motion are all relative to our own point of view. In case you have now settled back in your chair and have said to yourself, "Ah yes, I see the point, but the guy on the ground is correct," consider the appearance of your plane's motion to a third observer on the sun. Who does he say is at rest, and who is moving?

As a final exercise in relativistic ways of thinking, consider a motorcyclist cruising down a highway at 120 miles per hour. Behind him is another motorcycle, this one driven by a little old lady traveling at a mere 50 miles per hour. The first question is, who says those motorcycles are going at 120 mph and 50 mph? A fellow standing at the side of the road, watching the cyclists wheel by, would say that these figures are accurate. And, relative to his point of view, he would be correct. But let's shift our viewpoint to that of the little old lady on the 50-mph bike. She sees the bike in front of her pulling away at a rate of 70 mph, since that is how much faster he is going than she. Her viewpoint might be that her speed is zero and that of the front-runner is 70 mph. She also would note that the fellow by the side of the road passes her by at a speed of 50 mph in the other direction. There is, of course, a final point of view: that of the driver of the fast bike. If he thinks of himself as being at rest and of the others as in motion, he looks back to see granny moving at 70 mph, *but backwards,* and the bystander pulling away at 120 mph.

All of the difficulties concerning who is correct and who is wrong could be resolved if somewhere we could find a privileged point of view. It is tempting to choose the fellow at the side of the road as the privileged observer, but remember that with respect to an observer on the sun, he too is moving. Maybe we could solve all of our problems if we considered everything from the point of view of someone on the sun. But this won't work either, because with respect to the stars, the sun is not at rest! As you can see, the search could go on indefinitely.

Up until the present century, however, it was felt that there actually was a privileged viewpoint. It was held that electromagnetic waves (such as light) had to have a substance through which to travel, and it was thought that this substance was in a state of absolute rest. This material, called the ether, has been discussed in earlier chapters. The stationary ether would provide a unique framework with respect to which we could describe the motion of all objects. Such trivialities as whether an airplane moves above the earth or the earth moves beneath the plane would all be answered by comparing the motion of both as seen by an observer at rest with respect to the ether. The actual experiment is not so simple, however, because it is impossible to place an observer "at rest with respect to the ether." Remember, we cannot feel, taste, smell, hear, or even see the ether. But since it is the stuff through which light passes, we can use light itself to find the motion of the ether. An experiment to find its motion will tell us who is *really* at rest.

MICHELSON, MORLEY, AND THE SPEED OF LIGHT

Shout a word, and the sound spreads out from you at a speed of about 1100 ft/sec. We have mentioned this phenomenon previously, but we neglected to tell you that the speed with which the air or wind moves alters the speed of sound. As an example, consider the situation of Figure 23–

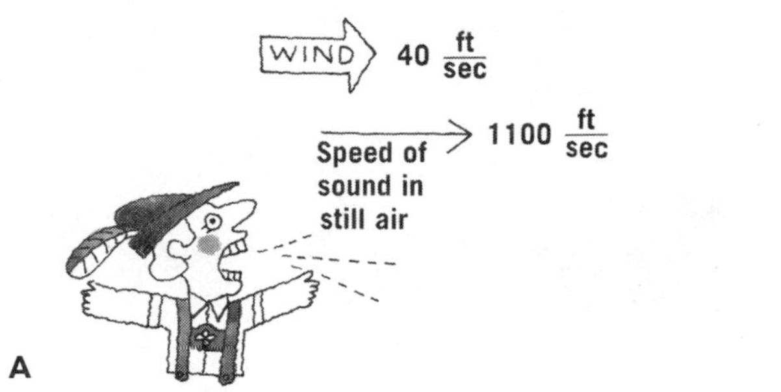

FIGURE 23–2. Note that the difference in wind direction affects the speed of sound measured by a person in the wind.

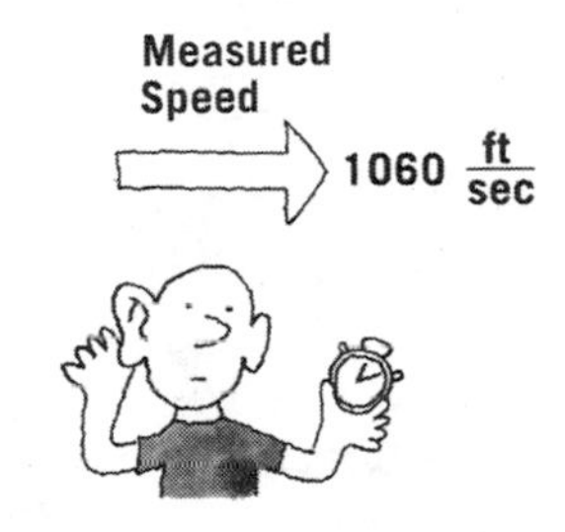

2*A*. A shout travels toward a listener who in some way measures the speed of the sound as it comes to him. His measurement will not be 1100 ft/sec, however, because as indicated in the drawing, there is a wind blowing in his direction at a speed of 40 ft/sec. The sound is carried along by the wind, which causes the experimenter to conclude that the speed of the sound is actually 1140 ft/sec. In Figure 23–2*B*, the situation is the same except that now the wind blows *away* from the listener at a speed of 40 ft/sec, so the measured speed would be 1060 ft/sec. Thus the motion of the material in which the sound wave travels changes the observed speed of the sound. But if this is the way that sound waves behave, do other waves act in the same fashion? And, in particular, what about light waves?

In 1883, Albert Michelson and E. W. Morley devised an experiment to measure the earth's speed through the "stationary" ether. They reasoned that as the earth moves through space, it must pass through this ether, and that this would cause an ether "wind" to blow across the earth (Fig. 23–3). Just like the sound wave in Figure 23–2, a beam of

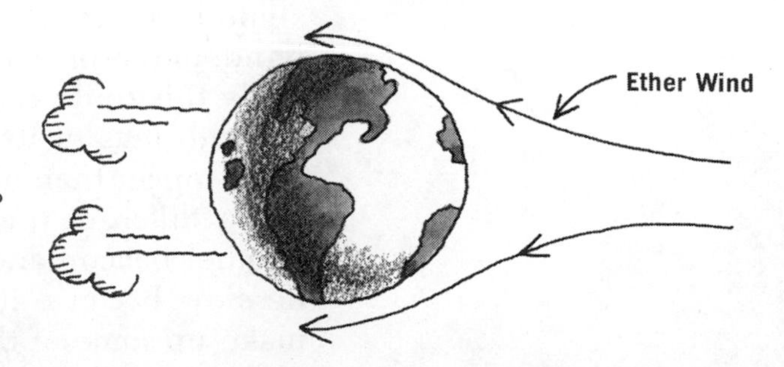

FIGURE 23–3. The motion of the earth should cause the ether to rush by as a wind.

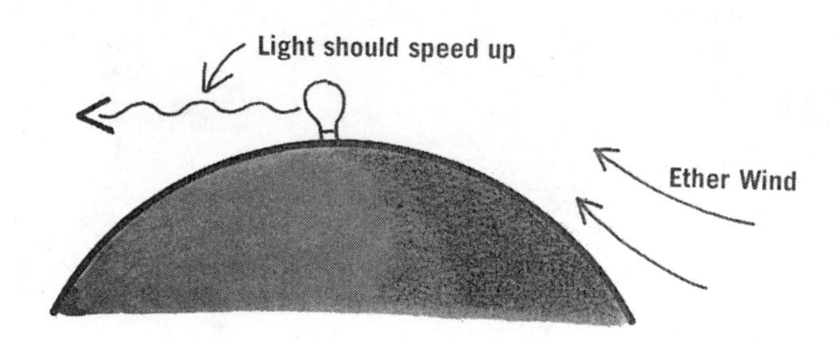

FIGURE 23–4. The ether wind should affect light speed just as air wind affects sound speed.

light sent against the ether wind would slow down, and a beam traveling with the wind would be speeded up (Fig. 23–4). Now consider what would happen if a beam of light were sent on a round trip to a mirror and back such that it first traveled "upwind" and returned "downwind." The time needed for such a trip would be slightly longer than if the ether wind were not blowing. But what about a beam of light traveling perpendicular to the wind on a round trip of the same length? A careful analysis of this trip would show that it also would require more time for its round trip than if the wind were not blowing. The extra amount of time required would not be as great when moving perpendicular to the wind as when moving parallel to it, however.

In order to better visualize this effect, consider the airplane race shown in Figure 23–5. The white plane is assigned a route which takes it on a path perpendicular to the wind, to the beacon and back to the start. Note that in order to fly this route the plane has to aim slightly into the wind on both legs of its journey. This causes its to-and-fro time to be longer than it would be on a calm day. The black plane has a different route: it must fly directly into the wind to another beacon and back. The first leg of its journey would be slow because it is going against the wind, but it would make up some of the lost time in flying back with the wind. If the total flying distance for each plane is the same, and

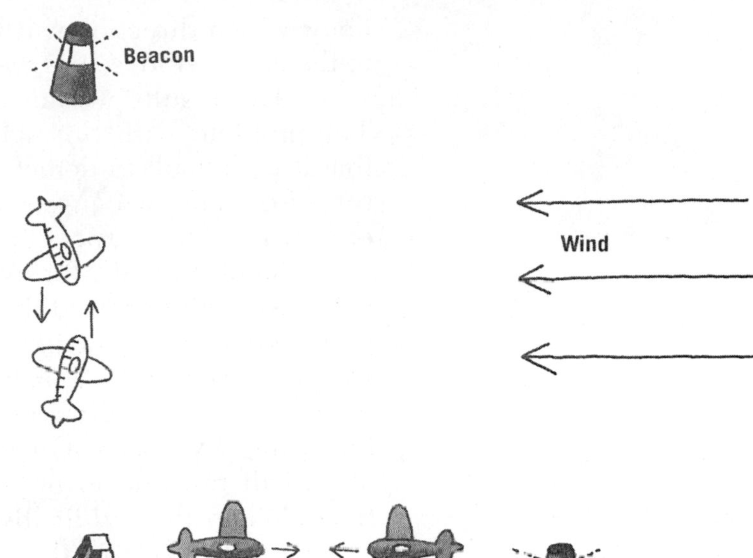

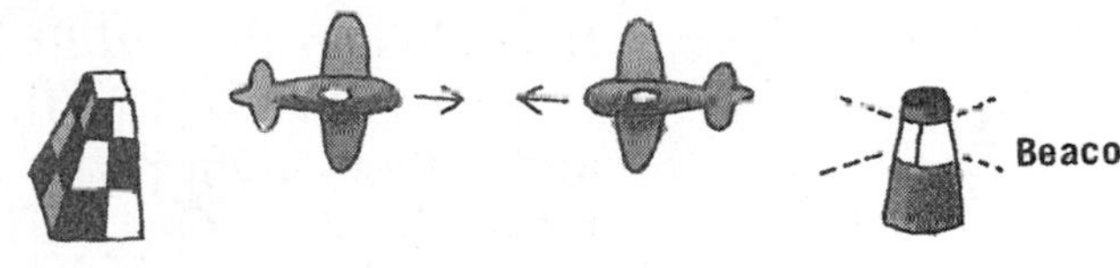

FIGURE 23–5. Both planes are identical and fly the same distance. The white one wins, however.

both planes have identical air speeds, which one would win? A careful analysis (or an actual race) shows that the white plane wins!

If there is an ether wind flowing by the earth, it should be possible to send one beam of light perpendicular to the wind and another parallel to it in a race similar to that of the planes (Fig. 23–6). Before the race there would be no way to

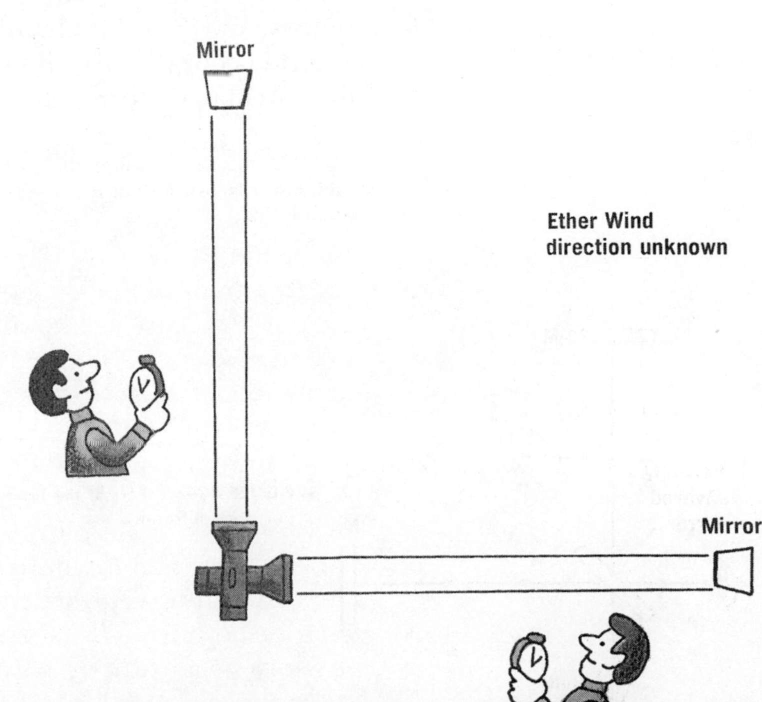

FIGURE 23–6. A light beam race similar to the plane race in Figure 23–5. Such an experiment should tell us the ether wind direction.

know which direction was parallel and which perpendicular to the ether wind, so we would only see which beam of light wins. The results would tell us the direction of the wind. The problem with this scheme is that light moves so fast that it is difficult to detect which of the two light beams returns first. The two guys with the stopwatches are dreaming, of course, if they expect to measure a time difference.

Michelson and Morley used a more ingenious method to measure this time difference. Their method, which depends upon the principle of interference of light waves, is shown in Figure 23–7. A beam of light leaves the bulb and travels to the partially silvered mirror, which breaks up the beam into two parts. One half goes toward mirror M_1, the other half toward mirror M_2. Both beams are reflected, retrace their paths and finally enter the eye of an observer.[1]

The two parts of the beam start their voyages in step, and if their flight paths are of the same length, they should return in step. Our earlier studies of light have indicated that two light waves arriving in step produce constructive interference and a bright spot is seen. In practice, the light must follow slightly different paths to arrive at different locations on the observer's retina, resulting in an alternating pattern of light and dark rings (Fig. 23–8).

Imagine our device placed in the ether wind. Slightly different travel times for the two paths should result, but we still will see a circular pattern. Suppose we start as in Figure 23–9A, with arm 1 luckily aligned with the wind, and we observe a pattern. Now the device is rotated until arm 2 is along the breeze. The two arms have, in effect, changed places, and a slight change in the observed circular pattern should occur (that is, the rings should move). This change in the circular pattern was expected to be small, as you might

[1]When passing through the partially silvered mirror for the second time, part of each beam is lost. Part of it does go to the eye, however, and that is the part we are considering.

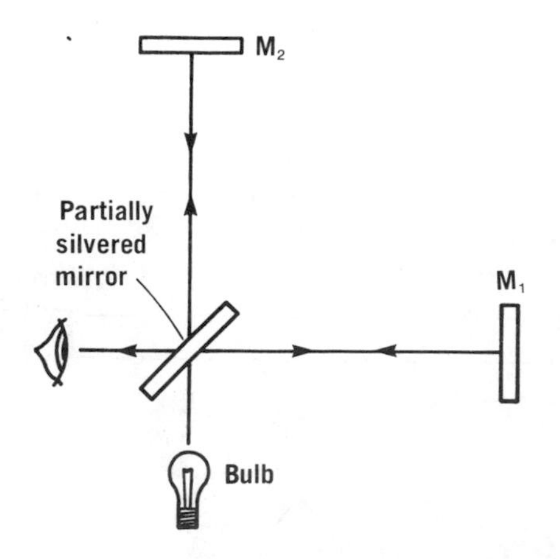

FIGURE 23–7. Part of the light from the bulb arrives at the eye from mirror M_1, and part from mirror M_2.

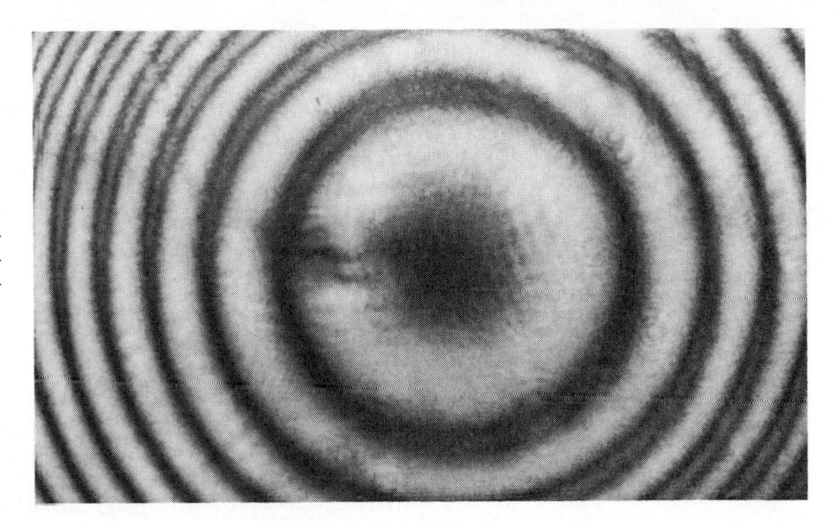

FIGURE 23–8. The interference pattern resulting from the combined beams in the Michelson-Morley experiment. There is destructive interference at the center.

suspect. But Michelson and Morley had predicted the amount of change they should expect to see, and they were convinced that their instrument was delicate enough to detect it. And so they tried . . . and tried . . . and tried. . . . Their results: nothing at all. Absolutely no change in the pattern was seen! Other experimenters repeated the experiment and met with the same result.

The scientific community had expected Michelson and Morley to verify the existence of the ether wind and, therefore, of the ether itself. The results, however, were conclusive: the ether wind did not show itself. This left two possibilities: either the ether moved right along with the earth, or it did not exist at all. The first of these was preposterous; to

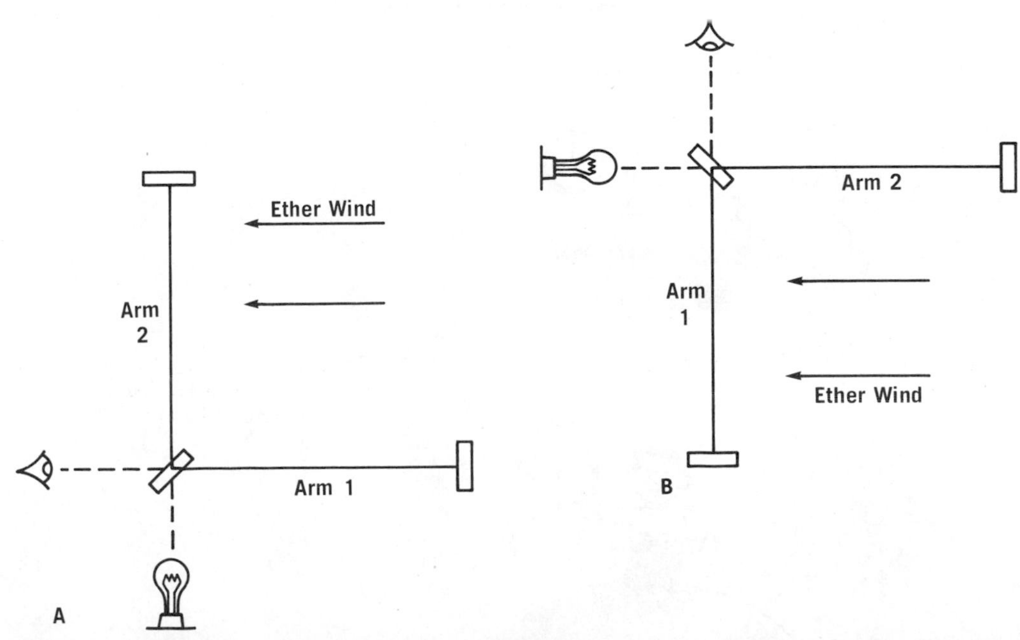

FIGURE 23–9. Between A and B the device has been rotated. Since the ether wind crosses it differently in B, the pattern should be changed.

think that the earth was so important that this universe-filling ether followed it in all its motions (rotating on its axis, revolving around the sun, following the sun's motion around the galaxy) was contrary to reason. The conclusion must be that ether does not exist! This discovery certainly puts a crimp in the plan to use ether as a privileged reference frame for comparing the motion of objects.

Where Do We Turn?

Many people devoted much time and effort attempting to find reasons why the Michelson-Morley experiment failed to detect an ether. Repeats of the experiment were performed in which many different orientations of the device were used and many different locations, such as mountain tops, were tried. The results were always negative. In fact, regardless of the relative motion between the source of light and the observer, it was found that the speed of light always remained the same.

There was reason to search for a way out of the Michelson-Morley dilemma other than scientists' reluctance to drop the ether theory: The result did great damage to other ideas about relativity. Consider the situation in Figure 23–10. A beam of light passes the earth, and an observer measures its speed as 186,000 miles/sec. Also measuring the speed of this beam is a fellow in a spaceship traveling at 93,000 miles/sec (half the speed of light).[2] Our old ideas about relativity say that the space traveler should find the speed of light to be 93,000 miles/sec relative to him. But

[2]An impossible speed by far for any of today's spaceships.

FIGURE 23–10. The experimenters measure the speed of light to be 186,000 miles/sec, even though the rocket is moving by the earth at 93,000 miles/sec.

does he? The Michelson-Morley experiment has indicated that whether you travel with, against, or perpendicular to a beam of light, the same speed of light is observed. And so the fellow in the spaceship, surprisingly, measures the speed of light to be 186,000 miles/sec also. In fact, regardless of how fast you go or in which direction the light travels, if you measure the speed of light, you always get the same answer—186,000 miles per second.

Einstein's Answer

Einstein took the results of the Michelson-Morley experiment at face value and, along with the results of the preceding section, simply(!) developed a new theory of relativity. The following starting points are called his postulates of special relativity:

(1) All the laws of physics are the same for all observers at rest with respect to one another or moving at a constant speed with respect to one another. (This is an expansion of Newtonian relativity to cover *all* the laws of physics.)
(2) The speed of light in free space is the same for all observers regardless of the motion of the source of light and the observer.

These postulates are simply an acceptance of the results of the Michelson-Morley experiment.

The mathematical manipulations performed by Einstein are beyond the scope of this text. We will, however, examine the results he obtained and inquire into some of the new revelations predicted by his theory.

SLOW CLOCKS AND FAST CLOCKS

The evidence that the speed of light is the same for all observers leads to a surprising prediction about the rate at which time passes for different observers. To indicate what happens and why it happens, consider a rocket ship traveling at a speed approaching the speed of light. In Figure 23–11A, an astronaut performs a simple experiment. He directs a beam of light from a light bulb at X straight down toward a spot on the floor, labeled Y. In principle, our astronaut could measure the speed of the light, because he can measure the distance from X to Y and he can measure the time required for the light to travel that distance. We know that his answer will be 186,000 miles/sec. In equation form the result is expressed as:

$$\frac{\text{Distance from X to Y}}{\text{Time for trip from X to Y}} = 186{,}000 \text{ miles/sec}$$

Albert Einstein (1879-1955)

Einstein was born in Germany and began his education in Munich. He did so poorly in school that it was feared that he was retarded. A teacher once told him, "You will never amount to anything, Einstein."

Einstein later finished school in Switzerland and then had trouble finding a job. With a little help from friends, he found a job at the Swiss patent office in Bern. His work there was such that he could complete his duties in four or five hours, and he was able to spend the rest of the day surreptitiously working on his theories and calculations. During his tenure in this job, in his early twenties, he published papers which explained the photoelectric effect and outlined his special theory of relativity.

Although many think of him only in terms of science, Einstein was vitally interested in world affairs. He enjoyed classical music, played the violin, and was of a deeply religious nature.

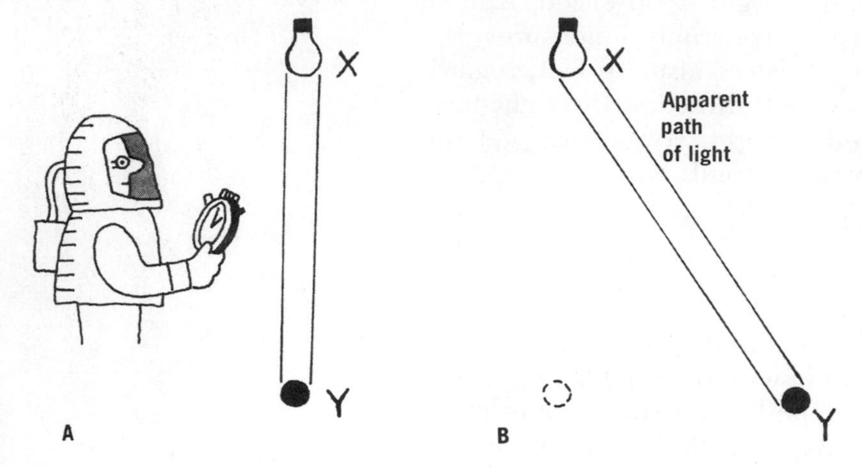

FIGURE 23–11. *A*, The astronaut measures the speed of the light beam from X to Y. *B*, The path of the light as seen by an observer on earth.

But what about an observer on the earth watching the same experiment? From his point of view, the light left the bulb and hit the spot, but by the time the beam had reached it, the spot had moved to the position shown in Figure 23–11*B*. This apparent motion of the spot is, of course, caused by the motion of the rocket ship. From the earthbound observer's point of view, the only way in which the light could leave the light bulb and hit point Y is for it to follow the path shown. Thus, he observes the light to travel a longer path in going from X to Y than does the observer in the rocket. Yet when our earthly experimenter proceeds to calculate the speed of light, he uses the same equation as the astronaut, and he has to get the same answer, 186,000 miles/sec. The equation he uses is:

$$\frac{\text{Distance from X to Y}}{\text{Time for trip from X to Y}} = 186{,}000 \text{ miles/sec}$$

Let's compare each term on the left side of the equation for both observers. The observer on the earth sees a greater distance from X to Y. However, when he divides this greater distance by the time required for the light to travel from X to Y, he gets the same answer as the moving observer. The only conclusion possible is that the earthbound observer must find that light takes a longer time to get from X to Y than does the astronaut. If the astronaut's clock (which, according to the earthbound observer, is in motion) says that three millionths of a second passes during the trip, the earthbound observer says that more time, perhaps five millionths of a second, passes on his clock.[3] A general statement

[3]How big would the rocket ship have to be for this much time to elapse?

applied to the situation at hand is: *moving clocks run slowly*. This effect also depends on the point of view. An observer comparing clocks with someone in a fast-moving car would observe the speeder's clock to be running more slowly. But the speeder in the car would deduce that he is at rest and that the speeder is the fellow outside the car. He perceives his clock to be running correctly and the clock belonging to the fellow at the side of the road to be slow. As far as people driving in cars are concerned, such differences are insignificant; the effect becomes important only at speeds approaching the speed of light.

How Time Flies

We have indicated in the previous section that moving clocks run slowly, but we have not looked at how slowly. Einstein developed an equation that answers this question by relating the time intervals according to different points of view. The equation is as follows:

$$T = \frac{T_0}{\sqrt{1 - V^2/C^2}}$$

Reprinted with permission of The Chicago Tribune Magazine/Sidney Harris.

in which V is the speed of the clock with respect to an observer, C is the speed of light (3×10^8 meters/second), T_0 is a time interval measured on a clock at rest with respect to the observer, and T is the same time interval on a clock moving at speed V. For example, let's assume that a clock whizzes past us at half the speed of light ($V = 1.5 \times 10^8$ meters/second). The equation says that when one minute passes on a clock at rest with respect to us (T_0 = one minute), T = 1.15 minutes. In other words, if a clock "ticks" once every minute when it is at rest with respect to us, it will "tick" every 1.15 minutes when traveling at half the speed of light: moving clocks run slowly.

Table 23–1 presents the results obtained for several different speeds of a moving clock. Rather than take our word for the answers, you might try verifying them for yourself. Note that at a speed of 10 per cent that of light (3×10^7 meters/sec), the moving clock takes 1.005 minutes to sweep out what we would see it sweep out in one minute if it were at rest. Now, 3×10^7 meters/sec is considerably faster than most of the speeds to which we are accustomed on earth, but the rate of the moving clock has not slowed all that much. As a result, it is not surprising that we do not observe this effect when we are jogging. Even at 90 per cent of the speed of light, the moving clock is running only about half as fast as a clock at rest, but note that it slows down very rapidly at percentages higher than this value.

HOW LONG IS A METER STICK?

If you pick up a meter stick in a laboratory and examine its length, the answer to the question posed in the heading

TABLE 23–1. RATE OF A MOVING CLOCK AT SEVERAL SPEEDS

Apparent Time	% of Speed of Light	Speed of Clock (m/sec)
1.005	10	3×10^7*
1.020	20	6×10^7
1.048	30	9×10^7
1.091	40	1.2×10^8
1.154	50	1.5×10^8
1.250	60	1.8×10^8
1.400	70	2.1×10^8
1.660	80	2.4×10^8
2.294	90	2.7×10^8
10.000	99.5	2.98×10^8
70.710	99.99	2.997×10^8

*Read across as follows: "A clock ticking once every minute at rest will tick once every 1.005 min at a speed 10 per cent the speed of light (3×10^7 m/sec)."

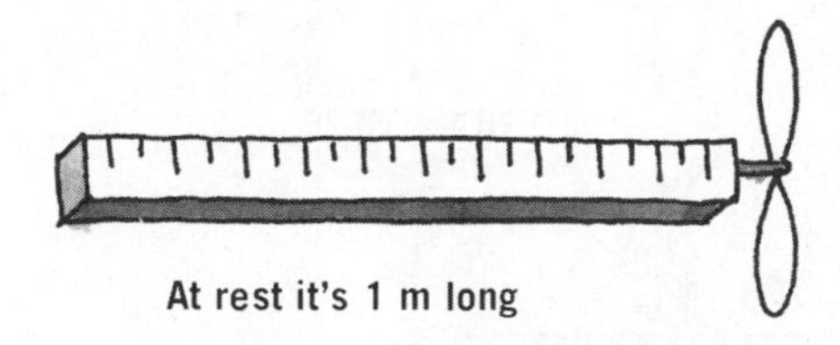

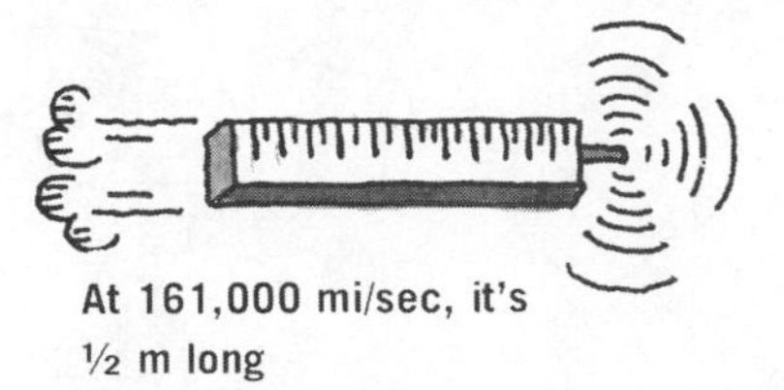

FIGURE 23–12. A meter stick equipped with an engine.

seems obvious. The answer that a meter stick actually is one meter long is a statement that depends on your point of view, however. The results of Einstein's theory of relativity predict that the length of an object appears to change if it moves relative to you. For example, suppose that a rather powerful engine is attached to the stick (Fig. 23–12), such that it is propelled through the air at a speed of 161,000 miles/sec. According to the theory of relativity, we would now observe that same meter stick to have shrunk to a length of only 50 centimeters, half its original length. The faster it goes, the more it shrinks. The speed of 161,000 miles/sec was selected in order to indicate the drastic apparent shrinkage that occurs at high speeds. If your roommate is nearby, ask him to run across the room carrying a stick. If you attempt to measure its length, you will not be able to detect any change; it does shrink, but the fantastic prediction of Einstein only becomes important and perceptible when the speeds are quite high. This is true not only for moving meter sticks but for most of the Einstein predictions. All of our old ideas such as the sanctity of the length of an object still hold at ordinary speeds. Troubles arise only when we observe objects at speeds approaching the speed of light.

Such results also differ depending upon your point of view. To illustrate, consider Figure 23–13. An observer in a spaceship holds a meter stick, and you, on the earth, hold one identical to it. You look at your stick and conclude that its length is one meter, and if his velocity is 161,000 miles/sec, you observe his stick to be one half meter long. But that's your point of view. From the pilot's point of view *you* are moving backward past *him*. As a result, when he looks at his meter stick, he says it is one meter long and it is yours that has shrunk to one half meter. It all depends on the point of view!

If you and the space traveler disagree on something so fundamental as the length of a little piece of wood, it seems inconceivable that you would agree on anything. But if you hold your meter sticks as in Figure 23–14, you will agree, because in this case, you are holding your sticks so that they are perpendicular to the direction of motion of the spacecraft (or to the motion of the earth, if we desire to satisfy the

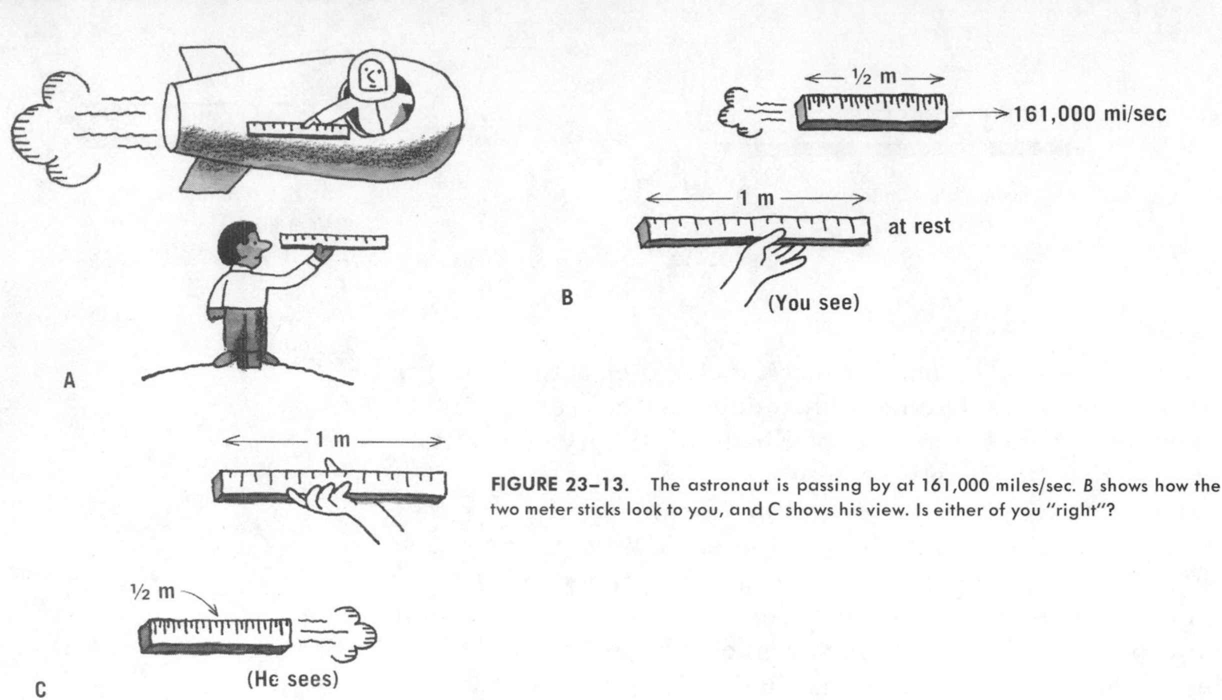

FIGURE 23–13. The astronaut is passing by at 161,000 miles/sec. *B* shows how the two meter sticks look to you, and C shows his view. Is either of you "right"?

pilot). In this situation, you both agree that the sticks are one meter long. *Lengths shrink only in a direction parallel to the direction of travel of the observer.* As an additional example, consider the hefty jogger of Figure 23–15A. At a rather fast jogging speed, near the speed of light, the distance across his middle would appear to shrink as in *B*, but his height would remain the same. The possibilities for exploitation of new diet fads are overwhelming.

The Long and the Short of Length Contractions

The expression that relates length of objects in motion to their length when at rest is given by:

$$L = L_0 \sqrt{1 - V^2/C^2}$$

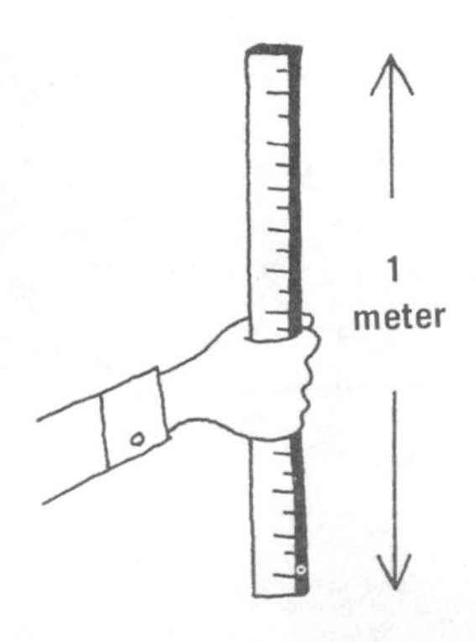

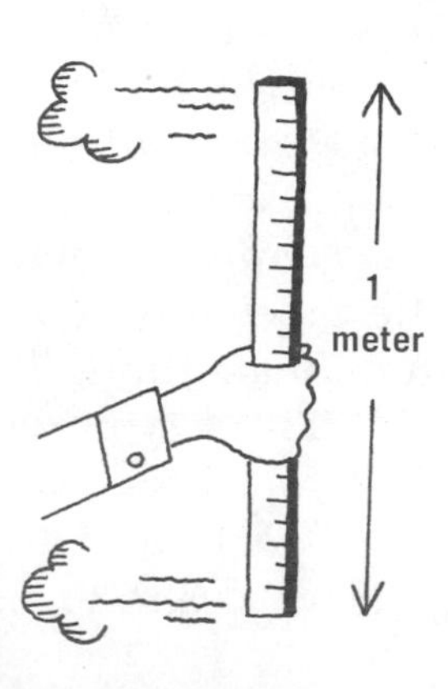

FIGURE 23–14. The meter stick contracts only in the direction of motion.

FIGURE 23–15. *A*, Jogger at ordinary speed; *B*, Jogger approaching speed of light.

In this equation V is the speed of the moving object with respect to an observer, C is the speed of light, L_0 is the length of the object at rest, and L is its length when in motion. Table 23–2 gives the apparent length of a moving object that is one meter long when at rest. The answers are given by the preceding equation, and it will be worth your time to verify a few for yourself.

MOVING MUONS

The disagreement on the passing of time and the length of objects must remain in the realm of interesting conjecture unless experimental evidence comes forth. A confirmation of the relativity predictions does occur for high-speed muons. The muon is a particle produced high in the atmosphere in collisions between cosmic rays and atoms in the air. The muon is radioactive and quickly decays into other particles. (The subjects of radioactivity and cosmic rays will be cov-

TABLE 23–2. APPARENT LENGTH OF A MOVING OBJECT WHICH IS ONE METER LONG AT REST

Apparent Length (meters)	% of Speed of Light
0.995	10
0.980	20
0.954	30
0.917	40
0.866	50
0.800	60
0.714	70
0.600	80
0.436	90
0.100	99.5
0.014	99.99

ered in later chapters on nuclear physics.) The half-life of muons is observed to be 1.5 microseconds (1.5×10^{-6} second) when the muons are at rest with respect to an observer. The half-life means that if an observer has, say, 1000 muons at one instant of time, then 1.5×10^{-6} seconds later, he will have only 500. The others have decayed into other particles.

To see what muons have to do with relativity, consider some muons approaching earth at 2.9×10^8 meters/second. If 1000 of these muons are observed at a height of 435 meters, their high speed will get them to the earth in 1.5 microseconds. This figure corresponds to the half-life of the muons, and thus we would expect to find 500 of the original muons reaching the earth. When the experiment is carried out, however, it is found that more than 500 of them reach the earth (about 840, actually). The reason? The muons were moving at 97 per cent of the speed of light; thus their "clocks" were running slower than ours. While 1.5 microseconds went by on the earth, only 0.38 microsecond passed for the muons. Thus, fewer than half of them disappeared by decay. The number of muons missing from the original 1000 corresponds exactly to the figure predicted by the special theory of relativity.

Now let's assume that instead of observing the high-speed muons from the earth, we ride along with them. From this point of view, we will not observe any change in the rate of passage of time. Instead, as we look down toward the earth, we see a difference in our *distance* from it. Instead of being 435 meters above the earth, we find this length shortened to about 111 meters. The time required to cover this distance is exactly 0.38 microsecond. So, if 1000 muons start out, we are not surprised that more than 500 remain at the earth's surface. Thus, even though there is a difference of opinion concerning which relativity effect occurs—time change or length change—the end result is the same for both points of view.

MASS IN MOTION

If the perception of length and time is affected by Einstein's relativity principle, can mass be far behind? The answer is no. Mass undergoes similar changes as the speed of a moving object approaches the speed of light. In this case, an object's mass is observed to increase as its speed increases. In fact, if an object's speed should reach that of light, its mass would become infinitely great.[4]

---

[4]What about our jogger of Figure 23–15 now?

In Chapter 1 we looked at the relation between force, mass, and acceleration:

$$\text{Force} = \text{Mass} \times \text{Acceleration}$$

From this equation we can deduce what it would mean if the mass of an object became infinitely large. As the equation indicates, if the mass becomes infinite, an infinite force would be required to accelerate it. Because it is difficult to imagine that we can find an infinitely large force to do the pushing, it is unreasonable to believe that we will be able to produce any more acceleration. If this is the case, its speed will no longer be able to be increased, and the object will travel at the speed attained: the speed of light. The conclusion, then, must be that there is an upper limit to the speed that an object can have. This upper limit is the speed of light.

Some Heavy Examples

According to Einstein, the relation between the mass of an object and its velocity is given by:

$$M = \frac{M_0}{\sqrt{1 - V^2/C^2}}$$

Here, V is the velocity of the object with respect to an observer, C is the speed of light, M_0 is the mass of the object when it is at rest, and M is its mass when in motion with respect to an observer. The mass of a one kilogram object at various speeds, as predicted by this equation, is given in Table 23–3.

TABLE 23–3. APPARENT MASS OF A MOVING OBJECT THAT HAS A MASS OF ONE KILOGRAM AT REST

Apparent Mass (kg)	% of Speed of Light
1.005	10
1.020	20
1.048	30
1.091	40
1.154	50
1.250	60
1.400	70
1.660	80
2.294	90
10.000	99.5
70.710	99.99

ENERGY AND MASS

In Chapter 2 we explored the concept of energy. One of the results of our investigation was the principle of the conservation of energy, which allows for many different forms of energy (such as potential, kinetic, and thermal) and allows for one form of these to be changed into another form. However, it does not allow energy to suddenly appear or disappear. Our final look at the changes wrought by Einstein's work will cause us to modify this idea.

From his work on relativity Einstein brought forth an equation that is probably the most widely known of any of his efforts. It states:

$$\text{Energy} = \text{Mass} \times (\text{speed of light})^2$$

or as often written: $E = MC^2$. This equation states that we can no longer consider mass and energy to be two completely independent entities. We must now incorporate into our ideas the possibility that mass can be converted into energy or, on the flip side, that energy can be converted into mass.

A piece of chalk about to be discarded has a mass of about one gram. If, in some way, this *entire* one gram object could be converted into energy, the amount created would be sufficient to light 10 million 100-watt light bulbs for a day. As we will see in Chapter 25, a calculation of this nature is not one of idle mathematical gymnastics. We will see that processes now underway (via nuclear reactors) and others under development utilize the conversion of mass into energy to provide electrical power. The humanitarian benefits are obvious, but unfortunately, the first use of this relationship was not peaceful. The atomic bombs exploded over Hiroshima and Nagasaki provided the first dramatic evidence of the equivalence of mass and energy.

A GOOD THEORY

The atomic bomb and energy-producing nuclear reactors are only two of a number of examples which verify Einstein's prediction that mass and energy are interchangeable. This prediction followed from the basic postulates of his theory of relativity. We have also seen that the time-change prediction is verified by experiment. These verifications serve as another good example of the scientific method. Einstein hypothesized a model of nature which accepts the result of the Michelson-Morley experiment. But at the same time, his model (or theory) resulted in new predictions. Through the years more and more of these predictions have been tested, and so far, all experimental evidence confirms his theory. Thus the theory is considered an extremely good one because it made many predictions which turned out to be right (and none which have proved wrong in 70 years of testing). This quality is what separates good theories from poor ones. Let's wind up our discussion of relativity by discussing one more prediction—a brain-bending one.

THE TWIN PARADOX

Moving clocks run slower than clocks at rest. So says the theory of relativity. But what is a clock? Is it simply the regularly ticking wristwatch variety, or can we expand our definition to include the regular beat of a heart? Is there something special about the muon that causes it to live longer when in motion, or will a simple cell in a human body also live longer? There is nothing restrictive about the theory of relativity that would prevent it from applying to any process including those listed above. All these "clocks" should slow down when in motion.

With this prospect in mind, let's examine an interesting possibility. Let's assume that a pair of twins reach their twentieth birthday. One of the twins, Goslo, is of the stay-at-home sort; his sister, Blasto, is adventurous. To celebrate her birthday, Blasto decides to go on a rocket ship voyage to a distant star some 20 light-years away. Traveling at very nearly the speed of light, she reaches the star and starts for home immediately. Because the star is 20 light-years away, and since Blasto travels at almost the speed of light, the trip naturally takes slightly more than forty years, according to Goslo's earthbound reckoning. Hastily downing his prune juice, the 60 year old Goslo rushes to the rocket to greet his sister. What he finds when he gets there depends on the point of view. According to Goslo's clock, his sister should be 60 years old, but if Blasto has existed in a high-speed environment and if moving clocks run slowly, all her bodily processes leading to aging should have been slowed. According to these moving clocks, very little time has passed. We would expect, then, that a young trim Blasto will alight from the rocket to greet her now arthritic twin.

There is another point of view. In Blasto's view of the trip, she could pretend that she was at rest all the time and that Goslo went sailing away from her on the earth and returned. Her picture says that *she* has aged by 40 years, and since Goslo is the one moving at a high speed, his clock should run slowly and it is he who should be youthful when they reunite.

Both of these viewpoints seem to be correct. Both obey the laws of the theory of special relativity, but obviously they cannot both be true. Herein lies the paradox. Which

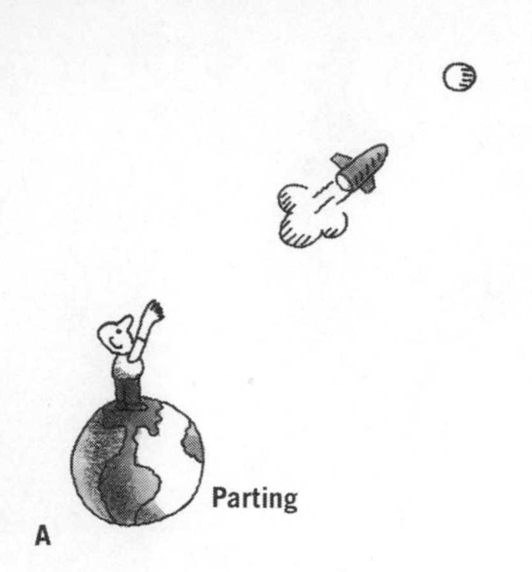

FIGURE 23–16. Are the twins the same age?

twin is correct? In order to answer this question, we must point out that the trips are not as symmetrical as we have led you to believe. Blasto has had to experience an acceleration in order to reach nearly the speed of light and then a deceleration and another acceleration to turn around for her voyage home. It is the twin who experiences these accelerations for whom time lags. As a result, when the twins reunite, Blasto is indeed the younger. That this is really the case has been verified by twin atomic clocks. One of the clocks was left behind as a reference while the other was flown around the world by commercial aircraft. Comparison of the clocks after the trip showed that the one which took the trip had "aged" less.[5]

OBJECTIVES

A relatively careful study of relativity should enable you to:

1. Distinguish between special and general relativity.
2. State the principle of Newtonian relativity, and use an everyday happening to give an example of its meaning.
3. Successfully argue that the earth is not a natural preferred point of view from which to consider motion.
4. Describe the Michelson-Morley experiment, stating its purpose and the result obtained.
5. State and explain each of the two postulates of the theory of special relativity.
6. Explain the effect of speed on the rate of clocks according to the theory of relativity.

---

[5]See Around-the-world atomic clocks: Predicted relativistic time gains. Science, *177*:166–168, 1972.

7. Explain the prediction by the theory of relativity concerning distances and lengths of objects.
8. Describe an experiment which verifies the time-change prediction of relativity.
9. Explain the predicted effect of speed on the mass of an object.
10. Explain how Einstein's theory altered the law of conservation of energy.
11. Use the history of the theory of special relativity to illustrate the scientific method.
12. Describe the twin paradox.

QUESTIONS—CHAPTER 23

1. It is said that as a child Einstein asked the question, "What would I see in a mirror if I carried it in my hands and ran at the speed of light?" What is your answer?

2. If the speed of light were 100 mph, discuss how our day-to-day lives would be changed.

3. A prism appears as shown in Figure 23–17 when it is at rest. What would happen to its apparent shape if it were set into motion at a speed approaching that of light?

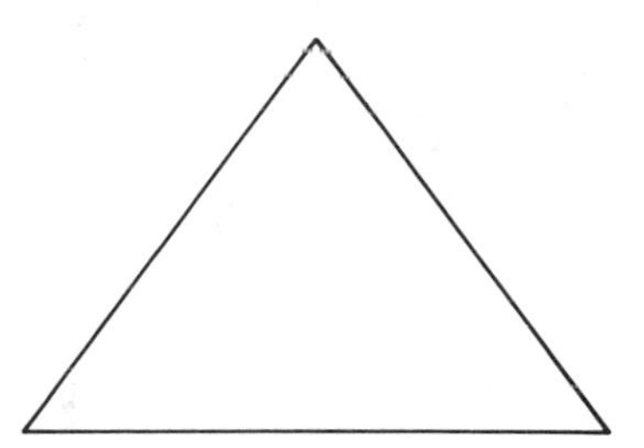

FIGURE 23–17. Prism for Question 3.

4. Suppose an astronaut travels at a speed near the speed of light. Does the astronaut's mass become greater or less while in motion? What will be his reaction when he opens his pay envelope and counts his flight pay?

5. Is the mass of an iron rod greater on a hot day than on a cold day? Why?

6. Which is larger, the mass of a spinning planet or of an otherwise identical but nonspinning planet?

7. If density is defined as mass divided by volume, is it conceivable that density of an object might not change when traveling near the speed of light?

8. A passenger at the rear of a train throws a baseball at 100 mph in the direction opposite to the motion of the train. If the train is traveling at 100 mph, where does the ball hit?

9. A bird flies at 10 mph through wind moving in his direction of travel at 5 mph. Both speeds are relative to the ground. What is the speed of the bird relative to the air? Repeat for the case in which the bird heads into the wind.

10. Atomic clocks are extremely accurate. Why was such accuracy needed for the around-the-world relativity experiment?

11. Why were there no TADI's included concerning special relativity? If you can't answer this question, *you* devise one and let us know.

12. If you studied the refraction of light, you found that light travels more slowly in glass than in air. Does this contradict the theory of relativity?

24

THE NUCLEUS AND RADIOACTIVITY

It was Einstein who derived the famous equation $E = mc^2$, which relates mass-loss in a nuclear reaction to energy-gain. (Reprinted with the permission of American Scientist/Sidney Harris.)

By now the planetary model of the atom (Fig. 24–1) has become quite familiar to us. We have used the model of a nucleus with tiny electrons orbiting it to describe a number of happenings in nature. Thus far, however, we have said almost nothing about the nucleus except that it is in the center, that it has a positive electrical charge, and that it is much more massive than the electrons. Yet it is nuclear transformations which are responsible for the energy of the sun and, therefore, for our life on earth. In this chapter, we will focus our attention on the nucleus, and we will find that it is quite a fascinating little speck.

CHEMICAL REACTIONS

Chlorine is a poisonous gas. Sodium is a metal which in its pure form can burst into flame upon contact with water. Yet the combination of these two elements—sodium chloride—is the common table salt we sprinkle on our eggs,

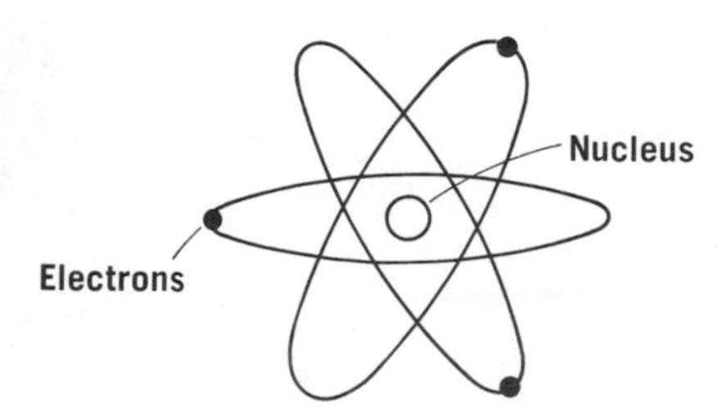

FIGURE 24–1. The planetary model of the atom.

beans, and bananas. Salt is nothing like either of the elements which make it up. Similarly, common water is made up of explosive hydrogen and life-sustaining oxygen. Observations such as these led men for centuries to spend their lives trying to produce gold by combining various other materials. What the alchemists, as these men are called, failed to understand is that all of the reactions they tried were chemical reactions (that is, processes in which two or more elements become bound together or broken apart). It should be obvious to you that no chemical reaction can produce gold (an element) where there was none before. Chemical reactions depend upon forces between entire atoms, and they involve only the electrons of the atoms. For example, an oxygen atom contains six electrons in its outermost orbit. This orbit is filled—completed—when it contains eight electrons. Each hydrogen atom contains one electron, and when two hydrogen atoms combine with one oxygen atom, the two electrons from the hydrogen atoms are shared by the oxygen atom. Thus the oxygen atom, in a sense, fills its outermost orbit by using the electrons of two hydrogen atoms, forming a water molecule (H_2O).

Water, then, is not an element but a substance composed of two elements. Gold, on the other hand, is one of the 92 naturally occurring elements. An atom of gold has 79 electrons. Because electrons are negatively charged and because an atom is normally without an overall electrical charge, the gold nucleus must contain enough positive charge to just compensate this negative charge. It is the protons in the nucleus which do this job; their number determines the nature of the element. Gold has 79 protons in its nucleus, and it keeps these 79 protons even when some of its electrons are temporarily removed from their orbits. Thus the alchemists, in order to produce gold, would have had to alter the nucleus of the atom.

The distinction between chemical processes and nuclear processes is an important one. In a chemical reaction, the nucleus of the atom is not disturbed. When sodium and chlorine combine, the nucleus of each of the atoms is not changed in any way. Such reactions occur between *atoms* and perhaps should have been called atomic reactions. Unfortunately, the term "atomic reactions" is often used interchangeably with the more appropriate term, nuclear reactions, to mean reactions which do involve a change in the nucleus of the atom.

THE DISTINCTION BETWEEN ELEMENTS

In Chapter 6 we considered the question of what makes an element what it is. We found that historically the elements were differentiated by their properties. Hydrogen, for example, is a very light element (the lightest), it is explosive, it reacts with various other elements in predictable ways, and it emits certain colors when heated. Any substance with this given set of properties is hydrogen. And each of the other elements has its own peculiar set of properties.

It was only in later years that it was found that an element's properties depend upon the arrangement of electrons in the atom's orbits and that this arrangement depends upon the number of electrons the neutral atom contains. Hydrogen, for example, has one electron; helium, two; lithium, three; beryllium, four; and so forth. At that point in time, then, one could say that any atom that had four electrons (when it was not electrically charged) was a beryllium atom. Still later it was found that this number of electrons depends upon the number of protons in the nucleus and that the number of protons is equal to the number of electrons. So now we can say that the number of protons in the nucleus of an atom identifies the element for us.

Figure 24–2 is a chart of the elements. The large number above each element's symbol is called the atomic

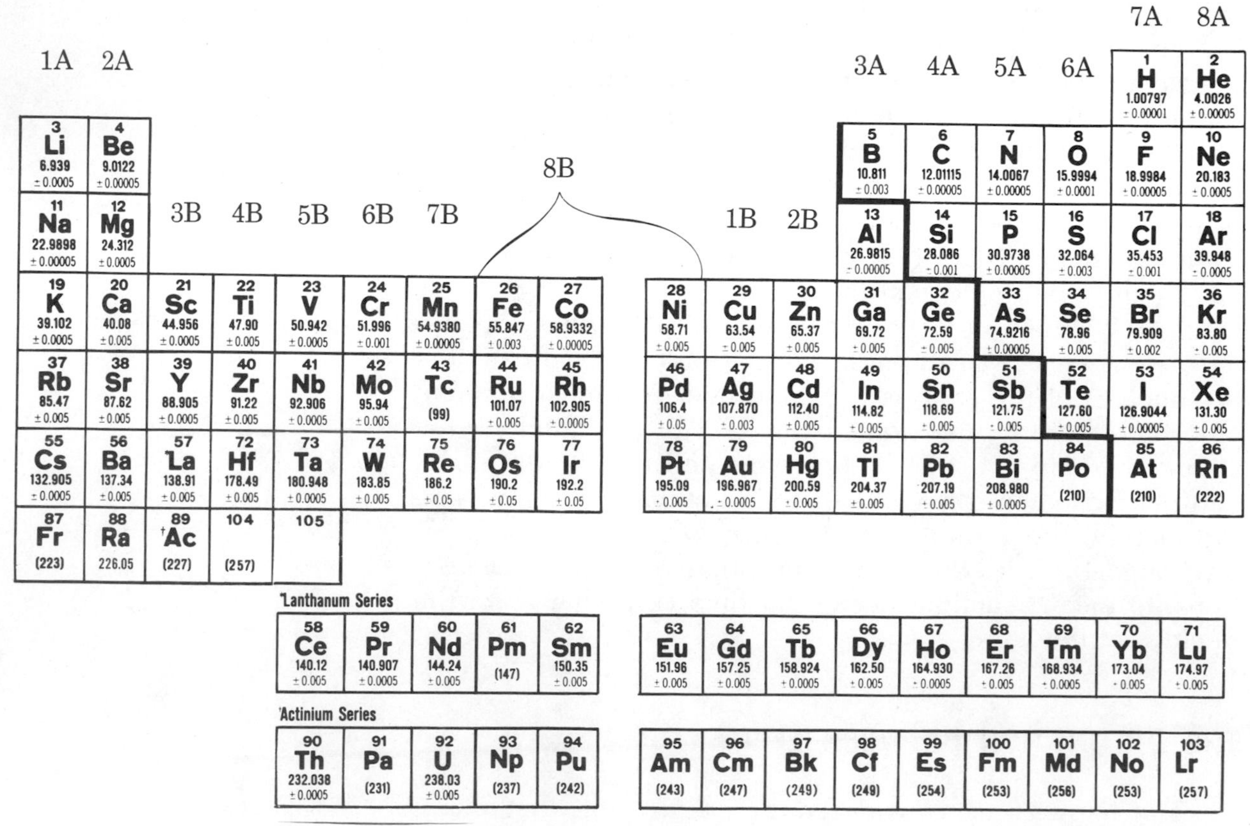

Atomic Weights are based on C^{12}—12.0000 and Conform to the 1961 Values

FIGURE 24–2. Chart of the elements.

number of the element and gives the number of protons in the nucleus of an atom of that element.

THE NUCLEUS

The nucleus contains protons, but that's not all. There is one other particle in there which is of some importance to us: the neutron. It is a neutral particle (that is, it has no electrical charge), and it is slightly more massive than the proton. Table 24–1 shows a comparison of the electron, proton and neutron in their relative mass, their charge, and their location in the atom.

Figure 24–3 shows the construction of the atoms of some elements, starting with the lightest. In the figure, *e* is an electron, *p* is a proton, and *n* a neutron. Note that some elements are pictured as having two forms. Although most hydrogen nuclei (*nōō-klē-ī*, plural of nucleus) contain only a proton, a small percentage of hydrogen atoms also contain a neutron in their nuclei (and an even smaller fraction have two neutrons). Likewise a nucleus of lithium may contain either three or four neutrons. Such differing forms of the same element are called *isotopes* of the element. In fact, it is the rule rather than the exception that there are a number of isotopes for each element. For example, calcium has six stable isotopes. All of these have the same number of protons (20), but various ones have 20, 22, 23, 24, 26, and 28 neutrons.

In Chapter 6, we discussed another number, in addition to the atomic number, that characterizes an element: the atomic weight. The atomic weight gives the relative weight of any atom as compared to a particular isotope of carbon. In Figure 24–2, this number appears below the symbol for the element. On the scale of atomic weights, the carbon isotope that has six protons and six neutrons is chosen as the standard for comparison and is assigned a weight of 12. Obviously, atomic weight reflects mainly the number of protons and neutrons in the nucleus. It might seem that the atomic weight of any element would be a whole number, but a quick glance at the table of elements shows that this is

TABLE 24–1. PARTICLES COMPOSING THE ATOM

Particle	Charge	Mass	Location
Electron	−1	1	Orbit around nucleus
Proton	+1	1836*	In nucleus
Neutron	0	1838	In nucleus

*This is the mass of the proton compared to the electron, whose mass is given as 1. The actual mass of the electron is 9×10^{-28} gram.

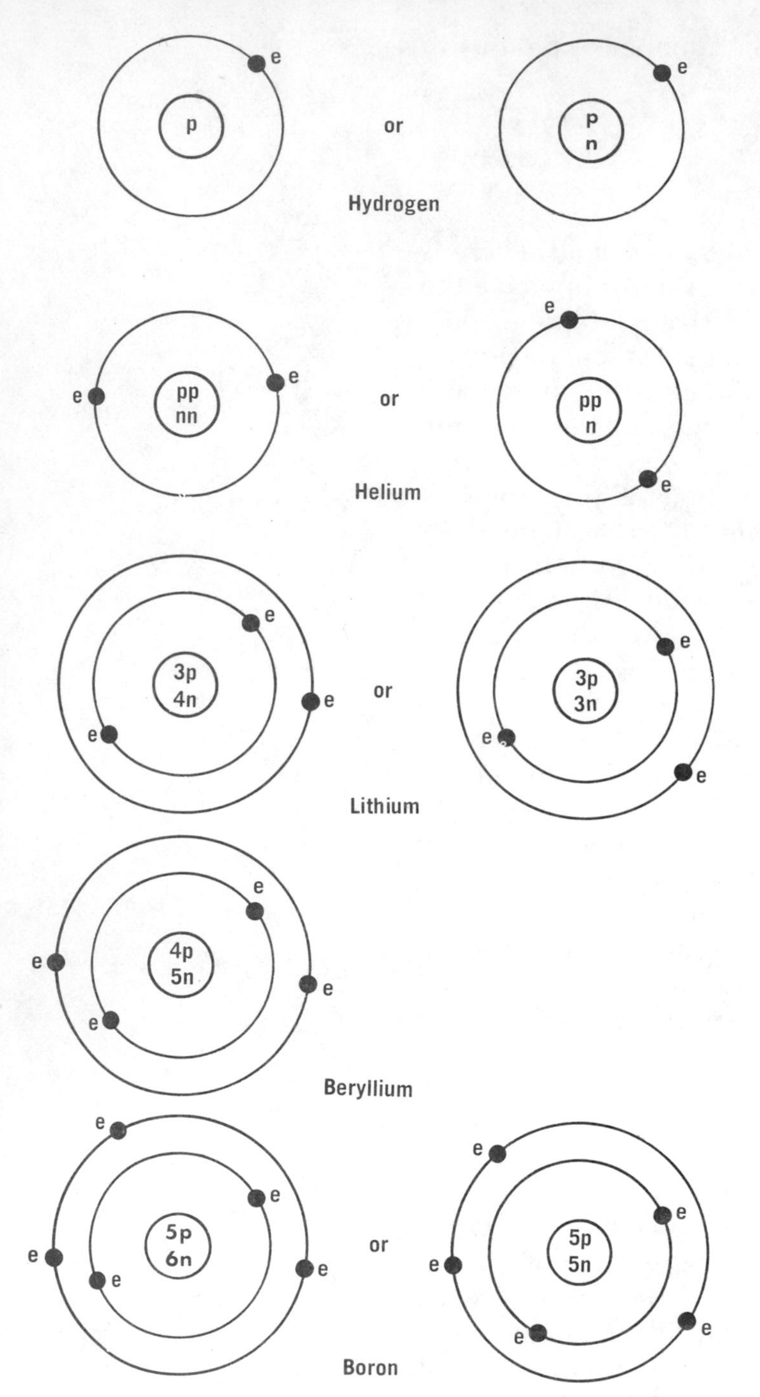

FIGURE 24–3. Note the number and type of nuclear particle for each isotope.

not the case. For example, the atomic weights of hydrogen and neon are 1.008 and 20.183, respectively. There are two reasons for this deviation from whole numbers: (1) When protons and neutrons are assembled to form heavy nuclei, the total mass of the nucleus is not equal to the sum of its parts, and (2) The atomic weight of an element given in the table is the *average* weight for the isotopes of that element.

As an example of (1), consider the formation of carbon by combining three nuclei of the helium isotope that has two protons and two neutrons. Since this combination would have six protons and six neutrons, it seems that the carbon isotope should have exactly three times the weight of the helium nucleus. But it doesn't—it weighs less. Einstein's equation, $E = mc^2$, tells us that the missing mass has been converted into energy, which is used to bind the combination together. Examples of how this energy can be recovered will appear shortly. As an example of (2), consider neon, which has three naturally occurring isotopes. Of these, about 90.92 per cent have a weight of 19.99, 0.25 per cent have a weight of 20.99, and 8.83 per cent have a weight of 21.99. The average weight (taking the percentages into account) is 20.183, which is the number given in the table of elements.

How Do We Know There Is a Nucleus?

Let's go no further before we examine the experimental background for believing that there is such a thing as a nucleus at all.

At the turn of the century, there was definite evidence that atoms exist (based in part upon the way chemicals always combine in definite proportions), but little was known of the make-up of the atom except that it contained small chunks of negative charge called electrons. One idea was that the atom was like a glob of jelly with the electrons scattered through it as fruit is scattered through fruit Jello (Fig. 24–4).

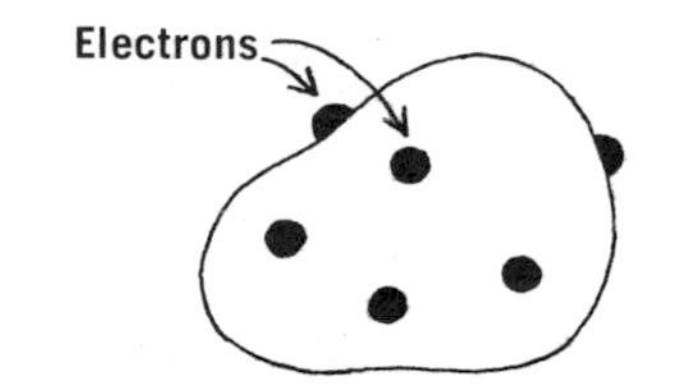

FIGURE 24–4. The "fruit Jello" model of the atom.

An important experiment, for which credit is generally given to Lord Ernest Rutherford (1871–1937), was performed in 1911, and from it the present idea of the nuclear atom developed. As we will see later, one of the particles resulting from radioactivity is called the alpha (α) particle.[1] Experiments were done with this particle of radiation by letting it pass through thin sheets of various materials and observing the deflection from its path as it went through the material. The following excerpt is from an essay by Rutherford in which he tells of the crucial experiment performed by a colleague and one of his students.

> In the early days I had observed the scattering of alpha particles, and Dr. Geiger in my tiny laboratory had examined it in detail. He found, in thin pieces of heavy metal, that the scattering was usually small, of the order of one degree. One day Geiger came to me and said, "Don't you think that young Marsden, whom I am training in radioactive methods, ought to begin a small research?" Now, I had thought that, too, so I said,

[1]The first letter of the Greek alphabet is alpha.

Ernest Rutherford (1871-1937)

This British physicist during his 66 years made not one, not two, but many major contributions to our knowledge about the world. Among his accomplishments were the formulation of the theory of radioactive decay, identification of alpha and beta particles and gamma rays, discovery of the atomic nucleus, and production of the first artificial transmutation of one element into another. He was an excellent teacher, and although quick-tempered, he had the habit (when things were going smoothly) of walking through the laboratory singing "Onward Christian Soldiers." When a physicist once said to Rutherford, "You are a lucky man... always on the crest of a wave!" he answered, "Well, I made the wave, didn't I?"

"Why not let him see if any alpha particles can be scattered through a large angle?" I may tell you in confidence that I did not believe that they would be, since we knew that the alpha particle was a very fast, massive particle, with a great deal of energy, and you could show that if the scattering was due to the accumulated effect of a number of small scatterings the chance of an alpha particle's being scattered backwards was very small. Then I remember two or three days later Geiger coming to me in great excitement and saying, "We have been able to get some of the alpha particles coming backwards...." It was quite the most incredible event that has ever happened to me in my life. It was almost as incredible as if you fired a 15-inch shell at a piece of tissue paper and it came back and hit you. On consideration, I realized that this scattering backwards must be the result of a single collision, and when I made calculations I saw that it was impossible to get anything of that order of magnitude unless you took a system in which the greater part of the mass of the atom was concentrated in a minute nucleus. It was then that I had the idea of an atom with a minute massive center carrying a charge.[2]

This scattering experiment indicated that not only is the greater part of the mass of the atom concentrated in the nucleus but also (from the way that the alpha particles were scattered) that the nucleus is charged *positively*. As a result, it was concluded that the positive charge in an atom is concentrated in a massive nucleus. This, obviously, was a death knell for the fruit Jello model of the atom.

take-along-do-it

Have a friend place either a saucer, a plate, a jar lid, or a few stacked coins under a large book, but don't let him tell you which of these he used (see Figure 24-5). The edges of the book should be high enough above the table so that you can roll marbles under it, but not so high that you can easily see what is underneath. Now without looking under the book, roll the marbles at the hidden object. By observing the results you should be able to determine the following: (1) that the object under the book is solid and fixed in position (as if you didn't already know this), and (2) the approximate size of the object. You should be able to decide whether it is a plate, saucer, jar lid, or stack of coins.

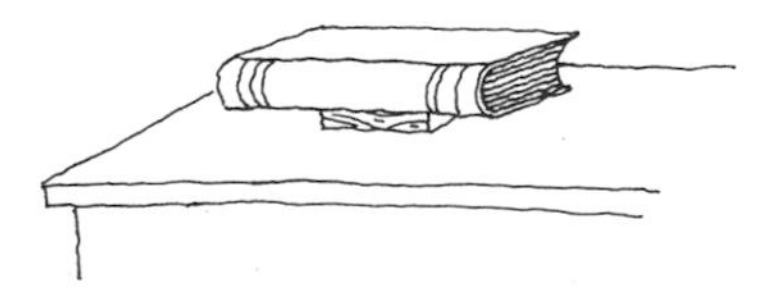

FIGURE 24–5. By rolling marbles, you can learn about what's under the book.

In a manner analogous to your TADI, further scattering experiments similar to those of Rutherford gave us the first ideas of the size of the nucleus. They showed that it is incredibly tiny, having a radius of about 10^{-15} meter. This is about 100,000 times less than the radius of an atom. Most of an atom, then, is actually empty space. Into this small nuclear speck is packed the mass of all the protons and neutrons. The result is nuclei having densities of about 10^9 tons per cubic inch! To illustrate just how small the nucleus is, suppose that it is blown up to the size of a BB (about 1/8 inch across). This BB would then be in the center of an atom the size of a football stadium. Look for the BB in Figure 24–6.

[2]From the essay (1936) "The Development of the Theory of Atomic Structure," by Lord Rutherford. In Needham, J. (ed.): *Background to Modern Science.* New York, Macmillan Company, 1940.

FIGURE 24–6. Find the BB in the stadium. (Courtesy of the Rose Bowl.)

Symbols for Nuclei

It will be convenient for us to have a symbolic way to represent nuclei and to show how many protons and neutrons are contained in an isotope under discussion. The following examples illustrate our method:

$^{1}_{1}$H This represents the hydrogen (H) nucleus, with its nuclear charge of +1 (the 1 at lower left). This nucleus, then, contains one proton and no neutrons, as shown in Figure 24–7*A*.

$^{2}_{1}$H The 2 in the upper left means that there are two nuclear particles, and since the lower left number tells us that one of these is a proton, there must be one neutron (Fig. 24–7*B*).

$^{14}_{6}$C Carbon, six positive charges (protons), eight (14 minus 6) neutrons (Fig. 24–7*C*).

Often we refer to an isotope by naming the element and stating the number of nuclear particles it contains. Thus the isotopes above are called "hydrogen-1," "hydrogen-2,"[3] and "carbon-14," respectively.

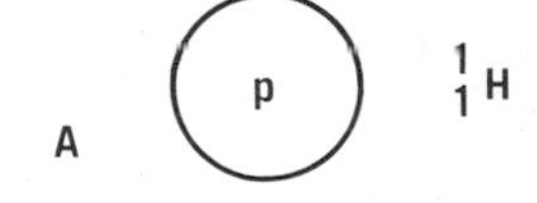

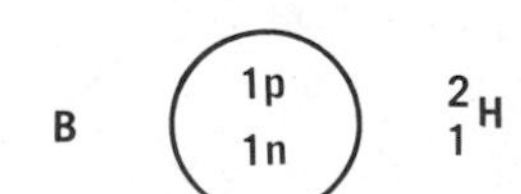

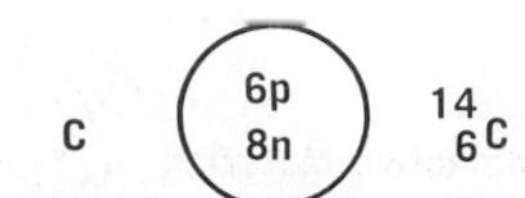

FIGURE 24–7. Three nuclei, showing what the symbols mean.

[3]Hydrogen-2 is more commonly referred to as deuterium. Hydrogen-3 is called tritium.

RADIOACTIVITY

Now that we have the vocabulary lessons out of the way, we can consider the nature of radioactivity. If you held a small amount of a radioactive substance, say bismuth-210, in your hand, you would have no way of knowing that it is radioactive. It would not be hot to the touch, nor would it glow or sparkle—that stuff is just for the movies. In fact, a real danger of radioactive substances lies in the fact that radiation is not perceived by the senses at all. You can't tell that a substance is radioactive by looking, feeling, tasting (God forbid!), or by using any other of the senses. But if you hold the bismuth-210 near a package of photographic film for a while (Fig. 24–8) and then develop the film, you will find that the film is exposed, just as if it had been exposed to light. Apparently something comes from the isotope, passes through the box that contains the film, and exposes the film. This "something" is the radiation from the radioactive substance.

Let's continue the imaginary experiment by placing some bobby pins on a fresh piece of film and again exposing it by holding the bismuth-210 above the pins (Fig. 24–9). Figure 24–10 was made in just that way. The picture is dark everywhere except for an outline of the metal objects. They blocked out the radiation and produced a shadow on the film. This is a characteristic of radiation: it can be shielded, or blocked off.

Types of Radiation

We can find out additional information about radiation by taking advantage of the fact that it can be shielded. We do this by putting the bismuth-210 into a heavy metal box with a single hole in it (Fig. 24–11) and placing a piece of film above the box. The shielding in this case would result in only a spot on the center of the film being exposed (because radiation leaving the box in other directions is blocked off). We arrange two metal plates, one on each side of the beam coming out of the box (Fig. 24–12*A*) and connect the plates

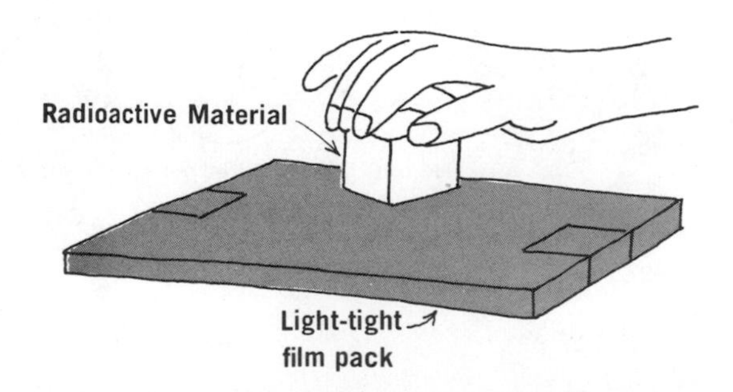

FIGURE 24–8. A radioactive material will cause fogging of photographic film.

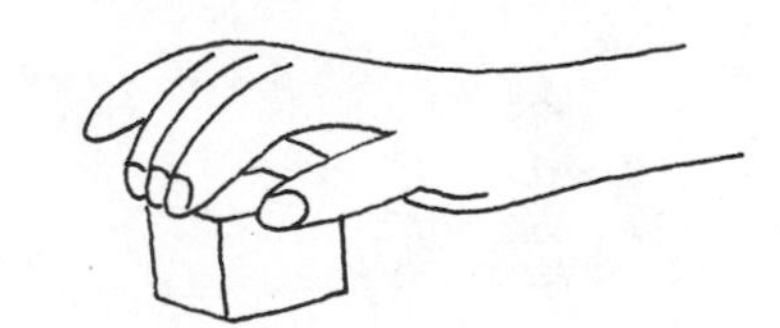

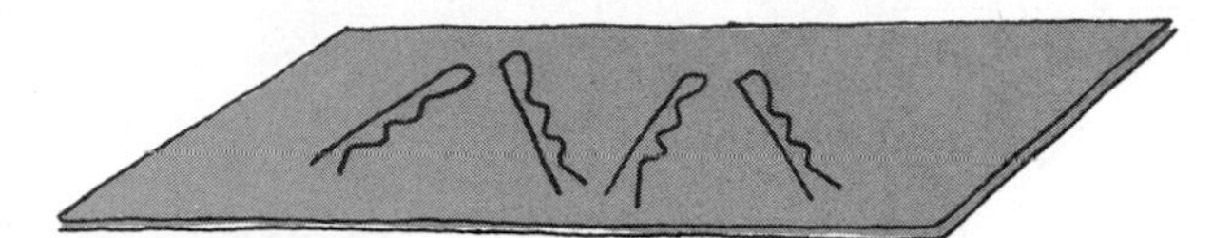

FIGURE 24–9. Bobby pins have been placed on the film below the radiation source.

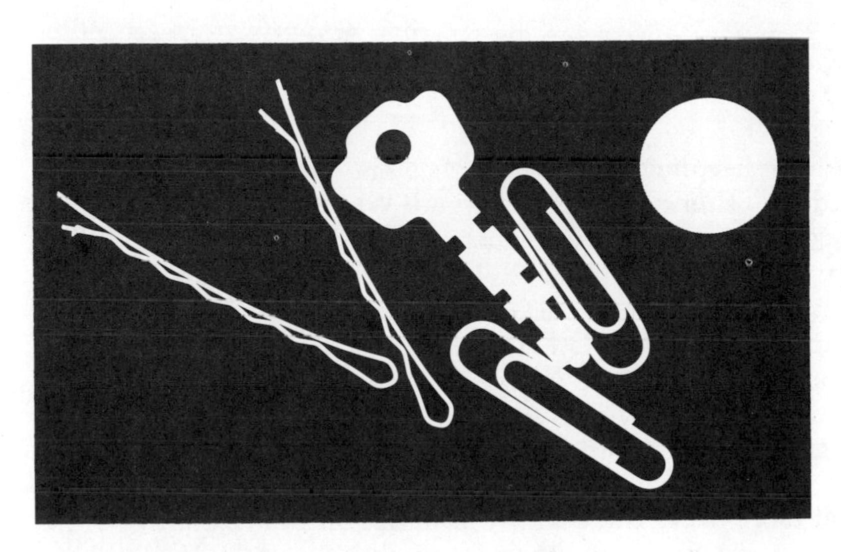

FIGURE 24–10. Radiation is stopped by the metal objects, and thus their shadows appear on the photographic film.

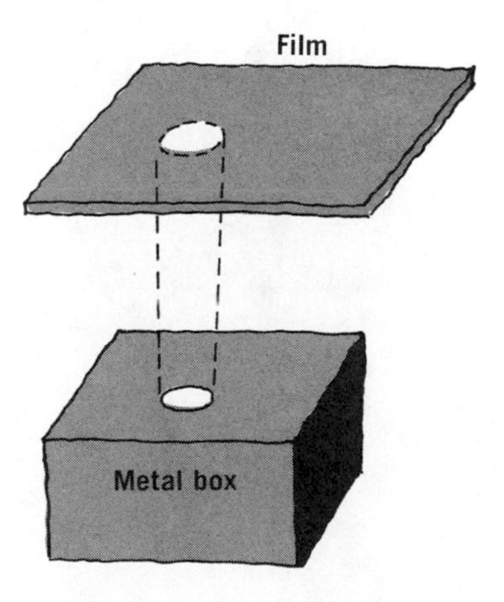

FIGURE 24–11. The Bi-210 is below the hole in the metal box.

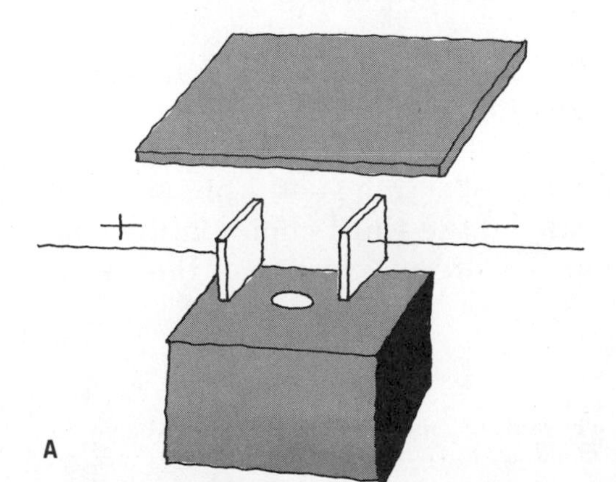

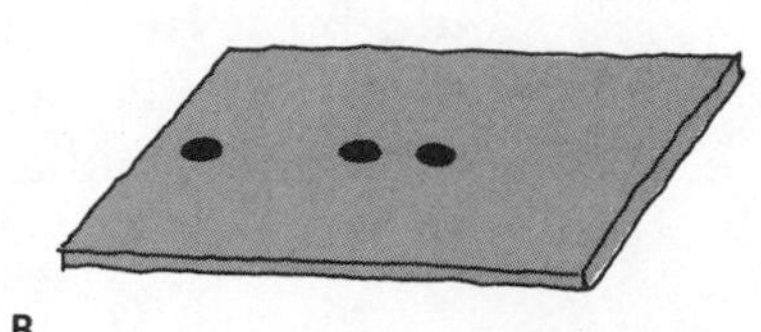

FIGURE 24–12. *A,* Set-up for analyzing radiation; *B,* The result.

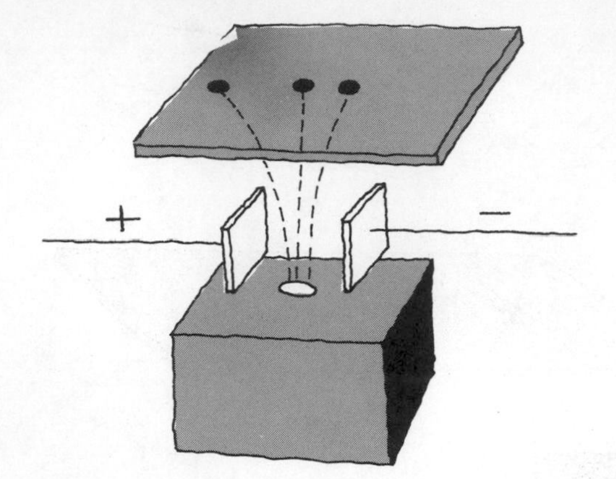

FIGURE 24–13. The radiation was bent as shown, resulting in Figure 24–12B.

to a high-voltage battery so that one is positively charged and the other negatively charged. When the film is exposed, it comes out with *three* exposed spots (Fig. 24–12*B*).[4]

Figure 24–13 shows what must have happened to the beam. Note that one part of the beam is deflected slightly toward the negative plate, which means that this type radiation must carry a positive charge. Another part bends quite a bit toward the positive plate; this radiation must be made up of negatively charged particles. The part going straight through the hole apparently has no charge. When it was first found that there were three kinds of radiation being emitted from radioactive materials, they were christened alpha (α), beta (β), and gamma (γ) rays, owing to a lack of knowledge of their true nature.

Consider the possible causes of the negative beam's bending more than the positive beam. Three possibilities present themselves:

(1) Perhaps each chunk of negative radiation is more highly charged than each chunk of positive radiation. This greater charge would cause a greater force on it and would result in more bending; and/or

(2) Perhaps the positive beam is moving faster than the negative beam. If this were the case, the electrical force would be less effective in deflecting the faster moving chunk; and/or

(3) Perhaps the positive beam was bent less because each chunk of the radiation is more massive. If this were the case, the electrical force exerted on the chunk would be less effective in deflecting such a big particle.

In fact, experiments have shown the third choice to be correct. The positive radiation is more massive than the negative.

---

[4]In order for this demonstration to work, the apparatus would have to be put in a vacuum, because some particles would not make it all the way through the air to the film.

With further experimentation involving placement of objects in the path of each of the beams to measure their penetrating power, we come to the conclusions of Table 24–2 concerning the three kinds of radiation. The table shows that none of the three radiations is new to us. The positive radiation is simply two protons and two neutrons bundled together—the same as the nucleus of helium. Although it has twice the amount of charge of the beta particle, because of its comparatively great mass the alpha particle deflected only slightly.

The neutral radiation is simply electromagnetic radiation, similar to radio waves, light, and ultraviolet rays but of greater energy than any of these. This accounts for its great penetrating power.

The negative radiation turns out to be ordinary electrons with their very small mass. Typical beta radiation will penetrate a book of the thickness of this one but will be stopped by greater thicknesses.

HOW RADIATION IS EMITTED

Gamma Radiation

Let's investigate each type of radiation and consider what happens when it is emitted. Gamma radiation is the easiest to explain, so we'll examine it first. Gamma rays are given off by a nucleus as the nucleus goes from a high-energy state to one of lower energy. If you have previously studied the emission of light from atoms (Chapter 17), you should recognize the terminology. Just as an atom emits a photon of light energy in order to get rid of extra energy, the nucleus uses the same method to dispose of any extra energy that it may have acquired as a result of some nuclear event. But since much more energy is generally involved in nuclear processes than in processes involving electron orbits, the photon emitted by the nucleus has more energy than a photon of visible light. Other than this loss of energy

TABLE 24–2. PROPERTIES OF ALPHA, BETA, AND GAMMA RAYS

Particle	Charge	Mass	Description	Penetrating Power
alpha	+2	7290	Helium nucleus	a piece of paper or a few inches of air
beta	−1	1	An electron	a sheet of steel
gamma	0	0	High energy electromagnetic wave	many feet of concrete

and a slight recoil,[5] emission of a gamma ray causes no change in the nucleus.

Alpha Particle Emission

Alpha emission is more interesting than gamma emission. Suppose you start with a small amount of pure uranium-238. You perform experiments to identify the radiation and find that alpha particles are emitted from this isotope. If you later do a chemical test of your sample (which originally was pure uranium), you will find that some *thorium* is present in the sample.[6] Alpha particles came out, and some of the uranium changed to thorium (Fig. 24–14). This can best be represented as a formula:

$$^{238}_{92}\mathrm{U} \rightarrow {}^{234}_{90}\mathrm{Th} + {}^{4}_{2}\mathrm{He}$$

This formula is a shorthand way of noting the change which occurs. The arrow indicates that the U-238 nucleus "decays"[7] into a Th-234 nucleus and an alpha particle. When this happens, the alpha particle is fired out of its original home. Note the following about the formula: (1) The total amount of charge on each side of the arrow is the same: $92 = 90 + 2$. (2) The total number of nuclear particles on each side of the arrow is the same: $238 = 234 + 4$. Knowing these facts, and knowing the number of protons and neutrons, you should be able to complete the following reaction:[8]

$$^{230}_{90}\mathrm{Th} \rightarrow {}^{4}_{2}\mathrm{He} + ?$$

---

[5]Recall the law of conservation of momentum from Chapter 2. It applies here too.

[6]Thorium, although it is about as abundant on earth as lead and more abundant than uranium, is not a household word. Pure thorium is a silvery-white metal which burns brilliantly when heated in air.

[7]The term "decay," when used in nuclear physics, has nothing to do with rot or corrosion but simply refers to a change in the nucleus produced by radioactivity.

[8]Answer: $^{226}_{88}\mathrm{Ra}$. You can calculate the two numbers from the others in the formula and then look for the symbol for element 88 in the table of elements. Ra is the symbol for radium.

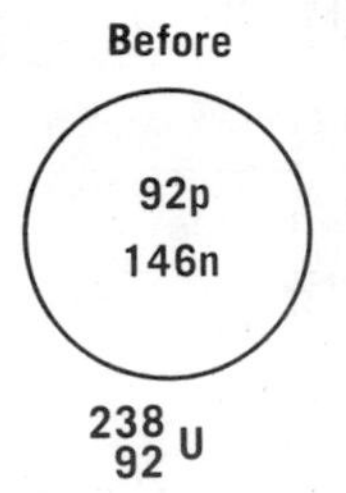

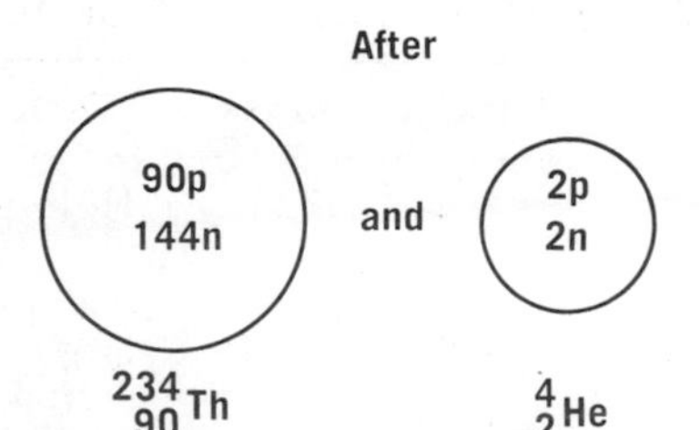

FIGURE 24–14. Uranium-238 emits an alpha particle, becoming thorium-234.

Alpha emission is the first example that we have seen of one element's changing into another. The reaction you completed above was actually a case of thorium decaying to radium. Later we will see more practical applications of this phenomenon.

Beta Particle Emission

The emission of beta radiation is the unusual one. Perhaps you noticed a paradox in an earlier discussion: having stated that the nucleus is composed of protons and neutrons, we indicated that electrons are emitted from the nucleus. And we were right in both cases: there are no electrons in the nucleus, but electrons *do* come out. This is possible because an electron can be created in the nucleus, and when this happens it is immediately ejected as a beta particle. The creation of the electron occurs when a neutron in the nucleus changes into a proton and an electron.[9] We can write this as a formula:

$$^{1}_{0}\mathrm{n} \rightarrow {}^{1}_{1}\mathrm{H} + {}^{0}_{-1}\mathrm{e}$$

The numbers written beside the neutron and proton (the hydrogen-1 nucleus is a proton) should come as no surprise, but perhaps the −1 and the 0 on the electron seem strange. The −1 simply indicates that the electron has a charge of negative one. The electron's mass is almost zero compared to the proton and the neutron, and it is not a nuclear particle; hence the 0. Note that again the charges on both sides of the arrow total zero and that the number of nuclear particles totals 1 on each side.

Platinum-197 is an example of an isotope which emits a beta particle. The formula for the event is as follows:[10]

$$^{197}_{78}\mathrm{Pt} \rightarrow {}^{0}_{-1}\mathrm{e} + {}^{197}_{79}\mathrm{Au}$$

Figure 24–15 represents this reaction. Note that the number of protons in the gold (Au) nucleus remaining after beta emission is one *greater* than in the original nucleus but that the number of nuclear particles stays the same. This should be no surprise when you remember that within the nucleus a neutron changed to a proton and an electron. The electron

[9]Verification that this reaction is more than mere hypothesis comes from observation of neutrons that have been freed from the nucleus. Such neutrons decay in this manner.

[10]Actually, every time an electron is emitted in a radioactive decay process, another particle, called a neutrino (or more strictly speaking, an antineutrino), is emitted with it. This particle is virtually undetectable, and we will ignore it for the present. We will examine its properties, however, in Chapter 27.

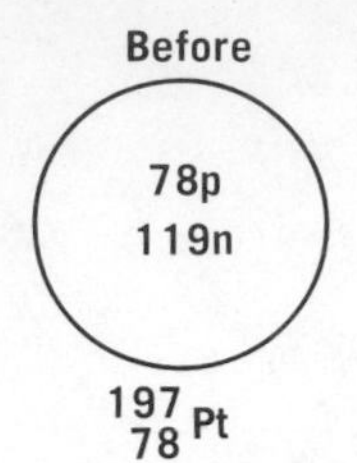

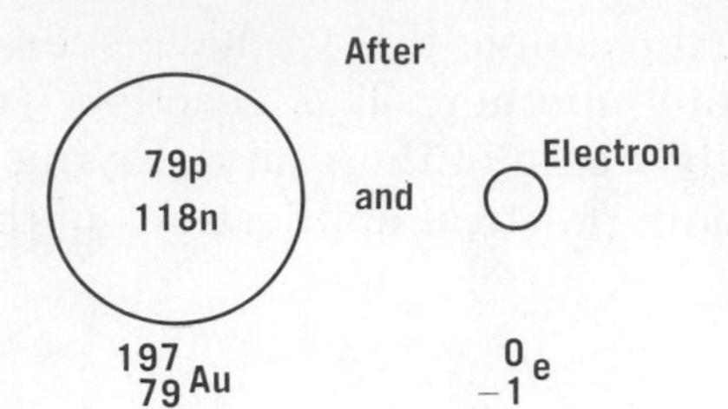

FIGURE 24–15. Platinum-197, upon emitting an electron (a beta particle), becomes gold-197.

comes barreling out and leaves behind a changed nucleus. Again, if you started with a quantity of pure platinum-197, you would eventually find some gold present. The alchemists' dream! The trouble is that platinum-197 does not occur in nature. We will show later how it is formed.

HALF-LIFE

Wherever you see a discussion of radioactive substances, you see the term "half-life." The meaning of this term is best explained with an example.

Suppose you have some pure sulfur-37 (which is a beta emitter that changes to chlorine-37 when the beta particle comes off). And suppose you are able to count the number of beta particles being emitted at any time and the number of sulfur nuclei in your sample at any time.[11] At exactly three o'clock you count the sulfur nuclei, of which there are exactly 5400 billion. During the next minute you count the number of beta particles emitted and obtain 700 billion. Thus at 3:01 you would have 4700 billion sulfur nuclei left (and 700 billion chlorine nuclei). You continue counting beta particles emitted and sulfur nuclei remaining, and you obtain the results shown in Table 24–3. The table also shows the number of chlorine atoms present at the end of each minute.

You probably skipped that table. It's natural to skip big lists of numbers, but in this case it's a mistake. The table has a lot to say. The first column is just the time of day, but notice the second column. It tells how many sulfur nuclei are present each minute. The number decreases continuously because the emission of each beta particle corresponds to the decay of one sulfur nucleus into one chlorine nucleus. But note how the numbers decrease—it is not a "regular" pattern.

Now look at the third column, the number of beta particles given off each minute. This number keeps decreasing

[11]The first of these measurements is possible with some accuracy; the second is not.

also. The reason is simple: fewer beta particles are emitted because fewer sulfur atoms are there to emit them.

Back to the second column. After five minutes the number of sulfur nuclei present is exactly half of the original number. And if you look at the number remaining after ten minutes, you see that half of that 2700 billion is left. In fact, if you choose any time on the chart, note the number of sulfur nuclei at that time, and compare it to the value listed for five minutes later, you will see that only half that number remains. (Compare the 3:02 figure with the figure for 3:07, for example.)

Based on the above data, we say that the half-life of sulfur-37 is five minutes. *The half-life of an isotope is defined to be the amount of time required for one half of the nuclei to undergo radioactive decay.* Half-lives of radioactive substances vary from very long times, such as 4½ billion years for uranium-238, to 10^{-21} second for lithium-5 ($^{5}_{3}Li$).

Surprisingly enough, there is absolutely nothing that one can do to change the half-life of a radioactive substance. Chemical reactions, temperature changes, pressure—you name it. We stated in an earlier section that chemical reactions occur without any participation by nuclei. Here we have the reverse situation: ordinary physical and chemical changes are not powerful enough to alter nuclear proceedings in any way.

TABLE 24-3. HALF-LIFE DATA

TIME	NUMBER OF $^{37}_{16}S$ NUCLEI EXISTING (IN BILLIONS)	NUMBER OF β'S EMITTED DURING THE MINUTE (IN BILLIONS)	NUMBER OF $^{37}_{17}Cl$ NUCLEI EXISTING (IN BILLIONS)
3:00	5400		0
		700	
3:01	4700		700
		610	
3:02	4090		1310
		530	
3:03	3560		1840
		460	
3:04	3100		2300
		400	
3:05	2700		2700
		350	
3:06	2350		3050
		305	
3:07	2045		3355
		265	
3:08	1780		3620
		230	
3:09	1550		3850
		200	
3:10	1350		4050

A Graph of the Data

It is often convenient to represent numerical data on a graph. In the case of half-life data, graphs are also instructive. In Figure 24–16, time is plotted on the horizontal axis, and the number of beta particles emitted during the previous minute is on the vertical axis. Thus if you compare Table 24–3 with the graph, you will find a point on the graph for each pair of figures in the first and third columns of the table.

A smooth line is drawn through the points, indicating that as time goes by, fewer beta particles are emitted each minute. But it shows more. Note the pattern of the line: as time goes by, it gets closer and closer to the bottom of the chart (the point of zero beta particles). It also flattens out more all the time. Will it ever get to the zero point? To answer this question, consider the following exercise. A person standing against one wall of a room walks halfway to a door on the opposite wall. After stopping and noting his position, he walks half the remaining distance to the door and stops. Then he does it again (see Fig. 24–17). Does he ever reach the door? Your answer may be that technically he does not but that for all practical purposes, after 10 or 20 "half-walks," he will be there. The answer is the same for radioactive decay: after 10 or 20 half-lives have passed (an hour or two for sulfur-37), the radioactivity of a substance is practically zero.

Although the half-life of sulfur-37 is indeed five minutes, the data in Table 24–3 are, of course, phony. First, if

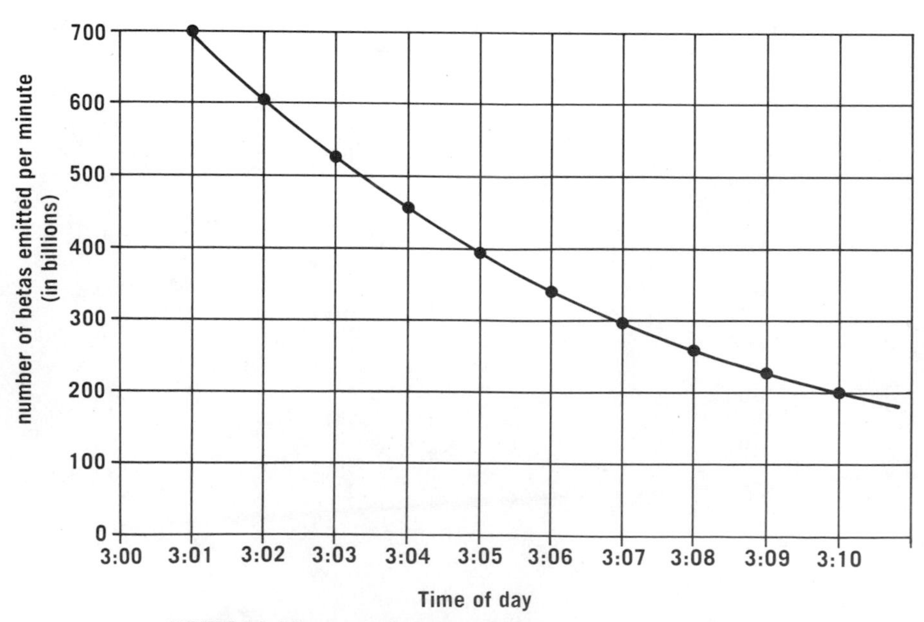

FIGURE 24–16. A graph showing the rate of radioactive decay.

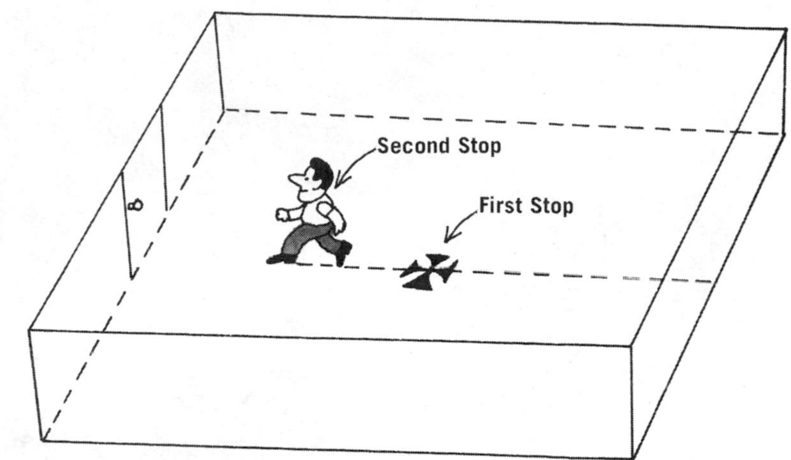

FIGURE 24–17. Imagine walking halfway to the door, stopping, walking half the remaining distance, and so on. . . . Do you reach the door?

you actually recorded such data you would not obtain numbers rounded off to billions. Secondly, these numbers were picked so that after each five minutes, they decreased by precisely one half. Actual measurements are not so exact. And thirdly, we cannot in practice count the number of nuclei present in a sample. How, then, can one determine half-life? Examine Column 3 more closely. Note that the same pattern holds here: if 610 billion beta particles are released between 3:01 and 3:02, half that number will be released between 3:06 and 3:07 (five minutes later). And again this is reasonable, because if only half as much radioactive sulfur is present at the later time, only half as much radiation will be emitted. We will see later that it is fairly easy to determine the point at which the amount of radiation from a substance decreases by one-half, and this information allows us to determine experimentally the half-life of the substance.

A Half-life Analogy

take-along-do-it

This TADI takes more time than most, but the time spent is worthwhile to help you understand the concept of half-life. To do this one you need a box of sugar cubes. With a pencil make a mark on one side of each of about 200 sugar cubes. Each of these cubes will represent the nucleus of a radioactive substance.

Now put the 200 marked cubes in a box and roll them out on a table, just as you would roll dice. Next, count and remove any cubes that fell marked-side up. These cubes represent nuclei which emitted radiation during the roll. They are no longer radioactive and thus do not participate in the rest of the action. Replace them with unmarked cubes. Unmarked cubes cannot land marked-side up and therefore cannot "decay." They represent stable nonradioactive nuclei.

Roll the mixture of "radioactive" and "nonradioactive" cubes again, and again count and remove the ones that decayed. Replace them with "stable" cubes. Continue rolling, counting, and replacing until you have completed 12 to 15 rolls. By then you should have only a few radioactive cubes left.

FIGURE 24–18. This is physics?

We are serious about your actually doing this TADI. Doing is better than reading. But we realize that perhaps one in 10,000 students won't do it, so we have supplied a table of data from one of our own trials (Table 24–4).

Note that the number of "radiations" tends to decrease with each roll. Why? (This is the same reason that real radiation decreases as time goes by.) About how many rolls were made before the number of radiations decreased to half the original number? (This corresponds to the half-life of a radioactive isotope.)

Because one side out of six on each cube was marked, you may have expected one sixth of the cubes on each roll to

TABLE 24–4. DATA FOR A DICE-ROLLING EXPERIMENT

Roll	Number of "Radiations"	Number of Cubes Remaining "Radioactive"
Before first		200
1	32	168
2	26	142
3	22	120
4	18	102
5	17	85
6	14	71
7	13	58
8	10	48
9	7	41
10	6	35
11	6	29
12	3	26

"decay." Why did it seldom (if ever) turn out to be exactly this value? Was it *almost* one sixth, at least while you still had quite a few radioactive cubes? This corresponds to the fact that radiation seems to happen *by chance*, at least as far as we can tell. We are unable to say at what moment a particular nucleus will emit its radiation, but we can make statements about the *probability* of emission, just as we can say that there is one chance in six of a cube's decaying in one roll.

Figure 24–19 is a graph of the data from this experiment. The line drawn shows the *expected* results. Note that the actual data do not always fall on the line. This is especially true at the lower end of the curve. When only a few occurrences are involved, prediction on the basis of probability is risky. After the eleventh roll, 29 radioactive cubes were left. One would have expected one sixth of these, or about five, to land marked-side up. But only three did. In the case of real nuclei, however, very large numbers are involved, and in these cases, prediction can be more accurate.

DETECTING RADIATION

Detection of radiation by allowing it to expose photographic film was one of the first methods for establishing the presence of these elusive rays. We've come a long way since then. Most of the newer methods depend upon the fact that as radiation passes through a gas (or any material) it ionizes some of the atoms of the gas (that is, it knocks out

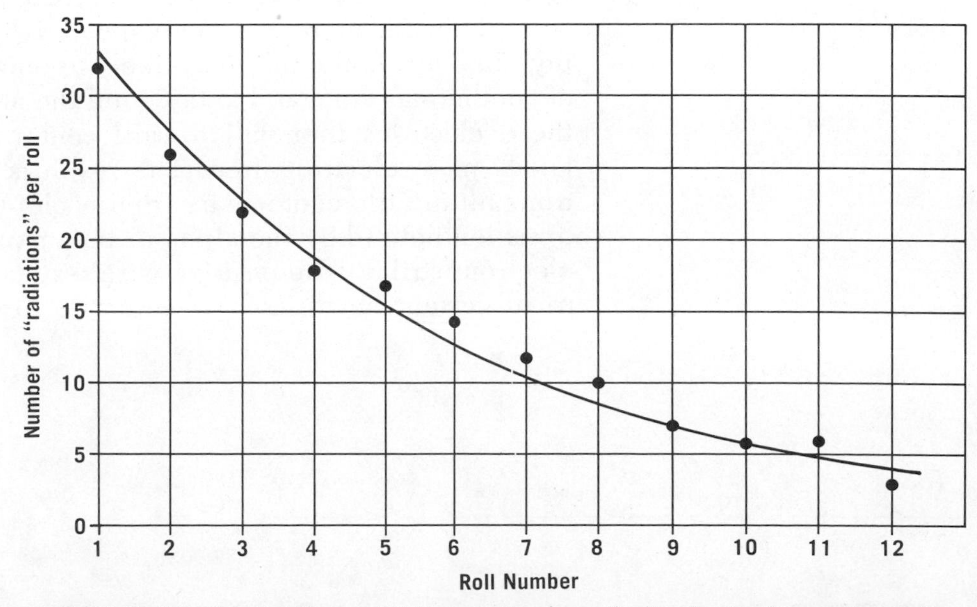

FIGURE 24–19. Note that the actual data are scattered near the line showing the *expected* results.

electrons from the atoms). To understand this effect, suppose that a beta particle, in passing through a gas, comes close to one of the electrons in an atom of the gas. Since both electrons have a negative charge, they will repel one another, and the fast moving beta particle may cause the electron to be knocked out of the atom. Thus in its trail, a beta particle leaves behind a line of free electrons and positively charged atoms (Fig. 24–20), and it is detection of this trail which allows us to conclude that a particle of radiation has passed by. In each of these collisions with electrons, the alpha or beta particle loses a slight amount of energy, slowing down a little bit, and finally the particle comes to rest.

The Geiger Counter

Figure 24–21 is a diagram of a Geiger tube. It consists of a metal tube with a wire down the center. The wire is held at one end by the glass or plastic that covers the entire tube except for the other end, which is covered by a very thin substance. Through this thin "window" the radiation we desire to detect is allowed to pass.

The Geiger tube is connected to a high-voltage electrical source, with the center wire connected to the positive and the outer metal tube to the negative terminal. The voltage between the two is typically 1000 volts, just slightly below the amount required to cause a spark to jump between them. The tube is filled with a gas (such as argon). When an alpha or beta particle passes through the gas, it leaves electrons and positive ions in its path. These charged particles respond to the pull of the high voltage: the electrons are pulled toward the center wire and the positive atoms toward the outer tube. As each electron heads for center, it gains more and more speed. It is, in a sense, "falling" in. During the fall, it is likely to encounter an electron of another gas atom and knock it off the atom, and as both of these electrons then fall toward center, each is likely to knock more electrons free. The result is a shower of electrons hitting the center wire from each of the original electrons left behind by the alpha or beta particle. This burst of electrons striking the positive wire results in a pulse of electrical current heading down the wire toward the electrical

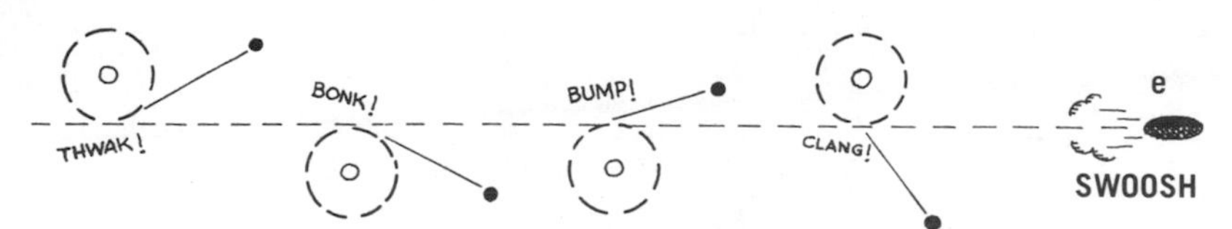

FIGURE 24–20. The beta particle dislodges electrons from atoms in its path.

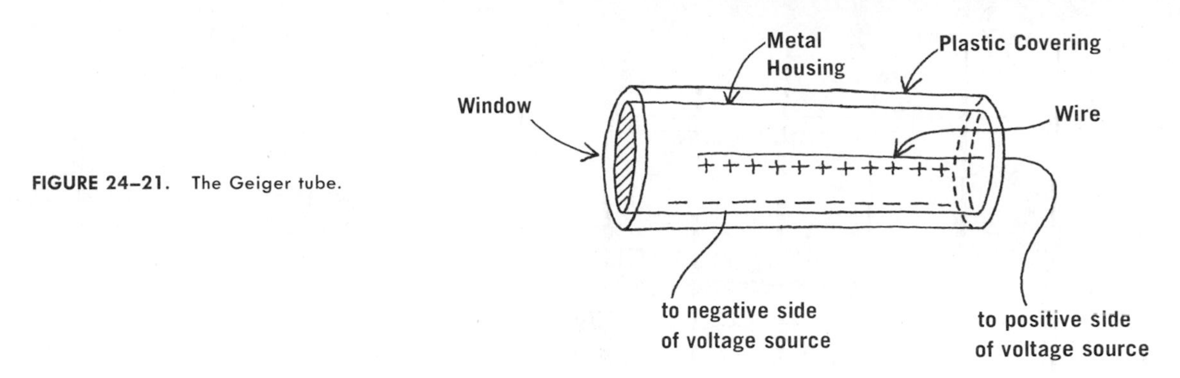

FIGURE 24–21. The Geiger tube.

power supply,[12] which contains an apparatus to detect and record pulses of current.

One method of recording the electrical pulse is simply to amplify it and send it on to a speaker, which then emits a single "click." This is the method used in Geiger counters designed to indicate roughly the amount of radiation present. Each alpha or beta particle passing through the tube results in a click. So "click, click, click, click, click" means that there is more radiation than does "click... click ... click."

A more sophisticated system is used when it is necessary to count the particles passing through the tube. In this case the clicking mechanism is replaced by a device that counts the pulses of current and hence the number of incoming particles. It is by the use of such a counting system that we are able to obtain half-life data such as were presented earlier in the chapter.

A gamma ray also knocks electrons from atoms but may travel a great distance before encountering an electron "head on," as it must to knock it free. Thus most gamma rays pass completely through the Geiger tube without being detected. Other methods are used when efficient gamma detection is desired.

The Cloud Chamber

Historically one of the most important devices used to detect alpha and beta particles was the cloud chamber devised by C.T.R. Wilson in 1912. A latter-day version of the chamber is shown in Figure 24–22. It is made by placing a metal pan, filled with alcohol, on a slab of dry ice.[13] The alcohol is drawn upward by a blotter to the upper edge of the chamber, where some of it evaporates and settles downward as it cools. A radioactive source, placed as shown,

[12]Meanwhile the positively charged nuclei go to the negative cylinder, where they pick up electrons to form neutral atoms again.

[13]Sometimes called "hot ice," this material is solid carbon dioxide.

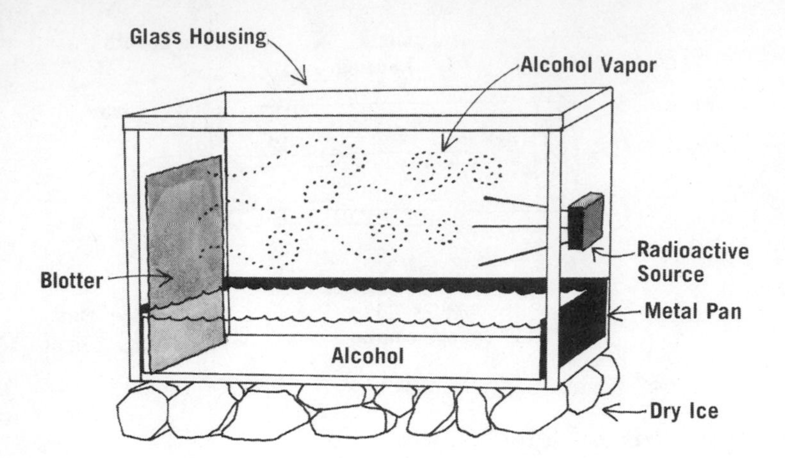

FIGURE 24–22. The cloud chamber allows us to see tracks left in the path of the radiation.

emits radiation, which leaves electrons and charged ions in its path as it streams through the air. In the earth's atmosphere, fog droplets are formed when water condenses on dust particles; in the cloud chamber the alcohol forms an alcohol fog as it condenses on the charges left in the wake of the radiation. Figure 24–23 shows that the fog or cloud records the path taken by the radiation.

ISOTOPES

Long and Short Lived Isotopes

There are 92 naturally occurring elements and more than 300 different isotopes found on earth. According to present theory of the life cycle of stars, many more isotopes than these 300 are formed in stars. Theory also holds that the earth was formed from the remains of supernova explosions (the explosion of most of the material of a star), and so we might expect to find many more than 300 isotopes on earth. But we don't. The explanation for this apparent paradox involves the half-lives of the isotopes. Those that are still here are the ones that have half-lives long enough so that they have not yet decayed to longer lived isotopes.

There are three groups of isotopes on earth that have half-lives of less than hundreds of millions of years, however: (1) A very few isotopes are constantly being formed in the upper atmosphere by cosmic rays. A practical application of this phenomenon will be discussed in Chapter 26. (2) A number of short-lived isotopes are produced by the radioactive decay of the naturally occurring long-lived isotopes. The example given earlier of the decay of uranium-238 to thorium-234 (by alpha emission) is a case in point. Uranium-238 has a half-life of 4.5×10^9 years, but thorium-234's half-life is only 24 days, so thorium-234 exists on earth at present solely because it is constantly being produced from uranium. (3) Many isotopes, and some entirely new elements, are man-made. Isn't anything sacred?

Man-made Isotopes

We have said that there are 92 naturally occurring elements. You also know that uranium has 92 protons. You might then conclude that these 92 elements range from hydrogen, with its single proton, to uranium with its 92 protons. But this isn't so. Element 43 is called technetium. It has no isotope that is not radioactive, no isotope which is formed

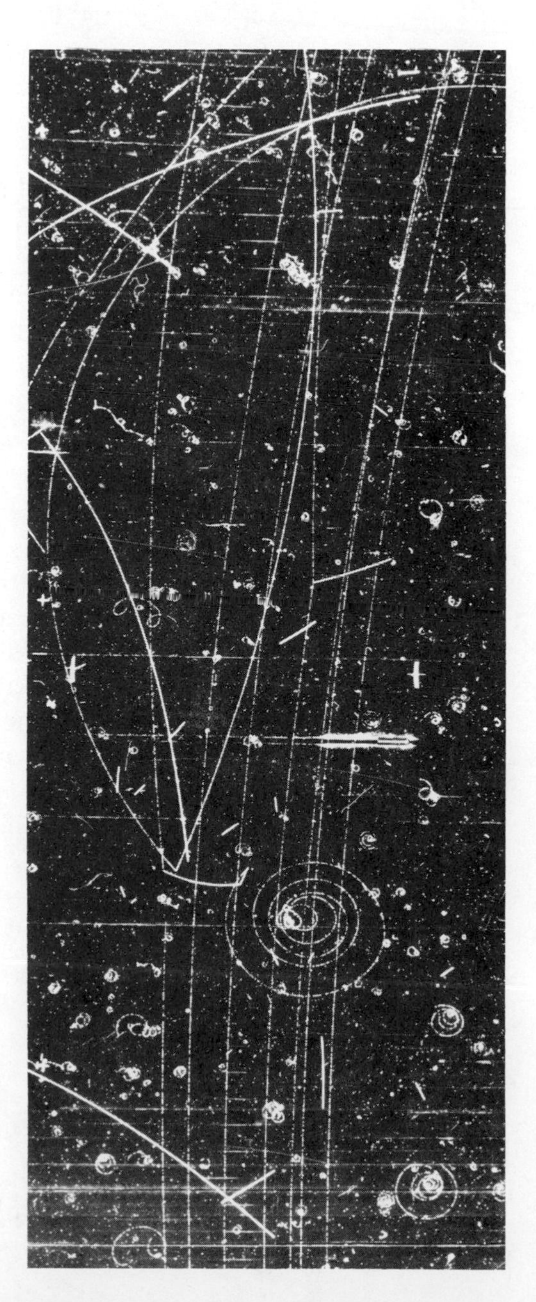

FIGURE 24–23. Fog droplets reveal the paths of the radiation. (By permission of University of California Lawrence Berkeley Laboratory.)

from radioactive decay of a naturally occurring element, and no isotope that has a half-life of more than 10^5 years. One hundred thousand years is but a short time compared to the age of the earth, so any technetium that may have once existed here has long since decayed to a stable isotope of some other element. Element 61, promethium, is in even worse shape. Its longest lived isotope has a half-life of 25 years, a short time even by man's reckoning.

So up to uranium there are only 90 naturally occurring elements. Let's look beyond uranium. Element 93 is neptunium, and it has an isotope, neptunium-237, with a half-life of 2.2×10^6 years. This isotope exists only in very small traces on earth. Likewise the next element on the list, plutonium, has a naturally occurring isotope with a half-life of 24,300 years. These isotopes are constantly being produced from uranium, as we will show. But the list ends with plutonium. There is no naturally occurring element beyond number 94. Enter man and his relentless researching.

In 1919 Ernest Rutherford, who produced the experimental evidence for the existence of the nucleus, did an experiment in which he allowed alpha particles to pass through nitrogen gas. He observed that some particles were emitted from the gas which were more penetrating than the alpha particles. By passing these new particles through an electric field he was able to determine that the particles were protons and that the following reaction had taken place:[14]

$$^{4}_{2}\text{He} + ^{14}_{7}\text{N} \rightarrow ^{17}_{8}\text{O} + ^{1}_{1}\text{H}$$

The formula, represented pictorially in Figure 24–24, simply means that the alpha particle ($^{4}_{2}$He) combines with the nitrogen nucleus, and the result is an oxygen-17 nucleus and an ejected proton. The important point to us is that oxygen is formed where there was none before; oxygen was created in this *nuclear reaction.* This was the first example of man's changing one element into another. Countless other similar experiments have been done since that time.

[14]Note that as before certain aspects of the formula must balance: the totals of the subscripts (representing the electric charge) and the superscripts (representing the mass) on each side of the arrow must be equal.

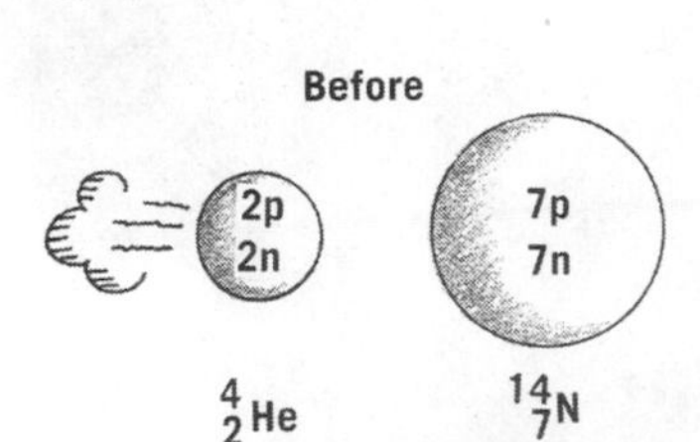

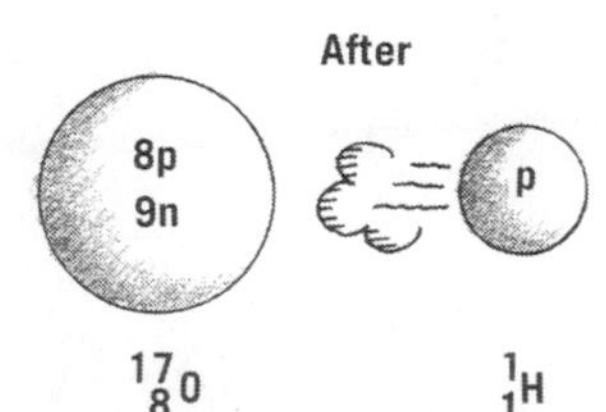

FIGURE 24–24. A nuclear reaction.

FIGURE 24–25. There is a strong repulsive force here because of the many positive charges in the uranium nucleus.

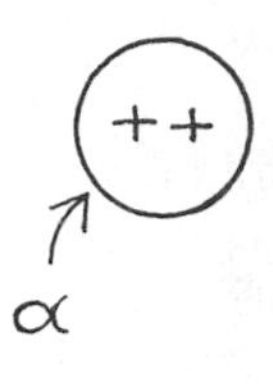

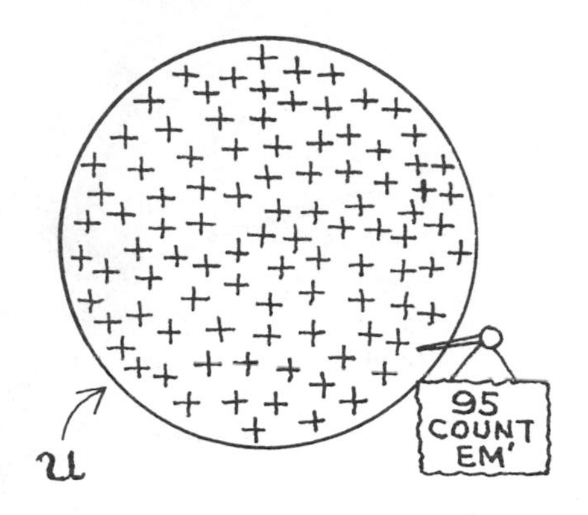

A reaction of particular value to the nuclear physicist is:

$$^{4}_{2}\text{He} + ^{9}_{4}\text{Be} \rightarrow ^{12}_{6}\text{C} + ^{1}_{0}\text{n}$$

The importance of this reaction (and others like it) is that a neutron is produced, and a neutron is a particularly good "bullet" with which to shoot other nuclei. To understand this concept, consider an alpha particle approaching a uranium-238 nucleus. As shown in Figure 24–25, a great electrical force of repulsion exists between the positively charged alpha particle and the nucleus because of the great amount of positive charge on the nucleus. Thus, in order to trigger a reaction, the alpha particle would have to be fired toward the uranium nucleus at a tremendous speed. There is no such repulsive force when a neutron approaches a nucleus, however, because the neutron has no charge. Thus the neutron is a good particle to use to cause reactions with heavy nuclei.

Consider the following take-along-*think*-it:

Form clay into a volcano-like hill as shown in Figure 24–26. Roll marbles toward the hill, trying to get one to fall into the center. The hill represents the way a nucleus appears to an approaching alpha particle. You'll find that most of your marbles (alpha particles) curve away from the hill or roll back toward you (Fig. 24–27). A real alpha particle does likewise, as shown in the results of Rutherford's alpha-scattering experiment.

Reduce the size of your clay hill and you have repre-

FIGURE 24–26. Roll a marble toward your volcano "nucleus."

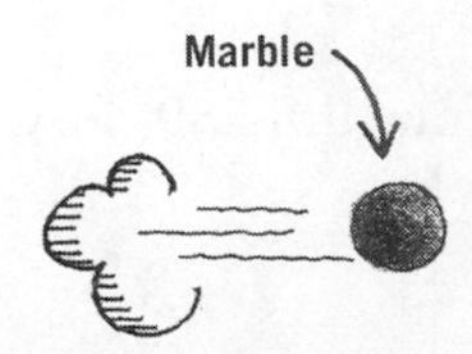

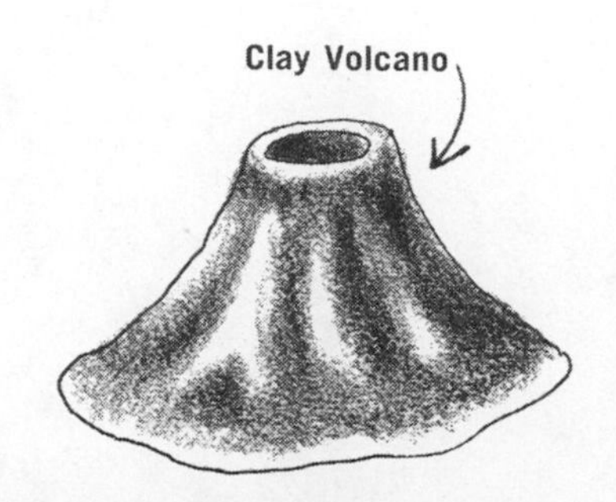

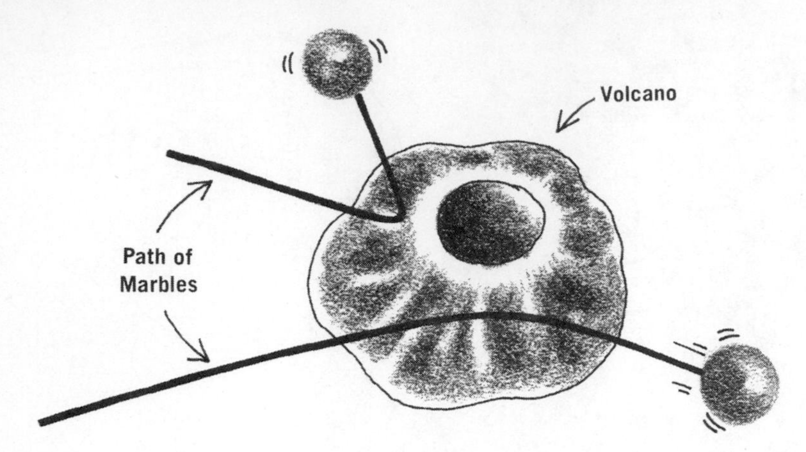

FIGURE 24–27. Most of the marbles will be deflected.

sented the nucleus as seen by a proton (which has less positive charge than the alpha particle). Then remove the hill entirely, leaving only a hole (drilled into the table) to represent the nucleus. This is how the nucleus appears to a neutron.

One of the many neutron-induced reactions which would have been of interest to the alchemists of old is the following:

$$^{1}_{0}n + ^{198}_{80}Hg \rightarrow ^{197}_{79}Au + ^{2}_{1}H$$

Hg is the symbol for mercury, the silvery metal used in thermometers, and Au stands for gold. The expense involved in producing gold in this manner, however, is so great that it is not worthwhile to do it except for experimental purposes in the laboratory. When used for the latter purpose, neutron bombardment has allowed us to learn much about the secrets of the nucleus. It has, in fact, allowed us to make new elements not found on earth.

BEYOND URANIUM

Uranium is the element with the greatest number of protons which exists in nature in any appreciable amount. Suppose that a neutron hits a uranium-238 nucleus. The reaction:

$$^{1}_{0}n + ^{238}_{92}U \rightarrow ^{239}_{92}U + \gamma$$

Uranium-239, however, is radioactive and emits a beta particle:

$$^{239}_{92}U \rightarrow ^{0}_{-1}e + ^{239}_{93}Np$$

And neptunium-239 also is beta-radioactive:

$$^{239}_{93}Np \rightarrow {}^{0}_{-1}e + {}^{239}_{94}Pu$$

Figure 24–28 represents pictorially this entire process.

Long before humans arrived on the scene with their experimentation, these reactions were occurring in nature. Neutrons are produced in the upper atmosphere by cosmic rays, which constantly bombard the earth from outer space, and (infrequent) collisions of these neutrons with uranium-238 nuclei produce (through the above reactions) very small traces of neptunium and plutonium in the natural environment.

In 1940, a team of researchers directed a beam of neutrons onto uranium; the products were plutonium and neptunium. These elements were thus the first man-made elements beyond uranium. By bombarding these heavy elements with neutrons and other particles, the list of man-made elements has been extended to include those shown in Table 24–5, each of which was named by the group of experimenters who isolated and identified the element.

Marie Curie (1867-1934)

Madame Curie, perhaps the most famous woman scientist, is the only person to win two Nobel prizes in science, one in physics (1903) and one in chemistry (1911). A daughter, Irene, won the 1935 Nobel prize in chemistry. Marie and her husband, Pierre, discovered the elements radium and polonium. She became the first women to teach at the Sorbonne in Paris. During World War I, she operated a mobile radiographic unit on the battlefields of France.

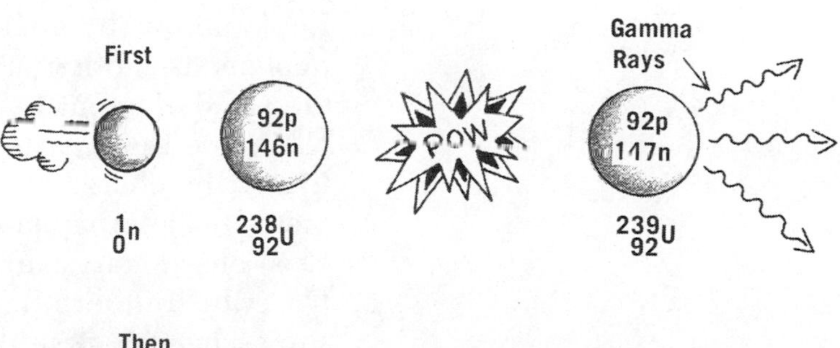

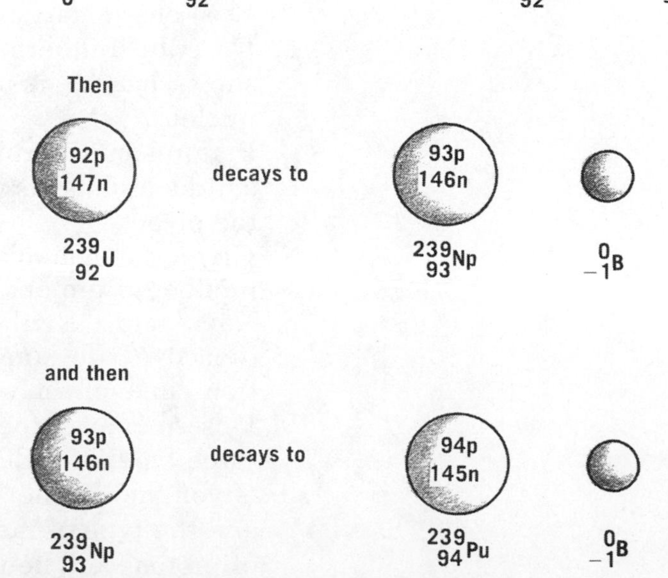

FIGURE 24–28. Carefully trace the process leading to formation of plutonium.

TABLE 24-5. MAN-MADE ELEMENTS BEYOND URANIUM

SYMBOL	NAME	NAMED FOR	YEAR
$_{93}Np$	Neptunium	the planet Neptune	1940
$_{94}Pu$	Plutonium	the planet Pluto	1940
$_{95}Am$	Americium	America	1944
$_{96}Cm$	Curium	Marie and Pierre Curie	1944
$_{97}Bk$	Berkelium	Berkeley, California	1949
$_{98}Cf$	Californium	(you guess)	1950
$_{99}Es$	Einsteinium	Albert Einstein	1952
$_{100}Fm$	Fermium	Enrico Fermi	1953
$_{101}Md$	Mendelevium	Dmitri Mendeleev*	1955
$_{102}No$	Nobelium	Alfred Nobel	1957
$_{103}Lw$	Lawrencium	Ernest Lawrence	1961
$_{104}Rf$**	Rutherfordium	Ernest Rutherford	1969
$_{104}Ku$	Kurchatovium	Igor Vasilievich Kurchatov	1964
$_{105}Ha$	Hahnium	Otto Hahn	1970
$_{105}Ns$	Nielsbohrium	Niels Bohr	1970

*He devised the periodic table of the elements in 1869.

**The priority of definite discovery of elements 104 and 105 is disputed between American and Russian scientists. The International Union of Pure and Applied Chemistry (IUPAC) is presently trying to decide which claims are more valid. We have listed the American name first in each case. Discovery of element 106 has recently been claimed by both groups, but each has agreed not to name it until priority is established.

OBJECTIVES

We hope that a study of Chapter 24 will enable you to:

1. Distinguish between chemical and nuclear reactions, giving an example of each.
2. Explain how the nucleus of one element differs from the nucleus of another and how it is similar to a nucleus of the same element.
3. Name the three principal particles making up the atom, listing the charge and location of each.
4. State the approximate relative mass of the three principal elementary particles.
5. Describe Rutherford's alpha-scattering experiment and show that its results confirmed the existence of the nucleus.
6. Perform an experiment in which the approximate size of a hidden object is determined by directing projectiles at the object.
7. Given the symbol for an isotope (such as $^{12}_{6}C$), state the number of protons and neutrons in the nucleus.
8. Name and describe the three major radiations from radioactive substances, stating the approximate penetration range of each and their relative masses.
9. Describe an experimental method of separating the three types of radiation.
10. Given the number of protons and neutrons in an isotope and the type of radiation emitted, calculate the number of protons and neutrons in the remaining nucleus.

11. Explain how electrons can be emitted from a nucleus even though they do not exist in the nucleus.
12. Define and give an example of the half-life of a radioactive substance.
13. Given the half-life of a substance and the amount existing at some time, calculate the amount existing at any whole number of half-lives later, and draw a graph of the amount existing versus time.
14. Perform a demonstration showing a process whose rate is analogous to the rate of decay of a radioactive substance.
15. Describe the operation of the Geiger counter and the cloud chamber.
16. Explain how some short-lived isotopes exist now, millions of years after the formation of the earth.
17. Describe a nuclear process in which a man-made isotope is produced.
18. Describe a demonstration which shows (by analogy) the difference between a nucleus as seen by an approaching alpha particle and an approaching neutron.
19. Name at least five elements beyond uranium in the periodic table.

QUESTIONS—CHAPTER 24

1. A friend mixes together a number of materials, heats the mixture, and claims that since the resulting substance is different from anything else known, he has discovered a new element. Is this possible? How do you know?

2. $^{12}_{6}C$ and $^{13}_{6}C$ both have six protons in the nucleus. $^{12}_{6}C$ and $^{11}_{5}B$ both have six neutrons. The first pair are both the same element, but the second pair are not. Why is the number of protons so much more important than the number of neutrons?

3. Alpha particles emitted with exactly the same energy as beta particles do not penetrate as far. Why?

4. If the atom were indeed like fruit Jello, what results would Geiger and Marsden have obtained in their alpha-scattering experiment?

5. Suppose that you start with 1.0 milligram (10^{-3} gram) of a pure radioactive substance and two hours later determine that 0.25 milligram of the substance remains. What is the half-life of the substance?

6. Indium-115 has a half-life of about 50 minutes. If you start with one gram of indium-115, how much would be left $2\frac{1}{2}$ hours later?

7. Why is the Geiger tube less likely to detect a passing gamma ray than a beta particle?

8. Complete the following radioactive decay formulas:

$$^{212}_{83}Bi \rightarrow ? + ^{4}_{2}He$$

$$^{12}_{5}B \rightarrow ? + ^{0}_{-1}e$$

$$? \rightarrow ^{234}_{90}Th + ^{4}_{2}He$$

$$? \rightarrow ^{14}_{7}N + ^{0}_{-1}e$$

9. Complete the following nuclear reactions:

$$? + {}^{14}_{7}N \rightarrow {}^{1}_{1}H + {}^{17}_{8}O$$

$${}^{7}_{3}Li + {}^{1}_{1}H \rightarrow {}^{4}_{2}He + ?$$

$${}^{27}_{13}Al + {}^{4}_{2}He \rightarrow ? + {}^{30}_{15}P$$

$${}^{1}_{0}n + ? \rightarrow {}^{4}_{2}He + {}^{7}_{3}Li$$

10. How many protons and how many neutrons are there in each of the following nuclei:

$${}^{23}_{11}Na,\ {}^{37}_{17}Cl,\ {}^{85}_{37}Rb,\ {}^{127}_{53}I,\ {}^{184}_{76}Os\ ?$$

11. Use an encyclopedia to identify the persons named in Table 24–5 with whom you are unfamiliar.

12. In Rutherford's experiment, assume that an alpha particle is headed directly toward the nucleus of an atom. Why does the alpha not make physical contact with the nucleus?

13. Rutherford used an alpha particle from a radioactive nucleus to bombard his targets. Since that day numerous devices have been constructed which accelerate nuclear bullets to very high speeds. Look up and discuss some of these devices. Some names you might look up: cyclotron, synchrocyclotron, linear accelerator, and betatron.

14. If an element has a half-life of two days, how much of an original sample of two pounds remains after eight days?

15. Why was the "fruit Jello" (or, as it is usually called by physicists, the "plum pudding") model of the atom abandoned?

16. The alpha particle has twice the charge of the beta particle. Why, then, does it deflect *less* than the beta particle when passing between electrically charged plates?

17. If film is kept in a box, alpha particles from a radioactive source outside the box cannot expose the film but beta particles can. Why?

18. A method of detecting radiation which was not mentioned in this chapter is based upon its effect on a charged electroscope (Chapter 12). What effect would you predict? Why? (Hint: It is not the fact that some radiation *hits* the electroscope which is important but that radiation passes through the air near it.)

19. Footnote 4 (p. 482) states that some particles would not reach the film if the apparatus were not put into a vacuum. What particles are being referred to?

Enrico Formi, pioneer of nuclear energy. (The Detroit Edison Company.)

25

NUCLEAR ENERGY

The term "nuclear energy" (and the misnomer "atomic energy," which means the same thing) evokes different feelings in different people. For some it means THE BOMB. Yet all of the energy ever used on earth can be traced back to its source—nuclear reactions in the interior of the sun. Without the nuclear energy of the sun we would have no coal, oil, gas, wind, or waterfalls to provide electricity for our television commercials. In this chapter we'll discuss both the bomb and the peaceful use of nuclear energy, the nuclear reactor.

THE SOURCE OF ENERGY

When a radioactive nucleus decays, the emitted particle is fired out of the nucleus at a high speed. It therefore possesses kinetic energy, and if the law of conservation of energy is not violated in this process, this energy must come from somewhere. To find its source let's examine what happens in a typical decay, such as the first one shown in the last chapter:

$$^{238}_{92}U \rightarrow {}^{234}_{90}Th + {}^{4}_{2}He$$

Imagine a process similar to the above, in which we slice a watermelon. Before slicing, we find the mass of our melon to be, say, 10 kilograms. We then cut off a small slice, being sure not to lose any seeds or juice, and measure the mass of the cut melon plus the slice. Result: 10 kilograms. But in the radioactive decay above, it doesn't work that way. Analogous to our melon we have a uranium nucleus with a certain mass which, upon decay, becomes thorium and an alpha particle. The uranium melon is, in effect, sliced. But the mass of the thorium nucleus and alpha particle when added together is *less* than the original uranium nucleus. It isn't much less: only 10^{-29} kilogram.[1] Even the scales of a gold dust speculator are not accurate enough to detect such a slight discrepancy, but this small amount of missing mass has profoundly altered the state of world politics.

In 1905 Einstein predicted, as a result of his theory of relativity, that mass can be changed into energy and vice-versa. (We introduced this idea in the chapter on relativity, and it was mentioned again in Chapter 24.) The missing 10^{-29} kg of mass actually goes out of existence in the reaction. It does not just disappear so that we cannot find it. It goes out of existence as mass and becomes energy. It is this effect which is responsible for the tremendous energy released in the nuclear bomb and the nuclear reactor. In both of these devices, however, more is involved than simple radioactive decay.

FISSION

In 1939, experiments were being carried out which involved bombarding the heavy elements with neutrons. It was found that the bombardment of uranium produced some confusing results. We stated previously that neptunium and plutonium were identified as products of the neutron–uranium-238 reaction, but other elements were found which could not so easily be explained. For example, the element barium (with 56 protons) was discovered among the prod-

[1]The mass of the original uranium nucleus was 4×10^{-25} kg.

FIGURE 25–1. The scale balances before and after cutting the watermelon. Nuclei don't behave like this.

FIGURE 25–2. The neutron causes the uranium nucleus to fission into two medium nuclei and three neutrons.

ucts. It was expected that only elements *beyond* uranium on the periodic table would be found, and certainly not elements with as few as 56 protons.

The inescapable conclusion was that some uranium nuclei were breaking up into two medium-sized nuclei. Further experiments showed the following to be a typical reaction of this type:

$$ {}^{1}_{0}n + {}^{235}_{92}U \rightarrow {}^{141}_{56}Ba + {}^{92}_{36}Kr + 3\ {}^{1}_{0}n $$

This reaction is illustrated in Figure 25–2. Uranium-235 is an isotope which makes up a small part (0.7 per cent) of all natural uranium, and when this isotope is struck by a neutron it breaks *(fissions)* into two medium-sized nuclei, releasing neutrons in the process.

An analogy in which the uranium nucleus is compared to a drop of water is often used to explain the fission process. In a drop of water, all the atoms that comprise it have energy, but this energy is not great enough to break up the drop. However, if energy is added to the drop such that it is set into vibration, it will undergo elongation and compression (Fig. 25–3*A*). Finally, if the amplitude of vibration becomes large enough, the drop may divide (Fig. 25–3*B*). In the uranium nucleus a similar process occurs. Energy is added when the neutron strikes the uranium nucleus, and the energy is absorbed by being distributed to all other neutrons and protons in the nucleus. A deformation occurs, as in the water drop, and a force of repulsion between the protons in each part encourages the separation to continue until a split, or fission, occurs. Freed in the "spray" are three neutrons. This release of neutrons was immediately seen by physicists as being crucially important. A neutron started the reaction, and the products of the reaction *included more neutrons.* If these neutrons could be made to initiate more reactions, the physicists reasoned, a chain reaction could be produced. As an example, imagine that two of the resulting neutrons cause splitting of other U-235 nuclei. Then from each of these two come three neutrons, two of which hap-

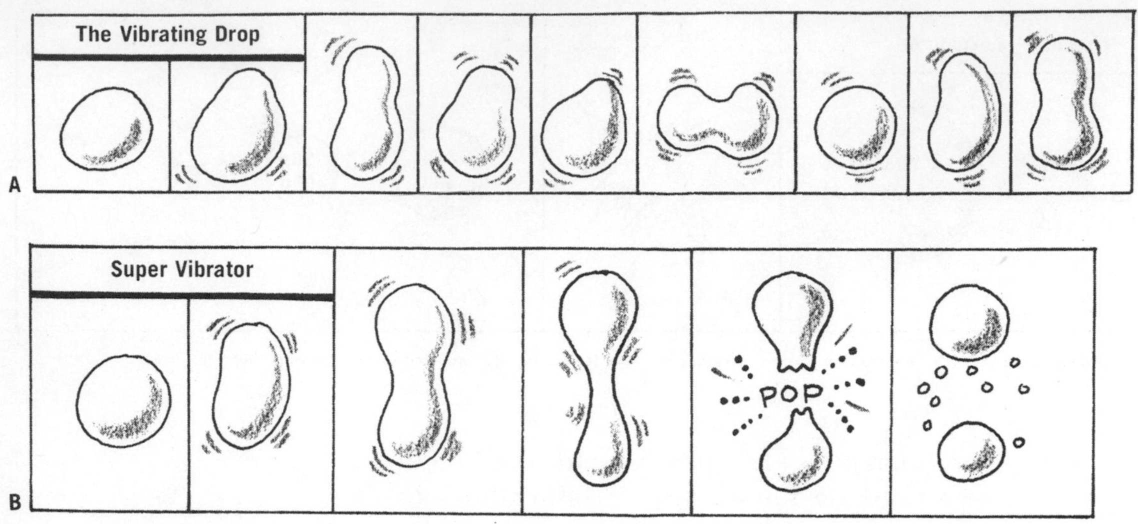

FIGURE 25–3. *A*, Suspended drop of water vibrating. If the amplitude gets large enough, *B* may result.

pen to split other U-235 nuclei. The reaction would grow as shown in Figure 25–4.

In actual practice the reaction given above is only one of several possible ways for uranium fission to occur. Many isotopes other than the two shown are produced as by-products of actual fission reactions. In some cases, three neutrons are produced, while in other cases only two may be produced. The average number of neutrons produced per fission turns out to be about 2.5. This value is sufficient to initiate or to continue a chain reaction.

A Mouse-Trap Chain Reaction

If it were reasonable to assume that the typical student has in his or her room 200 mousetraps and 201 corks, this

FIGURE 25–4. A single neutron starts the chain reaction, which then grows by leaps and bounds.

would be a good take-along-do-it. As it is, we'll settle for a take-along-think-it: Set the 200 mousetraps and (carefully!) place them as close together as possible on a table. Finally, a cork must be placed on *each* mousetrap in such a way that when the trap releases, the cork flies into the air. Now the experimenter can either wait for a mouse, or he can use the one remaining cork to start the reaction. He throws it to the center of the array. When it springs the trap there, it and the cork on that trap *both* fly up. What flies up, comes down, and soon four corks are flying, then eight, then . . . BOOM!

Energy Release

The energy of a fission reaction springs from the same sources as the energy of radioactive decay. As in the decay reaction, the amount of mass after the "event" is less than before it. In fission, about 10^{-26} kilogram per reaction is lost—about 1000 times more than in radioactive decay. This lost mass is converted into energy, some becoming kinetic energy of the fission products and some being carried away by gamma rays produced at the instant of fission.[2] These products zoom away from one another at high speeds. They are then slowed by collisions with nearby atoms, so that the energy is shared by the material as a whole and shows up as a higher temperature of the material.

The actual energy released in a single fission is small—about enough energy to lift a mosquito egg 1/1000 inch. But when we realize that the energy released in one chemical reaction associated with the burning of coal is only about one ten-millionth of this amount and that there are more than 10^{21} atoms in a gram of uranium,[3] we begin to appreciate the energy available in fission reactions.

Factors Affecting Fission

The principles of the fission chain reaction have been explained. The question arises as to why such reactions do not occur in the uranium found in the earth. Happily they do not—and will not. In order to cause a fission chain reaction, three factors have to be just right.

(1) Less than one per cent of natural uranium is U-235. The remainder, 99.3 per cent, is the (almost) nonfissionable U-238. Thus, of the neutrons produced by a fission (2.5 on the

---

[2]These were not shown in Figure 25–4 because gamma rays moving at the speed of light are too fast for an artist to draw.

[3]A gram of uranium-235 would take up about as much space as a drop of water and could lift a 100-car freight train 10 miles upon the fissioning of every nucleus in it.

average), some will be absorbed by the U-238 (producing neptunium and plutonium) and will not cause further fission. They are lost to the chain reaction. Figure 25–5 shows a neutron being lost in this way.

(2) It is found that slow-moving neutrons are more likely to cause fission of U-235 than fast neutrons. The neutrons produced in fission are emitted at high speeds and are four times more likely to be bounced away from a U-235 nucleus than to be captured for another fission. Thus fast neutrons are likely either to be absorbed by U-238 or to escape from the uranium altogether [see (3) below]. A slow-moving neutron, on the other hand, is 50 times more likely to cause fission than to bounce away from a U-235 nucleus.

(3) Nuclei occupy only a minute fraction of the total volume of a substance. Thus the likelihood of a neutron striking *any* nucleus is small; a neutron may pass billions and billions of nuclei without hitting one. Thus if the chunk of uranium is small, it is very likely that a neutron will completely escape without doing anything (Fig. 25–5).

Enrico Fermi (1901-1954)

The man for whom element 100 is named received his doctor's degree from the University of Pisa at the age of 21. In 1938 he and his family went from their home in Italy to Sweden to accept the Nobel prize for his demonstration of the existence of new radioactive elements produced by neutron bombardment and for his related discovery of nuclear reactions brought about by slow neutrons. The Italian Fascist press severely criticized him for not giving the Fascist salute when receiving the award and for not wearing the Fascist uniform. He and his family never returned to Italy but came to America to avoid persecution.

Following World War II, Fermi pioneered in research on high energy nuclear particles. He died of cancer only two weeks after being awarded $25,000 by the Atomic Energy Commission for work during the war. A very readable biography has been written by his wife, Laura: Atoms in the Family: My Life with Enrico Fermi.

THE FIRST CHAIN REACTION

In 1940, physicists were able to predict that a fission chain reaction could be produced, but the feat had never been accomplished. Enrico Fermi, an Italian physicist who emigrated to America to escape the political persecutions of pre–World War II Italy, set out to build the first nuclear reactor. (The location he chose was a squash court beneath the stands of the University of Chicago football field.) We will look at how each of the three fission factors was handled in this first reactor.

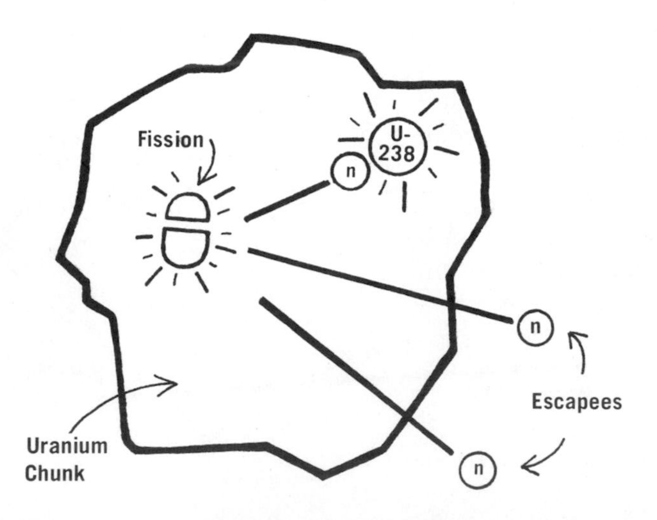

FIGURE 25–5. Three neutrons are produced in this particular fission, but none causes another fission.

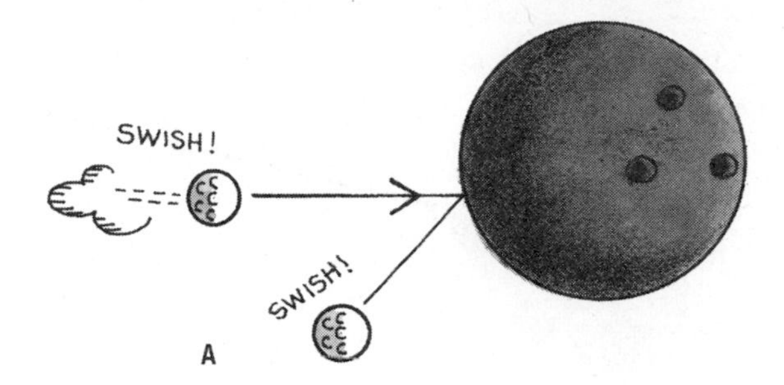

FIGURE 25–6. *A*, Golf ball hitting a bowling ball. It slows down more as a result of bouncing from a baseball, *B*,

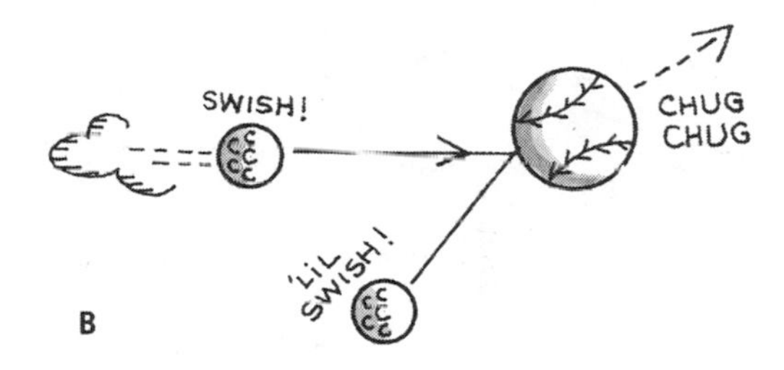

(1) No effort was made to increase the percentage of U-235 in the uranium. Natural pure uranium metal was used.

(2) The neutrons were slowed down. To understand how this can be done, consider the effect of golf balls being hit at a high speed past widely spaced bowling balls (Fig. 25–6). When a golf ball hits one of the massive bowling balls, it barely affects the bowling ball. The big ball may move very slightly, but most of the energy remains with the golf ball, and it rebounds with about the same speed that it had before the collision. But now change the conditions of the experiment: hit the golf ball past widely spaced baseballs. A baseball will be affected by collision: it will bounce away. The important point in this discussion, however, is that the golf ball will bounce back at a reduced speed. The closer that the mass of the struck ball is to the mass of the ball which strikes it, the more effective it will be in slowing down the fast-moving ball.[4]

The principles involved in the above thought-experiment are used in nuclear physics to slow down neutrons. Fermi chose to place bricks of graphite (carbon) between the chunks of uranium. Carbon nuclei are about twelve times more massive than neutrons, but after a number of collisions with carbon nuclei (Fig. 25–7), a neutron is slowed down sufficiently to increase the likelihood of fission with U-235. A material used in a reactor to slow down the

---

[4]These results follow from the laws of conservation of energy and conservation of momentum.

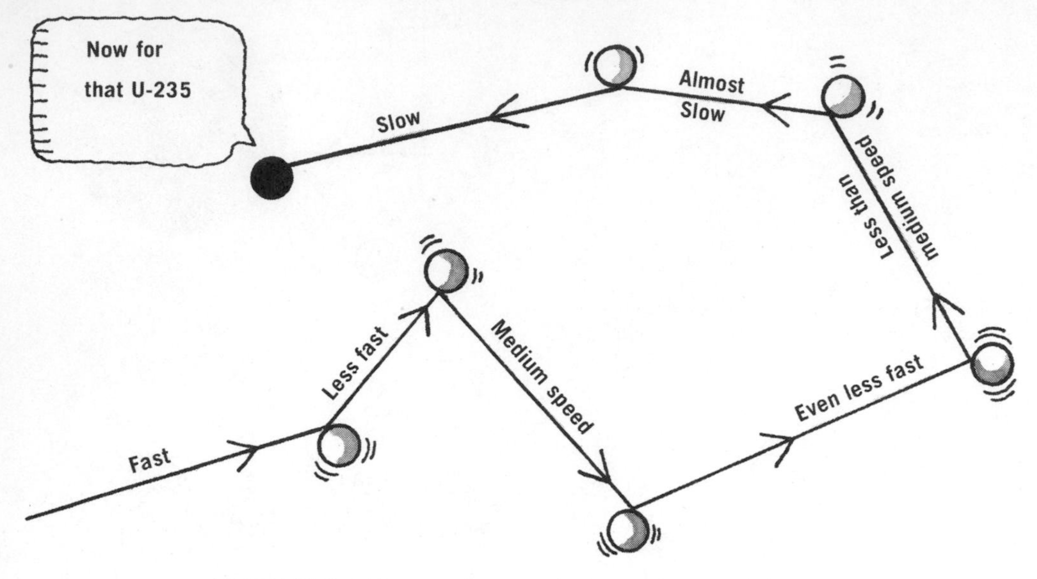

FIGURE 25–7. After many bounces, the neutron has slowed down.

neutrons is called a *moderator*. Figure 25–8, which shows some of the elements of Fermi's reactor, indicates that the carbon moderator composed the major portion of the reactor.

(3) The amount of uranium needed to prevent excessive loss of neutrons is not independent of the preceding two factors. Nor is it independent of the shape of the reactor, as shown by a comparison of Figures 25–9*A* and *B*. In *A*, the reactor (uranium plus moderator) is flattened into a wafer shape. A neutron produced in reactor *B* is much more likely to have to pass through a lot of uranium before escaping. Thus Fermi's reactor was constructed in a roughly spherical shape. Figure 25–10 is a drawing of that reactor in the squash court, based on descriptions by those present. The bottom half is composed of wood which surrounds and supports the graphite bricks and the enclosed uranium. Note the size of the reactor (24½ feet across and 19 feet high).

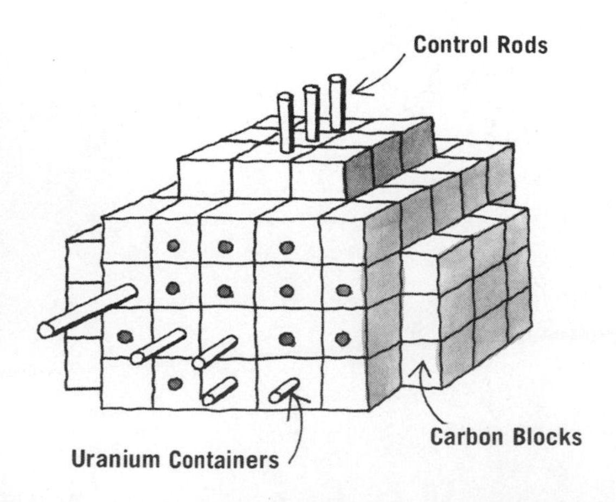

FIGURE 25–8. The world's first reactor contained these features.

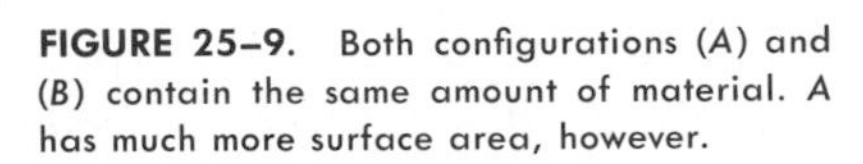
FIGURE 25–9. Both configurations (*A*) and (*B*) contain the same amount of material. *A* has much more surface area, however.

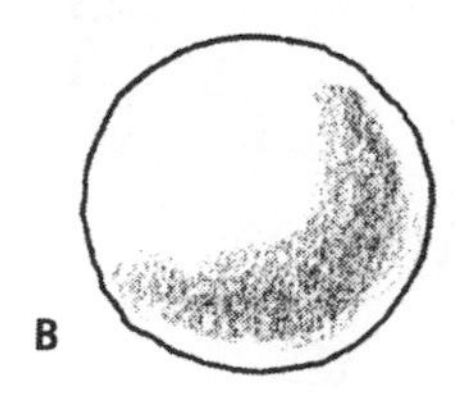

Control Rods

One other thing was needed for Fermi's reactor: a method of control. In order for a fission reaction to be self-sustaining, an average of exactly *one* neutron from each fission must cause another fission. If the number was slightly less than one, the reaction would slowly fizzle out. If slightly more than one, the reaction would continue to grow until it was uncontrollable and the heat produced would melt the reactor (and the floor under it).

Luckily, certain elements are very efficient in absorbing neutrons. One of these is the metal cadmium.[5] To control the squash court reactor, rods of cadmium were inserted in holes through the graphite bricks. Figure 25–11 illustrates the function of these control rods.

[5]Cadmium is a bluish-white metal which is so soft that it can easily be cut with a knife.

FIGURE 25–10. Owing to wartime secrecy, no photographs were taken of the first reactor. This sketch is based upon descriptions by the people present. (Courtesy of Argonne National Laboratory.)

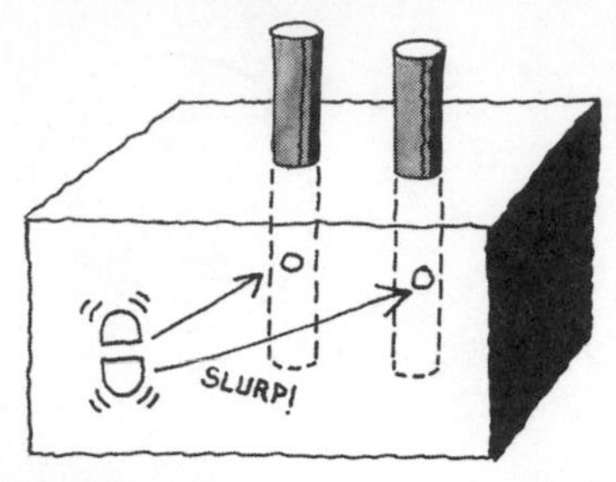

FIGURE 25–11. Control rods absorb neutrons.

On December 2, 1942 (almost one year after the attack on Pearl Harbor), Fermi's reactor was complete, and the experimenting team gathered to watch the control rods being pulled from the pile of graphite and uranium. One control rod was set so that it would slam into the reactor if a preset level of radiation were achieved. The rods were removed slowly, until at 3:25 PM the critical point was reached and the reaction became self-sustaining. The experimenters watched for 28 minutes before replacing the control rods, celebrating with Chianti wine, and leaving—hoping that they had been the first to see a self-sustaining nuclear reactor. Why "hoping"?—because it was feared that Germany might have progressed this far already and might be on the way to completing a nuclear bomb.

THE NUCLEAR BOMB

Background

Until the late 1930s almost all the important physicists lived and worked in Europe. Americans wishing to learn physics from the masters went there to study. But during the 1930s many scientists emigrated to America to escape the political persecutions of Hitler and Mussolini. Scientists with Jewish connections were especially affected. Among those arriving in America were Enrico Fermi, Albert Einstein, and Niels Bohr.

The question of whether the United States should have built—and then used—the atomic bomb is one that will probably never be resolved. The primary motivation in the early 1940s was the fear that Hitler was also working on the bomb. It was known that German scientists knew about uranium fission (since they discovered it) and that Germany had the "brains" to make the bomb. In addition, Germany had conquered Czechoslovakia and had stopped the sale of uranium from Czechoslovakian mines. The exiled physicists were especially concerned because they were aware of the terror and destruction that Hitler would spread if he succeeded in building a bomb. In the fall of 1939, Einstein wrote a letter to President Franklin Roosevelt, telling him about the possibility of the bomb and about his fear of Germany's intentions. In 1942, with the formation of the Manhattan Project, the United States began to work in earnest toward building the world's first nuclear bomb. The reactor under the football stands in Chicago was one of the early fruits of the secret project, and it proved that a fission chain reaction was indeed possible. Because of this successful experiment, the scientists were confident that a bomb could be built.

Problems

As you can see from the drawing of the first reactor, any bomb which was to be carried to its appointed place by airplane would have to be much smaller than such a reactor. To make the device smaller and still cause a chain reaction, the three conditions for fission were applied differently. To reduce weight and size, no moderator was used. But this change decreased the probability of a neutron's causing fission of a U-235 nucleus. To adjust for this factor, a much higher percentage of U-235 was used in the bomb than had been in the reactor.

Increasing the amount of U-235 in uranium, however, turned out to be a major problem. Remember that all isotopes of an element have almost identical chemical properties. If they did not, U-235 could simply be allowed to react with some chemical which does not affect U-238 and the two could then be separated. (This is how oxygen can be removed from the air, for example.) Up until 1942, the amount of U-235 that had been separated from U-238 was a few millionths of a gram. *Kilograms* were needed. The fact that the immense technological problems associated with this separation were solved in a few months is a tribute to the teams of men and women who worked on the project. By 1943, uranium enriched in U-235 was being produced in sufficient quantities.

Only one problem remained in the construction of the bomb. For conventional bombs made of TNT, the explosive can be put together ahead of time and ignited at the proper moment. If enough uranium-235 is put together, however, a chain reaction can be started by a stray neutron causing even a single fission.[6] The amount of fissionable material which is just enough to sustain a reaction is called the *critical mass*, and because of this feature, the uranium must be arranged in at least two chunks, each of which is less than the critical mass. At the appropriate time the chunks are fired together to form a greater-than-critical mass. In Figure 25–12,

[6] "Stray" neutrons are always present, being one of the particles produced by cosmic ray collisions (see Chap. 27).

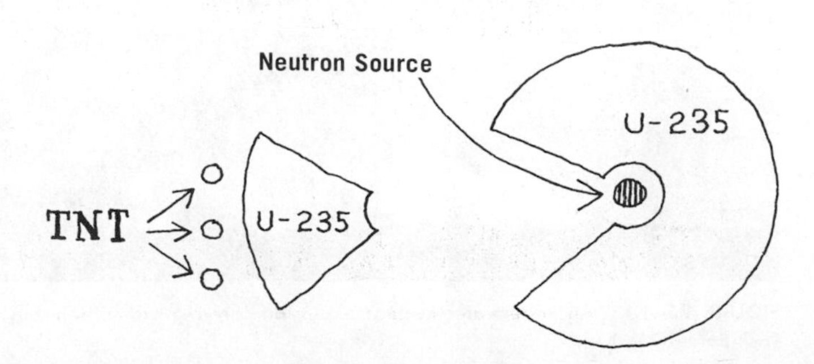

FIGURE 25–12. Possible configuration for a fission bomb. To detonate the bomb, the TNT is exploded, thereby slamming together the two pieces of uranium to form a greater-than-critical mass.

FIGURE 25–13. Nuclear weapon of the "Little Boy" type, the kind detonated over Hiroshima, Japan, during World War II. The bomb is 28 inches in diameter and 120 inches long. The first nuclear weapon ever detonated, it weighed about 9000 pounds and had a yield equivalent to approximately 20,000 tons of high explosive. (Los Alamos Scientific Laboratory.)

FIGURE 25–14. An underwater nuclear explosion carrying two million tons of water skyward. (Joint Task Force One.)

a conveniently located neutron source is provided so that the first fission is not left to a chance stray neutron.

The first nuclear bomb was detonated in a test explosion over the desert near Alamogordo, New Mexico, on July 16, 1945, only 2½ years after the first reactor was built. The bomb produced an explosive effect equivalent to nearly 20,000 tons of TNT.

On August 6, 1945, a nuclear bomb (weighing 4½ tons) was dropped on the Japanese city of Hiroshima. Three days later another bomb was dropped on Nagasaki. Two days after that, Japan surrendered.[7]

The Plutonium Fission Bomb

Uranium-235 is not the only fissionable material that can be used in reactors and bombs. Plutonium-239, the isotope which results after U-238 absorbs a neutron and emits two beta particles, can also be used. Remember that in a reactor with natural uranium as the fuel, a great deal of U-238 is available to absorb neutrons. The moderator prevents this effect from shutting down the reaction, but some neutrons are being absorbed by U-238 whenever a reactor is operating. Thus Pu-239 is always being produced. The scientists of the Manhattan Project used this process to produce plutonium. The test explosion in New Mexico and the second bomb dropped on Japan were both plutonium bombs.

Effects of Fission Bombs

The most obvious effect of "A-bombs" results, of course, from the explosion. It is this effect that is being compared to TNT explosions when an atomic bomb is rated as being a 50 kiloton bomb—its explosion does the same damage as 50,000 tons of TNT. The atomic bomb doesn't stop there, though. As indicated earlier, each fission results in the immediate release of gamma rays. In addition, visible light and infrared heat radiation are emitted from the tremendously hot explosion. All of these radiations travel at the speed of light, so that the area surrounding the bomb site is first affected by this radiation, which is intense enough to start fires some distance from the explosion.

[7]The interested reader is referred to a wealth of material discussing the "why" of the Manhattan Project. The list of books and articles on the subject is too long to include here; try the card catalog of your library. The controversy, however, resulted in the publication (by concerned scientists) of a journal which should be mentioned: *Bulletin of the Atomic Scientists*, published monthly since 1945, and now called *Science and Public Affairs.*

The blast itself is simply a pressure wave, similar to a sound wave but much more intense. Because it travels at the speed of sound, it is not felt until after the visible, gamma, and infrared radiation have passed. This occurs in exactly the same way and for the same reason that you see a lightning flash before you hear the thunder.

A third effect is due to the radioactive material produced. The barium and krypton isotopes mentioned earlier are only two of the roughly 200 fission products. Most are highly radioactive, having short half-lives, and most of these decay into isotopes which are also radioactive. In addition to these radioactive fission products, the tremendous number of neutrons released when the bomb explodes react with nuclei of the ground and air at the bomb site, in most cases producing more radioactive nuclei. Because of the high temperature (several million degrees at blast center), many of these radioactive atoms are in the gaseous state. Others are dust particles making up part of the mushroom cloud that forms after the explosion. Thus the radioactive by-products of the bomb end up in the upper atmosphere, to be spread by winds. This radiation may be detected downwind from the blast site for long periods of time following the explosion.

The amount of material produced by a bomb depends greatly upon where the bomb is exploded. If it is exploded high in the air, less material is produced because no earth or water is sucked into the fireball. An explosion at ground level, however, produces far greater amounts of this *radioactive fallout.*

Radioactive fallout is especially dangerous because some of the isotopes are absorbed by the body, which of course is unable to distinguish between a radioactive atom and a stable atom. For example, strontium-90 is deposited in bone. This isotope has a half-life of 28 years, and therefore when it is taken into the body it continues to emit radiation into the bones for the life of the person.

Table 25–1 shows the relative amounts of radiation our bodies receive from various sources, including fallout from nuclear bomb testing, medical x-rays, and natural radiation from our surroundings. Such figures, it must be emphasized, represent average values and may vary greatly from individual to individual. For example, the amount of radiation one receives from cosmic rays at several thousand feet elevation is *double* the amount received at sea level.

PEACEFUL FISSION

Although the history of the period is somewhat unsure, one of the first uses of fire was undoubtedly for warfare; it

TABLE 25–1. SOURCES OF RADIATION DOSE TO PERSONS IN THE UNITED STATES*

SOURCE	PER CENT (Approximate)
Man-made	
Diagnostic x-ray	40
Therapeutic x-ray	5
Radioactive fallout	5
Other	1
Total	51
Natural	
Radiation external to body	39
Radioisotopes inside body	10
Total	49

*Based on Radiation Protection Criteria and Standards: Their Basis and Use. Summary-Analysis of Hearings before The Special Subcommittee on Radiation of the Joint Committee on Atomic Energy, Congress of the United States. U.S. Government Printing Office, Washington, D.C., October, 1960.

must have terrorized those unfamiliar with it. We now consider fire a fairly peaceful happening. World War I provided the impetus for the development of the airplane; people of the time thought of the plane as an instrument of war. Of course, it still has military applications, but few people continue to think of it primarily as a weapon. The first use of nuclear energy was for war. There is reason to hope that in a relatively few years the term "atomic energy" will also lose the fearful, martial connotation many people still associate with it. The device which may justify this hope is the descendant of Enrico Fermi's experiment in the squash court: the nuclear reactor.

Production of Electrical Energy

The primary sources of electrical energy today are the energy of water as it falls over a dam and the energy from burning of fossil fuels. In the first of these processes, the water is made to turn turbines which are connected to electrical generators. Figure 25–15 illustrates the simple steps involved. Conversion of the chemical energy in fossil fuels (primarily oil and coal) into electrical energy is also a fairly simple procedure. The fuel is burned to convert water into high-pressure steam. As shown in Figure 25–16, the steam is then directed through a turbine (or turbines), which again is connected to electrical generators.

As previously stated, heat is produced in fission reactors when the high-energy fission fragments are slowed down in the surrounding material. When it was first operated, Fermi's squash court reactor produced heat at the rate of 1/2 watt. This is barely enough to light a small flashlight bulb.

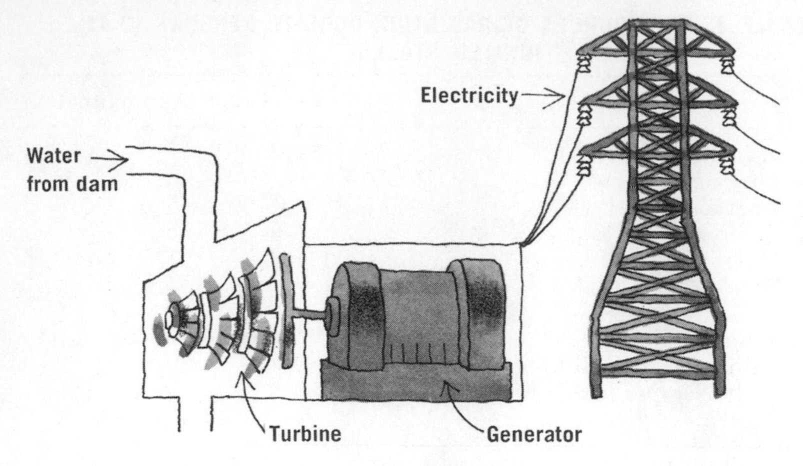

FIGURE 25–15. An electrical power plant that gets its energy from dammed water.

Later it was operated at a few hundred watts. Its power output was limited, however, because there was no way to remove heat from the core of the reactor, except by slow conduction to the surface. If allowed to operate at too high a rate, the reactor would have overheated (and melted). The modern power reactor must be constructed in a different way.

Figure 25–17 illustrates the essential parts of the core of a nuclear power reactor. Water serves as a coolant and runs between fuel elements that hold the slightly enriched uranium (shown in black). You may notice the absence of the graphite moderator; in this reactor the water itself serves as a moderator. Water can function in this way because each of its molecules contains two atoms of hydrogen, and hydrogen has only one proton in its nucleus. This means that its mass is comparable to the neutron, and therefore it will greatly slow down the neutron in a collision.

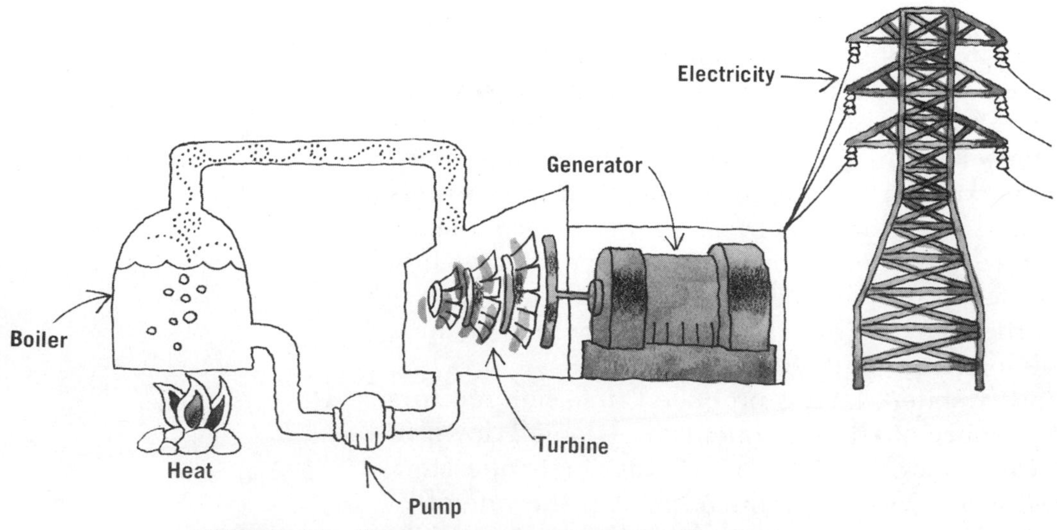

FIGURE 25–16. The electrical energy here comes from burning fossil fuels.

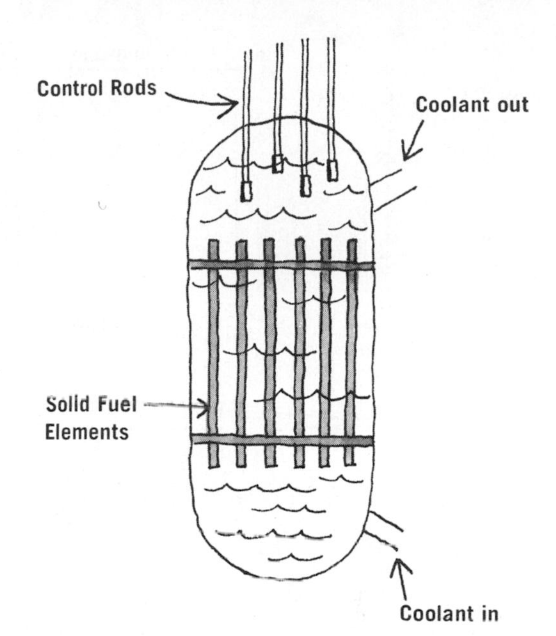

FIGURE 25–17. The core of a nuclear reactor.

Figure 25–18 shows how hot water is used to produce electricity. After leaving the reactor it goes to a steam generator. The water in the reactor's cooling loop is kept at a very high pressure—more than 2000 pounds per square inch. (This is equivalent to the pressure at a depth of one mile in the ocean.) Because of this great pressure the water can be heated to 600° Fahrenheit without boiling. The pressure in the second loop, however, is less, so that the water in it is boiled before being passed through the turbine. A comparison of Figures 25–18 and 25–16 shows that in the nuclear power reactor, the reactor core and the steam generator simply replace the fuel furnace as a method of producing steam.

The reactor described above is a *pressurized water* reactor. Roughly half of the power reactors built or being built in the United States are of this type. Most of the remaining are

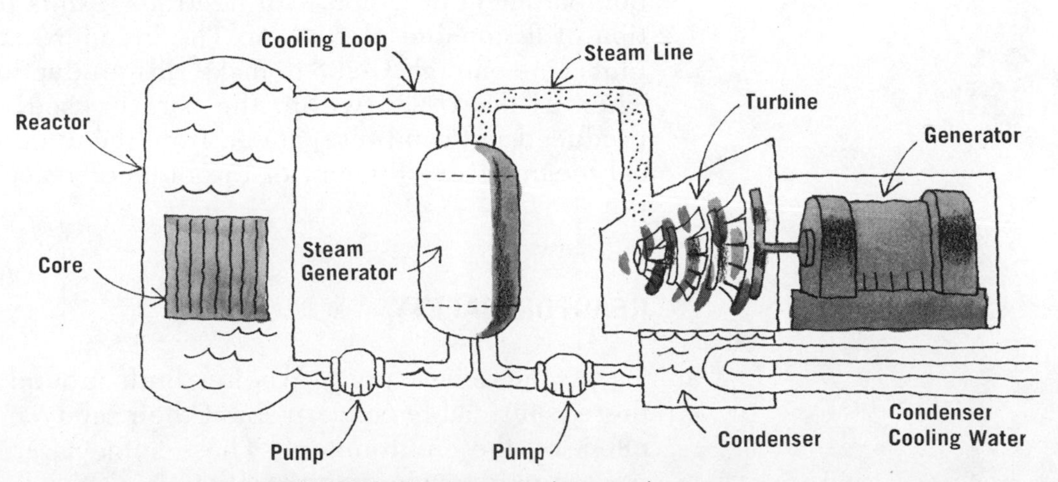

FIGURE 25–18. A pressurized-water nuclear reactor.

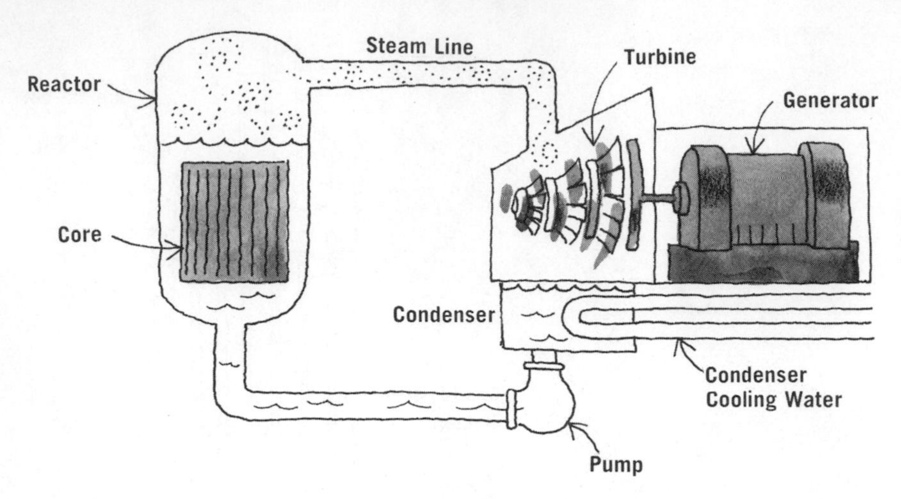

FIGURE 25–19. A boiling-water nuclear reactor. Note that no separate steam generator is used.

boiling-water types, shown schematically in Figure 25–19. In these reactors, the intermediate steam generator step is skipped, and the steam is produced within the reactor itself.

THE BREEDER REACTOR

Just as supplies of oil and coal are limited, so is the supply of U-235 for power reactors. Predictions of how long the supply will last vary greatly, depending upon estimates of undiscovered reserves, efficiency of mining, and number of reactors built. As a general statement, however, it is fairly safe to say that we will be running low on U-235 by the end of this century. A reactor which may extend the use of fission power well beyond that time is the breeder reactor.

This reactor, none of which have yet been built in the United States to function as full-scale electrical power producers, produces fuel as it "burns" fuel! This is not as fantastic as it may sound, for we have already discussed how bombardment of U-238 with neutrons results in the production of fissionable plutonium. The breeder reactor includes in its core enough U-238 to make this production significant. Then as the U-235 within the core is used up, Pu-239 is produced and can be separated from the uranium by chemical means. Development of the breeder reactor is continuing.

REACTOR SAFETY

As more reactors are being built around the country, there is justifiable concern about their safety and about their effect on the environment. These subjects are so vast that we can only touch upon them here.

One of the primary concerns of the public seems to be the fear of a nuclear explosion. Fortunately, the use of water as a moderator limits this possibility—and in an interesting way. As the temperature in a boiling water reactor increases, water is boiled faster. But this results in steam, rather than water, being present between the fuel elements. Steam, however, contains far fewer molecules in a given volume than does water, and since water molecules are necessary as a moderator, the number of neutrons which are moderated is reduced. The final result is less fission and a slowing down of the reaction. The same limiting feature occurs for the pressurized water reactor, although the pressure vessel may have to be ruptured before the water would change to steam and shut down the reaction.

A more real danger is the possibility of the water flow being interrupted so that the cooling of the reactor stops. As explained earlier, the production of steam would decrease the rate of fission, but heat would continue to be produced from the great amount of radioactive fission fragments within the core. The temperature could conceivably build to the point where the fuel elements melt, the bottom of the reactor melts, and the ground below the reactor melts. This possibility is referred to, appropriately enough, as the China syndrome.

In addition to the China syndrome, a regular non-nuclear explosion may occur because of the tremendous heat produced, causing radioactive materials to be spread through the area surrounding the power plant. To prohibit

FIGURE 25–20. The San Onofre nuclear generating station provides enough power to supply a city of well over one-half million people. A pressurized water nuclear reactor is housed in the steel containment sphere near the center of the picture. (Southern California Edison Company.)

such an event, all reactors are built with a back-up cooling system set to go into operation if the regular system fails.

Another concern about nuclear power plants is the everyday release of radioactive materials to the environment. It is apparently almost impossible to prevent small leaks of fission fragments from the fuel elements into the water of the cooling loop. Notice, though, that this water is recycled rather than being continuously released into the environment. In fact, in the pressurized water reactor, cooling water is two cycles away from the water that contains radioactive materials. (Refer again to Figure 25–18.) Nevertheless, slight leaks do develop, and some radioactivity is released into the stream.

There are also some radioactive materials released in a gaseous form from "smokestacks" at the reactor site. The total amount of radioactivity that any nuclear facility may release into the water or atmosphere is regulated by the Nuclear Regulatory Commission. Still, the question of what are safe limits is constantly debated.

The disposal of radioactive materials when the reactor core is replaced is another problem. This material contains such highly radioactive isotopes that it is literally *hot* and must be cooled for some time after removal from the reactor. It must then be stored in such a way that there is no chance that it will be released to the environment. At present, sealing it in deep salt mines seems to be the most promising solution.

A major concern about the proliferation of nuclear power plants is the danger of sabotage at reactor sites and the danger of nuclear fuel (or waste) being stolen during transport. One reason for fear is, of course, the fact that the materials could be used to make an atomic bomb. And aside from dangers due to its radioactivity, plutonium is one of the most poisonous materials known to man. It is frightening to imagine plutonium in the hands of terrorists.

A consequence not associated with radioactivity *per se* is the thermal effect on the cooling stream. So much stream water is required to cool the reactor's water after it has passed through the turbine that the river is measurably warmer downstream from a reactor. The heat affects living organisms within and adjacent to the river. In some cases the effect may be considered beneficial and in others detrimental, but in any case this factor must be taken into consideration in the location of nuclear power plants.

FUSION—BACK TO BOMBS

In our everyday language we refer to two types of nuclear bombs: atomic bombs (A-bombs) and hydrogen bombs (H-bombs). A better name for the first type is fission bombs. We have discussed these and also the use of the fission process to produce usable energy. The H-bomb uses another nuclear process, fusion, which may someday solve man's energy problems forever.

The term "fusion" refers to the fusing together of light nuclei to form a heavier nucleus. Two examples are:

$$^{2}_{1}H + ^{2}_{1}H \rightarrow ^{3}_{2}He + ^{1}_{0}n$$

$$^{2}_{1}H + ^{2}_{1}H \rightarrow ^{3}_{1}H + ^{1}_{1}H$$

Figure 25–21 illustrates these reactions. Hydrogen-2 (deuterium) is an isotope which makes up about 0.02 per cent of natural hydrogen, and hydrogen-3 (tritium) is an even rarer isotope.

Pick either of the two reactions; find the total mass before and compare to the total mass after. Guess what? In each case there is less mass after the reaction occurs than there was before. This mass becomes energy, just as did the disappearing mass in the fission reaction. In this case even more energy is produced per gram of fuel consumed. To get an idea of the amount of energy produced in fusion, consider the following. One gallon of ordinary water contains about 1/8 gram of deuterium. If each nucleus of this 1/8 gram undergoes fusion, the energy released is equivalent to that obtained by burning 300 gallons of gasoline. One gallon of water replacing three hundred gallons of gas! Wow, fill 'er up!

In order to trigger a fusion reaction, nuclei must be forced to unite. Since nuclei are all charged positively, they

FIGURE 25–21. The nuclei in each case fuse together to form a more massive nucleus.

repel one another and will fuse only if they are fired at each other at a high speed. Fusion is constantly taking place in stars. A star begins forming when the mutual gravitational force between chunks of matter and loose atoms in the vastness of space pulls them together. As they come closer together, the force becomes greater and they fall faster. Soon (in millions of years) they reach a speed great enough so that electrons are knocked free from nuclei. But the particles continue to fall. Finally, their speed becomes so great that collisions result in fusion. Fusion, of course, produces energy, and it is this energy which keeps the stars from continuing their collapse. Our sun is presently in a state of equilibrium between the contracting gravitational force and the expanding pressure due to the fusion reaction. (Approximately 650 million tons of hydrogen are being fused to helium *each second* in the sun.)

To induce fusion here on earth we must bring about temperatures and densities comparable to those in the sun, for the reacting nuclei must have enough energy to overcome their electrical repulsion. This is done in the H-bomb by first exploding a fission bomb. The fission process produces enough heat to start the fusion process.

The first fusion bomb was exploded in 1954. Fusion bombs are not limited by a critical mass factor[8] as are fission bombs. Thus H-bombs have been built that are well up into the megaton range. On the other hand, there is a lower limit to the size of a fusion bomb because it starts with a fission bomb.

CONTROLLED FUSION—THE ULTIMATE ENERGY SOURCE

Controlled fusion is called the ultimate energy source because of the availability of its fuel: *water.* At present it costs about four cents to extract the 1/8 gram of deuterium contained in a gallon of water. Such values would make the fuel costs of even an inefficient fusion reactor almost insignificant. Estimation of the number of gallons of water in the oceans will be left as an exercise for the student.

Economy and the ready availability of large amounts of fuel are certainly significant reasons for attempting to harness fusion, but another important factor is that very few radioactive by-products are formed during the fusion process. A typical fusion reaction fuses two hydrogen nuclei, producing safe nonradioactive helium. Anyone for a blimp ride?

---

[8]Since each of the chunks of uranium which are fired together in the fission bomb must be of less than critical mass, the total mass of uranium is limited to a few times the critical mass.

"Remember—it's better to light just one little thermonuclear power station than to curse the darkness." **(Reprinted with the permission of American Scientist/Sidney Harris.)**

What, then, are the problems? They stem from one factor: temperatures of hundreds of millions of degrees are required. At these temperatures the materials obviously cannot be kept in a glass jar. In fact, at these temperatures atoms do not exist; instead the nuclei and electrons fly around independently of one another (except for collisions). Such substances do not have properties like solids, liquids, or gases and are classified as a fourth state of matter, called *plasma* (see Chapter 8).

Because plasma consists of separate, fast-moving, charged particles, it is affected by magnetic fields. This feature is used to confine the plasma. Figure 25–22 shows one scheme which has shown promise of success. Magnetic fields set up by current in wires around the doughnut-shaped tube keep the plasma circling around inside the tube. The magnetic field serves to pinch the plasma in toward the center of the doughnut and thus keeps it away from the walls of the container (refer to the discussion of the magnetic bottle in Chapter 14). Using this mechanism, plasmas have been confined—for a few millionths of a sec-

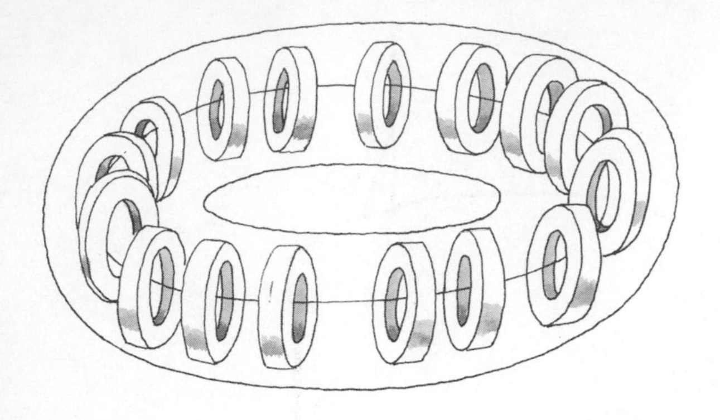

FIGURE 25–22. The principle of one type of "magnetic bottle."

ond. The problem is that the plasma is very unstable and easily leaks out of its confinement.

Another possibility for attaining the necessary high temperature involves the use of lasers. The plan is to drop pellets containing hydrogen-2 past a spot where the beams of a number of powerful lasers focus. If a high enough temperature can be reached—POW!

Many other schemes are being tried by a number of research teams throughout the world. When (and if) success will be achieved cannot be predicted, but there are hopes

FIGURE 25–23. A photo of fusion research under way at the US Atomic Energy Commission's Lawrence Livermore Laboratory. (Lawrence Radiation Laboratory.)

that by the 1980s or 1990s fusion power reactors will be under construction.

OBJECTIVES

Having studied this "bomb chapter" you should be able to:

1. Apply the law of conservation of energy (or mass-energy) to the production of energy in nuclear reactions.
2. Define nuclear fission and, given a fission reaction written in symbolic form, explain the reaction.
3. Describe a chain reaction produced by using mousetraps and corks, and state three similarities between this and a nuclear chain reaction.
4. List and explain three factors which must be controlled to produce a growing fission chain reaction.
5. Briefly relate the history of the development of the first fission reactor.
6. Explain the difference in the way the three principal fission factors are controlled in a bomb and in a reactor.
7. Define critical mass and explain how it is affected by the shape of the material.
8. List two effects of the fission bomb which do not occur in chemical explosions.
9. Compare the radiation dose received by a person due to radioactive fallout to that received from x-rays and natural radiation.
10. Sketch the major components of a power plant producing electrical energy from (a) water, (b) burning of fossil fuels, and (c) nuclear energy.
11. State the advantage of a breeder reactor over present nuclear reactors.
12. Explain the self-limiting feature which makes a nuclear explosion impossible in a water-cooled power reactor.
13. Discuss some safety- and environment-related problems of nuclear reactors.
14. Define nuclear fusion, and identify the source of energy in a fusion reaction.
15. Explain why fusion reactions are more difficult to control than fission reactions.

QUESTIONS—CHAPTER 25

1. A sample of radioactive material is always slightly warmer than its surroundings. What causes this?
2. Chain reactions are common in chemistry. When you set fire to a piece of paper, you excite a relatively few atoms to react with oxygen in the air. Explain in what sense the burning which follows is a chain reaction.

3. In a fission reaction 99 per cent of the neutrons are released at the same time that a fission occurs. The remaining 1 per cent are emitted somewhat later by fission products. Without these "delayed neutrons," control of a nuclear reactor would be impossible. Explain. (Hint: consider the time needed to move control rods.)

4. What causes the mushroom cloud to form after the explosion of a nuclear bomb?

5. From the periodic table find four different examples of pairs of elements that could be formed in a fission of uranium.

6. Your great-grandfather tells you that his physics book said that energy is conserved—period. Why is this statement incorrect?

7. Why would a uranium mine underwater be more likely to explode than one on land? (Assume that there are little cavities throughout the mineral through which water can circulate.) Actually, uranium concentrations in ores are so low as to make explosion impossible under any circumstances, but neglect this minor detail in your answer.

8. Do you think we should have dropped the atomic bomb on Japan? Defend your answer.

9. Why is an atomic bomb needed to start a hydrogen bomb?

10. What are your feelings about atomic power plants?

11. Describe how sources of energy such as coal, oil, water, wind, and tides can all be traced back to the sun as the original source.

12. Why is it impossible for a sustained chain reaction to occur in a very small piece of fissionable material?

13. In addition to fission and fusion as possible means for resolution of our energy problems, other means such as geothermal power, solar power, and tidal power will play an important role. Look up and discuss some of these energy sources.

14. The bomb is much smaller than a reactor. Why don't enough neutrons escape from the bomb to prevent a chain reaction?

15. Why would fusion reactors be far superior to fission reactors?

16. Why is it critical to the production of a chain reaction that neutrons be released when fission occurs?

17. Why was a major part of the scientific work on the Manhattan Project done by emigrants from Nazi Europe?

18. Note the locations of strontium (Sr) and calcium (Ca) on the chart of the elements. Use your knowledge of the chart (from Chapter 6) to explain why strontium is absorbed into bone (which contains quite a lot of calcium).

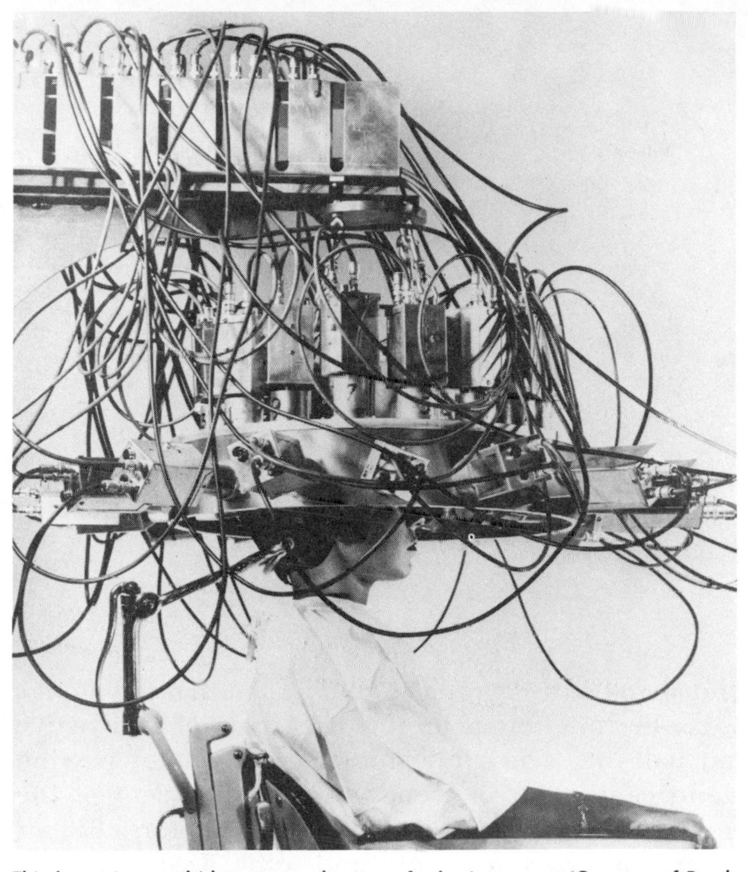

This beast is a multidetector used to scan for brain tumors. (Courtesy of Brookhaven National Laboratory.)

26 PEACEFUL USES OF THE NUCLEUS

Practical applications of nuclear physics are almost as widespread as sex (and a few are almost as interesting). Even a brief discussion of all of these possibilities would fill an entire book, and to keep such a book up to date would require a number of revisions each year. In this chapter we will present a few of the more interesting applications and a little of the nuclear theory supporting them.

GAUGING

A common use of radioactive materials employs a principle discussed briefly in Chapter 24: the absorption of radiation – in particular, of gamma rays. Figure 26-1 shows gamma photons hitting a board. Note that some photons are absorbed after passing through very little wood, others go farther before being stopped, and others make it all the way through. Suppose we draw a graph of the number of gamma rays passing through a certain thickness of the wood (Fig.

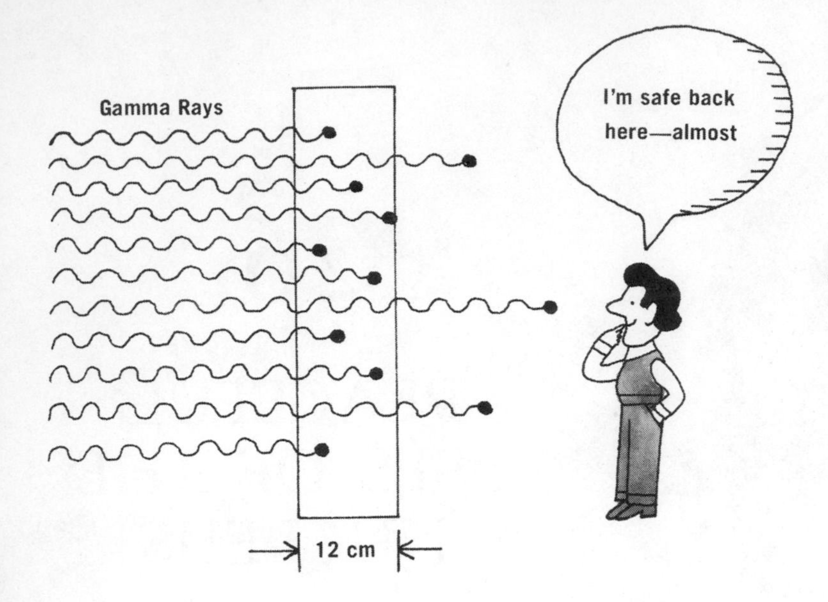

FIGURE 26–1. Some gamma rays penetrate more deeply than others into an absorber. The rate of absorption depends upon the number of rays, so the graph of Figure 26–2 results.

26–2). The shape of this graph should be familiar to you; it's the same as the one found in our discussion of radioactive decay and half-life. The graph indicates that after passing through one centimeter of wood, only 90 per cent of the gamma rays are left in the beam; two centimeters, 81 per cent (90 per cent of 90 per cent); and so on. Just as for the half-life graph, the line never reaches the bottom. No matter how thick the wood, *some* gamma rays will get through (although in negligible amounts with large thicknesses).

Figure 26–3 shows how this absorption principle is used to measure the thickness of a product on an assembly line. A radioactive isotope that emits gamma rays is positioned above the product as it rolls past a radiation detector. A graph similar to Figure 26–2 can be drawn for the material under test. The fraction of gamma rays that actually get through indicates the thickness of the sample.

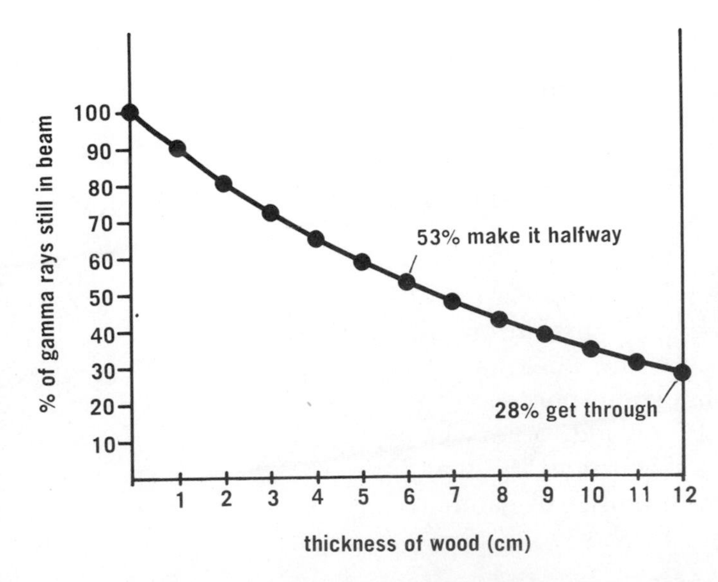

FIGURE 26–2. A graph showing the percentage of the original gamma rays present after passing through various thicknesses of absorber.

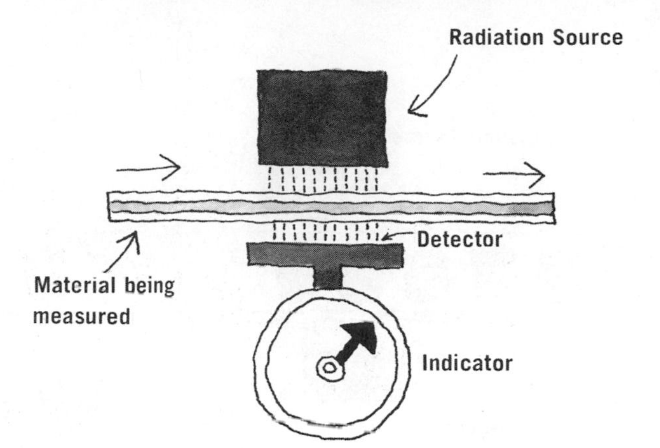

FIGURE 26–3. Principle of the thickness gauge.

Figure 26–4 shows a more elaborate set-up for sandpaper manufacturing. In actual applications, detection devices not only record but also control thickness. For example, the third indicator in the figure can be connected to a motor controlling a trap door on the sand container. If the indicator says, "Too thick!" the motor says, "O.K." and closes the door a little. Another common use of radiation in gauging is illustrated by Figure 26–5. Beverage cans pass between the radiation source and the detector. Any partially empty cans will be kicked out automatically by mechanical devices activated when the resulting overdose of radiation hits the detector (Fig. 26–5).

TRACING

Our second general category of applications uses the fact that in almost all respects a radioactive atom acts exactly

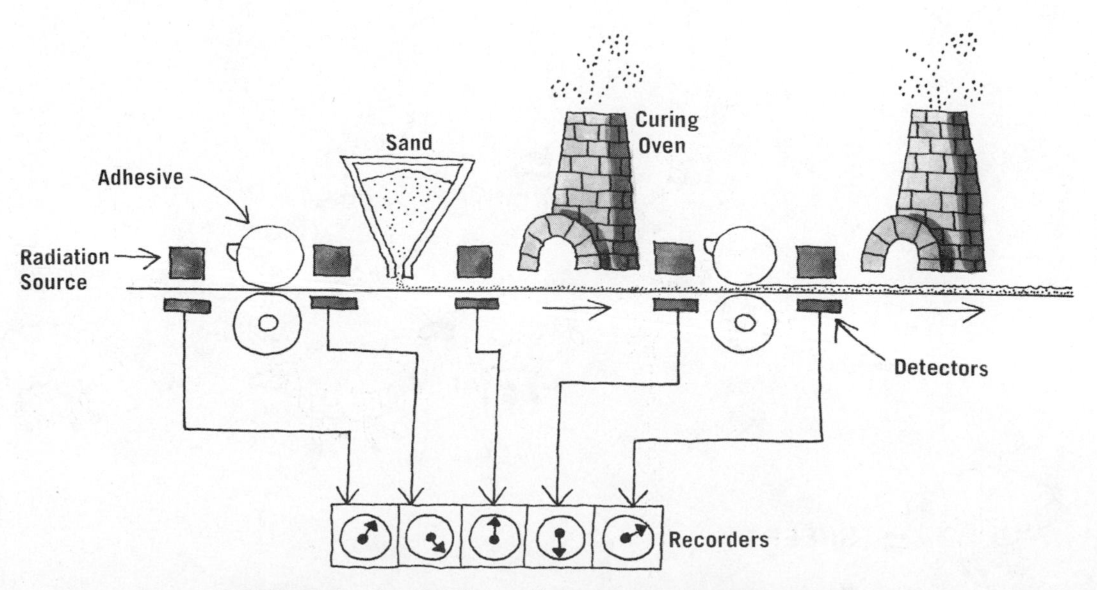

FIGURE 26–4. Use of multiple-radiation gauges with remote recorders.

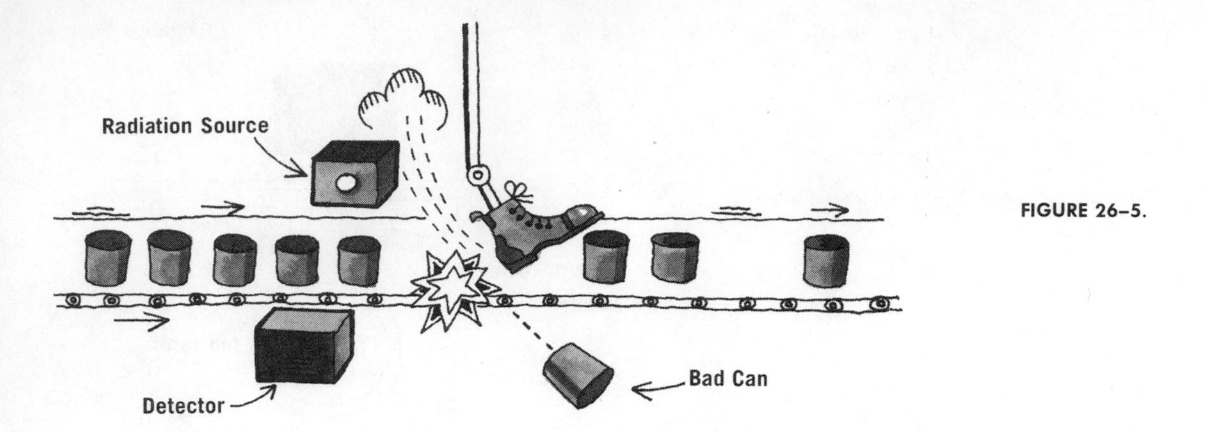

FIGURE 26–5.

like a nonradioactive atom. In one respect it is different: it emits radiation. Therefore, if a number of radioactive atoms are mixed with regular stable atoms of an element, we can use a radiation detector to trace their movement (Fig. 26–6).

One of the most valuable uses of radioactive tracers is in medicine. Iodine-131 is an artificially produced isotope of iodine; natural nonradioactive iodine is I-127. A certain amount of iodine is a necessary nutrient for our bodies and is obtained largely through the intake of iodized salt and seafood. The thyroid gland plays a major part in the distribution of iodine through the body. In order to evaluate the performance of the thyroid, a patient drinks a very small amount of radioactive sodium iodide. Two hours later the amount of iodine in the gland is determined by measuring the radiation coming from the neck area. Figure 26–7 shows a scan of the whole body of a patient four days after a dose of iodine-131 was administered.

The tracer technique is also useful in research. Suppose one wishes to determine the method of fertilization that will

FIGURE 26–6. Radioactive isotopes can be used as tags similar to the way one might tag a few sheep in a flock with bells.

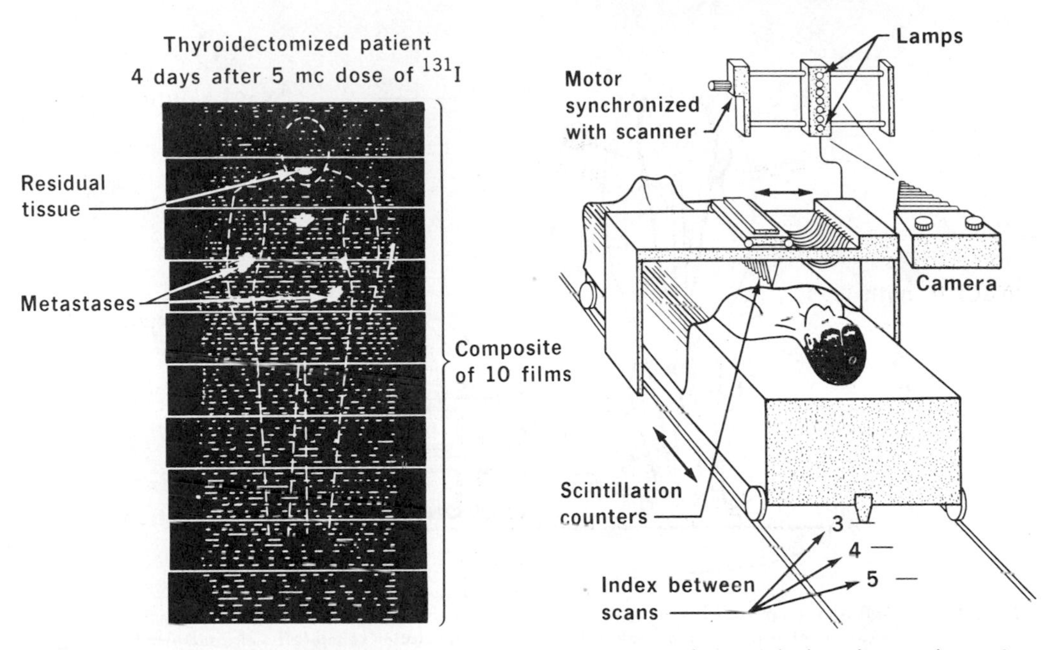

FIGURE 26–7. The drawing at right shows how the detector scans across the man's body to detect radiation from the radioactive iodine. The result is shown at left; note the four areas of intense radiation. (Courtesy of Lawrence Radiation Laboratory.)

provide maximum uptake into a plant. A certain material in the fertilizer, such as nitrogen, can be tagged with one of its radioactive isotopes. The fertilizer is sprayed on one group of plants. For another group it may be sprinkled on the ground; for a third group it may be raked into the soil. A geiger counter is then used to track the nitrogen through the plant.

Tracing techniques are employed as widely as man's ingenuity can devise. Present applications range from checking the adsorption of fluorine on teeth to checking contamination of food-processing equipment by cleansers to monitoring the deterioration inside an automobile engine.[1]

EFFECT OF RADIATION ON LIVING MATTER

It is general knowledge that radiation damages living things. This damage takes place at the cellular level within a body. The cell is a basic building block of life, and living objects grow when cells reproduce themselves by dividing in two. Within these cells are the complicated DNA molecules, whose structure determines the nature of the cell and thereby of the whole living body. (Figure 26–9 gives an idea of the complexity of the DNA molecule.)

---

[1]Radioactive pistons are used, and the oil is checked for radioactivity to determine the amount of wear on the pistons.

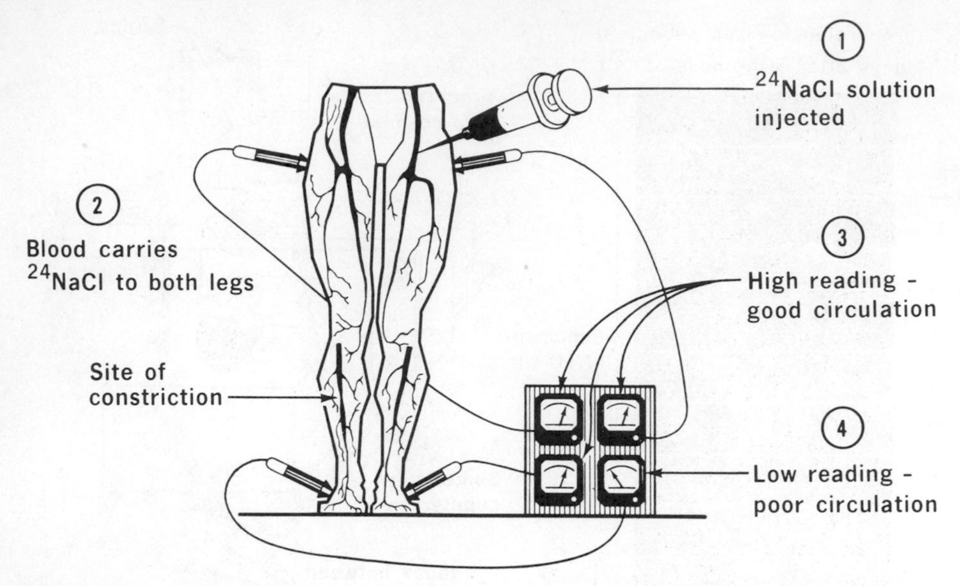

FIGURE 26–8. Salt containing radioactive sodium is injected into a vein in a leg. The time at which the radioisotope arrives at another part of the body is detected with a radiation counter. The elapsed time is a good indication of the presence or absence of constrictions in the circulatory system. (AEC drawing.)

When radiation strikes an atom of a DNA molecule it can jar an atom out of position, thus changing the structure of the molecule. If enough DNA molecules are altered, the cell may not be able to perform its function of reproducing by division. More radiation can completely kill the cell. Cells which are in the process of division are most susceptible to radiation damage. Thus the fast-growing fetus within its mother's body is particularly vulnerable. For this reason, pregnant women are not given routine x-rays. (X-rays, remember, are electromagnetic radiation like gamma rays.) The susceptibility of dividing cells to radiation damage makes radiation valuable in cancer control. Cancerous growth consists of rapidly dividing cells, so bombardment of the cells by x-rays or gamma rays is one of the common treatments for cancer.

Mutation

In cases where radiation damage is not severe enough to render a cell incapable of division, the damaged cell will pass on its newly acquired (due to the damage) characteristics to the two cells into which it divides. Since the damage may mean that these cells are incapable of performing their appointed function, their growth within a body can be serious. It is these rapidly multiplying cells which we call cancer. So radiation can both cause and cure cancer. Ironic!

When cells of a person's body are impaired, the harm is usually confined to that particular person. If your hand is burned by exposure to intense radiation, for example, the effect will not be passed on to your children. If a reproductive cell is involved, however, the effect is passed on to the next generation (if the damaged cell happens to be used in fertilization). Such changes from one generation to the next are called mutations.[2] Before humans started using radiation, about 99 per cent of all mutations resulted from causes other than radiation. In the last chapter we saw that today natural radiation accounts for about 50 per cent of the radiation received by the average person, the other 50 per cent being man-made. Thus man's use of radiation has doubled the number of radiation-induced mutations, resulting in about a one per cent increase in the total number of mutations from one generation to the next. For approximately every 100 mutations that occurred before we started using radiation, one more mutation is now caused by our use of this tool.

Figure 26–10 shows a number of potatoes. All are the same age – about 16 months. The difference is that the better

[2]Mutations also occur in vegetable life. Most of the peanuts grown in the Carolinas are of a type which was developed as a radiation-induced mutation of a less desirable strain of peanuts.

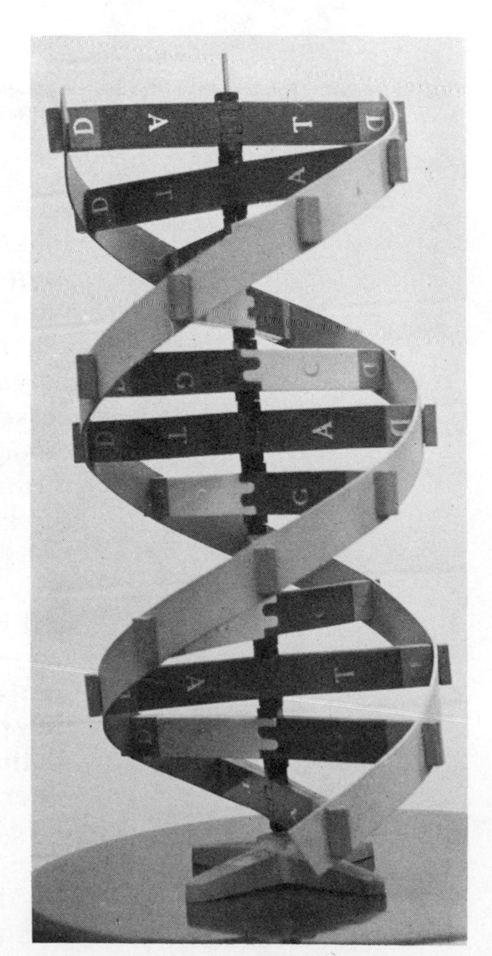

FIGURE 26–9. Vastly simplified model of the DNA molecule. The letters on the connecting links represent various molecular combinations.

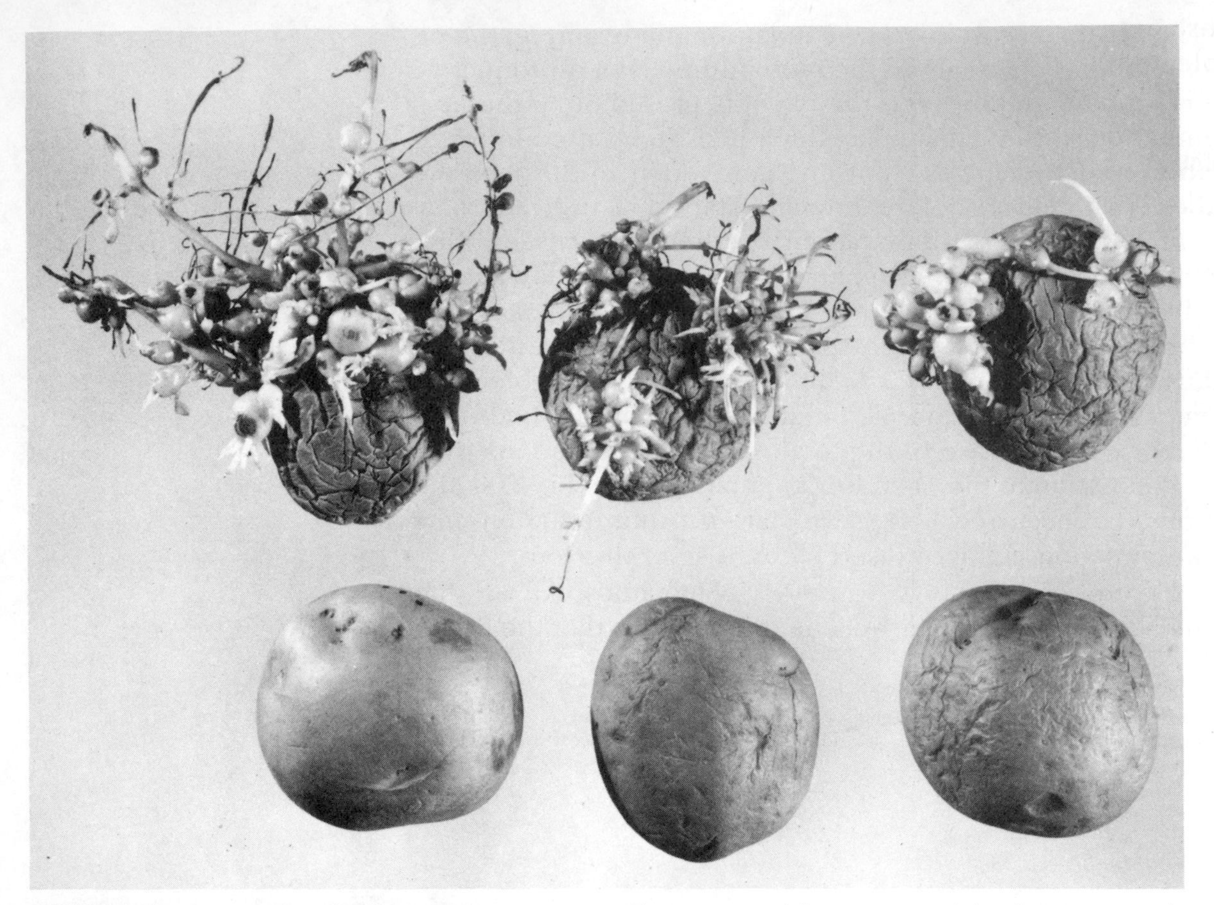

FIGURE 26–10. Potatoes 16 months after exposure to gamma rays. The one at upper left was not exposed; the others were exposed to varying dosages of gamma rays. (Courtesy of Brookhaven National Laboratory.)

looking potatoes were exposed to radiation after harvest. To understand how radiation accomplishes this minor miracle, we must look at the cause of food spoilage.

Foods begin spoiling almost from the moment they are harvested by the farmer, caught by the fisherman, or processed by the butcher. A primary source of spoilage is microorganisms existing within the food. As the food spoils, the microorganisms spread by cell division. This, then, is the key: radiation destroys microorganisms.

Radiation, of course, is only the latest method of food preservation. Others include drying, fermentation, canning, and freezing. Because of concern for safety and the fear of unknown effects of radiation, the US Food and Drug Administration has been very cautious about approving radiation processing of food. The result is that preservation by radiation has undergone far more intensive research than any other method. Practically all types of spoilage-prone food have undergone testing. In many cases neither the appearance nor the taste of the food was affected, and in no case has there been any evidence of adverse effects on people and animals eating the food. It is expected that within the next few years a number of irradiated foods will be available.

Nuclear Dirty Tricks

Radiation has found other uses in agriculture. By irradiating harvested wheat, for example, not only are insects within the wheat killed but also the eggs are destroyed. (Chemical poisons do not kill the eggs.) There is a more interesting application, however, which takes advantage of the fact that reproductive cells are susceptible to radiation damage.

A flying insect called the screwworm inhabits large areas of the southern United States and Mexico, and lays eggs in open wounds of livestock. The burrowing maggots often kill the animal. It was found that the screwworms produce a new generation about every three weeks and that if irradiated at the right stage, the insect could be sterilized with a fairly small dose of radiation. Based on this knowledge, a large screwworm-producing factory was set up, the insects were irradiated, and then starting in 1958 more than two billion sterile screwworms were released from airplanes over Florida, Georgia, and Alabama. Sterile flies are at no disadvantage in mating—they are sterile but not uninterested—so that because of the many unfruitful matings, fewer eggs were hatched for the following generation. After 18 months of releasing 50 million sterile insects weekly, the entire screwworm population was wiped out of the area.

NUCLEAR CLOCKS

Carbon-14

The earth is constantly being bombarded by what we call cosmic rays—nuclear particles and electromagnetic radiation from the far reaches of space. Most of this radiation, fortunately, is absorbed by the atmosphere before it reaches the surface of the earth. When neutrons from outer space enter the upper atmosphere and strike nitrogen nuclei, the following reaction is produced:

$$^{14}_{7}N + ^{1}_{0}n \rightarrow ^{14}_{6}C + ^{1}_{1}H$$

Carbon-14 is a radioactive isotope of carbon. Upon decay it emits a beta particle:

$$^{14}_{6}C \rightarrow ^{14}_{7}N + ^{0}_{-1}e$$

The half-life of carbon-14 is 5800 years. As we saw in an earlier chapter, this means that if we start with a quantity of carbon-14, half of it will have changed to nitrogen-14 in 5800 years. As far as we can tell, carbon-14 has been produced at a constant rate in the upper atmosphere for at least 50,000 years, and during this time it has just as regularly decayed to nitrogen.

take-along-do-it

Use a nail to punch a hole in the bottom of a large tin can—the fruit juice type is O.K. Hold the can beneath a faucet and adjust the water flow from the faucet to a fine constant stream. Although water flows from the hole at the bottom, you'll note that the level of the water in the can rises. As it does so, however, the flow of water leaving the can increases (owing to increased water pressure caused by greater depth of water). Unless the flow of water into the can is too great, an equilibrium point will be reached, at which the amount of water flowing out of the can each second exactly equals the amount flowing in each second. When this happens, the level of water in the can is constant (Fig. 26–11).

The relationship of the TADI with carbon-14 production and decay is as follows: water flowing into the can represents production of carbon-14, and water flowing out represents the carbon-14 changing to nitrogen. The analogy is appropriate because just as the rate of water leaving the can depends upon the amount which is inside, the rate of C-14 decay depends upon the amount which exists. So just as your water level reached a point where it remained constant, in our atmosphere there is a constant percentage of carbon-14 mixed in with the stable carbon-12 isotope.

The reason that we have focused on the production and decay of carbon-14 in the atmosphere (countless other isotopes are formed, too) is that carbon is one of the necessary constituents of life. Plants absorb carbon from the air, and it becomes part of them. Animals eat the carbon-containing plants, thereby making the carbon part of their bodies. But since living matter cannot distinguish between isotopes of an element, carbon-14 is absorbed right along with the normal carbon-12. In the same way that equilibrium is established in the atmosphere, there is equilibrium in living matter between absorption of carbon-14 and its radioactive decay. Thus all living matter contains a fixed percentage of carbon-14 within its cells and—of importance to us here—within its bones.

But all living matter finally dies, and the intake of carbon stops. Radioactive decay of carbon-14 within the body continues, however, making it possible for us to determine how long ago a once living organism died.[3] For example, if we find a kangaroo skull which has only half as much

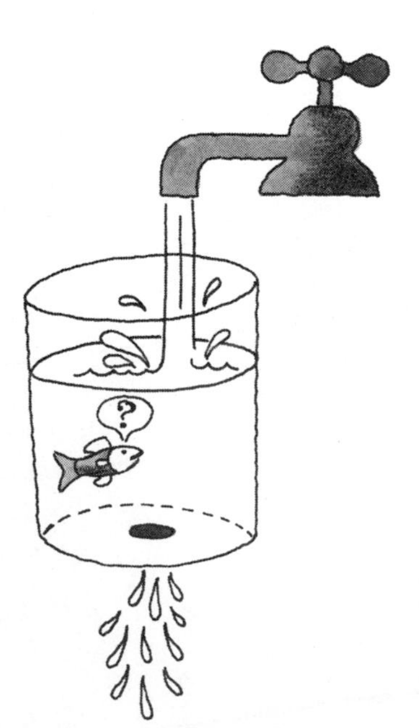

FIGURE 26–11. If the flow of water into the can is constant, the level will reach an equilibrium position, neither rising nor descending.

[3] If you suddenly turn off the water flowing into the can in the TADI, the amount in the can will begin to decrease. Appropriate measurement of the amount of water remaining could, in principle, allow one to calculate when the water was turned off.

carbon-14 as does a 1976 model kangaroo skull, we can conclude that the kangaroo died 5800 years ago.

In order for the carbon dating technique to be valid, it is necessary to assume that the percentage of carbon-14 in the atmosphere has not varied appreciably over the years. Significant variations, either upward or downward, in this percentage would alter the amount of carbon-14 available for intake into the body and would thus lead to inaccuracies in the calculated age of materials.

Using the carbon-14 dating process, samples of wood, charcoal, bone, and shell have been identified as having lived from 1000 to 50,000 years ago. This knowledge has helped us to reconstruct the history of living organisms—including man—during that time span. A particularly interesting example is the dating of the Dead Sea Scrolls. This group of manuscripts was first discovered by a shepherd boy in 1947. Translation showed them to be documents of a religious nature, including most of the books of the Old Testament. Because of their historical and religious significance, scholars desired to know their age. Carbon dating techniques applied to fragments of them and to the material in which they were wrapped established their age at about 1950 years.

Uranium Dating Techniques

The carbon-14 technique is not the only dating process relying on radioactivity. Carbon-14's application is limited to matter that lived within the last 50,000 years (because for matter which lived longer than 50,000 years ago, almost all of the carbon-14 has decayed by now). Other radioactive isotopes must be used for older specimens and for matter which was never alive.

One method, uranium dating, relies on the fact that uranium-235 (half-life, about 7.5×10^8 years) and uranium-238 (half-life about 5×10^9 years) decay eventually to lead-207 and lead-206, respectively.[4] By determining the percentages of lead-207 and lead-206 mixed intimately with uranium in a uranium-bearing rock, one can calculate how long ago the rock was formed. Using this method, the oldest rocks found on earth have been calculated to be about 3.5 billion years old, which is now believed to be a lower limit to the age of the earth.

ACTIVATION ANALYSIS

For centuries, a standard method of identifying the elements in a sample of material has been by chemical analy-

[4]Seven alpha emissions and four beta emissions change $^{235}_{92}U$ to $^{207}_{82}Pb$, and eight alpha and six beta emissions change $^{238}_{92}U$ to $^{206}_{82}Pb$.

sis, which involves testing a portion of the material for reactions with various chemicals. Because each element has its own characteristic chemical properties, one can determine the elements in a material by the results of a number of chemical reactions. A second method of analysis was discussed in Chapter 20. To use this technique, the investigator vaporizes the material to a gas, heats it until it emits light, and analyzes the spectrum of the light. Again, because each element emits its own characteristic set of electromagnetic wavelengths, the elements present can be determined (and from the relative intensities of the different spectra the relative abundances can be determined).

These methods are—and will continue to be—used. But now a third weapon has been added to the arsenal: neutron activation analysis. Both chemical and spectral analysis have the disadvantage that a fairly large sample of the material must be destroyed for the analysis. In addition, extremely small quantities of an element may go undetected by either method. Activation analysis has an advantage over the other two in both of these respects.

If the material under investigation is irradiated with neutrons (either in a reactor or by the use of some other source of neutrons), nuclei within the material will absorb neutrons and be changed to different isotopes. Most of these isotopes will be radioactive. For example, copper-65 absorbs a neutron to become copper-66, which is beta radioactive:

$$^{1}_{0}\mathrm{n} + ^{65}_{29}\mathrm{Cu} \rightarrow ^{66}_{29}\mathrm{Cu} \rightarrow ^{66}_{30}\mathrm{Zn} + ^{\;\,0}_{-1}\mathrm{e}$$

This formula would be of little value if we didn't know more about the radioactivity of copper-66. But we know that its half-life is 5.1 minutes, that the beta particles emitted have known energy limits, and that a gamma ray of known energy is released.[5]

Just as every element has its characteristic set of electromagnetic wavelengths (in the visible region of the spectrum) which it emits upon heating, the radiation from every radioactive isotope has definite properties. No two isotopes have identical radiation characteristics. By examining the radiation emitted by a substance after it has been exposed to neutron irradiation, one can detect extremely small traces of an element in the material. And usually it is not necessary to damage the object in order to run the test.

Neutron activation analysis is used routinely by a number of industries, but the following non-routine example of its use is more interesting. Napoleon died on the island of St. Helena, supposedly of natural causes. Through

---

[5]Beta energies: 2.63 and 1.59 million electron-volts. Gamma energy: 1.04 million electron-volts.

the ages, though, suspicion has existed that his death was not all that natural. After his death, his head was shaved and locks of his hair were sold as souvenirs. In 1961 a sample of this hair was submitted to neutron activation analysis. The result: unnatural quantities of arsenic in the hair! In addition, activation analysis is so sensitive that very small pieces of a single hair could be analyzed. Results showed that the arsenic was fed to him irregularly, and in fact the points in the hair with greatest arsenic concentration could be linked with the days on which records show that he was sickest. The tragedy is that by 1961 the statute of limitations prevented prosecution of the evildoer.

OBJECTIVES

From this chapter you should be able to learn enough about nuclear applications to be able to:

1. Describe the characteristic of radiation which permits its use in gauging.
2. List two manufacturing applications of radiation thickness gauging, and describe one example of how it is used to detect faulty products.
3. Explain the principle of tracing by radioactive isotopes, and list three applications.
4. Explain the distinction between radiation-induced cell damage which is passed on to the next generation and that which is not.
5. State quantitatively the approximate percentage increase in human mutations due to man-made radiation.
6. Describe two examples of radiation-induced mutations that are beneficial.
7. Explain the principle of carbon dating and uranium dating.
8. Perform a demonstration showing an analogy to the carbon-14 equilibrium established in living matter.
9. List two events which have been dated by nuclear dating processes.
10. Explain the principle of activation analysis, and relate one example of its use.

QUESTIONS—CHAPTER 26

1. Why is the shape of the graph in Figure 26–2 the same as for radioactive decay?
2. How could radioactive tracers be used to find the location of a leak in an underground pipe? Of sewage dumped into the ocean?
3. What are your feelings about food preservation by irradiation?
4. The technique of carbon dating relies on one important assumption.

Suppose that we find out that cosmic ray intensity was much greater 10,000 years ago. How would this affect present values of the ages of ancient samples of once-living matter?

5. Why is the carbon dating technique unable to provide accurate estimates of *very* old material?

6. A bone is 40,600 years old. What fraction of the original carbon-14 is present now ($1/8$, $1/16$, $1/32$, etc.)?

7. Which of the applications of radiation in this chapter do you find most interesting?

8. Use an encyclopedia to find a use of radiation not mentioned here.

9. Carbon dating was used on the Dead Sea Scrolls. Would it have worked if they had been tablets of stone? Explain.

By permission of the Atomic Energy Commission.

27

COSMIC RAYS AND ELEMENTARY PARTICLES

Every good detective story has, woven into its plot, an element of suspense. Who done it? Why? The answers are seldom forthcoming until the end. On the next to the last page, the ace crime fighter gathers the suspects together and slowly circles the room. He reflects on all the clues he has unearthed and pieces them together into a workable solution. Turning swiftly on his heels, he points a stabbing finger directly at the villain. All that remains is a tearful admission of guilt and the words "The End."

There have been very few classic elements of suspense in this text so far, but we now will explore some. The mysteries will be obvious; the questions many. A large number of clues will be gathered, and indeed, the only serious problem that will occur in our thriller is that we will provide very few answers. Perhaps what we need is an ace crime fighter to reduce the clues to their ultimate and obvious conclusion.

COSMIC RAYS

Discovery

The story of cosmic rays had its beginning with a simple observation with an electroscope. It will be remembered from Chapter 12 that an electroscope is a device used to de-

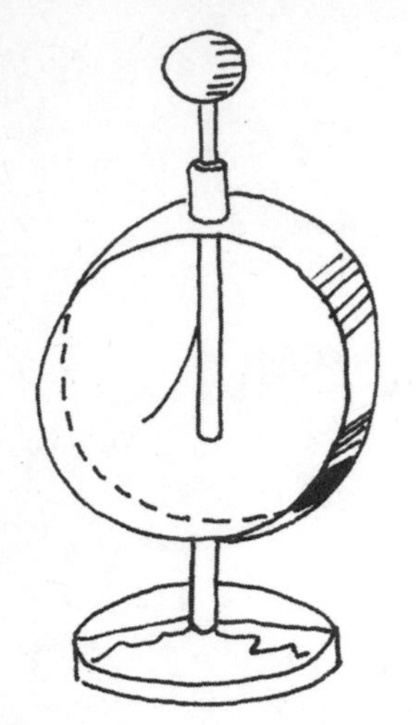

FIGURE 27–1. The position of the leaf indicates that the electroscope is charged.

tect the presence of electric charge. When charged, the leaves of the electroscope stand apart (Fig. 27–1). About 1900, it was noticed that if an electroscope was charged and left untouched overnight, morning would find the leaves either partially or totally collapsed. Charge was leaking off—but how? It was known that air will allow some leakage[1] but too much was observed to be accounted for in this way. Most scientists of the time felt that natural radioactivity present in surrounding objects was the culprit. Radiation coming from radioactive materials has enough energy to ionize some of the atoms of the air, and if many such free charges are present, some of them will move to the electroscope and neutralize a portion of the charge on it. In order to determine whether radioactivity was the cause of the leakage, experimenters surrounded the electroscope with shielding that was thick enough to substantially block out the effects of the strongest radiation source known at that time. The attempt proved unsuccessful.

In a supreme effort at shielding, an electroscope was taken out onto the ice of Lake Ontario in 1907. Here the electroscope was not in the presence of any possible sources of radioactivity but was instead surrounded by hundreds of feet of nonradioactive water. Observations showed that the discharge decreased slightly but did not vanish entirely, indicating that radioactivity of rocks and soil was only one of the culprits responsible for the electroscope's discharge. The remainder of the trouble must have been coming either from the atmosphere or from outer space.

Balloon flights then became the way of life for electroscopes. An Austrian scientist, Victor Hess, launched an instrument-laden balloon and showed that at low altitudes, the rate of discharge of the leaves slowed as the balloon rose and left behind the earth's radioactivity. But as the balloon soared above 2000 feet, the discharge began to increase again and became more rapid than it had been on earth! Upon analyzing his data, Hess concluded that the radiation must have its origin in outer space, so the name "cosmic rays" was coined to describe it. In 1936, Hess was awarded the Nobel prize for his solution to the puzzling problem.

In 1923, a series of experiments by Robert Millikan added further confirmation to the theory of the extraterrestrial origin of these cosmic rays. The experiments were performed at Lake Arrowhead and Lake Muir in California. These lakes are snow-fed, which at that time insured that no radioactive contamination was present in their water (with A-bomb fallout this could not be claimed today). The two lakes are several hundred feet above sea level, with Lake

[1]Air is a conductor of electricity, although a poor one. In general, the conductivity of air increases on humid days, when more water vapor is present in the air.

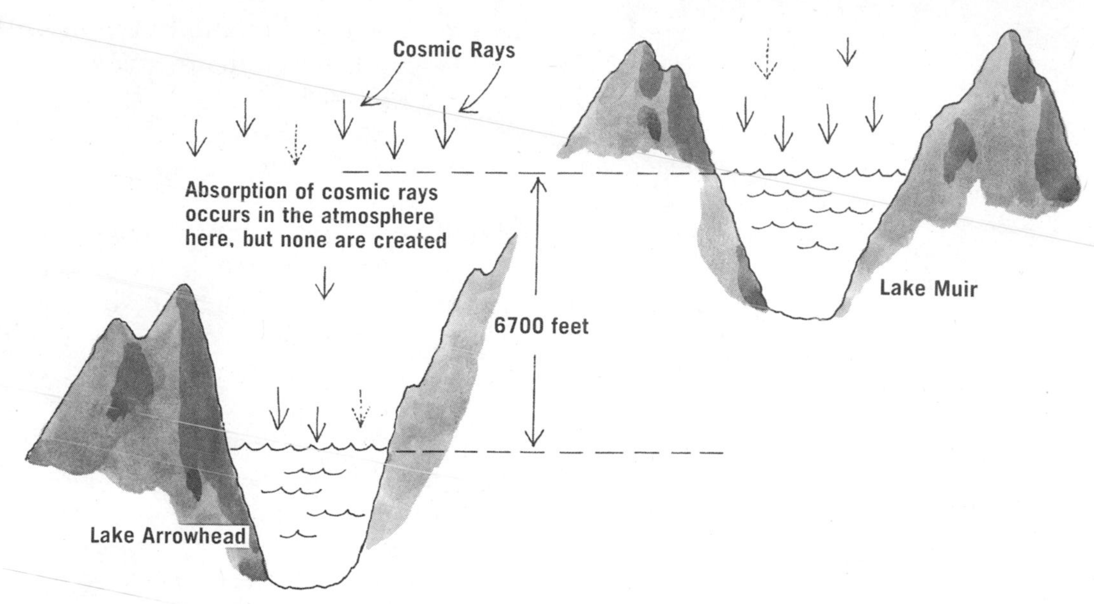

FIGURE 27–2. Cosmic rays are absorbed by the 6700 feet of air between the two lakes.

Muir being some 6700 feet higher than Lake Arrowhead. Millikan was able to show that the rate of discharge at the surface of Arrowhead was just enough less than at Lake Muir to be accounted for by absorption of cosmic rays in the 6700 feet of air between the levels of the two lakes (Fig. 27–2). His results proved that the source of cosmic rays must be extraterrestrial, because there was no source of radiation in the 6700 foot blanket of air which separates the lakes.

An additional piece of information surfaced from Millikan's work. Each of the lakes is surrounded by mountains that serve as a tube for a cosmic ray "telescope." The mountains prevent radiation from reaching the electroscope (Fig. 27–3) except from directly overhead. As the earth turns, this natural telescope points toward different parts of the sky, and it was found that regardless of the direction that the earth or the "telescope" pointed, the electroscope discharged at the same rate. Thus, the cosmic rays did not

FIGURE 27–3. The mountains ensure that cosmic rays reaching the electroscope come from a limited portion of the sky.

come from the sun, as some thought, but were instead coming in uniformly from all directions in space.

Detection

It was generally held by scientists that cosmic rays were gamma rays. This assumption was based largely on the fact that gamma rays have a greater range in air than do alpha or beta particles. The nature of cosmic rays could only be speculated on, however, until some method could be devised that would make them "visible." The answer was found through the use of a cloud chamber (which was described in Chapter 24). A magnetic field can be used in conjunction with a cloud chamber to bend the path of any charged particles passing through it, thus extending its use as a tool for analysis. By observing how the particles are bent, we can determine whether they have a positive or negative charge. Also, at a given energy it is found that massive particles are bent less than lighter ones if they both have the same charge. In addition, it is found that gamma rays and x-rays will not produce tracks at all. They ionize the air in the chamber, but the distance between ions—and therefore between drops of fog along the trail—is so great that a path cannot be discerned.

With the cloud chamber as a tool, the components of cosmic rays were soon determined. It was found that there are two different kinds, which came to be known as primaries and secondaries.

Types

Primaries. The primaries are those which actually come from outer space and strike the upper layers of the atmosphere. These are usually protons and alpha particles, but a large number of heavy nuclei such as iron and nitrogen have also been seen. Even in these days of machines that can give very high energies to nuclear particles, we are still unable to equal the energy possessed by the most energetic of these primaries. They have so much energy, in fact, that they travel almost at the speed of light. If one should strike the human body, it would pass through it, probably without hitting anything at all. And this is happening to you right now. Stay low!

Secondaries. The effect of primaries on the upper layer of the atmosphere is dramatic. A multiplying effect occurs which produces a large number of other particles and gamma rays: secondaries. For example, if we follow the path of a typical incoming primary, we would find that a probable

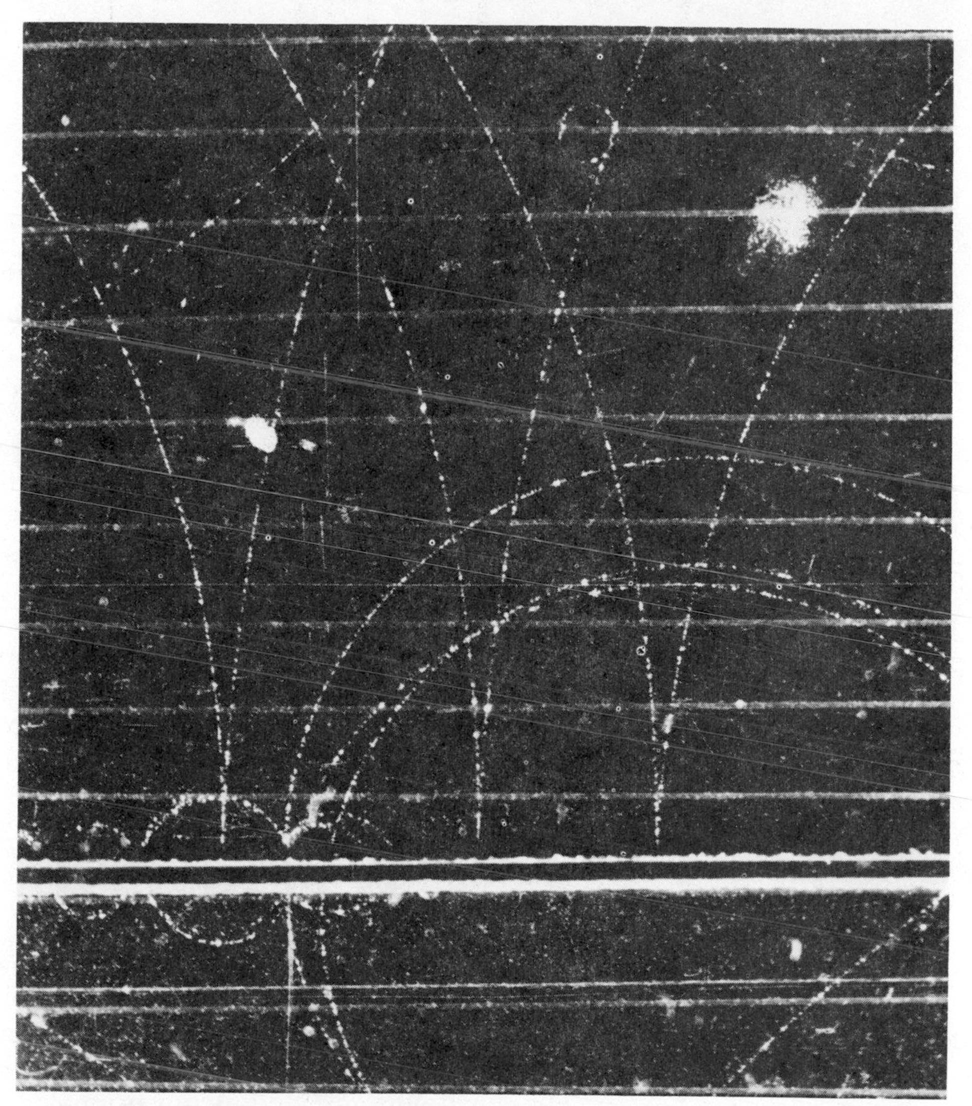

FIGURE 27–4. Cloud chamber photograph, showing tracks of submicroscopic particles. (University of California Lawrence Berkeley Laboratory.)

fate for it would be a collision with a molecule in the air. Like a bullet passing through a light bulb, the primary shatters the unfortunate air molecule into fragments, which also leave the scene of the collision with a very high speed. A number of them find unsuspecting air molecules with which to collide, and a multiplication effect somewhat like a chain reaction occurs. A few incoming primaries may lead to the production of a very large number of secondaries (Fig. 27–5). These are called "showers."

The incoming primary gives much of its energy to the fragments produced in a collision, which in turn lose much of their energy as they undergo similar collisions. A few high-energy specimens of both primaries and secondaries are present at sea level, however. The result of this production of secondaries and their consequent loss of energy is

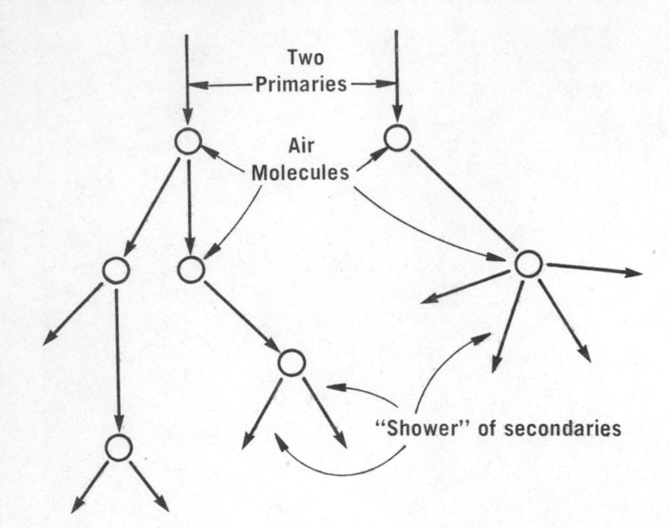

FIGURE 27–5. Incoming primary rays produce secondary rays.

that as we go upward from the surface of the earth, we find the number of cosmic rays increasing until we reach a height of about 15 miles. At this point, the number begins to decrease, until at the top of the atmosphere only primaries are present.

Cosmic Rays and the Earth's Magnetic Field

In 1927, evidence came to light which showed that there is a variation in the number of cosmic rays reaching sea level at different latitudes on the earth. If the intensity of cosmic rays is measured at various locations starting at the north pole and moving toward the equator, it is found that a gradual drop-off starts to occur at 45 degrees north latitude. At the equator, the intensity decreases by about ten per cent of its maximum value, which occurs at the pole. The same results are obtained when starting from the south pole. It had long been suspected that this might be the case even before careful measurements were made. Cosmic rays are charged particles and, as such, are deflected by magnetic fields. The field of the earth is sufficiently strong to deflect many low-energy primaries, as Figure 27–6 shows. A primary entering the vicinity of the earth and heading toward a point in the higher latitudes is relatively unaffected by the field of the earth because it is traveling more or less along one of the lines of the earth's magnetic field and, therefore, is not bent appreciably. However, at the equator, the primaries enter at right angles to the magnetic lines, and the deflecting force is greater. In fact, many of the low-energy particles are deflected enough so that they miss the earth completely.

Van Allen Belts. With the advent of orbiting satellites, a most astounding discovery was made by James Van Allen of the University of Iowa. Instruments carried aloft in spacecraft demonstrated that two belts exist high in the atmosphere, in which many charged particles are trapped. A cosmic ray particle moving like the one shown in Figure 27–7 will be trapped and will become a part of these Van Allen belts. The force on the particle due to the earth's magnetic field causes it to move in a spiral. As the particle gets closer to the north pole of the earth, the field gets stronger, and the spiral gets smaller and smaller (the magnetic bottle effect of Chapter 14 again). Finally, the particle turns back and retraces its path to the south pole, where the process repeats itself. A proton trapped in this way may spiral from pole to pole once every few seconds and may remain trapped in this way for several hundred years. Figure 27–8 indicates that there are two belts in which the trapping process occurs. The inner belt is about 2000 miles above the earth's surface, and the second is about 10,000 miles above the earth.

For the most part, these trapped particles go unnoticed by us in our day-to-day lives. However, they occasionally make their presence known when large numbers of them are spilled out over the poles. The spilled particles lose energy in collisions with air molecules (Fig. 27–9), which explains the beautiful auroras that light the northern and southern skies.

FIGURE 27–6. Cosmic rays are deflected by the earth's magnetic field.

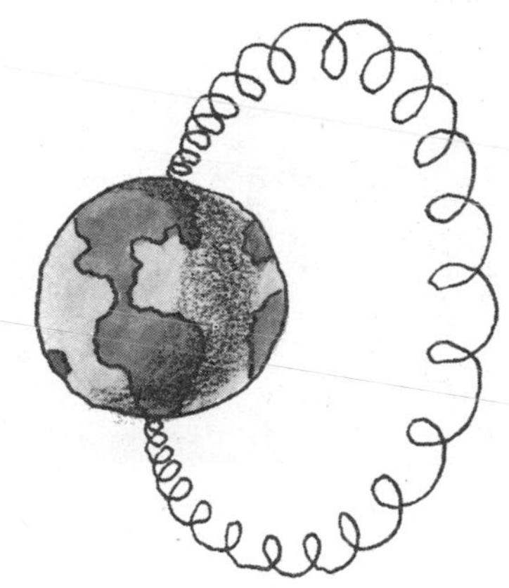

FIGURE 27–7. Cosmic rays spiral around in the magnetic field and become trapped, forming the Van Allen belts.

Where Do You Come From, Cosmic Ray?

The experiment by Millikan, in which the mountains near Lakes Arrowhead and Muir served as a cosmic ray "telescope," proved that there is no preferred direction from which these rays come. Since Millikan's experiment, more sophisticated methods for scanning the sky have been developed, but the results are not changed. The particles apparently reach earth in a uniform manner from all directions.

It is thought that many of the low-energy rays emanate from the sun, and that external effects such as the magnetic field of the earth and perhaps also of other planets subsequently mix up their direction of travel such that they appear to come equally from all directions. However, scientists believe that it is inconceivable that the more energetic primaries are produced by the sun. Their origin is unexplained. There exist in space exploding stars called novae and much brighter ones called supernovae, which are energetic enough to be possible sources. Perhaps they are responsible, or perhaps pulsars are the source. A pulsar is a star that varies in brightness at a very fast rate, and the great amount of energy it emits thrusts it into the forefront as a possible source. Perhaps far out in the vast reaches of space there is something we simply have not discovered. Perhaps

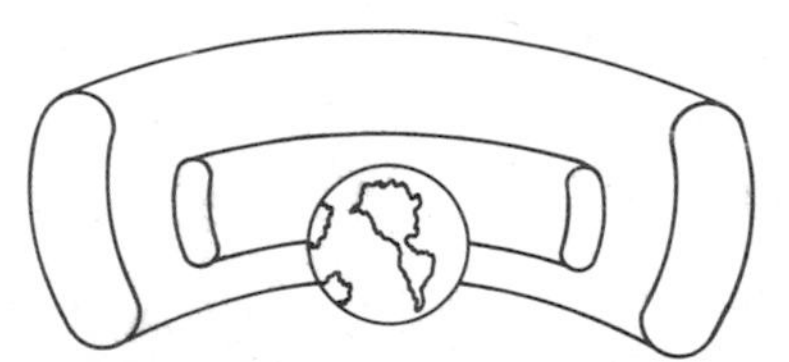

FIGURE 27–8. There are two Van Allen belts.

FIGURE 27–9. When cosmic rays collide in great numbers with air molecules, the aurora borealis results.

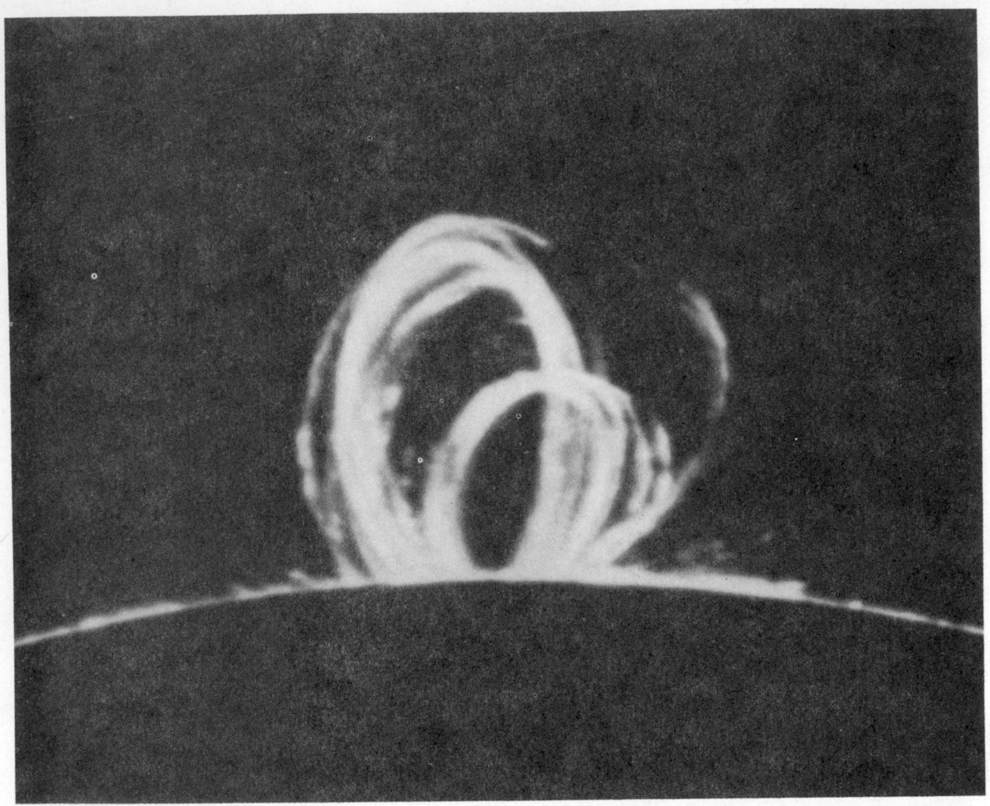

FIGURE 27–10. A solar flare. The gas is concentrated along magnetic field lines. (Sacramento Peak Observatory. From Smith, E., and Jacobs, K.: *Introductory Astronomy and Astrophysics.* Philadelphia, W. B. Saunders Co., 1973.)

the discoveries have already been made but no one has found the proper connections. Perhaps. . . .

Obviously, the verdict is not in yet, but the study of cosmic rays has added much to our knowledge. Examination of cloud chamber photographs has revealed new information about particles found inside the nucleus of the atom; these so-called elementary particles will be discussed in future sections. Before moving to this topic, however, let's look at an interesting use of cosmic rays.

Search for the Tomb of Chephren

In 1966 cosmic rays were used in a most unusual fashion. At that time, a pyramid project was established by the United Arab Republic and the United States. A group under the leadership of Luis Alvarez of the United States took up the task of "exploring" what is referred to as the second pyramid of Giza. The reason for the study can best be explained by considering a cutaway view of the Great Pyramid of Cheops (Fig. 27–11). The architects outdid themselves in their design of the interior of this pyramid. Shown at U is an underground chamber; G is the Grand Gallery leading to

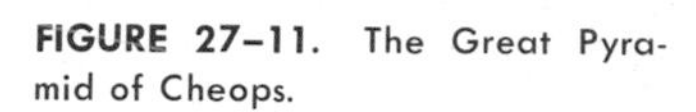
FIGURE 27–11. The Great Pyramid of Cheops.

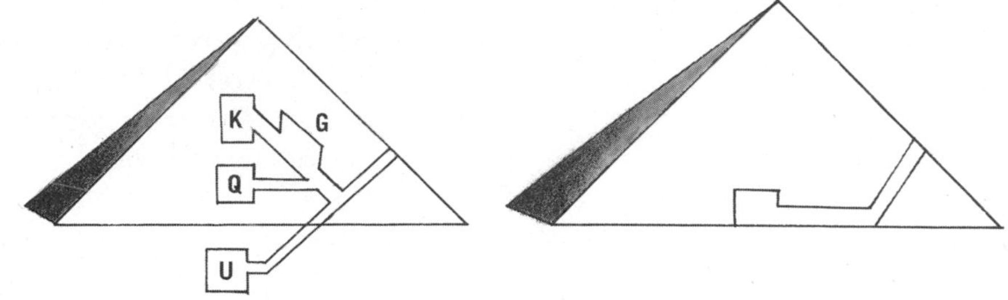

FIGURE 27–12. The pyramid of Chephren, son of Cheops.

the King's Chamber at K, and below the King's Chamber, at position Q, is the Queen's Chamber. Cheops was succeeded on the throne by his son Chephren. The pyramid of Chephren has been examined by archeologists, and the result of their efforts is shown in Figure 27–12. One small chamber, known as the "Belzoni" chamber, was located near the center of the base of the pyramid. Considering the elaborate design of his father's pyramid, it seemed likely that other chambers existed, and exploration of this possibility was the task given to the pyramid project group.

To examine the pyramid they placed elaborate cosmic ray detectors in the Belzoni chamber. Pointing the detectors in all directions,[2] they hoped to see changes in the number of cosmic rays detected. From the known shape and dimensions of the pyramid, they were able to predict how many cosmic rays would be absorbed by the pyramid, and hence, they could determine how many should come from any given direction. As the cosmic rays pass through the limestone blocks of the pyramid, many of them are absorbed, but if a room is in the path of the ray, more particles than expected will be able to get through and reach the detectors. Even a room as small as five feet on a side should have been "seen." In effect, the investigators were x-raying the pyramid.

At this writing, an appreciable amount of the pyramid has been scanned, but no rooms have been located. Even though it has not been successful in finding a room, the accuracy of the method has been demonstrated, and similar techniques could be used to determine the thickness of rock layers above underground tunnels, mines, and caves.

ELEMENTARY PARTICLES

The Positron

The electron, the proton, the neutron—are these all there are? They are certainly the most important and the

[2]More elaborate cosmic ray detectors than electroscopes are now used. They are able to indicate the direction from which the detected rays are coming.

most easily observed of all the nuclear particles. It was in 1932, however, that new kinds of particles began to appear on the scene. The first of this new breed was the *positron*, discovered by Carl Anderson. Since its discovery, the proliferation of particles has reached astounding proportions. The positron is identical to the electron in all respects except one: the electron has a negative charge; the positron has a positive charge. The discovery was made by Anderson while he was examining a photograph of particles passing through a cloud chamber. The photographs were taken with the cloud chamber placed in a magnetic field, which caused the particles to be bent into a circular path. Anderson noted that one of the particles behaved in an unusual fashion. It appeared to be an electron because the degree of curvature of its path was the same as an electron's would be, but it was bent in the opposite direction from that expected for an electron.

Since the initial discovery, the positron has been observed in a number of ways. Perhaps the most common way that a positron is produced in nature is by a process called *pair production*. In this process a gamma ray traveling through space passes close by the nucleus of an atom. (This is a very bad thing to do if you are a gamma ray.) If it passes close enough to the nucleus, it will have to pass through a very strong electric field. The electric field of the nucleus is, in fact, strong enough to completely destroy the gamma ray. But there is more to the process than merely a simple bit of destruction. A gamma ray carries energy, which cannot just disappear. Observations reveal that something is produced in place of the gamma ray—in fact, *two* somethings. Two particles suddenly appear, each picking up half the energy of the original gamma ray; one of these is an electron, the other is a positron. This process is important in the sense that a new particle appears from the ashes, but it carries additional significance: a gamma ray (pure energy) has been transformed into *mass*. This is a striking confirmation of Einstein's prediction that energy and mass are not separate entities but that transformations can take place between them.

Antimatter

In all respects except one, the electron and the positron are alike—the difference is in their charge. It has become customary to refer to one of them as a particle and to the other as its antiparticle, with the positron taken to be the antiparticle half of the team. Perhaps if the electron has an antiparticle, then so do our old friends the proton and the neutron. In fact, they do—they are called simply the antiproton and the antineutron.

The antiproton can be produced in exactly the same fashion as described for the positron. The only difference is that the gamma ray that is destroyed must have a much higher energy than is needed for positron production. The process seems never to happen freely in nature, but it can be made to occur in a laboratory. The antiproton, as its name should now suggest, is exactly like the proton except that it has a negative charge. The difference between a neutron and an antineutron is much more subtle than for our other two sets of twins. Since the neutron has no charge at all, its counterpart cannot very well have an opposite charge. But there is a difference, as we shall see later.

We now have three antiparticles—the positron, the antiproton, and the antineutron. We pictured the atom as consisting of a collection of protons and neutrons in the nucleus, with electrons orbiting about. Could it be that there are atoms composed of antiparticles? We have all the necessary pieces to construct such antimatter. Before this question can be answered, let's consider one more piece of information about the positron.

The positron was discovered in 1932. But why had it not been seen before? The reason is that the positron simply does not have a very long life expectancy when allowed to wander around freely in the air. If a positron is created, it moves through space and very quickly encounters an electron. They attract one another, being of opposite charge, and begin to spin around like two kids holding hands and whirling one another around (Fig. 27–13*A*). This spinning dance does not last very long because soon after they come together, they destroy one another (Fig. 27–13*B*). In the process in which the positron was produced, a gamma ray was destroyed and the positron and electron created; now the process goes the other way. When the positron and electron disappear, two or three gamma rays appear to replace them. Likewise, when a proton and an antiproton meet, they annihilate one another, producing gamma rays. And we promised to show that neutrons and antineutrons differ. You

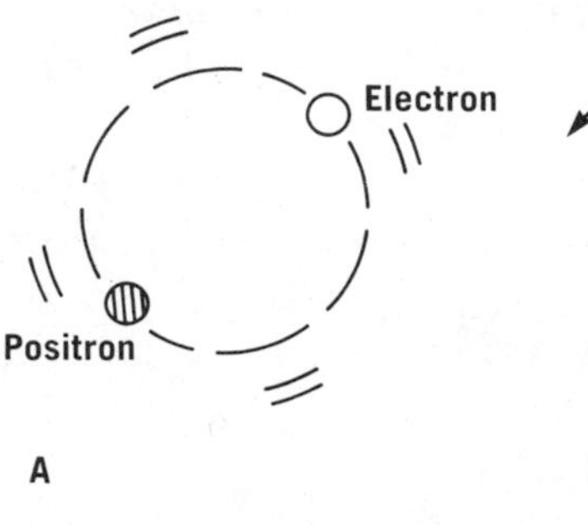

B

FIGURE 27–13. *A*, The electron and positron, having opposite charge, will orbit one another, but when they collide, both are annihilated, releasing gamma rays (*B*).

guessed it—they annihilate (and of course, two regular neutrons do not annihilate one another).

So . . . perhaps somewhere out in the universe is a solar system composed of planets made completely of antimatter. Walking the antimatter grassy plains of an antimatter earth are antimatter people. What would these people be like? Do they sterilize objects with a septic? Would they put freeze in their cars to winterize them? When things grow old and decrepit do they become quated? Do they wait for us to contact them in eager cipation? If we do make contact, will everything thereafter be climactic for them? One point is not speculation, however: if such a world does exist, it would not be advisable to go there to visit. As soon as our spaceship landed, we would find ourselves involved in a gigantic matter-antimatter annihilation. Gone—in a puff of gamma rays.

Antimatter in Siberia? Some scientists believe that a less dramatic version of an encounter with antimatter has already occurred right here on earth. In 1908 there was a gigantic explosion in a remote region of Siberia which leveled trees within a distance of 20 miles, and the shock waves from it were detected around the world. Many theories have been presented in an attempt to explain this event. It could have been a giant meteor, but many scientists speculate that it could have been produced by a lesser meteor of antimatter.[3]

There is still more speculation of a possible major role of antimatter in the universe. Astronomers observe many very distant objects (quasars) which give off a fantastic amount of energy—more energy than can possibly be obtained from normal fusion reactions taking place in stars. Perhaps—just perhaps—these objects are not really objects, as such. Perhaps all that energy is coming from a collision between an antimatter galaxy and a regular galaxy.

The Neutrino

Of all the elementary particles, one of the most difficult—if not the most difficult—to detect is the neutrino. This particle was first proposed by Wolfgang Pauli in 1930 to explain some puzzling features of beta decay, and it was named the neutrino (little neutral one) by Enrico Fermi shortly thereafter. To appreciate the difficulties of detecting it, consider some of its properties: it is electrically neutral, it has no rest mass, and it practically never interacts with mat-

---

[3]In 1973, a team of University of Texas physicists offered another even more dramatic possibility. They suggest that a black hole (see Chapter 3) may have struck the earth. A black hole as small as a grain of sand might be devastating enough to produce the chaos seen.

ter. As an example of the last characteristic, if a beam of 10^{12} neutrinos should be fired at the earth, all but about one of these would pass directly through without noticing the earth's presence at all.

Because of its unusual properties, it might be surprising to you that anyone would even suspect the existence of such a particle. But if it were not for the elusive neutrino, two of the most sacred principles of physics, the conservation of energy and the conservation of momentum, would be violated. To see why this is so, let's return to the subject of beta decay, which was introduced in Chapter 24. Any beta decay process is suitable to demonstrate the dimensions of the problem, but let's choose a particularly simple one—a neutron decaying into a proton and an electron as follows:

$$^{1}_{0}n \rightarrow {}^{1}_{1}H + {}^{0}_{-1}e$$

In this process, we know the mass of all the particles involved, and therefore it is not difficult to see that the mass of the neutron is greater than the sum of the masses of the proton and electron. Such situations have occurred before, and in those cases, we found that the missing mass had been converted into energy, which was given to the products of the reaction (in this case to the proton and electron). The difficulty that arises here is that the electron and proton do carry away some of the energy, but only rarely do they carry away as much as they should. The conservation of energy is a principle built on hallowed ground, and physicists were not going to give it up without a fight. In order to preserve the conservation of energy principle, Pauli predicted that there was another particle—the neutrino—coming off with the proton and electron and carrying with it the necessary energy to insure agreement with the sacred conservation law.

There was additional evidence to indicate the need for such a particle, because momentum also was not being conserved in beta decay processes. For example, when a neutron at rest decays, the proton and electron should zip away in opposite directions in order to conserve momentum, but they rarely did. Instead, they moved away in a V as shown in Figure 27–14A. Again the neutrino solves our problem, if

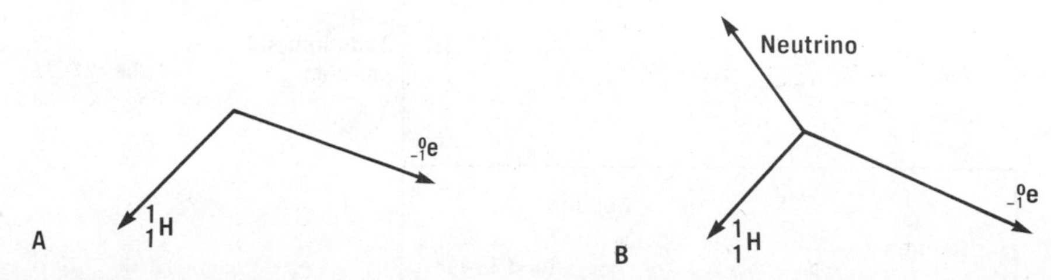

FIGURE 27–14. A, The observed motion of the proton and electron after beta decay. B, In reality a neutrino must also move off.

we assume that it carries momentum as it zips away along a path as in *B*.

If the neutrino exists, the beta decay reaction should be written as follows:

$$^{1}_{0}\text{n} \rightarrow {}^{1}_{1}\text{H} + {}^{0}_{-1}\text{e} + \nu$$

where ν (the greek letter nu) is the symbol for the neutrino. It produces some satisfaction to write the reaction in this way, but it's a somewhat unsatisfactory satisfaction until the neutrino is actually detected. In order to detect the neutrino, a large number of them are needed because they interact so weakly with other particles. Such a source is provided by a nuclear reactor, because many fission fragments are beta-radioactive and therefore emit neutrinos. The process sought was one in which a neutrino strikes a proton to produce a neutron and a positron as follows:

$$\nu + {}^{1}_{1}\text{H} \rightarrow {}^{1}_{0}\text{n} + {}^{0}_{1}\text{e}$$

In 1956, C. Cowan and F. Reines were able to detect this event by the scheme indicated in Figure 27–15. The reaction occurs in a cadmium-salt solution, which emits a flash of light as the atoms of the solution are ionized by the positron. The positron then will shortly encounter an electron, and the two annihilate one another, producing two gamma rays. These gamma rays pass through the liquid layers surrounding the cadmium-salt solution, which signal the presence of the gammas by emitting flashes of light. While all this is going on, the neutron has been wandering around and is finally captured by cadmium, at which point three gamma rays are emitted. These three gammas also produce flashes of light as they pass through the liquid layers that sandwich the cadmium-salt solution. In short, every neutrino collision is signaled by flashes of light, which occur at accurately known time intervals. The detection of these flashes verified that the process occurs as predicted and confirmed the existence of the neutrino. Trapping the neutrino

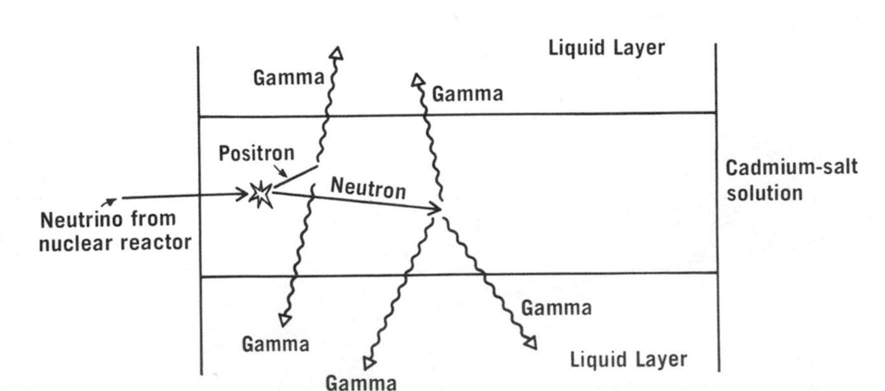

FIGURE 27–15. The scheme for detecting the neutrino.

in this indirect way was satisfying even though its existence had been hypothesized some 20 years earlier.

Other Particles

Economists worry about inflation; since 1932 physicists have had to worry about inflation also, but of another kind. In the last few sections of this text, we have increased our original three particles (electron, proton, and neutron) by adding three antiparticles and the neutrino. Six years after discovering the positron, Carl Anderson and a co-worker discovered two more new particles, one positive and one negative, which have come to be known as muons. Their mass is intermediate between that of the electron and proton. In 1947, another group of particles, called pions, with masses in the same range as the muons was discovered. Both groups were discovered from cloud chamber photographs of cosmic ray collisions with air molecules. Since then the known particles have grown in number and complexity. There are *many* more particles known now than were mentioned in this chapter. The situation has often been described as an embarrassment of riches. We have a large number of particles but little understanding of the part they play in the design of nature. Perhaps when a few of the questions raised by the presence of these particles are answered, new insight into how our world works will come forth. If so, the next generation of college students may look to these particles to find the answers to questions not even dreamed of today. Such is physics.

OBJECTIVES

Careful study of this cosmic chapter should enable you to:

1. Explain how an electroscope acts as a detector of cosmic rays.
2. Trace historically the steps taken to determine the source of what were eventually found to be cosmic rays.
3. Explain how the presence of mountains allowed scientists to determine directions from which cosmic rays came.
4. Name and describe the two classifications of cosmic rays.
5. Relate the rate of electroscope discharge to height above the earth, and explain the relationship.
6. Describe the interaction between cosmic rays and the earth's magnetic field that results in the Van Allen belts.
7. State three possible sources of cosmic rays.

8. Explain how cosmic rays are being used in research on ancient pyramids.
9. Define antimatter, positron, antiproton, and antineutron.
10. Describe electron-positron formation and relate it to Einstein's theory of relativity.
11. Speculate on the existence of an "anti-universe."
12. Describe a theory explaining the gigantic 1908 explosion in Siberia.
13. Explain why neutrinos were hypothesized and why they are difficult to detect.
14. Describe briefly the experiment in which neutrinos were detected.

QUESTIONS—CHAPTER 27

1. Devise a system wherein two geiger counters can be used together to determine the direction of cosmic rays.

2. Why can a gamma ray photon produce an electron-positron pair, but an ultraviolet photon cannot?

3. If gamma rays have momentum, why is it necessary that at least two gamma rays be produced in an annihilation?

4. If we could visit an antimatter world, carrying with us any equipment that we choose, could we detect that we were not in a normal-matter environment?

5. It has often been said that rocket ships of the future might run on matter-antimatter propulsion. How would they work? In what kind of container would you keep your antimatter fuel?

6. Pions can now be produced in the laboratory by allowing protons to strike protons or neutrons. Typical reactions are shown below.

$$p + p \rightarrow p + n + \pi^+$$
$$p + n \rightarrow p + p + \pi^-$$

   Why must the incoming proton have a high energy in both cases?

7. Pick any beta decay process and show that the neutrino must have zero charge.

8. Why, as one goes from sea level to great heights, does the number of cosmic rays first increase and then decrease?

9. If you have not studied Chapter 23, read the section entitled "Moving Muons" in that chapter to learn of another use of cosmic rays. What was this use?

10. There is a certain minimum energy that a gamma ray can have if it is to be destroyed to form the electron-positron pair. Why won't a low-energy gamma ray play a part in pair-production? Why is an even more energetic gamma ray needed to produce the proton-antiproton pair?

B.C. by permission of John Hart and Field Enterprises, Inc.

Appendix 1

POWERS OF TEN

As you proceed through the chapters of this book, you encounter discussions of distances ranging from the incredibly small to the incredibly large. Time periods and masses of objects cover a similarly extensive range. Examples: Are you aware that at the site where your classroom now stands, a dinosaur may have stood 1,000,000,000,000,000 seconds ago, give or take half a day? On the other hand, the electron in a hydrogen atom circles the proton, making one round trip every 0.000,000,000,000,001 second.

Consider distance: The distance from the earth to the nearest star (beyond our own sun) is a mere 100,000,000,000,000,000 yards, while the average distance between the air molecules in the air around you is 0.000,000,001 yard. As for mass—the sun has a mass of about 1,000,000,000,000,000,000,000,000,000,000,000 grams, and the ink required to make the dot over an "i" in this text has a mass of about 0.000,001 gram.

These examples have been used for two reasons: first, to show that the scope of physics is enormous, and secondly, to point out that reading numbers of this size is awfully tedious. To simplify the task of writing large and small numbers, a method called powers-of-ten notation was developed. It's exceedingly easy to use and will make these large and small numbers more meaningful and comprehensible.

We know that $10 \times 10 = 100$, or more efficiently, $10^2 = 100$. Also, $10 \times 10 \times 10 = 1000$, or writing this another way, $10^3 = 1000$. If we carry this process on and on, we can construct a table as shown below.

TABLE A1-1. POSITIVE POWERS OF TEN

$10^1 = 10$
$10^2 = 100$
$10^3 = 1000$
$10^4 = 10,000$
$10^5 = 100,000$
$10^6 = 1,000,000$
ETC.

Table A1–1 could be continued as far as we wish to go, but more important is the pattern developing: $10^2 = 100$; the numeral "2" on the left side of the equation corresponds to the two zeros in the 100. Likewise, 10^6 is ten raised to the sixth power and is the same as the number one followed by six zeros. With the above facts in mind, let's return to the example of the dinosaur mentioned previously. If you count the zeros in the number representing the time since he walked the earth, you will find fifteen of them. So isn't it a lot simpler to say that he strolled in your back yard 10^{15} seconds ago? The earth-to-star distance translates to 10^{17} yards and the mass of the sun to 10^{33} grams.

The problem of writing small numbers can be solved just as easily. The number 0.1 can be written as 10^{-1}, which is read as "ten to the negative one power." If we multiply 0.1×0.1, we get 0.01. The number 0.01 can be written in powers-of-ten notation as 10^{-2}. (See Table A1–2.) Again, a pattern emerges. The number 10^{-6} is equal to 0.000001, which has five zeros between the decimal and the one. There is always one less zero than the negative power of ten; that is, the number 10^{-14} has 13 zeros between the decimal and the one.

Referring to the amount of time required for one round trip of the electron around the proton in a hydrogen atom, we find a time of 0.000,000,000,000,001 second. There are 14 zeros in this number, and it may be written in scientific notation as 10^{-15} second. Convince yourself that the separation between air molecules is 10^{-9} yard and that the mass of ink in the dot of the i is 10^{-6} gram.

As a slight extension of this process, consider the number 6×10^3. This is the same as 6 multiplied by 10^3, which equals 6 times 1000, or 6000. Note that the answer 6000 is the same as six followed by three zeros. Check your comprehension with the following set of numbers:

$$7 \times 10^1 = 70$$

$$3 \times 10^2 = 300$$

$$5 \times 10^6 = 5{,}000{,}000$$

Very small numbers are often written as follows: 6×10^{-3}. This is equivalent to 6×0.001, or 0.006. The same reasoning is applied here as to the situation above. Try these:

$$7 \times 10^{-1} = 0.7$$

$$3 \times 10^{-2} = 0.03$$

$$5 \times 10^{-6} = 0.000{,}005$$

TABLE A1–2. NEGATIVE POWERS OF TEN

$0.1 = 10^{-1}$
$0.01 = 10^{-2}$
$0.001 = 10^{-3}$
$0.0001 = 10^{-4}$
ETC.

For practice, the following set of problems will be of value to you.

1. Man first walked on the earth about 10^{13} seconds ago. Write this time value without powers-of-ten notation.
2. The diameter of the nucleus of the atom is about 10^{-13} centimeter. Express this number without the aid of powers-of-ten notation.
3. The length of a football field is 100 yards. How is this expressed in powers of ten?
4. The radius of the earth is 4000 miles. Find this distance in feet, round off your answer, and express it in powers-of-ten notation.
5. A certain molecule has a mass of 0.000,000,000,000,000,000,-000,000,008 kilogram. Express in powers of ten.

Appendix 2

MEASUREMENT

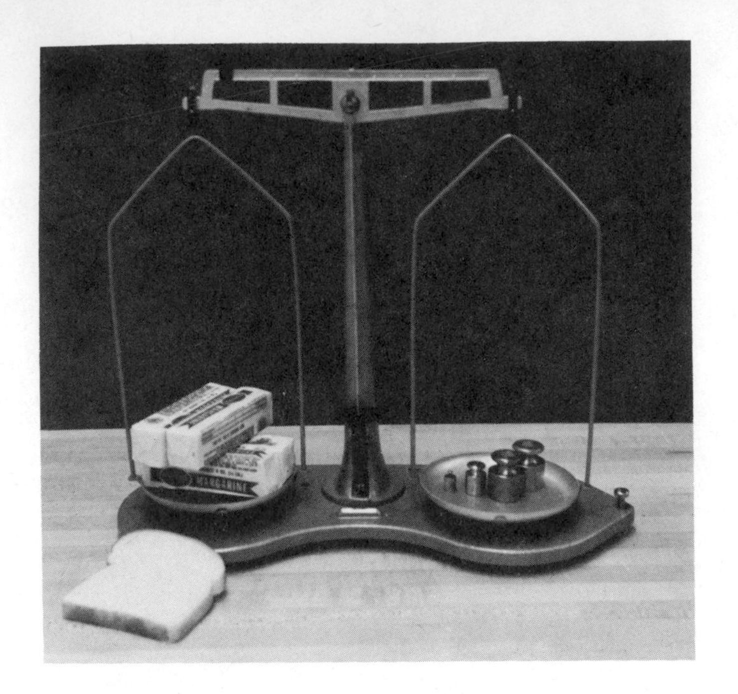

LENGTH (or GIVE HIM 2.54 CENTIMETERS AND HE'LL TAKE 1.6 KILOMETERS)

The realization that a standard for the measurement of length is necessary appears early in recorded history. The Biblical references to the "cubit"—a unit of length—are familiar. In more modern times (if 1120 AD can be called modern) King Henry of England decreed that the length of the yard should be precisely equal to the distance from the tip of his nose to the end of his outstretched arm. This perishable standard was later translated into the length of an iron rod kept by the Exchequer in London. In 1742, an alert public servant noticed that this rod had been broken and welded together so poorly that it drooped when held extended. As is true in all good bureaucracies, repairs were quickly made—in 1760! Then in 1834, Parliament House was burned in the great fire of London, and our standard of length became a puddle of liquid metal. Reliable copies and fond memories remained, however, and suitable duplicates soon appeared.

Not to be outdone, the French adopted as their original standard of length the length of the royal foot of King Louis XIV. This standard prevailed until 1790 when one ten-millionth of the distance from the equator to the pole became the legal standard and was given the name "meter." Ink was barely dry on documents proclaiming the new standard when another public servant discovered an error in the measurement of the equator-to-pole distance.

Fraught with difficulties though they were, these systems have nevertheless survived until today in terms of measurements that can be easily and accurately reproduced. The advantages of the metric system (using the French meter) have caused it to become the prevailing system in most countries and within scientific circles everywhere. A recent National Bureau of Standards report indicates that only eleven nations retain the "English" system of yards, feet, and inches. They are Barbados, Burma, Gambia, Liberia, Muscat and Oman, Naurua, Sierra Leone, Southern Yemen, the United States, Tonga, and Trinidad.* No editorial comment seems necessary.

The metric system uses the meter as the basic unit of length. The meter is divided into 100 equal parts called *cent*imeters, just as our dollar is divided into 100 *cents.* The centimeter is further divided into 10 equal parts called *mil*limeters. (Our cent is divided into 10 *mils,* but they're worthless.) The only other metric unit of length in everyday use is the kilometer, which is 1000 meters.

TABLE A2-1. METRIC AND ENGLISH UNITS OF LENGTH

1 meter = 39.37 inches	1 inch = 2.54 centimeters
1 kilometer = 0.6214 mile	1 mile = 1.609 kilometers

MASS

A favorite trick in television skits is to load a table with dishes and then to jerk a cloth out from under them without causing any to fall to the floor. A not-so-favorite happening occurs when you are forced to slam on the brakes of your car quickly. You find yourself thrown forward toward the instrument panel. Both of these occurrences are due to a property of all objects called inertia. The name "inertia" is well chosen, because the word implies a resistance to change. The dishes do not move because initially they were stationary, and since they possess the property of inertia, they prefer to remain stationary. The driver of the car also has inertia, and since initially he is moving forward, he resists a change in his motion and subsequently is thrown forward as the car beneath him stops.

It is necessary to describe the property of inertia in numerical terms just as it was necessary to describe length numerically. We describe the amount of inertia an object has in terms of a quantity called its mass. Scientists use a unit called the gram to measure mass and, hence, inertia. A larger unit called the kilogram is equal to 1000 grams (*kilo* = 1000). Objects in our environment cover wide ranges in mass. The sun has a mass of about 10^{30} kilograms, whereas an electron has a mass of only 10^{-30} kilogram. A nickel is about 5 grams and a house about 10^5 kilograms.

Measuring Mass

Chapter 3 contains a discussion of the difference between mass and weight. Further difference is shown in the way the quantities are measured. Weight can be measured in a manner with which we are all familiar: spring scales. The common bathroom scale is an

*Note that even the British no longer use the English system.

example—the more you weigh, the more the spring inside is distorted and the more the scale reads.

Mass, however, is measured differently. The figure on page 562 shows a common laboratory balance. This device relies on the fact that if the mass of two objects is the same, the force of gravity acts equally on both of them. On the right side of the balance are several pieces of metal of known mass. They total 454 grams. Since the balance does not tilt one way or the other with this mass on the right side, we know that the object on the left also has a mass of 454 grams.

A primary difference between results from using a balance and a spring scale is that the scale will show a different reading for the weight of an object at different points on the earth. Thus four sticks of margarine may weigh one pound in New Orleans but 0.98 pound on top of Pike's Peak. And though the force of gravity on the margarine varies in different locations, it also varies for the metal masses, so the balance remains the same. The margarine will have a mass of 454 grams anywhere.

INDEX

Numbers followed by a (t) indicate tables.

OUR DUMB CENTURY

the ONION®

Presents

OUR

100 Years of Headlines

DUMB

from America's

CENTURY

Finest News Source

RUNNING PRESS
PHILADELPHIA • LONDON

A Running Press® Miniature Edition™

Printed in China

Library of Congress Control Number: 2003110721

ISBN 978-0-7624-1866-4

This book may be ordered by mail from the publisher. Please include $1.00 for postage and handling.
But try your bookstore first!

Running Press Book Publishers
2300 Chestnut Street
Philadelphia, PA 19103-4371

Visit us on the web!
www.runningpress.com
www.theonion.com

INTRODUCTION

To correct the problem of bulkiness found in the original *Our Dumb Century*, *The Onion* introduces this Miniature Edition. It's small enough to fit in your medicine cabinet along with all of your other books and, best of all, can be found for purchase near a bookstore's cash registers, requiring no lengthy walk past shelves with frightening labels like "Literature."

Enjoy!

1900–1929

A NATION TURNS ITS CRANK

THE ONION Monday, January 1, 1900

DEATH-BY-CORSET RATES STABILIZE AT ONE IN SIX.

GROWING USE OF DR. SCHEIDT'S PATENTED SAFETY CORSET.

Ladies Breathe Slightly Less Painful Sigh of Relief.

The Republic's health-officials are reporting an overall seven percent fewer deaths by corset this year, a statistic that will no doubt gladden the hearts of those dutiful ladies who unselfishly risk sacrifice upon the altar of Beauty everyday so that they may appear comely and supple to their husbands and suitors.

This reduction in corset-related fatalities is credited to the growing popularity of Dr. Ezra Scheidt's Safety and Wellness Corset for the Improvement and Uplifting of the Pleasing Female Form. The supporting under-garment received a patent last year for its unique styling and construction. With this corset, the waist is bound to only seventeen inches, but it still dramatically enhances the figure and beautifies its wearer, giving her an alluring pouter-pigeon appearance which will assuredly turn the heads of gentle-men.

THE ONION Monday, January 1, 1900

VATICAN CONDEMNS 'RHYTHM METHOD.'

Pontiff Excoriates Infidels for Fiendish 'Sin by the Calendar.'

RELEASES PAPAL EDICT OUTLINING FORBIDDEN FAMILY PRACTICES.

ITALIANS IN ATTENDANCE VOW TO PEOPLE THE EARTH.

VATICAN CITY, DEC. 31.—In a proclamation rife with the righteous anger befitting a clergyman of his most exalted station, His Holiness the Pope, Bishop of Rome, Vicar of Christ, Leader of all Christians of the

World, lashed out at those with the effrontery to use the calendar to defeat God's purpose of peopling the globe.

"This prideful 'rhythm method' is a blasphemy before God," Leo XIII declared to a crowd of as yet non-American-emigrant Italians at St. Peter's Basilica. "Those who would count the days He created against the ripeness of their goodwives are indulging in the darkest brand of sorcery."

The reactions of lay civilization have been varied, as most prominent Americans are expected to seek the counsel of their bishops for clarification of the edict.

THE ONION Tuesday, May 20, 1902

REPUBLIC'S NEGROES STILL WAITING FOR 40 ACRES, MULE.

AGRICULTURE SECRETARY FAULTS LOW MULE, ACREAGE AVAILABILITY AT PRESENT.

WASHINGTON, MAY 19.—Nearly 30 years after the end of Reconstruction, freed-men and free-born throughout the land's Negro populace experienced to-day yet another delay in their long-promised allocation of forty acres and a mule.

Citing "prohibitive federal budget inadequacies," U.S. Agriculture Secretary James Wilson announced that "the United States presently cannot afford to allocate forty acres or mules to any citizen, particularly those of dark-skinned, African persuasion, at this time."

Wilson urged all Negroes to remain forth-right and stead-fast in their patience. He addressed Negroes directly by adding, "Please do not commit any acts of violence or impropriety against our nation's white women while waiting."

TO-DAY'S LYNCHINGS.

10:30 A.M., RED SEAVER AT THE TREVERTON TOWN SQUARE.

12 NOON, POKEY WILLIS AT THE HORBART FARM, BEAUMONT. LUNCHEON TO FOLLOW.

12:30, LONNIE LITTLE AT THE CRESTWOOD VILLAGE FESTIVAL.

4:30 THE BROTHERS BROWN, ON THE WILMUTH PROPERTY, ARLINGTON

THE ONION Tuesday, May 20, 1902

CONGRESS REDUCES WORK-WEEK TO 135 HOURS.

Captains of Industry Scarlet with Rage.

In a move that resounds like thunder throughout our great Republic, a Congress made up of influential Progressive Republicans has voted "Yea" on legislation that would cut the work-week to a scant 135 hours and hobble fair Dame Industry with many other insidious and crippling reforms.

THE ONION Thursday, August 6, 1903

ZULUS ATTACK LONDON!

KING HELD AT SPEARPOINT.

LONDON, AUG. 5.—This quondam glorious metropolis was in abject shambles this evening, the sad victim of a cruel and villainous attack by members of Rhodesia's Zulu tribes.

Horrified women clutched their young children bosomward and watched helplessly as the combined forces of the once-proud British Army, joined by dozens of amateur soldiery culled from the ranks of Scotland Yard, proved no match for the thousands of uluating warriors.

THE ONION Friday, December 18, 1903

SCIENCE CONQUERS SKY WITH WRIGHT BROS. FLYING MACHINE;

HEAVEN EXPEDITION SLATED FOR NEXT YEAR.

Historic Meeting Between Roosevelt and St. Peter Planned; President Preparing Speech.

Citizens of Republic Eager to Visit Passed Relatives.

WILL DEATH BECOME A THING OF THE PAST?

WASHINGTON, D.C., DEC. 17.—The serene kingdom of heaven will be the domain of man next year, thanks to the scientific prowess of the Wright brothers of Ohio, who verified man's mastery of the skies at 9 a.m. today with their historic flight above the dunes at Kitty Hawk, North Carolina.

Possessing at last a heavier-than-air craft capable of lifting itself, and listing where directed by the able pilot, man-kind will now be able to journey to God's fair abode, where the moral and philosophical conundrae which vex Earth dwellers may be explained by the Creator himself.

The expedition is slated for June of next year. It is anticipated that when the craft reaches the kingdom, cherubim will sound majestic trumpets heralding man's return to the Creator.

THE ONION Sunday, April 22, 1906

EARTH-QUAKE MARKS LEAST GAY DAY IN SAN FRANCISCO HISTORY

'QUEEN CITY ON THE PACIFIC' LIES IN RUINS.

Garment District Still Flaming.

SAN FRANCISCO, CALIFORNIA, APRIL 18.— Gaiety was not to be had on the streets of the city this morning, as it lies in ruins from what fair citizens are calling the most outlandish earth-quake in known history.

"This is not a gay day for San Francisco," the honorable Mayor Eugene Schmitz said. At precisely 5:12 a.m., the local time, the quake shook the city to its conservative core, rendering it no more than rubble, ash, and faggots.

City authorities fear that of the thousands of fair citizens trapped under collapsed homes, some 700 will never come out.

THE ONION Sunday, April 22, 1906

SHOULD U.S. SET LIMITS ON INDIAN SLAUGHTER?

CONGRESS-MEN SEARCH FOR SPORTING ANSWER.

Some Naturists Fear Indian May Become Scarce if Hunted Wantonly.

Bureau of Savage Affairs Chair-man Gives Speech.

WASHINGTON, D.C., APRIL 21.—Congressmen rose to speak on all sides of the issue of the unremitting Indian slaughter as the House of Representatives debated whether to set limits on the pastime enjoyed by so many.

See The Onion's newest comical-sequential story panel, "Little Fat-Pants and his Billy-Goat, Juniper."

Kindly turn to page 14.

THE ONION Tuesday, April 16, 1912

WORLD'S LARGEST METAPHOR HITS ICE-BERG

TITANIC, REPRESENTATION OF MAN'S HUBRIS, SINKS IN NORTH ATLANTIC

1,500 DEAD IN SYMBOLIC TRAGEDY

NEW YORK, APRIL 15.—Officials of the White Star Line have confirmed the sinking, during her maiden voyage, of the R.M.S Titanic, the world's largest symbol of man's mortality and vulnerability.

First reports of the calamity were received Monday at the telegraph office of the White Star Line, which owns the nautical archetype.

INDIFFERENCE

It is believed that at this time that upwards of 1,500 passengers aboard the metaphor may have perished in the imperturbable liquid immensity that, irrespective of man's self-congratulatory "progress," blankets most of the globe in its awful dark silence. Seven hundred more passengers survived to objectify human insignificance in the face of the colossal placidity of the universe.

THE ONION Friday, February 19, 1915

600,000 KILLED IN 4-INCH ADVANCE ON WESTERN FRONT

HEROIC SOLDIERS PAY ULTIMATE PRICE TO MAKE PATCH OF MUD SAFE FOR DEMOCRACY

'These 4 Inches of Mud No Longer Bow Before the Whip of Tyranny,' Says British Command

PLESSIER, FRANCE, FEB. 18.—Allied High Command is jubilant today after last night's triumphant four-inch advance into enemy territory at the cost of a mere 600,000 troops. The strategically vital one-third of a foot is already being fortified by British forces against the expected German counter-assault.

Strategically Significant

Allied strategists consider this advance extremely significant. "After this bold offensive, the allied armies are only 931,720,458 inches from Berlin," said Lt. Gen. Thomas Fuller, former leader of the Second Army Corps, which was eliminated in the fighting. "My men fought valiantly for at least five minutes. Thank God we claimed the mud, and their deaths were not in vain."

The mud had been seized by the Kaiser two days before, despite a valiant defense at the cost of 450,000 soldiers. It is expected to remain safely in Allied hands for two or three days, British Army Intelligence officials said.

THE ONION Thursday, November 8, 1917

PRETENTIOUS, GOATEED COFFEEHOUSE TYPES SEIZE POWER IN RUSSIA

STUDENTS IN BERETS AND TURTLENECK SWEATERS STAGE ARMED COUP D'ÉTAT

REVOLUTION FUELED BY TRIPLE ESPRESSOS

WINTER PALACE LOOTED FOR POETRY ANTHOLOGIES

If the Huns Had Their Way

TEUTON—"I will make short work of vis 'Fraulein' Liberty."

LADY LIBERTY—"Oh, red-blooded American men between the ages of 18 and 35, please save me!"

Ways to Guard Against the Spanish Influenza:

Wear a flimsy cotton mask; it will filter the influenza bug from the air you breathe.

Read the entire book of Numbers as loudly as possible.

Bury your family alive.

Wear a cast-iron collar ringed with lit candles. This will ward off the invisible evil spirits bringing this pestilence.

Advise your children that the infected corpses cluttering the gutters may not be used as play-dolls.

Scrape skin with sharp edge of sandstone; cut off lips, eyelids, nose, fingernails, hair; cover body in paste made from rock salt, borax, and wood alcohol; wrap body in burlap; sing hymns.

If you must attend mass bond rallies, try not to inhale or exhale.

Persuade Congress to declare war on the hated Spaniards, purveyors of this wicked and fearsome malady, so that they may receive a soundly deserved punishment.

THE ONION Sunday, June 29, 1919

PRESIDENT WILSON CALLS FOR CREATION OF USELESS WORLD GOVERNING BODY

New 'League of Nations' to Offer Gift Shop, Guided Tours

SNACK BAR 'A POSSIBILITY,' WILSON SAYS

WASHINGTON, D.C., JUNE 28.—President Woodrow Wilson today called for the creation of a "League of Nations," a world governing organization with limited powers and an unlimited selection of souvenir world flags and postcards, available for purchase in its lower-level gift shop.

"I am fully committed," President Wilson told members of the press corps assembled at the White House, "to making the new League of Nations an extremely powerful, effective world tourist attraction, one which will draw visitors from around the globe. With the most terrible war in human history finally behind us, with 22 million dead from the most grisly, bloody fighting the world has ever seen, it is vital that the peoples of the world have a place where they may gather to purchase miniature wax statuettes of famous landmarks with national slogans on them."

PRESIDENT WILSON ACCIDENTALLY DEPORTED

Our report on page 31.

THE ONION Wednesday, October 29, 1919

GANGSTERS PASS 18TH AMENDMENT

'Lucky' Luciano Casts Deciding Vote to Make Alcohol Illegal

Mob to Take over $500 Million Liquor Industry

'Sinful Drink' Rightfully Condemned, Says Al Capone

CHICAGO, OCTOBER 28.—At a special meeting at its Chicago headquarters, the United Organized Criminals of America voted in favor of passing the 18th Amendment, outlawing alcohol in the United States, effective immediately.

Gangsters present were elated by the amendment's passage, which they called "long overdue."

"Alcohol is a depraved and immoral substance which has brought ruin to countless lives," U.O.C.A. president Alphonse "Scarface" Capone said. "Our government is wrong to have tolerated its manufacture for so many years. The only proper place for such a substance is in the hands of unsavory mobsters."

'Root Of Evil'

New York U.O.C.A. member Dutch Schultz agreed. "As Carrie Nation and so many braves leaders of this nation's temperance movement have argued, alcohol is the root of nearly all the evil that surrounds us. Crime, sexual promiscuity, and decreased church attendence can all be traced to it," Schultz said.

THE ONION Thursday, February 2, 1922

INDIA'S NATIONALIST LEADER PUMMELED SENSELESS BY PRACTITIONERS OF BRITISH 'VIOLENCE' MOVEMENT

Mohandas Gandhi Given 'Non-Passive Head-Bashing'

Agitator Mohandas K. Gandhi, who reportedly received a non-peaceful beating from his jailers yesterday.

Two New Literary Journals Founded

T.S. Eliot's 'The Criterion' and Mr. and Mrs. DeWitt's 'Reader's Digest'

Which Will Endure?

(a noted scholar's viewpoint, page 9)

Unattractive Dresses
Available in several shapeless, unalluring styles
Slocombe's Department Store
11 Canal Street, New York

THE ONION Tuesday, December 18, 1923

DADAIST MOVEMENT ENDS; 'VICTORY,' CRY DADAISTS

Dadaists Drench Selves with Glue in Celebration

'Make Angry Love to the Fish!' Cry Triumphant Artistes

ZURICH, SWITZERLAND, DEC. 17.—The revolutionary "anti-art" movement known as "Dada" officially came to an end today.

According to poets, painters, and sculptors identified throughout the art world as Dadaists, the increasing chaos and alienation of the modern world made the movement no longer tenable within intellectual circles.

"At last, Dada has achieved its ultimate victory over itself," said Dadaist poet and essayist Tristan Tzara, who spoke from inside a dank, fetid cabinet constructed of rotting cheese. "This day shall be remembered as Dada's greatest hour: the death of Dada."

SCOPES MONKEY TRIAL RAISES TROUBLING QUESTION:

IS SCIENCE BEING TAUGHT IN OUR SCHOOLS?

DAYTON, TENN., JULY 19.—The trial of high-school biology teacher John Scopes took another controversial turn today with the introduction by defense attorney Clarence Darrow of a surprise witness: Cornelius the intelligent chimpanzee-man, a strikingly anthropomorphic simian claiming to originate from thousands of years in Earth's future.

Darrow hopes that the presence of the chimp, a member of the "intellectual class" of a future society ruled by apes, will illustrate to the jury the merit of the biological theory of evolution, the teaching of which now stands in question under Tennessee law.

THE ONION Tuesday, August 23, 1927

BILLIONAIRES BUY U.S. FROM MILLIONAIRES

Future of Nation in Yet Wealthier Hands

The assembled billionaires plan to take control of all aspects of American society, including the passage of laws, levying of taxes, and all else that makes up the greater structure of daily life. The government, they say, will be left in place, but its function will, of course, be regulated by the richest of American citizens, and it will seem a democracy only upon its most superficial face.

Tuesday,
April 17, 1928

HEELS KICKED UP ACROSS NATION; PRESIDENT CALLS FOR CALM

Frolicking, Cavorting Levels at 30-Year High

MILLIONS BUYING ON CREDIT

CHARLESTON-DANCING, JALOPY-RIDING ALL THE RAGE IN '28

WASHINGTON, D.C., APR. 16.—The United States is currently in the midst of the greatest era of sustained prosperity since the Prancing '00s, and, as a result, heels are being kicked up across the land.

Among the evidence of heel-kicking: the rising popularity of marathon dancing; a 42 percent increase in the manufacture and consumption of outlawed liquor in 1927 alone, and a quadrupling in rouge-wearing and hair-bobbing since 1920. A dramatic rise in the number of college-age men who wear crew-neck athletic pullovers decorated with hand-lettered "slang" terms such as "Oh, you, kid!" and "Go easy, Mabel!" is being cited as proof of the rise in spirits.

THE ONION Tuesday, October 22, 1929

Stock Market Invincible

'Buy, Buy, Buy!' Experts Advise

Wall Street, Spirits Soaring

U.S. Enjoying Embarrassment of Riches

Solid-Gold Plumbing Fixtures Best-Selling Consumer Product of 1929

Even Immigrants Enjoying Measure of Comfort

Citizens Urged to Put Everything They Have into Stock Market

NEW YORK, OCT. 21.—Leading economic indicators of the day, such as numbers of raccoon-skin coats in the process of being paid off on "easy credit" and the highly inflated value of major common stocks being over-sold on margin, point to one thing, say experts: prosperity for the U.S.A.

Average Americans are getting rich by buying stocks with money they don't even have.

"I bought several shares of Westinghouse stock on margin and have a net worth of over $30,000," said New York City teacher Sid Grossman. "I'm going to retire on that money."

And stockbrokers are advising everyone to buy in big while the going is good.

THE ONION Tuesday, October 29, 1929

Pencils for Sale

Stock Market Crashes; Debacle Linked to Jews, Negroes, Catholics, Anarchists, Foreigners, Women Voters

MILLIONS THRUST INTO DESPERATE POVERTY

WALL STREET FAT CATS BLAMELESS, SAY FINANCIAL EXPERTS

NEW YORK, OCT. 28.—Although a formal inquiry has not yet been initiated, our nation's political and economic experts have revealed damning evidence tying Monday's collapse of the New York Stock Exchange to Jews, Negroes, Catholics, Anarchists, Foreigners and Women Voters.

In what would be a dramatic reversal of

previous laissez-faire economic policy, it is rumored that the Hoover Administration may advocate legislation ensuring that all future share transactions are carried out with the complete exclusion of Jews, Catholics, Negroes, anarchists, foreigners, and women voters.

Of one point, all Wall Street investors are certain: They themselves are blameless.

1929-1946

DUST, DESPAIR AND DEATH: THOSE WERE THE DAYS

Hoover Hopes to Restore Faith in Nation's Banks with Free-Toaster Offer

Bread-Toasting Device to Ease Economic Hardship

'Open an Account Today, Have Toast Tomorrow,' Says President

Bread Not Included

After months of criticism from political rivals for his inaction during the great economic crisis which grips our nation, President Hoover introduced a bold initiative on Wednesday, urging banks to offer a free toaster and up to five other "premium" gift items to customers who face the loss of their life savings.

"I will work closely with the nation's banks to make the dispensation of these premium gifts, including the new Toastmaster Model 1-C-1 bread-toaster, a priority on our nation's road to recovery," Hoover said in a meeting with the nation's leading banking officials this week.

"The American people are facing hard times," the president said. "Nearly one in four is out of work. Countless families have been made penniless by the collapse of banks and the nation's financial infrasturcture. What better way to renew our citizens' faith in our banking system than with the free gift of an attractive, cast-aluminum three-slice toaster?"

THE ONION Wednesday, December 6, 1933

Stalin Announces Five-Year 'Everybody Dies' Plan

50 Million Russians to Perish by 1939

Over Two Million New Jobs to Be Created in Grave-Digging

MOSCOW, DEC. 5.—Speaking before the Supreme Soviet yesterday, Premier Josef Stalin unveiled a bold new five-year "Everybody Dies" plan for the U.S.S.R.

If successful, Stalin said, more than 50 million Soviet citizens will die by 1939.

"This is a plan that can work," Stalin said. "Whether by government-backed murder, war, starvation, or disease, the people of the Soviet Union are very good at dying."

According to Stalin, the plan consists of three parts: mass famine through collectivization of agriculture, purging of all suspected enemies of the state, and gross mismanagement of resources by a bloated and corrupt government bureaucracy.

Hollywood Careers Destroyed Today:

Bow, Clara
Brooks, Louise
Fitzgerald, F. Scott
Gilbert, John
Haines, William
Keaton, Buster

THE ONION Saturday, December 30, 1933

President Confronts Depression with 'Big Deal' Plan

'BIG DEAL, I'M RICH!' ROOSEVELT SAYS

In a special statement to Congress Friday, President Roosevelt unveiled the "Big Deal," his plan to remedy America's financial crisis. "Big Deal," the president said. "I'm rich!"

A Comprehensive Plan

President Roosevelt is urging Congress to launch emergency legislation and rally the nation back to health with programs under the umbrella title, "The Big Deal." "I expect the plan to fail, which is why it is called the Big Deal," he said. "My family will always have money no matter what happens to anyone else, so: Big Deal."

As for the rest of the nation, Roosevelt said, "I'll try to pass some emergency banking legislation or something. But basically, Big Deal. See if I care."

Said Roosevelt, "Do I need to worry about about every last uneducated urchin on the street? No, I do not. In fact, to them I say, "Big Deal!"

THE ONION Monday, September 9, 1935

Bumper Crop of '35: Dust

OKLAHOMA, KANSAS REPORT RECORD HARVESTS

Legislators Debate Dust Tariff for Upper Atmosphere

An Oklahoma dust-bowl farmer surveys this season's spectacular yield of fine, airborne dirt.

Dirt farmers from Arkansas to the Dakotas are vomiting topsoil today, as an unexpected bumper crop of finely particulated dirt continues to set records as the highest-yield harvest in American dust-farming history. Locals are just trying to keep up, as the region formerly known as the Great Plains continues to soar eastward at up to 40 miles per hour without reprieve.

The storm of high-quality, American-grown dirt is the Midwest's largest ever agricultural windfall, and it can be seen from as far away as Washington, D.C., as the skies are blackened with the fast-moving clouds of dust.

"There can be no complaining about the financial woes of our nation when there is such a surplus of arable land in the country," said Franklin Olsen of the U.S. Airborne Dirt Planning Office.

SOULLESS CULTURAL WASTELAND 'ON THE GROW' IN SOUTHERN CALIFORNIA DESERT

Los Angeles to Be Hellish Megalopolis by 1950

LOS ANGELES, CALIF.—The soulless cultural wasteland in the California desert, considered to be one of the bleakest and most God-forsaken stretches of uninhabitable scorched earth in the nation, is "on the grow," West Coast sources say, as the burgeoning city of Los Angeles continues its cancerous expansion.

Originally a tiny villa called Los Diablos, a coastal settlement of no distinction save for its capacity for heartlessness, the boomtown is now bigger than ever.

Despite its lack of any life-sustaining natural resources, the city, which has no reason to exist, has all the earmarks of a truly spectacular, soulless cultural wasteland on the rise.

Tourist-Friendly Dystopia

Thanks to its policy of draining every conceivable water source within a hundreds of miles via a massive network of pipes, as well as the Chamber of Commerce's approval of a name-change to the more tourist-friendly "Los Angeles," the up-and-coming wasteland shows every sign of ballooning into a full-scale dystopia.

THE ONION Sunday, September 3, 1939

HEADLINE CONTINUED ON PAGE 2

Hitler Invades Britain, Belgium, Denmark, Norway, Netherlands, France, Greece, Yugoslavia, Hungary, Luxembourg

French Surrender After Valiant Ten-Minute Struggle

BORDEAUX.—A French military spokesman offered surrender yesterday after a fierce 10-minute struggle left as many as two-thirds of the French fighting forces dispersed on foot or in hiding.

German armies marching into the country had scarcely lifted the barrels of their guns before French forces still holding ground surrendered, dropped their weapons, and raised their hands.

At 7:44 a.m. Saturday, German Panzer and stormtrooper units crossed the Alsace River and entered French territory, where they were met with furious retreats of the French armies. German supreme commander Wilhelm Keitel was surprised and pleased to see, upon overrunning fortifications along the river, a completed, notarized, and legally binding document of military submission atop a table on the French command headquarters.

THE ONION Monday, December 8, 1941

DASTARDLY JAPS ATTACK COLONIALLY OCCUPIED U.S. NON-STATE

Congress Declares War After Sneak Attack on U.S. Imperial Holding

FDR: 'We Conquered the Hawaiians First'

French Surrender

THE ONION Monday, December 8, 1941

PEARL HARBOR, DEC. 7—The Empire of Japan launched a villainous attack on democracy, freedom, and decency Sunday morning, when her military forces bombed the tiny island paradise of Hawaii, which U.S. forces rightfully subjugated over a hundred years ago.

Japanese planes also attacked the U.S. territories of Guam, Wake Island, and Clark Field in the Philippines, all areas rightfully considered U.S. brown-people holdings.

The U.S. Senate voted unanimously to declare war against the Japanese Empire, which they ruled can have no claim on the territories, since "America conquered the ignorant savages upon these lands first."

Ladies, Negroes Momentarily Useful

"With America's fine young men off at war, millions of females and coloreds are no longer without function," War Production Board Chairman Donald Nelson said. "Under these extreme wartime conditions, these otherwise useless members of society are temporarily not completely worthless."

Women, who, in times of peace, sit at home and bide their time with such activities as cooking, giving birth, and washing dishes until U.S. husbands return home from work at the end of the day are now finding employment in factories, where they manufacture the heavy weaponry their men use on the fighting fronts of the Pacific.

Plot to Assassinate Hitler Fails When He Misses TNT-Filled Piñata

LEIPZIG.—A group of top German military officers, fearing that their Führer is leading them down a path of military foolishness and self-destruction, failed in an attempt on his life last week.

During an impromptu "south of the border" theme party thrown at Hitler's mountain villa, the German leader repeatedly swung at, but failed to strike, an explosive piñata fashioned from a 15-pound block of TNT cleverly shaped and colored to resemble a festive burro.

According to sources within Germany, those loyal to Hitler first suspected that the crepe-paper party item was an explosive device when Field Marshal Klaus von Stauffenberg jumped behind a massive stone fireplace after spinning the Führer around for the fourth time in preparation for his swatting at the frilly, donkey-shaped piñata.

Upon inspection, the colorful object was found to contain enough high explosives to kill not only the blind-folded, stick-wielding Hitler, but also every high-ranking Nazi official in attendance at the Mexican-themed fiesta.

THE ONION Sunday, November 21, 1943

Loose Lip Sinks Ship

613 Sailors Killed in South Pacific by Careless Talk in Michigan

Part-Time Ford Employee Arrested

DEARBORN, MICHIGAN—The War Department reported yesterday that the sinking of the U.S.S. *Saginaw*, which was torpedoed by the Japanese 100 miles east of Guam last Wednesday, was the result of the careless talk of Raymond Fowlie, a part-time second-shift bolt sorter at Ford's Dearborn Assembly Works.

"Mr. Fowlie gave the Japs the clues they needed to send 613 of our boys to their deaths," said Special Agent John Smith of the Federal Bureau of Investigation who took Fowlie into custody late yesterday. "While visiting his neighborhood tavern, Fowlie indulged in alcoholic beverages which loosened his lip to the point where he discussed his sensitive bolt-sorting work more loudly and candidly than is permitted by War Department authority."

INSIDE:

MARIJUANA USE UP AMONG LOUIS ARMSTRONG

See story, page 8

THE ONION Thursday, June 8, 1944

War Rationing Board Restricts Nylon Use to Armed Forces, J. Edgar Hoover Only

Frilly Panties and Garters Especially Needed Now, FBI Director Says

FBI Director J. Edgar Hoover

WASHINGTON, D.C.—In an effort to conserve limited supplies of badly needed nylon for use in the war, access to the material has been reserved solely for the Armed Forces and FBI Director J. Edgar Hoover.

Sale of nylon and goods containing nylon, such as fishing line and women's hosiery, ended in retail stores this week, as the distribution of all nylon was diverted to meet the military needs of the Army, Navy, and Air Force, as well as the confidential needs of Hoover.

General Forrester of the Air Force reports that his branch of the military alone needs a weekly supply of 32 tons of nylon material for use in parachutes, ropes, straps, and collapsible awnings. J. Edgar Hoover, who has requested an allotment of 22 pounds of nylon, has made no explanation as to its pupose other than to say that it is of a highly sensitive nature.

THE ONION Wednesday, March 14, 1945

Nothing Going On in New Mexico

Top Physicists 'Just Camping,' Say Army Intelligence Officials

ALAMOGORDO, N.M.—Atomic Energy Commission officials announced yesterday that nothing whatsoever is going on in the New Mexico desert, where dozens of the world's leading physicists are gathered to enjoy relaxing, non-suspicious camping.

"Oppenheimer, Teller and the other physicists are very good friends, and they decided it would be nice to get away and go camping and fishing together," General Leslie Groves said. "So they all headed to New Mexico for a few weeks where they will be relaxing and not testing anything even remotely resembling an incredibly destructive new weapon."

Suspicions were aroused Saturday when a flash of blinding light, followed by a mile-high column of smoke, was spotted by motorists several miles from the physicist's campsite.

In response to the sighting, the War Department released a statement explaining that the flash was merely the result of physicist Robert Oppenheimer adding too much lighter fluid to the campfire.

THE ONION Friday, August 10, 1945

Nagasaki Bombed 'Just for the Hell of it'

SECOND A-BOMB WOULD HAVE JUST SAT AROUND ANYWAY, SAY GENERALS

FRENCH SURRENDER

GUAM.—The world's most destructive weapon, the atomic bomb, was used for a second time against nearly defeated Japan Thursday, in a raid against the industrial city of Nagasaki.

Military officials explained that the bomb, which vaporized thousands of innocent Japanese civilians, charred thousands more with third-degree burns over their entire bodies, and obliterated the city with an explosive force equal to a half-million tons of TNT, was dropped "just for the hell of it."

"We still have one device remaining after our display of force in Hiroshima the other day," said B-29 pilot Col. Paul Tibbets, "and because of the wonderful success of that bombing, we were aware that Japan was already moving toward surrender. If we were going to drop the other bomb, we had to do it quickly, before peace made it inadvisable."

Family Unit Gazes Happily Into Glorious, Shining Future

Husband, Wife, Children Joyfully Spellbound by Wonderful Things to Come

Even Greater Happiness Apparently Possible

PROSPECT CORNERS, OHIO—A nuclear family unit stood rigidly on the manicured front lawn of their beautiful four-bedroom suburban home Tuesday, gazing unblinkingly at some undetermined fixed point in the distance, their expressions a mixture of industrious can-do spirit and satisfaction with a job well done.

Dick Larson, dressed in a snappy pin-striped suit, and Annie Larson, dressed in a lovely house-dress and heels and a conservative pearl necklace, lovingly held the hands of their two bright and well-behaved children, daughter Sally and son Chip, as they looked bravely toward the horizon. The well-respected church going family seemed to radiate positivity and get-up-and-go as they gazed into the glorious future of our great and inordinately wealthy nation.

An uncommon scene? Hardly! They might just as well be the John Q. Anytown, U.S.A, family!

1946-1963

THE SWELL YEARS

THE ONION Thursday, February 20, 1947

'Tele-Vision' Promises Mass Enrichment of Mankind

'Drama and Learning Box' Will Make Schools Obsolete by 1970

New Device to Provide High-Minded Alternative to Mindless Drivel Found on Radio

Imagine, if you will, touring the ancient ruins of the Roman Empire, watching a demonstration on the fine art of sculpture, or having the theories of velocity explained to you by a doctor of physics—all in the comfort of your own den after supper! Imagine, ladies, learning the newest crochet patterns and discovering the latest stain-removal tips—all while ironing the day's laundry! The amazing tele-vision, the most promising invention of the century, will make all this possible.

Antlike Conformity Now Affordable

Row upon Row of Identical Box-Like Homes Replace Ugly Long Island Prairie

New 'Levittown' Homes to Be as Interchangeable as Inhabitants

HEMPSTEAD, N.Y.—In what is sure to be a boon to the nation's millions of growing families, maverick developer William J. Levitt announced Tuesday that he is preparing to unveil a new type of suburban housing development on a vast parcel of Long Island real estate. The new community, consisting of thousands of identical structures built on identical 60 x 100-foot lots and accessed by a geometrically precise grid of two-way streets is being called "Levittown."

"You're a lucky fellow, Mr. Homeowner," Levitt said at a ground-breaking ceremony. "As an American citizen, you have the right to live among people whose age, income, number of children, habits, topics of conversation, modes of dress, possessions and religious beliefs are identical to yours."

War-Weary Jews Establish Homeland Between Syria, Lebanon, Jordan, Egypt

'In Israel, Our People Will Finally Have Safety and Peace,' Says Ben-Gurion

Jordan Welcomes New Neighbors with Celebratory Gunfire, Rock Throwing

JERUSALEM, ISRAEL.—After more than 2,000 years of wandering and persecution, including six million deaths at the hand of Nazi Germany, the Jewish people finally established a homeland Monday, a place of safety and peace, nestled between Syria, Lebanon, Jordan and Egypt.

"No longer will the Jewish race live in a constant state of fear and endangerment, its very existence threatened at every turn by hostile outsiders," said David Ben-Gurion, the new nation's first prime minister, addressing a jubilant crowd of Zionists at Jerusalem's Western Wall. "Here in Israel, we are safe, far away from those who seek to destroy us."

For two millennia, the Jewish people have wandered without a home, facing an endless series of hostile enemies. With the establishment of a sovereign Jewish state in the Middle East, Israeli officials believe this 2,000-year ordeal has at last come to an end.

"Israel is the land of milk and honey," Ben-Gurion said. "The only gunfire we shall hear is that which lingers in our minds from troubled times long past."

THE ONION Sunday, October 2, 1949

Toxic Levels of Self-Involvement Found in Many Post-War Babies

70 Percent of Infants' First Word: 'Me'

BOSTON, MASS.—According to an article published in the latest *New England Journal of Medicine*, toxic levels of self-involvement have been discovered in nearly 80 percent of children born in the post-war "baby boom."

The article's findings were based on blood samples taken from 25,000 babies born between 1945 and 1948, whose blood contained dangerously high levels of self-involvement, some 40 times higher than those of Americans born in the previous generation.

"Ordinarily, blood absorbs many substances present in a person's body, including nutrients, salt, cholesterol and urea. But rarely does it actually absorb the person himself," *New England Journal of Medicine* Managing Editor Dr. Malcolm Causewell said.

"Curiously, though in the case of these post-war babies, that is exactly what happened, creating a strange condition called 'Self-Absorption.'"

THE ONION Thursday, November 6, 1952

A National Hygiene Crisis:

Boy-Girl Malt-Sharing at All-Time High

Wanton, Promiscuous Exchange of Saliva Rampant at Nation's Soda Fountains

Unsafe Practice Could Lead to Social Diseases, Childbirth

Malted Treats a Breeding Ground for Dangerous Microbes, Sperm

WASHINGTON D.C.,—In what could represent the most serious public-health crisis since the influenza epidemic of 1918, Department of Health officials announced Wednesday that opposite-sex malt-sharing among U.S. young people reached an all-time high in the first half of 1952

The study found that more than 80 percent of the American youths between the ages of 16 and 21 have shared malts with members of the opposite sex. In 67 percent of cases, the teens were using the same straw.

"This kind of reckless behavior on the part of young people is inexcusable," U.S. Hygiene Secretary O. Russell Shea said.

THE ONION Sunday, March 8, 1953

New Medical Report Finds Heavy Petting Linked to Communism

Touching Selves, Others in Impure Manner Exposed as Bolshevik Plot

Nation's Clergy Urge Immediate Scrubbing-Down of All U.S. Teens

PROVIDENCE, R.I.—A Brown University medical study released Saturday demonstrates a link between intimate touching connecting teens and the spread of Communism.

"Our nation's teen-agers are in closer proximity to each other than ever before," study director Dr. Milton Lambert said.

"Not only is this dirty petting exposing our children to germs and head lice, but we now have evidence that it is also putting them at risk of becoming brainwashed agents of the Kremlin."

The study found that the heated, fevered exchange of such bodily fluids as sweat and saliva creates conditions ideal for the development of Communist microbes.

"These Red cells spread quickly into the brain, reducing the victim to an unwitting footservant of the Bolsheviks," Lambert said.

To conduct the study, high-school students were asked to "neck" in a controlled laboratory setting while being monitored for galvanic skin response and heart rate.

THE ONION Wednesday, August 12, 1953

Pentagon Develops A-Bomb-Resistant Desk

Schoolchildren Now Safe from Atomic Blast

Desks Will Be Standard Issue for All U.S. Public Schools by 1955

WASHINGTON, D.C.—In a scientific breakthrough with far-reaching implications in light of current tensions between the U.S. and Russia, U.S. Secretary of Defense Charles Wilson announced Tuesday that Civil Defense Authority scientists have successfully developed a school desk that is impervious to atomic attack.

"Should the U.S.S.R. choose to drop the atom bomb on the United States, our nation's children would merely have to hide under their school desks, and they would be safe," Wilson said. "Everything in the surrounding 10-mile radius, including hospitals, homes, grocery stores and loved ones, would be instantly incinerated, vaporized in three-millionths of a second from the searing heat of an atomic blast. But the children would be unharmed."

If an A-bomb were detected entering a school's air space, Wilson said, the school's principal would sound a special nuclear-attack siren.

Rosa Parks to Take Cab

'Screw This Bus Shit,' Says Montgomery, Alabama, Commuter

MONTGOMERY, ALA.—Tired of being asked to give up her seat to white passengers on segregated buses, negro commuter Rosa Parks decided to begin taking a cab to work late this week.

"Screw this bus shit," the 42-year-old seamstress said in an impromptu statement to fellow Montgomery bus commuters Friday.

"The white passengers and the Montgomery buses can take their segregated, negroes-giving-up-their-seats Jim Crow horseshit and shove it up their cracker butts. I'll take a cab."

Although Parks will pay approximately $3 more each day for transprotation to her and from her job at Millon's Clothiers, she said it was "abolutely worth it" to avoid what she called "the bullshit seating rules" on Montgomery buses.

THE ONION Monday, November 21, 1955

Defense Department to Reinforce Nation's Brassieres

Angora Sweaters Nearing Breaking Point

Congress Calls for Emergency Loosening of Strained Tops of Monroe, Mansfield, Turner

THE ONION Monday, November 21, 1955

WASHINGTON, D.C.—A federal panel warned Friday that the approximately 14 million pink, fuzzy sweaters in the U.S. are approaching critical mass and will likely reach a breaking point by mid-July.

"America's angora sweaters just keep getting tighter and tighter," said Dr. Edgar O. Retzloff of the Federal Bosom Containment Task Force, established by President Eisenhower to examine the crisis. "At a certain point, they are going to burst."

"It is unreasonable to expect these garments to maintain their buxom payloads for much longer without defense-contractor reinforcement," he said.

THE ONION Friday, July 27, 1956

Supreme Court Upholds Mississippi Law Requiring Negro Voters to Be White

WASHINGTON, D.C.—The Supreme Court Thursday upheld a Mississippi judge's decision against a Jackson, Miss., voter charged with willfully breaking a state law requiring negro voters to be white.

The case involved Marcus Jefferson, a Mississippi voting hopeful who, in November 1952, was deemed "not white enough" to vote in the state.

"Mr. Jefferson is technically a negro, and therefore did undeniably violate the law by trying to vote," Chief Justice Warren wrote in the majority opinion.

Warren cited Amendment XV, Section 1 of the Constitution which states that "the right to vote shall not be denied to a citizen of any race, color or creed who has white skin."

THE ONION Monday, June 24, 1957

Status Quo Maintained for 2,000th Consecutive Day

With the nation's housewives, and nearly everyone else, in agreement, the U.S. achieved a milestone in stasis yesterday, reaching its 2,000th consecutive day of societal status quo.

Order and sameness are the rules of the day across our land, as national power structures remain safely entrenched and bold new ideas remain at an all-time low.

"From our collective love of enormous automobiles to our fear and mistrust of outsiders, to our shared desire to own attractive new patio furniture, America has not changed in three wonderful years," said Brad Edwards, 33, a Canoga Park, Calif., advertising executive. "I, like other citizens, am extremely satisfied with the way things are."

THE ONION Friday, July 19, 1957

EVERYTHING'S NIFTY

Swellness in America at All-Time High

All Americans Pleased as Punch

Ike Says 'Okey-Dokey' to American Prosperity

In a nationally televised address Thursday night, President Eisenhower announced that everything is hunky-dory in the U.S.

"My fellow Americans, I stand before you tonight to tell you that everything is swell, real swell," the president said in his one-hour address. "Our kids are getting A's on all their tests at school, discontentment is at a 30-year low, and U.S. wives are cooking more delicious meatloaf than at any point in American history. We, as a nation, are more than okey-dokey; we are super-duper okey-dokey."

According to Eisenhower nearly every major economic and social indicator points upward. Among the more notable statistics: Industrial production is up 20 percent in the past 12 months, unemployment is at just 2 percent, the divorce rate is at just 4 percent, teeth are 57 percent whiter, and U.S. soft drinks have 46 percent more pep.

National politeness indicators are also at a four-decade high with 33 percent more people saying, "Thank you, dear," and "Gee, you're welcome, ma'am."

THE ONION Wednesday, November 9, 1960

JFK to Lead Nation in Good-Natured Game of Touch Football

HYANNISPORT, MASS.—Speaking to supporters, President-elect Kennedy pledged Tuesday night to lead the nation's 180 million citizens in a good-natured game of touch football on the lawn of his family's estate Sunday.

Kennedy said the spirited yet casual game will symbolize America's innocence and bright future and be not so much a competition as a metaphor for American spirit and teamwork.

Everyone will have a chance to participate, he said, even the ladies. His young wife Jackie, looking both demure and athletic in one of her husband's old sport shirts, will join the women of America as they cheer from the sidelines.

The president-elect pledged to wear comfortable chinos and a faded Harvard sweatshirt during the game.

Eisenhower Warns of Military-Industrial-Oedipal Complex

Nation Increasingly Jealous of Relationship Between Pentagon, Big Business

'America Must Become Aware of Its Deep-Seated National Feelings,' Ike Says

WASHINGTON, D.C.—President Eisenhower, in his farewell address to the nation, warned Friday of the "growing, unwarranted influence of the military-industrial-Oedipal complex," which he said threatens to "undermine the most primal foundations" of society.

"It is, perhaps, natural," Eisenhower

said, "that we Americans find ourselves snuggled so deeply in the protective, maternal bosom of the defense-contracting industry that we come to be jealous of—and at times, even loathe—the paternalistic military machine to which it is wedded. Nevertheless, it is vital for the future of our American family that we come to see the armed forces not as a rival for the affections of General Electric or McDonnell-Douglas, but as a stern yet loving hand, capable of imparting great wisdom and moving us beyond the era of international adolescence."

Eisenhower appealed directly to the American people, asking them to demand regular, private, tri-weekly "sessions" with President Kennedy over the next four years.

THE ONION Monday, April 17, 1961

OUTER SPACE FALLS TO COMMUNISTS

Russian Man Is First in Orbit

Will a Statue of Lenin Look Down on Us from the Moon?

Alarmed Legislators Slash Education Budget to Fund Space Race

At 10 p.m. EST Sunday, the Russians launched a man into orbit around the Earth, claiming the heavens in the name of

Communism.

"The Russians have captured space," Pentagon spokesman John Powell said in a somber official statement. "The national-security implications of this event are chilling, as the United States is now surrounded by 360 degrees of Communist-occupied vacuum."

In a mission lasting nearly two hours, Russian communaut and vanquisher of millions of square miles of outer space Yuri Gagarin orbited once around the Earth at an altitude of 200 miles. It is unknown to which planets he may have directed his craft to plant the flag of the Union of Soviet Socialist Republics.

THE ONION Monday, October 29, 1962

ALLEN FUNT LETS PRESIDENT IN ON HILARIOUS 'CUBAN MISSILE CRISIS' PRANK

Relieved Kennedy Laughs Heartily after Learning of Elaborate International Gag

JFK a Good Sport, Says Host of TV's *Candid Camera*

WASHINGTON, D.C.—A period of intense national tittering came to a hilarious end early Monday morning, when notorious prankster and *Candid Camera* host Allen Funt ended his most elaborate gag to date by letting President Kennedy in on the "Cuban Missile Crisis" that has kept America in stitches since last week.

Following a briefing by the "Joint Chiefs of Staff" (in actuality, professional actors) about the latest developments in the crisis, Kennedy was surprised by a loud knock at the Oval Office door. Funt, who has passed himself off as Nikita Khrushchev for the last two weeks in a series of increasingly implausible televised speeches, walked unimpeded through a phalanx of "Secret Service agents" and approached the president.

"Didn't mean to "rush in" like that, Comrade Presidentski," a straight faced, Russian-accented Funt told the president, "but I am havink somethink important vith vich to be telink you."

THE ONION Friday, November 22, 1962

KENNEDY SLAIN BY CIA, MAFIA, CASTRO, LBJ, TEAMSTERS, FREEMASONS

President Shot 129 Times from 43 Different Angles

DALLAS, TEX.—President Kennedy was assassinated Friday by operatives of the CIA, the Giancana crime syndicate, Fidel Castro, Vice President Johnson, the Freemasons and the Teamsters as he rode through downtown Dallas in a motorcade.

According to eyewitnesses, Kennedy's limousine had just entered Dealey Plaza when the president was struck 129 times in the head, chest, abdomen, arms, legs, hands, feet, back and face by gunfire.

THE ONION Friday, November 22, 1962

Preliminary reports indicate that hitmen for the Giancana crime syndicate fired from a nearby grassy knoll, CIA agents fired from an office building slightly off the parade route, Cuban nationals fired from an overpass overlooking Dealey Plaza, an elite hit squad working for Teamsters President Jimmy Hoffa fired from perches atop an oak tree, a 'lone nut' fired from the Texas Book Depository, a shadow-government sharp-shooting team fired from behind a wooden fence, a consortium of jealous husbands fired from an estimated 13 sites on the sidewalk along the route, a hitman working for Johnson fired from a sewer grate over which the limousine passed, and Texas Gov. John Connally lunged at the president from within the limousine.

1963-1981

PEACE, LOVE AND OTHER BULLSHIT

Malcolm X: 'I Also Have a Dream'

'I Have a Dream That One Day Little Black Children Will Beat the Living Crap out of Little White Children'

HARLEM, N.Y.—Responding to the stirring speech delivered by Dr. Martin Luther King at last year's March on Washington, fellow negro-rights crusader Malcolm X announced Sunday that, like King, he also has a vision for the future of negroes in America, albeit a slightly different one.

"I also have a dream," Malcolm X said to the congregation of Holy Mercy Evangelical Church in Harlem. "I have a dream that, one day, a little white child and a little black child will sit together in peace and harmony. I have a dream that little white child will then steal that little black child's lollipop and laugh at the little black child, just as the white man has been degrading, stealing from and abusing the black man for the last 400 years."

The Black Muslim leader paused, the congregation confused as to his intended message. "And that little black child will cry, as all black men and women have, over the cruelty and greed of the white man. But my friends, my dream does not end there."

"Brothers and sisters I have a dream that he will beat that little white boy like a red-headed stepchild until that son of a bitch soils his drawers."

THE ONION Saturday, April 4, 1965

Sanford, Son Killed in Watts Rioting

Aunt Esther Missing

LOS ANGELES—Last week's Watts riots, in which 34 people were killed, 4,000 injured, and millions of dollars in property destroyed, have claimed two more victims, as local junk dealer Fred G. Sanford and his son Lamont are now listed among the dead.

The Sanfords' bodies were uncovered under a heap of burned furniture, the body of a pick-up truck and other scrap-pile items the businessmen collected at their 134th Street residence.

Witnesses reported that, during the riot, Fred Sanford stood amid the frenzy of looting, rampaging citizens, clutching his chest and yelling, "This is the big one!"

Damage to Sanford's junk is estimated at nearly $80.

Female Orgasm Discovered

BALTIMORE, MD.—Scientists at Johns Hopkins University Medical School announced Monday that they have discovered what may be a sexual reflex in women: the mythical, long-rumored female orgasm.

According to Johns Hopkins' Dr. Randolph Stolper, the discovery was made when a Baltimore-area woman, "Jane," was admitted to the hospital after reporting feelings of intense pleasure during sex with her husband.

"Sex, an act science has long defined as the forcing of the man's penis into the woman's vagina for the singular purpose of male ejaculation and orgasm," Stolper said, "did not appear to cause Jane any of the agonizing pain females typically experience during the procedure. Appropriately concerned by this abnormality, she consulted her physician."

Jane, in fact, reported actually feeling sensations of pleasure similar to those typically associated with the male sexual experience.

"While many invasive tests are still necessary before we can attempt to determine what caused this phenomenon, it would appear that this woman experienced what can be termed a 'female orgasm,'" Stolper said. "The entire medical community is mystified."

THE ONION Saturday, June 10, 1967

UC-BERKELEY STUDENTS PROTEST NOT BEING TOLD ABOUT LATEST PROTEST BY FRIENDS

BERKELEY, CALIF.—A coalition of left-wing students at the University of California at Berkeley held an impromptu rally Friday night to protest "the unconscionable failure of our classmates to tell us about a really great protest that happened earlier in the day."

Waving hand-drawn signs with such slogans as, "Why Didn't You Call Us," and, "We Should Have Been There," the crowd of approximately 1,000 students congregated in Sproul Plaza, then marched across campus chanting, "Hey, hey, ho, ho, if there's a protest, we ought to know." The impassioned demonstration was met with force by campus and city police, who dispersed the students with tear gas and clubs.

The protest the students had missed, the "Feminist Anti-War Make-Love-for-Civil-Rights-Free-Speech Peace-In," included police, rousing free-speech rhetoric and music.

The protest featured a surprise performance by the popular San Francisco rock group Moby Grape, which only added to the neglected students' sense of outrage.

"Everyone who was anyone was at that protest," campus radical leader Mario Savio said.

THE ONION Tuesday, June 13, 1967

PENTAGON WORKING TO HARNESS FLOWER POWER FOR DESTRUCTIVE NEW WEAPON

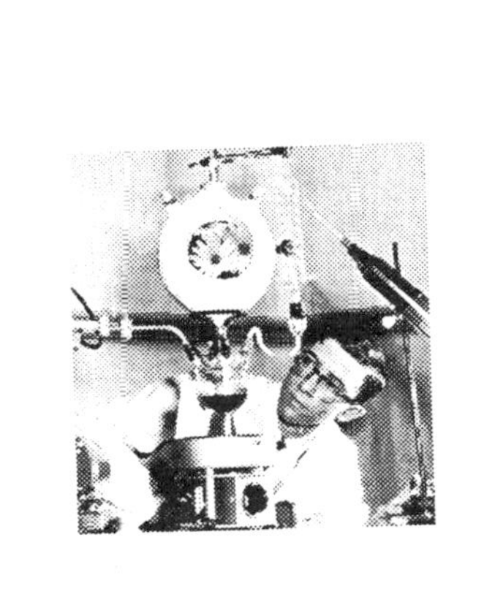

THE ONION Tuesday, January 30, 1968

JOHNSON DEPLOYS 20,000 BODY BAGS TO VIETNAM

Bag Escalation Part of U.S. Policy of 'Corpse Containment'

Total Military Commitment in Vietnam up to 60,000 Bags

Army helicopters deliver crucial body-bag reinforcements to the 33rd Infantry near Puc Tho.

WASHINGTON D.C.—President Johnson stepped up his commitment to the war in Vietnam Monday, deploying an additional 20,000 body bags to military zones throughout the Southeast Asian nation.

"If the U.S. is to contain the threat of rotting, fetid corpses in Vietnam, we need to send more bags," Johnson said. "Only by zipping up these dead bodies in quality bags of durable, rubberized-canvas construction can we ensure that the stench of deteriorating flesh does not spread beyond the 16th parallel and into the southernmost regions of Vietnam.

THE ONION Monday, July 21, 1969

HOLY SHIT

MAN WALKS ON FUCKING MOON

NEIL ARMSTRONG'S HISTORIC FIRST WORDS ON MOON:

'HOLY LIVING FUCK'

THE MOON—Jesus fucking Christ.

The distant, lonely, and mysterious satellite that has fascinated mankind since the dawn of time is distant and lonely no more.

At 4:17 p.m. EST yesterday, astronauts

Neil Armstrong and Edwin E. Aldrin, Jr. touched down on the Sea of Tranquility in the lunar module Eagle and radioed back to Earth the historic report: "Jesus fucking Christ, Houston. We're on the fucking moon."

Armstrong and Aldrin then made final technical and psychological preparations for the un-fucking-believeableness of the next phase of the operation, the moon walk.

As two billion spellbound earthlings watched on television, Armstrong slowly descended the four steps leading out of the module, paused, and took one small but epoch-making step onto the soft, virgin soil.

"Holy living fuck....Are you fucking believing this? Over," Armstrong radioed back to NASA headquarters nearly 250,000 miles away.

Rain at Woodstock Results in Slightly Dirtier Hippies

A sudden downpour at 3 p.m. Sunday during the third day of the Woodstock festival reportedly resulted in even dirtier hippies. Mud slicks quickly formed on the tramped fields of Max Yasgur's farm, giving concert-goers the chance to cover their already-grimy hair and filthy bodies in yet more dirt and mud.

"Whoo-hoo!" yelled one dirty hippie, Kevin "Tree-Branch" Roberts, as he threw himself into a 15-square-foot puddle of mud created by the sudden shower. He was immediately joined by seven others. Upon walking away from the puddle, a slight increase could be detected in the amount of grime on Roberts' body.

"Somehow, the rain, which could have given the hippies an opportunity to cleanse themselves," noted Jim Fingerton, a policeman assigned to the festival, "only resulted in an ever thicker husk of crusted putridity upon their loathsome bodies."

In an effort to increase the skin area available to cover with filth, many of the hippies removed their clothing before rolling in the mud, emerging up to 14 percent dirtier.

Puke Orange, Pea Green, Mustard Yellow Adopted as New National Colors

WASHINGTON, D.C.—A federally appointed panel of interior decorators, assigned the task of giving the American flag a more up-to-date look for the '70s, chose puke orange, pea green and mustard yellow as the new national colors Monday.

"These bright, vibrant colors really say, 'Welcome to the '70s, America,'" panel chairman Leslie Lurman said. "And the new design of the flag is much more 'with it.'"

"Red, white and blue served America well," Lurman added, "but, as we fast approach our Bicentennial, we need to revamp America's image for a new age of good taste and impeccable fashion sense."

The panel is recommending that the flag's dated-looking stars be replaced by exaggeratedly rounded, bathmat-style daisies; that the stripes be made slightly more modern and psychedelic; and that all the flags be made of hemp.

"Everything from bean-bag chairs to polyester dresses is available in these popular attention-getting colors," Lurman said. "Stomach-flu avocado, concrete gray and rancid-orange-juice tangerine are also good choices, but, in the end, the most dynamic colors won out. Puke orange, pea green and mustard yellow are truly the colors of our time."

Bumper-Sticker Industry Applauds *Roe v. Wade* Decision

WASHINGTON, D.C.—The nation's bumper-sticker lobbyists are hailing the Supreme Court's *Roe v. Wade* decision Monday as a major victory for bumper-sticker sales.

Said Bumper-Sticker Manufacturers of America President Karl Steinholz: "This Supreme Court decision is a triumph for the entire bumper-sticker industry. But it is also a victory for women, as they now have the freedom to look at two bumper stickers and decide which choice is right for them."

Chief Justice Warren Burger, in the majority opinion, wrote, "This court has deemed abortion a constitutional right.

Now, a woman, before making her very personal decision, must first look deep within the bumpers of the cars in front of her at traffic stops."

Steinholz, who sells bumper stickers to stores nationwide, said he is somewhat conflicted regarding the complex issue. "I am really undecided about abortion," he said. "On the one hand, the nation's lawmakers should definitely keep their laws off a woman's body. But, on the other hand, it's important to remember than an aborted fetus is a child, not a choice."

Steinholz noted that both "It's a Child, Not a Choice" and "Keep Your Laws Off My Body" bumper stickers are available at stores everywhere.

THE ONION Saturday, April 8, 1974

U.S. SEVERS TIES WITH SYMBIONIA FOLLOWING PATTY HEARST KIDNAPPING

Symbionese Leaders Outraged

U.S. Stands to Lose $168 in Export Revenue

WASHINGTON, D.C.—Responding to the Symbionese Liberation Army's kidnapping and subsequent brainwashing of publishing heiress Patricia Hearst, the U.S. officially severed trade with the nation of Symbionia Friday.

President Nixon made the formal announcement from the White House. "The United States will, as of today, sever all diplomatic relations with the people and government of Symbionia," he said.

Following the announcement, Nixon ordered Warren M. Reed, the U.S. ambassador to Symbionia, to return to the U.S. immediately. The president also deployed seven Army C-130 transports to airlift any Americans from Symbionese territory.

"I will always reflect fondly upon the Symbionese people, culture and way of life," Reed said of Symbionia, a nation consisting of a small apartment on San Francisco's Golden Gate Avenue. "But after this recent incident, once I pass through customs and say my final good-byes, I will never go back."

NATIONAL MOOD RING GREEN

We Are Sensitive Right Now, Say Ring Experts

WASHINGTON, D.C.—The U.S. Department of Health, Education and Welfare reported Friday that the national mood ring is currently green.

"America has cooled down, mellowed out and gotten more in touch with its feelings," said HEW Secretary Forrest Matthews. "We are a little sensitive right now."

The news of the green ring marks the most assuring American mood change since the National Science Foundation's Magic 8-

Ball indicated "yes" in February.

Department of Karma and Astrology officials warn that, while the green mood is promising, serious obstacles lie ahead for future American mellowness.

"The U.S., a Cancer, is still feeling fragile after the Vietnam War and Watergate," U.S. Astrologer General Phyllis Conroy said. "If we are not given adequate time to get in touch with our feelings, the ring could turn amber-green or perhaps even amber in a matter of months."

"This is a time for the nation to relax," Secretary of Karma and Astrology Marvin Frisch said, "and maybe do some yoga and deep-breathing."

President Calls for Calm Following Nipple Sighting on Farrah Fawcett Poster

WASHINGTON, D.C.—In an address to the American people Wednesday, President Carter called for calm and an end to the panicked rioting that has swept the nation since Monday's sighting of an erect nipple on a poster of *Charlie's Angels* star Farrah Fawcett.

"Level heads must prevail," Carter said during the speech, televised live from the Oval Office. "I urge the American public, especially young men and boys, to remain calm, even in the face of such an obviously aroused nipple."

"I've never seen anything like this, and I was here during Watts," said Lt. Mike Weber of the Los Angeles Police Department, standing before the burning wreckage of a poster shop torched by an out-of-control mob after supplies of the Fawcett poster had run out. "Fathers are attacking their own sons for this poster. I can't say I blame them, though—that chick looks like she's ready to go."

Also affected are the nation's schools: Record truancy rates are being reported, the result of male students home to masturbate while viewing the Fawcett poster.

Carter Offers Ayatollah 'Helpful Energy-Saving Tips' in Exchange for Hostages

WASHINGTON, D.C.—President Carter broke his vow not to negotiate with terrorists Monday, offering the Ayatollah Khomeini what he described as "helpful energy-saving tips" in exchange for the safe return of the 62 U.S. hostages in Iran.

"With this deal, I am appealing to both the Iranians' sense of common decency and their sense of practicality," Carter said. "After all, what sort of person could hold another human being hostage for months on end, brutally beating him until he has lost his very will to live? And, furthermore, what sort of person could turn down the chance to cut electric bills by more than $25 a month and reduce over-all household heat-loss by over 40 percent?"

Carter's Energy-Saving Tips

Turn off your mechanical bull when not in use.

Illuminate the floor of your discotheque using reflected sunlight.

Don't fast-forward your 8-track tape to find the song you want. If you are patient, the song will come around again.

While driving to work, carpool with colleagues to save gasoline. Also refrain from exciting-but-inefficient car chases set to bongo music.

Campaign '80

Jimmy Carter

"Let's Talk Better Mileage"

Ronald Reagan

"Kill the Bastards"

ANN ARBOR, MICH.—In a nationally televised debate Thursday night on the University of Michigan campus, President

Carter and Republican presidential candidate Ronald Reagan outlined their positions on foreign and domestic policy issues.

Public opinion polls show a 76 percent approval rating for Reagan's "Kill the Bastards" compared to only 11 percent for Carter's "Better Mileage" platform. The "Kill the Bastards" plan rates even higher in polls than Reagan's highly popular policies while governor of California, including his "Nuke the Bastards" plan, his "Kill the Foreigners" plan, and his recent "Just Kill 'Em All" plan.

THE ONION Wenesday, January 21, 1980

Reagan May Have Been Elected, Doesn't Recall

40th President 'Not Entirely Sure' If He Swore to Uphold Constitution

WASHINGTON, D.C.—Ronald Wilson Reagan, sworn in Tuesday as the nation's 40th president, said he "can't recall" defeating Jimmy Carter in November's general election.

"As to whether I defeated Mr. Carter in last year's presidential election, I have no memory of that," the new president told reporters following his inaugural address at the U.S. Capitol. "Whether I did or did not win, I unfortunately cannot say with 100 percent certainty."

"President Reagan is a very busy man who does a lot of things in the course of a day," said Reagan aide Michael Deaver in defense of the new president. "And, as such, he cannot be expected to remember every last detail of his activities or promises."

In his inaugural address, Reagan said that, in addition to being unable to recall details of last November's election, "I cannot recall a new morning in America. I have no memory of a public mandate to reform our government. And I am afraid I have no recollection of our nation's renewed determination, pride and strength."

1981-2000

A NATION FINDS ITS REMOTE

THE ONION Thursday, July 30, 1981

CIA Unveils Cheaper, More Powerful Form of Cocaine

'Crack' Expected to Make Drug Addiction More Accessible to Inner-City Poor

LANGLEY, VA.—The CIA unveiled its latest mind-altering drug Wednesday: "crack," a concentrated, smokable form of cocaine.

Cheap and highly addictive, crack is expected to revolutionize drug addiction in America's blighted urban ghettos, CIA officials said.

"Our pharmacologists worked long and

hard on this new drug, and, judging by the response from our test subjects, it was worth the wait," CIA Director William Casey said. "We expect America's minorities to be thoroughly ravaged by crack."

A multimedia presentation at CIA headquarters, attended by the mayors of nearly every major U.S. city, explained the differences between crack and conventional powdered cocaine.

"The crack high is much more intense, yet also much shorter, than that of normal cocaine," said Casey, standing in front of a large video screen depicting an inner-city youth smoking crack out of a glass pipe.

MIT Scientists Have Three Sides of Rubik's Cube Complete

CAMBRIDGE, MASS.—A team of MIT physicists announced Sunday that it has completed the vital third side of the Rubik's Cube, bringing it ever closer to solving the puzzle.

The scientists credited the breakthrough to a shift in their approach to the multi-colored cube. Previously, they had focused their efforts on the center-most square on each side of the puzzle. But team member Dr. Jim Barstow suggested the group concentrate on the four outer corners while closely monitoring the position of each square on the reverse side of the cube during each turn.

"It was the classic example of being too close to a problem to see the solution," Barstow said. "Only by stepping back and rethinking our whole way of looking at the cube could we ultimately move forward."

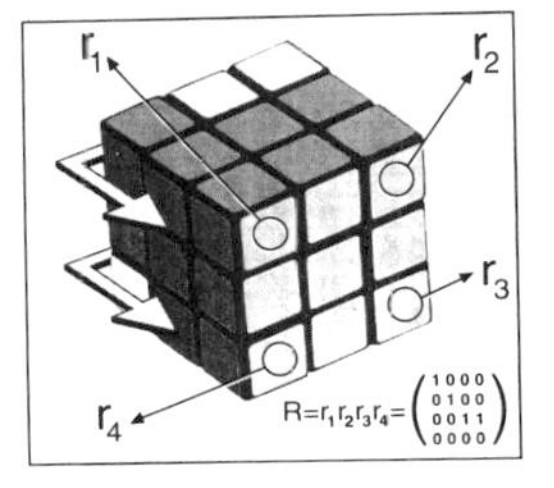

Your Legs: Are They Warm Enough?

A serious lack of attention is being paid to the warmth of Americans' legs, warns a federal legwear study released Friday.

According to the National Leg-Warmth Council report, each day in the U.S., more than 50 million lower limbs are inadequately protected from the elements—a condition researchers say can easily be remedied with the use of colorfully striped, knitted calf-coverings.

"We have long known that 'leg warmers' are vital to dancers for keeping their Spandexed legs warm, preventing potentially dangerous lower-leg chills," NLWC Chair Adrienne Bosch said. "But we are now finding that leg warmers are extremely useful for the general public, as well. They can be worn during activities ranging from club dancing to roller skating, to just hanging around the house."

The study found that women, whose typically smaller frames make them more susceptible to cold calves, should wear leg warmers, not only with "work-out gear," but also with blue jeans, denim skirts, sweater dresses and nylon short-shorts.

For extra warmth, the addition of a color coordinated head- and wristband set was also strongly advised.

Donald Trump Announces Plans to Gold-Plate Atlantic City

ATLANTIC CITY, N.J.—Real-estate mogul Donald Trump unveiled plans Monday to gold-plate Atlantic City, a $900 billion project scheduled for completion in Fall 1988.

"Atlantic City will out-gold the Aztecs. Atlantic City will out-gold King Midas. And I will bring you that gold," said Trump, speaking from his gold-plated Trump Towers in Manhattan.

According to Trump mistress and project coordinator Marla Maples, a fleet of silver helicopters will pour liquid gold over all of Atlantic City starting this spring.

Said Atlantic City Tourism Director Stanley Schmidt, “We’re very enthusiastic about Mr. Trump’s plan for Atlantic City. We have many tourist attractions here already, but, as exciting and dynamic as those attractions are, they will be even more spectacular when covered with solid gold.”

Trump said he intends to pay for the project in cash.

Dynamic New Soviet Leader Not on Brink of Death

MOSCOW—Mikhail Gorbachev, 54, was named the USSR's premier Friday, ushering in a bold new era of Soviet leadership not on the brink of death.

Gorbachev, who had served as Deputy of the Supreme Soviet since 1970, is expected to live for at least another five years.

"Unlike such premiers as Brezhnev, Andropov and Chernenko, Mr. Gorbachev is not about to keel over at any moment," said Samuel Hilliard, a Duke University professor of Russian history. "Perhaps even more impressive, Mr. Gorbachev does not constantly cough up blood. The Soviet government should be greatly invigorated by this dynamic man who was born in the 20th century."

In his first speech before the Supreme Soviet as premier, Gorbachev pledged a break from the policies of his predecessors.

THE ONION Tuesday, August 13, 1985

CDC: New 'AIDS' Disease Could Put Nation at Increased Risk of Gala Celebrity Benefits

Epidemiologists Predict 10,000 New Cases of Lavish Hollywood Fundraisers in Next 10 Years

ATLANTA—According to an alarming report released Monday by the Centers for Disease Control, a deadly new disease known as Acquired Immune Deficiency Syndrome, or "AIDS," could lead to an exponential increase in gala celebrity-benefit events over the next 10 years.

"In 1984, there were six reported cases of AIDS fundraisers in the U.S.," CDC chief

Lawrence Preston said. "Expect these figures to soar over the course of the next decade. By 1995, we estimate that there will be some 10,000 new cases of lavish AIDS benefits, including 7,500 in Los Angeles County alone."

Preston said side-effects of the new disease will take many forms, including glamorous black-tie dinners, fun runs, charity walks, and art auctions.

"Cancer is on its way out," a prominent Hollywood agent said. "Everyone I know is inspired by people with AIDS. Everyone who's anyone in Hollywood is concerned about this exciting new disease."

Mr. T Releases 'Pity List '86'

LOS ANGELES—At a packed press conference at the Four Seasons Hotel Sunday, actor and former bodyguard Mr. T released his official 1986 Pity List.

"Ladies and gentlemen of the press," Mr. T said, "I pity a great many fools this year."

Pausing to don his reading glasses, the hot-tempered star of *Rocky III* and *The A-Team* then began slowly, methodically reciting the list.

Another 49 fools from previous years were singled out for a "lifetime pity" designation, including those who ridiculed his 35 pounds of gold chains or made fun of his mohawk-style haircut.

THE ONION Monday, January 27, 1986

Horrible Asshole Richer Than Ever

NEW YORK—Horrible asshole Brad Thorstad, known throughout the New York metropolitan area as an insufferable, mean-spirited prick whose enormous wealth has enabled him to disregard the basic human dignity of everyone with whom he comes into contact, is even richer today, thanks to the bullish stock market, Wall Street sources reported Tuesday.

It is expected that Thorstad, an awful person who already has more money than any human being could possibly need, will reap untold profits by reinvesting his earnings.

"My watch cost more than you'll make in your entire life," Thorstad said to a waitress at the Russian Tea Room. "Get me another one of these highballs, you stupid bitch."

Thorstad has, sources report, never afforded another human being an ounce of respect in his life. He is said to be incapable of any form of love, except for the inhuman lust for money that feeds his black heart. Described as "that evil, evil fucker" by family members, domestic servants and virtually every other person who encounters him, Thorstad possesses an all-consuming assholitude that is widely attributed to his vast wealth.

THE ONION Tuesday, December 29, 1987

God Kills Oral Roberts For Fundraising Shortfall

Televangelist Called Home By Lord As Warned

TULSA, OK—Angered by what He called His disciple's "shameful failure to raise sufficient funds for My goodly earthly works," God struck down prominent televangelist Oral Roberts Monday.

Roberts, who had received numerous warnings from God that he would be "called home" if his fundraising goal was not met, was killed when a large thunderbolt from the heavens struck him during a taping of his weekly television show.

The 67-year-old Roberts died instantly from the heavenly bolt. He was $211,000 short of God's $3 million fundraising goal at the time of his death.

"I said I would smite Oral, and I kept My word," God told reporters. "While $2,789,000 is an impressive sum of money for one man to raise in the name of the Lord, I made it very clear that I would accept no less than $3 million."

Roberts had been aware for several months that his life was in danger, repeatedly warning his followers that he would be killed by God if donations did not reach the target amount.

THE ONION Saturday, March 25, 1989

Bush Decries Exxon Valdez Spillage of 'Precious, Precious Oil'

Beloved Crude Lost As Tanker Runs Aground Off Alaska Coast

WASHINGTON, D.C.—In a highly charged White House press conference Friday, President Bush lashed out against Exxon's supertanker spill off the Alaska coast, decrying the company's "shocking lack of respect for our planet's greatest natural resource: precious, precious oil."

"What has happened there in Alaska is a tragic, tragic waste of the fossil fuel most dear to my heart," the visibly grieving president said. "We've got to do whatever we can to make sure every drop of the beautiful black crude is recovered."

The Exxon Valdez, a crude-oil carrier in the Alaska-West Coast trade, ran aground in shallow coastal waters off Alaska's Prince William Sound some time Thursday night. Though the cause of the accident is still unknown, reliable workers have been summoned from as far as 2,000 miles away in a desperate attempt to save the oil, which spilled out of the supertanker's badly ruptured hull at a rate of 500 gallons per second. "The large quantities of dead waterfowl and aquatic life in the spill region can be tracked down and wrung out for excess oil residue," Bush said.

Berlin Wall Destroyed In Doritos-Sponsored Super Bowl Halftime Spectacular

EAST BERLIN—Amid fireworks, stirring John Philip Sousa marching music and soaring Blue Angels jets, the Berlin Wall was torn down Sunday in a thrilling climax to "Doritos Presents A Super Salute To Freedom," the Super Bowl XXIV halftime show.

"Ladies and gentlemen, let's hear it for democracy!" the announcer exclaimed, as the wall that separated East and West

Berlin for nearly 30 years fell. "Let there be freedom. And enjoy Doritos. Munch all you want. We'll make more!"

More than 300,000 people gathered at the Berlin Wall for the halftime event, which symbolically ended the Cold War, entertained over a billion football fans worldwide and coincided with Doritos' launch of its new "Cool Ranch" flavor.

As mobs of East Germans pounded at the wall with sledgehammers, Frito-Lay CEO Richard Teller announced to a TV audience, "Today, we crunch our way into an exciting new age. After decades of isolation and oppression, Germany can bite into the great taste of liberty at last."

THE ONION Friday, January 18, 1991

CNN Deploys Troops To Iraq

ATLANTA—CNN deployed more than 50,000 ground troops to the Persian Gulf Thursday in anticipation of a military showdown with Iraq.

Speaking from CNN Command Post Central in Atlanta, Major General Wolf Blitzer said the cable network can no longer allow Saddam Hussein's aggression to go unchecked.

"We must send a message, loud and clear, that Iraq's continued military presence in Kuwait will not be tolerated. CNN is drawing a line in the sand," said Blitzer, flanked by President Turner and Chief of Staff Peter Arnett. "Although this network maintains hope that a peaceful solution can be reached, it is fully prepared to go to war if Saddam refuses to budge."

The CNN Joint Chiefs have already commissioned the composition of new theme music for the war, and Arnett said "Operation Showdown In The Gulf" troops are expected to be ready to attack key Iraqi strongholds every half-hour, on the hour.

Blitzer said CNN is also working closely with NBC News, whose Tom Brokaw pledged full support for his cable-based allies.

Supreme Court Nominee Clarence Thomas: 'The Ass-Slapping Was Never Done In An Inappropriate Manner'

WASHINGTON, DC—Testifying before a senate committee in Day Six of his confirmation hearings Wednesday, Supreme Court nominee Clarence Thomas insisted that his frequent slapping of Anita Hill's ass over the course of many months was "never done in an inappropriate manner."

"The striking of Miss Hill's ass with my open hand was never more than an expression of mutual professional respect," Thomas told the committee.

Thomas went on to state that the numerous times he lifted Hill into the air, slinging

THE ONION

This book has been bound using handcraft methods and Smyth-sewn to ensure durability.

The dust jacket was designed by
THE ONION.

The interior was designed by
DUSTIN SUMMERS.

The text was excerpted from
OUR DUMB CENTURY by **THE ONION.** www.theonion.com

The text was written by
SCOTT DIKKERS and the Staff of **THE ONION.**

The text was edited by
ALISON TRULOCK.

The text was set in Excelsior
and Anzeigen Grotesk.

Michael Jordan Shot For His Sneakers

CHICAGO—Chicago Bulls star Michael Jordan is in critical condition following a Tuesday street-corner mugging in which he was shot for his Air Jordans.

The assault occurred outside Chicago Stadium at approximately 11:15 p.m., shortly after the end of a Bulls-Pistons game. According to one witness, Jordan was approached by a group of teens who demanded he remove his Nikes. When he refused, one of the teens shot Jordan in the chest and ran off with the $110 shoes.

that someone in Washington cares," he continued. "And the women of this country need to know that someone in Washington is willing to cup their breasts in both hands and give them a gentle squeeze."

Voters are praising Clinton's new vision for America.

"I met with President Clinton privately to discuss my economic situation, and he really seemed to understand how I feel," said Jennifer Langer, a 22-year-old nursing student from Little Rock. "That massage really helped."

Added Langer, "President Clinton was right: Going through tough financial times can make you really tense."

New President Feels Nation's Pain, Breasts

WASHINGTON, DC—In his inaugural address to the nation Wednesday, President Bill Clinton delivered a message of compassion, telling Americans that he is fully committed to feeling their pain, and in the case of attractive young women, their breasts.

"As your president, I feel your pain," Clinton said from a podium overlooking a crowd of supporters and nubile campaign volunteers. "I would also very much like to feel your firm, plump breasts."

"The American people need to know

her over his shoulder and "noogie-ing" her ass while making growling noises, were always "a sign of my approval of her fine work, devoid of any sexual overtones."

"Ours was a cordial, relaxed office environment, and my conduct toward her was intended to be friendly," Thomas added. "I am sorry Miss Hill misinterpreted my actions."

Thomas assured the committee there was no improper subtext to his frequent pants-droppings, which were often accompanied by instructions to "kiss the birdie if you like having a job."

Mail-Related Killings This Week
55% Killed by disgruntled postal workers
34% Killed by Unabomber
11% Killed by Publishers Clearinghouse spokesperson Ed McMahon

THE ONION Thursday, August 10, 1995

O.J. Finds Killer

Los Angeles—After months of tireless searching, a triumphant O.J. Simpson announced Wednesday that he has found the killer of Nicole Brown Simpson and Ron Goldman.

Simpson, who had been on a personal crusade to uncover the truth about the double homocide ever since it occurred on June 13 of last year, personally delivered 27-year-old Hermosa Beach auto mechanic Stuart Rogers to LAPD headquarters, telling law enforcement officials that he had "found your man."

"At last, justice is served," Simpson told

reporters. "This should prove to everyone that they rushed to judgment when they saw me panic-stricken and fleeing police in my Bronco, and when they heard that my blood-soaked glove and traces of my DNA had been found at the scene of the crime, and when they found out that I had a long history of abusing my wife and suspected she was having an affair with Mr. Goldman. The fact that I have found this man Stuart Rogers, clears up the confusion once and for all."

"Mr. Simpson is a model citizen who single-handedly tracked down the lowlife who perpetrated this great tragedy upon his family," said Police Chief Williams.

"He is to be commended for solving this case."

THE ONION Sunday, May 26, 1996

Oprah Secedes From U.S., Forms Independent Nation Of Cheesecake-Eating Housewives

Twenty million American housewives, led by daytime talk-show host Oprah Winfrey, seceded from the U.S. Saturday, forming the independent nation of Ugogirl.

Ugogirl, a nurturing, supportive republic, will be subject to a set of laws and provisions which prioritize access to healthy, low-fat recipes, creative home-decorating tips and inspirational stories of personal triumph over adversity. The new nation will have its own congress, constitution and national anthem, Whitney Houston's rendition of "I'm Every Woman."

"We women just weren't being heard," said Ugogirlian Nancy Gordon, a former

housewife from Charleston, SC, who helped unanimously elect Winfrey the fledgling nation's President for Life. "It was time to do something about it."

Though situated within the U.S., Ugogirl will be recognized by the U.N. as a "sovereign nation with attitude and sass."

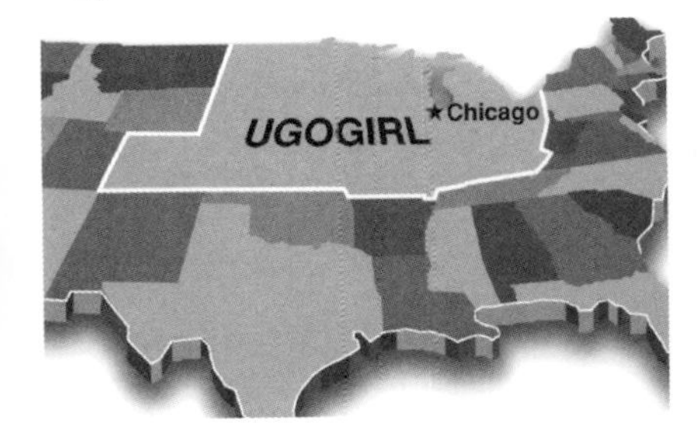

Philip Morris Denies 'Marlboro Babies' Cartoon Targets Minors

LOUISVILLE, KY—Responding to anti-smoking lobbyist's charges Monday, Philip Morris vehemently denied that its new "Marlboro Babies" cartoon characters, featured prominently in recent print ads, target children.

"Once again, the tobacco industry is being forced to defend itself against ridiculous, paranoid accusations," Philip Morris CEO Steven Hatchins said. "The Marlboro

Babies and their lovable sidekick, Puff the puppy, in no way encourage children to smoke."

Philip Morris introduced the cuddly smoking kids last month following the success of its "Filter King & The Nico-Teens" billboard campaign, a marketing move the tobacco giant also denies was aimed at minors.

"We intend adults, and adults alone, to enjoy the antics and adventures of this rag-tag group of mischievous, chain-smoking moppets," Hatchins said. "Remember, kids: Only in the make-believe land of Smokadonia can someone under 18 get Marlboro-brand cigarettes from an older friend or go to a vending machine in an empty hotel lobby and buy some for themselves."

THE ONION Saturday, January 10, 1998

Drugs Win Drug War

WASHINGTON, DC—After nearly 30 years of combat, the U.S. has lost the drug war.

Drug Czar Barry McCaffrey delivered the U.S.'s unconditional surrender in a brief statement Friday. "Drugs, after a long, hard battle, you have defeated us," he said. "Despite all our efforts, the United States has proven no match for the awesome power of the illegal high."

"In retrospect," McCaffrey added, "this was not a winnable war."

McCaffrey then handed over power to *High Times* magazine editor Steven Hager, who will now head the new U.S. Office of Drug Policy, replacing the now-defunct DEA.

"We must all get behind drugs now," outgoing DEA Chief Thomas Constantine said. "I recommend we all get really, really baked."

With the defeat, drugs will begin a full-scale occupation of the vanquished U.S. Massive quantities of crack, heroin, PCP, LSD, marijuana and other drugs will flood the nation legally, saving America's estimated 75 million drug users billions of dollars on their yearly drug budgets.

Christian Right Ascends To Heaven

TULSA, OK—At the stroke of midnight, Jan. 1, 2000, the clouds opened above the Bible Belt and a golden staircase appeared for all born-again Christians who do not bear the Mark of the Beast to ascend into Heaven and enjoy Everlasting Salvation.

Night turned to day as Jesus Christ appeared at the top of the staircase in a blinding white sun-beam to select only 1,000 believers for ascension into Heaven, as outlined in the Book of Revelation.

"Follow Me," the bearded unkempt Jew

told His assembled flock as He unrolled a papyrus scroll bearing a list of names. The list was a veritable Who's Who of the Christian Right. "Pat Buchanan, Bob Dornan, Jerry Falwell, Fred Phelps, Ralph Reed, Trent Lott . . ." Jesus read on, as those names followed Him into the clouds.

Millionaire cable-TV executive and right-wing politician Pat Robertson smiled gleefully as he slowly climbed the stairs. "I've been waiting for this moment all my life," he said, his three-piece suit shimmering in the beatific glare.

PHOTO CREDITS

p. 21, "Little Fat Pants":
Maria Schneider/Onion Graphics

p. 27, "If the Huns Had Their Way":
Maria Schneider/Onion Graphics

p. 35, "Unattractive Dresses":
Dover Clip Art

p. 65, steelworker; p. 83, homes:
Lambert/Archive Photos

p. 153, Rubik's Cube:
Mike Loew/Onion Graphics

p. 179, “Mail-Related Killings”:
Mike Loew/Onion Graphics

p. 183, Ugogirl:
Mike Loew/Onion Graphics

All other photos: **ARCHIVE PHOTOS**